ELEMENTS OF Statistics

ELEMENTS OF Statistics

F. Daly D. J. Hand
M. C. Jones A. D. Lunn K. J. McConway

An imprint of Pearson Education

Harlow, England · London · New York · Reading, Massachusetts · San Francisco
Toronto · Don Mills, Ontario · Sydney · Tokyo · Singapore · Hong Kong · Seoul
Taipei · Cape Town · Madrid · Mexico City · Amsterdam · Munich · Paris · Milan

The M246 Course Team

Authors	Fergus Daly David J. Hand Chris Jones Daniel Lunn Kevin McConway
External Assessor	Professor P. J. Diggle, *University of Lancaster, UK*
Consultants	Dr Colin Chalmers, *DataStat Consultants, London, UK* Dr David Jerwood, *University of Bradford, UK* Professor Clifford Lunneborg, *University of Washington, Seattle, USA* Dr Tony Dusoir, *Mole Software, Co. Antrim, N. Ireland*
Readers	Shirley Hitchcock Gillian Iossif Jann Allison Shelley Channon Trevor Lambert Sue Luton Christopher Minchella Nora Dykins Carole Pace
Course Manager	Janet Findlay
Publishing Editor	Jenny Kwee Brown
Text Composition	Nicky Kempton Sharon Powell Karen Lemmon Liz Lack Jeannie Dean Maybelle Acquah James Campbell
Designer	Diane Mole
Artist	Sue Dobson
Cover Design	Martin Brazier
BBC	John Jaworski Michael Peet Andrew Law Anne-Marie Gallen
Academic Computing Service	Andrew Bertie

The course team would like to thank Jennifer Harding for her work on the course during the early stages of production. They would also like to express their appreciation to Alison Cadle for her help.

Prentice Hall is an imprint of **Pearson Education Limited**

Visit us on the World Wide Web at:
http://www.pearsoneduc.com

First edition 1995.

This book is a component of the Open University course, M246 *Elements of Statistics*. Details of this and other Open University courses are available from the Central Enquiry Service, The Open University, PO Box 200, Milton Keynes, MK7 6YZ, United Kingdom. Telephone (0908) 653231.

Edited, designed and typeset by The Open University using the Open University TEX system. Produced by Pearson Education Asia Pte Ltd. Printed in Singapore (KKP)

British Library Cataloguing in Publication Data

A catalogue record for this book is available from the British Library.

ISBN 0-201-42278-6

10 9 8 7 6 5
04 03 02 01 00

1.1

Contents

Preface

Statistics long ago escaped from its origins as a summary term for tables of numbers describing quantitative features of 'the State'. Nowadays it is a fully-fledged scientific discipline playing a central role in science, industry and commerce. It is the discipline of discovering patterns and structure in data. Because modern statistical techniques allow one to discern relationships which are concealed from the unaided brain, they have been likened to the telescope or the microscope, that allow one to see things invisible to the naked eye. The aim of this book is to provide an introduction to some of these statistical tools.

This book serves as the core component of the Open University course M246 *Elements of Statistics.* Its objective is to provide a solid introduction to the basic concepts and methods of modern statistics.

Its orientation is that it is *problem* and *data* driven. That is, we begin with the attitude that statistics is about solving problems. We collect data relevant to those problems and we analyse those data to answer the problems. To this end, the book is heavily illustrated with a large number of real data sets arising from real problems. The data sets are small, to make them comprehensible to you, the reader, and manageable in the text, but they are real nonetheless.

The book is also *computer-based.* The computer is the essential tool in modern statistical data analysis and we assume that you will have such a tool. The computer and, to a lesser extent, the calculator permit us to forget about the low-level mechanics of the arithmetic, and focus on higher-level issues: what we want to know and how we can answer the questions using statistical methods. In addition to this, you can use your computer to perform simulations which demonstrate statistical concepts in strikingly convincing ways.

Given its central role in an Open University course, the book is aimed at those who will be studying with little or no conventional lecture support: it is intended primarily for lone study. That explains why there are so many worked examples and exercises, and why solutions are given for all of the exercises.

So much for what the book *is.* It is perhaps also appropriate to say a few words about what it is *not.*

It is not a book on mathematical statistics. Although we attempt to give sound justifications for the concepts and methods we describe, we have not hesitated to side-step any lengthy mathematics which we felt would have obstructed the focus on the statistical content. In this vein, we have avoided exercises which test mathematical skills and understanding rather than statistical skills and understanding.

Similarly there are, of course, many topics which, regrettably, we have had to omit for reasons of space. Perhaps chief amongst these is our lack of treatment of experimental design and sampling methods. A truly comprehensive book at this level would be vast, given that statistics is now a very extensive discipline, so something had to be omitted.

The text comprises fourteen chapters which should be studied sequentially. Each of the first twelve chapters consists of teaching material interspersed with examples and exercises and fully accounted illustrative data sets (including references to their source); each chapter ends with a summary, listing the main points that were studied. *Chapter 13* uses a case study approach to revise most of the main methods and techniques taught in the course. *Chapter 14* provides a look forward to problems which require a more advanced approach, and includes a brief description of how those approaches may be applied.

Readers are encouraged to work through the exercises in the text, rather than merely to read them. Many exercises by their nature involve the use of a calculator or a computer, or may require reference to sets of statistical tables. In such cases a marginal icon is used to suggest whichever of these may be the more appropriate: but, of course, calculators and statistical computer packages differ greatly in their competences and readers may find that they prefer to do most of the exercises on a computer or they might manage everything with a hand-held machine. Our suggestions, if followed, would ensure skilled practical use of all three.

All the exercises have full solutions at the end of the book (they constitute nearly one-fifth of the whole text). Where the exercises are computer-based, the solutions provided include no particular command syntax or expected output format. We might suggest as part of the preamble to an exercise on comparing two means that you 'explore the facilities of your computer' which is another way of saying 'learn how your computer does t-tests' before you embark on the exercise proper.

In each chapter there are copious illustrative diagrams in addition to examples, exercises and tables of data since we believe that in a statistical context as well as in others, a picture is often worth a thousand words and, frequently, quite a few numbers as well.

A word about printed accuracy and precision: often in the course of an extended calculation we have shown intermediate results and it is helpful to see these. However, just as you would pursue the computation yourself, subsequent calculations are made to the fullest attainable background accuracy (usually 12 or 15 significant figures, not all of which are printed) and this is reflected in the final result. So you will see instances such as

$$\frac{7 \times 37.75}{16.01} = 16.50$$

when out of context the right-hand side is 16.5053 and should be printed (surely?) 16.51. Actually, the denominator is 16.012 764... and hence the printed result.

It is easy to get hung up on displayed precision, but you should not. As you will see, results are printed sometimes to two decimal places or to four, or sometimes to the nearest integer, depending on what they are ... there is no rule for this (how could there be?): we have used whatever seems the most appropriate precision at the time.

Finally, many of the standard results and formulas given in the text may be proved (or confirmed, at least) using calculus. If you know about differentiation and integration, and if you enjoy practising those skills, you can do this. As a notational aid, we have used the integration symbol $\int$ to show when we are interested in 'the area under a curve'. However, the point about most of these results is that indeed they are *standard*: they are there to be used, not checked. You do not need to use the techniques of the calculus to understand this book.

This text, whilst standing alone as an introduction to the wide application of the discipline of statistics, forms only part of an Open University undergraduate course which itself forms part of a degree profile. Other components of the course include a statistical software package with associated manuals and guides (including a detailed guide to the solutions of the computer-based exercises), a series of eight television programmes under the general title *Statistics in Action*, an audio tape offering an introduction to computers and statistical computing, a booklet of extra problems and exercises, and any number of support booklets (study guides, and so on). The course is examined through coursework during the academic year and by an end-of-year examination paper.

If you would like further information about studying with the Open University, either as an undergraduate pursuing a study programme that may extend over several years, or as an associate student studying courses independently of the university degree structure, you should write to the Central Enquiry Service, PO Box 200, The Open University, Milton Keynes, MK7 6YZ, United Kingdom.

Finally, we hope that some of the excitement we have found in analysing data and answering research questions will be apparent from the following pages. And we hope that you, too, will begin to appreciate the excitement of this queen of sciences.

Fergus Daly
David J. Hand
Chris Jones
Daniel Lunn
Kevin J. McConway

Department of Statistics, The Open University, Milton Keynes, MK7 6AA, United Kingdom.

Acknowledgements

Grateful acknowledgement is made to the following sources for permission to reproduce the photographs in this book.

Figure 4.1 Archives de l'Academie des Sciences, Paris; *Figure 5.5(a)* and *Figure 5.5(b)* Mansell Collection; *Figure 5.6* University College London, Archives; *Figure 8.10* Iowa State University, Statistical Laboratory; *Figure 8.11* Hulton Deutsch Picture Collection; *Figure 8.12* University of California, Public Information Office; *Figure 8.13* The Granger Collection; *Figure 10.14* Godfrey Argent and The Royal Society; *p. 434* University College London, Library; *p. 441* University College London, Library.

Acknowledgments

Chapter 1
Data

An excellent approach to achieving an initial understanding of a set of data is to summarize the data in graphical form. In this chapter several diagrammatic representations are described. A data set may also be summarized numerically, and some common measures are described. The chapter ends with a look forward to the rest of the course.

Chambers English Dictionary defines the word data as follows.

> **data**, *dātă*, ..., *n.pl.* ...facts given, from which others may be inferred: — *sing.* **da'tum**(q.v.) .... [L. *data*, things given, pa.p. neut. pl. of *dare*, to give.]

You might prefer the definition given in the *Shorter Oxford English Dictionary*.

> **data**, things given or granted; something known or assumed as fact, and made the basis of reasoning or calculation.

Data arise in many spheres of human activity and in all sorts of different contexts in the natural world about us. The science of statistics may be described as exploring, analysing and summarizing data; designing or choosing appropriate ways of collecting data and extracting information from them; and communicating that information. Statistics also involves constructing and testing models for describing chance phenomena. These models can be used as a basis for making inferences and drawing conclusions and, finally, perhaps for making decisions. The data themselves may arise in the natural course of things (for example, as meteorological records) or, commonly, they may be collected by survey or experiment.

Later in the course, beginning in *Chapter 2*, we shall look at models for data. However, we shall begin here by examining several different data sets and describing some of their features.

Depending on the way they are expressed (perhaps as a mere list or in a complicated table), very large data sets can be difficult to appreciate without some initial consolidation (perhaps as a series of simpler tables or in a diagrammatic form). The same applies to smaller data sets, whose main message may become evident only after some procedure of sorting and summarizing.

It is now relatively easy to use a statistical computer package to explore data and acquire some intuitive 'feel' for them. This means that you can approach statistics as a numerical detective rather than as a theoretician who may be required to take on board difficult assumptions and preconceptions.

Indeed, you would have to take on trust the direction and validity of any theoretical approach unless guided by the structure or 'shape' of your data. This is reassuring in that the most important and informative place to start is the logical one, namely with the data themselves, and the computer will make your task both possible and relatively quick.

However, you must take care not to be misled into thinking that computers have made statistical theory redundant. Far from it, you will find the computer can only lead you to see where theory is needed to underpin a common-sense approach or, perhaps, to reach an informed decision. It cannot replace such theory and it is incapable of informed reasoning.

However, if you are to gain real understanding and expertise, your first steps are best directed towards learning to use your computer to explore data, and to obtain some tentative inferences from them.

The technology explosion of recent years has made relatively cheap and powerful computers available to all of us. Furthermore, it has brought about an information explosion which has revolutionized our whole environment. Information pours in from the media, advertisements, government agencies and a host of other sources and, in order to survive, we must learn to make rational choices based on some kind of summary and analysis of it. We need to learn to select the relevant and discard the irrelevant, to sift out what is interesting, to have some kind of appreciation of the accuracy and reliability of both our information and our conclusions, and to produce succinct summaries which can be interpreted clearly and quickly.

Our methods for summarizing data will involve the computer in producing graphical displays as well as numerical calculations. You will see how a preliminary pictorial analysis of your data can, and indeed should, influence your entire approach to choosing a valid, reliable method.

But we shall begin with the data themselves. In this course, except where it is necessary to make a particular theoretical point, all of the data sets used are genuine; none are artificial, contrived or 'adjusted' in any way.

Statistics exists as an academic and intellectual discipline precisely because real investigations need to be carried out. Simple questions, and difficult ones, about matters which affect our lives need to be answered; information needs to be processed; and decisions need to be made. The idea of the statistician as a detective has already been introduced. 'Finding things out' is fun: this is the challenge of real data.

1.1 Data and questions

The data sets you will meet in this section are very different from each other in both structure and character. By the time you have reached the end of this chapter, you will have carried out a preliminary investigation of each, identified important questions about them and made a good deal of progress with some of the answers. As you work through the course developing statistical expertise, these data sets will be revisited and different questions addressed.

There are eight data sets here. Do not spend too long with them at this early stage; you should spend just long enough to see how they are presented and think about the questions that arise. However, if you think you have identified something interesting or unusual about any one of them, make a note of your idea for later reference.

Example 1.1 *USA workforce*

The first data set comprises the figures published by the US Labor Department for the composition of its workforce in 1986. It shows the average numbers over the year of male and female workers in the various different employment categories and is typical of the kind of data published by government departments.

Table 1.1 Average composition of the USA workforce during 1986

Type of employment	Male (millions)	Female (millions)
Professional	15.00	11.60
Industrial	12.90	4.45
Craftsmen	12.30	1.25
Sales	6.90	6.45
Service	5.80	9.60
Clerical	3.50	14.30
Agricultural	2.90	0.65

In spite of this being a small and fairly straightforward data set, it is not easy to develop an intuitive 'feel' for the numbers and their relationships with each other when they are displayed as a table. What is the most meaningful and appealing way to show the information? How best can you compare the male and female workforces in each category? Or is the important question perhaps a comparison between the total number of employees in each of the seven categories? ■

Example 1.2 *Opinion polls*

If you found the data displayed in Table 1.1 awkward to interpret, or even if you did not, you should find the layout of the next data set thoroughly confusing. On 19 December 1989, the *Guardian* newspaper published the following table of opinion poll figures.

Table 1.2 A newspaper summary of voting intentions

RECENT OPINION POLLS

Fieldwork	Poll	Sample	Con	Lab	L. Dem	SDP	Grn
10–11.11	ICM/*Guardian*	1416	36	49	6	3	3
8–13.11	NOP	1628	34	46	9	†	6
22–23.11	Harris/*Obsvr*	1037	36	47	9	1	4
23–24.11	MORI/*S. Times*	1068	37	51	4	3	4
24–25.11	ICM/*S. Corr*	1460	38	48	3	4	5
24–26.11	ASL*	801	37	44	4	3	8
29.11–4.12	NOP	1760	38	45	7	5	
1–4.12	Gallup/*D. Tele*	950	37.5	43.5	9	4	
8–9.12	ICM/*Guardian*	1333	37	49	4	3	4
12–14.12	Harris/*Obsvr*	978	39	46	6	1	5
Guardian average of last 5			38	46	6	5	3
General election June 1987			43	32	Alliance 23		

Where two or more polls were sampling at the same time, only one average is given. *Telephone poll. † NOP no longer records a separate figure for the SDP but includes them in Others; average figures are adjusted accordingly.

Presumably the column labelled Fieldwork gives the dates during which the polls in the second column were carrying out sampling. The precise meaning of most of the footnote is anybody's guess and what is the '*Guardian* average of last 5'? How was it calculated?

When displaying tables of numbers there is much to be said for clarity and simplicity. Tables should not be complicated and Table 1.3, which follows, has the information most readers would want.

Table 1.3 A simpler summary of voting intentions

DECEMBER OPINION POLLS

Poll	Number of people questioned	Percentages			
		Con	Lab	L. Dem	Others
NOP	1760	38	45	7	10
Gallup/*D. Tele*	950	37.5	43.5	9	10
ICM/*Guardian*	1333	37	49	4	10
Harris/*Obsvr*	978	39	46	6	9

How reliable are such polls? Do they really give an accurate reflection of party support? What are the likely errors? You can see that there is enormous variation in sample size between the different polls, with NOP's sample almost double that of Gallup/*Daily Telegraph.* Does the way of drawing the sample matter, and how do we know that the sample is typical of the voting population as a whole? ■

Example 1.3 Infants with SIRDS

This data set comprises recorded birth weights of 50 infants who displayed severe idiopathic respiratory distress syndrome (SIRDS). This is a serious condition which can result in death. The data appear below in Table 1.4.

Table 1.4 Birth weights (in kg) of infants with severe idiopathic respiratory distress syndrome

1.050*	2.500*	1.890*	1.760	2.830
1.175*	1.030*	1.940*	1.930	1.410
1.230*	1.100*	2.200*	2.015	1.715
1.310*	1.185*	2.270*	2.090	1.720
1.500*	1.225*	2.440*	2.600	2.040
1.600*	1.262*	2.560*	2.700	2.200
1.720*	1.295*	2.730*	2.950	2.400
1.750*	1.300*	1.130	3.160	2.550
1.770*	1.550*	1.575	3.400	2.570
2.275*	1.820*	1.680	3.640	3.005

* child died

van Vliet, P.K. and Gupta, J.M. (1973) Sodium bicarbonate in idiopathic respiratory distress syndrome. *Arch. Diseases in Childhood*, **48**, 249–255.

At first glance, there seems little that one can deduce from these data. The babies vary in weight between 1.03 kg and 3.64 kg. Notice, however, that some of the children died. Surely the important question concerns early identification of children displaying SIRDS who are at risk of dying. Do the children split into two identifiable groups? Is it possible to relate the chances of eventual survival to birth weight? ■

Example 1.4 Runners

The next data set comes from 22 of the competitors in an annual championship run, the Tyneside Great North Run. Blood samples were taken from eleven runners before and after the run, and also from another eleven runners who collapsed near the end of the race. The measurements are plasma β endorphin concentrations in pmol/litre. Unless you have had medical training you are unlikely to know precisely what constitutes a plasma β endorphin concentration, much less what the units of measurement mean. This is a common experience even among expert statisticians working with data from specialist experiments, and usually gives little cause for concern. What matters is that some physical attribute can be measured, and the measurement value is important to the experimenter. The statistician is prepared to accept that running may have an effect upon the blood, and will ask for clarification of medical questions as and when the need arises. The data are given in Table 1.5.

The letter β is the Greek lower-case letter beta, pronounced 'beeta'.

Table 1.5 Blood plasma β endorphin concentration (pmol/l)

Normal runner before race	Same runner after race	Collapsed runner after race
4.3	29.6	66
4.6	25.1	72
5.2	15.5	79
5.2	29.6	84
6.6	24.1	102
7.2	37.8	110
8.4	20.2	123
9.0	21.9	144
10.4	14.2	162
14.0	34.6	169
17.8	46.2	414

Dale, G., Fleetwood, J.A., Weddell, A., Ellis, R.D. and Sainsbury, J.R.C. (1987) Beta-endorphin: a factor in 'fun run' collapse? *British Medical Journal*, **294**, 1004.

You can see immediately that there is a difference in β endorphin concentration before and after a race, and you do not need to be a statistician to see that collapsed runners have very high β endorphin concentrations compared with those who finished the race. But what is the relationship between initial and final β endorphin concentrations? What is a typical finishing concentration? What is a typical concentration for a collapsed runner? How do the dispersions of data values compare?

The table raises other questions. The eleven normal runners (in the first two columns) have been sorted according to increasing pre-race endorphin levels. This may or may not help make any differences in the post-race levels more immediately evident. Is this kind of initial sorting necessary, or even common, in statistical practice? The data on the collapsed runners have also been sorted. The neat table design relies in part on the fact that there were eleven collapsed runners measured, just as there were eleven finishers, but the two groups are independent of each other. There does not seem to be any particularly obvious reason why the two numbers should not have been different. Is it necessary to the statistical design of this experiment that the numbers should have been the same? ■

Example 1.5 Cirrhosis and alcoholism

These data, quoted for several countries in Europe and elsewhere, show the average alcohol consumption in litres per person per year and the death rate per 100 000 of the population from cirrhosis and alcoholism. It would seem obvious that the two are not unrelated to each other, but what is the relationship and is it a strong one? How can the strength of such a relationship be measured? Is it possible to assess the effect on alcohol-related deaths of taxes on alcohol, or of laws that aim to reduce the national alcohol consumption?

The data are given in Table 1.6.

Osborn, J.F. (1979) *Statistical exercises in medical research.* Blackwell Scientific Publications, Oxford, p. 44.

Table 1.6 Average alcohol consumption and death rate

Country	Alcohol consumption (l/person/year)	Cirrhosis & alcoholism (death rate/100,000)
France	24.7	46.1
Italy	15.2	23.6
W. Germany	12.3	23.7
Austria	10.9	7.0
Belgium	10.8	12.3
USA	9.9	14.2
Canada	8.3	7.4
England & Wales	7.2	3.0
Sweden	6.6	7.2
Japan	5.8	10.6
Netherlands	5.7	3.7
Ireland	5.6	3.4
Norway	4.2	4.3
Finland	3.9	3.6
Israel	3.1	5.4

France has a noticeably higher average annual individual alcohol consumption than the others; the figure is more than double that of third-placed West Germany. The French alcohol-related death rate is just under double that of the next highest. Should the figures for France be regarded as atypical? If so, how should they be handled when the data are analysed? ■

Example 1.6 Body and brain weights for animals

The next data set comprises average body and brain weights for 28 kinds of animal, some of them extinct. The data are given in Table 1.7.

Jerison, H.J. (1973) *Evolution of the brain and intelligence.* Academic Press, New York.

These data raise interesting questions about their collection and the use of the word 'average'. Presumably some estimates may be based on very small samples, while others may be more precise. On what sampling experiment are the figures for *Diplodocus*, *Triceratops* and other extinct animals based? The three-decimal-place 'accuracy' given throughout the table here is extraordinary (and certainly needs justification).

Table 1.7 Average body and brain weights for animals

Species	Body weight (kg)	Brain weight (g)
Mountain Beaver	1.350	8.100
Cow	465.000	423.000
Grey Wolf	36.330	119.500
Goat	27.660	115.000
Guinea Pig	1.040	5.500
Diplodocus	11700.000	50.000
Asian Elephant	2547.000	4603.000
Donkey	187.100	419.000
Horse	521.000	655.000
Potar Monkey	10.000	115.000
Cat	3.300	25.600
Giraffe	529.000	680.000
Gorilla	207.000	406.000
Human	62.000	1320.000
African Elephant	6654.000	5712.000
Triceratops	9400.000	70.000
Rhesus Monkey	6.800	179.000
Kangaroo	35.000	56.000
Hamster	0.120	1.000
Mouse	0.023	0.400
Rabbit	2.500	12.100
Sheep	55.500	175.000
Jaguar	100.000	157.000
Chimpanzee	52.160	440.000
Brachiosaurus	87000.000	154.500
Rat	0.280	1.900
Mole	0.122	3.000
Pig	192.000	180.000

Once again it would seem obvious that the two variables, body weight and brain weight, are linked: but what is the relationship between them and how strong is it? Can the strength of the relationship be measured? Is a larger brain really required to govern a larger body? These data give rise to a common problem in data analysis which experienced practical analysts would notice as soon as they look at such data. Can you identify the difficulty? You will see it immediately in Figure 1.14. ■

Example 1.7 Stock-market averages

Table 1.8 lists the annual highs and lows for the Dow Jones industrial average on the New York stock-market from 1954 to 1985.

Table 1.8 Dow Jones industrial averages

Year	High	Low	Year	High	Low
1954	404	280	1970	842	631
1955	488	388	1971	951	798
1956	521	462	1972	1036	889
1957	521	420	1973	1052	788
1958	584	437	1974	892	578
1959	679	474	1975	882	632
1960	685	566	1976	1015	859
1961	735	610	1977	1000	801
1962	726	536	1978	908	742
1963	767	647	1979	898	797
1964	892	766	1980	1000	759
1965	969	841	1981	1024	824
1966	995	744	1982	1071	777
1967	943	786	1983	1287	1027
1968	985	825	1984	1287	1087
1969	969	770	1985	1360	1185

The World Almanac and Book of Facts 1985. Pharos Books, New York.

These data are known as *time series* because they give values measured at a series of times (i.e. at consecutive times). Look down the columns for annual lows in chronological order. Starting with 1954, you see a steady increase followed by a decrease which does not go as low as the 1954 value before it starts to increase again. After a little fluctuation, it increases to the 1968 value, drops again, hits a high in 1972, 1976, and so on. Notice that the values show an overall increase over the 32-year period. The aim, of course, is to use these data to predict highs and lows of the Dow Jones average for 1986 and subsequent years. ■

Example 1.8 Surgical removal of tattoos

The final data set in this section is different from the others in that the data are not numeric. So far you have only seen numeric data in the form of measurements or counts. However, there is no reason why data should not be verbal or textual. Table 1.9 comprises clinical data from 55 patients who have had forearm tattoos removed by two different surgical methods. Their tattoos were of large, medium or small size, either deep or at moderate depth. The final result is scored from 1 to 4, where 1 represents a poor removal and 4 represents an excellent result. In Table 1.9 the two methods of removal are denoted A and B. The sex of the patient is also shown.

Lunn, A.D. and McNeil, D.R. (1988) *The SPIDA manual.* Statistical Computing Laboratory, Sydney.

Table 1.9 Surgical removal of tattoos

Method	Sex	Size	Depth	Score	Method	Sex	Size	Depth	Score
A	M	large	deep	1	B	M	medium	moderate	2
A	M	large	moderate	1	B	M	large	moderate	1
B	F	small	deep	1	A	M	medium	deep	2
B	M	small	moderate	4	B	M	large	deep	3
B	F	large	deep	3	A	F	large	moderate	1
B	M	medium	moderate	4	B	F	medium	deep	2
B	M	medium	deep	4	A	F	medium	deep	1
A	M	large	deep	1	A	M	medium	moderate	3
A	M	large	moderate	4	B	M	large	moderate	3
A	M	small	moderate	4	A	M	medium	deep	1
A	M	large	deep	1	A	F	small	deep	2
A	M	large	moderate	4	A	M	large	moderate	2
A	F	small	moderate	3	B	M	large	deep	2
B	M	large	deep	3	B	M	medium	moderate	4
B	M	large	deep	2	B	M	medium	deep	1
B	F	medium	moderate	2	B	F	medium	moderate	3
B	M	large	deep	1	B	M	large	moderate	2
B	F	medium	deep	1	B	M	large	moderate	2
B	F	small	moderate	3	B	M	large	moderate	4
A	F	small	moderate	4	B	M	small	deep	4
B	M	large	deep	2	B	M	large	moderate	3
A	M	medium	moderate	4	B	M	large	deep	2
B	M	large	deep	4	B	M	large	deep	3
B	M	large	moderate	4	A	M	large	moderate	4
A	M	large	deep	4	A	M	large	deep	2
B	M	medium	moderate	3	B	M	medium	deep	1
A	M	large	deep	1	A	M	small	deep	2
B	M	large	moderate	4					

What are the relative merits of the two methods of tattoo removal? Is one method simply better, or does it depend upon the size or depth of the tattoo? ■

As the course develops, there will be suggestions from time to time that you use your computer to produce graphs and perform calculations on these and other data sets.

In the next section, some different diagrammatic representations of data are described.

1.2 Graphical displays

The data set in Example 1.8 in Section 1.1 comprised non-numeric or categorical data. Such data often appear in newspaper reports and are usually represented as one or other of two types of graphical display, one type being called a **pie chart** and the other a **bar chart**; these are arguably the graphical displays most familiar to the general public, and are certainly ones that you will have seen before.

1.2.1 Pie charts

Suppose we count the numbers of large, medium and small tattoos from the data in Table 1.9: there were 30 large tattoos, 16 of medium size and 9 small tattoos. A pie chart display of these data is shown in Figure 1.1.

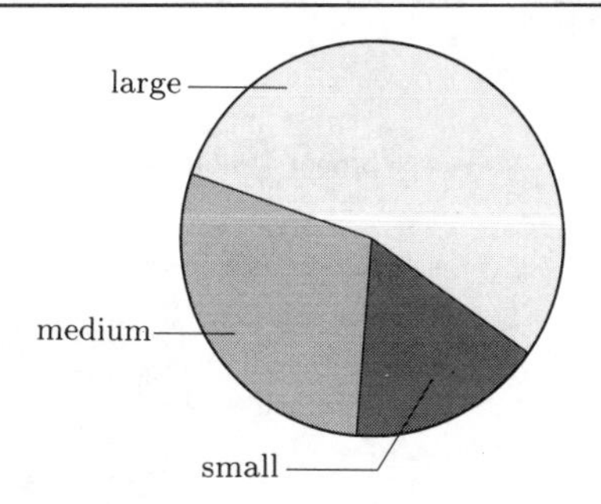

Figure 1.1 Tattoo sizes

This is an easy display to construct because the size of each 'slice' is proportional to the angle it subtends at the centre, which in turn is proportional to the count in each category. So, to construct Figure 1.1, you simply draw a circle and draw in radii making angles of

$$360^\circ \times \frac{30}{30+16+9} = 196^\circ,$$

$$360^\circ \times \frac{16}{30+16+9} = 105^\circ,$$

$$360^\circ \times \frac{9}{30+16+9} = 59^\circ,$$

to represent the counts of large, medium and small tattoos respectively. Then shade the three sectors in order to distinguish them from each other.

At first sight the pie chart seems to fulfil the basic requirements of a good statistical display in that it is informative, easy to construct, visually appealing and readily assimilated by a non-expert.

Pie charts can be useful when all you want the reader to notice is that there were more large than medium size tattoos, and more medium than small tattoos. In conveying a good impression of the relative magnitudes of the differences, pie charts have some limitations. They are also only useful for displaying a limited number of categories. Figure 1.2 shows a pie chart of the number of nuclear power stations in countries where nuclear power is used.

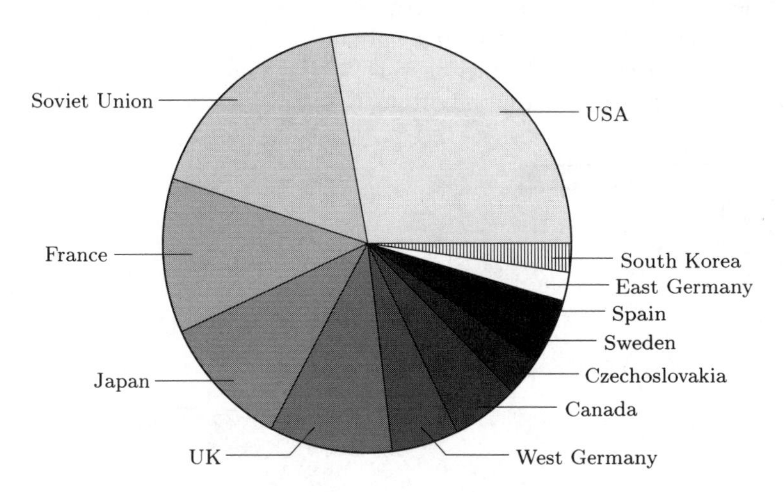

Figure 1.2 Nuclear power stations (a pie chart)

In this diagram the names of the countries listed are those that pertained at the time the data were collected. Since then, geographical borders and political circumstances have altered.

It is not so easy to extract meaningful information from this more detailed diagram. You can pick out the main users of nuclear power, and that is about all. When trying to construct pie charts with too many categories, a common ploy of the graphic designer is to produce a chart which displays the main contributors and lumps together the smaller ones.

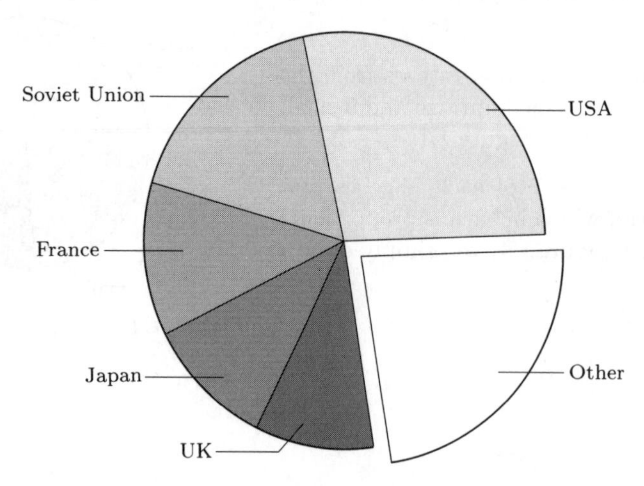

Figure 1.3 Nuclear power stations, smaller groups consolidated

Figure 1.3 shows how this can be done, but it is still unsatisfactory. Information has been lost: for instance, the reader cannot now compare France's nuclear power provision with that of West Germany.

Pie charts are of limited use in that they are able to give an immediate visual impression of proportion for a small number of categories, say, four or five categories at most. However, the pie chart is the most common method used by the news media: you can find an example of a pie chart in the business section of a daily newspaper on most days of the week.

1.2.2 Bar charts

A better way of displaying the data on nuclear power stations is by constructing a rectangular bar for each country, the length of which is proportional to the count. Bars are drawn separated from each other and, since in this context order does not matter, in order of decreasing size from top to bottom.

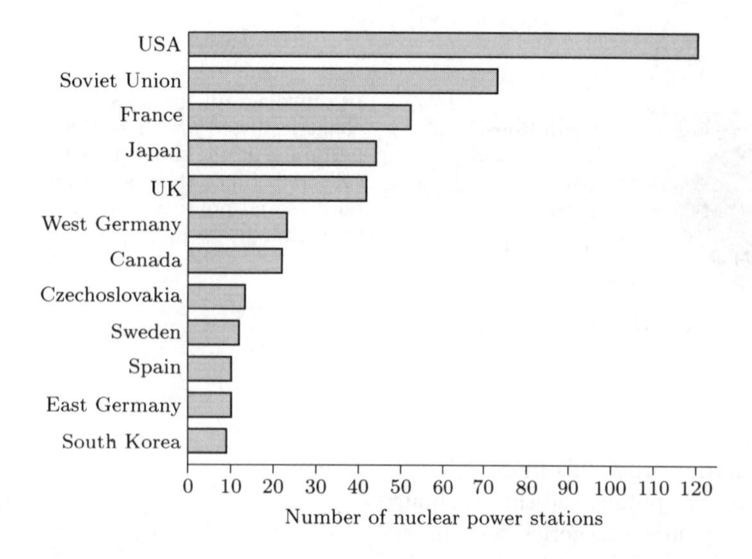

Figure 1.4 Nuclear power stations (a bar chart)

The display in Figure 1.4 is called a bar chart. The bars may be drawn vertically or horizontally according to preference and convenience. Those in Figure 1.4 have been drawn horizontally because of the lengths of the names of some of the countries. Had the bars been drawn vertically, the names of the countries would not have fitted along the horizontal axis unless the bars were drawn far apart or the names were printed vertically. The former would make comparison difficult, while the latter would make the names difficult to read. However, it is conventional to draw the bars vertically whenever possible and Figure 1.5 shows a bar chart of the effectiveness of tattoo removal, using the data in Table 1.9.

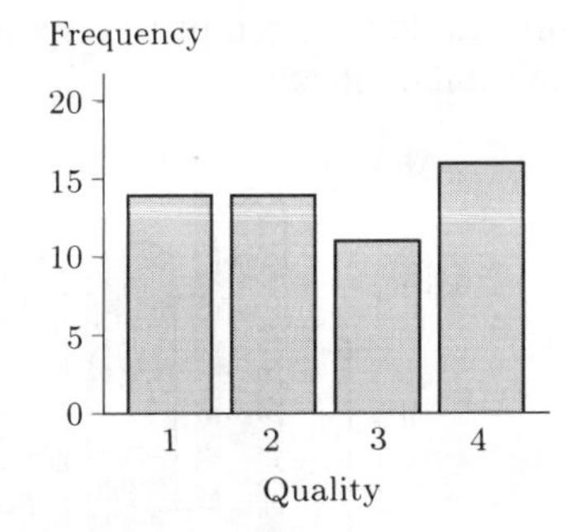

Figure 1.5 Quality assessment, surgical removal of 55 tattoos

Sometimes order is important. The quality of tattoo removal was given a score from 1 to 4, and this ordering has been preserved along the quality (horizontal) axis. The vertical axis shows the reported frequency for each assessment. Figure 1.6 shows the same bar chart in three dimensions. You can see that it is quite difficult to discern the corresponding frequency value for each bar.

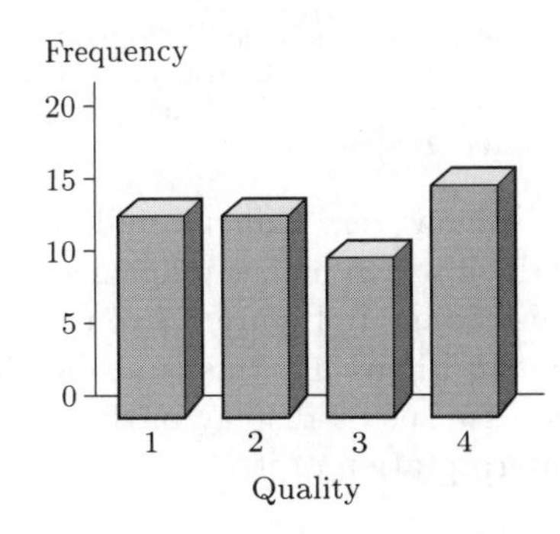

Figure 1.6 A three-dimensional bar chart

This kind of three-dimensional bar chart is commonly used as a television graphic for showing data such as the results from an opinion poll on the popularity of the main political parties. Viewers do not necessarily realize that they are supposed to use the back edge of the bar to determine its height. If you want to be able to interpret this kind of graphic properly, you need to be aware of how misleading it can be.

The danger of using three-dimensional effects is really brought home when two data sets are displayed on the same bar chart. Look at the data from Table 1.1, for example. Without perspective effects, we have a clear, informative display that lets us compare the patterns of male and female employment in the USA. We simply plot the bars side by side and distinguish the sexes by shading. The bar chart is shown in Figure 1.7.

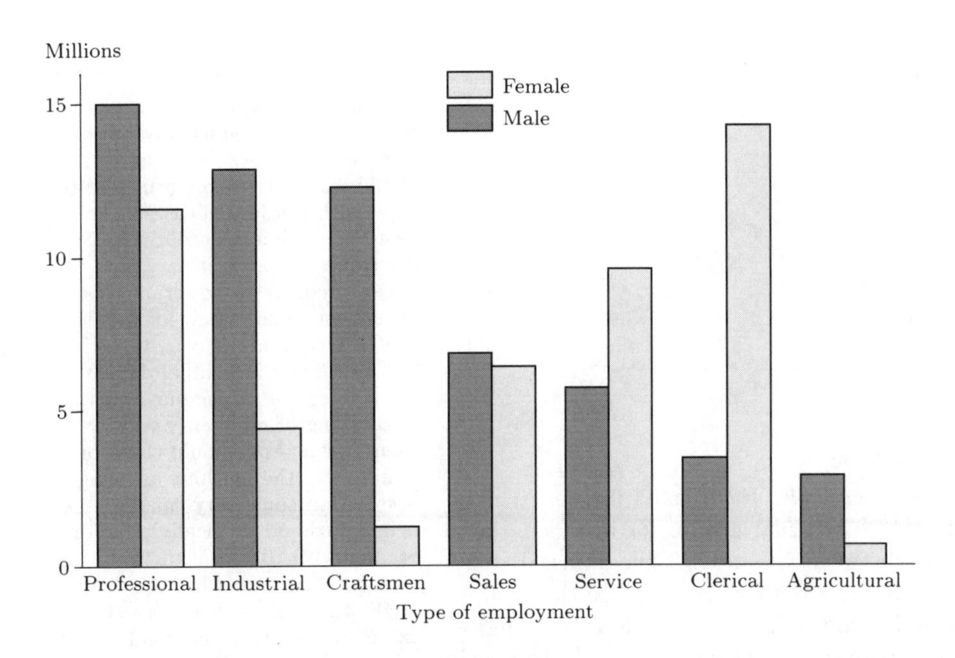

Figure 1.7 USA workforce data, 1986 averages

Figure 1.8 is an attempt to display the same information by means of a three-dimensional effect.

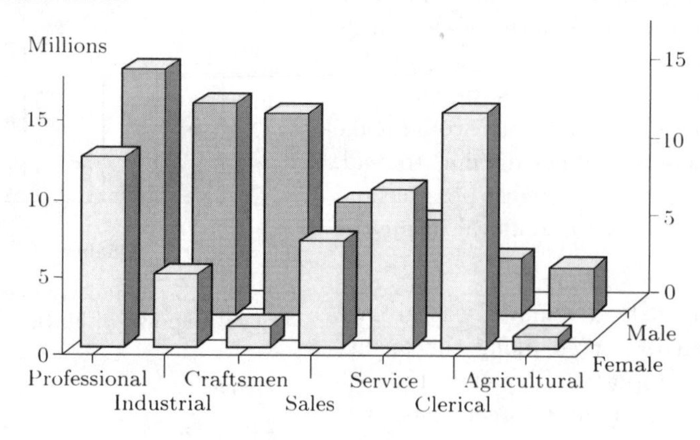

Figure 1.8

It is now very difficult indeed to identify values. Some blocks are partly hidden which makes judgement difficult. The display for *Sales* is particularly misleading; in Figure 1.7 you can see that the bars are almost the same height, but in Figure 1.8 this is less obvious. Try to remember this effect next time you see such a display in the press and try to be a little more critical in your interpretation of it.

1.2.3 Histograms

In Section 1.1 we looked briefly at the set of birth weights (given in Table 1.4) of infants with severe idiopathic respiratory distress syndrome. The list of weights is in itself not very informative, partly because there are so many weights listed. Suppose, however, that the weights are grouped as shown in Table 1.10.

Table 1.10 Birth weights (kg)

Group	Birth weight (kg)	Frequency
1	1.0–1.2	6
2	1.2–1.4	6
3	1.4–1.6	4
4	1.6–1.8	8
5	1.8–2.0	4
6	2.0–2.2	3
7	2.2–2.4	4
8	2.4–2.6	6
9	2.6–2.8	3
10	2.8–3.0	2
11	3.0–3.2	2
12	3.2–3.4	0
13	3.4–3.6	1
14	3.6–3.8	1

Such a table is called a *grouped frequency table*. Each listed frequency gives the number of individuals falling into a particular group: for instance, there were six children with birth weights between 1.0 and 1.2 kilograms. It may occur to you that there is an ambiguity over borderlines between the groups. Into which group, for example, should a value of 2.2 go? Should it be included in Group 6 or Group 7? Provided you are consistent with your rule over such borderlines, it really does not matter.

For large sets of data, a computer would normally be used to obtain a grouped frequency table, and it would normally be programmed to cope with borderline cases. The user would simply accept its decisions. Even so, it may not always manage to be scrupulously consistent. The reason for this is that the computer's way of storing and retrieving individual data items does not mimic our own way of thinking of numbers and writing them down. You should think of numbers in the machine as being *at best* within some very small neighbourhood (typically a factor of about 1 ± 10^{-15}) of their intended value, or what the user might think of as their correct value. However, in the great scheme of data analysis, the problem is negligible.

In fact, among the 50 infants there were two with a recorded birth weight of 2.2 kg and both have been allocated to Group 7. The infant weighing 2.4 kg has been allocated to Group 8. The rule followed here was that borderline cases were allocated to the higher of the two possible groups.

With the data structured like this, certain characteristics can be seen even though some information has been lost. There seems to be an indication that there are two groupings divided somewhere around 2 kg or, perhaps, three groupings divided somewhere around 1.5 kg and 2 kg. But the pattern is far from clear and needs a helpful picture, such as a bar chart. The categories are ordered, and notice also that the groups are contiguous (1.0–1.2, 1.2–1.4, and so on). This reflects the fact that here the variable of interest (birth weight) is not a count but a measurement. In this kind of situation, the bars of the bar chart are drawn without gaps between them. This is shown in Figure 1.9. This kind of bar chart, of *continuous data* which has been put into a limited number of distinct groups or classes, is called a *histogram*. In this example, the 50 data items were allocated to groups of width 0.2 kg: there were 14 groups. The classification was quite arbitrary. If the group classifications were narrower, there would have been more groups each containing fewer observations; if the classifications had been wider, there would have been fewer groups with more observations in each group. The question of an optimal classification is an interesting one, and surprisingly complex.

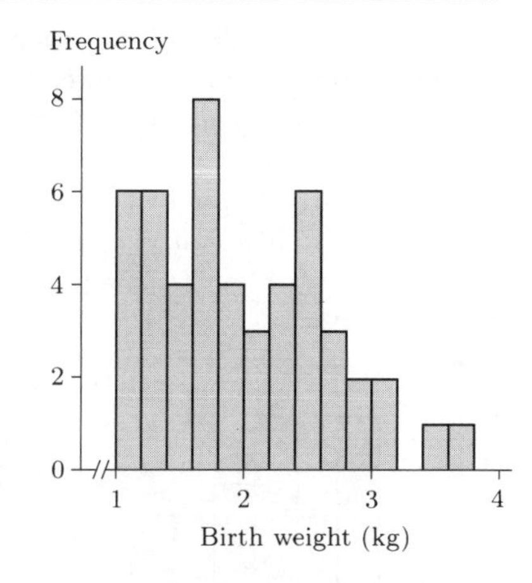

Figure 1.9 Birth weights of infants with SIRDS (kg)

The distinction between 'counting' and 'measuring' is quite an important one. In later chapters we will be concerned with formulating different models to express the sort of variation that typically occurs in different sampling contexts, and it matters that the model should be appropriate to the type of data. Data arising from measurements (height, weight, temperature and so on) are called *continuous* data. Those arising from counts (family size, hospital admissions, monthly launches of a lifeboat) are called *discrete*.

How many groups should you choose for a histogram? Too few and you will not have a picture of the shape, too many and the display will be too fragmented to show an overall shape. When these data were introduced in Example 1.3, the questions posed were: do the children split into two identifiable groups? Is it possible to relate the chances of eventual survival to birth weight? We are not, as yet, in a position to answer these questions, but we can see that the birth weights might split into two or three 'clumps'. On the other hand, can we be sure that this is no more than a consequence of the way in which the borderlines for the groups were chosen? Suppose, for example, we had decided to make the intervals of width 0.3 kg instead of 0.2 kg. We would have had fewer groups, with Group 1 containing birth weights from 1.0 to 1.3 kg, Group 2 containing birth weights from 1.3 to 1.6 kg, and so on, producing the histogram in Figure 1.10.

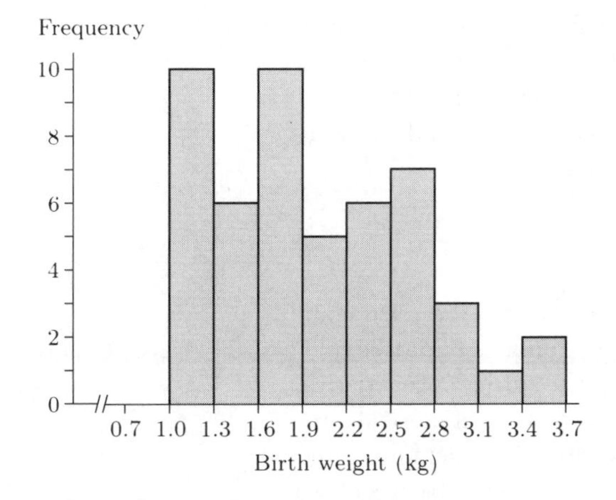

Figure 1.10 Birth weights, 0.3 kg group widths

This looks quite different, but then this is not surprising as the whole display has been compressed into fewer bars. The basic shape remains similar and you might be tempted to conclude that the choice of grouping does not really matter. But suppose we retain groupings of width 0.3 kg and choose a different starting point. Suppose we make Group 1 go from 0.8 to 1.1 kg, Group 2 from 1.1 to 1.4 kg, and so on. The resulting histogram is shown in Figure 1.11(a).

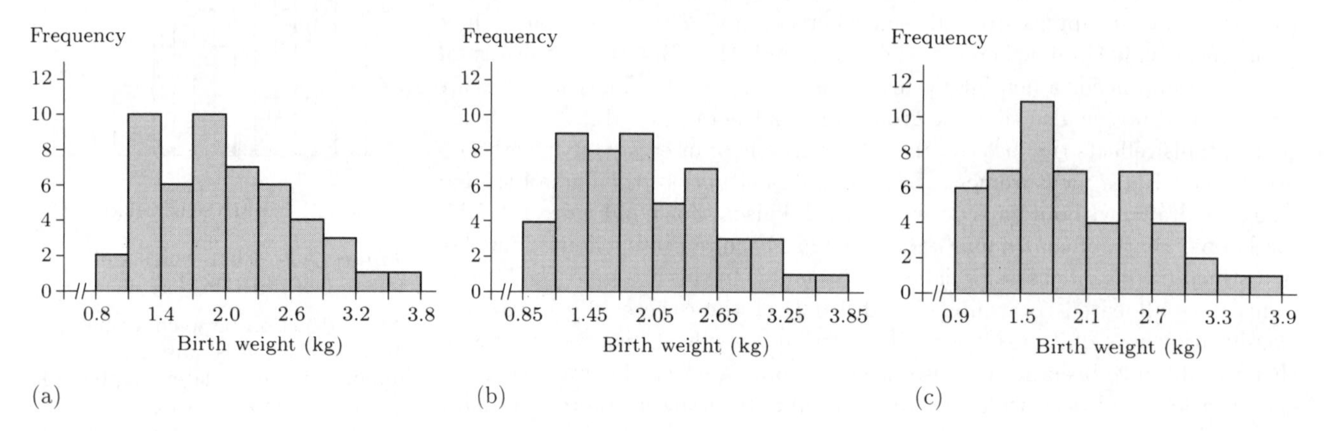

Figure 1.11

Figure 1.11 shows three different histograms. Figures 1.11(b) and 1.11(c) show histograms in which the first group was started at 0.85 kg and 0.9 kg respectively. However, only Figure 1.11(c) gives a good indication that the data are split into two 'clumps'. What you have seen is a series of visual displays of a data set which warn you against trying to reach firm conclusions from histograms. It is important to realize that they often produce only a vague impression of the data—nothing more. One of the problems here is that we have only 50 data values, which is not really enough to make a clear pattern evident. However, the histograms all convey one very important message indeed: the data do not appear in a single, concentrated clump. It is a fundamental principle in modern practical data analysis that all investigations should begin, wherever possible, with one or more suitable diagrams of the data. Such displays should certainly show overall patterns or trends, and should also be capable of isolating unexpected features which might otherwise be missed. Clearly it is a good idea to look at the way frequencies of data such as the birth weights are distributed and, given that any statistical computer package will quickly produce a histogram for you, comparatively little effort is required. This makes the histogram a valuable analytic tool and, in spite of some disadvantages, you will find that you use it a great deal.

1.2.4 Scatter plots

In recent years, graphical displays have come into prominence because computers have made them quick and easy to produce. Techniques of data exploration have been developed which have revolutionized the subject of statistics, and today no serious data analyst would carry out a formal numerical procedure without first inspecting the data by eye. Nowhere is this demonstrated more forcibly than in the way a *scatter plot* reveals a relationship between two variables. Look at the scatter plot in Figure 1.12 of the data on cirrhosis and alcoholism from Table 1.6.

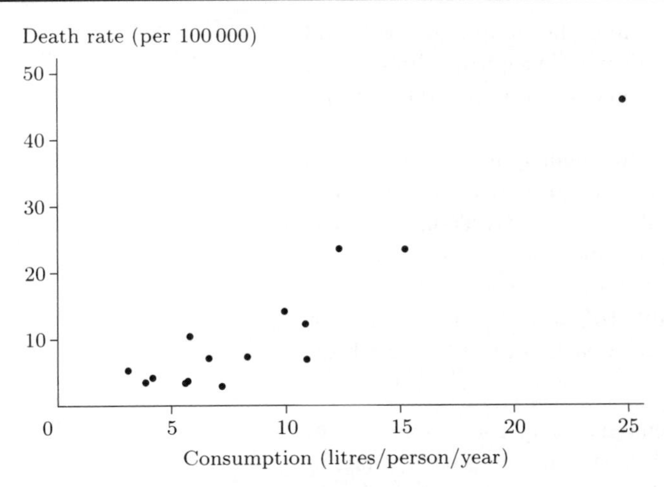

Figure 1.12 Alcohol-related deaths and consumption

In a scatter plot, one variable is plotted on the horizontal axis and the other on the vertical axis. Each data item corresponds to a point in two-dimensional space. For example, the average annual individual consumption of alcohol in France for the time over which the data were collected was 24.7 litres per person per year, and the death rate per hundred thousand of the population through cirrhosis and alcoholism was 46.1. In this diagram consumption is plotted along the horizontal axis and death rate is plotted up the vertical axis. The data point at the co-ordinate (24.7,46.1) corresponds to France.

Is there a strong relationship between the two variables? In other words, do the points appear to fit fairly 'tightly' about a straight line or a curve? It is fairly obvious that there is, although the picture is not made any easier to see since most of the points are concentrated in the bottom left-hand corner. There is one point that is a long way from the others and the size of the diagram relative to the page is dictated by the available space into which it must fit. We remarked upon this point, corresponding to France, when we first looked at the data, but seeing it here really does put into perspective the magnitude of the difference between France and the other countries. The best way to look for a general relationship between death rate and consumption of alcohol is to spread out the points representing the more conventional drinking habits of other countries by leaving France, an extreme case, out of the plot.

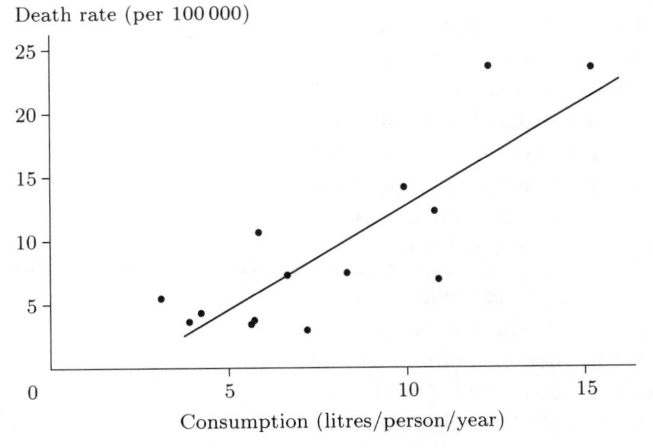

Figure 1.13 Alcohol-related deaths and consumption, excluding France

The picture in Figure 1.13 is now much clearer and shows up a general (and hardly surprising) rule that the incidence of death through alcohol-related disease is strongly linked to average alcohol consumption, the relationship being plausibly linear. A 'linear' relationship means that you could draw a straight line through the points which would fit them quite well, and this has been done in Figure 1.13. Of course, one would not expect the points to sit precisely on the line but to be scattered about it tightly enough for the relationship to show. In this case you could conclude that, given the average alcohol consumption in any country not included among those on the scatter plot, you would be fairly confident of being able to use your straight line for providing a reasonable estimate of the national death rate due to cirrhosis and alcoholism.

It is worth mentioning at this stage that demonstrating that some sort of association exists is not the same thing as demonstrating causation, that, in this case, alcohol use 'causes' (or makes more likely) cirrhosis or an early death. For example, if cirrhosis were stress-related, so might be alcohol consumption, and hence the apparent relationship. It also should be noted that these data were averaged over large populations and (whatever may be inferred from them) they say nothing about the consequences for an individual of alcohol use. The idea of association between variables, and ways of interpreting that association, are dealt with in *Chapter 11*.

In leaving out France, that data point was treated as an extreme case. It corresponded to data values so atypical, and so far removed from the others, that we were wary of using them to draw general conclusions.

'Extreme', 'unrepresentative', 'atypical' or possibly 'rogue' observations in sets of data are all sometimes called *outliers*. It is important to recognize that, while one would wish to eliminate from a statistical analysis data points which were erroneous (wrongly recorded, perhaps, or observed when background circumstances had profoundly altered), data points which appear 'surprising' are not necessarily 'wrong'. The identification of outliers, and what to do with them, is a research question of great interest to the statistician. Once a possible outlier has been identified, it should be closely inspected and its apparently aberrant behaviour accounted for. If it is to be excluded from the analysis there must be sound reasons for its exclusion. Only then can the data analyst be happy about discarding it. An example will illustrate the point.

In our discussion of the data on body and brain weights for animals in Example 1.6, we conjectured a strong relationship between these weights on the grounds that a large body might well need a large brain to run it properly. At that stage a 'difficulty' with the data was also anticipated. It would be useful to look at a scatter plot, but you will see the difficulty if you try to produce one. Did you spot the problem when it was first mentioned? There are many very small weights such as those for the hamster and the mouse which simply will not show up properly, if displayed on the same plot as, say, those for the elephants! Figure 1.14 shows the difficulty very clearly.

Now, this often happens and the usual way of getting round the problem is to **transform** the data in such a way as to spread out the points with very small values of either variable, and to pull closer together the points with very large values for either variable. The objective is to reduce the spread in the large values relative to the spread in the small values. (In later chapters, some

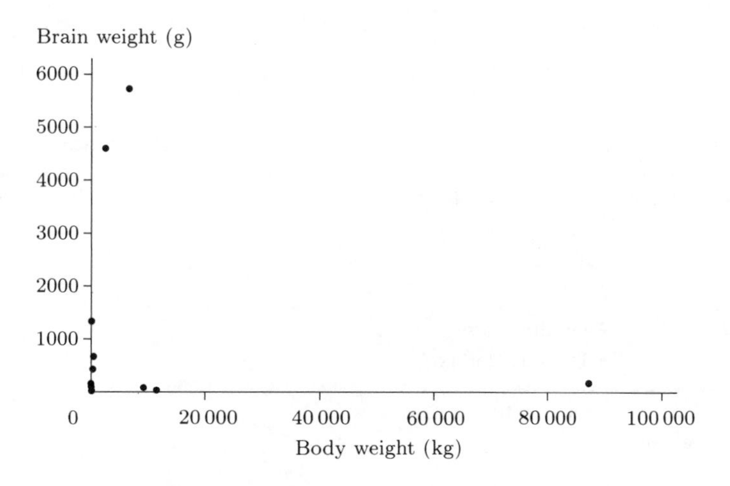

Figure 1.14 Body weight and brain weight

attention is given to how we tell the need for a transformation and the way in which we decide which transformation to use.) In this case it can be done by plotting the logarithm of brain weight against the logarithm of body weight. The log transformation compresses the large values but stretches the small ones. Notice that simply treating the large values as outliers and removing them would not solve the problem because the tight clumping of points close to the origin would still remain. The scatter plot which results from a log transformation is shown in Figure 1.15.

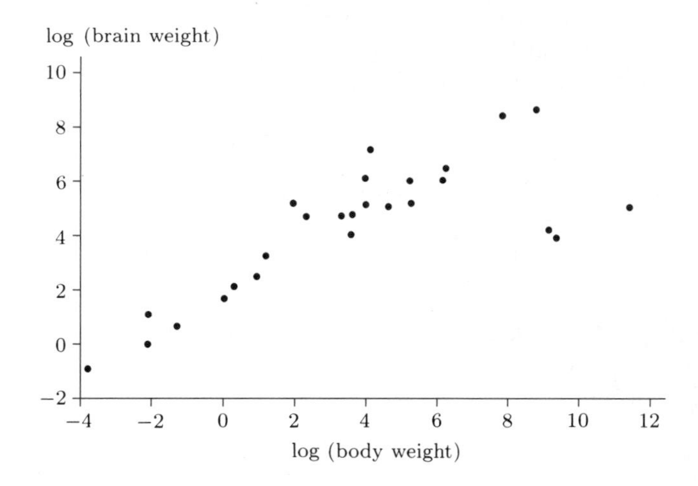

Figure 1.15 Body and brain weights after a log transformation

The plot immediately reveals three apparent outliers to the right of the main band of points, and when you discover the animals to which they correspond you will not be surprised. They are easily identified by the stratagem of labelling the animals with the first letters of the names of their species and plotting the letters in place of the points. This scatter plot is shown in Figure 1.16.

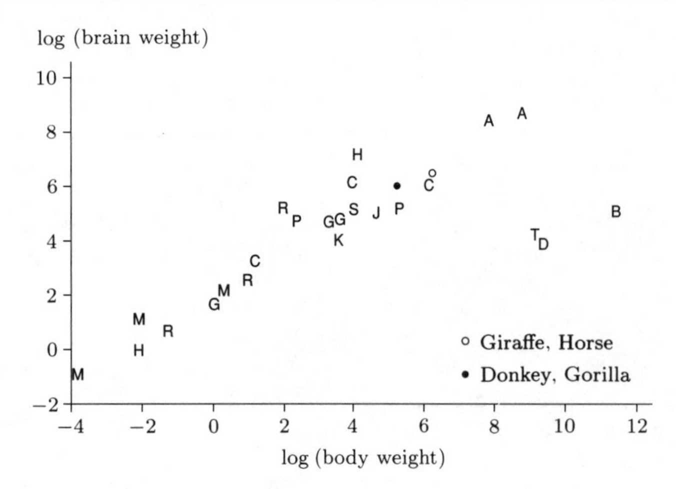

Figure 1.16 Scatter plot, labelled points

The outlying points are B, D and T which correspond to *Brachiosaurus*, *Diplodocus* and *Triceratops*. Excluding these three species, there is a convincing linear relationship, although the human, the mole and the rhesus monkey all appear to have exceptionally high brain weight to body weight ratios.

The scatter plots you have just seen have revealed simple straight line relationships. Much more complicated patterns can emerge. Let us look at a scatter plot of the data on stock-market averages in Table 1.8 by plotting the annual lows against their year. With time series, one might expect a value at a particular time point to depend strongly upon the value at the time point immediately preceding it; that is, one might expect the lowest value of the Dow Jones average for 1984 to be linked to the lowest value of the Dow Jones average for 1983. It is conventional to represent this linking on a graph of a time series by joining adjacent time points with straight lines, and this convention is adopted in Figure 1.17.

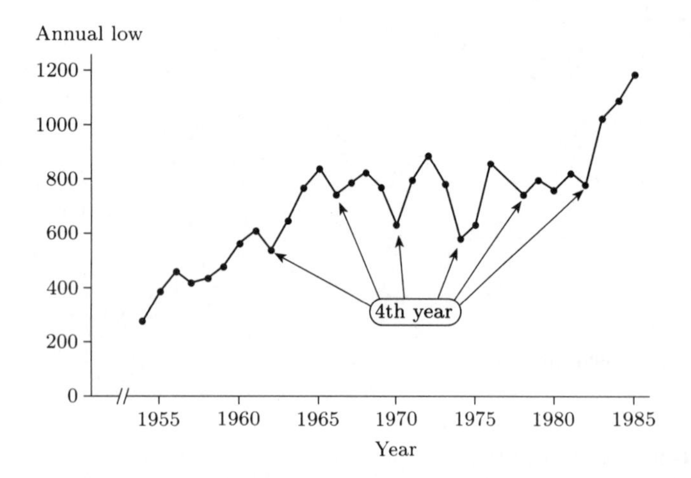

Figure 1.17 Dow Jones lows, 1954 to 1985

The overall increasing trend was remarked earlier, when looking at the values themselves, but Figure 1.17 brings out an extra, interesting feature. If you look carefully, you can see that superimposed upon the steady increase is a kind of cyclic behaviour. Clearly there is a strong relationship between successive points because each point shows an increase over the previous year, except for a sharp drop which seems to occur every four years. Occasionally there is a small drop after three years, but, apart from some muddled behaviour in the late 1970s, a large drop inevitably follows. Why? What can this pattern mean? We can only speculate that the four-year cycle may have something to do with the frequency of presidential elections in the USA. The unusual performance in the late 1970s may well be a reflection of the market's lack of confidence in the Carter administration.

Once such a pattern has been observed, the very brave among us may use it for predicting future stock-market behaviour. Many methods, both crude and sophisticated, exist for doing this, and there is a ready market for computer programs which attempt to predict stock-market behaviour. With one popular method using these data, the predicted low for 1986 is close to 1100, and for 1992 it is close to 1278. While it is not unreasonable to use these data to predict the 1986 value, you should be very wary of extrapolating for predictions as far in advance of the data as 1992.

In fact, the 1986 low was 1185 and 1992 low was 3172!

There is a further useful graphical display which will be described in Section 1.4. It is called a *boxplot* and, along with the scatter plot, is amongst the most useful and informative of visual summaries of statistical data. However, you need first to have read Section 1.3, which deals with numerical summaries of data sets.

The following exercises are designed to help you become acquainted with the graphical (and other) facilities offered by your computer.

Exercise 1.1

(a) Obtain frequency tables for the SIRDS birth weight data with the following group classifications.

(i) 1.0–1.2, 1.2–1.4, ...

(ii) 1.0–1.3, 1.3–1.6, ...

(iii) 0.8–1.1, 1.1–1.4, ...

(iv) 0.85–1.15, 1.15–1.45, ...

(v) 0.9–1.2, 1.2–1.5, ...

(b) Plot histograms for the SIRDS birth weight data with the same group classifications as in part (a).

Exercise 1.2

(a) Plot the variable death rate against alcohol consumption for all fifteen countries, including France, in the data set recording alcohol-related deaths and average individual alcohol consumption.

(b) Plot these data points again, excluding France from the data set.

Exercise 1.3

(a) For the data set recording body and brain weights for different species of animal, sort the species according to decreasing brain weight to body weight ratio.

(b) Obtain a scatter plot of brain weight against body weight (i) before and (ii) after a log transformation.

1.3 Numerical summaries

Histograms provide a quick way of looking at data sets, but they lose sight of individual observations. However, we may often want to summarize the data in numerical terms; for example, we could use a number to summarize the general level of the values and, perhaps, another number to indicate how spread out or dispersed they are.

1.3.1 The median, the mean and the mode

Everyone professes to understand what is meant by the term 'average', in that it should be representative of a group of objects. The objects may well be numbers from, say, a batch or *sample* of measurements, in which case the average should be a number which in some way characterizes the batch as a whole. For example, the statement 'a typical adult female in Britain is 160 cm tall' would be understood by most people who heard it. Now, not all adult females in Britain are the same height: there is considerable variation. To state that a 'typical' height is 160 cm is to ignore the variation and summarize the distribution of heights with a single number. Even so, it may be all that is needed to answer certain questions. (For example, is a typical adult female smaller than a typical adult male?)

But how should this representative value be chosen? Should it be a typical member of the group or should it be some representative measure which can be calculated from the collection of individual data values? There are no straightforward answers to these questions. In fact, two different ways of expressing a representative value are commonly used in statistics, namely the *median* and the *mean*. The choice of which of these provides the better representative numerical summary is fairly arbitrary and is based entirely upon the nature of the data themselves, or the particular preference of the data analyst, or the use to which the summary statement is to be put. The median describes the central value of a set of data.

A third representative measure, the *mode*, is sometimes used. It is a less common summary of a set of data, because it is not uniquely defined. However, as we shall see, it is a useful term when describing some general characteristics of a set of data.

The sample median

The **median** of a sample of data with an odd number of data values is defined to be the middle value of the data set when the values are placed in order. If the sample size is even, then the median is defined to be half-way between the two middle values.

Example 1.9 Beta endorphin concentration (collapsed runners)

The final column of Table 1.5 listed the β endorphin concentration levels for 11 runners who collapsed towards the end of the Great North Run. The observations were already sorted. They were

66 72 79 84 102 110 123 144 162 169 414.

Eleven is an odd number, so the middle value of the data set is the sixth value (five either side). So, in this case, the sample median is 110 pmol/l. ■

Example 1.10 Birth weights, infants with SIRDS

The data in Table 1.4 are the birth weights (in kg) of 50 infants suffering from severe idiopathic respiratory distress syndrome. There are two groups of infants: those who survived the condition (there were 23 of these) and those who, unfortunately, did not. The data have not been sorted, and it is not an entirely trivial exercise to do this by hand (though it is a task that a computer can handle very easily). The sample size is even: the sample median is defined to be the midpoint of the 25th and 26th observations. That is to say, it is obtained by splitting the difference between 1.82 (the 25th value) and 1.89 (the 26th value). This is

$$\tfrac{1}{2}(1.82 + 1.89) = 1.855\,\text{kg}.$$ ■

Incidentally, the question of whether or not there are identifiable differences between the two groups of infants has already been raised. We explored this, not with any great success, using histograms. The next exercise continues this exploration.

Exercise 1.4

Find the median birth weight for the infants who survived, and for those who did not. (Sort the data by hand, rather than using your computer. For data sets much larger than about 30, you will appreciate that hand-sorting becomes impractical.)

Exercise 1.5

The first two columns of Table 1.5 give the blood plasma β endorphin concentrations of 11 runners before and after the race (successfully completed). There is a marked difference between these concentrations. The data are reproduced in Table 1.11 below with the 'After – Before' difference shown.

Table 1.11 Differences in pre- and post-race β endorphin concentration levels

Before	4.3	4.6	5.2	5.2	6.6	7.2	8.4	9.0	10.4	14.0	17.8
After	29.6	25.1	15.5	29.6	24.1	37.8	20.2	21.9	14.2	34.6	46.2
Difference	25.3	20.5	10.3	24.4	17.5	30.6	11.8	12.9	3.8	20.6	28.4

Find the median of the 'After – Before' differences given in Table 1.11.

Exercise 1.6

The annual snowfall (in inches) in Buffalo, New York, USA was recorded for the 63 years from 1910 to 1972. These data are listed in Table 1.12.

Parzen, E. (1979) Nonparametric statistical data modelling. *J. American Statistical Association*, **74**, 105–31.

Table 1.12 Annual snowfall in Buffalo, NY, 1910–1972 (inches)

126.4	82.4	78.1	51.1	90.9	76.2	104.5	87.4	110.5	25.0	69.3	53.5
39.8	63.6	46.7	72.9	79.7	83.6	80.7	60.3	79.0	74.4	49.6	54.7
71.8	49.1	103.9	51.6	82.4	83.6	77.8	79.3	89.6	85.5	58.0	120.7
110.5	65.4	39.9	40.1	88.7	71.4	83.0	55.9	89.9	84.8	105.2	113.7
124.7	114.5	115.6	102.4	101.4	89.8	71.5	70.9	98.3	55.5	66.1	78.4
120.5	97.0	110.0									

Use your computer to find the median annual snowfall over this period.

The second representative measure defined in this course for a collection of data is the *sample mean.* This is simply what most individuals would understand by the word 'average': all the items in the data list are added together, giving the *sample total.* This number is divided by the number of items (the sample size).

The sample mean

The **mean** of a sample is the arithmetic average of the data list, obtained by adding together all of the data values and dividing this total by the number of items in the sample.

Denoting the n items in a data set $x_1, x_2, \ldots, x_n$, then the sample size is n, and the sample mean is given by

$$\overline{x} = \frac{x_1 + x_2 + \cdots + x_n}{n} = \frac{1}{n}\sum_{i=1}^{n} x_i.$$

The symbol $\overline{x}$ denoting the sample mean is read 'x-bar'.

Example 1.9 continued

From the figures in Table 1.5, the mean β endorphin concentration of collapsed runners is

$$\overline{x} = \frac{66 + 72 + 79 + 84 + 102 + 110 + 123 + 144 + 162 + 169 + 414}{11}$$
$$= 138.64,$$

where the units of measurement are pmol/l. ■

Exercise 1.7

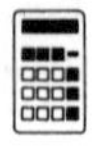

Use your calculator to find the mean birth weight of infants who survived SIRDS, and of those who died. What was the mean birth weight for the complete sample of 50 infants?

Exercise 1.8

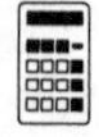

Find the mean of the 'After – Before' differences given in Table 1.11.

Exercise 1.9

Use your computer to find the mean annual snowfall in Buffalo, New York, during the years 1910 to 1972 (see Table 1.12).

Two plausible measures have been defined for describing a typical or representative value for a sample of data. Which measure should be chosen in a statement of that typical value? In the examples we have looked at in this section, there has been little to choose between the two. Are there principles that should be followed? It all depends on the data that we are trying to summarize, and our aim in summarizing them.

To a large extent deciding between using the sample mean and the sample median depends on how the data are distributed. If their distribution appears to be regular and concentrated in the middle of their range, the mean is usually used. It is the easier to compute because no sorting is involved, and as you will see later, it is the easier to use for drawing inferences about the population from which the sample has been taken. (Notice the use of the word *range* here. This is a statement of the extent of the values observed in a sample, as in '...the observed weights ranged from a minimum of 1.03 kg to a maximum of 3.64 kg'. It need not be an exact statement: '...the range of observed weights was from 1 kg to about 4 kg'. However, in Subsection 1.3.2 we shall see the word 'range' used in a technical sense, as a measure of dispersion in data. This often happens in statistics: a familiar word is given a technical meaning. Terms you will come across later in the course include *expect*, *likelihood*, *confidence*, *estimator*, *significant*. But we would not wish this to preclude normal English usage of such words. It will usually be clear from the context when the technical sense is intended.)

With the help of a computer, neither the sample mean nor the sample median is easier to calculate than the other; but a computer is not always ready to hand.

If, however, the data are irregularly distributed with apparent outliers present, then the sample median is usually preferred in quoting a typical value, since it is less sensitive to such irregularities. You can see this by looking again at the data on collapsed runners in Table 1.5. The mean endorphin concentration is 138.6 pmol/l, whereas the median concentration is 110. The large discrepancy is due to the outlier with an endorphin concentration of 414. Excluding this outlier brings the mean down to 111.1 while the median decreases to 106. From this we see that the median is more stable than the mean in the sense that outliers exert less influence upon it. The word *resistant* is sometimes used to describe measures which are insensitive to outliers. The median is said to be a resistant measure, whereas the mean is not resistant.

The data in Table 1.1 were usefully summarized in Figure 1.7. The variable recorded here is 'type of employment' (professional, industrial, clerical, and so on) so the data are categorical and not amenable to ordering. In this context the notion of 'mean type of employment' or 'median type of employment' is not a sensible one. For any data set, a third representative measure sometimes used is the **mode**, and it describes the most frequently occurring observation. Thus, for males in employment in the USA during 1986, the *modal* type of employment was 'professional'; while, for females, the modal type of employment was 'clerical'.

The word **mode** can also reasonably be applied to numerical data, referring again to the most frequently occurring observation. But there is a problem of definition. For the birth weight data in Table 1.4, there were two duplicates:

two of the infants weighed 1.72 kg, and another two weighed 2.20 kg. So there would appear to be two modes, and yet to report either one of them as a representative weight is to make a great deal of an arithmetic accident. If the data are classified into groups, then we can see from Figures 1.9 to 1.11 that even the definition of a 'modal group' will depend on the definition of borderlines (and on what to do with borderline cases). The number of histogram peaks as well as their locations can alter.

However, it often happens that a collection of data presents a very clear picture of an underlying pattern, and one which would be robust against changes in group definition. In such a case it is common to identify as modes not just the most frequently occurring observation (the highest peak) but every peak.

Here are two examples. Figure 1.18 shows a histogram of chest measurements (in inches) of a sample of 5732 Scottish soldiers. This data set is explored and discussed in some detail later in the course; for the moment, simply observe that there is an evident single mode at around 40 inches. The data are said to be *unimodal*. Figure 1.19 shows a histogram of waiting times, varying from about 40 minutes to about 110 minutes. In fact, these are waiting times between the starts of successive eruptions of the Old Faithful geyser in the Yellowstone National Park, Wyoming, USA, during August, 1985. Observe the two modes. These data are said to be *bimodal*.

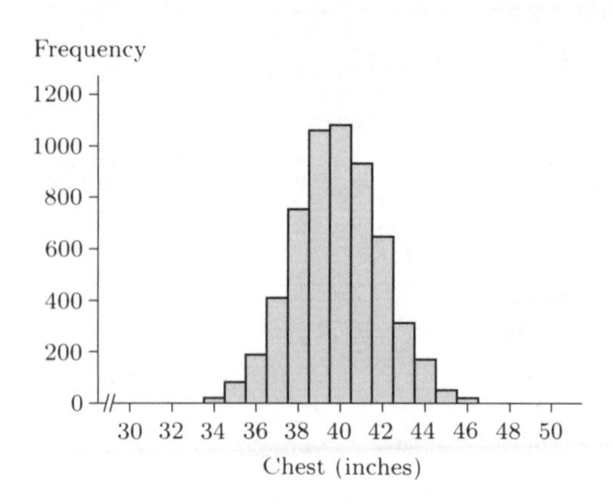

Figure 1.18 Chest measurements (inches)

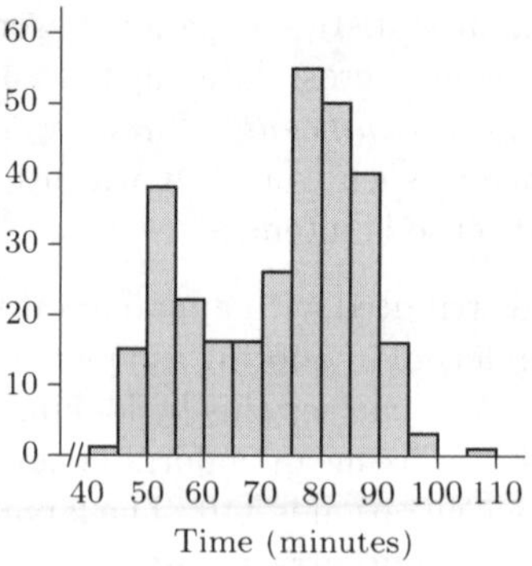

Figure 1.19 Waiting times (minutes)

Sometimes data sets may exhibit three modes (*trimodal*) or many modes (*multimodal*). You should be wary of too precise a description. Both the data sets in Figures 1.18 and 1.19 were based on large samples, and their message is unambiguous. As you will see later in the course, smaller data sets can give rise to very jagged histograms indeed, and any message about one or more preferred observations is consequently very unclear.

1.3.2 Measures of spread

During the above discussion of suitable numerical summaries for a typical value, you may have noticed that it was not possible to make any kind of decision about the relative merits of the sample mean and median without introducing the notion of the extent of variation of the data. In practice, this means that the amount of information contained in these measures, when taken in isolation, is not sufficient to describe the appearance of the data and that a more informative numerical summary is needed. In other words, we need some measure of the spread of observations if we are to be happy about replacing a data set by a few summary numbers.

The *range*, taken here to mean the difference between the smallest and largest data values, is certainly the simplest measure of spread, but it can be misleading. The range of β endorphin concentrations for collapsed runners is $414 - 66 = 348$, suggesting a fairly wide spread. However, omitting the value 414 reduces the range to $169 - 66 = 103$. This sensitivity to a single data value suggests that this is not a very reliable measure; a much more modest assessment of dispersion may be more appropriate. By its very nature, the range is always going to give prominence to outliers and therefore cannot sensibly be used in this way.

This example indicates that we need an alternative to the range as a measure of spread, and one which is not over-influenced by the presence of a few extreme values. An alternative measure is the *interquartile range*: this is the difference between summary measures known as the *lower* and *upper quartiles*.

The sample median has been defined already as in some sense representative of a set of data: it is the middle value of the observations. If the number of observations is odd, this middle value is easily identified; if the number is even, the median is defined to be the midpoint of two middle values.

It would be convenient to express this wordy definition in a concise symbolic form, and this is easy to do. Any data sample of size n can be written as a list of numbers

$$x_1, x_2, x_3, \ldots, x_n.$$

In order to calculate the sample median it is necessary to sort the data list into order, and in any case it is often informative to do this. The sorted list can then be written

$$x_{(1)}, x_{(2)}, x_{(3)}, \ldots, x_{(n)},$$

where $x_{(1)}$ is the smallest value in the original list (the minimum) and $x_{(n)}$ is the largest (the maximum). Each successive item in the ordered list is greater than or equal to the previous item. For instance, the list of six data items

7, 1, 3, 6, 3, 7

may be ordered as

1, 3, 3, 6, 7, 7

with $x_{(1)} = 1, \quad x_{(2)} = x_{(3)} = 3, \quad x_{(4)} = 6, \quad x_{(5)} = x_{(6)} = 7.$

In any such ordered list, the sample median m may be defined to be the number

$$m = x_{(\frac{1}{2}(n+1))},$$

as long as the right-hand side is correctly interpreted. If the sample size n is odd, then the number $\frac{1}{2}(n+1)$ will be an integer, and there is no problem of definition. (For instance, if $n = 27$ then $\frac{1}{2}(n+1) = 14$, and the sample median is $m = x_{(14)}$, the middle value, with thirteen data items either side of it.) If the sample size n is even (as in the example above, where six numbers are listed) then the number $\frac{1}{2}(n+1)$ will not be an integer but will have a fractional part equal to $\frac{1}{2}$. (Such numbers are sometimes called 'half-integer'.) For instance, if $n = 6$ then the sample median is

$$m = x_{(\frac{1}{2}(n+1))} = x_{(3\frac{1}{2})}.$$

If the number $x_{(3\frac{1}{2})}$ is interpreted as the number 'half-way between $x_{(3)}$ and $x_{(4)}$' then you can see that the wordy definition survives intact. It is a very obvious interpretation to make, and can be extended. As well as expressing through the sample median a representative value for a set of data, it is useful to be able to say something about the dispersion in the data set through the **lower quartile** (roughly, one-quarter of the way into the data set) and the **upper quartile** (approximately three-quarters of the way through the data set).

Sample quartiles

If a data set $x_1, x_2, \ldots, x_n$ is re-ordered as $x_{(i)}, i = 1, 2, \ldots, n$, where

$$x_{(1)} \leq x_{(2)} \leq \ldots \leq x_{(n)},$$

then the **lower sample quartile** is defined by

$$q_L = x_{(\frac{1}{4}(n+1))},$$

and the **upper sample quartile** is defined by

$$q_U = x_{(\frac{3}{4}(n+1))}.$$

The lower and upper sample quartiles are sometimes called the first and third sample quartiles. The median is the second sample quartile. Other definitions are possible, and you may even be familiar with some of them. Some practitioners use

$q_L = x_{(\frac{1}{4}n+\frac{1}{2})}, \quad q_U = x_{(\frac{3}{4}n+\frac{1}{2})};$

others use

$q_L = x_{(\frac{1}{4}n+\frac{3}{4})}, \quad q_U = x_{(\frac{3}{4}n+\frac{1}{4})}.$

Still others insist that the lower and upper quartiles be defined in such a way that they are identified uniquely with actual sample items. Almost all definitions reduce to the same thing when it comes to identifying the sample median.

Example 1.11 Quartiles for the SIRDS data

For the 23 infants who survived SIRDS (Table 1.4), the ordered birth weights are given in Solution 1.4. The first quartile is

$$q_L = x_{(\frac{1}{4}(23+1))} = x_{(6)} = 1.720\,\text{kg};$$

the third quartile is

$$q_U = x_{(\frac{3}{4}(23+1))} = x_{(18)} = 2.830\,\text{kg}. \quad ■$$

Example 1.12 Quartiles when the sample size is awkward

For the ordered list $1, 3, 3, 6, 7, 7$ $(n = 6)$ the lower quartile is given by

$$q_L = x_{(\frac{1}{4}(n+1))} = x_{(\frac{7}{4})} = x_{(1\frac{3}{4})}.$$

In other words, the number q_L is given by the number three-quarters of the way between $x_{(1)} = 1$ and $x_{(2)} = 3$. Their difference is 2: so $q_L = x_{(1)} + \frac{3}{4}(x_{(2)} - x_{(1)}) = 1 + \frac{3}{4}(2) = 2.5$. The upper quartile is given by

$$q_U = x_{(\frac{3}{4}(n+1))} = x_{(\frac{21}{4})} = x_{(5\frac{1}{4})}.$$

So q_U is the number one-quarter of the way between $x_{(5)} = 7$ and $x_{(6)} = 7$. This is just the number 7 itself. ■

Exercise 1.10

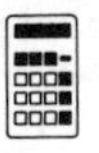

(a) Find the lower and upper quartiles for the birth weight data on those children who died of the condition. (See Solution 1.4 for the ordered data.)

(b) Find the median, lower and upper quartiles for the data in Table 1.13, which give the percentage of silica found in each of 22 chondrites meteors. (The data are ordered.)

Good, I.J. and Gaskins, R.A. (1980) Density estimation and bump-hunting by the penalized likelihood method exemplified by scattering and meteorite data. *J. American Statistical Association*, **75**, 42–56.

Table 1.13 Silica content of chondrites meteors

20.77	22.56	22.71	22.99	26.39	27.08	27.32	27.33
27.57	27.81	28.69	29.36	30.25	31.89	32.88	33.23
33.28	33.40	33.52	33.83	33.95	34.82		

A simple measure of dispersion, the interquartile range, is given by the difference $q_U - q_L$.

The interquartile range

The dispersion in a data set may be simply expressed through the **interquartile range**, which is the difference between the upper and lower quartiles, $q_U - q_L$.

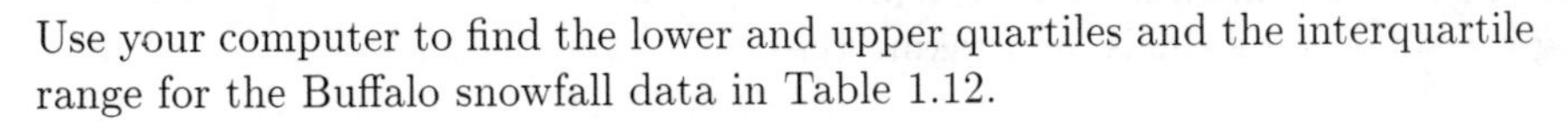

Exercise 1.11

Use your computer to find the lower and upper quartiles and the interquartile range for the Buffalo snowfall data in Table 1.12.

The interquartile range is a useful measure of dispersion in the data and it has the excellent property of not being too sensitive to outlying data values. However, like the median it does suffer from the disadvantage that its computation requires sorting the data. This can be very time-consuming for large samples. Another measure that is easier to compute and, as you will find in later chapters, has good statistical properties is the *standard deviation.*

The standard deviation is defined in terms of the differences between the data values and their mean. These differences $(x_i - \overline{x})$, which can be positive or negative, are called **residuals**.

Example 1.13 *Calculating residuals*

The mean difference in β endorphin concentration for the 11 runners sampled who completed the Great North Run in Example 1.4 is 18.74 pmol/l (to two decimal places). The eleven residuals are given in the following table.

Difference, x_i	25.3	20.5	10.3	24.4	17.5	30.6	11.8	12.9	3.8	20.6	28.4
Mean, $\overline{x}$	18.74	18.74	18.74	18.74	18.74	18.74	18.74	18.74	18.74	18.74	18.74
Residual, $x_i - \overline{x}$	6.56	1.76	−8.44	5.66	−1.24	11.86	−6.94	−5.84	−14.94	1.86	9.66

■

If x_i is a data value ($i = 1, 2, \ldots, n$, where n is the sample size) then the ith residual can be written

$$r_i = x_i - \overline{x}.$$

These residuals all contribute to an overall measure of dispersion in the data. Large negative and large positive values both indicate observations far removed from the sample mean. In some way they need to be combined into a single number.

There is not much point in averaging them: positive residuals will cancel out negative ones. In fact their sum is zero, since

$$\sum_{i=1}^{n} r_i = \sum_{i=1}^{n} (x_i - \overline{x}) = \sum_{i=1}^{n} x_i - n\overline{x} = 0;$$

and so, therefore, is their average. What is important is the magnitude of each residual, the absolute difference $|x_i - \overline{x}|$. The absolute residuals could be added together and averaged, but this measure (known as the *mean absolute deviation*) does not possess very convenient mathematical properties. Another way of eliminating minus signs is to square the residuals and average them. This leads to a measure of dispersion known as the *sample standard deviation.*

The sample standard deviation

A measure of the dispersion in a sample

$$x_1, x_2, \ldots, x_n$$

with sample mean $\overline{x}$ is given by the **sample standard deviation** s, where s is obtained by averaging the squared residuals, and taking the square root of that average:

$$s = \sqrt{\frac{r_1^2 + r_2^2 + \cdots + r_n^2}{n-1}} = \sqrt{\frac{\sum (x_i - \overline{x})^2}{n-1}}. \qquad (1.1)$$

There are two important points you should note about this definition. First, you should remember to take the square root of the average. The reason for this is that the residuals are measured in the same units as the data, and so their squares are measured in the squares of those units. After averaging, it is necessary to take the square root, so that the standard deviation is measured in the same units as the data.

Second, although there are n terms contributing to the sum in the numerator, the divisor is not the sample size n, but $n - 1$.

The reason for this surprising amendment will become clear in *Chapter 6* of the course. Either average (dividing conventionally by n or dividing by $n - 1$) has useful statistical properties, but these properties are subtly different. The definition at (1.1) will be used in this course.

Example 1.13 continued

The sum of the squared residuals for the eleven β endorphin concentration differences is

$$\begin{aligned}&\sum(x_i - \overline{x})^2\\ &= 6.56^2 + 1.76^2 + (-8.44)^2 + \cdots + (9.66)^2\\ &= 43.03 + 3.10 + 71.23 + \cdots + 93.32\\ &= 693.85;\end{aligned}$$

Notice that a negative residual contributes a positive value to the calculation of the standard deviation. This is because it is squared.

so the sample standard deviation of the differences is

$$s = \sqrt{\frac{693.85}{10}} = 8.33. \quad \blacksquare$$

Even for relatively small samples the arithmetic is rather awkward if done by hand. Fortunately, it is now common for calculators to have a 'standard deviation' button, and all that is required is to key in the data.

Exercise 1.12

Use your calculator for each of the following calculations.

(a) Confirm the sample standard deviation for the 11 differences listed in Example 1.13.

(b) Calculate the standard deviation for the 22 silica percentages given in Table 1.13.

(c) Calculate the standard deviation for the β endorphin concentrations of the 11 collapsed runners. (See Table 1.5.)

Exercise 1.13

Use your computer for each of these calculations.

(a) Compute the standard deviation for the birth weights of all 50 infants in the SIRDS data set. (See Table 1.4.)

(b) Find the standard deviation for the annual snowfall in Buffalo, NY. (See Table 1.12.)

In later chapters you will find that the main use of the standard deviation lies in making inferences about the population from which the sample is drawn. Its most serious disadvantage, like the mean, results from its sensitivity to outliers. In Exercise 1.12 you calculated a standard deviation of 98.0 for the data on collapsed runners. Try doing the calculation again, but this time omit the outlier at 414. You will find a drastic reduction in the standard deviation to 37.39, a reduction by a factor of almost three!

Which, then, should you prefer as a measure of spread: range, interquartile range or standard deviation? For exploring and summarizing dispersion in data values, the interquartile range is safer, especially when outliers are present. For inferential calculations, which you will meet in subsequent chapters of the course, the standard deviation is used, possibly with extreme values removed. The range should only be used as a check on calculations. Clearly the mean must lie between the smallest and largest data values, somewhere near the middle if the data are reasonably symmetric; and the standard deviation, which can never exceed the range, is usually close to about one-quarter of it.

Exercise 1.14

When hunting insects, bats emit high-frequency sounds and pick up echoes of their prey. The data given in Table 1.14 are bat-to-prey detection distances (i.e. distances at which the bat first detects the insect) measured in centimetres.

Griffin, D.R., Webster, F.A. and Michael, C.R. (1960) The echo location of flying insects by bats. *Animal Behaviour*, **8**, 141–154.

Table 1.14 Bat-to-prey detection distances (cm)

62 52 68 23 34 45 27 42 83 56 40

Calculate the median, interquartile range, mean and standard deviation of the sample.

Exercise 1.15

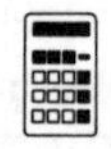

The following data are taken from the 1941 Canadian Census and comprise the sizes of completed families (numbers of children) born to a sample of Protestant mothers in Ontario aged 45–54 and married at age 15–19. The data are split into two groups according to how many years of formal education the mothers had received.

Keyfitz, N. (1953) A factorial arrangement of comparisons of family size. *American J. Sociology*, **53**, 470–480.

Table 1.15 Family size: mothers married aged 15–19

Mother educated for 6 years or less
14 13 4 14 10 2 13 5 0 0 13 3 9 2 10 11 13 5 14
Mother educated for 7 years or more
0 4 0 2 3 3 0 4 7 1 9 4 3 2 3 2 16 6 0 13 6 6 5 9 10 5 4 3 3 5 2 3 5 15 5

Find the median, interquartile range, mean and standard deviation of each of the groups of mothers. Comment on these measures for the two groups.

For future reference, the square of the sample standard deviation in a sample of data is known as the **sample variance**.

The sample variance

The **sample variance** of a data sample $x_1, x_2, \ldots, x_n$ is given by

$$s^2 = \frac{\sum(x_i - \overline{x})^2}{n-1},$$

where $\overline{x}$ is the sample mean.

Finally, many practitioners find it convenient and useful to characterize a data sample in terms of the **five-figure summary**.

The five-figure summary

For a sample of data, the **five-figure summary** lists, in order,

- the sample minimum, $x_{(1)}$;
- the lower quartile, q_L;
- the sample median, m;
- the upper quartile, q_U;
- the sample maximum, $x_{(n)}$.

For instance, the five-figure summary for the snowfall data in Table 1.12 can be written

$$(25, 63.6, 79.7, 98.3, 126.4).$$

For the silica data in Table 1.13 it is

$$(20.77, 26.91, 29.03, 33.31, 34.82).$$

1.4 Graphical displays and numerical summaries

Bearing in mind the advice that we should always start a data exploration by looking at a graphical display of the data, we have displays such as bar charts and histograms to guide us when looking at data sets which involve only one variable. But the problem with these displays is first, that they can have too much detail, and second, that they are not very useful for comparing two or more samples. Now that we know how to calculate useful numerical summaries of the data, it would be useful to have available a graphical display showing the summary statistics in a visually appealing and interpretable way: a simple method for doing this is by drawing a **boxplot**.

A boxplot is designed to depict, as clearly as possible, the median, the quartiles, the range of the data and any outliers which may be present. It gives a clear picture of all of these features and, as you will see, allows a quick comparison of data sets.

A boxplot is simple to construct. It will be useful to follow the construction through with an example, and we shall use the 'collapsed runners' data in Table 1.5. The eleven β endorphin concentrations recorded were

$$66, 72, 79, 84, 102, 110, 123, 144, 162, 169, 414.$$

It will also be useful to have the five-figure summary for these data, given by

$$(x_{(1)}, q_L, m, q_U, x_{(11)}) = (66, 79, 110, 162, 414).$$

First, a convenient horizontal scale is drawn, covering the extent of the data: say, in this case, a scale from 0 to 500. Against this scale, the three quartiles (the first quartile q_L, the second quartile m, and the third quartile q_U) are drawn in as shown in Figure 1.20.

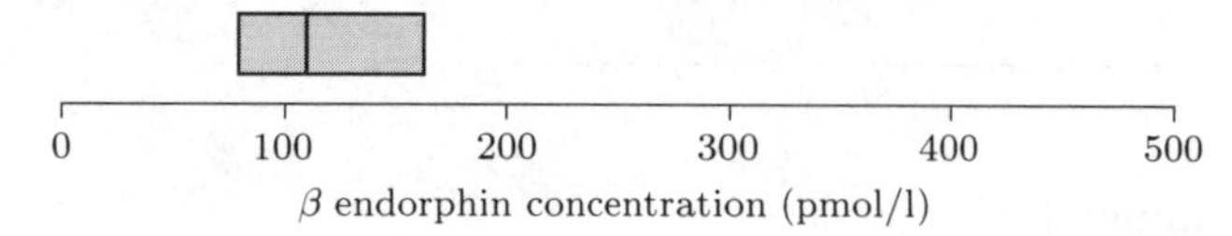

Figure 1.20 Drawing a boxplot

The rectangle is the 'box', with edges defined by the lower and upper quartiles; the median is shown by a vertical line appropriately located in the box.

Next, calculate the interquartile range. In this case, $162 - 79 = 83$. 'Whiskers' are then constructed, extending to the furthest observation within one i.q.r. (interquartile range) either side of the box. In this case,

$$q_U + \text{i.q.r} = 162 + 83 \\ = 245,$$

and so the right-hand whisker will extend as far as the observation 169, the highest observation not exceeding 245. The number 169 is called the **upper adjacent value**. Similarly,

$$q_L - \text{i.q.r.} = 79 - 83 \\ = -4,$$

and so the left-hand whisker extends all the way to 66, the lowest observation. So, in this case, the **lower adjacent value** is the same as the sample minimum. This is shown in Figure 1.21.

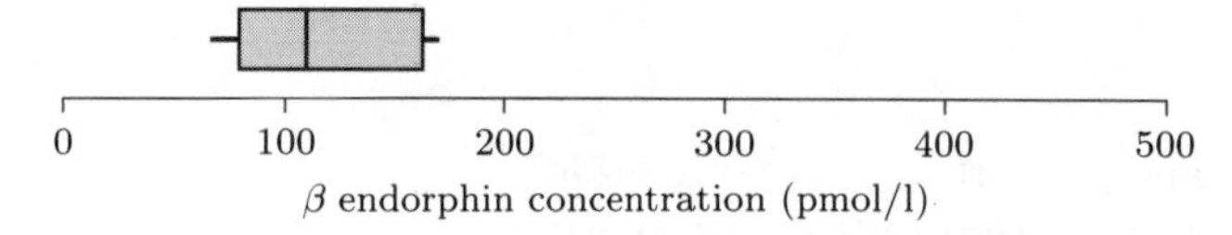

Figure 1.21 Drawing a boxplot, continued

Finally, any observations not covered by the whiskers are marked in as separate items. (They may in some circumstances be deemed outliers, or at least worth special attention.) In this case, the only observation not covered by the whiskers is that to the extreme right, the maximum observation of 414. This is shown in Figure 1.22, which is the completed boxplot.

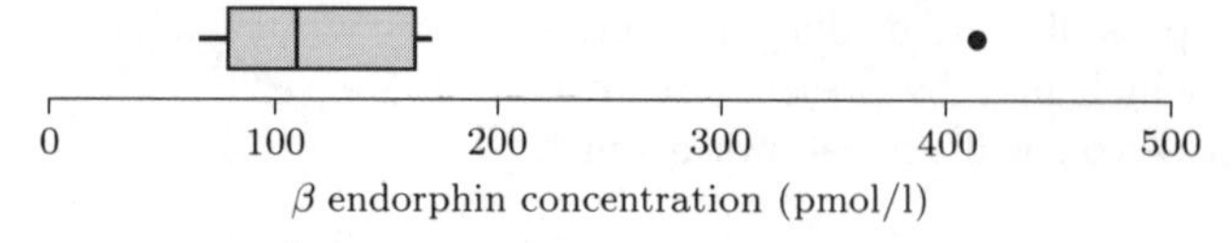

Figure 1.22 The completed boxplot

Again, you should be aware that boxplot construction is an area where there are no clear rules. All boxplots show the three quartiles, but the conventions defining the whiskers vary from text to text and from one computer package to another. The whiskers may extend up to one and a half or even two interquartile ranges either side of the box; and some approaches distinguish moderate and severe outliers with different symbols. The approach adopted here is the simplest approach that shows all that needs to be illustrated.

You can see how a boxplot gives a quick visual assessment of the data. The length of the box shows the interquartile range and the lengths of the whiskers relative to the length of the box give an idea of how stretched out the rest of the values are. The unusually large value is clearly shown in this case and the median gives an assessment of the centre. If the sample median is to be used as an estimate of an unknown population value, then the boxplot gives an overall feel for the precision of such an estimate. Some kind of assessment of symmetry is possible, since symmetric data will produce a boxplot which is symmetric about the median. However, it should be borne in mind that this particular data set has only 11 values, and this is too small a number to infer anything definite about any underlying structure. You should now make sure you understand boxplots by constructing a couple for yourself.

Exercise 1.16

Using a pencil and ruler, construct a boxplot (a) for the silica data in Table 1.13 and (b) for the snowfall data in Table 1.12. Use the five-figure summaries given at the end of Section 1.3.

Boxplots are particularly useful when used for quick comparisons. When the data on birth weights of children exhibiting severe idiopathic respiratory distress syndrome were introduced in Table 1.4 the question asked was: is it possible to relate survival to birth weight? We are now in a good position to make some headway. Figure 1.23 shows comparative boxplots of the two groups of birth weights.

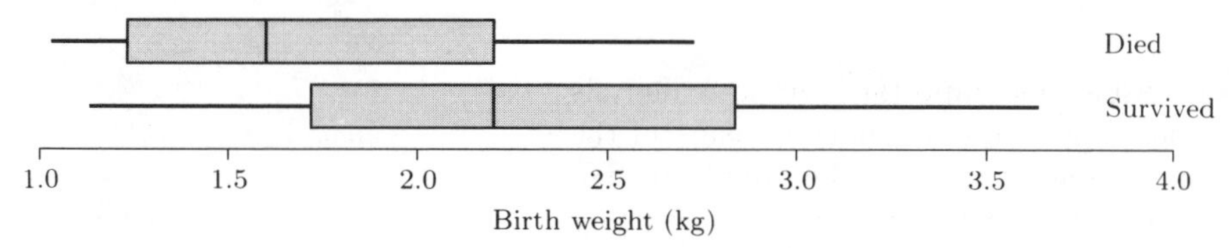

Figure 1.23 Comparative boxplots, two groups

You can see immediately that the median birth weight of children who died is less than the lower quartile of birth weights of children who survived. The picture immediately gives a compact, quickly assimilated summary of the data, suggesting that children who survive and children who do not may typically have different birth weights.

Using boxplots you have seen how summary statistics can be represented by graphical displays which not only give some feel for the way the data are dispersed but also make it possible visually to compare two or more samples of data.

1.4.1 Data and symmetry

Figure 1.24 shows a bar chart of some of the data of Table 1.15: it shows the numbers of children born to the 35 Protestant mothers in Ontario who had at least seven years' education.

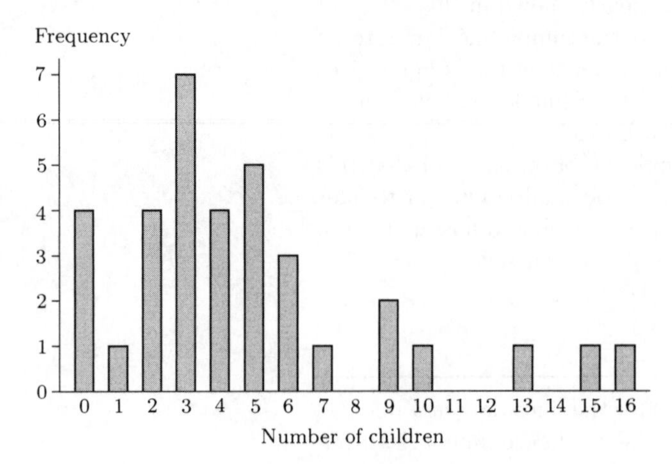

Figure 1.24 Bar chart, family size

The bar chart shows a marked lack of symmetry. The corresponding boxplot is shown in Figure 1.25.

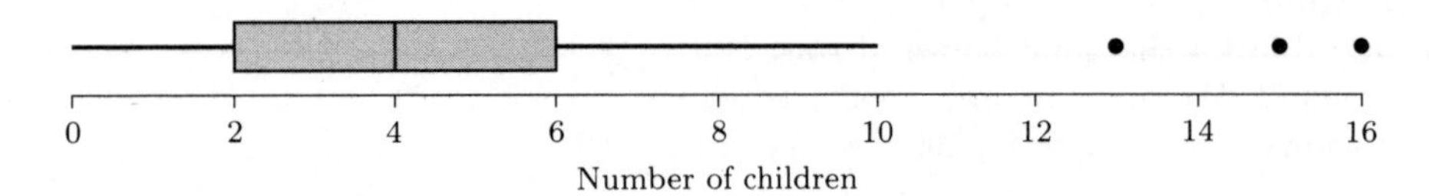

Figure 1.25 Boxplot, family size

Detection of lack of symmetry is of considerable importance in data analysis and inference. This is because the most important summary measure of the data is the typical or central value in the context of which the sample median and the sample mean were introduced. Now, when the data are roughly symmetrically distributed, all ambiguity is removed because the median and the mean will nearly coincide. When the data are very far from symmetric, not only will these measures not coincide but we may even be pressed to decide whether any summary measure of this kind is appropriate.

A measure of symmetry would clearly be both meaningful and useful.

The generally accepted measure is the **sample skewness**, defined as follows.

The sample skewness

The **sample skewness** of a data sample $x_1, x_2, \ldots, x_n$ is given by

$$\frac{1}{n-1}\sum_{i=1}^{n}\left(\frac{x_i-\overline{x}}{s}\right)^3,$$

where $\overline{x}$ is the sample mean and s is the sample standard deviation.

Notice in this formula the term $(x_i - \overline{x})^3$: observations greater than the sample mean contribute positive terms to the sum, while observations less than the sample mean will contribute negative terms. Perfectly symmetric data would have a skewness of 0. The data of Figures 1.24 and 1.25 have a sample skewness of 1.28—the data are said to be *positively skewed.*

Exercise 1.17

Use your computer to calculate the sample skewness for the family size data for the first group of 19 mothers, who had six or less years' education.

In the case of the first group of mothers the sample skewness is negative: the data are said to be *negatively skewed.* In this case, the asymmetry is rather slight. However, the consequences of asymmetry are important for subsequent analyses of the data, as we shall see as the course develops.

The following exercise is a computer exercise to reinforce some of the ideas you have met so far.

Exercise 1.18

Data were taken from an experiment on three groups of mice. The measurements are amounts of nitrogen-bound bovine serum albumen produced by normal mice treated with a placebo (i.e. an inert substance), alloxan-diabetic mice treated with a placebo, and alloxan-diabetic mice treated with insulin. The data are given in Table 1.16.

Table 1.16 Nitrogen-based BSA for three groups of diabetic mice

Normal	Alloxan-diabetic	Insulin treatment
156	391	82
282	46	100
197	469	98
297	86	150
116	174	243
127	133	68
119	13	228
29	499	131
253	168	73
122	62	18
349	127	20
110	276	100
143	176	72
64	146	133
26	108	465
86	276	40
122	50	46
455	73	34
655		44
14		

Dolkart, R.E., Halpern, B. and Perlman, J. (1971) Comparison of antibody responses in normal and alloxan diabetic mice. *Diabetes,* **20**, 162-167.

(a) Summarize the three groups in terms of their five-figure summaries.

(b) Calculate the mean and standard deviation for each group.

(c) Calculate the sample skewness for each group.

(d) Obtain a comparative boxplot for the three groups. Are any differences apparent between the three treatments?

1.5 A look ahead

The examples in this chapter have been drawn from a wide variety of areas, the diversity of which serves to illustrate the unlimited range of applicability of statistical methods. Statistical ideas, however, are not merely used to address questions from a broad range of areas, they are also used to address a multitude of different types of problem.

One important division of problem types is into formulation, estimation and testing situations. In the first, we are searching for patterns or structures in the data, and you have seen some of that in this chapter. This is done in the hope that such patterns or structures might lead to simplified explanation and improved understanding of what is going on, thereby helping the formulation of hypotheses and theories. Sometimes the word **exploratory** is used to describe statistical methods with this aim in mind because they are being used to explore the data to see what can be discovered. Most of the graphical methods you have met in this chapter could be described as exploratory.

Example 1.14 *Hypothesizing a linear relationship*

We used an exploratory method to investigate the relationship between alcohol consumption and the death rate from cirrhosis and alcoholism (Table 1.6). It is clear from Figures 1.12 and 1.13 that some pattern does exist. We might hypothesize a linear relationship of the form

$$\text{Death rate } = \alpha + \beta\,(\text{Alcohol consumption}),$$

The letter α is the Greek lower-case letter alpha.

where α and β are constants (to be determined numerically by a method which is described in *Chapter 10*). With this as a basic relationship, a government might then wish to examine its policy of alcohol taxation. ■

Example 1.15 *Old Faithful geyser, August 1978*

The Old Faithful geyser in Yellowstone National Park has been much observed: we saw in Figure 1.19 the results of monitoring in 1985. Some years earlier, observations were carried out with the objective of deciding whether a prediction rule could be formulated so that the Park Rangers would be able to tell visitors when the next eruption would be likely to occur, based on the duration of the previous eruption. The durations of eruptions and the times to next eruption were recorded between 6 am and midnight from 1 to 8 August 1978. The data are given in Table 1.17.

Table 1.17 Eruptions of Old Faithful geyser, August 1978

Day 1		Day 2		Day 3		Day 4		Day 5		Day 6		Day 7		Day 8	
D	I	D	I	D	I	D	I	D	I	D	I	D	I	D	I
4.4	74	4.3	76	4.5	71	4.0	71	4.0	67	1.8	53	3.5	50	4.2	73
3.9	70	1.7	54	3.9	78	3.7	69	2.3	65	4.6	70	2.0	87	4.4	73
4.0	64	3.9	76	4.4	80	3.7	63	4.4	77	3.5	69	4.3	40	4.1	70
4.0	72	3.7	65	2.3	51	4.3	64	4.1	72	4.0	66	1.8	76	4.1	84
3.5	76	3.1	54	3.8	82	3.6	82	4.3	79	3.7	79	4.1	57	4.0	71
4.1	80	4.0	86	1.9	49	3.8	68	3.3	73	1.2	48	1.8	71	4.1	79
2.3	48	1.7	40	4.6	80	3.8	71	2.0	53	4.6	90	4.7	70	2.7	58
4.7	88	4.1	87	1.8	43	3.8	71	4.3	69	1.7	49	4.2	69	4.6	73
1.7	53	1.8	49	4.7	83	2.5	63	2.9	53	4.0	78	3.9	72	1.9	59
4.9	71	3.2	76	1.8	49	4.5	79	4.6	78	1.8	52	4.3	51	4.5	76
1.7	56	1.9	51	4.6	75	4.1	66	1.9	55	4.1	79	1.8	84	2.0	49
4.6	69	4.6	77	1.9	47	3.7	75	3.6	67	1.9	49	4.5	43	4.8	75
3.4	72	2.0	49	3.5	78	3.8	56	3.7	68	4.6	75			4.1	75
						3.4	83	3.7	73	2.9	75				

D = duration of eruption, I = time interval to next eruption (minutes)

A scatter plot of 'time interval to next eruption' against 'duration of eruption' is given in Figure 1.26. ■

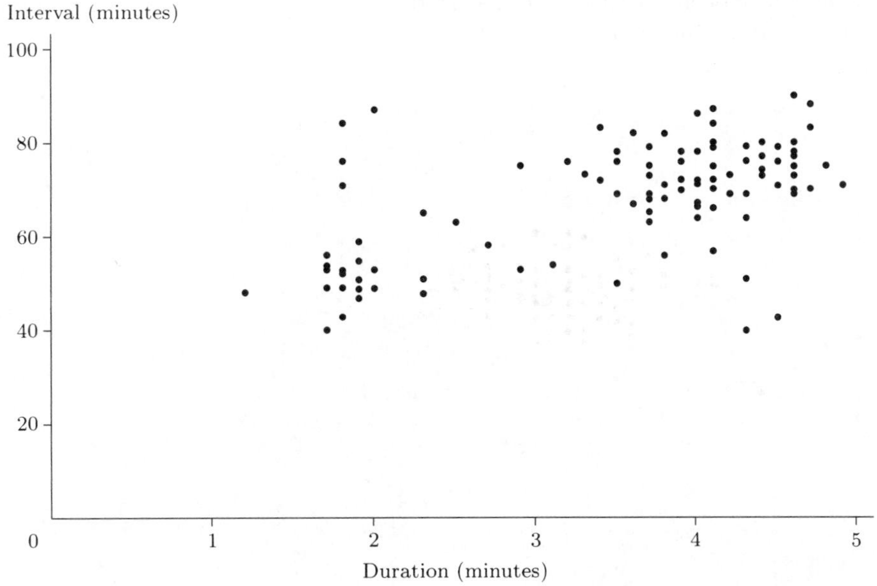

Interval (minutes)
100
80
60
40
20
0
1
2
3
4
5
Duration (minutes)

Figure 1.26 Eruptions of Old Faithful geyser, August 1978

The scatter plot clearly shows that, if the duration is large, the interval to the next eruption is also likely to be large, and if the duration is small the interval is likely to be small. However, the situation is not entirely clear cut because the plot also shows some outlying points which are far from the others, notably in the bottom right-hand and top left-hand corners of the plot. Two statisticians from Bell Laboratories in the United States, Lorraine Denby and Daryl Pregibon, were intrigued by this curious distribution of the points and explored it by replacing the points on the plot with symbols showing the days on which the measurements were taken. This is shown in Figure 1.27, and you can see that Day 7 accounts for the anomalous outlying points.

Interval (minutes)
100
80
60
40
20
0
1
2
3
4
5
Duration (minutes)
Day 1
Day 2
Day 3
Day 4
Day 5
Day 6
Day 7
Day 8

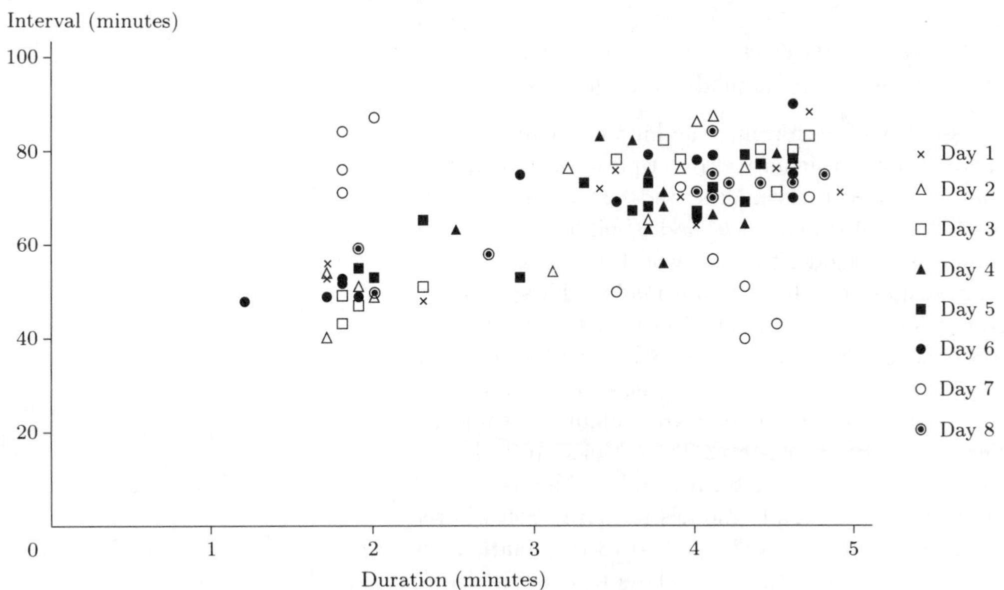

Figure 1.27 Eruptions of Old Faithful geyser, August 1978, day by day

This led Denby and Pregibon to check that the results for Day 7 had been recorded correctly. They found that a mistake had indeed been made, and that the recorded intervals for Day 7 should have been matched with the listed duration following the one with which they had originally been matched. This led to the revised scatter plot shown in Figure 1.28.

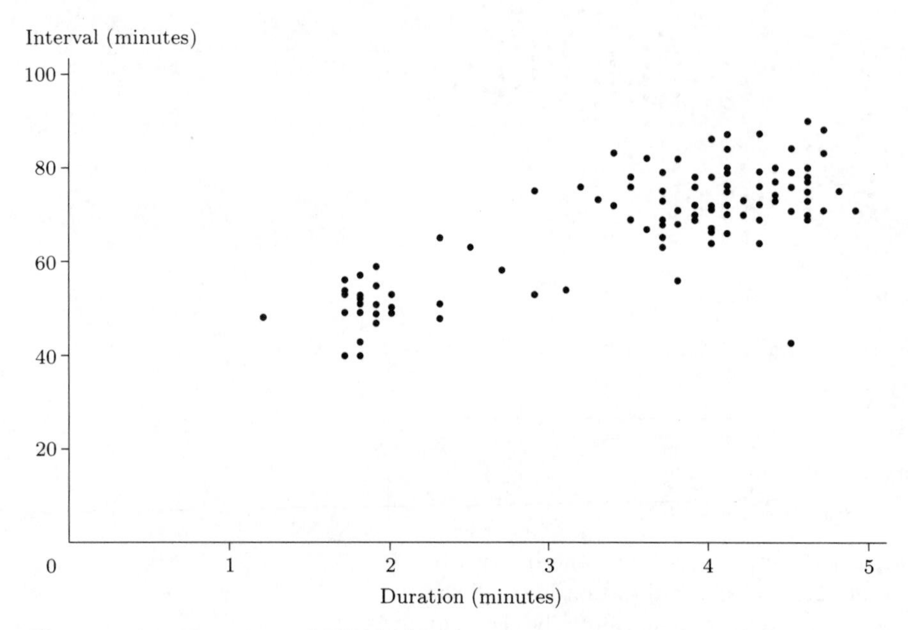

Figure 1.28 Eruptions of Old Faithful geyser, August 1978, revised plot

Now the points follow a fairly well-defined pattern (except for one!), and predictions can be made.

Another type of problem concerns testing. Here, statistical methods are used to determine the possible truth of a hypothesis or theory by comparing it with a set of data. In effect we are asking if it is likely that the observed data arose from a structure such as the one hypothesized or if it is, in fact, extremely unlikely. Later chapters will include formal methods for such tests.

Before we have even the possibility of performing the kind of comparison involved in a test, we need to have a mathematical description or explanation of the data. In other words, we need to formulate and test a *model* which explains how the data arose. The first step in such model-building is to decide what is relevant. It is impossible to model the real world in all its awesome complexity and simplifying assumptions have to be made. These may be along the lines that certain things are irrelevant (Newton, for example, in formulating his laws of gravitation, decided that the colour and material of gravitating bodies were irrelevant—only distance and mass were important), or certain aspects of the data may be too complex to explain (in Table 1.8 and Figure 1.17, we might decide that short-term fluctuations are too complex to model and focus our attention on long-term trends). The next three chapters will be about constructing probability models for data. You will see how to formulate common-sense assumptions which lead to tidy mathematical descriptions of the way in which the data arise. These will enable you to quantify such things as the tests mentioned in the preceding paragraph.

Of course, modelling does not end with formulating a plausible mathematical description of the way data arise. We also need to test the validity of the resulting model. The results of such a test may lead us to accept the adequacy of a model; we may decide that the patterns in the data are explained successfully. But is the test we have used powerful enough? A weak test is one which would pass an inadequate model and clearly would not be of much use. A powerful test is more discriminating and acceptance will require a closer fit between model and data.

Alternatively, the results of a test may lead us to conclude that the model is inadequate: it does not explain successfully the patterns in the data. We say that the test has *rejected* the model and we have to try some other model. Sensitive use of statistical methods can lead to ideas about how a rejected model may be modified to improve the fit.

Example 1.16 Incidence of Down's syndrome births, Australia, 1942–1952

Table 1.18 shows the incidence of Down's syndrome babies born to Australian mothers from 1942 to 1952. The total number of births is also given. The mothers are grouped into seven categories, by age.

Table 1.18 Down's syndrome births

Age group of mother	Number of Down's syndrome births	Total number of births
under 20	15	35 555
20–24	128	207 931
25–29	208	253 450
30–34	194	170 970
35–39	297	86 046
40–44	240	24 498
45 and over	37	1 707

Moran, P.A.P. (1974) Are there two maternal age groups in Down's syndrome? *British Journal of Psychiatry*, **124**, 453–455.

The objective is to model the risk of a child born with the syndrome given the mother's age at the birth.

Suppose we plot the proportion of Down's syndrome births for each age group against the midpoint of the age group; that is, we shall plot the number of Down's syndrome births divided by the total number in each age group against each interval midpoint, the midpoints being taken to be 17.5, 22.5, 27.5, 32.5, 37.5, 42.5, 47.5. The plot is shown in Figure 1.29.

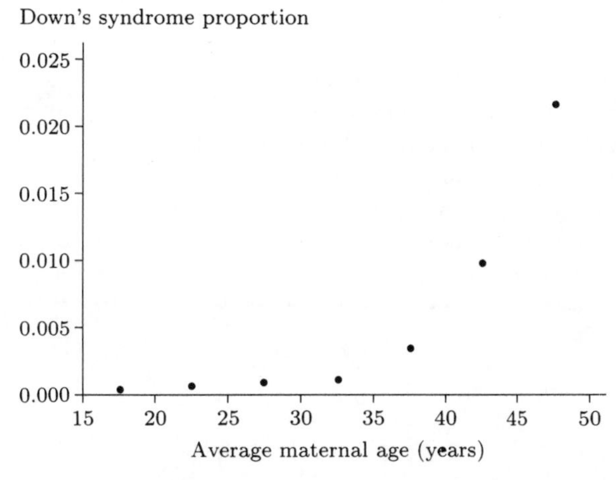

Figure 1.29 Proportion of Down's syndrome births against age of mother

You can see how the risk increases with age, but the proportions increase so rapidly at the higher ages that we need to use a transformation of the data if we are to see what is really happening. In Figure 1.15 we obtained a clear scatter plot for the data in Table 1.7 by taking logarithms. In this case, we want to spread out the small proportions and compress the larger ones whilst leaving the ages evenly spread, so we take logarithms of the proportions only.

In fact, the appropriate transformation is to plot the variable

$$\log\left(\frac{\text{proportion}}{1 - \text{proportion}}\right)$$

against age. You need not worry about the arithmetic. The transformed plot is shown in Figure 1.30.

You are not expected at this stage of the course to be able to select an appropriate transformation yourself, or to recognize why some particular transformation might lead to a more informative view of the data.

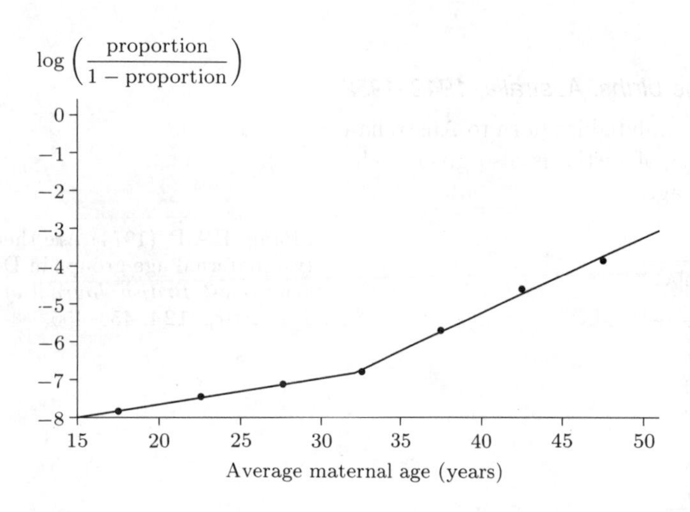

Figure 1.30 Down's syndrome plot, after transformation

The plot shows two groups of points, each approximately linear but with different slopes, the changeover point being the midpoint of the 30–35 age interval. In fact the fit is almost perfect.

The model gives the proportion of Down's syndrome births as a function of the maternal age and enables us to estimate the relative risks for different ages. ■

Sometimes models are very complex and it is difficult to see what is going on. In such cases, one method which is sometimes adopted to gain understanding is simulation. A computer program is written which acts as if the model were true, and generates data from it. Thus large numbers of simulated random samples may be produced and compared. In this way different models may be tried and the consistency of their performance may be assessed. You will meet computer simulation throughout the text, beginning towards the end of *Chapter 2*.

Summary

1. An important first stage in any assessment of a collection of data, preceding any numerical analysis, is to represent the data, if possible, in some informative diagrammatic way. Useful graphical representations include pie charts, bar charts, histograms, scatter plots and boxplots. Possibly, transformations may be useful to aid the representation of the data.

2. Most diagrammatic representations have some disadvantages; in particular, histograms are very sensitive to the choice of origin and the widths of the class interval.

3. The design of boxplots varies with analysts' preferences and between computer packages. In this course a boxplot exhibits the lower adjacent value, the lower quartile, the sample median, the upper quartile and the upper adjacent value. The quartiles are spanned by the 'box', the adjacent values by 'whiskers'. Other extreme values are marked individually and may be called 'outliers'.

4. Numerical summaries for a set of data include the sample median

$$m = x_{(\frac{1}{2}(n+1))},$$

where n is the sample size; the lower quartile

$$q_L = x_{(\frac{1}{4}(n+1))};$$

the upper quartile

$$q_U = x_{(\frac{3}{4}(n+1))};$$

the minimum value $x_{(1)}$; and the maximum value $x_{(n)}$. The distance $x_{(n)} - x_{(1)}$ is sometimes called the range of the sample; the distance $q_U - q_L$ is called the interquartile range.

The data summary $(x_{(1)}, q_L, m, q_U, x_{(n)})$ is called the five-figure summary.

5. Other summary measures include the sample mean

$$\overline{x} = \frac{1}{n}(x_1 + x_2 + \cdots + x_n) = \frac{1}{n}\sum_{i=1}^{n} x_i;$$

the sample standard deviation (a measure of dispersion)

$$s = \sqrt{\frac{1}{n-1}\left((x_1 - \overline{x})^2 + (x_2 - \overline{x})^2 + \cdots + (x_n - \overline{x})^2\right)}$$

$$= \sqrt{\frac{1}{n-1}\sum (x_i - \overline{x})^2};$$

and the sample skewness (a measure of asymmetry)

$$\frac{1}{n-1}\sum\left(\frac{x_i - \overline{x}}{s}\right)^3.$$

The quantities $(x_i - \overline{x})$ are called residuals; the number s^2 is known as the sample variance.

Chapter 2
Models for Data I

A very common type of statistical experiment involves taking measurements of an attribute on some sample from a population (for example, height, weight, pressure, income, and so on). What is often observed is that the measurements are not all the same, but vary. A mathematical model describing this variability (that is, what sorts of measurements are 'likely' and which are 'unlikely') is called a probability model. In this chapter we make a start on the exercise of modelling, and some of the language of probability is introduced.

You have already seen some examples of data sets in *Chapter 1*. Here is one more. This example is about the results of a botanical experiment.

Example 2.1 Leaf lengths

Table 2.1 shows the lengths (in cm) of 100 leaves from an ornamental bush measured by a researcher who wished to obtain an idea of average leaf length. The bush was of particular interest because it was being grown in a new environment.

Data provided by D.J. Hand, The Open University.

Table 2.1 Leaf lengths (cm)

1.6	1.9	2.2	2.1	2.2	1.0	0.8	0.6	1.1	2.2	1.3	1.0	1.1
0.8	1.4	2.2	2.1	1.3	1.0	1.3	1.1	2.1	1.1	1.1	1.0	0.9
1.3	2.3	1.3	1.0	1.0	1.3	1.3	1.5	2.4	1.0	1.0	1.3	1.1
1.3	1.3	0.9	1.0	1.4	2.3	0.9	1.4	1.3	1.2	1.5	2.6	2.7
1.6	1.0	0.7	1.7	0.8	1.3	1.4	1.3	1.5	0.6	0.5	0.4	2.7
1.6	1.1	0.9	1.3	0.5	1.6	1.2	1.1	0.9	1.2	1.2	1.3	1.4
1.4	0.5	0.4	0.5	0.6	0.5	0.5	1.5	0.5	0.5	0.4	2.5	1.6
1.5	2.0	1.4	1.2	1.6	1.4	1.6	0.3	0.3				

In this example, leaf length is called a **random variable** since it *varies* from leaf to leaf and in a way that is unpredictable, i.e. *random*. In the other examples you have examined so far, in *Chapter 1*, other quantities are random variables. In the example on infants with severe idiopathic respiratory distress syndrome (*Chapter 1*, Example 1.3), birth weight (in kg) was a random variable. In the marathon runners example (*Chapter 1*, Example 1.4), blood plasma β endorphin concentration (in pmol/l) was a random variable. ■

A random variable may take any value from a set of possible values, but some may be more likely than others to occur. If another leaf were to be measured from the same bush, we should not be too surprised if its length took a value between 0.3 cm (the smallest value from the 100 the researcher measured) and 2.7 cm (the largest value amongst the 100). However, it is clear from the histogram of the 100 leaf lengths in Figure 2.1, that not many leaves have lengths greater than 2.4 cm. Only five of those measured were as long as that. On the other hand, many of the leaves from the sample have lengths between 0.8 cm and 2.0 cm (68 from the 100, including the three that were 0.8 cm long).

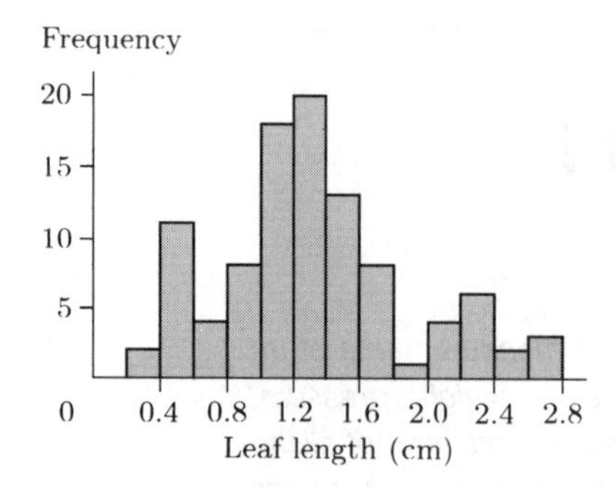

Figure 2.1 Leaf lengths

In this diagram, leaves whose recorded lengths were exactly 0.4 cm (say) have been allocated to the class interval 0.4–0.6 cm.

If another leaf were to be taken from the same bush and measured, we might feel that it was more likely to be between 0.8 cm and 2.0 cm long than it was to be longer than 2.4 cm. Putting this another way, we might say that the **chance**, or **probability**, of obtaining a leaf between 0.8 cm and 2.0 cm long is greater than the chance, or probability, of obtaining a leaf longer than 2.4 cm. On the same sort of experimental basis, we might feel that there was no chance at all of the bush having any leaves longer than 10 cm. This chapter explores the notions of *random variable* and *probability* in some detail, showing how they are related to each other and how they can be used to shed light on natural phenomena.

In statistics, as in any other science, we observe a natural phenomenon and then try to formulate a **model** to describe it. The models we use, as in many other sciences, are mathematical models. However, as also in other sciences, we often cannot model the awesome complexity of a natural phenomenon in all its detail, and so we make simplifying assumptions. We try to identify the important influences and relationships in the system we are studying. Then we try to represent those features with sufficient accuracy so that our conclusions or predictions are good enough for the purpose for which we intend to use them.

Statistical techniques are used to produce models in several ways, some more formal than others. You have already seen in *Chapter 1*, examples of statistics being used to find structures or patterns in data. In this chapter we are primarily concerned with probability models: models for how likely it is that particular events will occur.

We shall begin in Section 2.1 by addressing the notions of *sample* and of *population*. A population is the set of all possible objects from which the measurements might be taken—all the leaves on the bush in Example 2.1—while a sample is the set on which measurements are actually taken (the leaves that are actually measured). Section 2.1 discusses why one might draw a sample from a population, rather than measuring all of the objects in the population.

The concept of probability has already been informally introduced. Probability is an abstract concept that can be used to model random phenomena in the real world. These models can become fairly intricate, as the systems they represent become more and more complex. We shall begin rather simply by formulating models for a restricted class of random phenomena where all that is of interest to the researcher is whether or not a particular event happened. For instance, an insurance underwriter renewing a motorist's annual cover will primarily be interested in whether or not a claim has been made; a quality inspector, sampling items from a production line, will primarily be concerned with whether or not a sampled item is defective; an archer in whether or not her next arrow hits the target; a chess master in whether or not he wins his next game.

You can probably appreciate that a model expressing only two possible outcomes (Yes–No, Win–Lose, Hit–Miss, Present–Absent) may not be quite adequate in all these examples: the insurance underwriter will be concerned about the *size* of any claim made against issued policies, or the *number* of claims; the archer in her actual *score*. Different, more elaborate, models would need to be developed to meet these requirements. We shall look at such models later in the course.

In Section 2.2 this concept of probability is extended from situations where there are just two possibilities (something occurring or not) to situations where more than two outcomes are possible (for example, rolling a die in a board game where the possible outcomes are $1, 2, 3, 4, 5$ and 6) and then to situations where any possible value within a certain range (a continuum) might occur.

The leaf length example is an example of the latter. The data listed in Table 2.1 suggest that the leaf length measurements were only recorded to the nearest tenth of a centimetre; however, in practice, leaf length is a *continuous random variable*, not restricted in nature to exact millimetre separation. In other words, leaves do not just come in two (or six or twenty) different lengths, but can have any length between some minimum and some maximum. This kind of consideration is intrinsic to the formulation of probability models for observed variation. The idea of determining the probability, for some random variable, that a value within a certain range will be observed, is fundamental to much of statistics and will recur throughout this course.

In Section 2.3 two particular models are discussed: the *Bernoulli probability distribution* and the *binomial probability distribution*. The first of these is used to describe situations in which the random variable can take only two possible values (corresponding to a response of Yes or No in an opinion poll, success or failure in an examination, and so on). It is a matter of chance which of the two outcomes is observed. The second distribution is concerned with collections of such **binary** outcomes and looks at *the total number* of outcomes of one kind (for example, the total number of Yeses in an opinion poll, or the number of students in a class who pass the examination or the total number of defective items in a quality sample.)

The word binary is derived from the Latin word *binarius*; cf. the **binary system** of counting, which uses only the two digits 0 and 1.

In order to build a mathematical model for a random variable, it is useful to match outcomes to numbers. If you are measuring lengths, scores, times and so on, then the necessary identification is naturally made. In the case of outcomes like Yes–No, Hit–Miss, Pass–Fail, On–Off, then it is usual to match one outcome to the number 1 and the other to the number 0. You would then obtain the total number of Yeses in your poll (say) in a very natural way by adding together all the 1s and 0s. Bernoulli random variables and binomial random variables are *integer-valued*: the possible values observed are restricted to integers (whole numbers). Such random variables are called discrete.

In Section 2.4 we take a first look at a probability model for a particular class of *continuous* random variable (such as leaf length). The model is known as the *normal distribution*. *Chapter 5* is devoted to discussing the properties and importance of the normal distribution. The section in this chapter is short and you need not spend too much time on it. It contains no exercises.

In Section 2.5 the notions of *simulation* and the use of a computer to model a random phenomenon are briefly introduced: simulation is discussed again in more detail in *Chapter 4* and throughout later chapters of the course.

The chapter ends with a series of exercises on the material so far, which you are recommended to attempt using a computer. (It is not really feasible to attempt them any other way.) The computer exercises in the course should be regarded as essential: one learns by doing and not just by reading about doing.

2.1 Random variables and probability

2.1.1 Samples and populations

In the leaf length experiment described in the introduction to this chapter, the researcher did not completely denude the bush in his investigation. In fact, he took the measurements of only 100 leaves. In statistical terminology, we say that he measured a **sample** of 100 leaves from the entire **population** of leaves on the bush. The word *population* thus refers to all of the leaves and the word *sample* to some subset of them. In the example on infants with severe idiopathic respiratory distress syndrome in *Chapter 1* (Example 1.3), a sample of 50 from the population of all infants with this condition was examined.

The leaf length researcher went on to calculate the average leaf length for the sample of 100 leaves, obtaining a sample mean of 1.276 cm. However, we know that leaf length is a random variable. If, for example, instead of one of the leaves in his sample, he had plucked a different leaf, it would probably have had a different length. This, of course, is true for all of the leaves in his sample. The implication is that if he had taken a different sample of 100 leaves his calculated average would have been different. This raises the obvious question: if his results vary according to the random choice of leaves in the sample, of what value are the results? Presumably he really hoped to say something about the average length of the population of leaves and not just of an arbitrarily chosen sample. How can the sample average, which varies from sample to sample, tell us anything about the average length of all the leaves on the bush?

These questions find their answers in the discipline of statistics. Statistics enables us to take just a sample of values and from these deduce conclusions about the entire population; conclusions whose validity or accuracy we can assess. It tells us *how* the sample should be taken so that valid and accurate conclusions can be drawn. (For example, it would clearly be incorrect for the researcher to pick large leaves deliberately—the average of a sample of 100 large leaves would produce an average larger than the average of the entire population of leaves. Somehow he should strive to produce a *representative* sample, or at least a sample for which he can say how likely it is to be representative and how likely it is to be extreme.) Finally, (and perhaps most importantly) statistics shows us how to compute numerical measures of the reliability of our conclusions.

Sometimes working with a sample is a necessity because it is impossible to take a **census**, that is, it is impossible to use the entire population. For example, this is clearly the case if the population is the population of humans on the earth. Indeed, in this example it is arguable that the population is not even well-defined since people are being born and others are dying even as you read this. In other cases, even if it is possible in principle to use the entire population, it may not be realistic to consider doing so. The savings in money or time arising from working with a relatively small sample as opposed to an entire population may make the difference between permitting a question to be answered and leaving it unanswered. Of course, the use of a sample can only be considered if we have some way of determining how accurately the sample results reflect population characteristics. In other words, the use of a sample can only be considered because of the existence of statistical techniques.

Example 2.2 *Social surveys*

A very important application of statistical sampling is in social investigation. In this area many different kinds of samples are drawn for many different purposes. They include surveys to explore the demographic characteristics of a set of people, that is, their social environment, economic functioning, health, opinions and activities. Many such surveys are conducted for governmental planning purposes.

Market research is also an important user of social survey techniques. Here, a sample of people from the population being studied is questioned about their purchases in some sector. You can see here how important it is to draw the sample properly—to make sure that it is chosen from the population one wishes to draw conclusions about, and in such a way that it permits conclusions to be drawn. For example, in studying the relative consumption of different brands of chocolate bars we would not want to restrict ourselves to households with no children. Since children are major consumers, such a sampling strategy would grossly distort the results. Similarly, in conducting a public opinion poll on the desirability of restricting access by private cars to a town centre, it would be important that the opinions of all those likely to be affected were represented and that the relative numbers of those likely to be affected could be estimated.

Other major types of sample-based social surveys are audience research, readership, family expenditure, nutrition, education, social mobility, leisure, and studies of criminal and deviant behaviour.

A census is based on a complete enumeration of the entire population of interest and so no issues of inference from a sample to the population arise. Notice, however, that other statistical considerations will arise with census data. We shall still have to condense the data, summarize them, and find ways of displaying and representing them so that they are accessible and comprehensible. ■

Exercise 2.1

A researcher studying reading habits stands outside a London railway station and approaches people at random as they leave the station. She asks them which newspapers they read. What are the inadequacies of this way of drawing the sample? What differences might you expect between samples drawn at 9 am on a weekday and at 7 pm on a Saturday?

The point being made in Exercise 2.1 is that in drawing a sample it is essential that it should represent the population being studied. In a study of road traffic accidents, you might decide to examine a sample of 50 accidents. However, unless you wanted only to make inferences about motorway accidents, these 50 should not just be motorway accidents but should be drawn from the entire population of different types of road accident.

2.1.2 Probability

Example 2.3 *Street lamp bulbs*

Suppose that a local council suspects that the latest consignment of sodium tubes for its street lamps is of poor quality, with an unacceptably large proportion of them being faulty and not working. To investigate this, a researcher examines a sample of bulbs from this consignment waiting in the warehouse.

This example is fairly typical of a particular problem of quality control: the estimation of 'percentage defectives' in a batch of supplied items. In practice, quality procedures are much more rigidly prescribed than is suggested by the sequential testing approach described in what follows, both for the supplier and end-user. However, we shall be mainly concerned with the essential idea of testing a random sample from a population (the population being, in this case, the recent consignment of bulbs).

A 'random sample' here is strictly one where no item in the population is any more likely than any other to be incorporated into the sample. In this simplified description of events, bulbs less accessible than others are probably less likely to be sampled! Hence the need, in practice, for strict quality procedures.

The researcher enters the warehouse and randomly chooses bulbs. The first bulb works. Amongst the first ten bulbs he examines, he finds that only one does not work. Amongst the first fifty, nine do not work. Amongst the first 100, fifteen do not work. The researcher could go on, examining all the bulbs in the warehouse, and find the exact proportion of bulbs which do not work. Alternatively, he may use statistical arguments to support a claim that the proportion of faulty bulbs within a sample of a certain size provides a sufficiently accurate estimate of the overall proportion that is faulty.

As the researcher increases his sample size he will observe an interesting phenomenon. For small samples, quite substantial deviations between the proportions of faulty bulbs for different sample sizes are observed. Suppose that the first two bulbs he examined were functioning, then he came across a faulty one so that the sequence of the observed proportions of defectives initially was $0/1, 0/2$ and $1/3$. These may be regarded as successive estimates of the unknown overall proportion of defectives and the last of these three estimates is quite different from the first two. As the sample gets larger, however, the deviations between successive estimates tend to become smaller. Suppose for the 98th, 99th, and 100th bulbs that he tested the estimates were $14/98, 15/99$ and $15/100$. These proportions are much closer to one another. The proportion of bulbs observed to be faulty seems to be settling down towards some constant value (about 0.15). ■

We can **simulate** this phenomenon ourselves, that is, we can produce a simplified version or model of this situation which behaves in the same way. Later on you will be encouraged to perform such simulations using a computer, but for this first exploration the results will be more striking if we carry it out for some physical system. The simulation tools that we shall use are dice and coins. In many ways these provide ideal models because they strip away much of the complexity so that we can focus on the essentials.

Exercise 2.2

Using a die from a family board game, carry out a sequence of 30 rolls with a view to obtaining either a three or a six. Each time that a three-spot face or a six-spot face occurs record a 1, to represent success. Each time that some other face of the die occurs record a 0, to represent failure. In this way you will generate a sequence of 1s and 0s. At any stage you will not be able to predict whether the next number will be a 1 or a 0—so the next number is a random variable.

By adding the 1s and 0s as you go along you can calculate the total number of successes after each roll. At each roll you can also calculate the proportion of rolls so far that have produced a three- or six-spot face. Letting P denote this proportion, plot a graph of P, on the vertical axis, against the number of rolls, on the horizontal axis. Join together successive points to produce a jagged line running from left to right across the page.

Not everybody has easy access to a six-sided die—some people will not possess a board game. You can get round the problem this way. Obtain a six-sided pencil or ball-point pen and identify particularly two sides of the six (to match the 'three' and the 'six' of the exercise). These could be the side impressed with the maker's name and the next side round; or, in the case of a ball-point pen, the side with the maker's name and the side with the little hole in the casing.

Your sequence in Exercise 2.2 might have begun

0 0 1 1 0 0 0 0 0 0 1 1,

which would give you a sequence of points to plot as shown in Table 2.2.

Table 2.2 The first 12 rolls

Roll number	1	2	3	4	5	6	7	8	9	10	11	12
Observed result	0	0	1	1	0	0	0	0	0	0	1	1
Total so far	0	0	1	2	2	2	2	2	2	2	3	4
Proportion (P)	0/1	0/2	1/3	2/4	2/5	2/6	2/7	2/8	2/9	2/10	3/11	4/12

An example of the kind of graph you might have obtained is given in Figure 2.2. The thirty rolls which produced this graph gave the following results.

0 0 1 1 0 0 0 0 0 0 1 1 0 0 1 0 0 0 0 1 0 0 1 0 1 0 0 1 0 0

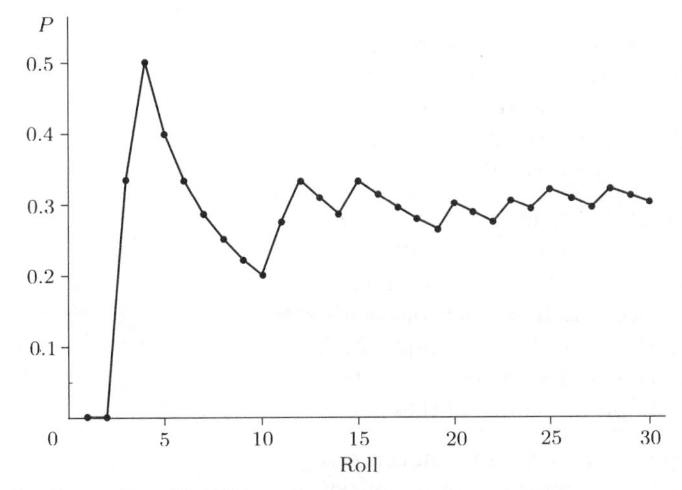

Figure 2.2 Proportion P, thirty rolls of a die

The phenomenon described above is apparent in this graph. As the number of rolls increases, the sequence of estimates given by successive values of P becomes less variable. The sequence appears to be tending to a particular constant value, in this case about $\frac{1}{3}$. Did you notice something like this happening for your sequence of estimates?

The sequence of rolls that resulted in the graph in Figure 2.2 was taken further—up to 500 rolls of the die. The points in Figure 2.3 show the proportions obtained. The phenomenon is even more marked here.

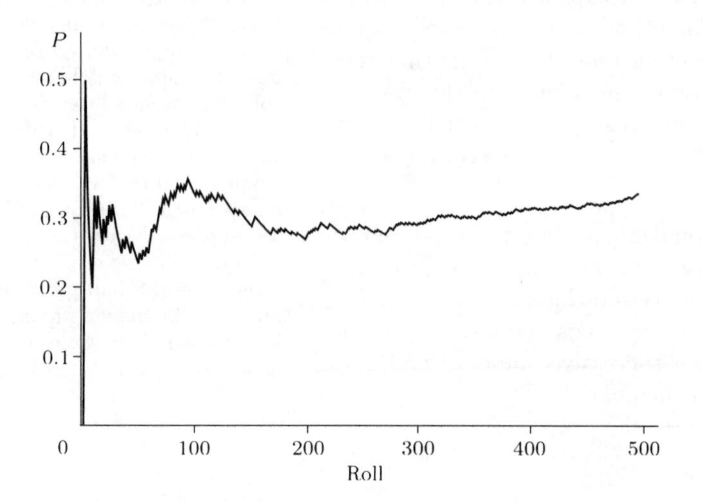

Figure 2.3 Proportion P, 500 rolls of a die

Exercise 2.3

Take a coin and carry out a sequence of 40 tosses, recording 1 for Heads and 0 for Tails. Again, the outcome of each toss is a random variable. Plot the proportion of tosses producing Heads against the number of tosses. What do you observe about the proportion of Heads as the number of tosses increases?

In Exercise 2.3, you might have observed behaviour similar to that of Exercise 2.2, but with the points defining the jagged line converging within a narrow region close to $\frac{1}{2}$ instead of $\frac{1}{3}$. However, what is readily apparent from as many as 500 repetitions of an experiment (Figure 2.3) will not necessarily be evident in as few as 30 repetitions (Figure 2.2) or even 40.

The four examples described so far (leaf lengths, quality testing of street lamp bulbs, the die and coin experiments) have some common features. In each example a number of cases is observed in which recorded measurements on a random variable may differ from case to case. In the leaf measuring example, the length of the leaves differed. In the street lamp example, bulbs may or may not be faulty. In the die-rolling and coin-tossing examples, there were two possible outcomes and each repetition produced one of these.

The leaf length example is different from the other three because, in this case, the measured random variable is continuous. For the moment, let us consider only the other three. Each example consists of a sequence of observations on a binary random variable (two possible outcomes only). Using 0s and 1s to label the different outcomes in each case, the results for any one of the examples could be recorded as a sequence of 0s and 1s, such as

0 0 1 1 1 1 0 . . . .

The term **Bernoulli trial** is used to describe a single statistical experiment resulting in one of only two possible outcomes. Each of these examples therefore consists of a *sequence of Bernoulli trials.* The outcome of each trial is uncertain (it was not possible to predict beforehand what it would be). We have also noticed, for the bulb, die and coin examples at least, a tendency for the observed proportions to settle down towards some constant value.

The value towards which the proportion is tending as the number of trials increases is called the **probability** of the particular outcome of interest.

In the introduction to this chapter it was remarked that probability models are used to represent the important aspects of random phenomena, and that it is often necessary in the construction of such models to make simplifying assumptions. This is true for the cases under discussion now. It is often assumed that a coin is *fair*—that there is no reason to suppose that Heads is more or less likely than Tails. So we might construct a probability model for the outcome of a coin-tossing experiment by starting with the assumption that the probability that Heads will appear is the same as the probability that Tails will appear—both being equal to exactly $\frac{1}{2}$.

Similarly, for the die example, we might assume that each face has the same probability of appearing. That is, we might choose to *model* the outcomes of rolls of a real die by those of a perfect die in which the appearance of each face is equiprobable. If we do this, the probability that a particular one of the six faces appears is $\frac{1}{6}$, and so the probability that one of the two faces showing three or six spots will appear is exactly $\frac{2}{6}$, or $\frac{1}{3}$.

Two other terms sometimes used in statistics are **sample frequency**, to describe the number of times an outcome occurs, and **sample relative frequency**, to describe the proportion of times an outcome occurs in a particular experiment. In a coin-tossing experiment in which a coin is tossed 100 times, coming up Heads 47 times, we would say that the sample frequency (or just 'frequency') of Heads was 47 and the sample relative frequency of Heads was $\frac{47}{100} = 0.47$.

In the examples above, the modelling assumption is made that the probability that a three- or six-spot face will appear in a roll of a perfect die is $\frac{1}{3}$; and that the probability that Heads will appear in a toss of a perfect coin is $\frac{1}{2}$. The bulb-testing example is different. We cannot deduce, from any kind of theoretical argument based on symmetry, a numerical value for the probability that a bulb will be faulty. This probability has to be estimated from data—in Example 2.3, this probability estimate was the sample relative frequency of failures, 0.15 (based on 15 observed failures in 100 trials). We might say, on the basis of this experiment, that we believe the proportion of faulty bulbs in the latest consignment to be about 0.15.

Exercise 2.4

In an experiment to explore the issue of whether people are generally more helpful to females than to males, eight students approached people and asked if they could change a 5p coin. Altogether 100 people were approached by the male students and 105 by the female students. The results of the experiment are displayed in Table 2.3.

Sissons, M. (1981) Race, sex and helping behaviour. *British Journal of Social Psychology*, **20**, 285–292.

The sum of 5p may seem very small indeed to the modern reader, and prompt the question why change might be required for such a sum. At the time the experiment was carried out, a local telephone call could be made from public telephone boxes for as little as 2p.

Table 2.3 Helping behaviour

Gender of student	Help given	Help not given
Male	71	29
Female	89	16

(a) Use these data to estimate the probability that a male is given help under these circumstances.

(b) What would you estimate this probability to be for a female?

(c) Do the results of the experiment support the notion that people are more helpful to females?

In Exercise 2.4 you were asked to estimate two probabilities. A large part of this course is concerned with estimation—methods for obtaining estimates and assessing how reliable these estimates are. You were also asked to comment on the meaning of the results of the experiment. Later in the course (*Chapter 8*) you will see formal ways of doing this—so that you can actually quantify the extent to which experimental results support a research hypothesis.

Now let us take a slightly more complicated example. In a Bernoulli trial there are only two possible outcomes: what if we were to consider, say, each of the six faces of a die as distinct possible outcomes? Suppose the die is rolled. What is the probability that the face with a single spot appears, what is the probability that the face with two spots appears and so on?

We could carry out an experiment analogous to those above—rolling the die a large number of times and calculating the relative frequency with which each of the faces came up. This would enable us to directly estimate the probability that each face occurs. Alternatively, since there is no reason to suppose that any one of the six faces is favoured more than any of the others, we could argue that each face is equally likely to appear—that the probabilities are equal for each face. That is, our *model* would be that each face has a probability of $\frac{1}{6}$ of appearing. This would mean that about $\frac{1}{6}$ of the time we should expect the face with one spot to come up, about $\frac{1}{6}$ of the time we should expect the face with two spots to come up and so on. As we rolled the die more and more times, so, under our model, we should expect the proportion of times each face appears to get closer and closer to $\frac{1}{6}$.

Example 2.4 *A simulated die-rolling experiment*

Here is a simulation of an experiment based on the assumption of a perfect die. The simulation was achieved using a computer to mimic the rolls and also to record the outcomes. The die was rolled a large number of times, and the observed frequencies of 1s, 2s, ..., 6s (six possible outcomes) recorded after 30, 300 and 3000 rolls. At each stage the sample relative frequency for each outcome was calculated.

Table 2.4 Simulation results for a perfect die

	Outcome	1	2	3	4	5	6
30 rolls	Frequency	5	3	6	4	5	7
	Relative frequency	0.1667	0.1000	0.2000	0.1333	0.1667	0.2333
300 rolls	Frequency	55	54	56	52	46	37
	Relative frequency	0.1833	0.1800	0.1867	0.1733	0.1533	0.1233
3000 rolls	Frequency	506	508	497	525	467	497
	Relative frequency	0.1687	0.1693	0.1657	0.1750	0.1557	0.1657

■

It is apparent from the observed frequencies that the computer model is capable of reflecting the random variation inherent in the physical process. Although we might 'expect' 50 observations of each of the six equiprobable outcomes after 300 rolls of the die, we would nevertheless be somewhat surprised to see exactly that: actually, there were only 37 6s. However, after a large number of rolls, the observed relative frequency for each of the possible outcomes seems to be showing a tendency to approach the theoretical value of $\frac{1}{6} = 0.1667$. This tendency is shown graphically in Figure 2.4, summarizing the findings in Table 2.4.

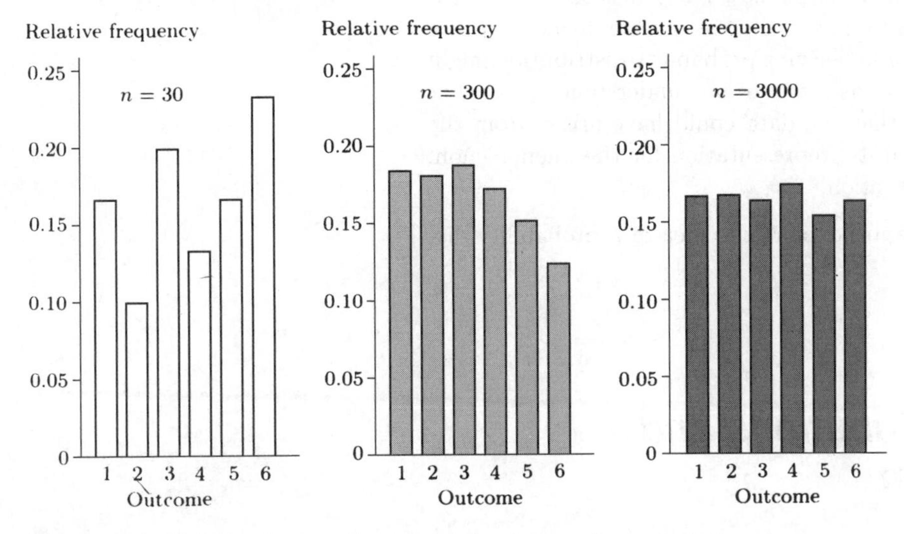

Figure 2.4 Relative frequencies from a simulation of a perfect die
(a) 30 rolls (b) 300 rolls (c) 3000 rolls

Assuming a perfect die and applying a symmetry argument, Figure 2.5 illustrates the conclusions of a very extended experiment involving many rolls of the die.

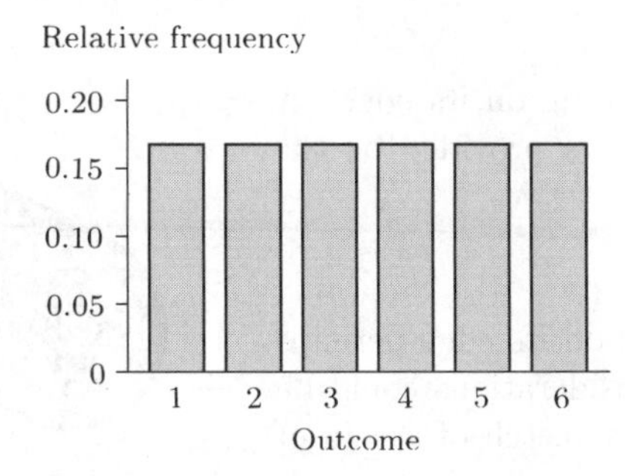

Figure 2.5 The perfect die

Put another way, the diagram in Figure 2.5 shows the theoretical **probability distribution** over the six possible outcomes $(1, 2, \ldots, 6)$ for a die that is assumed to be perfect.

It will be useful if at this stage some simple notation is introduced. In this course, the probability that a statistical experiment results in a particular outcome will be denoted $P(\cdot)$. For instance, the earlier statement that, in tossing a perfect coin, Heads and Tails are equally likely so each outcome has probability $\frac{1}{2}$, would be written

$$P(\text{Heads}) = P(\text{Tails}) = \tfrac{1}{2}.$$

The theoretical probability distribution for the perfect die may be written

$$P(1) = P(2) = P(3) = P(4) = P(5) = P(6) = \tfrac{1}{6}.$$

In particular, referring to the original experiment (Exercise 2.2),

$$P(3 \text{ or } 6) = \tfrac{1}{6} + \tfrac{1}{6} = \tfrac{1}{3}.$$

This concept of a probability distribution is a very important and fundamental one in statistics. It tells us how likely it is that different outcomes (or, in general, *events*) will occur. For example, you will see later how the idea can be used to work out for a random sample how likely it is that a particular average will occur. Conversely, we might express our confidence, given a sample average, that some particular underlying probability distribution might have generated the data. This serves as a test of our understanding of what is going on. If it is very unlikely that the data could have arisen from the model that we believed to be a realistic representation for the phenomenon, then maybe our beliefs need modification.

The next three sections of this chapter explore the idea of a probability distribution in more detail.

2.2 Probability distributions and describing them

In Section 2.1 the notation $P(\cdot)$ was introduced to stand for the probability of a particular outcome in a statistical experiment or, in general, for the probability of any event of interest. In the case of a perfect coin, for instance, a theoretical probability model for the two possible outcomes is given by

$$P(\text{Heads}) = P(\text{Tails}) = \tfrac{1}{2};$$

but that model precludes the possibility of the coin landing on its edge. A different model, incorporating a third possible outcome, is provided by the probabilities

$$P(\text{Heads}) = P(\text{Tails}) = 0.4995, \quad P(\text{Edge}) = 0.001.$$

This model suggests that, on average, a coin might land on its edge perhaps once in a thousand tosses. Presumably, engineering considerations could be brought to bear on the problem. Here in Figure 2.6 is a sketch of a coin of rather unusual shape. You can see that the coin is exceptionally thick. Were

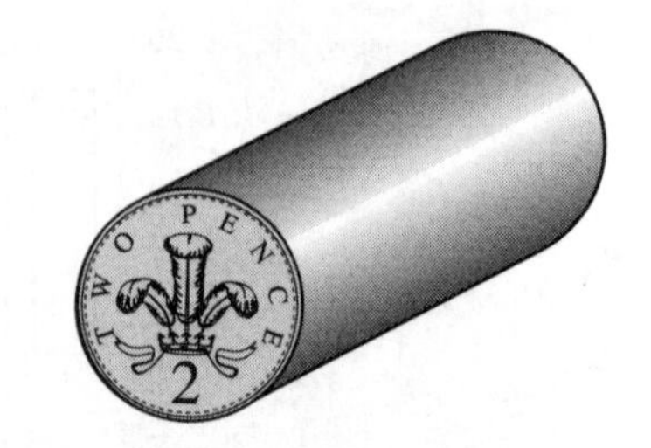

Figure 2.6 An unusual coin

this coin to be tossed, then its chances of landing on its edge would be high and the probability of coming down Heads or Tails correspondingly somewhat reduced.

In the next subsection we shall explore ways of expressing such probability statements in mathematical language. We have already made a start on this: associated with any Bernoulli trial (just two outcomes) there has often been defined a corresponding random variable taking just two possible values, 0 and 1. For instance, in a quality-testing context we might score 1 for a working item and 0 for an item that is defective in some way. A doctor completing a diagnostic check-list on a patient might assign the score 1 if a symptom is present and 0 if it is not. (It does not matter which way round the association is made, only that the two possible outcomes should be distinguished in this way.)

2.2.1 Using mathematical language

In the introduction to this chapter, it was said that in order to build a mathematical model for a random variable, it is useful to match outcomes to numbers. This kind of identification is naturally made if the random variable of interest is a count or a measurement and, in the case of a Bernoulli trial, as we have just discussed, we invent a correspondence between the trial outcomes and the numbers 0 and 1.

The reason for making a point of this is that we can then use the idea of a **mathematical function** (or, simply, *function*). If you know anything at all about functions, you will remember that the essential feature is this: starting with a number, a function is a process that alters that number to produce a second (usually different) number. Here are some examples.

Example 2.5 *Mathematical functions*

(a) Most calculators include keys with labels such as $\sqrt{x}$ (the square root function), $1/x$ (the reciprocal function) and usually several others. If you key in a number to the calculator display and then press one of these 'function' keys, the calculator display alters. The display shows you 'what has happened to' the number you started with, that is, the result of processing it according to the function key you used. In mathematical language, we might denote the process of finding the square root of a number by the function $\text{sqrt}(\cdot)$ and write

$$\text{sqrt}(25) = 5.$$

Similarly, $\text{sqrt}(81) = 9$ and $\text{sqrt}(3) = 1.732\,050\,808$ (depending on how many digits your calculator is capable of displaying).

(b) Depending on the function you choose, some starting numbers are allowable and others are not. If you try to evaluate the square root of the number -3 (a negative number), your calculator will respond in such a way as to indicate that this was a meaningless question: negative numbers do not possess square roots.

It is possible to define a counting system in which square roots of negative numbers exist, but it is a system more elaborate than this course requires.

(c) In most systems for raising revenues through taxation on income, there is a set threshold income below which an individual will pay no tax. Above that threshold, the difference between actual income and the set threshold is called taxable income. A fraction of this taxable income is paid to the revenue collectors. In a country where the threshold is, say, £3000 and where the tax rate on taxable income is 25% (one-quarter), an individual earning £2800 will pay no tax. This may be expressed in mathematical language as

$$\text{tax}(2800) = 0.$$

Somebody earning £3800 is liable to pay one-quarter of the taxable income to the tax collectors: that is,

$$\text{tax}(3800) = \tfrac{1}{4}(3800 - 3000) = 200,$$

a tax bill of £200. The mathematical function tax$(\cdot)$ may be expressed as

$$\text{tax}(x) = \begin{cases} 0 & 0 \le x \le 3000 \\ \frac{1}{4}(x - 3000) & x > 3000, \end{cases}$$

where the 'starting number' x is an individual's total income (in £). The function tax$(\cdot)$ gives the tax (in £) that must be paid. The need for the curly bracket in the definition of the function tax$(\cdot)$ is a consequence of the fact that there are two formulas for calculating individuals' tax liability, depending on whether or not their income exceeds the threshold. The function tax$(\cdot)$ is not defined for negative x (an income less than zero, which has no interpretation). ■

The full mathematical description of a function is more complicated than is suggested by the foregoing, but we now know enough about functions to return to a probability context. The point about probability statements of the kind

$$P(\text{Heads}) = \tfrac{1}{2} \quad \text{or} \quad P(\text{faulty}) = 0.15$$

is that, whilst they are clear and unambiguous, the starting element for $P(\cdot)$ is not a number but an event. ('The result of the next toss of the coin will be Heads'; or 'the next bulb will be faulty'.) It follows that $P(\cdot)$ is not a 'function' in the sense that the notion of function has been described. On the other hand, the notation $P(\cdot)$ is a very useful general shorthand.

Associating a particular outcome of a statistical experiment with a number enables us to express the probability distribution for the range of possible outcomes in terms of mathematical functions. If, in the usual way, we were to match the outcome Heads to the number 1 and the outcome Tails to the number 0, we could write

$$p(x) = \tfrac{1}{2}, \quad x = 0, 1,$$

to suggest that the random variable X denoting the outcome of the experiment takes the value 0 or 1 with equal probability. The function $p(\cdot)$ possesses the properties of a mathematical function: starting with a number x (describing an event), the function produces a second number $p(x)$ (the probability of that event).

In such a case, the function $p(x)$ is called the **probability function** for the random variable X, and the following probability statements are equivalent:

$$P(\text{Tails}) = P(\text{Heads}) = \tfrac{1}{2};$$
$$P(X = 0) = P(X = 1) = \tfrac{1}{2};$$
$$p(x) = \tfrac{1}{2}, \quad x = 0, 1.$$

Here the distinction between $P(\cdot)$ and $p(\cdot)$ is exemplified: $P(\cdot)$ describes the probability of an event, while $p(\cdot)$ is the corresponding methematical function.

However, only the third statement uses a mathematical function and this function now defines our probability model for the perfect coin. To a great extent these subtle distinctions are unimportant, but, in general, it would be more difficult to make progress with the idea of a probability model and (more importantly) with the applications of the model, if the notion of the probability function $p(\cdot)$ was lacking.

Notice the convention that an upper-case letter (e.g. X) is used for the label of the random variable under consideration, while the corresponding lower-case letter (x) is used as representative of the possible values it might take.

In the case of the consignment of bulbs (Example 2.3), the proportion faulty was estimated to be 0.15 based on a sample of 100. A proposed probability model, scoring 1 for a faulty bulb and 0 for one that 'passed', could be expressed as

$$p(x) = \begin{cases} 0.85 & x = 0 \\ 0.15 & x = 1, \end{cases}$$

where the random variable X takes the value 0 or 1 as each bulb tested passes or fails the quality test.

For the die-rolling experiment, there is already a natural correspondence between the outcome of a roll ('the face that came up') and a number $(1, 2, 3, 4,$ 5 or 6). So, if the random variable N denotes the outcome of a single roll of a fair die, we could write down the probability function for N as

$$p(n) = \tfrac{1}{6}, \quad n = 1, 2, 3, 4, 5, 6.$$

Notice again the convention that an upper-case letter is used for the random variable under consideration, while the corresponding lower-case letter is used to represent the possible values it might take. The list of possible values is called the **range** of the random variable.

Compare this usage of the word 'range' with its use in a sampling context: the range of a random sample is the difference between the maximum sample value obtained and the minimum value.

By convention, the notation lower-case $p(\cdot)$ is often used for the probability function, but other letters of the alphabet will also be used. The only letter you should avoid using to stand for anything other than the phrase 'the probability of' is $P(\cdot)$.

There are two extensions of this idea of associating a particular outcome with a number. They occur where the random variable can, in principle, take an unlimited number of different values—not just two, or six, for example.

First, the range of observations might be restricted to exact integer counts, but where there is no necessary maximum (as there is in the case of a score on a die). Some board games involving dice require that a 6 is rolled before a player can start. The number of rolls necessary to achieve a 6 is a random variable—it could happen first time or possibly not for several rolls—and in fact, theoretically, there is no assurance of success by any particular roll. The random variable describing the number of rolls necessary to start playing the game has an infinite range $1, 2, 3, 4, \ldots$. (Extremely high values are unlikely, but they are not impossible.)

Secondly, in the leaf length example (Example 2.1) it was mentioned that although observations were recorded to the nearest millimetre, it is not a law of nature that leaves only come in exact millimetre lengths—in fact, the range of observable leaf lengths constitutes a continuum between some minimum and maximum values. (In this case, observations at either extreme were seen to be less likely than intermediate values; when the time comes to formulate a *continuous* probability model for leaf length, this is the sort of feature that it must reflect.)

We shall return to these points. First, let us consider another example involving die rolling—one where not all the outcomes have the same probability.

Example 2.6 Crooked dice

Certain crooked dice players use dice called *Tops*. These are dice for which it is not the case that all six faces are marked with a different number of spots—some of the faces are duplicated on opposite sides of the die. Since the players cannot see round corners, no one player will recognize that the die is numbered incorrectly—unless he picks it up, of course, and the cheat makes sure this never happens by switching the *Top* in and out of the game as required. A particular kind of *Top* is like a normal die except that it has the two-spot face replaced by another five-spot face. This is called a *Double-Five*. Figure 2.7 shows the probability distribution for the face recorded when such a die is rolled. A five-spot face has twice as much chance of appearing as any of the other four possible faces—no roll of the die will result in the outcome 2. Our probability model assumes that the die has been perfectly engineered. The probability function for W, the outcome observed when a *Double-Five* is rolled, may be written as follows.

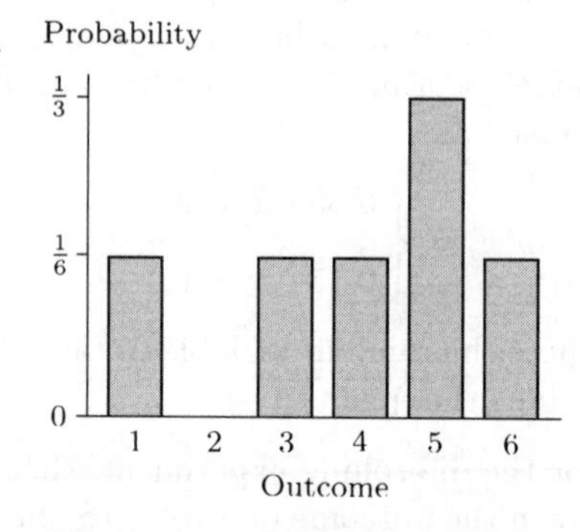

Figure 2.7 The probability distribution for a *Double-Five*

$$p(w) = \begin{cases} \frac{1}{6} & w = 1, 3, 4, 6 \\ \frac{1}{3} & w = 5. \end{cases}$$

The range of W is $\{1, 3, 4, 5, 6\}$. ■

Now let us consider a case in which the random variable is continuous and can take any value from within a particular interval. We could use the histogram of the sample of leaf lengths in Figure 2.1 to provide estimates of the probabilities for the leaf population that the leaf lengths lie within certain intervals. For instance, 68% of our sample lies in the interval 0.8–2.0, and we might estimate the probability that a randomly plucked leaf would have length between 0.8 cm and 2.0 cm to be 0.68. Similarly, we might estimate the probability that a randomly plucked leaf would have length less than 1 cm to be 0.25.

Of course, had we used a different sample of size 100, we should not have expected the histogram to have precisely the same shape as that in Figure 2.1—in just the same way, as we remarked at the time, that a different sample would be expected to give rise to a slightly different average. This would mean that our estimates of the probabilities would vary from sample to sample. However, if we used a larger sample size, say, samples of 1000 leaves, then we should expect the shapes of histograms from several such samples to be very similar. This is just a generalization of the fact, noted in Example 2.4, that the proportion of dice rolls producing different outcomes does not vary greatly from sample to sample if large samples are used. Figure 2.8 illustrates summary histograms for three different samples of leaf lengths. All the samples were of size 1000.

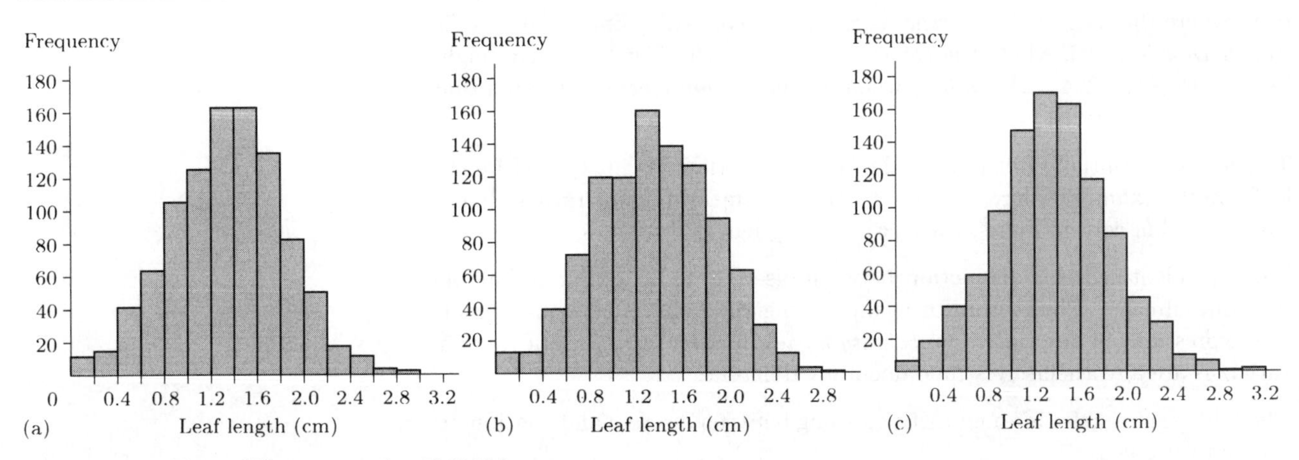

Figure 2.8 Three different samples of 1000 leaves

The three samples of leaf lengths were in fact generated by computer from one probability model, the details of which are not essential to make the point. (Notice that the model evidently permits extremely small leaves! Maybe the model needs refining.) The point is that the histograms are very similar to one another. Sample relative frequency estimates of the probability that a randomly selected leaf will be between 0.8 cm and 2.0 cm were obtained for each sample. They were $\frac{805}{1000} = 0.805$ (sample (a)), 0.798 (sample (b)) and 0.801 (sample (c)). This suggests that a large sample is not only very suggestive of the overall 'shape' of the variation that might be observed in measurements on a random variable, but provides good estimates of relevant probabilities. A large enough sample will result in a histogram which is effectively a smooth curve (see Figure 2.9), providing a very accurate picture of the population probability distribution. As in the cases of coins and dice, a mathematical expression of the curve—a function, in other words—could be used to provide a simplified *model* for the probability distribution of leaf lengths.

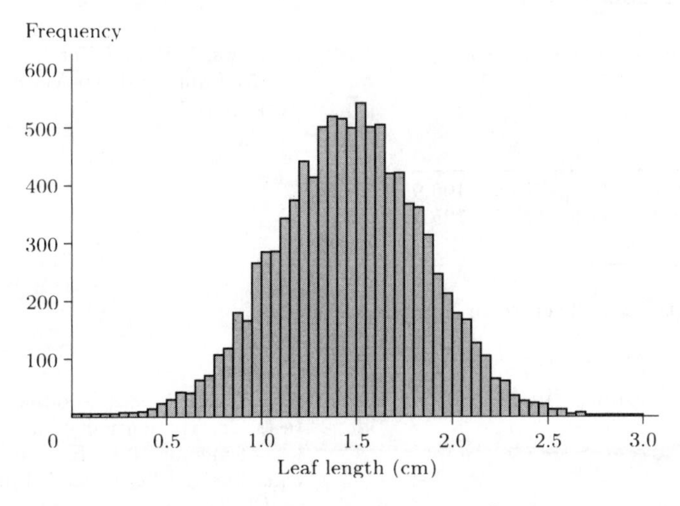

Figure 2.9 A histogram based on a very large sample

We have now seen examples of two essentially different kinds of random variable. The first, illustrated by the coin, die and bulb examples, takes only integer values $0, 1, 2, \ldots$ and so on. The range might be finite, as for a Bernoulli trial, where the range is $\{0, 1\}$, and as for the outcome of a single roll of a die (not a *Double-Five*), where the range is $\{1, 2, 3, 4, 5, 6\}$. Or the range might be infinite ($\{1, 2, 3, 4, \ldots\}$) as it is when listing the number of attempts that might be necessary to join a board game.

The second essentially different kind of random variable is illustrated by the leaf lengths example. Here, the observation is a measurement rather than a count—within certain limits, leaves can be any length.

We have identified this distinction between discrete and continuous random variables already. The reason for again making the point is that mathematical expressions of descriptive *probability models* for the two types of random variables are also themselves fundamentally different.

The following exercise is about distinguishing between discrete and continuous random variables.

Exercise 2.5

(a) Table 2.5 gives the numbers of yeast cells found in each of 400 very small squares on a microscope slide when a liquid was spread over it. The first row gives the number n of yeast cells observed in a square and the second row gives the number of squares containing n cells. (For instance, 213 of the 400 squares contained no yeast cells at all; no square contained more than 5 cells.)

'Student' (1906) On the error of counting with a haemocytometer. *Biometrika*, **5**, 351–360. 'Student' was the pseudonym of William Sealy Gosset (1876–1937). Gosset pursued his research whilst working for the Guinness brewery company, which prohibited its employees from publishing their work—hence the need to write under a pen name. The probability distribution bearing the name 'Student' is a very important statistical model, which we shall meet in *Chapter 7*.

Table 2.5 Yeast cells on a microscope slide

Cells in a square, n	0	1	2	3	4	5
Frequency	213	128	37	18	3	1

Evidently, the number of cells per square is a random variable taking (in this experiment) observed values between 0 and 5. Would you model this variation using a discrete or continuous probability model?

(b) Table 2.6 gives the lengths (in mm) of the jawbones of 23 kangaroos of the *Macropus Giganteus* species.

Andrews, D.F. and Herzberg, A.M. (1985) *Data.* Springer Verlag, New York, p. 311.

Table 2.6 Jawbone lengths (mm)

108.6	115.8	113.1	109.0	117.5	90.1	108.4	114.9	106.9
124.0	134.5	117.9	130.9	144.3	133.9	136.1	137.7	125.3
129.3	153.9	153.0	152.6	154.7				

Would you choose to model jawbone length as a discrete or continuous random variable?

(c) The following is a list of 20 operational lifetimes (in hours) for components.

Angus, J.E. (1982) Goodness-of-fit tests for exponentiality based on a loss-of-memory type functional equation. *J. Statistical Planning and Inference*, **6**, 241–251.

Table 2.7 Operational lifetimes (hours)

6278	3113	5236	11584	12628	7725	8604	14266	6125	9350
3212	9003	3523	12888	9460	13431	17809	2812	11825	2398

What sort of model would you adopt for the variation in lifetimes?

(d) Daily rainfall (in mm) was recorded over a 47-year period in Turramurra, Sydney, Australia. For each year, the wettest day was identified, that is, that having the greatest rainfall. Table 2.8 shows the rainfall recorded for the 47 annual maxima.

Rayner, J.C.W. and Best, D.J. (1989) *Smooth tests of goodness of fit.* Oxford University Press, p. 146.

Table 2.8 Annual maxima, daily rainfall (mm)

1468	909	841	475	846	452	3830	1397	556	978	1715
747	909	2002	1331	1227	2543	2649	1781	1717	2718	584
1859	1138	2675	1872	1359	1544	1372	1334	955	1849	719
1737	1389	681	1565	701	994	1188	962	1564	1800	580
1106	880	850								

What sort of random variable would you use to model the variation evident here?

Incidentally, the seventh observation of 3830 mm seems very exceptional, even in a location as wet as Turramurra would appear to be! In fact, it turns out that a fairly standard probability model fits these data quite well without ignoring that observation as an outlier, or assuming a misprint. Without local knowledge, it is difficult to say whether or not the units of measurement (mm) have been incorrectly recorded or whether perhaps there is confusion over reporting what was actually measured. However, the fact that something has gone wrong in the recording of these data is clear: weather records give the *annual* average rainfall in Sydney as about 47 inches or about 1200 mm! The *Encyclopedia Britannica* (15th edition, 1985) reports that the heaviest *daily* rainfall then recorded anywhere in the world was 1870 mm at Cilaos, La Reunion in the Indian Ocean, during 15–16 March 1952.

This kind of situation where the data units are wrong, or omitted, or where it is not entirely clear what was being recorded, occasionally happens in statistics.

(e) Data are given in Table 2.9 for library book usage. A sample of 122 books was selected and for each of them the number of times they had been borrowed in the preceding twelve months was counted. (For instance, of the 122 books sampled, 65 had been borrowed just once in the preceding year.)

Burrell, Q.L. and Cane, V.R. (1982) The analysis of library data. *J. Royal Statistical Society, Series A*, **145**, 439–471. The authors collected the data from several libraries. These data are from one of the sections of the Wishart Library in Cambridge.

Table 2.9 Library book usage

Loans per year	1	2	3	4	5	6	7
Frequency	65	26	12	10	5	3	1

The number of loans is a random variable. Is the random variable discrete or continuous?

Here is a further example.

Example 2.7 Memory recall times

In a study of memory recall times, a series of stimulus words was shown to a subject on a computer screen. For each word, the subject was instructed to recall either a pleasant or an unpleasant memory associated with that word. Successful recall of a memory was indicated by the subject pressing a bar

Dunn, G. and Master, D. (1982) Latency models: the statistical analysis of response times. *Psychological Medicine*, **12**, 659–665.

on the computer keyboard. Table 2.10 shows the recall times (in seconds) for twenty pleasant and twenty unpleasant memories. The random variable 'recall time' is an example of a continuous random variable—in principle, any value within a certain range could arise.

Table 2.10 Memory recall times (seconds)

Pleasant memory	Unpleasant memory
1.07	1.45
1.17	1.67
1.22	1.90
1.42	2.02
1.63	2.32
1.98	2.35
2.12	2.43
2.32	2.47
2.56	2.57
2.70	3.33
2.93	3.87
2.97	4.33
3.03	5.35
3.15	5.72
3.22	6.48
3.42	6.90
4.63	8.68
4.70	9.47
5.55	10.00
6.17	10.93

Of key interest in this study was whether pleasant memories could be recalled more easily and quickly than unpleasant ones. Figure 2.10 shows a comparative boxplot for these two samples and we can see that although the smaller values of recall time for the unpleasant memories have values similar to the smaller values for the pleasant memories, the larger values tend to be substantially greater for the unpleasant memories. Both distributions are skewed. We shall explore more formal techniques for examining problems like this later in the course.

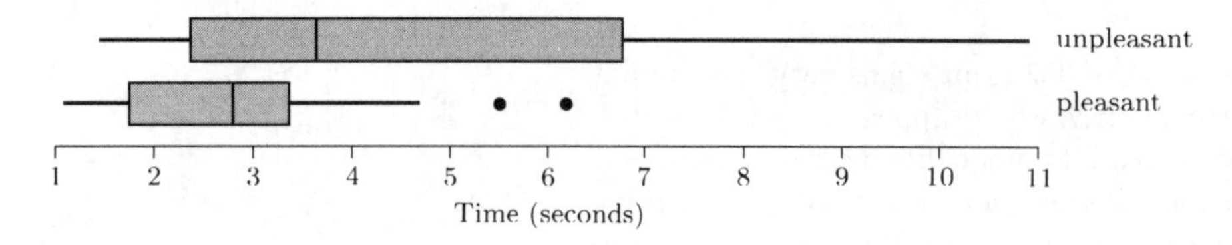

Figure 2.10 Boxplot for memory recall times ■

2.2.2 Modelling random variables

In the rest of this chapter and in the two that follow, we are going to be mainly concerned with two activities. The first is *constructing probability models* for some simple random phenomena (such as we have already done for the outcome of rolls of fair and crooked dice). The second activity is *using the models* to calculate probabilities and discover other features of the random variables for which the models were constructed. (Later in the course we shall *test* the models against the original data in order to assess the quality of the fit.)

Denoting by X the score on a perfect die when it is rolled, we proposed on symmetry grounds (no outcome more likely than any other) the probability model

$$p(x) = \tfrac{1}{6}, \quad x = 1, 2, 3, 4, 5, 6.$$

If the model is a good one, then we can write down the probabilities of various events. For example,

$$P(3 \text{ or } 6) = p(3) + p(6) = \tfrac{2}{6} = \tfrac{1}{3};$$

and

$$P(\text{an even number}) = P(2 \text{ or } 4 \text{ or } 6) = p(2) + p(4) + p(6) = \tfrac{3}{6} = \tfrac{1}{2}.$$

Obtaining an even number is just as likely as obtaining an odd number (assuming the model to be a good one). The probability of scoring between 2 and 5 inclusive (say) is

$$P(2 \leq X \leq 5) = p(2) + p(3) + p(4) + p(5) = \tfrac{4}{6} = \tfrac{2}{3}.$$

Something rather different is going to be needed in the case of a probability model for a continuous random variable, such as that illustrated by the leaf lengths. Earlier on (Figure 2.9) you saw what might happen to a summary histogram as the sample size becomes large. Eventually the shape of the histogram becomes less jagged, suggesting that a smooth curve might provide an adequate model for the frequency distribution of the random variable (see Figure 2.11).

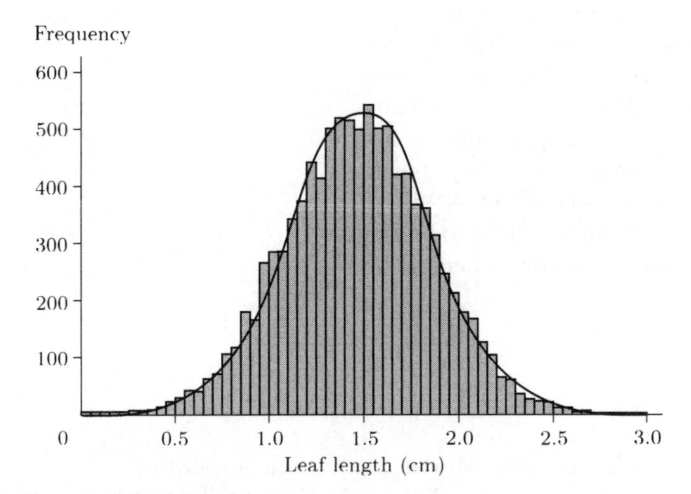

Figure 2.11 A smooth curve fitted to a histogram

The smooth curve could be described by a mathematical function—a formula, in other words. If the curve is scaled so that the total area under the curve is 1, then, if we wish to know the probability that a randomly plucked leaf will be between 1.0 cm and 1.5 cm, we need simply to find the area beneath this curve between $x = 1.0$ and 1.5. This is shown in Figure 2.12.

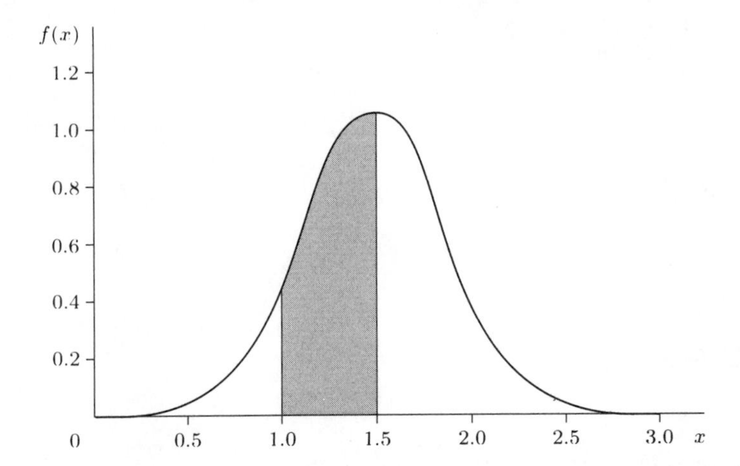

Figure 2.12 Leaf lengths, theoretical probability distribution

The area of the shaded region is equal to the probability required. Such curves are called **probability density functions**. They provide the fundamental modelling apparatus for continuous random variables, just as probability functions (such as those represented in Figures 2.5 and 2.7) provide models for discrete random variables.

The mathematical technique used to calculate the area under a given curve between a left-hand limit and a right-hand limit (see Figure 2.12) is called

integration. You might be familiar with this procedure. If you wished to (and were able to), you could use this approach on every occasion that you needed to calculate a probability for a continuous random variable. However, as the course progresses and as we explore many different data sets arising from many different random contexts, it will become apparent to you that only a relatively small number of basic density functions is necessary to represent adequately the different kinds of random variation most often observed. Further, for each of these basic models, the important mathematical results have already been worked out and are relatively well-known (amongst statisticians, at least). It is therefore not necessary to work out probabilities from scratch each time using integration: all that is required is to implement a known result in the particular context of interest or concern to you. This is the approach that will be adopted in this course, though it will be convenient to use the *notation* of integration, so that it is clear what area is being found between what limits under what curve.

Here is an example.

Example 2.8 Traffic data

The data shown in Table 2.11 are the time intervals (in seconds) between the first 51 consecutive vehicles passing a particular point on the Kwinana Freeway in Perth, Western Australia, after 9.44 am on a particular day. A histogram for these data is given in Figure 2.13. The data are recorded as integers, but a good theoretical model would be continuous. (The gaps between consecutive vehicles will, in reality, have lasted fractional seconds.)

Data provided by Professor Toby Lewis, Centre for Statistics, University of East Anglia.

Table 2.11 Gaps in traffic, Kwinana Freeway (seconds)

5	8	2	1	8	2	3	5	1	3	1	7	3	3	4	3	4
4	5	2	10	1	5	1	6	14	3	8	5	6	2	5	1	12
6	2	3	2	1	6	7	2	2	4	2	1	1	2	16	2	

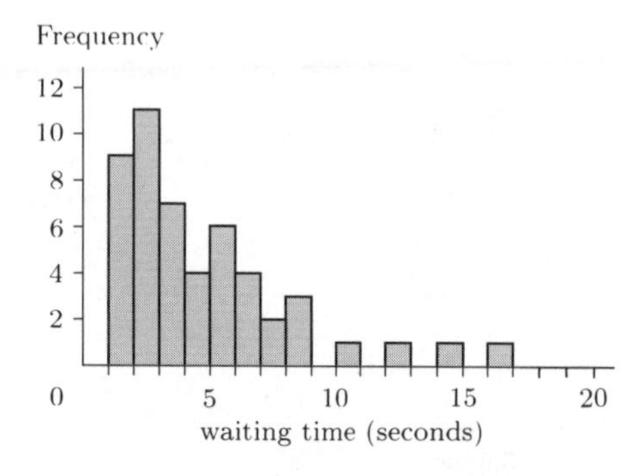

Figure 2.13 Histogram of the Kwinana Freeway traffic data

You can see from Figure 2.13 that the random variation exhibited by these waiting times has quite different characteristics to the variation in the leaf lengths in Figure 2.1. While, for the leaves, intermediate lengths were the most frequent, with shorter and longer measurements occurring less frequently, it appears for these waiting times that shorter gaps occur rather more often than longer ones—the data are very skewed. Although times up to 16 seconds were recorded, almost three-quarters of the waiting times were of 5 seconds' duration, or less.

Time delays during which nothing happens are often called 'waiting times' in statistics.

One possible model for the variation in the Kwinana Freeway waiting times is given by the **triangular density** shown in Figure 2.14. The density function depicted in Figure 2.14 is non-zero only for values along the horizontal axis between 0 seconds and 20 seconds. It is a rather arbitrary model, based on the fact that none of the observed waiting times exceeded 20 seconds. The graph's shape captures the essential skewness apparent from the data. More elaborate models are possible and perhaps even desirable, but this one at least allows us to make a start on the exercise of modelling. (For instance, the model says that gaps longer than 20 seconds between consecutive vehicles are *impossible*! No law of nature decrees that this should be so and, if observation was continued for a long enough period, it is virtually certain that such an extended gap would eventually be recorded. This is a limitation of the model, but we shall assume that such long gaps are so infrequent that this is not a serious limitation. We are merely learning early that statistical models purporting to represent the important aspects of observed variation may have inadequacies in other areas. In this case, one would need to construct a better model.)

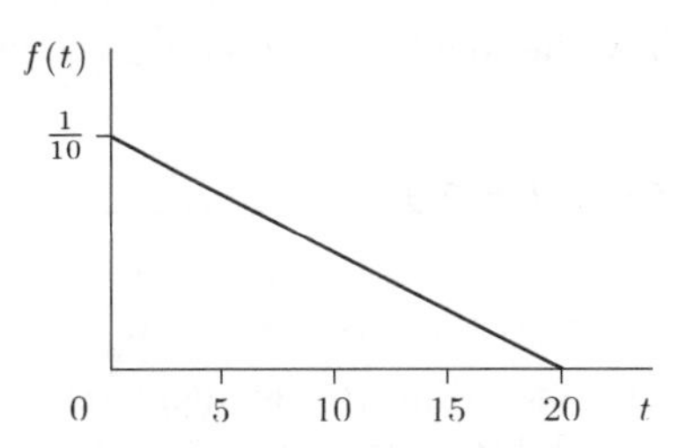

Figure 2.14 A triangular probability density function

You can confirm that the area under the probability density function is 1 by using the formula for the area of a triangle:

$$\tfrac{1}{2} \times \text{base} \times \text{height} = \tfrac{1}{2} \times 20 \times \tfrac{1}{10} = 1.$$

The total area under the density function has to equal 1: this is a property of all probability density functions. This constraint is sufficient to specify the triangular density function completely. Denoting by T the random variable which is the waiting time (in seconds) between consecutive vehicles, then the variation in T may be modelled by the density function

$$f(t) = \frac{20-t}{200}, \quad 0 \le t \le 20.$$

This is the only mathematical function which has the triangular shape shown in Figure 2.14; the function is non-zero between $t = 0$ and $t = 20$, and the total area under the curve is 1.

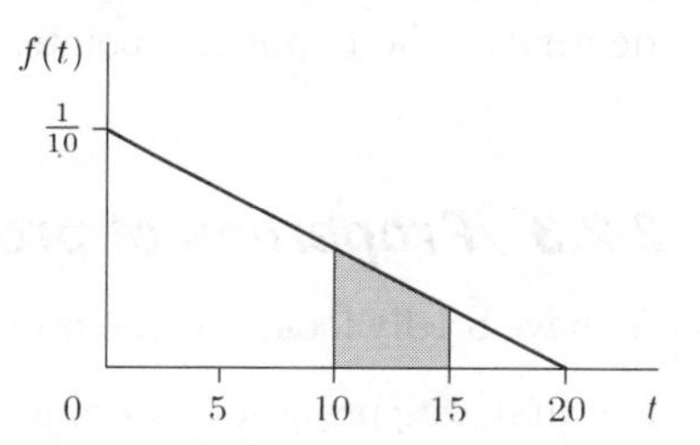

Figure 2.15 The probability $P(10 \le T \le 15)$

Now we can use the *model* rather than the *data* to estimate the proportion of waiting times that are, say, between 10 seconds and 15 seconds. This particular proportion is given by the shaded area in Figure 2.15. This area could be found formally, using integration. It is given by

$$P(10 \le T \le 15) = \int_{10}^{15} f(t)\,dt = \int_{10}^{15} \frac{20-t}{200}\,dt$$

and read as 'the integral of the function $f(t)$ between $t = 10$ and $t = 15$'. Actually, as integrals go, this is not a particularly difficult one and if you know how to integrate you might want to try it. Alternatively, the area of the shaded region in Figure 2.15 may be calculated geometrically. It is

$$\begin{aligned}
\text{area} &= (\text{average height}) \times (\text{width}) \\
&= \tfrac{1}{2}\,(\text{long side} + \text{short side}) \times (\text{width}) \\
&= \tfrac{1}{2}(f(10) + f(15)) \times (\text{width}) \\
&= \tfrac{1}{2}\left(\frac{20-10}{200} + \frac{20-15}{200}\right) \times (15-10) \\
&= \tfrac{1}{2}(0.05 + 0.025) \times 5 \\
&= 0.1875.
\end{aligned}$$

So the model says that less than 20% of traffic gaps will be between 10 seconds' and 15 seconds' duration. ■

Try the following exercise. (Use geometrical arguments.)

Exercise 2.6

(a) According to the model suggested in Example 2.8, what proportion of gaps exceeds 5 seconds?

(b) According to the suggested model, can you find a formula for the probability that a gap is not longer than t, for some general value of t?

In your solution to Exercise 2.6(b), you used geometrical arguments to find a formula for the probability $P(T \leq t)$ for the triangular density defined over the interval $0 \leq t \leq 20$. A different approach is to use integration and say

$$P(T \leq t) = \int_0^t f(v)\,dv = \int_0^t \frac{20-v}{200}\,dv.$$

This would give you the same answer. However, most of the results we shall use in this course are standard and do not need to be worked out from first principles. If you cannot remember a particular formula, then all you need do is turn the pages to find it.

This section finishes with a summary of the ideas introduced so far, thus identifying the important points.

2.2.3 Properties of probability models

We have briefly looked at two probability models for discrete random variables.

The first was proposed in order to describe the variation that would be observed when a six-sided die is rolled. Under the assumption that the die is perfectly fair, the corresponding probability function is

$$p(x) = \tfrac{1}{6}, \quad x = 1, 2, 3, 4, 5, 6,$$

where the random variable X denotes the number of spots on the face that appears.

The second probability model was used to describe the variation in outcome when a *Double-Five* is rolled. Denoting the outcome of that experiment by the random variable Y, the corresponding probability function is

$$p(y) = \begin{cases} \frac{1}{6} & y = 1, 3, 4, 6 \\ \frac{1}{3} & y = 5 \end{cases}$$

and the range of Y is $\{1, 3, 4, 5, 6\}$.

For continuous random variables, we needed a different mathematical entity, reflecting the fact that the range of observable values is continuous. We saw that in this case we can represent the variation that might be observed through a probability density function $f(\cdot)$, and calculate the probability that a particular value is observed between a lower limit x_1 and an upper limit x_2 as

$$P(x_1 \leq X \leq x_2) = \int_{x_1}^{x_2} f(x)\,dx,$$

where the expression on the right-hand side means the area under the curve $f(x)$ between $x = x_1$ and $x = x_2$. In the traffic example (Example 2.8), we used the triangular density function

$$f(x) = \frac{20 - x}{200}, \quad 0 \leq x \leq 20$$

as a continuous probability model for traffic waiting times (in seconds).

Different models will be appropriate in different contexts: for example, data suggest that a model very different to the triangular density would be appropriate to represent variation in leaf lengths, at least for one particular bush of one particular species.

The expression 'probability density function' is often abbreviated to p.d.f. and we may summarize the situation so far as follows.

Probability density function

For a *continuous* random variable X, observed variation may be modelled through the **probability density function** (or simply the **density function**) $f(x)$. This is often abbreviated to p.d.f. Then the probability that an observation on X lies between limits x_1 and x_2 may be calculated as the area under the curve of $f(x)$ between $x = x_1$ and $x = x_2$. This is written in mathematical notation as the integral

$$P(x_1 \leq X \leq x_2) = \int_{x_1}^{x_2} f(x)\, dx.$$

The p.d.f. is defined for all values of x in the range of X.

What we have so far loosely called the probability function $p(x)$ describing the probability distribution of a discrete random variable X is sometimes called a **probability mass function**, which clearly distinguishes it from the probability density function $f(x)$ for a continuous random variable. The term is often abbreviated to p.m.f.

Probability mass function

The probability function for a *discrete* random variable is usually called the **probability mass function** (or simply the **mass function**) of the random variable. This is often abbreviated to p.m.f. For a discrete random variable X, the probability mass function gives the probability distribution of X:

$$p(x) = P(X = x).$$

The p.m.f. is defined for all values of x in the range of X.

During this discussion of probability mass functions and probability density functions some important properties of probability models have been identified. First, a probability (whether of taking a particular value, or of observing a value within a particular interval) is a number representing a long-term proportion for an event; and that event will either never happen, sometimes happen, or always happen. Second, of all the possible outcomes (all the values of x in the range of X) one will certainly happen. Thus we are led to these properties of probability functions.

For a discrete random variable X with probability mass function $p(x)$,

$$0 \leq p(x) \leq 1$$

for all x in the range of X, and

$$\sum p(x) = 1,$$

where the summation is over all x in the range of X.

For a continuous random variable Y with probability density function $f(y)$, the p.d.f. cannot be negative,

$$f(y) \geq 0$$

and

$$\int f(y)\,dy = 1,$$

where the integration is over the whole range of possible values of the random variable Y. (The total area under the curve $f(y)$ over the entire range of Y is equal to 1.)

Using the appropriate probability function, we can calculate the probability that a random variable will lie in any given interval by summing probabilities (or by integration in the continuous case). For the sort of standard probability density function that will be considered in this course, it will not be necessary to go through the process of integration, since most results are well-documented.

It is often useful to know the probability that the random variable will not exceed some specified value. Suppose, for example, that we wish to know the probability that a random variable X, which only takes non-negative values, will not exceed some specified value x. That is, we want to find the probability $P(X \leq x)$.

If X is discrete, this probability may be obtained from the p.m.f. $p(x)$ by summing appropriate terms: so, for a random variable X with range $0, 1, 2, \ldots$, this probability may be written

$$F(x) = P(X \leq x) = \sum_{j=0}^{x} p(j) = p(0) + p(1) + \cdots + p(x).$$

On the other hand, if X is continuous, then we would have

$$F(x) = P(X \leq x) = \int_0^x f(w)\,dw$$

(the lower limit of the integral being, in general, the lowest observable value of X, not necessarily zero).

In either case, the function $F(x) = P(X \leq x)$ is called the **cumulative distribution function** of the random variable X.

The cumulative distribution function

The **cumulative distribution function** $F(x)$ of a random variable X is the probability that X takes a value less than or equal to x:

$$F(x) = P(X \leq x),$$

where x is in the range of X.

Often the abbreviation c.d.f. is used, or the simpler term **distribution function**.

The notation $p(\cdot)$ is standard for probability mass functions; so is the notation $f(\cdot)$ for probability density functions. In either case, $F(\cdot)$ is standard notation for the cumulative distribution function.

In the following example, the c.d.f. of a discrete random variable is obtained.

Example 2.9 *The c.d.f. for a fair die*

For a fair die, the probability mass function is given by

$$P(X = x) = p(x) = \tfrac{1}{6}, \quad x = 1, 2, 3, 4, 5, 6,$$

and the cumulative distribution function is defined by $F(x) = P(X \leq x)$. For instance,

$$F(3) = P(X \leq 3) = P(X = 1 \text{ or } 2 \text{ or } 3) = \tfrac{3}{6} = \tfrac{1}{2}.$$

The p.m.f. $p(x)$ and the c.d.f. $F(x)$ for the random variable X may conveniently be set out in a table as shown in Table 2.12. ■

Table 2.12 The probability distribution for a fair die

x	1	2	3	4	5	6
$p(x)$	$\frac{1}{6}$	$\frac{1}{6}$	$\frac{1}{6}$	$\frac{1}{6}$	$\frac{1}{6}$	$\frac{1}{6}$
$F(x)$	$\frac{1}{6}$	$\frac{1}{3}$	$\frac{1}{2}$	$\frac{2}{3}$	$\frac{5}{6}$	1

Exercise 2.7

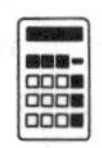

If the random variable Y taking values $y = 1, 3, 4, 5, 6$ is used to model the outcome of a roll of a perfectly engineered *Double-Five*, and if the c.d.f. of Y is written $F(y)$, construct a table like Table 2.12 to represent the probability distribution of the random variable Y.

Exercise 2.8

You have already found in Exercise 2.6(b) the c.d.f. of T, the random variable describing the waiting time (in seconds) between consecutive vehicles using the Kwinana Freeway in Perth. It was

$$F(t) = P(T \leq t) = \frac{40t - t^2}{400}, \quad 0 \leq t \leq 20.$$

Use the c.d.f. to calculate the probabilities

(a) $P(T \leq 10)$;

(b) $P(T > 5)$.

(Again, you should notice that the supposition that these calculations are useful is based on the assumption that $F(t)$ as given above provides a good model for the probability distribution of the time intervals between cars.)

The probability function for a random variable tells us all there is to know about the probability that a random variable will take particular values or fall in a particular interval of the range. If, for a given random variable, we knew this function, then life would be straightforward—questions about likely outcomes could be answered simply by reading off values from this function or by using appropriate formulas. Unfortunately, life is not so straightforward. In general, the probability function will not be known. This means that to make headway in answering questions about probabilities we shall have to estimate the function. Histograms and bar charts based on samples of data provide one general way of estimating probability functions. In some cases we may have more information on which to base our model for the process generating the values of the random variable.

In the street lamp example (Example 2.3) there were only two possible outcomes (faulty or not faulty) so that the probability distribution had only two peaks (0.15 and 0.85).

In the coin-tossing example (Exercise 2.3) which is very similar, we might go further and postulate that the coin is fair—it favours neither Heads nor Tails—so $P(\text{Heads}) = P(\text{Tails})$. Since, as we have seen, the sum of the probabilities of all possible outcomes is 1, i.e. $P(\text{Heads}) + P(\text{Tails}) = 1$, we must have $P(\text{Heads}) = P(\text{Tails}) = \frac{1}{2}$. This means that the two non-zero peaks of the corresponding probability distribution in the coin-tossing example must have equal heights. If we do not wish to make the assumption that the coin is fair, then the situation is, from a mathematical viewpoint, identical to the bulb-testing example.

In both these cases we shall have to estimate the probabilities from a sample, that is, from the data. In other situations, such as the leaf length example (Example 2.1) or the blood plasma β endorphin concentration data from *Chapter 1*, Example 1.4, we might also have some reason (perhaps based on theory or on extensive experience with similar situations) for believing that the probability function will have a particular form. In the rest of this chapter, and in *Chapters 3* and *4*, particular forms which occur frequently and have very widespread application are described.

The probability models associated with the bulb-testing example, the coin-tossing example and the first die-rolling example (with two outcomes: either the three- or six-spot face, or one of the other four) all used Bernoulli trials: this illustrates an important feature of such situations. This is that the probability function in question will typically be a member of a *family* of similar functions. In these three examples, each function consisted of just two peaks, but the heights of the peaks differed between the examples. We can *index* the family by the heights of the peaks. In fact, in each case the height of only one of the peaks needs to be given since the other can be found by subtraction from 1. This indexing value is called a *parameter* of the distribution. The value of the parameter tells us exactly which member of a family of distributions is being discussed.

2.3 The Bernoulli and binomial probability models

2.3.1 The Bernoulli probability model

In Section 2.2 we met three examples which had the same basic structure: the bulb-testing example, the coin-tossing example and the first die-rolling example (with only two possible outcomes). In each case we can define a random variable taking just the values 1 or 0. In fact, we have already done this: for the coin example the value 1 is taken if Heads appears and for the die example the value 1 is taken if a three- or a six-spot face appears. For the bulb-testing example, we defined the random variable as taking the value 1 if a bulb is faulty and 0 otherwise.

The probability distributions associated with these examples thus all belong to the same **family**, in that they have the same basic form. However, they differ in a specific way—a way which is **indexed**, that is, determined, by the value of a **parameter**. This parameter happens to be the probability that the random variable will take the value 1. The other probability in each case—the probability that the random variable will take the value 0—can be obtained by subtraction from 1, since we know that $p(0) + p(1) = 1$.

Thus, once we know that the probability distribution consists of just two non-zero peaks and the height of one of the peaks (the value of the indexing parameter), the probability mass function is completely known. Figure 2.16 shows the three probability distributions for the three examples.

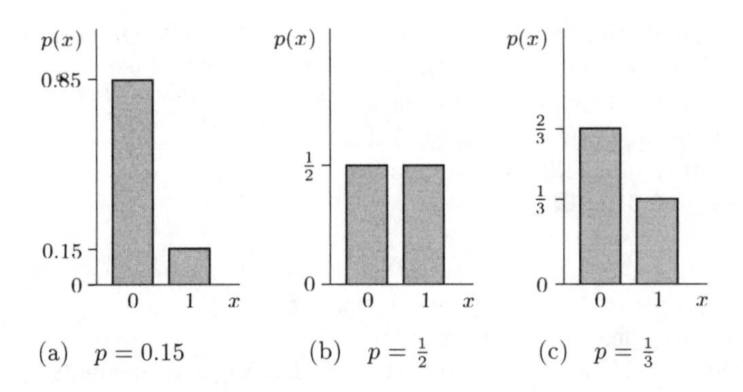

Figure 2.16 Same family; different parameters

Random variables such as those in these three examples, which can take only two possible values, are called **Bernoulli random variables** (after James Bernoulli, 1654–1705). This matches our previous notation: each observation of such a random variable, taking one of the two possible outcomes, is termed a *Bernoulli trial.* Suppose we have a Bernoulli random variable X, which can take the values 0 or 1, and suppose $P(X = 1) = p$ and $P(X = 0) = 1 - p$. Then this probability function, with two non-zero peaks, is said to be that of a *Bernoulli distribution with parameter* p. Thus the coin, die and bulb examples involve Bernoulli distributions with parameters $\frac{1}{2}$, $\frac{1}{3}$ and (we estimated) 0.15, respectively.

The Bernoulli distribution can be written simply as $P(X = 1) = p$, $P(X = 0) = 1 - p$; but, for reasons which will become apparent later, we normally write the probability mass function rather more mathematically as

$$p(x) = p^x(1-p)^{1-x}, \quad x = 0, 1.$$

This means that for $x = 0$,

$$p(0) = P(X = 0) = p^0(1-p)^{1-0} = 1 - p;$$

The value of any non-negative number raised to the power 0 is 1.

and for $x = 1$,

$$p(1) = P(X = 1) = p^1(1-p)^{1-1} = p.$$

The Bernoulli probability model

The discrete random variable X with range $\{0, 1\}$ and probability mass function

$$p(x) = p^x(1-p)^{1-x}, \quad x = 0, 1,$$

is said to follow a **Bernoulli distribution with parameter p**,

$$0 < p < 1.$$

This is written $X \sim \text{Bernoulli}(p)$.

The symbol '$\sim$' is read 'is distributed as' (or, simply, 'is').

The model may be applied to a Bernoulli trial where the probability of obtaining the outcome 1 is equal to p.

***Example 2.10** Blood groups*

In a sample of people living in London who suffered from peptic ulcer, 911 had blood group O and 579 had blood group A. If we pick one of these people at random then the probability is 911/(911+579) that this person will have blood group O. Within this sample, the random variable taking the value 1 if the person is group O and 0 if the person is group A thus follows a Bernoulli distribution with parameter $p = 911/(911 + 579) = 0.61$. ■

Woolf, B. (1955) On estimating the relation between blood group and disease. *Annals of Human Genetics*, **19**, 251–253.

***Example 2.11** Roulette wheels*

The number of compartments in which the ball may come to rest on roulette wheels differs on wheels in different parts of the world. In Monte Carlo the wheels have 37 compartments. These compartments are numbered from 0 to 36. If we define a random variable X as taking the value 1 if the ball stops in the '19' compartment and 0 otherwise, then, assuming the wheel is fair, this random variable follows a Bernoulli distribution with parameter $\frac{1}{37}$. The probability mass function is given by

In Las Vegas there are 38 compartments numbered $0, 1, \ldots, 36, 00$.

$$p(x) = \left(\tfrac{1}{37}\right)^x \left(\tfrac{36}{37}\right)^{1-x}, \quad x = 0, 1.$$

A different question is: at what number will the ball come to rest? Assuming the wheel is fair, this number is equally likely to be any in the range $0, 1, 2, \ldots, 36$. The random variable N denoting this score has probability mass function

$$p(n) = \tfrac{1}{37}, \quad n = 0, 1, 2, \ldots, 36.$$

The physical situation is the same, but a different question has been asked and so a different model has been constructed to answer it. ■

Exercise 2.9

In a study in the Norwegian county of Sogn and Fjordane of all 6503 women aged between 35 and 49, 591 of them were found to have passed menopause.

(a) If we were to choose one of these women at random, what is the probability that she would have passed the menopause?

(b) Define a suitable random variable X to indicate whether a randomly chosen woman has passed the menopause, and write down its probability mass function.

Keiding, N. (1991) Age-specific incidence and prevalence: a statistical perspective. *J. Royal Statistical Society, Series A*, **154**, 371–412.

Exercise 2.10

Telephone surveys are surveys conducted over the telephone. These are easier and cheaper than sending out interviewers, but not all households have telephones so the coverage will be incomplete. In 1986, 78% of British households had telephones. For a randomly chosen British household, write down the probability distribution of the random variable X which takes the value 1 if the household does have a telephone and 0 if it does not.

2.3.2 The binomial probability model

Often situations are encountered in which there is not just a single Bernoulli trial, but a *set* of such trials. For example, in evaluating a new drug we would not wish to base conclusions on the reaction of a single patient. Instead, we would treat several patients and look at the proportion for whom the treatment was beneficial. (Then we might go on to compare this proportion with the proportion for whom the usual treatment was successful, to see whether or not the new treatment is better—there is more on this in later chapters.) We would be using 'the proportion successful' as an estimate of the probability that a randomly selected patient would respond favourably to the drug. Note that there is an implicit assumption in this, namely that the fact that one patient has responded favourably should not affect the probability that another will also respond favourably. If this assumption is true—that the outcome of one trial does not influence the probabilities of the outcomes of another trial—then the trials are said to be **independent**. The assumption that the trials are independent seems a natural one in the case of different patients responding to treatment, but independence is not always so obvious (nor can it always be assumed).

If we let X take the value 1 when a trial is successful and 0 when it fails, then the total number successful in the sample of patients is the sum of the X values, and the proportion successful is the sum of the X values divided by the total number of patients.

Example 2.12 Headache relief

In a small sample of eight patients, five responded successfully to a treatment to relieve headache (five scores of 1); the other three failed to respond (three scores of 0). The proportion of patients in the sample who showed a positive response is

$$\frac{(5 \times 1) + (3 \times 0)}{8} = \frac{5}{8}.$$

This number provides an estimate for the proportion of successful responses in the 'population' of headache sufferers given the treatment. ■

There are many contexts, not just medical ones, that can be regarded, statistically speaking, as a collection of independent Bernoulli trials and where what is of interest is the total number of 1s (indicating, for example, positive responses, Heads, or Yeses). This number is a random variable.

Exercise 2.11

An experiment is performed consisting of a sequence of 15 independent Bernoulli trials. If a trial is successful (the interpretation of 'success' depends on the context) a score of 1 is recorded; otherwise, a score of 0 is recorded. The total 'score' for the experiment, Y, a random variable, is obtained by adding together all 15 1s and 0s.

What is the range of the random variable Y?

The situation where what is of interest is the total number of successful trials in a set of independent Bernoulli trials is a very common one in statistics. The probability model for this random variable is one of the standard models that we shall be using in the course. Its definition follows.

> If a set of n independent Bernoulli trials each has identical probability of success, p, then the random variable Y, defined as the total number of successes over all the trials, is said to follow a **binomial distribution with parameters n and p**. This is written $Y \sim B(n,p)$.

Sometimes the expression 'Y is a binomial random variable' or 'Y is binomially distributed' or, simply, 'Y is binomial' is used.

Before going on to examine properties of the binomial probability distribution, we shall first explore different situations where it might or might not provide an adequate model.

Example 2.13 Coin tossing

Suppose a (fair) coin is tossed three times. Each time it may come up Heads or Tails, and each of these outcomes has probability $\frac{1}{2}$. It is reasonable to assume that the trials are independent—that the fact that it came up Heads (or Tails) the first time does not influence the chance of it doing so at the second (or any subsequent) trial. These three trials thus satisfy the conditions for the total number of Heads to follow a binomial distribution with parameters $n = 3$ and $p = \frac{1}{2}$.

This means that the probability distribution for the total number of Heads (either 0, 1, 2 or 3) is binomial $B(3, \frac{1}{2})$. ■

Example 2.14 Political attitudes in one household

In a survey of political attitude, we might ask all adults in a household how they intend to vote in a two-party system. There are two possible outcomes. Let us define, for each adult in the household, a random variable X, taking the value 1 if they say they intend to vote RED and 0 if they say BLUE. Then we could add up these random variables to produce a new one, Y, the total number of adults who say they intend to vote RED. However, the X scores would not be independent since adults within a household are likely to influence the way the others vote, so the random variable Y would not follow a binomial distribution. ■

Exercise 2.12

Half of the married people in the United Kingdom are female and half male. Thus, if a married person is chosen at random, the probability of picking a female is $\frac{1}{2}$. If 100 married people are randomly picked, the distribution of the number of females will follow the binomial distribution $B(100, \frac{1}{2})$. However, suppose that 100 married people are picked by choosing 50 different people at random, along with their spouse. Does the number of females now follow a binomial distribution?

Example 2.15 Visual perception

In an experiment on visual perception, it was necessary to create a screen of small squares ($\frac{1}{12}$ inch by $\frac{1}{12}$ inch) and, at random, colour the squares black or white. The size of the screen was 27 inches by 40 inches, so there were 155 520 small squares to colour. A computer was used to make the decision according to a very long sequence of Bernoulli trials with predetermined probabilities

Laner, S., Morris, P. and Oldfield, R.C. (1957) A random pattern screen. *Quarterly Journal of Experimental Psychology*, **9**, 105–108.

$$P(\text{Black}) = 0.29, \quad P(\text{White}) = 0.71.$$

After this was done, and before performing the experiment, the screen was sampled to see whether the colouring algorithm had operated successfully.

A total of 1000 larger squares ($\frac{1}{3}$ inch by $\frac{1}{3}$ inch) each containing 16 of the small squares were randomly selected and the number of black small squares was counted in each case. The results should have been consistent with 1000 observations from the binomial probability distribution $B(16, 0.29)$. The results are shown in Table 2.13.

If the computer algorithm for the random colouring was correct, then the data in Table 2.13 give us our first glimpse of a random sample from a binomial distribution. Figure 2.17 gives the corresponding bar chart for these data. You can see that the sample is unimodal and moderately skewed.

Table 2.13 Counts on a random screen pattern

Number of black squares	Frequency
0	2
1	28
2	93
3	159
4	184
5	195
6	171
7	92
8	45
9	24
10	6
11	1
12	0
13	0
14	0
15	0
16	0

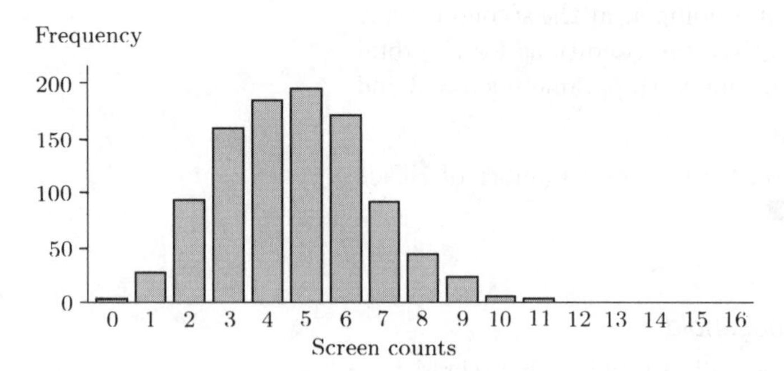

Figure 2.17 Counts on a random screen pattern ■

The statistical procedures for testing the quality of the fit of a hypothesized probability model to a particular data set will be discussed in *Chapter 9*. Indeed, at this stage we have only discussed circumstances under which the binomial model might be assumed to hold, and the model itself has not yet been developed. That is, we have not yet discovered how to calculate the probability that a particular total will be obtained if we observe data assumed to arise from a binomial distribution.

To illustrate the calculations, let us use the following example. Suppose that we are interested in how likely people are to visit their doctor. We might be interested in this as members of a team reporting to a local health authority on the demand for health treatment. To investigate this, we could draw a random sample of people and ask each person if they have visited their doctor in the last year. We shall assume that whether or not one person has visited the doctor implies nothing about the probability that any other person will have visited the doctor—the observations are independent. Let us also suppose that, unknown to us, 1 in 3 people will answer Yes to the question, so the probability is $\frac{1}{3}$ that a randomly chosen person will answer Yes.

In reality, estimating this probability would be the purpose, or one of the purposes, of such an investigation.

We shall go through the calculation in two stages. First, suppose that three people are asked the question: have you visited your doctor in the past twelve months? The probability that the first answers Yes is $\frac{1}{3}$. The probability that the second answers Yes is also $\frac{1}{3}$. The probability that the third gives a No response is $\frac{2}{3}$. An assumption of the binomial model is that these responses are independent. What then is the overall probability that the question elicits the responses

Yes Yes No

in that order? The easiest way to think about this is to calculate the *proportion* of occasions in repeated samples that the first three responses would take precisely that form. Only in one-third of such samples would the first response be Yes; and only in one-third of *those* would the second also be Yes—so only in one-ninth of repeated samples would the first two responses be Yes:

$$P(\text{Yes Yes}) = P(\text{Yes}) \times P(\text{Yes}) = \tfrac{1}{3} \times \tfrac{1}{3} = \tfrac{1}{9}.$$

Continuing the process, you can see that only in two-thirds of *those* samples would the third response be No:

$$P(\text{Yes Yes No}) = P(\text{Yes}) \times P(\text{Yes}) \times P(\text{No}) = \tfrac{1}{3} \times \tfrac{1}{3} \times \tfrac{2}{3} = \tfrac{2}{27}.$$

We can apply the same sort of argument to any particular sequence. For example, the probability that the sequence Yes No No Yes Yes would result after asking the question of five people is

$$\tfrac{1}{3} \times \tfrac{2}{3} \times \tfrac{2}{3} \times \tfrac{1}{3} \times \tfrac{1}{3} = \tfrac{4}{243}.$$

Now we move to the second stage of the calculation. In pursuing this survey, we are not actually interested in recording the order in which the responses occurred, only in counting the different responses. What, for instance, is the probability that amongst the first three people questioned, two say Yes and one says No? There are exactly three different ways this could happen and their probabilities are as follows.

$$P(\text{Yes Yes No}) = \tfrac{1}{3} \times \tfrac{1}{3} \times \tfrac{2}{3} = \tfrac{2}{27}$$

$$P(\text{Yes No Yes}) = \tfrac{1}{3} \times \tfrac{2}{3} \times \tfrac{1}{3} = \tfrac{2}{27}$$

$$P(\text{No Yes Yes}) = \tfrac{2}{3} \times \tfrac{1}{3} \times \tfrac{1}{3} = \tfrac{2}{27}$$

If we now disregard the order of response, we find that the probability of achieving two Yes responses and one No after questioning these three people is

$$P(\text{Yes Yes No}) + P(\text{Yes No Yes}) + P(\text{No Yes Yes})$$
$$= \tfrac{2}{27} + \tfrac{2}{27} + \tfrac{2}{27} = 3 \times \tfrac{2}{27}.$$

Fortunately, it is not necessary to list all the possible sequences of responses in order to establish, say, the probability that in a sample of ten people questioned seven responded Yes; or that more than twelve responded Yes in a sample of twenty. There is a formula that makes such efforts unnecessary. It states generally that the number of different ways of ordering x Yeses (and $n - x$ Noes) in a sequence of n Yes–No responses is given by

$$\binom{n}{x} = \frac{n!}{x!\,(n-x)!},$$

for x taking any value from 0 (all Noes) to n (all Yeses).

The notation $x!$ is a shorthand for the number $1 \times 2 \times 3 \times \cdots \times x$. You might also have seen the notation

$$\binom{n}{x} = \frac{n!}{x!\,(n-x)!} = {}^nC_x.$$

The number $x!$ is read 'x factorial'. The number $\binom{n}{x}$ is read 'n C x'.

So, for instance, the number of different ways of obtaining 7 Yeses (and 3 Noes) in a sample of ten people questioned is

$$\frac{10!}{7!\,3!} = \frac{1 \times 2 \times 3 \times 4 \times 5 \times 6 \times 7 \times 8 \times 9 \times 10}{(1 \times 2 \times 3 \times 4 \times 5 \times 6 \times 7)(1 \times 2 \times 3)} = 120.$$

Each one of these 120 ways occurs with probability $\left(\frac{1}{3}\right)^7 \left(\frac{2}{3}\right)^3$; for instance,

$$P(\text{Yes Yes Yes No No Yes Yes Yes No Yes})$$
$$= \tfrac{1}{3} \times \tfrac{1}{3} \times \tfrac{1}{3} \times \tfrac{2}{3} \times \tfrac{2}{3} \times \tfrac{1}{3} \times \tfrac{1}{3} \times \tfrac{1}{3} \times \tfrac{2}{3} \times \tfrac{1}{3} = \left(\tfrac{1}{3}\right)^7 \left(\tfrac{2}{3}\right)^3.$$

Overall, then, the probability of obtaining 7 Yes responses and 3 No responses in a sample of size ten is

$$\frac{10!}{7!\,3!}\left(\tfrac{1}{3}\right)^7\left(\tfrac{2}{3}\right)^3 = 120\times\left(\tfrac{1}{3}\right)^7\times\left(\tfrac{2}{3}\right)^3 = 120\times\frac{2^3}{3^{10}} = 0.016.$$

This is the probability that a binomial random variable with parameters $n = 10$ (the sample size) and $p = \frac{1}{3}$ (the probability of obtaining a Yes response) will take the value 7. That is, if $Y \sim B\left(10, \frac{1}{3}\right)$, then

$$P(Y=7) = \binom{10}{7}\left(\tfrac{1}{3}\right)^7\left(\tfrac{2}{3}\right)^3;$$

or, more generally, if $Y \sim B\left(10, \frac{1}{3}\right)$, then

$$P(Y=y) = \binom{10}{y}\left(\tfrac{1}{3}\right)^y\left(\tfrac{2}{3}\right)^{10-y}, \quad y = 0, 1, \ldots, 10.$$

The number 0! is *defined* to be 1. Setting y equal to 0 in this formula gives

$$P(Y=0) = \binom{10}{0}\left(\tfrac{1}{3}\right)^0\left(\tfrac{2}{3}\right)^{10} = \frac{10!}{0!\,10!}\left(\tfrac{1}{3}\right)^0\left(\tfrac{2}{3}\right)^{10} = 1\times 1\times\left(\tfrac{2}{3}\right)^{10},$$

which is the probability that all ten responses are No.

The binomial probability model may be summarized as follows.

The binomial probability model

The random variable X follows a binomial distribution with parameters n and p if it has probability mass function

$$p(x) = \binom{n}{x} p^x (1-p)^{n-x}, \quad x = 0, 1, 2, \ldots, n,$$

where $0 < p < 1$, and

$$\binom{n}{x} = \frac{n!}{x!\,(n-x)!}, \quad x! = 1\times 2\times\cdots\times x, \quad 0! = 1.$$

This is written $X \sim B(n, p)$ and provides a probability model for the total number of successes in a sequence of n independent Bernoulli trials, in which the probability of success in a single trial is p.

Exercise 2.13

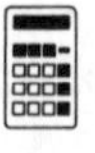

Suppose that a study is undertaken to compare the safety and efficacy of two antidepressant drugs. Eighteen patients are each randomly allocated to one of three groups, six to a group. The first group is treated with Drug A and the second with Drug B. Patients in the third group are treated with a placebo (a substance which is inert as far as antidepressive effects go, but which is given to the patients in the same way as the treatments being studied, so that the analysis can be controlled for any natural remission).

One of the problems associated with studies of this sort is that patients occasionally 'drop out': they cease treatment before the study is completed. This might be for reasons unrelated to their course of treatment, or because they suffer from side-effects, or it might be because they perceive no beneficial effect from their treatment. The phenomenon is consequently a complicating feature in a statistical analysis of the results of such studies.

A previous study suggests that the proportion of patients in placebo groups who drop out might be about 14%.

Dunbar, G.C., Cohn, J.B., Fabre, L.F., Feighner, J.P., Fieve, R.R., Mendels, J. and Shrivastava, R.K. (1991) A comparison of paroxetine, imipramine and placebo in depressed outpatients. *British Journal of Psychiatry*, **159**, 394–398.

(a) Using this estimate for the value of the parameter p in a binomial model, calculate for the placebo group in the present study
   (i) the probability that all six patients drop out;
   (ii) the probability that none of the six drop out;
   (iii) the probability that exactly two drop out.

(b) An assumption of the binomial model is that of 'independence from trial to trial'. Interpret this assumption in the context of the study, and comment on whether you believe that, in this case, it is a reasonable assumption.

Example 2.16 *Multiple choice examination scores*

One of the components of assessment of the statistics students at a certain British university is a multiple choice examination consisting of twenty questions. For each question the correct answer is one of five possible options. Students indicate the one of the five they believe to be correct. Sometimes some of the students give the impression that they have gone through the paper guessing answers at random.

Since such a student guesses answers at random, an answer to any particular question is independent of an answer to any other question. Moreover, since there are five possible answers to each question and since the selection is made at random, the probability of picking the correct one is $\frac{1}{5}$ for each question. Thus the answers of a student who guesses at random form a sequence of twenty independent Bernoulli trials, each with probability of success $\frac{1}{5}$, i.e. 0.2. So the total number of correct answers given to the twenty questions is a random variable following a binomial distribution with parameters $n = 20$ and $p = 0.2$. That is, letting the random variable T denote the total number of correct answers, $T \sim B(20, 0.2)$.

In such a situation we might like to know the proportion of students who would score more than the pass mark if they guessed in the way described above. For example, if the pass mark was 10, the probability that a student adopting a random guessing procedure such as that described above would score less than 10 (and so fail the examination) is given by

$$\begin{aligned} &P(T < 10) \\ &= P(T \leq 9) = P(T = 0 \text{ or } 1 \text{ or } \ldots \text{ or } 9) = \sum_{t=0}^{9} \binom{20}{t} 0.2^t 0.8^{20-t}. \end{aligned}$$

Here, we have summed the probabilities of all the scores less than 10.

Table 2.14 gives the probability mass function $P(T = t) = p(t)$ and the cumulative distribution function $P(T \leq t) = F(t)$ for the binomial random variable $T \sim B(20, 0.2)$. For example,

$$\begin{aligned} &P(T = 3) \\ &= \binom{20}{3} (0.2)^3(0.8)^{17} = \frac{20!}{3!\,17!}(0.2)^3(0.8)^{17} = 1140(0.2)^3(0.8)^{17} = 0.2054. \end{aligned}$$

The probabilities are given correct to four decimal places, but notice that, because of rounding effects, the numbers in the right-hand column are not all exactly equal to the 'running sum' of the probabilities in the middle column. For instance, the table gives

$$P(T = 0) = 0.0115, \quad P(T = 1) = 0.0576;$$

adding gives

$$P(T = 0) + P(T = 1) = 0.0691;$$

but

$$P(T \leq 1) = 0.0692 \quad \text{(to 4 decimal places).}$$

This kind of discrepancy occurs occasionally in probability calculations involving rounding and *can be ignored.*

If the pass mark is set at 10, the probability that a student who randomly guesses his answers passes may be found as follows:

$$P(T \geq 10) = 1 - P(T \leq 9) = 1 - 0.9974 = 0.0026. \quad ■$$

Table 2.14 The probability distribution of $T \sim B(20, 0.2)$

t	$P(T = t)$	$P(T \leq t)$
0	0.0115	0.0115
1	0.0576	0.0692
2	0.1369	0.2061
3	0.2054	0.4114
4	0.2182	0.6296
5	0.1746	0.8042
6	0.1091	0.9133
7	0.0545	0.9679
8	0.0222	0.9900
9	0.0074	0.9974
10	0.0020	0.9994
11	0.0005	0.9999
12	0.0001	1
13	0	1
14	0	1
15	0	1
16	0	1
17	0	1
18	0	1
19	0	1
20	0	1

In general, the c.d.f. of a binomial random variable $Y \sim B(n, p)$ is given by

$$P(Y \leq y) = \sum_{j=0}^{y} \binom{n}{j} p^j (1 - p)^{n-j}.$$

Unfortunately, this does not have a convenient mathematical form that is quick and easy to evaluate. (If you are doing a statistical analysis using a statistical computer package, then the computer will probably do the calculations for you and produce a probability value directly. This is the easiest way of obtaining particular values of $p(y)$ and $F(y)$ when Y is binomial. Otherwise, you can sum the individual probabilities using a calculator.)

In Figure 2.18 are diagrams of four typical binomial probability distributions. They show

(a) the number dropping out in the placebo group (Exercise 2.13): $B(6, 0.14)$,

(b) the number of correct answers guessed (Example 2.16): $B(20, 0.2)$,

(c) the number of faulty bulbs in a box of ten (Example 2.3): $B(10, 0.15)$,

(d) the number of 5s in eight rolls of a *Double-Five* (Example 2.6): $B\left(8, \frac{1}{3}\right)$.

Exercise 2.14

(a) If $V \sim B(8, 0.3)$, find the probability $P(V = 2)$.

(b) If $W \sim B(12, 0.5)$, find the probability $P(W = 8)$.

(c) If $X \sim B(6, 0.8)$, find the probability $P(X > 4)$.

(d) If $Y \sim B\left(6, \frac{1}{3}\right)$, find the probability $P(Y \leq 2)$.

(e) If $Z \sim B\left(10, \frac{1}{4}\right)$, find the probability $P(Z \leq 7)$.

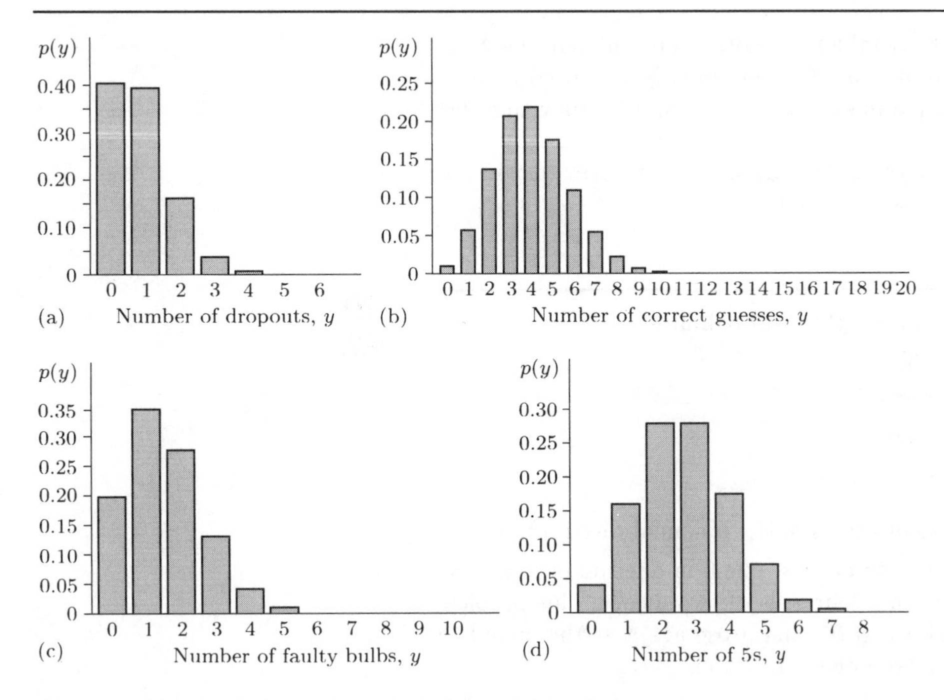

Figure 2.18 Typical members of the binomial family

Exercise 2.15

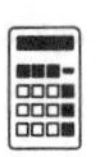

The science of genetics includes the study of hereditable physical features such as eye and hair colour, bone structure, blood group or unusual skin conditions. The Augustinian monk Gregor Mendel (1822–1884) investigated, in some experiments on pea plants, what happened when plants with round yellow peas of a certain kind were crossed. The offspring were not all round and yellow—some were wrinkled and others green, according to the following combinations and in the proportions shown. (These proportions are based on theory. However, observed experimental frequencies were very close to the theoretical predictions.)

$P(\text{round and yellow}) = \frac{9}{16}$
$P(\text{round and green}) = \frac{3}{16}$
$P(\text{wrinkled and yellow}) = \frac{3}{16}$
$P(\text{wrinkled and green}) = \frac{1}{16}$

Round yellow peas are crossed and produce eight offspring peas.

(a) What is the distribution of the number of wrinkled yellow peas amongst the offspring?

(b) What is the probability that all eight offspring are wrinkled yellow peas?

(c) What is the probability that there are no wrinkled green peas amongst the offspring?

Calculation of binomial probabilities using the formula can involve a surprisingly large number of key strokes on your calculator. This is especially true when you are calculating more than one probability in order to calculate the probability that a particular random variable takes one of several possible values. A set of statistical tables is another option, but these are not necessarily very easy to read and usually contain only a limited number of possibilities for the binomial parameter p. (For example, they may list binomial probabilities for n from 2 to 20 and for p from 0.1 to 0.9 by steps of 0.1 or possibly 0.05. It may be that the problem of interest to you has p equal to 0.37.) An option is to use a computer.

Exercise 2.16

Use your computer to calculate directly these probabilities.

(a) $P(X = 4)$ when $X \sim B(10, 0.3)$;

(b) $P(X = 6)$ when $X \sim B(20, 0.17)$;

(c) $P(X \leq 8)$ when $X \sim B(11, 0.33)$;

(d) $P(X \geq 8)$ when $X \sim B(20, \frac{1}{3})$.

Parts (e) and (f) illustrate two applications of the binomial distribution.

(e) Which is more probable: to obtain at least one 6 in one roll of four dice, or at least one double-6 in twenty-four rolls of two dice? (This problem was discussed by Cardano (1501–1576) and arose again at the gambling table when, in 1654, de Méré proposed it to Pascal.)

(f) On any given day of the year (365 days), past experience suggests that the probability of rain is about 0.3. What is the probability that in a whole year it rains at least 100 times?

2.4 The normal distribution

In this section we shall briefly consider a probability distribution that can provide a useful model for all sorts of widely disparate random phenomena. The model is continuous. First, let us look at some data sets.

Below are presented the histograms for several data sets arising from a number of quite different sources.

Example 2.17 *Heights*

This data set from a study of osteoporosis gives the heights (in cm) of a sample of 351 elderly women randomly selected from the community. (For example, three women amongst the 351 were 145 cm tall—that is, between 144.5 cm and 145.5 cm.)

Data provided by D.J. Hand, The Open University. Osteoporosis is a disease affecting the bones, and can cause a reduction in height.

Table 2.15 Heights of 351 elderly women (cm)

Height	142	143	144	145	146	147	148	149	150	151	152	153	154
Frequency	1	0	0	3	1	4	2	1	6	6	12	17	11
Height	155	156	157	158	159	160	161	162	163	164	165	166	
Frequency	21	20	20	31	17	21	20	18	30	17	18	11	
Height	167	168	169	170	171	172	173	174	175	176	177	178	
Frequency	7	6	8	11	3	0	3	1	0	1	1	2	

The histogram summarizing these data is shown in Figure 2.19.

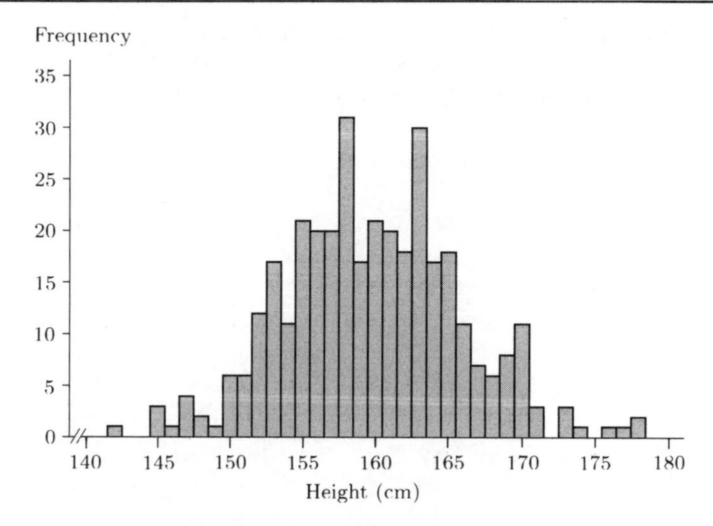

Figure 2.19 Heights of 351 elderly women ■

Example 2.18 Nicotine levels

This data set lists measurements of blood plasma nicotine levels (in nanograms per millilitre) for 55 smokers, in a study of cigarette smoking habits.

Data provided by D.J. Hand, The Open University.

1 ng $\equiv 10^{-9}$ g

Table 2.16 Blood plasma nicotine levels for 55 smokers (ng/ml)

123	311	242	474	375	449	419	185	33	564	256	242
312	179	456	232	389	429	309	269	274	274	157	348
384	274	179	306	260	346	106	468	597	233	304	448
182	527	155	347	74	471	260	213	346	304	256	233
227	607	464	469	209	314	456					

The histogram summarizing these data is shown in Figure 2.20.

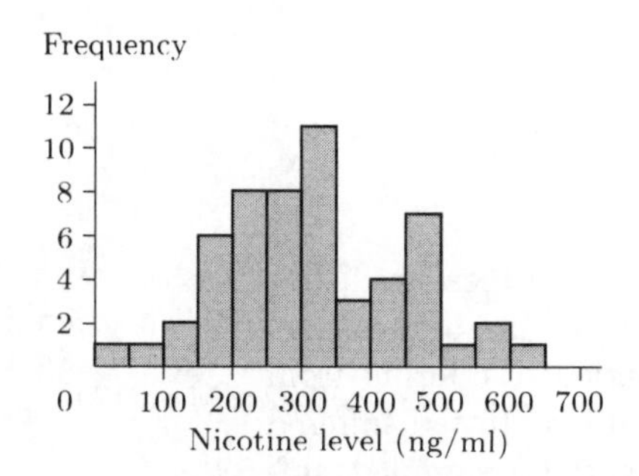

Figure 2.20 Blood plasma nicotine levels for 55 smokers ■

Example 2.19 Chest measurements

This data set (see Table 2.17) gives the chest measurements (in inches) of 5732 Scottish soldiers.

Stigler, S.M. (1986) *The History of Statistics—The Measurement of Uncertainty before 1900.* Belknap Press of Harvard University Press, p. 208.

These data have an interesting history. The Belgian mathematician Adolphe Quetelet (1796–1874) was interested in probability and social statistics, and, amongst other things, in the fitting of statistical models to data. The chest measurements data were extracted from the *Edinburgh Medical and Surgical Journal* (1817) and were first given erroneously. For instance, Quetelet's summary found 5738 soldiers (not 5732); his total for the number of soldiers with chest measurements between $33\frac{1}{2}$ inches and $34\frac{1}{2}$ inches was 18 (not 19). Stigler writes: 'Although the errors have no important bearing on the explanation [of Quetelet's method for fitting a statistical model], they do exemplify Quetelet's tendency to calculate somewhat hastily, without checking his work'.

Table 2.17 Chest measurements of 5732 Scottish soldiers (inches)

Measurement	Frequency
33	3
34	19
35	81
36	189
37	409
38	753
39	1062
40	1082
41	935
42	646
43	313
44	168
45	50
46	18
47	3
48	1

The data are summarized in the histogram in Figure 2.21.

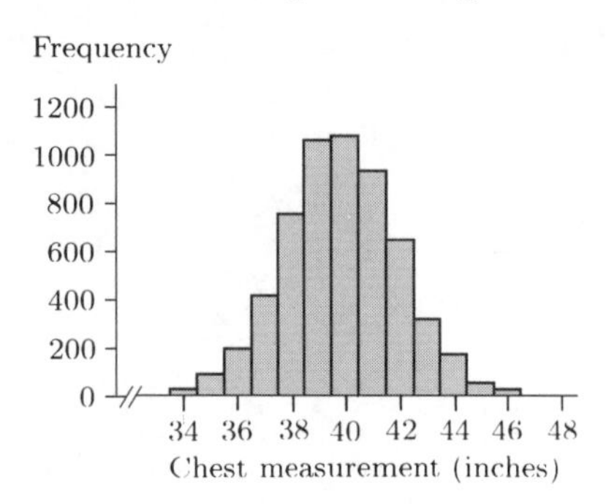

Figure 2.21 Chest measurements of 5732 Scottish soldiers ■

You can see that all the histograms in these examples have a very similar basic shape. If we discount random variation, it seems as if they each come from populations that have a single maximum value—there is just one hump. In other words, the distributions are unimodal. Furthermore, all of the histograms are roughly symmetrical. Of course, the data sets have different sample means and sample standard deviations—but this is to be expected since they measure different quantities on different scales.

Notice that some of the diagrams are more jagged than others: the sample size is a very important consideration when assessing the reliability of a histogram as an indicator of shape.

Here is another example.

Example 2.20 *Hepatitis*

In the following data sets the same random variable is being measured, but on two different sets of individuals. The variable is the logarithm of measures of ornithine carbonyltransferase (a liver enzyme) in patients suffering from two different forms of hepatitis. In Table 2.18, the 57 patients tested suffered from acute viral hepatitis; in Table 2.19, there were 40 patients sampled, suffering from aggressive chronic hepatitis. The investigators were interested in a comparison of the two groups of patient—specifically, whether or not it was possible to distinguish between the patient groups on the basis of measurements of the enzyme.

Albert, A. and Harris, E.K. (1987) *Multivariate interpretation of clinical laboratory data.* Marcel-Dekker, Inc., New York.

Table 2.18 Enzyme measurements, acute viral hepatitis (log measure)

2.66	2.38	2.37	2.31	2.50	1.96	2.85	2.68	1.76	2.36	2.56	2.09
2.85	2.67	2.37	2.40	2.79	1.82	3.00	2.50	2.36	2.48	2.60	2.42
2.51	2.51	2.80	2.50	2.57	2.54	2.53	2.78	2.07	2.35	2.98	2.31
2.45	2.75	2.56	2.50	3.00	2.94	2.46	2.83	3.61	2.99	2.78	3.02
2.93	2.78	2.57	2.62	2.71	2.18	3.21	2.86	2.51			

Table 2.19 Enzyme measurements, aggressive chronic hepatitis (log measure)

3.01	2.99	2.60	2.47	3.04	1.92	2.17	2.33	2.07	2.30	2.56	2.11
3.32	2.21	1.71	2.60	2.79	2.71	2.64	2.52	2.21	2.58	2.40	2.45
3.18	2.84	2.84	2.31	2.71	2.47	2.72	3.71	2.73	3.69	3.40	2.77
2.28	2.84	2.80	3.02								

Figure 2.22 shows the histograms for the two hepatitis data sets.

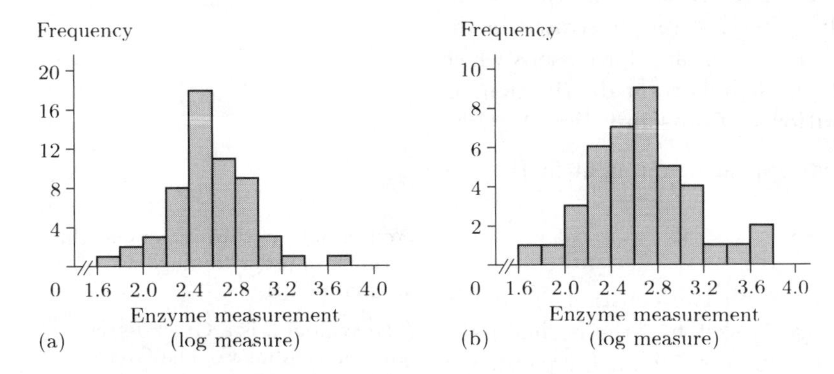

Figure 2.22 (a) Acute viral hepatitis (b) Aggressive chronic hepatitis

Again, these two histograms show the same basic shape; and since they are measurements on the same variable, they can be compared directly. (See the comparative boxplot in Figure 2.23.)

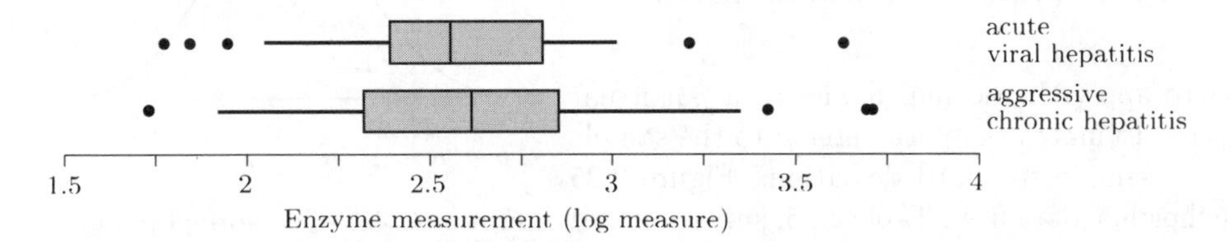

Figure 2.23 Comparison of the two sets of liver enzyme measurements on hepatitis patients

Figure 2.23 suggests very clearly that, while the two sample means are not very different from each other, the measurements on the liver enzyme are more variable for the group of patients suffering from aggressive chronic hepatitis than for those suffering from acute viral hepatitis. ■

The similarity of all the histogram shapes that we have seen in this section suggests a family of probability distributions, all possessing the essential characteristic that they are symmetric about a central point at which the distribution peaks. The only differences between members of the family are the location of that central point, and the dispersion, or degree of variability about it that is exhibited.

We have come across the idea of a *family* of probability distributions already. The Bernoulli distribution can be used to model the outcome of a Bernoulli trial: this is a family of probability distributions indexed by the parameter p, the probability of success. The binomial family has two parameters: n, the total number of trials, and p, the probability of success at any one trial. The context may differ (faulty light bulbs in a consignment, correct guesses in an examination), but the statistical circumstances are the same (count the successes in a sequence of independent trials) and the probability calculations involve the same kind of arithmetic procedures.

Similar considerations apply to the examples in this section (except that the variation observed is continuous rather than discrete, as in the case of the Bernoulli and binomial distributions). The general shape illustrated in the preceding histograms arises frequently in practice. In fact, for reasons which are explained in *Chapter 5*, it is one of the most important distributions in statistics. It is called the **normal distribution** or **Gaussian distribution**.

The normal probability density function may appear daunting at first:

$$f(x) = \frac{1}{\sigma\sqrt{2\pi}} \exp\left[-\frac{1}{2}\left(\frac{x-\mu}{\sigma}\right)^2\right], \quad -\infty < x < \infty.$$

You do not need to remember this formula in order to make use of the normal model.

You can see that the density function depends on two constants μ and σ: these are the parameters of the normal family. A sketch of the normal p.d.f. is given in Figure 2.24. The normal family is symmetric about the parameter μ and is such that observations less than about $\mu - 3\sigma$ or more than about $\mu + 3\sigma$ are rather unlikely. (Theoretically, the model permits any value of the random variable, negative or positive. It may seem illogical then to apply it, say, to chest measurements or to the amount of an enzyme present in a biological sample. Remember, however, statistical models (like most mathematical models) are not intended to be *exact* representations of the natural world, just *sufficiently good* ones for conclusions drawn from the model to be sufficiently accurate.)

The symbol μ is a Greek letter pronounced 'mew'. The Greek letter σ is pronounced 'sigma' (it is the lower-case version of the upper-case Greek letter $\sum$).

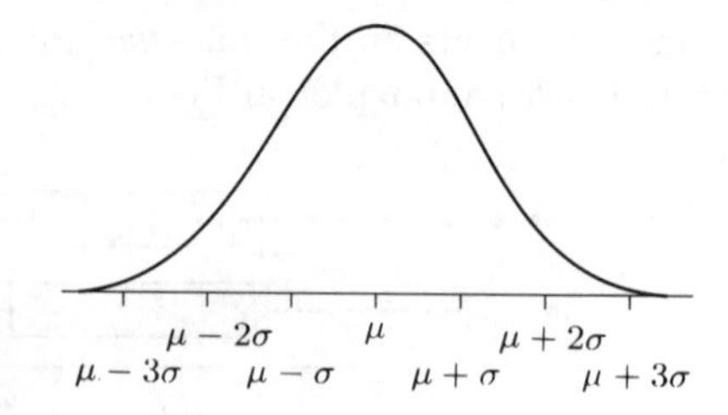

Figure 2.24 The normal p.d.f.

We shall see that in attempting to apply the normal model to a particular random phenomenon, it makes sense to match the parameter μ to the sample mean, and the parameter σ to the sample standard deviation. Figure 2.25 shows again the histogram of the 'heights' data from Table 2.15; superimposed on the histogram is a theoretical normal model with μ set equal to 159.8 (the sample mean for the data) and σ to 6.0 (the sample standard deviation).

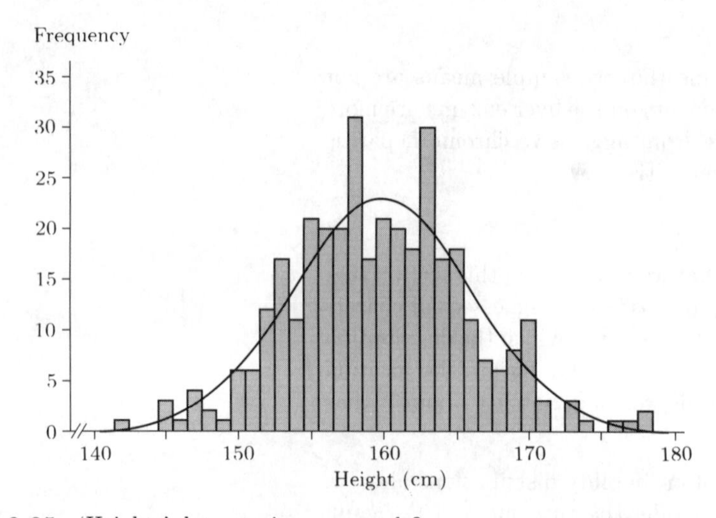

Figure 2.25 'Heights' data: trying a normal fit

The diagram suggests that the normal model provides a good 'fit' to the data—there will be much more to say about the quality of fit of a statistical model to data in the rest of the course.

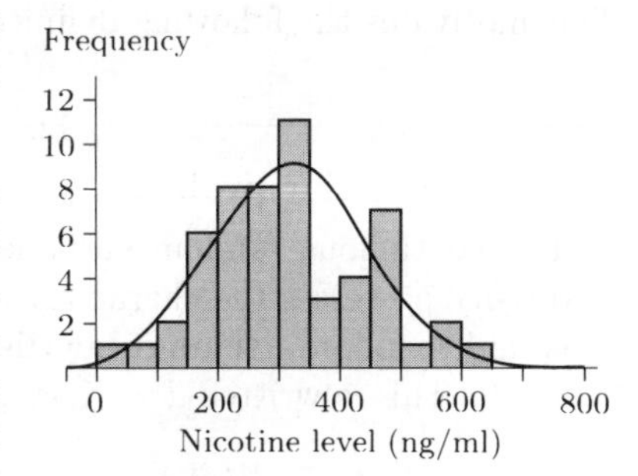

Figure 2.26 'Plasma' data: trying a normal fit

Figure 2.26 repeats the histogram of blood plasma nicotine levels from Table 2.16; superimposed on the histogram is a normal curve, where $\mu = 314.9$ (the sample mean for the data) and $\sigma = 131.2$ (the sample standard deviation). The fit is not quite as good, since the data are perhaps slightly skewed but it may still be a *usable* model for practical purposes (which is what models are for). In Figure 2.27, the histogram for the Scottish chest measurements data from Table 2.17 is reproduced; superimposed on the histogram is a fitted normal curve, with μ set equal to 39.85 and σ equal to 2.07.

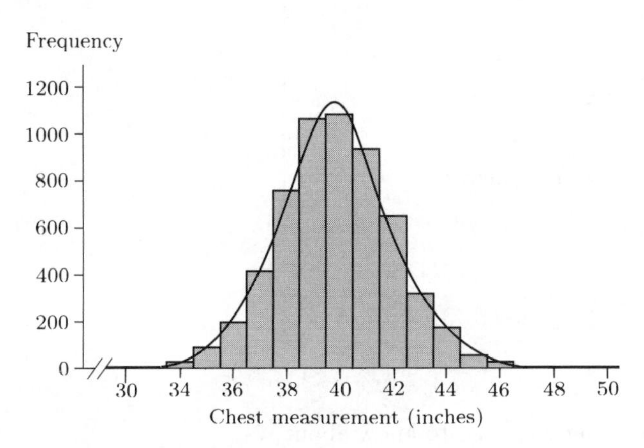

Figure 2.27 'Chest measurements' data: trying a normal fit

Here, the fit seems excellent, and a normal model with these parameters should provide a very good model for practical purposes.

We shall return to the normal distribution in *Chapter 5*. The normal distribution is an example of a continuous distribution, like the triangular density function introduced in Section 2.2. At that stage, some rather drawn-out calculations were performed based on geometrical arguments. The point was made then that an easier approach is simply to use existing formulas and results.

Now, with the word 'parameter' in our vocabulary, we can summarize all we need to know about this distribution. The triangular density that was adopted to model the Kwinana Freeway traffic data was characterized by the assumed maximum observable waiting time, $t = 20$ (measured in seconds). For the 'area under the curve' between $t = 0$ and $t = 20$ to be 1, it followed necessarily that the corresponding density was

$$f(t) = \frac{20 - t}{200}, \quad 0 \leq t \leq 20$$

(though you were spared the details of the argument that leads to this result).

This triangular density is just one of a whole *family* of triangular densities, whose *indexing parameter* can conveniently be taken to be the assumed maximum observable waiting time (or length, or weight, or whatever quantity is appropriate to the random variable under consideration).

This motivates the following definition.

The triangular probability model

The continuous random variable T with range $0 \leq t \leq \theta$, and whose density possesses the characteristic skewed shape shown in Figure 2.28, is said to follow a **triangular distribution with parameter θ**, where $\theta > 0$. This is written $T \sim \text{Triangular}(\theta)$.

The Greek letter θ is pronounced 'theta', and is often used as the label for the indexing parameter of probability distributions.

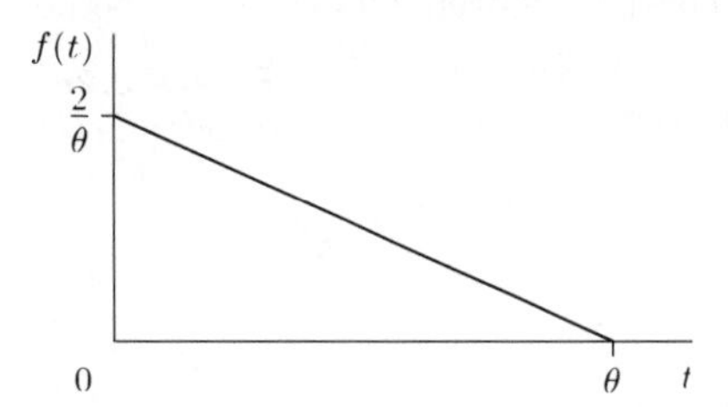

Figure 2.28 The density function $f(t)$ when $T \sim \text{Triangular}(\theta)$

The c.d.f. of T is given by

$$F(t) = P(T \leq t) = 1 - \left(1 - \frac{t}{\theta}\right)^2, \quad 0 \leq t \leq \theta.$$

This is an example of 'a standard result': everything you need to know about the triangular probability model is summarized in the box. It is the last line that permits probability statements to be made. If the model is a good one, these statements will be a useful aid to understanding the random variation observed.

Notice, incidentally, that the p.d.f. $f(t)$ is not given in the description of the random variable T. This is because it is not needed in the calculation of probabilities (unless you want to go through the mathematical exercise of integration each time you need to find a probability).

Now, if ever you decide that the triangular model with some given parameter provides an adequate representation of random variation in some context, all you need to do in order to calculate probabilities is to use the given formula for the c.d.f. You might have to look it up, if you have forgotten it; or perhaps your computer knows about the triangular model, in which case you will not have to remember the formula at all. This approach is standard in this course. A typical scenario for the statistician might be as follows.

A random context (such as coin tossing, bulb testing, measuring traffic waiting times) will be described, and observations on some associated random variable will be collected in order to obtain some idea of the inherent random variation. A probability model might then be deduced on mathematical grounds, as was done in the case of the binomial probability model, where the choice followed as a necessary consequence of the assumption that the Bernoulli trials were independent. In another case, the data might suggest a useful model (a normal model for heights, or a triangular model for the duration of gaps in traffic). In either case, the model might be found on further scrutiny to be faulty, in which case it needs further development.

Later in the course we shall return to the question of model fitting and model testing. In the final section of this chapter, no additional material is taught. You will be encouraged to try some further exercises consolidating the material of the chapter and to practise using your computer.

2.5 Some further exercises

As the course develops, you will be encouraged to make frequent use of your computer, pausing in your reading to examine some property of a data set, or to carry out a statistical test, the theory of which you have read about, but the practice of which would be tedious with pencil and paper. (Indeed, with some larger data sets it would be impossible to undertake a pencil analysis and to finish it within a reasonable time. With some others, it would be a possible but not an intelligent undertaking.) A sensible approach would be to perform some arithmetic calculations on your calculator, which may well include simple or even some fairly advanced statistical functions, and to use your computer for the larger data sets. You were encouraged to use both approaches in the exercises at the end of Section 2.3.

Exercise 2.17

At the time of writing this section, one of the authors had a library of 5152 books—too many for his shelves—so many of them were stored in boxes. He intended to move house and hoped that in his new house he would be able to have enough shelving for all his books. He considered that the books currently on his shelves possessed the same distribution of widths as his entire collection and so took a sample of 100 books from his shelves to estimate the shape of this distribution. The data are shown in Table 2.20.

Data supplied by K.J. McConway, The Open University.

Table 2.20 Widths of 100 books (mm)

16	13	10	19	31	10	26	20	17	8	15	24	32	16	9	16
12	22	16	28	17	17	27	16	32	24	20	28	25	23	6	24
10	30	29	18	14	19	26	17	18	13	34	8	35	28	11	12
29	40	14	20	13	11	47	13	17	17	18	35	11	19	11	26
11	22	12	6	7	16	21	10	14	17	14	6	20	11	13	22
18	5	12	10	14	11	10	8	15	14	11	9	11	18	18	48
14	19	25	17												

(a) Plot a histogram of this sample, and calculate its mean and standard deviation. Using your calculated mean, obtain an estimate in metres of the total length of shelf space the author needed.

(b) Comment on any similarities or differences between your sample histogram and the bell-shaped distribution of Section 2.4.

In Section 2.1 you conducted some simulations using a die and a coin. Simulation can be a very powerful tool to investigate natural phenomena and is normally done using a computer. For example, suppose we wish to explore the statistical properties of an engineering production line. We may believe that the objects produced each have a probability p of being faulty, and that whether or not an object is faulty is independent of whether any other object is faulty. We might be interested in how often all six objects in a group of six will be in working order (perhaps the objects are packed in boxes of six). We could try to explore this mathematically and indeed the problem is not very hard. Following the description (in Section 2.3) of the binomial distribution, we know that the proportion of boxes containing six working components out of six is $(1-p)^6$. Alternatively, since we have a well-defined mathematical model, we could program a computer to behave in the same way as we believe the production line behaves. We could get the computer to generate random 0s and 1s so that a proportion p of them were 1s and the others 0s, and interpret 1 as corresponding to a faulty product. Then we would look through the sequence of 0s and 1s generated by the computer and grouped in sixes, and see how often a set of six is all 0s.

This is an example of simulation. It is a rather artificial example in that it is easy to work out the answer algebraically, as has been done. As the course progresses, we shall discover more realistic situations and reasons for using simulation approaches. To get used to the idea of generating data on your computer, try the following exercises.

Exercise 2.18

(a) Using the Bernoulli random number generator of your computer, generate 10 values from a Bernoulli distribution with parameter $p = 0.2$, and calculate the total number of 1s amongst the 10 values.

Now use the binomial random number generator to generate a single value from a binomial distribution $B(10,0.2)$.

What is the relationship between the populations from which these two numbers have come?

(b) Again using the binomial random number generator, generate 10 values from a $B(20,0.5)$ distribution, 100 values from the same $B(20,0.5)$ distribution and then 500 values from the same $B(20,0.5)$ distribution.

In each of the three cases produce a bar chart of the samples.

Comment on the differences between the three bar charts in terms of how jagged and irregular they are.

(c) Using the binomial random number generator, generate 500 values from
(i) a binomial distribution with parameters $n = 10$ and $p = 0.1$;
(ii) a binomial distribution with parameters $n = 10$ and $p = 0.3$;
(iii) a binomial distribution with parameters $n = 10$ and $p = 0.5$.

In each case, produce a bar chart for the 500 values.

How do you think the parameter p affects the shape of the binomial distribution?

(d) Using the binomial random number generator, generate 500 values from
(i) a binomial distribution with parameters $n = 10$ and $p = 0.2$;
(ii) a binomial distribution with parameters $n = 30$ and $p = 0.2$;
(iii) a binomial distribution with parameters $n = 50$ and $p = 0.2$.

In each case produce a bar chart of the 500 values.

What effect do you think the parameter n has on the shape of the binomial distribution?

The next exercise is about the simulation of an industrial process.

Exercise 2.19

Fuses for use in domestic electrical appliances are packaged in bags of eight for sale to customers. The probability that any fuse is defective is 0.012.

A retailer buys a box of 100 of these bags direct from the manufacturer. Use your computer to 'open the bags' and count the number of defective fuses in each one. How many bags contained no defectives? How many contained just one? Complete a frequency table like the one in Table 2.21.

Table 2.21 Defective fuses

Number of defective fuses	Frequency
0	
1	
2	
3	
4	
5	
6	
7	
8	
Total	100

The next exercise is about generating the results from simulated rolls of a die.

Exercise 2.20

(a) Use your computer to roll a fair six-sided die 6 times; compare the sample relative frequencies for each of the six different outcomes with the theoretical probabilities.

(b) Now use your computer to roll the die 600 times and repeat the calculations.

Exercise 2.21

A researcher is exploring the theory that after early injury to the left half of the brain, the brain reorganizes itself to compensate for the loss. She hypothesizes that one manifestation of such reorganization would be that people who would otherwise have been right-handed would, after such damage and reorganization, become left-handed. To test this hypothesis she wishes to compare the proportion of left-handed people in the population of the United Kingdom with the proportion of left-handed people in a sample of brain-damaged people.

Data were provided by Dr S.L. Channon, Middlesex Hospital, University College London.

She knows that in the United Kingdom population, around 10% of the people are left-handed.

In a sample of 33 people who were brain-damaged at birth, 8 were found to be left-handed. Eight out of 33 represents a proportion of about 24%.

This suggests that there may be some basis for the researcher's theory. However, we should explore the random variation that might be observed in samples from the wider population. If a count as high as 8 turns out to be not unusual, then, on the evidence of this experiment at least, there may be no foundation for her belief.

Using the binomial random number generator on your computer, generate 100 samples from a binomial distribution with parameters $n = 33$ and $p = 0.1$, i.e. 10%, and see what fraction of these samples produce a count of 8 or more. This fraction gives you an indication of how often you might expect to obtain as many as 8 out of 33 left-handed people, if the proportion of left-handed people in the overall population from which the sample was drawn was only 10%. If this fraction is very small then the researcher can feel confident that there is some substance to her hypothesis. If the fraction is large then such an outcome (at least 8 out of 33) could easily have arisen by chance.

This is an example of a statistical *hypothesis test*; hypothesis testing is described more formally in depth later in the course.

If your computer incorporates functions or procedures for the triangular density Triangular(θ), you can use the following exercise to investigate the command syntax. Otherwise, use your calculator.

Exercise 2.22

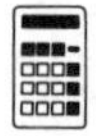

(a) If the continuous random variable V follows a triangular distribution with parameter 60, i.e. $V \sim$ Triangular(60), find

(i) $P(V \leq 20)$;

(ii) $P(V > 40)$;

(iii) $P(20 \leq V \leq 40)$.

(b) The following table gives the waiting times, that is, the gap (to the nearest half-hour) between consecutive admissions at an intensive care unit. (The data are read across the rows.)

Cox, D.R. and Snell, E.J. (1981) *Applied Statistics—Principles and Examples.* Chapman and Hall, London, p. 53. The data were collected by Dr A. Barr, Oxford Regional Hospital Board.

Table 2.22 Waiting times between admissions (hours)

6.0	102.0	59.0	122.0	45.0	7.0	42.0	5.5	28.5	80.0
8.0	62.0	34.0	41.5	98.0	4.5	52.0	144.0	9.5	33.5
77.0	51.0	47.0	2.0	1.0	87.0	0.0	54.0	76.0	0.0
96.0	63.0	6.5	23.5	21.0	96.0	28.5	171.0	116.5	23.0

(i) Obtain a histogram for these data.

(ii) Comment on the feasibility of fitting a triangular model to these data, and decide on a sensible value for the indexing parameter θ.

(iii) Use your model to determine the long-term proportion of waiting times that would exceed 100 hours. Compare this with the corresponding estimate of that proportion, based on these data.

Summary

1. A random variable taking integer values only is called discrete. The probability mass function of a random variable X

$$p(x) = P(X = x),$$

for $x = 0, 1, 2, \ldots,$ (or some other subset of the integers) describes the probability distribution of X.

2. The random variable X follows a Bernoulli distribution with parameter p if it has probability mass function

$$p(x) = p^x(1-p)^{1-x}, \quad x = 0, 1,$$

where $0 < p < 1$.

This is written $X \sim \text{Bernoulli}(p)$.

3. The random variable X follows a binomial distribution with parameters n and p if it has probability mass function

$$p(x) = \binom{n}{x} p^x(1-p)^{n-x}, \quad x = 0, 1, 2, \ldots, n,$$

where $0 < p < 1$ and

$$\binom{n}{x} = \frac{n!}{x!\,(n-x!)}, \quad x! = 1 \times 1 \times 2 \times \cdots \times x, \quad 0! = 1.$$

This is written $X \sim B(n, p)$, and provides a probability model for the total number of successes in a sequence of n independent Bernoulli trials.

4. A random variable X is called continuous if it can take any real value between some lower limit (a, say) and some upper limit (b). Its probability distribution is given in terms of the probability density function

$$f(x), \quad a \leq x \leq b.$$

The cumulative distribution function of X is given by

$$P(X \leq x) = F(x) = \int_a^x f(w)\,dw, \quad a \leq x \leq b,$$

where the integral notation means the 'area under the curve $f(w)$ between the limits $w = a$ and $w = x$'.

5. The continuous random variable W is said to follow a triangular distribution with parameter θ if it has cumulative distribution function

$$F(w) = P(W \leq w) = 1 - \left(1 - \frac{w}{\theta}\right)^2, \quad 0 \leq w \leq \theta.$$

This is written $W \sim \text{Triangular}(\theta)$.

Chapter 3
Models for Data II

In this chapter, more models for random variation within a population are described and characteristics of the models are identified. In particular, the notions of population moments and population quantiles are introduced.

In *Chapter 1*, some straightforward ways of examining data were described. The methods were applied to samples of data to shed light on particular questions. The techniques introduced included a variety of graphical methods for picturing the data, and some numerical quantities for summarizing the data in different ways—such as the sample mean and sample standard deviation. Very often, however, the sample itself is not of any special interest. What we really want to do is make some statement about the entire population from which the sample is drawn. Much of statistics is concerned with making inferences from a sample to the population: how to estimate population parameters from samples, how to evaluate the confidence we should have in the results, how to design experiments or draw samples so that we can obtain the most accurate estimates, and so on.

In *Chapter 2*, probability models were introduced. In particular, probability distributions were introduced as models for random variation. Some sections of that chapter focused on general properties (such as probability mass functions, probability density functions and cumulative distribution functions), while others concentrated on specific models and their properties. The distinction between discrete and continuous random variables was drawn. The ideas were illustrated by the Bernoulli and binomial distributions (both discrete), and the normal distribution (continuous). This chapter continues in the same vein: some sections introduce general properties of population models, while others deal with particular models.

General properties are introduced in Sections 3.1, 3.2 and 3.5. In Section 3.1, population analogues of the sample mean and sample variance introduced in *Chapter 1* are discussed. *Chapter 1* also introduced sample quartiles and the sample median. There are again corresponding population measures and these are discussed in Section 3.5. Just as sample statistics can be used to describe aspects of the shape of the distribution of values in the sample, so the corresponding population parameters describe features of the population models.

Section 3.2 concerns a fundamental concept, introduced briefly in *Chapter 2*, namely the concept of independence. The idea of independence—that the outcome of one trial (measurement, score, observation, and so on) does not influence the outcome of another—is vitally important. A great many statistical techniques assume that observations are independent (for example, that the outcome of two tosses of a coin are independent, or that the responses of two patients in a test of a new medicine are independent). If this assumption turns out to be false then more sophisticated statistical techniques have to be devised.

The two particular models developed in this chapter are the *geometric distribution* and the *uniform distribution*. The former is outlined in Section 3.3 and arises, like the binomial distribution, from a statistical experiment involving a sequence of independent trials each of which has a binary outcome (either success or failure). Here, however, interest focuses not on the total number of successes in a sequence of trials, but on the number of trials needed until the first success is obtained.

The uniform distribution has already been used in *Chapter 2*, page 62 as a model for the outcomes observed when a perfect die is rolled. In Section 3.4, the uniform distribution is discussed in more detail.

3.1 Population means and variances

In *Chapter 1*, we saw how certain important characteristics of samples of data can be encapsulated by various numerical summaries. Examples included the sample mean, the sample median and the sample standard deviation. If a bar chart or histogram, as appropriate, were used to represent the variation in the data, then these numerical summaries could be seen to be describing various aspects of the shape of the bar chart or histogram. For example, the sample mean and median tell us about the 'location' of the sample in certain senses; the 'spread' or 'dispersion' of the data can be measured by the sample standard deviation, or alternatively by the sample interquartile range; and the sample skewness measures asymmetry of the data.

Given a sample of values, we know how to compute numerical descriptors such as those mentioned above: you had some practice at doing so in *Chapter 1*. However, if we were to assume a probability model as adequate for the variation within a population, how might we define and calculate similar numerical summaries for the population such as the *population mean* and *population standard deviation*? These are sometimes required for a statistical analysis.

Here is an example of the sort of context within which we might need to perform a comparative test, and for which we therefore first need a model.

Example 3.1 Origins of the Etruscan empire

The origins of the Etruscan empire remain something of a mystery to anthropologists. A particular question is whether Etruscans were native Italians or immigrants from elsewhere. In an anthropometric study, observations on the maximum head breadth (measured in mm) were taken on 84 skulls of Etruscan males. These data were compared with the same skull dimensions for a sample of 70 modern Italian males. The data are summarized in Table 3.1.

The Etruscan urban civilization reached its zenith in about the sixth century BC.

Table 3.1 Maximum head breadth (mm)

84 Etruscan skulls										
141	148	132	138	154	142	150	146	155	158	150
140	147	148	144	150	149	145	149	158	143	141
144	144	126	140	144	142	141	140	145	135	147
146	141	136	140	146	142	137	148	154	137	139
143	140	131	143	141	149	148	135	148	152	143
144	141	143	147	146	150	132	142	142	143	153
149	146	149	138	142	149	142	137	134	144	146
147	140	142	140	137	152	145				
70 modern Italian skulls										
133	138	130	138	134	127	128	138	136	131	126
120	124	132	132	125	139	127	133	136	121	131
125	130	129	125	136	131	132	127	129	132	116
134	125	128	139	132	130	132	128	139	135	133
128	130	130	143	144	137	140	136	135	126	139
131	133	138	133	137	140	130	137	134	130	148
135	138	135	138							

Barnicot, N.A. and Brothwell, D.R. (1959) The evaluation of metrical data in the comparison of ancient and modern bones. In Wolstenholme, G.E.W. and O'Connor, C.M. (eds) *Medical Biology and Etruscan Origins.* Little, Brown and Co., USA.

The statistical procedures for such a comparison will be described in *Chapter 8*. (In fact, a simple comparative boxplot as shown in Figure 3.1 suggests marked differences between the Etruscan skulls and those of modern Italian males.)

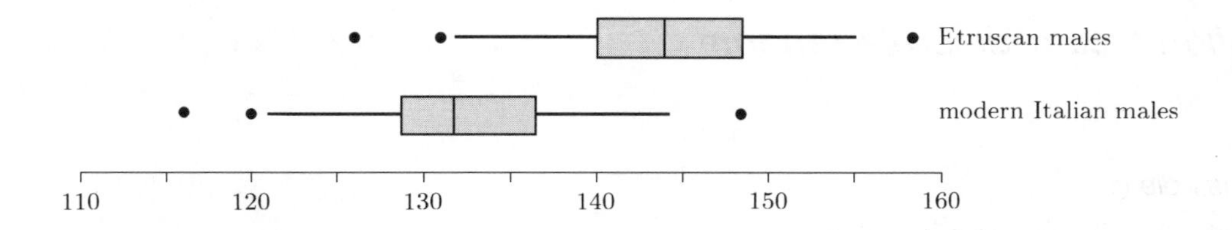

Figure 3.1 Comparative boxplot, two skull samples

Histograms for the two skull samples are shown in Figure 3.2. These suggest that in either case a normal model as described in *Chapter 2*, Section 2.4, might be adequate for the purposes of a comparative test, but with different indexing parameters in the two cases.

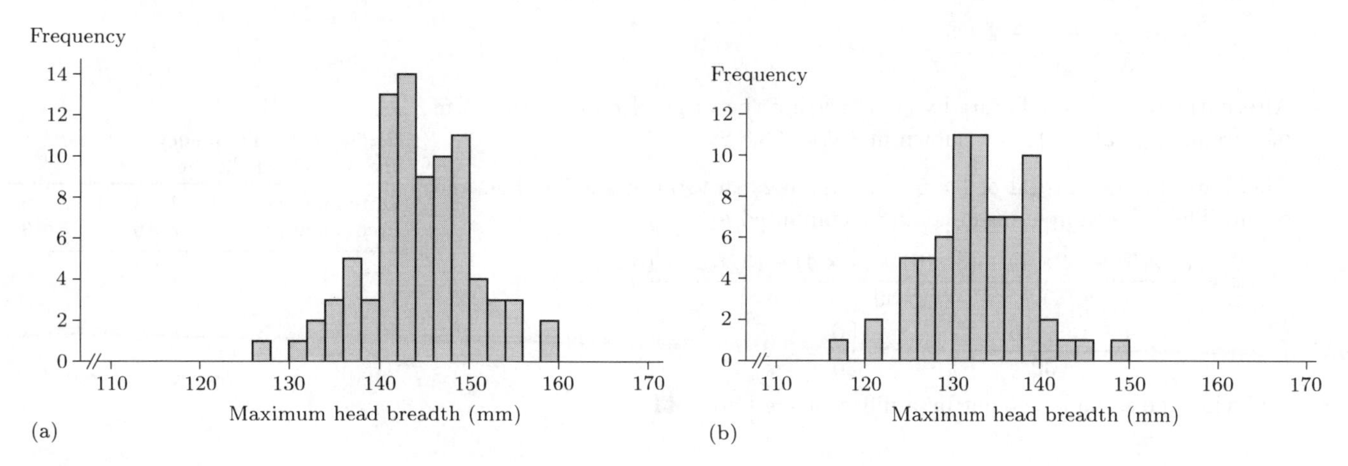

Figure 3.2
Histograms, two skull samples: (a) 84 Etruscan males (b) 70 modern Italian males

The two usual sample summaries for location and dispersion are the sample mean and sample standard deviation. For these data, we have the following summary statistics:

84 Etruscan skulls: $\overline{x} = 143.8$, $s_x = 6.0$;
70 Italian skulls: $\overline{y} = 132.4$, $s_y = 5.7$.

Notice the use of subscripts to distinguish between the two samples.

These values suggest that the skulls of the modern Italian male are, on average, both narrower and slightly less variable than those of the ancient Etruscan. If the breadth of Etruscan skulls may be adequately modelled by a normally distributed random variable X and the breadth of modern Italian skulls by a normally distributed random variable Y, then differences in the indexing parameters should reflect the differences observed in the two samples. ■

In this section, we shall be concerned chiefly with two of the main numerical summaries of population probability models, the population analogues of the sample *mean* and sample *variance* (or sample *standard deviation*). Other important concepts, such as the population median and other quantiles, will be discussed in Section 3.5.

Let us begin with a look at how the notion of the sample mean can be developed to help us define its population counterpart, the population mean.

3.1.1 The population mean: discrete random variables

Example 3.2 *Rolls of a fair die*

The outcomes of 30 rolls of a fair die are given in Table 3.2.

Table 3.2 30 rolls of a fair die

4	3	2	5	6	3	2	6	4	3	6	4	3	2	5
3	1	5	2	1	4	3	1	2	1	1	2	5	2	5

One way of calculating the sample mean for the 30 rolls (if you had to do this by hand) is simply to obtain the sample total and divide by the sample size:

$$\overline{x} = \frac{4+3+2+\cdots+2+5}{30} = \frac{96}{30} = 3.2.$$

Alternatively, one could start by summarizing the original data in the form of a frequency table. This is shown in Table 3.3.

Table 3.3 Frequency table for the 30 rolls of the die

Outcome (j)	1	2	3	4	5	6
Frequency (f_j)	5	7	6	4	5	3

There are five 1s (a total of $1 \times 5 = 5$), seven 2s (a total of $2 \times 7 = 14$), and so on. Then the sample mean could be computed as

$$\begin{aligned}\overline{x} &= \frac{(1 \times 5) + (2 \times 7) + (3 \times 6) + (4 \times 4) + (5 \times 5) + (6 \times 3)}{30} \\ &= \frac{5 + 14 + 18 + 16 + 25 + 18}{30} = \frac{96}{30} = 3.2,\end{aligned}$$

achieving the same result but by a different method. ■

This example shows that, given a sample of data from a discrete distribution, there are two equivalent ways of calculating the sample mean.If $x_1, x_2, \ldots, x_{30}$

denote the values in the sample in Example 3.2, in the first method we obtain the sample mean by adding all the values together and dividing by 30. In this case we are using the familiar definition of the sample mean,

$$\overline{x} = \frac{1}{n}\sum_{i=1}^{n} x_i. \tag{3.1}$$

The idea behind the second approach is to count how many of each particular outcome there are in the sample in order to obtain a frequency table. If we denote the number of occurrences of outcome j in the sample by f_j (for instance, $f_2 = 7$, in Example 3.2), the contribution made to the total by each of the outcomes j is $j \times f_j$ (for example, $2 \times 7 = 14$). Adding all these up and then dividing by n (the sample size, equal to 30 in Example 3.2), we obtain the sample mean in the form

$$\overline{x} = \frac{1}{n}\sum_{j} j f_j. \tag{3.2}$$

Here, the sum is over all possible outcomes j $(1, 2, \ldots, 6$, for the die). The two formulas (3.1) and (3.2) for $\overline{x}$ give the same answer.

However, in *Chapter 2*, Subsection 2.1.2, you saw how a discrete probability $p(j)$ could be defined by taking the limiting value of the sample relative frequency of the observation j as the sample size n gets larger and larger. By writing $\overline{x}$ in the form (3.2), or perhaps even more clearly in the form

$$\overline{x} = \sum_{j} j \frac{f_j}{n},$$

the same idea can be employed here to produce a definition of the population mean, at least for a discrete random variable. By simply replacing the sample relative frequency f_j/n by its limiting value, the probability $p(j)$, we obtain the following definition.

We shall discuss the case of continuous distributions in Subsection 3.1.2.

For a discrete random variable taking the value j with probability $p(j)$, the **population mean** is given by

$$\mu = \sum_{j} j p(j),$$

where the sum is over the set of possible observed values, that is, over the range of the random variable.

The lower-case Greek letter μ is often used to denote a population mean.

Example 3.3 *The average score when a fair die is rolled*

For a die that is assumed to be fair, each of the six possible outcomes $j = 1, 2, \ldots, 6$ occurs with probability $p(j) = \frac{1}{6}$. The mean outcome in a single roll of a fair die is therefore

$$\begin{aligned} \mu &= \sum_{j} j p(j) \\ &= \left(1 \times \tfrac{1}{6}\right) + \left(2 \times \tfrac{1}{6}\right) + \left(3 \times \tfrac{1}{6}\right) + \left(4 \times \tfrac{1}{6}\right) + \left(5 \times \tfrac{1}{6}\right) + \left(6 \times \tfrac{1}{6}\right) \\ &= \tfrac{1}{6}(1 + 2 + \cdots + 6) = 3.5. \quad \blacksquare \end{aligned}$$

Notice that it is not necessary for the mean of a random variable X to be in the range of X: 3.5 is not a possible outcome when a die is rolled. As another example, the 'average family size' in Britain is not an integer.

Exercise 3.1

In *Chapter 2*, you were also introduced to certain unfair dice called *Tops* and you saw that, for a *Double-Five*, the probability distribution for the outcome of a single roll is as given in Table 3.4.

What is the mean of this distribution? How does it compare with the mean outcome for a fair die?

Table 3.4 A *Double-Five*

j	1	3	4	5	6
$p(j)$	$\frac{1}{6}$	$\frac{1}{6}$	$\frac{1}{6}$	$\frac{1}{3}$	$\frac{1}{6}$

One very important application of statistics is to epidemiology—the study of health and illness in human populations. Many different models have been developed for the transmission of infectious diseases, a few of them simple, but most of them rather complicated. In small communities (for instance, families and schools) one variable of interest is the total number of people who catch a disease, given that initially one member of the community becomes infected. In a family of 4, say, that number could be $1, 2, 3$ or 4. This number is a random variable because epidemic dynamics are, to a great extent, a matter of chance—whether or not you catch your brother's cold, for instance, is not a predetermined event.

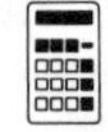

Exercise 3.2

One model for a certain disease within a particular family of 6 gave the probability distribution for the number X, who eventually suffer from the disease, as shown in Table 3.5.

What is the mean of the distribution of X?

Table 3.5 A family of 6

x	1	2	3	4	5	6
$p(x)$	$\frac{3}{90}$	$\frac{8}{90}$	$\frac{15}{90}$	$\frac{20}{90}$	$\frac{24}{90}$	$\frac{20}{90}$

An alternative terminology for the mean of a random variable X is the **expected value** of X or simply the **expectation** of X. This is written $E(X)$ as an alternative to μ.

We can summarize the foregoing as follows.

The mean of a discrete random variable

For a discrete random variable X with probability mass function

$$P(X = x) = p(x)$$

over a specified range, the **mean of *X*** or the **expected value of *X*** or the **expectation of *X*** is given by

$$\mu = E(X) = \sum_X x p(x), \qquad (3.3)$$

where the sum is taken over the range of X.

The notation μ for the mean of a random variable X is sometimes modified to include the subscript X, that is, μ_X. This notation is particularly useful where a model involves more than one variable, as the means of different random variables cannot then be confused. However, in this course the subscript will not usually be included except where it is necessary to avoid ambiguity.

Example 3.4 Using the $E(\cdot)$ notation

If X denotes the score on a single roll of a fair die (Example 3.3), then the expected value of X is

$$E(X) = \tfrac{1}{6}(1+2+3+4+5+6) = 3.5.$$

On the other hand, if the random variable Y denotes the score on a single roll of a *Double-Five* (Exercise 3.1), then the expected value of Y is

$$E(Y) = 4.$$

For the family of 6 (Exercise 3.2), the expected number $E(Z)$ of family members who eventually suffer from the disease is

$$E(Z) = 4.3.$$

Occasionally, the phrase 'expected value' or 'expected number' is more natural than 'mean'. But notice that 'the value you would expect to get' is not usually a valid interpretation: you could not actually have 4.3 ill people in a family, because the number 4.3 is not an integer. ■

The examples above are for probability distributions where the probability mass function could be specified exactly. Earlier, the idea of indexing distributions by means of some unspecified quantity or quantities called the parameter(s) of the model was introduced. It will be particularly useful if some simple link can be established between the indexing parameter(s) and some summary measure, such as the mean: this would aid the interpretation of model parameters.

Let us take a look, then, at the first 'parametric family' of discrete distributions to which you were introduced in *Chapter 2*, Section 2.3. This was the Bernoulli family of models: each distribution in this family allows only the two possible outcomes 0 or 1, and the probability mass function of a Bernoulli distribution is $p(1) = p$, $p(0) = 1 - p$. (Here, p is the indexing parameter.) The population mean of any Bernoulli distribution can therefore be found in terms of p. In fact,

$$\begin{aligned}\mu &= \sum_{j=0}^{1} jp(j) \\ &= 0 \times p(0) + 1 \times p(1) = 0 \times (1-p) + 1 \times p = p. \qquad (3.4)\end{aligned}$$

That is, the population mean of a Bernoulli distribution is the parameter p used to index this family of models.

Exercise 3.3

(a) What is the mean score resulting from a toss of a fair coin, if we score 1 for Heads and 0 for Tails?

(b) Suppose that a random variable is defined to take the value 1 when a fair die shows a 3 or a 6, and 0 otherwise. What is the mean value for this Bernoulli distribution?

The second discrete probability distribution introduced in *Chapter 2* was the binomial distribution or family. If the random variable X has a binomial distribution $B(n,p)$ then the probability mass function of X is

$$p(x) = \binom{n}{x} p^x (1-p)^{n-x}, \quad x = 0, 1, 2, \ldots, n.$$

So, using the definition in (3.3), the mean of X (or the expected value of X) is given by the sum

$$E(X) = \sum_{x=0}^{n} x \binom{n}{x} p^x (1-p)^{n-x}.$$

In fact, this calculation is not quite as unwieldy as it looks, and a small amount of algebraic manipulation would give us the answer we need. However, no algebra is necessary if we think about what the binomial random variable X represents. It is the number of successes in a series of n trials, where the probability of a successful outcome at each trial is p. So the problem posed is this: what is the expected number of successes in such a series of n trials? No formal mathematics is required to provide an answer to this question (though it could be used to provide a formal proof). If, for instance, 100 trials are performed and for each the probability of success is $\frac{1}{4}$, then the expected number of successes is $100 \times \frac{1}{4} = 25$ and the expected number of failures is $100 \times \frac{3}{4} = 75$. In general, the mean of a binomial random variable X indexed by the two parameters n and p is given by the product

$$E(X) = np. \tag{3.5}$$

Notice, as in the case of a fair die, that the mean may be some value not in the range of X (for instance, take $n = 100$ and $p = \frac{1}{3}$; the number $np = 33\frac{1}{3}$ is not an integer). Nevertheless, the mean or expectation is a statement about the 'long-term' average number of successes in sequences of Bernoulli trials.

Example 3.5 *The mean of a binomial random variable: two methods of calculation*

If X is binomial $B(4, 0.4)$, then its probability mass function is given by

$$p(x) = \binom{4}{x} (0.4)^x (0.6)^{4-x}, \quad x = 0, 1, 2, 3, 4.$$

The individual probabilities are as follows.

x	0	1	2	3	4
$p(x)$	0.1296	0.3456	0.3456	0.1536	0.0256

The mean of X is then

$$\begin{aligned} E(X) &= \sum_{x=0}^{4} x p(x) \\ &= (0 \times 0.1296) + (1 \times 0.3456) + \cdots + (4 \times 0.0256) \\ &= 0 + 0.3456 + 0.6912 + 0.4608 + 0.1024 = 1.6. \end{aligned}$$

This is much more easily obtained using (3.5):

$$E(X) = np = 4 \times 0.4 = 1.6.$$

This may be shown graphically as in Figure 3.3. The point $x = 1.6$ (the mean of the distribution) is shown as an arrowhead on a sketch of the probability mass function of X.

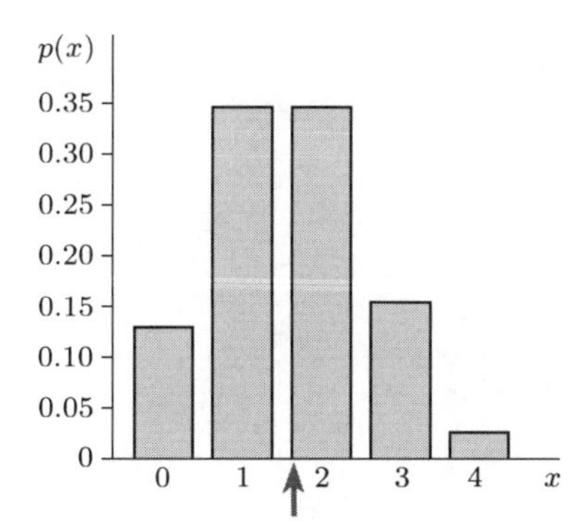

Figure 3.3 The mean of X when $X \sim B(4, 0.4)$ ■

The mean has the following physical interpretation (rather idealized): imagine lead weights of mass 0.1296, 0.3456, 0.3456, 0.1536 and 0.0256 units placed at equal intervals on a thin plank of zero mass. If it is represented by the horizontal axis in Figure 3.3, the plank will balance at the point indicated by the arrowhead—at a point just to the right of the midpoint between the two largest weights.

Example 3.6 *Spores of the fungus Sordaria*

The spores of the fungus *Sordaria* are produced in chains of eight. Any chain may break (at any of the seven joints) and the spores are thus projected in chainlets varying in length from one to eight. It turns out that it is reasonable to model the breakages occurring at the joints as independent; and each joint has the same probability p of breaking.

The probability of a conjunction of independent events is found from the product of the probabilities of the component events. This result was used in our derivation of the binomial probability distribution in *Chapter 2*, Subsection 2.3.2.

So, for instance, an original chain of eight might survive unbroken with probability $(1-p)^7$. However, it might break at all seven joints, so that eight chainlets all of length one are projected. This occurs with probability p^7. Chainlets of length one are called 'singletons'.

The number of singletons, X, produced by a chain of eight spores is a random variable taking values $0, 1, 2, \ldots, 8$. (Actually, a count of seven is not possible, for the eighth spore would itself necessarily be a singleton too.) The probability distribution of X is given in Table 3.6. (The distribution derives directly from a complete enumeration of possible cases; you should not bother to check these results.)

Table 3.6 The probability distribution of the number of singletons, X

x	$p(x)$
0	$(1-p)^4(1+2p-p^2-p^3)$
1	$2p(1-p)^4(1+4p+p^2)$
2	$p^2(1-p)^3(3+12p-5p^2)$
3	$4p^3(1-p)^3(1+4p)$
4	$5p^4(1-p)^2(1+2p)$
5	$6(1-p)^2p^5$
6	$7(1-p)p^6$
7	0
8	p^7

In this case there is no particularly obvious way of deducing the average number of singletons produced, other than by applying the formula. This gives

$$\begin{aligned} E(X) &= \sum_{x=0}^{8} x p(x) \\ &= 0 \times (1-p)^4(1+2p-p^2-p^3) + \cdots + 8 \times p^7 \\ &= 2p(3p+1), \end{aligned}$$

after considerable simplification. For instance, if p is 0.8 (the original chain is very fragile) the expected number of resulting singletons is 5.44. If p is as low as 0.1 (the original chain is robust) the expected number of resulting singletons is only 0.26.

The simplification in the last line of the calculation of the mean $E(X)$ involved quite a lot of algebraic manipulation, which you should not attempt to verify! A computer running an algebra program was used for these calculations.

However, in this case there is no immediate interpretation of the mean in terms of the indexing parameter p.

These arithmetic calculations are illustrated in Figure 3.4. Notice the interesting multimodal nature of the probability distribution of the number of singletons when p is 0.8.

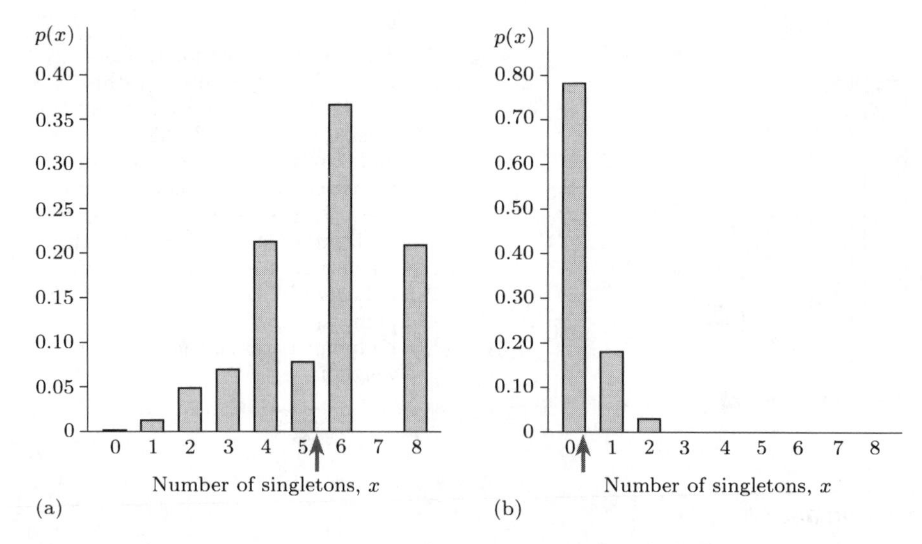

Figure 3.4 The probability distribution of the number of singletons (a) for $p = 0.8$; (b) for $p = 0.1$. ■

Exercise 3.4

A chainlet of length four is called a 'quad'. The original chain will result in two quads being projected, for instance, only if the middle joint breaks and the other six do not: the probability of this is $p(1-p)^6$. The probability distribution of the number of quads projected is given in Table 3.7. Again, do not worry about the algebraic details.

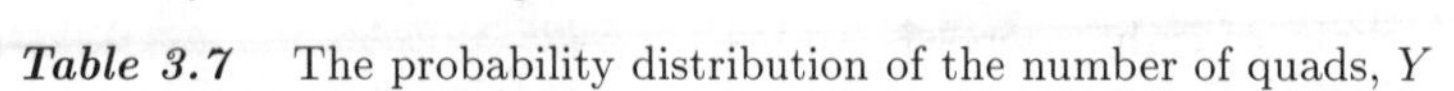

Table 3.7 The probability distribution of the number of quads, Y

y	$p(y)$
0	$1 - p(1-p)^2\left(3p^4 - 2p^3 + 11p^2 - 4p + 1\right)$
1	$p^3(1-p)^2\left(2p^2 + 2p + 5\right)$
2	$p(1-p)^6$

Find the expected value of Y when

(a) $p = 0.1$; (b) $p = 0.4$; (c) $p = 0.6$; (d) $p = 0.8$.

3.1.2 The population mean: continuous random variables

The variation that might be observed in measurements on a discrete random variable is expressed through its probability mass function, and you have seen how to use the p.m.f. to calculate the mean or expected value of a discrete random variable. Similarly, variation observed in measurements on a continuous random variable may be expressed by writing down a probability density function. If this density function is a good model then we should be able to use it not just for forecasts about the likelihood of different future measurements but, as in the case of discrete random variables, to provide information about the long-term average in repeated measurements.

As in the case of the mean of a discrete random variable, the value of the expectation $\mu = E(X)$ of a continuous random variable X has a physical interpretation. It is the point about which a physical model of the density would balance, if such a model were constructed (say, of tin plate). Here are some examples.

Example 3.7 Means of continuous random variables

(a) In *Chapter 2*, Example 2.8, the triangular density Triangular(20) was used as a model for the waiting time (in seconds) between vehicles, for traffic using the Kwinana Freeway in Perth, Western Australia. The density is shown in Figure 3.5. Also shown is the point at which a tin triangle with these dimensions would balance. This point occurs one-third of the way along the base of the triangle, at the point $t = 6\frac{2}{3}$. This is the mean of the triangular distribution with parameter 20.

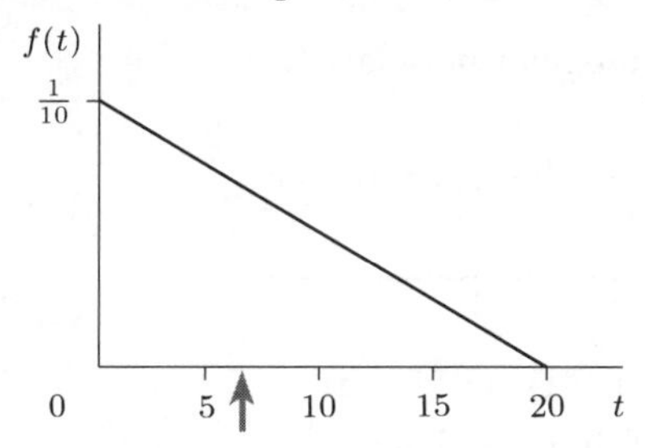

Figure 3.5 The triangular density Triangular(20), showing the mean $\mu = 6\frac{2}{3}$

(b) The density of the random variable $X \sim N(\mu, \sigma^2)$ is shown in Figure 3.6. (Observations on X much below $\mu - 3\sigma$ or much above $\mu + 3\sigma$ are possible, but unlikely.) The normal density is symmetric: if it were possible to construct a tin-plate model of the density, the model would balance at the point $x = \mu$.

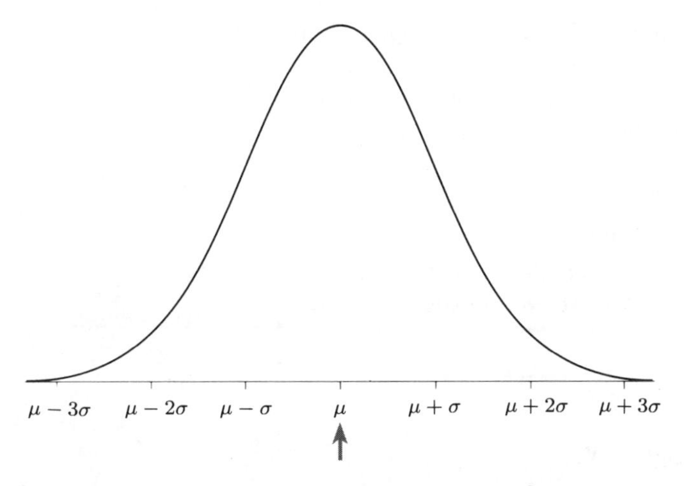

Figure 3.6 The p.d.f. of the normal distribution $N(\mu, \sigma^2)$

So, in fact, one of the two indexing parameters for the normal family is the mean of the distribution. Observations are no more likely to be above the central value μ than below it. The long-term observed average is bound to be μ itself. ■

In the following exercise you are asked to simulate random observations from the triangular family, and use your findings to suggest a general formula for the mean of the triangular distribution Triangular(θ).

Exercise 3.5

(a) Generate a random sample of size 10 from the triangular distribution with parameter $\theta = 20$. List the elements of your sample and calculate the sample mean.

(b) Find the mean of a random sample of size 1000 from the triangular distribution with parameter $\theta = 20$.

Repeat the sampling procedure a further nine times and list the ten sample means you obtained. (You should find that your list offers supporting evidence for the result stated in Example 3.7(a): that the mean of the triangular distribution Triangular(20) is $6\frac{2}{3}$.)

(c) Find the mean of random samples of size 1000 from triangular populations with parameters (i) $\theta = 30$; (ii) $\theta = 300$; (iii) $\theta = 600$.

(d) Use your results in this exercise to hazard a guess at the mean of the continuous random variable T when $T \sim$ Triangular(θ).

Most probability density functions possess neither the symmetry of the normal density function nor the rather convenient geometrical form of the triangular density function. How, in general, is the mean of a continuous random variable calculated?

The result (3.3) tells us that for a discrete random variable X with probability mass function $p(x)$, the mean of X is given by the formula

$$\mu = E(X) = \sum_X x p(x),$$

where the summation is taken over the range of X. This represents an average of the different values that X may take, according to their chance of occurrence. The definition of the mean of a continuous random variable is analogous to that of a discrete random variable.

The mean of a continuous random variable

For a continuous random variable X with probability density function $f(x)$ over a specified range, the **mean of X** or the **expected value of X** is given by

$$\mu = E(X) = \int_X x f(x)\,dx, \qquad (3.6)$$

where the integral is taken over the range of X.

As you can see, the technique of integration is required for the calculation of the mean of a continuous random variable. If you are familiar with the technique, you might like to confirm for yourself some of the standard results that follow now and in the rest of the course. However, these are standard results and you would not be expected to obtain them from first principles.

If you are not familiar with the technique of integration, this will not interfere with your understanding of the results.

Example 3.8 *The mean of the triangular distribution*

(a) You saw in *Chapter 2*, Subsection 2.2.2, that the p.d.f. of the continuous random variable $T \sim$ Triangular(20) is given by

$$f(t) = \frac{20-t}{200}, \quad 0 \le t \le 20.$$

Using (3.6), the mean of T is

$$\begin{aligned}\mu = E(T) &= \int_T t f(t)\, dt \\ &= \int_0^{20} t\left(\frac{20-t}{200}\right) dt \\ &= \tfrac{1}{200}\int_0^{20} t(20-t)\, dt \\ &= \tfrac{1}{200}\int_0^{20} \left(20t - t^2\right) dt \\ &= \tfrac{1}{200}\left[10t^2 - \tfrac{1}{3}t^3\right]_0^{20} \\ &= \tfrac{1}{200}\left(4000 - \tfrac{8000}{3}\right) \\ &= \tfrac{20}{3} \\ &= 6\tfrac{2}{3}.\end{aligned}$$

(b) In Exercise 3.5(d) you guessed the mean of the triangular distribution Triangular(θ). Now your guess can be confirmed (or otherwise). The p.d.f. of the continuous random variable $T \sim$ Triangular(θ) is given by

$$f(t) = \frac{2(\theta - t)}{\theta^2}, \quad 0 \le t \le \theta.$$

So the mean of the distribution is given by

$$\begin{aligned}\mu = E(T) &= \int_T t f(t)\, dt \\ &= \int_0^{\theta} t\left(\frac{2(\theta-t)}{\theta^2}\right) dt \\ &= \frac{2}{\theta^2}\int_0^{\theta} \left(\theta t - t^2\right) dt \\ &= \frac{2}{\theta^2}\left[\tfrac{1}{2}\theta t^2 - \tfrac{1}{3}t^3\right]_0^{\theta} \\ &= \frac{2}{\theta^2}\left[\tfrac{1}{2}\theta^3 - \tfrac{1}{3}\theta^3\right] \\ &= \frac{2}{\theta^2}\left[\tfrac{1}{6}\theta^3\right] \\ &= \tfrac{1}{3}\theta. \quad ■\end{aligned}$$

The c.d.f. of T,

$$F(t) = 1 - \left(1 - \frac{t}{\theta}\right)^2, \quad 0 \le t \le \theta,$$

was given in *Chapter 2*, Section 2.4. Just as a c.d.f. is obtained from a p.d.f. by integration, so a p.d.f. is obtained from a c.d.f. by the inverse operation, known as differentiation. If you are familiar with this technique, you can check that the triangular p.d.f. is as stated. However, it is used here only as a means to an end: to find the mean of the triangular distribution.

***Example 3.9** The mean of the normal distribution*

Were you to calculate the mean of the normal distribution $N(\mu, \sigma^2)$ directly from (3.6),

$$E(X) = \int_X x f(x)\, dx = \int_{-\infty}^{\infty} \frac{x}{\sigma\sqrt{2\pi}} \exp\left[-\tfrac{1}{2}\left(\frac{x-\mu}{\sigma}\right)^2\right] dx,$$

you would find that the parameter σ vanished and you were left with $E(X) = \mu$. ■

Now try the following exercise. (Notice that the two results you need to answer the question are given to you: you do not need to obtain them from first principles.)

Exercise 3.6

For the triangular distribution with parameter θ, the c.d.f. is given by

$$F(t) = 1 - \left(1 - \frac{t}{\theta}\right)^2, \quad 0 \le t \le \theta,$$

and the mean of the distribution is $\mu = \frac{1}{3}\theta$.

It is assumed in a traffic monitoring context that the triangular density provides an adequate model for the waiting time between successive vehicles, but the value of the parameter θ is unknown. (It could be one of the aims of a sampling experiment to estimate the value of θ: see *Chapter 6* for more on the topic of estimation.)

What proportion of waiting times is longer than average?

3.1.3 The population variance

In Subsection 3.1.2 the idea of a sample mean was extended to the mean of a theoretical model for the observed variation, which was denoted by μ or $E(X)$ (the expected value of X). Now we require a measure of dispersion for a population, analogous to the sample variance or sample standard deviation. The following example illustrates a typical context in which knowledge of a population variance is essential to answering a scientific question.

***Example 3.10** Measuring intelligence*

A psychologist assessing intellectual ability decides to use the revised Wechsler Adult Intelligence Scale—WAIS-R—to measure IQ. She finds that one subject has a score of 110. This is above the population mean of 100. But how far above the mean is it? Should she expect many people to score as high as this, or is the difference of 10 points a large difference? To answer this question she needs to know something about the spread or dispersion of IQ scores in the population, and she might, for example, choose to measure this spread using the population analogue of the sample standard deviation encountered in *Chapter 1*. ■

As defined in *Chapter 1*, the sample variance is given by

$$s^2 = \frac{1}{n-1} \sum_{i=1}^{n} (x_i - \overline{x})^2.$$

This measure of dispersion gives the *average squared deviation* of each item in the sample from the sample mean, with the small distinction that the average is obtained through division by $n-1$ rather than by n. The analogous measure for a probability model is the *expected squared deviation* of a random variable X from the mean of X. This may be written using the 'expectation' notation $E(\cdot)$ as

$$E[(X-\mu)^2].$$

So now we require not simply 'the expected value of X', but 'the expected value of a function of X'. We need to calculate the value of that function $(x-\mu)^2$ for each value of x in the range of X, and then average the squared deviations obtained over the probability distribution of X.

Example 3.2 continued

In Example 3.2 we looked at the results of 30 rolls of a fair die. The sample mean was found to be $\overline{x} = 3.2$. For this sample, the sum of squared deviations from the mean is given by

$$\begin{aligned}
&\sum_{i=1}^{30}(x_i - \overline{x})^2 \\
&= (1-3.2)^2 + (1-3.2)^2 + (1-3.2)^2 + (1-3.2)^2 + (1-3.2)^2 \\
&\quad + (2-3.2)^2 + (2-3.2)^2 + (2-3.2)^2 + (2-3.2)^2 + (2-3.2)^2 + (2-3.2)^2 + (2-3.2)^2 \\
&\quad \vdots \\
&\quad + (6-3.2)^2 + (6-3.2)^2 + (6-3.2)^2.
\end{aligned}$$

This can more conveniently be written as

$$\begin{aligned}
&\sum_{x=1}^{6}(x - \overline{x})^2 f_x \\
&= (1-3.2)^2 \times 5 + (2-3.2)^2 \times 7 + (3-3.2)^2 \times 6 + \cdots + (6-3.2)^2 \times 3 \\
&= 76.8;
\end{aligned}$$

and so the sample variance (dividing by $n-1=29$) is

$$s^2 = \frac{76.8}{29} = 2.65.$$

However, a *theoretical probability model* for the outcome of rolls of a fair die is provided by the random variable X with probability mass function

$$P(X=x) = p(x) = \tfrac{1}{6}, \quad x = 1, 2, \ldots, 6.$$

The mean of X is given by

$$\mu = E(X) = \tfrac{1}{6}(1+2+3+4+5+6) = 3.5.$$

The expected value of $(X - \mu)^2$ is found by averaging the values obtained for $(x - \mu)^2$ over the probability distribution of X:

$$\begin{aligned} &E[(X-\mu)^2] \\ &= (1-3.5)^2 \times \tfrac{1}{6} + (2-3.5)^2 \times \tfrac{1}{6} + (3-3.5)^2 \times \tfrac{1}{6} + \cdots + (6-3.5)^2 \times \tfrac{1}{6} \\ &= \frac{6.25}{6} + \frac{2.25}{6} + \frac{0.25}{6} + \cdots + \frac{6.25}{6} \\ &= 2.92. \end{aligned}$$

So we see that our sample of 30 rolls was, on that occasion, a little less variable than theory would have suggested. ■

For discrete probability distributions, then, the variance is given by (3.7).

The variance of a discrete random variable

For a discrete random variable X with probability mass function

$$P(X = x) = p(x)$$

over a specified range, with mean $\mu = E(X)$, the **variance of X** is given by

$$\sigma^2 = V(X) = E[(X-\mu)^2] = \sum_X (x-\mu)^2 p(x), \qquad (3.7)$$

where the sum is taken over the range of X.

It is a common statistical convention to denote the variance of a random variable by σ^2. The alternative notation $V(X)$ will also often be used.

The **standard deviation of X** is given by the square root of the variance:

$$SD(X) = \sqrt{V(X)} = \sigma = \sqrt{\sum_X (x-\mu)^2 p(x)}.$$

For instance, the standard deviation in the outcome of rolls of a fair die is $\sqrt{2.92} = 1.71$.

Exercise 3.7

What is the variance of the outcome of throwing a *Double-Five*? How does this compare with the variance for a fair die?

We can work out the variance of any Bernoulli distribution with parameter p. Recall that the possible outcomes are just 0 and 1; so, from the definition,

$$\sigma^2 = (0-\mu)^2 p(0) + (1-\mu)^2 p(1).$$

Since $p(0) = 1 - p$, $p(1) = p$ and, from (3.4), the mean is p, it follows that

$$\begin{aligned}\sigma^2 &= (-p)^2(1-p) + (1-p)^2 p \\ &= p(1-p)(p+1-p) \\ &= p(1-p).\end{aligned}$$

Calculation of the variance of the binomial distribution $B(n, p)$ will be postponed to *Chapter 4*, Section 4.3.

We can move from the case of discrete distributions to that of continuous distributions just as we did for means. In short, we replace the p.m.f. by a p.d.f., and the sum by an integral.

The variance of a continuous random variable

For a continuous random variable X with probability density function $f(x)$ over a specified range and with mean $\mu = E(X)$, the **variance of *X*** is given by

$$\sigma^2 = V(X) = E\left[(X-\mu)^2\right] = \int_X (x-\mu)^2 f(x)\,dx, \tag{3.8}$$

where the integral is taken over the range of X.

The formula for the variance of a random variable will shortly be rewritten in a way that leads to some easement of the algebra. However, for common models it is not usually necessary to go through the algebra at all—the results are standard and well-known. At this stage, it is useful to notice that regardless of whether the random variable X is discrete or continuous, its variance can be written as

$$V(X) = E[(X-\mu)^2]. \tag{3.9}$$

The normal probability distribution is discussed in more detail in *Chapter 5*. However, just as the parameter μ indexing the normal distribution is the mean of the normal distribution, so the second parameter σ^2 indexing the normal distribution is the variance of the distribution (and σ is its standard deviation).

Now recall the question posed in Example 3.10: is a score of 110 on an IQ test, where the population average score is 100, unusually high? If the underlying probability distribution of WAIS-R scores were known, this question could be answered.

In fact, the IQ scale is *designed* to take account of the variability in responses from within the population in such a way that the resulting scores are normally distributed with mean $\mu = 100$ and standard deviation $\sigma = 15$ (variance 225).

We can plot the corresponding normal density and identify the location of the particular score of interest, 110. This is shown in Figure 3.7.

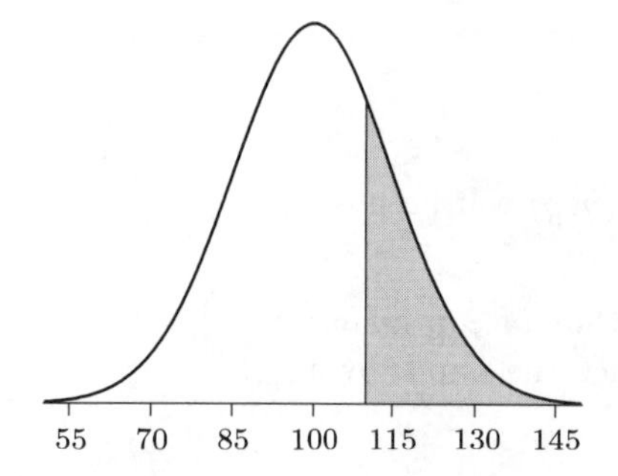

Figure 3.7 An observed score of 110 when $X \sim N(100, 225)$

The observed score is higher than average; nevertheless it looks as though a considerable minority of the population (the proportion represented by the shaded area in the diagram) would score as well or better. You will see in *Chapter 5* how to compute this proportion. It is given by the following integral:

$$P(X \geq 110) = \int_{110}^{\infty} \frac{1}{15\sqrt{2\pi}} \exp\left[-\frac{1}{2}\left(\frac{x-100}{15}\right)^2\right] dx.$$

In practice, evaluation of the integral is unnecessary since, for the normal distribution, such probabilities can be obtained using tables or a computer package. This is discussed further in *Chapter 5*. (The area of the shaded region in Figure 3.7 is 0.252: more than a quarter of the population would score 110 or more on the WAIS-R intelligence test.)

3.2 Independence of random variables

In *Chapter 2*, Subsection 2.3.1, the Bernoulli distribution was introduced as a model for a particular class of random variable. In *Chapter 2*, Subsection 2.3.2, we proceeded to the binomial distribution which is a model for an extended class of random variable obtained by adding together a number of separate Bernoulli random variables. However, the components of the sum all had to involve the same value of the parameter p, and also the different Bernoulli random variables had to be *independent*. In such a case, the components of the sum are described as **independent identically distributed** (i.i.d.) random variables. This notion of independence recurs through so much of statistics that the whole of this fairly short section will be devoted to it. The idea is a particularly important one for *Chapters 6* and *11*.

If the random variable X represents the number showing on a rolled red die, and Y represents the number showing on a rolled white die, the value x taken by X has no bearing on the value y taken by Y, and vice versa. Likewise, knowledge of the number of newspapers sold in a particular newsagent's shop in Wrexham one morning will not tell us much about the number of babies born in a maternity hospital in Luton that day, and vice versa.

However, Seal (1964) describes an experiment in which frogs' skull lengths X and skull breadths Y were measured. One might reasonably expect a large reading on one variable to indicate a similarly large reading on the other, though there will still be random variation evident in both variables. A sample of paired measurements will yield information on the nature and degree of the association between the two variables.

Seal, H.L. (1964) *Multivariate Statistical Analysis for Biologists.* Methuen, London, p. 106.

There is a clear distinction between the first two of these three situations and the third. In the first two, the random variables have no bearing on each other; conversely, in the frogs' skulls example, they are, in some way, related to one another. The distinction is to do with the degree of dependence between them. In this section, we are concerned with random variables of the former type that do not have any effect on each other. Dependent random variables, like skull length and skull breadth, will be dealt with in later chapters (particularly in *Chapter 11*).

A particular property of independent random variables was used in the derivation of the binomial probability distribution in *Chapter 2*, Section 2.3. There, each component in a sequence of Bernoulli trials is assumed to be independent of every other component. We used the result that the probability of obtaining a particular sequence of responses is found by multiplying together the individual probabilities for each response. Thus, for instance, if p is the probability of a Yes response, then

$$P(\text{Yes Yes No}) = P(\text{Yes})P(\text{Yes})P(\text{No}) = p^2(1-p)$$

and

$$P(\text{No No No}) = (1-p)^3.$$

If the assumption of independence breaks down, then this approach fails too. In an extreme case, the second and third respondents might simply repeat the previous response, so that

$$P(\text{Yes Yes Yes}) = p$$

$$P(\text{No No No}) = 1-p,$$

and these are the only two possible sequences of responses. Here, the degree of dependence between the responses is very high.

The idea that the probability of a conjunction of independent events is found by multiplying together the individual probabilities for each component event may be extended naturally to independent random variables in the following way.

Independence of random variables

If X and Y are discrete random variables that are independent, then the probability that X takes the value x and Y takes the value y, simultaneously, is given by

$$P(X = x, Y = y) = P(X = x)P(Y = y), \qquad (3.10)$$

for all x in the range of X and all y in the range of Y.

The comma on the left-hand side of the expression may be read as 'and'.

Example 3.11 Scores on two different dice

If X is the score when a fair die is rolled and Y the score when a *Double-Five* is rolled, and you throw one after the other, the outcomes are clearly independent (for by what mechanical influence could the scores be related?). The probability of scoring two 5s is

$$P(X=5, Y=5) = P(X=5)P(Y=5) = \tfrac{1}{6} \times \tfrac{2}{6} = \tfrac{1}{18}. \quad \blacksquare$$

Example 3.12 Leaves of Indian creeper plants

The leaves of Indian creeper plants *Pharbitis nil* can be variegated or unvariegated and, at the same time, faded or unfaded. Are the two characteristics independent? In one experiment, plants of a particular type were crossed. They were such that the offspring plants would have leaves that were variegated with probability $\frac{1}{4}$ (and unvariegated with probability $\frac{3}{4}$); and also the leaves would be faded with probability $\frac{1}{4}$ (and unfaded with probability $\frac{3}{4}$). If the theory of independence is correct, then one would expect to observe unvariegated unfaded leaves in $\frac{3}{4} \times \frac{3}{4} = \frac{9}{16}$ of the offspring plants, unvariegated faded leaves in $\frac{3}{4} \times \frac{1}{4} = \frac{3}{16}$ of them, and so on.

Bailey, N.T.J. (1961) *Mathematical Theory of Genetic Linkage.* Clarendon Press, Oxford, p. 41.

The seeds of *Pharbitis nil* may be roasted and used as a purgative.

As it turned out, of 290 offspring plants observed, 187 had unvariegated unfaded leaves, 35 had unvariegated faded leaves, 37 had variegated unfaded leaves and 31 had variegated faded leaves.

The observed sample relative frequencies occurred in the ratios

$$\tfrac{187}{290} : \tfrac{35}{290} : \tfrac{37}{290} : \tfrac{31}{290} = 0.64 : 0.12 : 0.13 : 0.11.$$

These differ somewhat from the forecast proportions of

$$\tfrac{9}{16} : \tfrac{3}{16} : \tfrac{3}{16} : \tfrac{1}{16} = 0.56 : 0.19 : 0.19 : 0.06.$$

In fact, even allowing for random variation in the observed experimental results, the theory of independence is resoundingly rejected on the basis of the results. ■

We shall explore how to test this sort of theory in *Chapter 9*.

When discussing two or more random variables some reduction in notation can be usefully achieved by using the notation $p(x)$ for a probability mass function, and distinguishing between p.m.f.s by including the names of the random variables as subscripts. Thus we write

$$P(X=x) = p_X(x),$$
$$P(Y=y) = p_Y(y).$$

In the same way, we write

$$P(X=x, Y=y) = p_{X,Y}(x,y).$$

The function $p_{X,Y}(x,y)$ is called the **joint probability mass function** for the random variables X and Y. Then condition (3.10) for independence between random variables X and Y can be rewritten as

$$p_{X,Y}(x,y) = p_X(x)p_Y(y), \qquad (3.11)$$

for all x in the range of X and all y in the range of Y.

Exercise 3.8

The entries in Table 3.8 are values of the joint probability mass function $p_{X,Y}(x, y)$ of random variables X and Y. The random variable X takes the values 0, 1 and 2, and Y takes the values -1 or 1. The columns of the table correspond to the random variable X with p.m.f. $p_X(0) = 0.4$, $p_X(1) = 0.4$ and $p_X(2) = 0.2$. The rows are associated with the random variable Y with p.m.f. $p_Y(-1) = 0.3$ and $p_Y(1) = 0.7$. Notice that the separate probabilities for X are found by adding within columns, and the separate probabilities for Y by adding within rows.

Are X and Y independent random variables?

Table 3.8 A joint p.m.f.

		x		
		0	1	2
y	-1	0.12	0.10	0.08
	1	0.28	0.30	0.12

The analogue for continuous random variables of the joint probability mass function for discrete random variables is the **joint probability density function**. A direct analogue of (3.11) for continuous random variables involves the p.d.f.s $f_X(x)$ and $f_Y(y)$ of two random variables X and Y and their joint p.d.f. $f_{X,Y}(x, y)$. However, the idea of independence is more easily interpreted if it is expressed in terms of probabilities of events. For example, the following result involves the tail probabilities $P(X \leq x)$, $P(Y \leq y)$ and $P(X \leq x, Y \leq y)$, the probability of the conjunction of two events.

The subject of dependent random variables is discussed in detail in *Chapter 11*.

If X and Y are continuous random variables that are independent, then

$$P(X \leq x, Y \leq y) = P(X \leq x)P(Y \leq y), \tag{3.12}$$

for all x in the range of X and all y in the range of Y.

It also follows from independence that, for example,

$$P(X > x, Y > y) = P(X > x)P(Y > y).$$

That concludes our preliminary discussion of independence, though it is a topic to which we shall return as the course develops. You should be aware that sometimes in statistical experimentation an assumption of independence is wrongly made when, in fact, there is a subtle dependence between responses, and so the conclusions of the experiment are, or may be, flawed. It is one of the roles of the statistician to exhibit these dependencies, where they occur, and control for them.

3.3 The geometric probability model

In *Chapter 2*, Subsection 2.1.2, you learned about the Bernoulli trial, a statistical experiment where exactly one of only two possible outcomes occurs. The main features of the model were described in *Chapter 2* and are as follows. The outcomes are usually something like Success–Failure, Yes–No, On–Off, Male–Female. In order to standardize the associated Bernoulli distribution so that it becomes a useful statistical model with wide application, it is usual to identify one of the outcomes with the number 1 and the other with the number 0. Then, for example, a sequence of twenty engineering trials run on identical pieces of equipment as part of a quality control exercise, resulting in the outcomes Pass, Pass, Fail, ..., Fail, Fail could be written more easily as

$$X_1 = 1, \quad X_2 = 1, \quad X_3 = 0, \quad \ldots, \quad X_{19} = 0, \quad X_{20} = 0.$$

In this case the number 1 was used as a label to indicate that a piece of equipment had passed the quality test. The numbers could easily have been arranged the other way round: a score of 1 for a piece of defective equipment.

The Bernoulli distribution has associated with it a single parameter, the number p. This number is a probability (so $0 \leq p \leq 1$): it represents the probability that any single trial results in the outcome 1.

In this section we shall consider two particular examples based on the Bernoulli model.

Example 3.13 The sex of consecutive children

The pattern of boys and girls in a family is one that has received a lot of scientific attention, both biological and statistical, for nearly three centuries. In 1710 the philosopher John Arbuthnot, having examined parish records and noted that for 82 consecutive years more boys than girls had been christened and (reasonably) deducing that for 82 years more boys than girls had been born, proposed that sex determination was not a simple matter of chance, akin to the result of the toss of a fair coin. (Actually, he proposed a degree of divine intervention: that while the coin was in mid-air, God temporarily suspended the laws of chance.) The philosopher and probabilist Nicholas Bernoulli (1687–1759) had also noted the imbalance in the sexes, and commented that sex determination was like rolling a 35-sided die, with 18 faces marked 'boy' and 17 'girl'.

The biologists, statisticians, genealogists, philosophers, social scientists and demographers who have all at one time or another made their various investigations in this area, would have found their research greatly hampered had they not developed various models (and an easy notation) for the ideas and theories involved. The easiest model is to assume independence from child to child, write 1 for a son and 0 for a daughter, use Bernoulli's estimate for the probability of a boy of $p = 18/35 = 0.514$ and write

$$P(X = 0) = 0.486, \quad P(X = 1) = 0.514,$$

where X is the random variable denoting the sex of a child.

Actually, nearly all the very considerable mass of data collected on the sex of children in families suggests that the Bernoulli model would be a very bad model to adopt, for at least two reasons. First, one theory suggests that the 'boy' probability p, even if it averages 0.514 over a very large population, is probably not the same from family to family, with some couples seeming to have a preponderance for boys, and others for girls, greater than would be suggested simply by sampling variation on the binomial distribution. Second, and even more tantalizingly, some statistical analyses seem to show that the independence assumption of the Bernoulli model breaks down: that is, that Nature has a kind of memory, and the sex of a previous child affects to some degree the probability distribution for the sex of a subsequent child. Nevertheless, for the rest of this chapter, we shall assume that a Bernoulli model does provide, for our purposes, an adequate fit to the observed numbers of the sexes in families. ■

Statistical model testing is one very important way of testing biological and other scientific theories: if what actually happens is wildly different from what one might reasonably expect to happen if the scientific theory were true, then that scientific theory probably does not hold. It is the application of the science of statistics that helps one decide how different 'different' can reasonably be if due to chance alone. These tests are called tests of 'goodness of fit' and are dealt with in *Chapter 9*.

Example 3.14 *Silicon chips*

The manufacture of silicon chips is an extremely sensitive operation, requiring engineering accuracies many orders of magnitude greater than those required in most other manufacturing contexts, and a working environment that is clinically 'clean'. (At the time of writing, only the production of compact discs requires higher standards.) In the early days of chip technology, most chips were 'defective'—they did not work properly (or often, they did not work at all). Either they were dirty (a mote 0.5 microns across can cause havoc on a circuit board where the tracks carrying current are only 0.3 microns across); or not all the connections were correctly made. At all the stages of slicing, lapping, etching, cleaning and polishing involved in the manufacture of a chip, defective units are identified and removed. Even so, possibly as many as one chip in twenty is faulty.

1 micron = 10^{-6}m

During the manufacturing process, there are probably 'runs of rough', intervals during which nothing seems to go very well and the product defective rate is rather high. These periods will alternate with intervals where things go rather better than average. However, we shall assume for the rest of this chapter that chip quality can be regarded as invariant and independent from chip to chip, unrealistic though this might seem. ■

We have just seen that the Bernoulli model does not fit exactly in either Example 3.13 or Example 3.14, and yet we are going to go on and apply it in what follows! This is the way things often are in statistics (and in science, in general): these models are merely supposed to be *adequate* reflections of reality, not perfect ones. For many purposes, the Bernoulli model is adequate for both these situations.

3.3.1 The geometric distribution

After studying *Chapter 2*, you can answer the following types of questions. The following questions relate to Example 3.13. In families of five children, what proportion of families have all boys (take p equal to 0.514)? What is the probability that in a family of four children, all the children will be girls? In what proportion of families of three children do the boys outnumber the girls? The next question relates to Example 3.14. If silicon chips are boxed in sealed batches of one hundred, what is the probability that a purchaser of a box will find he has bought more than ten defectives?

All these questions may be answered by reference to the appropriate binomial distribution.

But now consider the following extensions to these problems.

Example 3.13 continued

One of the factors complicating the development of a satisfactory statistical model for family size and structure is that parents often impose their own 'stopping rules' for family limitation, depending on the number or distribution of boys and girls obtained so far. For instance, among completed two-child families in a recent issue of *Who's Who in America* there was a striking excess of boy-girl and girl-boy sets, more than would be suggested by a simple binomial model: parents (apparently) prefer to contrive their families to include at least one of each sex. Family limitation rules may be more extreme than this: for instance, 'keep going until the first son is born, then stop'. Under this rule, completed families (M for a son, F for a daughter) would show the structure M, FM, FFM, FFFM, ... The number of children in a completed family of this type is evidently a random variable: what is its probability distribution? ■

Example 3.14 continued

A quality inspector at a silicon chip factory introduces a new quality test. At random times he will sample completed chips from the assembly line. He makes a note of the number of chips sampled up to and including the first defective he finds. If this number reaches or exceeds some predetermined tolerance limit, then he will assume that factory procedures are running efficiently. Otherwise (defectives are occurring too frequently) the production process is stopped for assessment and readjustment. ■

You will have noticed that in both these examples the same type of random variable is being counted: essentially, the number of trials from the start of a sequence to the first success. Notice that the trial at which that success occurs is included in the count.

In this context, success denotes the identification of a defective chip.

The assumptions of a sequence of Bernoulli trials are that the outcomes of successive trials are independent, and that the probability of success remains the same from trial to trial. If these twin assumptions hold, and if the number of trials to the first success is denoted by N, then we can say

$N = 1$ if the first trial is a success,

$N = 2$ if the first trial is a failure, the second a success,

$N = 3$ if the first two trials are failures, the third a success,

and so on.

Exercise 3.9

Evidently, the number N is a random variable: it is impossible at the start of the sequence to forecast with certainty the number of the trial at which success will first occur. Assume independence and that the probability of success at any trial is p, $0 < p < 1$, in the following.

(a) Write down the probability $P(N = 1)$.

(b) Write down the probability $P(N = 2)$.

(c) Write down the probability $P(N = 3)$.

(d) Using your answers to parts (a), (b) and (c), try to find a general formula for the probability $P(N = n)$.

(e) State the range of possible values for the random variable N.

The results of Exercise 3.9 lead to the following definition. Like the Bernoulli, the binomial and the normal distributions, the probability model whose definition follows is one of the 'standard' probability models. It is the third to be associated with the fundamental notion of a Bernoulli trial.

The first two are the Bernoulli probability distribution and the binomial probability distribution.

The geometric distribution

If in a sequence of independent trials the probability of success is constant from trial to trial and equal to p, $0 < p < 1$, then the number of trials up to and including the first success is a random variable N, with probability function given by

$$P(N = n) = p_N(n) = q^{n-1}p, \quad n = 1, 2, \ldots, \tag{3.13}$$

where $q = 1 - p$.

The random variable N is said to follow a **geometric distribution with parameter p** and this is written $N \sim G(p)$.

It is very common in the context of Bernoulli trials to write the probability of failure as q, where $q = 1 - p$.

The reason for the name 'geometric' is that the sequence of probabilities $P(N = 1), P(N = 2), \ldots$, form a geometric progression: each term is a constant multiple (in this case, q) of the preceding term. That multiple is less than 1 (since it is a probability), so successive terms of the probability function of N become smaller and smaller. This is illustrated in Figure 3.8. In (a), the parameter p is equal to 0.8—that is quite high: you would not have to wait long for the first successful trial. In (b), the parameter p is much lower, with the probability of success equal to only 0.3. In this case, you could find you have to wait for quite a time for the first success to occur.

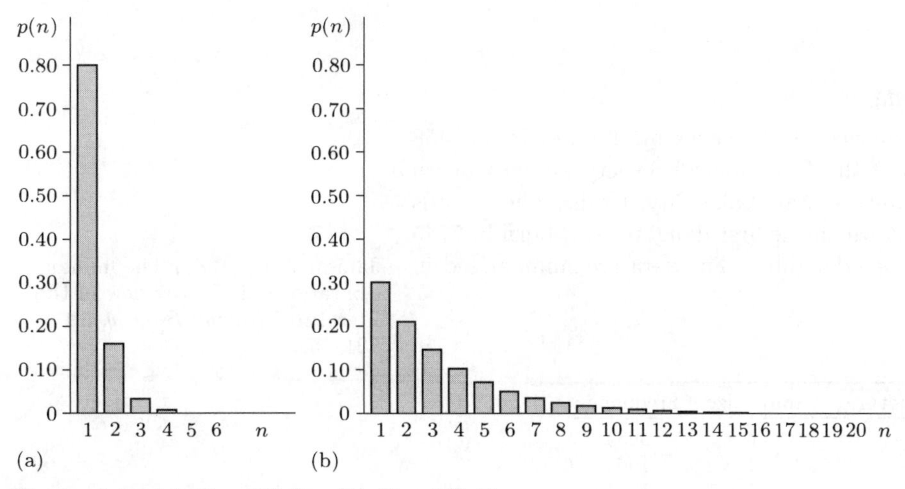

Figure 3.8 (a) $N \sim G(0.8)$ (b) $N \sim G(0.3)$

Notice that whatever the value of p, the most likely value of N is 1.

Example 3.15 Family size with a stopping rule

One of the difficulties in the collection and analysis of data on family structures is in the definition of a 'completed' family. Parents do not always know whether or not more children will appear. Table 3.9 contains simulated data on family size (the number of children) for 1000 completed families under a hypothetical rule of 'stop after the first son', where p is taken to be 18/35 (Bernoulli's estimate for the probability of a boy). The table gives the observed frequency for each family size. Also shown are the theoretical frequencies, obtained by multiplying the probabilities $p_N(n)$ by 1000. For instance, the probability that a family is of size 3 is

$$p_N(3) = P(\text{FFM}) = \left(\tfrac{17}{35}\right)^2\left(\tfrac{18}{35}\right) = 0.121\,329,$$

and multiplying this by 1000 gives the theoretical frequency for families of size 3 in a sample of 1000.

Table 3.9 Simulated family size with a stopping rule

Family structure	Family size	Observed frequency (simulated)	Theoretical frequency
M	1	508	514.3
FM	2	255	249.8
FFM	3	138	121.3
FFFM	4	53	58.9
FFFFM	5	24	28.6
FFFFFM	6	12	13.9
FFFFFFM	7	4	6.8
FFFFFFFM	8	3	3.3
FFFFFFFFM	9	3	1.6
	≥ 10	0	1.5

In this case the geometric 'fit' to the simulated data seems quite good; but this does not prove anything, since the data were generated in the first place from a precisely stated geometric model. ■

The next example is necessarily sparse, but gives data on birth order taken from an investigation reported in 1963.

Example 3.16 Salt Lake City data

Details were obtained on the sequence of the sexes of children in 116 458 families recorded in the archives of the Genealogical Society of the Church of Jesus Christ of Latter Day Saints at Salt Lake City, Utah. The records were examined to find the stage at which the first daughter was born in 7745 families where there was at least one daughter. The data are summarized in Table 3.10.

James, W.H. (1987) The human sex ratio. Part I: A review of the literature. *Human Biology*, **59**, 721–752.

Table 3.10 First daughter

First daughter	Family structure	Family size	Frequency
Firstborn	F	1	3684
Secondborn	MF	2	1964
Thirdborn	MMF	3	1011
Fourthborn	MMMF	4	549
Later than fourth		≥ 5	537

Here, the data are really too sparse to provide evidence for or against a geometric model; but it is worth noting that successive frequencies are in the ratio

$$\tfrac{1964}{3684} = 0.53, \quad \tfrac{1011}{1964} = 0.51, \quad \tfrac{549}{1011} = 0.54;$$

all approximately one-half. The frequencies themselves form a geometric progression (roughly speaking). ■

The geometric probability function $p_N(n) = q^{n-1}p$ is a much simpler formula than, for instance, the binomial probability function, and nobody would bother to publish tables listing the value of $p_N(n)$ for different values of n and p. The exercise that follows is quite straightforward: it only requires a few key-presses on your calculator.

Exercise 3.10

(a) Suppose that Nicholas Bernoulli was right, and his hypothesis that the sex of children may be modelled as independent rolls of a 35-sided die, with 18 faces marked 'boy' and the other 17 marked 'girl', was a correct one. Under a stopping rule 'stop after the first son', what proportion of completed families comprise at least four children?

(b) Suppose that the proportion of defective chips leaving a chip factory's assembly line is only 0.012: manufacturing standards are quite high. A quality inspector collects daily a random sample. He examines sampled chips in the order in which they came off the assembly line until he finds one that is defective. He decides that he will halt production if fewer than six chips need examining. What proportion of his daily visits result in a halt in production?

As we have seen, there is no convenient formula for the sum

$$P(X \le x) = \sum_{j=0}^{x} P(X = j) = \sum_{j=0}^{x} \binom{n}{j} p^j (1-p)^{n-j},$$

when X is binomial $B(n,p)$; so its calculation could be very time-consuming (and error-prone). If these probabilities exist in printed form, the problem reduces to one of using the tables correctly; alternatively, they may be obtained more or less directly from your computer.

In the case of the geometric random variable N, it is possible to deduce a formula for the probability $P(N \le n)$ by writing down successive individual probabilities and then finding their sum. Actually, it is rather easier to return to the original context where the model arose and argue straight from there. An argument 'from first principles' goes as follows.

Remember that we are interested in counting the number of trials necessary to record the first success. If we are rather unlucky, the sequence could go on for a long time. If we start with 7 failures in the first 7 trials, for instance, then we know that N has to be at least 8 (that is, more than 7). The probability of recording 7 failures in 7 trials is q^7, so $P(N > 7) = q^7$. If after 20 trials we have recorded 20 failures, then we know that N has to be more than 20. The probability of recording 20 failures in 20 trials is q^{20}, so $P(N > 20) = q^{20}$. In general, the probability of recording n failures in n trials is q^n, so

$$P(N > n) = q^n.$$

It follows that the cumulative distribution function $F_N(n)$ for the geometric random variable N is given by

$$F_N(n) = P(N \le n) = 1 - P(N > n) = 1 - q^n, \qquad (3.14)$$

where $n = 1, 2, 3, \ldots$.

The argument that led to (3.14), the statement of the c.d.f. of the random variable N, is quite watertight; but you might like to see the following mathematical check. The distribution function $F_N(n)$ can be obtained directly from the probability function $p_N(n)$ by summing probabilities:

$$P(N \le n) = P(N = 1) + P(N = 2) + \cdots + P(N = n),$$

so

$$P(N \le n) = p + qp + q^2p + q^3p + \cdots + q^{n-1}p. \qquad (3.15)$$

The terms in this series form a geometric progression (sometimes abbreviated to g.p.). Perhaps you already know a formula for the sum of the first n terms in a geometric progression. If not, we can proceed in this case as follows.

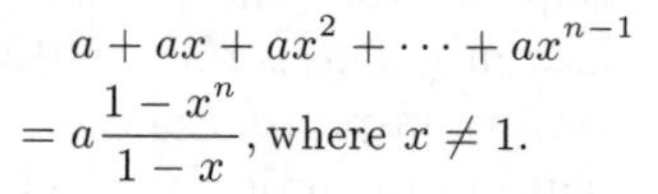

$a + ax + ax^2 + \cdots + ax^{n-1}$
$= a\dfrac{1 - x^n}{1 - x}$, where $x \ne 1$.

Multiplying both sides of (3.15) by q, we obtain

$$qP(N \le n) = qp + q^2p + q^3p + \cdots + q^np. \qquad (3.16)$$

Subtracting (3.16) from (3.15), we obtain (since most terms vanish) the identity

$$(1 - q)P(N \le n) = p - q^np.$$

So replacing $(1 - q)$ by p and dividing by p, we have

$$P(N \le n) = 1 - q^n, \quad n = 1, 2, 3, \ldots. \qquad (3.17)$$

This is the answer that was obtained in (3.14) by a more direct argument.

Exercise 3.11

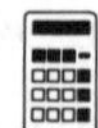

The proportion of defective products in a battery factory is 0.02. A quality control inspector tests batteries drawn at random from the assembly line. What is the probability that he will have to examine more than 20 to obtain a faulty one? What is the probability that he will have to examine at least 50?

3.3.2 The mean and variance of the geometric distribution

As with all models for random variables, two useful measures exist: one to give an idea of what value to expect, and the other to suggest how far away actual values might be from that expected value. These are the mean and standard deviation of the probability distribution.

Exercise 3.12

Without doing any arithmetic, try to answer the following questions. (Just jot down your first reaction.)

(a) The probability that a fair coin shows Heads when it is tossed is $\frac{1}{2}$. How many times, on average, do you think a coin would need to be tossed to come up Heads?

(b) The probability that a *Double-Five* comes up showing 5 when it is rolled is $\frac{2}{6} = \frac{1}{3}$. On average, how many times do you think a *Double-Five* would need to be rolled to show a 5?

(c) One car in six on British roads is white. If you stand by the side of the road and start counting, how many cars, on average, do you think you would have to count to record your first white one?

(d) The probability that a car starts first time in the morning is p, where $0 < p < 1$. Assuming that attempts to start it can be modelled as a sequence of Bernoulli trials, guess in terms of p the average number of attempts necessary to get it going each morning. (The assumption that a Bernoulli model will be a useful one is probably not reasonable in this case: most cars just need 'encouraging'. If it does not start first time, then it will almost certainly start at the second or third; and if not then, then not at all.)

What did you write down in Exercise 3.12(d)? Intuitively, you might feel that the mean of the geometric distribution with parameter p is the reciprocal of p, $1/p$. In this case your intuition is not misplaced. A proof of this result is as follows.

The mean of a geometric random variable N with parameter p is given by

$$E(N) = \sum_{n=1}^{\infty} n p_N(n) = 1p + 2qp + 3q^2p + 4q^3p + \cdots. \qquad (3.18)$$

This series for $E(N)$ is not itself a geometric progression, since terms are not obtained from the preceding term by multiplying by a constant factor—the coefficients $1, 2, 3, \ldots$ are complicating features. However, proceeding as before, let us try multiplying both sides by q. This gives

$$qE(N) = 1qp + 2q^2p + 3q^3p + 4q^4p + \cdots.$$

If we now subtract this expression from (3.18), on the left-hand side we have $E(N) - qE(N) = pE(N)$; while on the right-hand side, terms almost vanish as they did before, but not quite. We are left with

$$\begin{aligned} pE(N) &= 1p + (2-1)qp + (3-2)q^2p + (4-3)q^3p + \cdots \\ &= p + qp + q^2p + q^3p + \cdots, \end{aligned}$$

and the series on the right-hand side we know sums to 1: it is just a list of all the terms of the geometric probability mass function, so it *must* sum to 1.

Or use the result: for $|x| < 1$, $a + ax + ax^2 + ax^3 + \cdots = \dfrac{a}{1-x}$.

Hence we have $pE(N) = 1$.

Dividing by p gives the final result:

$$E(N) = \frac{1}{p}. \qquad (3.19)$$

The less likely an event is, the longer one should expect to wait for it to happen.

For a probability distribution as *skewed* (i.e. as asymmetric—see Figure 3.8) as the geometric distribution, knowledge of the value of the standard deviation is not as useful as it is in the case of the normal distribution, where (as you will see in *Chapter 5*) all probability statements may be made in terms of the number of standard deviations an observation is from its expected value. However, the following result will be useful for future work: we shall see in *Chapter 5* that knowledge of the variance of a random variable can be put to other uses. This result for the variance of the geometric random variable $N \sim G(p)$ is included without proof:

$$V(N) = \frac{q}{p^2}. \qquad (3.20)$$

So the standard deviation of N is $SD(N) = \sqrt{q}/p$.

The next exercise summarizes the work of this section.

Exercise 3.13

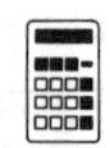

In some board games, progress round the board is dictated by the score from a roll of a six-sided die. In some games, you cannot start playing until you have obtained your first six (and then you move accordingly). If you score some other number, you have to wait until your next turn and then make another attempt.

(a) What is the probability that you can start playing with your first roll of the die?

(b) What is the probability that you can start playing only at your second roll? At your third?

(c) What is the probability that you will need at least six rolls to get started?

(d) Find the expected number of rolls required to get you started, and calculate the standard deviation.

3.4 Two models for uniformity

3.4.1 The discrete uniform probability distribution

In *Chapter 2*, page 54 we considered the theoretical 'perfect die', which when rolled would land displaying any one of its six faces with equal probability. At any given roll of the die, the outcome is a random variable—one cannot forecast precisely what will happen. The probability mass function for the random variable (X, say) is given by

$$p(x) = \tfrac{1}{6}, \quad x = 1, 2, \ldots, 6.$$

This is an example of a random variable following a *discrete uniform distribution*. The list of possible values that X can take (the range of X) is given as a set of integers with stated lower and upper limits; and no possible value is more probable than any other possible value.

Example 3.17 Digit frequencies

Most computers include a 'random number generator' designed to print out in no predictable order but with equal likelihood the digits $0, 1, 2, \ldots, 9$ for as long as the user requires. (Many printed books of statistical tables contain at least a page of such random digits. A list of random digits is given in Table A1.) The way the computer generates successive digits is to follow some complicated rule involving earlier digits (and possibly the date or time as well, if the computer has an internal clock). So what is printed out is not random (in the sense that if you knew the rule you could predict the sequence exactly), but merely indistinguishable from random, or at least similar in certain key respects to the output of a random device such as a ten-sided die.

In an experiment using four different computer programs, the following digit frequencies in sequences of 1000 digits were observed.

Table 3.11 Digit frequencies (four programs)

Digit	0	1	2	3	4	5	6	7	8	9
SC v.1.09	92	107	85	85	109	95	104	95	113	115
GW-Basic v.3.23	85	110	91	95	106	110	92	106	101	104
Spida v.5.50	110	94	86	97	101	94	113	133	84	88
Minitab v.7.20	112	93	96	87	108	84	103	120	111	86

Data provided by F. Daly, The Open University. The Minitab data were supplied by K.J. McConway, The Open University.

The four bar charts in Figure 3.9 show the sample relative frequencies for each of the ten digits and for the four programs. The sample relative frequencies may be compared with the theoretical proportions: $\frac{1}{10}$ in each case. There is some evidence of variability, as one must expect. However, none of the four programs manifests serious departures from the theoretical uniform distribution.

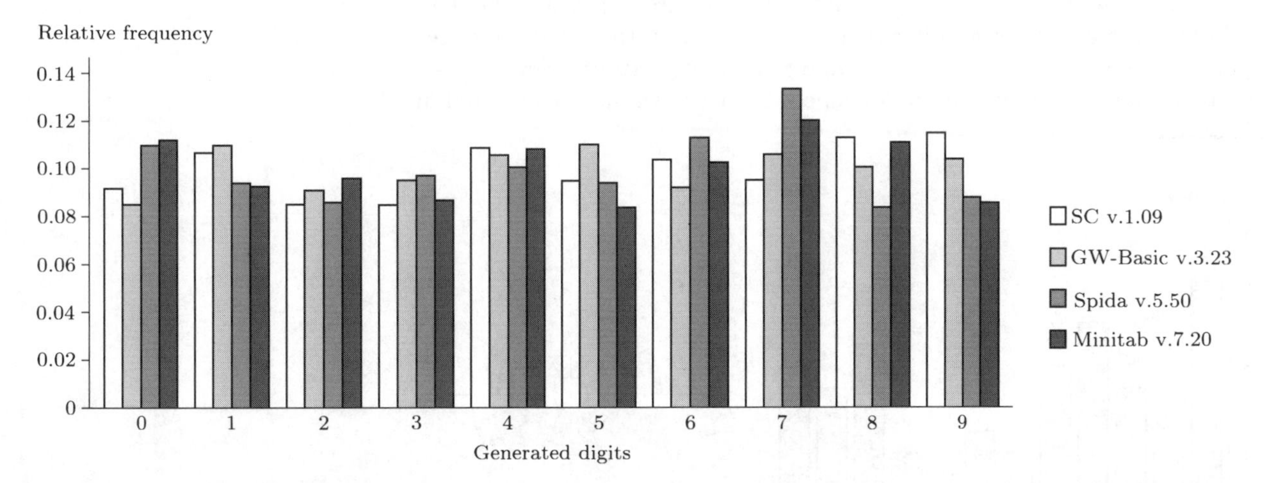

Figure 3.9 Random number generation: relative frequencies for four computer programs

A different, faulty, program gave the frequencies listed in Table 3.12.

Table 3.12 Digit frequencies (a faulty program)

Digit	0	1	2	3	4	5	6	7	8	9
Frequency	100	100	100	100	100	100	100	100	100	100

So this program resulted in sample relative frequencies of exactly $\frac{100}{1000} = \frac{1}{10}$ for each of the ten digits $0, 1, 2, \ldots, 9$. The uniform 'fit' is perfect!

This example demonstrates that there is more to the generation of random digits than that each should occur with equal likelihood. The faulty program was generating the sequence

0 1 2 3 4 5 6 7 8 9 0 1 ... 8 9,

and so gave a perfectly uniform distribution of digits. So a 'good' uniform fit does *not* imply that the random number generator is satisfactory. A 'bad' fit on the other hand might suggest that the generator is unsatisfactory. ■

Example 3.18 Month of death of royal descendants

The data in Table 3.13 give the month of death (January = 1, February = 2, ..., December = 12) for 82 descendants of Queen Victoria who died of natural causes.

Table 3.13 Month of death, royal descendants

Month	1	2	3	4	5	6	7	8	9	10	11	12
Frequency	13	4	7	10	8	4	5	3	4	9	7	8

A straight test of uniformity here would not reflect the fact that, leap year or not, February is a short month; but at a first glance the data certainly seem to suggest that the summer months (6, 7, 8, 9) are less likely to include a death than winter months. ■

(A formal test of uniformity for the data in Table 3.13 should really reflect the fact that the months of January and December are adjacent. The data would be represented not as a bar chart or histogram in the usual sense (see Figure 3.10(a)) but as a circular histogram (see Figure 3.10(b)). Several tests for circular uniformity have been developed, but they will not be explored in this course.)

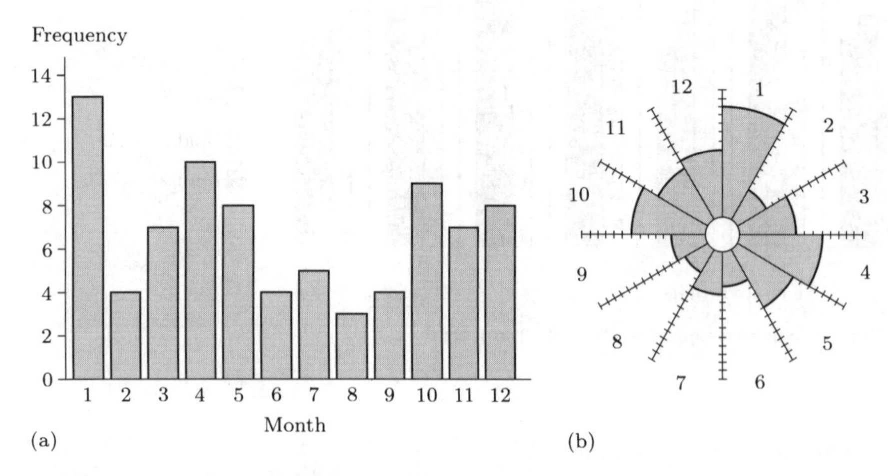

Figure 3.10 Representations of the death data

A definition of the discrete uniform distribution is as follows. Notice the phrase '*a* definition' rather than '*the* definition'. Here, the stated range of the discrete random variable X is $1, 2, \ldots, n$. Elsewhere, you might see an alternative definition where the list of possible values includes zero: $0, 1, 2, \ldots, m$, say. The essential feature is that each of the possible values occurs with equal probability.

The discrete uniform distribution

The random variable X is said to follow a **discrete uniform distribution** if it has probability mass function

$$p(x) = \frac{1}{n}, \quad x = 1, 2, \ldots, n. \tag{3.21}$$

Again, there is a whole family of discrete uniform probability distributions: the indexing parameter in this case is n, the maximum attainable value.

The c.d.f. of X is found from

$$\begin{aligned} P(X \leq x) &= p(1) + p(2) + \cdots + p(x) \\ &= \frac{1}{n} + \frac{1}{n} + \cdots + \frac{1}{n} \\ &= \frac{x}{n}, \quad x = 1, 2, \ldots, n. \end{aligned}$$

That is,

$$F(x) = \frac{x}{n}, \quad x = 1, 2, \ldots, n. \tag{3.22}$$

The mean of the random variable X following a discrete uniform distribution with parameter n is given by

$$\begin{aligned} E(X) &= \sum_{j=1}^{n} j p(j) \\ &= \sum_{j=1}^{n} j \left(\frac{1}{n}\right) \\ &= \frac{1}{n}(1 + 2 + 3 + \cdots + n) \\ &= \frac{1}{n}\left(\tfrac{1}{2}n(n+1)\right) \\ &= \tfrac{1}{2}(n+1) \end{aligned}$$

(which is, as you might expect, the middle of the range of X).

Here, the result

$$1 + 2 + \cdots + n = \tfrac{1}{2}n(n+1)$$

is used. You can see this by writing

$$S = 1 + 2 + \cdots + n;$$

then write the right-hand side last to first, as

$$S = n + (n-1) + \cdots + 1.$$

Now sum each side of the two expressions, term by term. This gives

$$\begin{aligned} 2S &= (n+1) + (n+1) \\ &\quad + \cdots + (n+1) \\ &= n(n+1). \end{aligned}$$

Dividing by 2 gives

$$S = \tfrac{1}{2}n(n+1).$$

The variance of X is given by

$$V(X) = \frac{n^2 - 1}{12}.$$

This result is obtained by a straightforward but time-consuming application of the variance formula, and you need not bother with the details.

Exercise 3.14

Generate 1200 rolls of a fair six-sided die.

(a) Obtain a frequency table for your data.

(b) Plot a bar chart of your data.

(c) Evaluate the sample mean and sample standard deviation, and compare these two statistics with the corresponding population **moments**.

The word 'moment' is often used as a technical description of expectations of a random variable. For a positive integer r, $E(X^r)$ is called the rth raw moment of X; $E((X-\mu)^r)$ is called the rth central moment of X, where $\mu = E(X)$. In this course we shall occasionally loosely refer to 'the first two moments of X', to mean the mean and variance (or possibly the mean and standard deviation) of X.

Essentially, that is all that needs to be said about the discrete uniform distribution. There is a continuous analogue that we shall now examine.

3.4.2 The continuous uniform probability distribution

Here are some simple examples of situations for which a useful statistical model has not yet been developed.

Example 3.19 *Admissions to an intensive care unit*

In *Chapter 2*, Table 2.22, data were given on the waiting time (in hours) between the first 41 admissions to an intensive care unit. The full data set lists 254 admissions over a period of about 13 months. Information is given on the date and time of day of admission. It would be interesting, and helpful to planners, to explore whether admissions were more frequent at some times of day than others, or whether any time of admission is as likely as any other. The times of admission over two-hour intervals are summarized in Table 3.14.

Cox, D.R. and Snell, E.J. (1981) *Applied Statistics—Principles and Examples.* Chapman and Hall, London, p. 53. The data were collected by Dr. A. Barr, Oxford Regional Hospital Board.

Actually, the data are sufficiently sparse over the early hours 4 am–10 am (and dense over the early evening period 4 pm–6 pm) to provide strong statistical evidence that frequency of admission does depend on the time of day. We shall look in *Chapter 9* at statistical procedures enabling this sort of evidence to be examined. The important point is that an essential requirement for such procedures is the formulation of a model to describe (possibly rather badly) the variation observed. ■

Table 3.14 Time of day of 254 admissions to an intensive care unit

Time interval	Frequency
midnight–2 am	14
2 am–4 am	17
4 am–6 am	5
6 am–8 am	8
8 am–10 am	5
10 am–midday	25
midday–2 pm	31
2 pm–4 pm	30
4 pm–6 pm	36
6 pm–8 pm	29
8 pm–10 pm	31
10 pm–midnight	23

Example 3.20 *Faulty cable*

Faults in underground television cable cause degradation, or even complete loss, of the signal. When this happens, the cable needs to be repaired. In the absence of any indication of where the fault might be, the repair company has to search the cable until the fault is located—this is just as likely to be near the end, near the beginning or in the middle of the cable. The distance searched to locate the fault is a random variable, and a factor in the cost of the repair. ■

Example 3.21 *Green-haired Martians*

Suppose you were required to guess the unknown proportion p of inhabitants of Mars with green hair (assuming you knew that they had hair). One way of expressing your (assumed) total ignorance on this matter might be to say '$p = \frac{1}{2}$', on the principle that as far as you are concerned a Martian's hair is as likely as not to be green.

Another way to express your ignorance is to say that the proportion p could be anywhere between 0 and 1, and that you have no evidence that any value is any more likely than any other. Then you express your guess in terms of a probability distribution, rather than giving any particular value. ■

Example 3.21, though somewhat contrived, actually raises some very important questions. Much of the science of statistics that we shall be exploring in this course is directed at the problem of 'estimation': making a guess, and preferably a good one, at the value of some model parameter. This parameter is a constant; the problem is that we do not know its value.

An entirely different approach is to express your uncertainty about the value of such descriptive parameters by giving them a probability distribution which says 'this value is credible but unlikely, this value is (to me) the most credible, this one is not at all plausible, ... ' and so on. The data you collect then allow you to modify or update this probability distribution. The whole question of this kind of inference, called **Bayesian inference**, is a very interesting one, but unfortunately it is beyond the scope of this course.

The central idea in these examples has been that of 'no preferred value'. The probability model that permits left- and right-hand bounds to be stated, and that carries the sense that between those bounds there is no preferred value, is known as the *continuous uniform distribution.* Since there is no preferred value in the range of a continuous uniform random variable X, the height of the p.d.f. of X on a sketch must be constant over the range. For instance, suppose that X can take values between a and b ($a < b$); then a sketch of its p.d.f. will look like that in Figure 3.11.

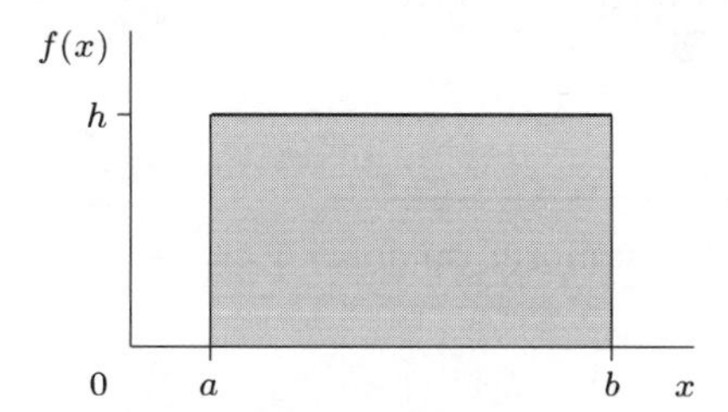

Figure 3.11 No preferred value between a and b

The p.d.f. is of the form $f(x) = h$, for $a \leq x \leq b$. Since the total area under the p.d.f. must be 1, the area of the rectangle in Figure 3.11 must be 1, so $(b-a)h = 1$. It follows that $h = 1/(b-a)$. Hence the following definition can be made.

The continuous uniform distribution

The continuous random variable X, equally likely to take any value between two stated bounds a and b ($a < b$), is said to be **uniformly distributed over the interval $a \leq x \leq b$** and has probability density function

$$f(x) = \frac{1}{b-a}, \quad a \leq x \leq b. \tag{3.23}$$

This is written $X \sim U(a, b)$.

A sketch of the density is shown in Figure 3.12.

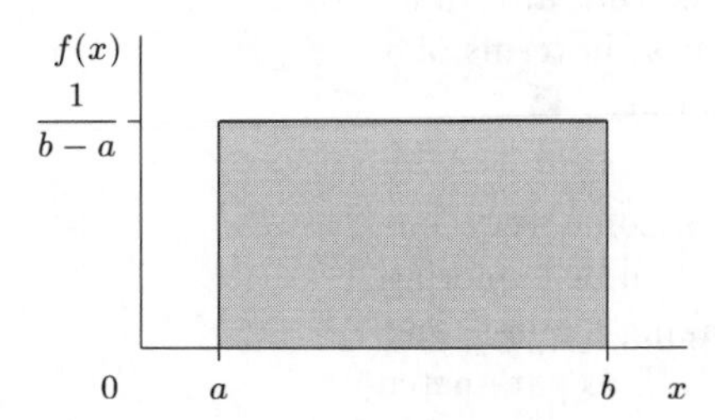

Figure 3.12 The uniform density, $U(a, b)$

In Figure 3.12 it is assumed that the bounds a and b are both positive. This is not a necessary condition. Depending on the situation being modelled, one bound or both bounds could be negative. The only constraint is that a should be less than b.

Example 3.19 continued

In this example, one possibility (in the absence of data) is that times of admission might be independent of the time of day. A model for the time T of admission reflecting this suggestion is $T \sim U(0, 24)$ (using the 24-hour clock). Once suggested, a model can be examined for the adequacy of its fit to a data set. (In this case, a uniform model provides a very bad fit.) ■

Example 3.21 continued

In this case the unknown proportion p of green-haired Martians is given a uniform probability distribution with $a = 0$ and $b = 1$. ■

Exercise 3.15

Use a sketch of the p.d.f. of X to find the following for the uniform random variable $X \sim U(a, b)$:

(a) the mean (use a symmetry argument);

(b) the cumulative distribution function.

The mean of the uniform distribution $U(a, b)$ is given by

$$E(X) = \mu = \frac{a+b}{2} \tag{3.24}$$

(in other words, the midpoint of the range) and the cumulative distribution function is given by

$$F(x) = \frac{x-a}{b-a}, \quad a \le x \le b. \tag{3.25}$$

Calculation of the variance involves an exercise in integration. It is given by

$$V(X) = \sigma^2 = \frac{(b-a)^2}{12}. \tag{3.26}$$

Notice that the value of the variance depends only on the difference $(b - a)$ and not on the actual values of a and b.

3.4.3 The standard continuous uniform distribution

One special case of the uniform distribution is the member of the family that starts at 0 and ends at 1; this has applications to Bayesian inference and to simulation (amongst others).

The probability density function for the uniform distribution on $(0, 1)$ is given (setting $a = 0$, $b = 1$ in (3.23)) by

$$f(x) = 1, \quad 0 \leq x \leq 1. \tag{3.27}$$

The uniform distribution $U(0, 1)$ is known as the **standard uniform distribution**. Its graph is shown in Figure 3.13.

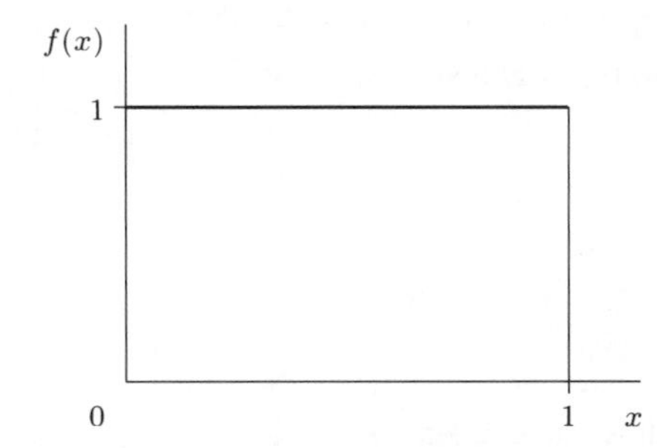

Figure 3.13 The uniform density, $U(0, 1)$

The cumulative distribution function for the standard uniform distribution is particularly easy to find: the probability $P(X \leq x)$ turns out to be x itself. Formally,

$$P(X \leq x) = F(x) = \int_0^x f(v)\, dv = \int_0^x 1\, dv = [v]_0^x = x. \tag{3.28}$$

However a geometrical argument like that used in Exercise 3.15(b) is easier!

The mean of the standard continuous uniform random variable X is

$$E(X) = \mu = \tfrac{1}{2}.$$

Exercise 3.16

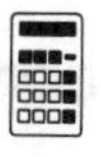

Write down the variance of the standard uniform random variable $X \sim U(0, 1)$, and hence calculate its standard deviation.

This section ends with one final example.

Example 3.22 *Traffic wardens*

Many parking zones in city centres have notices that read: *Waiting limited to 60 minutes. Return prohibited within 1 hour.* One way that traffic wardens keep track of which cars are parked where, and when, is to note the position of the valve dust-cap on one (or, in some cases, all) of the four wheels. A typical record is shown in Figure 3.14.

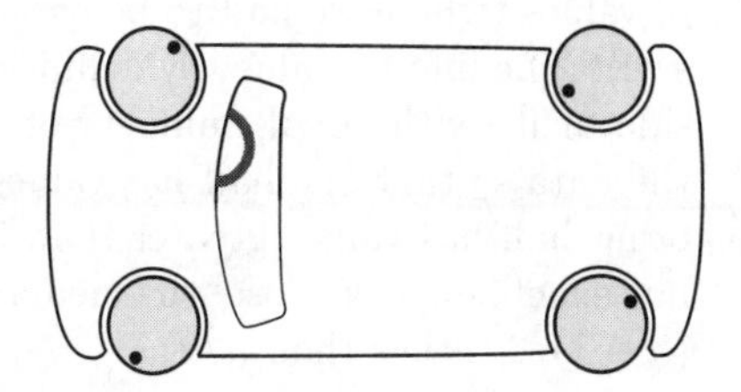

Figure 3.14 Record of dust-cap positions

A motorist accused of overstaying in a parking zone, claimed that he had been away and returned only after a proper interval as the law required. Counsel for the authority taking him to court produced evidence that, if this was so, then the rather unlikely event had occurred that all four valve dust-caps had ended up in identical locations about the wheel (at least, to within recording variation) as they had been earlier. Expert statistical advisors suggested that after a car journey of anything more than a negligible distance, even given the four starting orientations, there would at the end of the journey be no orientation for the dust-caps more likely than any other. It may also be assumed that even after a short journey, and given the four starting orientations, the final orientations of the four dust-caps are independent. (In cornering, the wheels travel different distances; differential axles preventing skidding.) Assuming that the four dust-cap positions had been recorded to an accuracy no greater than one twelfth of a turn (like hours on a clock-face—accuracy rather greater than this is not difficult to attain) then the motorist's version of events meant that an event with probability $(1/12)^4 = 0.000\,048$ had occurred. The magistrates found these odds (more than 20 000 to 1 against) less than credible. ■

3.5 Population quantiles

In Sections 3.3 and 3.4, three particular probability models have been described: the geometric distribution, and both discrete and continuous uniform distributions. To finish the chapter, let us, in this section, consider a topic generally applicable to any probability distribution. As in Section 3.1, where the population analogues of the sample mean, variance and standard deviation were described, here we shall consider the population analogues of some more quantities you met in *Chapter 1*, specifically the sample median and sample quartiles (and the sample interquartile range). More generally, it will prove to be useful to introduce a new idea—which covers population medians and quartiles—namely that of *population quantiles*.

'Quantiles' and 'quartiles' are two separate words with closely related but distinct meanings.

This is another instance where it is best to treat discrete and continuous situations separately. It is slightly more straightforward to deal with the continuous case, so that is where we shall start.

3.5.1 Quantiles for continuous probability distributions

The median

In *Chapter 1*, the sample median was introduced as the middle value of a set of values (where we had to be careful about what to do with samples of even size). The middle value, by definition, splits the data into two portions either side of it, with equal numbers of data points in each. The idea is to split the data so that one half has values smaller than the sample median and the other half has values greater than it (although for samples of odd size this is not exact, since we 'lose' the median value itself so that the subsets are of size $(n-1)/2$ rather than $n/2$).

In the modelling context, there is a natural analogue of this notion of splitting the data in half: we should like to define our population median m, say, to split the distribution into two halves in the sense that, if the random variable concerned is denoted by X and its median by m, $P(X \le m) = \frac{1}{2}$ and $P(X \ge m) = \frac{1}{2}$. In terms of the c.d.f., this may be written $F(m) = 1 - F(m) = \frac{1}{2}$.

> For a continuous random variable, the **population median** is the value x which is the solution to the equation
>
> $$F(x) = \tfrac{1}{2};$$
>
> this solution is denoted $x = m$.

Often the qualifying adjective 'population' is dropped, and we shall speak simply of the median of a random variable.

***Example 3.23** Salary distributions*

Distributions of salaries are typically skewed to the right (positively skewed) since relatively few people earn large salaries. As a consequence, the median is in many ways a better summary statistic for salary surveys than the mean. The median tells us the midpoint of the salary distribution within a population, in the sense that half of the people earn less and half earn more. ■

Exercise 3.17

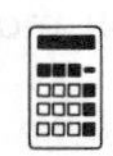

(a) In a simulation experiment, a computer is programmed to generate pseudo-random numbers from the continuous uniform distribution on $U(a, b)$. What is the median of this distribution?

(b) Theory suggests that a particular random variable X has p.d.f.

$$f(x) = 3x^2, \quad 0 < x < 1.$$

(i) Sketch the density function of X.

(ii) The mean of X is $\mu = \frac{3}{4}$. Show this on your sketch of the density of X.

(iii) The c.d.f. of X is given by

$$\begin{aligned} F(x) &= P(X \le x) \\ &= \int_0^x f(v)\,dv = \int_0^x 3v^2\,dv = \left[v^3\right]_0^x = x^3, \quad 0 \le x \le 1. \end{aligned}$$

Calculate the median m of the random variable X and show m on your sketch.

Quartiles

The general idea in the definition of sample quartiles in *Chapter 1*, Section 1.3 was to choose values which split the data into proportions of one-quarter and three-quarters. There are, of course, two ways to do the latter, and consequently there are two quartiles: a lower quartile designed to have (approximately) a quarter of the data with values smaller than it and three-quarters with larger values, and an upper quartile for which three-quarters of the data have values below it and a quarter above it. Their difference gives the sample interquartile range. To define population quartiles for continuous distributions, 'proportions' are replaced by 'probabilities', as follows. For example, to obtain the lower population quartile, denoted by q_L, the defining requirement is that $P(X \leq q_L) = \frac{1}{4}$.

For a continuous random variable, the **lower quartile** is the value x which is the solution to the equation

$$F(x) = \tfrac{1}{4};$$

the solution is denoted $x = q_L$.

Similarly, the **upper quartile** is the value x which is the solution to the equation

$$F(x) = \tfrac{3}{4},$$

and this solution is denoted $x = q_U$.

The **interquartile range** is $q_U - q_L$.

Just as the sample interquartile range is a measure of the spread of samples, so the population interquartile range, like the standard deviation, is a measure of the dispersion of population models.

Example 3.24 Childhood growth

In studies of growth of children, it is often of interest to show the position of a particular child relative to the overall distribution of heights of children of that age. The upper and lower quartiles for the distribution of heights at each age provide useful information. The lower quartile is the height such that 25% of children of that age are shorter; and the upper quartile is the height such that 75% are shorter (only 25% are taller). The median is, of course, the height such that 50% are shorter and 50% are taller.

The interquartile range of the heights of the children for any particular age covers the range of values taken by the 'middle 50%' of children of that age. It is the difference between the heights of the tallest and shortest children left after excluding the shortest 25% and the tallest 25%. ■

Exercise 3.18

Find the interquartile range for the distribution with p.d.f.

$$f(x) = 3x^2, \quad 0 < x < 1.$$

Quantiles

There is a simple pattern to the definitions of continuous population medians and quartiles that we have given above. Each is defined as a point x, say, such that $F(x)$ equals $\frac{1}{2}$ or $\frac{1}{4}$ or $\frac{3}{4}$. The idea can be generalized to cover *population quantiles* as follows.

For a continuous random variable with c.d.f. $F(\cdot)$, the **α-quantile** is the value x which is the solution to the equation

$$F(x) = \alpha, \quad 0 < \alpha < 1;$$

this value is denoted q_α.

So, in particular, the median is $m = q_{0.5}$, the lower quartile is $q_L = q_{0.25}$ and the upper quartile is $q_U = q_{0.75}$. The terms **percentile** and **percentage point** are synonymous with 'quantile', and are often used when α is expressed as a percentage. Some other special cases of quantiles also have more specialized names. For example, if α is an integer multiple of 0.1, the corresponding quantiles are sometimes called *deciles*. There is good reason for generalizing to population α-quantiles and that is because some values of α other than $\frac{1}{4}$, $\frac{1}{2}$ and $\frac{3}{4}$ will be made much use of as the course develops. From *Chapter 4* onwards, you will often find that interest centres on the 'extremes' of distributions, and that quantiles associated with values of α like 0.9, 0.95, 0.99 and, at the other extreme, 0.1, 0.05, and so on, may be important.

A diagram should help clarify the idea of a population quantile. Figure 3.15 shows a graph of a typical continuous c.d.f. $F(x)$. For a given value a, say, on the horizontal axis, you could evaluate (or read off the graph) the corresponding point $F(a)$ (a probability, $0 < F(a) < 1$) on the vertical axis.

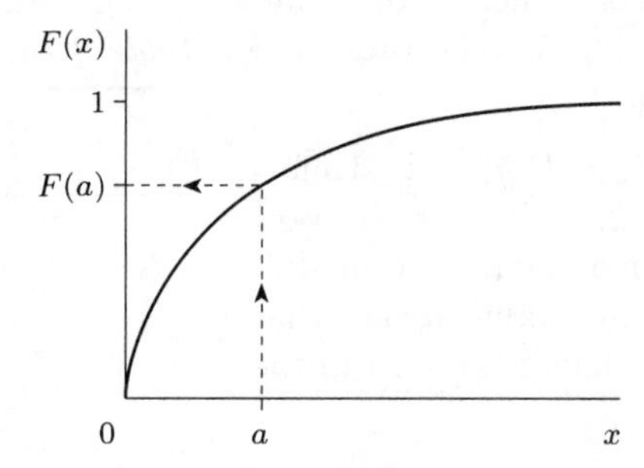

Figure 3.15 Calculating $F(a)$ for a given a

In Figure 3.16 the starting point is a number α $(0 < \alpha < 1)$ on the vertical axis. The corresponding point on the horixontal axis is the α-quantile of X, the value q_α.

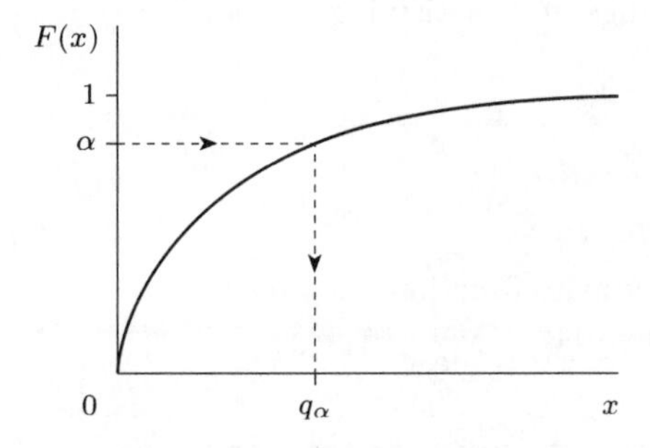

Figure 3.16 Calculating q_α for a given α

(Expressed mathematically, the α-quantile is given by $q_\alpha = F^{-1}(\alpha)$, where $F^{-1}(\cdot)$ is the inverse function of $F(\cdot)$.)

3.5.2 Quantiles for discrete probability distributions

Quantiles for continuous distributions have been defined readily and unambiguously. The equation $F(x) = \alpha$ is solved for x. In all of our examples, and in all the cases you will encounter in this course, this equation has a unique solution.

Unfortunately, when quantiles for discrete distributions are considered, this simplicity and lack of ambiguity disappears. Quantiles for discrete distributions are not as important as those for continuous ones, so in this course we shall not dwell on them, but shall simply illustrate some of the problems and how they might be overcome.

Let us begin with the example of the uniform distribution on the integers $1, 2, \ldots, 6$: the model adopted for the score X when a fair die is rolled. The c.d.f. of X is given by $F(x) = x/6$, tabulated in Table 3.15.

Table 3.15 A fair die

x	1	2	3	4	5	6
$F(x)$	$\frac{1}{6}$	$\frac{1}{3}$	$\frac{1}{2}$	$\frac{2}{3}$	$\frac{5}{6}$	1

Using the above method for finding the median of X, we need to solve the equation $F(m) = \frac{1}{2}$ for m. This yields $m = 3$.

If the median is regarded as an indication of the 'centre' of the distribution then this is unsatisfactory. If we reversed the distribution, starting at the score of 6, and applied the same process to find the median, then we would obtain a value of 4. Given the symmetry of the distribution, this is not very appealing. An attractive property of a symmetric distribution would be that it has the same 'centre' from whichever end you look at it!

An extra problem is illustrated by the probability mass function of a random variable Y which is uniform on the integers $1, 2, \ldots, 5$. The tabulated cumulative distribution function is given in Table 3.16.

Table 3.16

y	1	2	3	4	5
$F(y)$	$\frac{1}{5}$	$\frac{2}{5}$	$\frac{3}{5}$	$\frac{4}{5}$	1

Again, to find the median we need to solve the equation $F(y) = \frac{1}{2}$. Unfortunately, this equation is not satisfied for any $y = 1, 2, \ldots, 5$. At the very least this requires the definition of the median to be modified. One modification is to redefine the population median as the minimum value m such that $F(m) \geq \frac{1}{2}$. Here, this would yield the solution $m = 3$, which is at least in the middle of the range.

Similar problems arise with other quantiles. The definition we shall use in this course follows.

> For a discrete random variable X with c.d.f. $F(x)$, the **α-quantile** q_α is defined to be the minimum value of x in the range of X satisfying
>
> $F(x) \geq \alpha.$

Example 3.25 Quartiles of the binomial distribution

If X is binomial $B(6, 0.6)$, then X has the probability mass function $p(x)$ and cumulative distribution function $F(x)$ given in Table 3.17.

Table 3.17

x	0	1	2	3	4	5	6
$p(x)$	0.004 096	0.036 864	0.138 240	0.276 480	0.311 040	0.186 624	0.046 656
$F(x)$	0.004 096	0.040 960	0.179 200	0.455 680	0.766 720	0.953 344	1

To find the median of X we need to find the minimum value of x such that $F(x) \geq 0.5$. Since $F(3) = 0.455\,680$ (less than 0.5) and $F(4) = 0.766\,720$ (greater than 0.5), $x = 4$ is the least value satisfying $F(x) \geq 0.5$. Hence the median m is 4. Similarly, since $F(2) < 0.25 \leq F(3)$, the lower quartile of X is $q_L = 3$. Finally, since $F(3) < 0.75 \leq F(4)$, the upper quartile of X is $q_U = 4$. Notice that in this example the median and the upper quartile are the same. ■

Exercise 3.19

(a) Find the median of the binomial distribution $B(10, 0.5)$.

(b) Find the median of the binomial distribution $B(17, 0.7)$.

(c) Find the upper quartile of the binomial distribution $B(2, 0.5)$.

(d) Find the interquartile range of the binomial distribution $B(19, 0.25)$.

(e) Find the 0.85-quantile of the binomial distribution $B(15, 0.4)$.

3.5.3 Population modes

For discrete probability models, the mode, if there is just one, is the value that has the highest probability of occurring (much as the sample mode is the value that occurs the highest proportion of times in a sample). For continuous population models, interest simply shifts to maxima: the maximum, if there is only one, of the p.d.f. rather than the p.m.f. Also—and this notion is especially useful for continuous distributions—distributions can have one or more maxima and thus may be multimodal. You need not take away much more from this short subsection than the idea that it makes sense to talk of, for example, bimodal probability distributions. An example of a bimodal probability density function is shown in Figure 3.17.

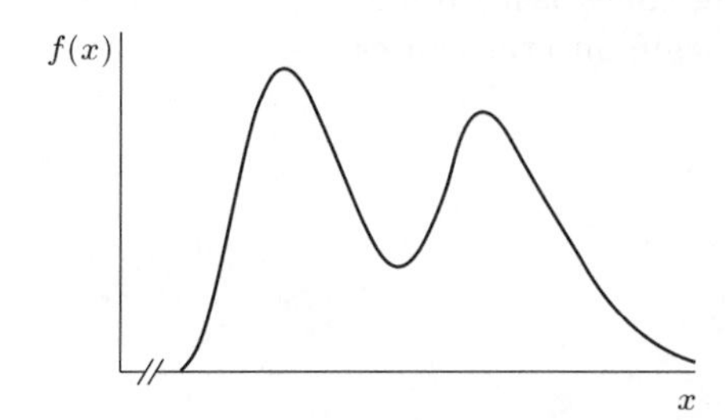

Figure 3.17 A bimodal density function

The Old Faithful geyser at Yellowstone National Park, Wyoming, USA, was observed from 1–15 August 1985. During that time, data were collected on the duration of eruptions and the waiting time between the starts of successive eruptions. There are 299 waiting times (in minutes), and these are listed in Table 3.18.

Azzalini, A. and Bowman, A.W. (1990) A look at some data on the Old Faithful geyser. *Applied Statistics*, **39**, 357–366.

Table 3.18 Waiting times (minutes) between eruptions, Old Faithful geyser

80	71	57	80	75	77	60	86	77	56	81	50	89	54	90	73	60	83
65	82	84	54	85	58	79	57	88	68	76	78	74	85	75	65	76	58
91	50	87	48	93	54	86	53	78	52	83	60	87	49	80	60	92	43
89	60	84	69	74	71	108	50	77	57	80	61	82	48	81	73	62	79
54	80	73	81	62	81	71	79	81	74	59	81	66	87	53	80	50	87
51	82	58	81	49	92	50	88	62	93	56	89	51	79	58	82	52	88
52	78	69	75	77	53	80	55	87	53	85	61	93	54	76	80	81	59
86	78	71	77	76	94	75	50	83	82	72	77	75	65	79	72	78	77
79	75	78	64	80	49	88	54	85	51	96	50	80	78	81	72	75	78
87	69	55	83	49	82	57	84	57	84	73	78	57	79	57	90	62	87
78	52	98	48	78	79	65	84	50	83	60	80	50	88	50	84	74	76
65	89	49	88	51	78	85	65	75	77	69	92	68	87	61	81	55	93
53	84	70	73	93	50	87	77	74	72	82	74	80	49	91	53	86	49
79	89	87	76	59	80	89	45	93	72	71	54	79	74	65	78	57	87
72	84	47	84	57	87	68	86	75	73	53	82	93	77	54	96	48	89
63	84	76	62	83	50	85	78	78	81	78	76	74	81	66	84	48	93
47	87	51	78	54	87	52	85	58	88	79							

Figure 3.18 shows a histogram of the geyser waiting time data. Notice that there are two very pronounced modes.

These data were referred to in *Chapter 1*, Figure 1.19.

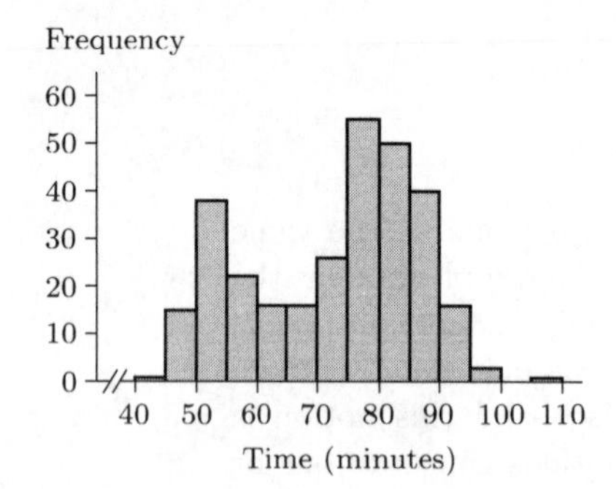

Figure 3.18 Waiting times between eruptions, Old Faithful geyser

The histogram suggests that a good model for the variation in waiting times should also be bimodal. This is important: the bimodality may indicate that waiting times are essentially of two characters. 'Short' waiting times last a little under an hour (with some variation); 'long' waiting times last around 75–80 minutes (with some variation). Research can then begin into the causes of the interesting phenomenon observed.

Summary

1. The mean and variance of a discrete integer-valued random variable X, with probability mass function $p(x)$, are given by

$$E(X) = \mu = \sum xp(x),$$
$$V(X) = \sigma^2 = E((X-\mu)^2) = \sum (x-\mu)^2 p(x),$$

where the summations are taken over all x in the range of X.

2. The mean and variance of a continuous random variable X, with probability density function $f(x)$, are given by

$$E(X) = \mu = \int x f(x)\, dx,$$

$$V(X) = \sigma^2 = E((X-\mu)^2) = \int (x-\mu)^2 f(x)\, dx,$$

where the integrations extend over the range of X.

3. Two discrete random variables X and Y are said to be independent if and only if

$$P(X = x, Y = y) = P(X = x)P(Y = y),$$

for all x in the range of X and all y in the range of Y.

Two continuous random variables X and Y are said to be independent if, for example,

$$P(X \le x, Y \le y) = P(X \le x)P(Y \le y),$$

for all x in the range of X and all y in the range of Y.

4. In a sequence of independent Bernoulli trials indexed by the parameter p, $0 < p < 1$, the number N of trials from success to next success is a random variable following a geometric distribution with parameter p, written $N \sim G(p)$.

The random variable N has probability function

$$p_N(n) = q^{n-1}p, \quad n = 1, 2, 3, \ldots,$$

where $q = 1 - p$.

The cumulative distribution function is

$$F_N(n) = 1 - q^n, \quad n = 1, 2, 3, \ldots.$$

The first two moments of N are $E(N) = 1/p$ and $V(N) = q/p^2$.

5. The random variable X following a discrete uniform probability distribution with parameter n has probability mass function

$$p(x) = \frac{1}{n}, \quad x = 1, 2, \ldots, n.$$

The mean is $\dfrac{n+1}{2}$ and the variance is $\dfrac{n^2-1}{12}$.

6. When a continuous random variable X is constrained to take values between stated limits a and b $(a < b)$, but within those bounds no value of x is any more likely than any other, then the random variable X is said to follow a continuous uniform distribution on the interval $a \le x \le b$, with probability density function

$$f(x) = \frac{1}{b-a}, \quad a \le x \le b.$$

This is written $X \sim U(a, b)$.

The mean of X is $\frac{a+b}{2}$ and the variance of X is $\frac{(b-a)^2}{12}$.

7. If X is uniformly distributed on the interval $0 \leq x \leq 1$, then this is written $X \sim U(0,1)$ and X is said to follow the standard uniform distribution.

The random variable X has probability density function

$$f(x) = 1, \quad 0 \leq x \leq 1.$$

The cumulative distribution function is

$$F(x) = x, \quad 0 \leq x \leq 1.$$

8. The α-quantile of a continuous random variable X with cumulative distribution function $F(x)$ is defined to be the solution of the equation

$$\alpha = F(x),$$

written $x = q_\alpha$. In particular, the first (lower) population quartile q_L, the median m and the third (upper) quartile q_U are, respectively, the quantiles $q_{0.25}$, $q_{0.50}$ and $q_{0.75}$.

9. The α-quantile for a discrete integer-valued random variable X with cumulative distribution function $F(x)$ is defined to be the minimum value of x in the range of X such that $F(x) \geq \alpha$.

Chapter 4
Models for Data III

This is the last of the three chapters concerned with fundamental probability models; two new models for variation are described. The notion of 'a random sample' is discussed in some detail: the point is made that in repeated experiments the sample mean, which is an important summary statistic, is itself a random variable.

This chapter continues the development of fundamental probability models for variation, and our discussion of the properties of those models. Following the pattern developed in *Chapters 2* and *3*, the chapter deals both with specific models for variation and the more general properties of such models.

Two new probability models are developed in this chapter. In *Chapter 2* the notion of a Bernoulli trial was introduced, and a probability model for the number of successful trials in a set or sequence of trials was developed. This model was the binomial probability distribution, which applies where the results of different trials may be assumed to be independent and where the probability of success is assumed to remain the same from trial to trial. In Section 4.1, we study the sort of experiment where the number of trials involved could be very large indeed, but the chance of 'success' in any one trial is very small; the random variable of interest is still the total count of successes. But in this case, one where 'something rather unlikely to happen has many opportunities to do so', the actual count distribution, which you know to be binomial, turns out to have a very useful approximate form which is a discrete probability model of considerable usefulness in its own right. The result is known as *Poisson's approximation for the occurrence of rare events*, and the ensuing model is the *Poisson distribution.*

The Poisson model is explored in depth in *Chapter 12*.

The topic of Section 4.4, the *exponential distribution*, is a continuous probability model of wide application. The situation modelled is this. Some events happen every now and then, but they are not precisely forecastable. Some of the nastier ones are floods, domestic accidents, lightning strikes, air crashes and earthquakes; a less unpleasant example might be an unexpected cheque in the post. If the average rate at which such haphazard events occur is more or less constant (data suggest that major earthquakes world-wide, for instance, occur roughly once every 14 months or so—not that this helps you forecast the next one) then it turns out that something can be said about the probability distribution of the time lag, or 'waiting time', between consecutive occurrences.

It was noted in *Chapter 2* that it is in the nature of a random sample that if you draw on two different occasions a random sample of the same size from the same population, then you (probably) get different individual sample observations and therefore (probably) different sample means and variances. And yet, on both occasions, the sample was drawn to obtain information about precisely the same population. So, in repeated experiments, the sample mean is itself a *random variable*. In this chapter we shall investigate simple properties of the sample mean. For instance, we followed in *Chapter 2* a botanical experiment in which a researcher drew at random a sample of 100 leaves from an ornamental bush, and measured their lengths. In the sample the leaves ranged in length from 0.3 cm to 2.7 cm; the sample mean was 1.276 cm. A different experiment would almost certainly have led to different results, and yet the sample mean clearly provides some sort of assessment, an *estimate*, of the average length of all the leaves on the bush. This was one of the aims of the experiment. Our first look at properties of the sample mean will give us an idea of the accuracy (in other words, the usefulness) of this estimate.

Example 2.1

There are occasional computer exercises dispersed throughout this chapter; Section 4.5 is entirely devoted to exercises in which nothing new is introduced, but where some ideas of the preceding sections are revisited. As far as the sort of material studied in this course is concerned, the computer has four main uses. You have by now seen examples of some of these: (i) as a very fast calculator; (ii) for the speedy and efficient analysis of statistical data; (iii) using its graphics capabilities to produce enlightening diagrammatic representations of data. The fourth role of a computer is as a *simulator*. There are many uses for computer simulation. In one application, the computer mimics the drawing of a random sample from some specified probability distribution. This is particularly useful for comparing statistical models with one another and with real data. You saw some examples of this in *Chapter 2*, and the idea is continued in this chapter.

4.1 Rare events

Figure 4.1 Siméon Denis Poisson

This section introduces a probability distribution that is one of the most useful for the statistical modeller. It is called the *Poisson distribution* after the Frenchman Siméon Denis Poisson (1781–1840) who, in 1837, introduced it in a scientific text on the subject of jury verdicts in criminal trials. In the early nineteenth century in France a twelve-person jury could return a 'guilty' verdict by a majority of seven to five. Poisson's work followed that of Condorcet and Laplace in regarding individual jurors' decisions as the results of tosses of a weighted coin. These researchers were concerned with the effects of jury size and the definition of 'majority' on the chance of reaching a correct verdict.

For more on the history of the Poisson distribution, and much else besides, see Stigler, S. M. (1986) *The History of Statistics—The Measurement of Uncertainty before 1900*, published by the Belknap Press of Harvard University Press.

The Poisson distribution is introduced here as a convenient approximation to the binomial probability distribution, one which provides good accuracy under certain circumstances.

4.1.1 Two examples

We shall start by looking at two different situations where the Bernoulli trial, or a sequence of trials, might be a useful statistical model, answering some questions using the distributions and techniques available to us so far.

A **Bernoulli process** is the name given to a *sequence of Bernoulli trials* in which

(a) trials are independent;

(b) the probability of success is the same from trial to trial.

For a Bernoulli process the idea of trials occurring in order, one after the other, is crucial. Departures from the modelling assumptions of the Bernoulli process include cases where the result of one trial might influence the probabilities of the results of later trials. In *Chapter 12* we shall consider some statistical tests for examining this kind of dependence.

Example 4.1 Cycle usage

At the time of writing, people tend to commute from their homes to their places of work either by car or by public transport (buses, underground trains and railway trains); some work within walking distance from home; and some, if the distance is not too great, cycle. One particular disadvantage of cycling (and one probably sufficient to put many commuters off the whole idea of cycling to work) is the possibility of arriving utterly drenched, having been caught in a downpour. This is indubitably a very unpleasant experience; but actually its frequency has been estimated at only 15 times a year 'for a regular cyclist'. ■

Ballantine, R. (1975) *Richard's Bicycle Book.* Pan, Great Britain.

It is not entirely clear from the reference to it quite how this value has been calculated; if we wish to use it for forecasting the likelihood of such an eventuality, then we need to make some further assumptions. Suppose a cyclist uses her bicycle regularly for travelling to and from work and, occasionally, for shopping and social visits. Her own assessment of cycle usage (which is rather rough) works out at 50 journeys a month of a non-negligible distance. This gives (at twelve months a year) an estimated proportion of rides on which a downpour occurs of

$$p = \frac{15}{12 \times 50} = \frac{15}{600} = \frac{1}{40}.$$

600 journeys a year, during 15 of which it rains.

Unpleasant and complicating to one's existence as the experience is, it therefore occurs (on average) only once in 40 rides. If we assume that these events occur independently of previous experiences and with constant probability, then it follows that the Bernoulli process, a sequence of independent Bernoulli trials, could be a useful model (with a trial corresponding to a single journey, and a success—a score of 1—to getting caught in a downpour during that journey). Then the number of times per month that this occurs has a binomial distribution $B(50, 1/40)$. (The mean of this distribution, $np = 1.25$, is equal to $15/12$, the average number of soakings per month calculated from Richard's figure.)

It was stated without proof in *Chapter 3*, page 102 that the mean of the binomial distribution $B(n, p)$ is the product np.

(The Bernoulli process is not as good a statistical model as one might develop. A better model would probably include some sort of mechanism that allows the parameter p to vary with the time of the year, and perhaps even permits some sort of dependence from trial to trial, so that if you got wet cycling to work then you are rather more likely than otherwise to get wet cycling home as well since it is a wet day. However, it is as good a model for the phenomenon as we are at the moment in a position to construct.)

Exercise 4.1

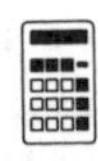

Assuming the cyclist's own estimate of '50 journeys a month' is accurate, what over the next month is the probability

(a) that she never gets wet;

(b) that she gets wet twice;

(c) that she gets wet at least four times?

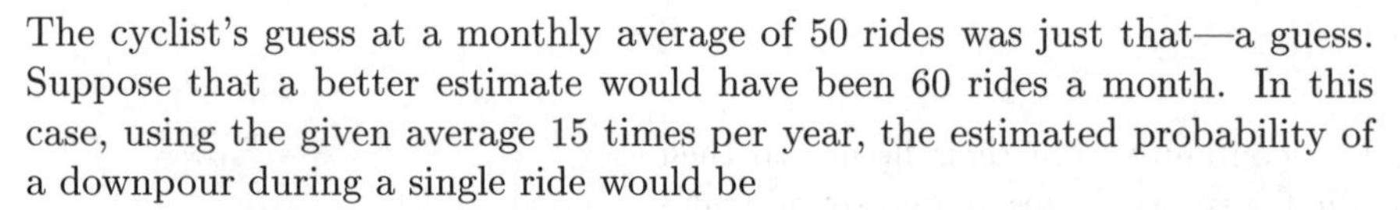

The cyclist's guess at a monthly average of 50 rides was just that—a guess. Suppose that a better estimate would have been 60 rides a month. In this case, using the given average 15 times per year, the estimated probability of a downpour during a single ride would be

$$p = \frac{15}{12 \times 60} = \frac{15}{720} = \frac{1}{48}.$$

Then the assumed probability distribution for the number of drenchings per month is $B(60, 1/48)$ (which has mean 1.25, as before).

Exercise 4.2

The table following (Table 4.1) is half completed: it gives certain probabilities for the binomial distribution $B(50, 1/40)$. From it, you could read off all your answers to Exercise 4.1. Complete the table by computing the corresponding probabilities for the binomial distribution $B(60, 1/48)$.

Table 4.1 Probabilities, monthly downpours while cycling

	$P(X=0)$	$P(X=1)$	$P(X=2)$	$P(X=3)$	$P(X \geq 4)$
$B(50, 1/40)$	0.2820	0.3615	0.2271	0.0932	0.0362
$B(60, 1/48)$					

You will have noticed from Exercise 4.2 that the probabilities calculated for the cyclist but based on the higher estimate for cycle usage are not very different from those calculated for Exercise 4.1. To two decimal places, these values are identical.

Example 4.2 Typographical errors

One of the necessary stages in getting a book published is the careful and attentive checking of the proofs, the first version of the printed pages. Usually there are typographical errors, such as tihs, that need identifying and correcting. With a good publisher and printer such errors, even in the first proofs, are fairly rare. Experience with one publisher of children's books (for scientific texts the problem is a little more complicated) suggests an average of 3.6 errors per page at proof stage. ■

The number of words per page is, of course, not constant from page to page; if it was (and equal to 320, say, taking an estimate from a random children's paperback), then the estimated probability that a word was mistyped would be

Of course, there may be other errors besides mistyped words. This is a simplified description of the problem.

$$p = \frac{3.6}{320} = 0.01125,$$

or a little over 1%; and assuming independence from word to word, the number of errors per page would follow a binomial distribution with parameters $n = 320$, $p = 0.01125$.

We shall perform explicitly the binomial calculations involved in finding the first few terms in the probability distribution of the number X of errors per page of proofs, based on a page length of 320 words. For instance,

$$p_X(0) = (1 - 0.01125)^{320} = 0.98875^{320} = 0.0268;$$

and

$$p_X(1) = 320(0.01125)(0.98875)^{319} = 0.0975.$$

For the next term, the binomial formula gives

$$p_X(2) = \binom{320}{2}(0.01125)^2(0.98875)^{318}.$$

If you try to find this value using the factorial notation, writing

$$\binom{320}{2} = \frac{320!}{2!318!},$$

you will probably find that your calculator fails: most calculators cannot cope with the number 320!. However, that number can be rewritten

$$\binom{320}{2} = \frac{320!}{2!318!} = \frac{320 \times 319 \times 318!}{2!318!} = \frac{320 \times 319}{1 \times 2},$$

so the probability becomes

$$p_X(2) = \frac{320 \times 319}{1 \times 2}(0.01125)^2(0.98875)^{318} = 0.1769.$$

These calculations are included only to demonstrate what is actually happening. You should find that your computer will give explicit binomial calculations without baulking at $n = 320$ (though there probably is some sort of upper limit to what it will do). You can certainly use your computer to complete Exercise 4.3.

Further terms in the sequence of binomial probabilities can be calculated similarly:

$$p_X(3) = \frac{320 \times 319 \times 318}{1 \times 2 \times 3}(0.01125)^3(0.98875)^{317} = 0.2134.$$

Actually, the page counts are a little variable: turning to a different page, the estimate of the word count per page would almost certainly have been different—360, perhaps.

Exercise 4.3

(a) Write down the probability distribution of X, the number of errors per page, based on an estimate of 360 words per page and an average of 3.6 errors per page.

(b) Calculate the probabilities $p_X(0)$, $p_X(1)$, $p_X(2)$, $p_X(3)$ in this case, find the probability that there are more than three errors on a page, and comment on any differences in the results obtained between your model and the earlier binomial $B(320, 0.01125)$ 'guess'.

In all these calculations, the binomial distribution has been used and the probability p has been 'rather small'. In Example 4.1, the results also suggested that for values of n not too different from one another, and with a corresponding estimate for the parameter p of

$$p = \frac{15}{12 \times n} = \frac{1.25}{n},$$

probability calculations for the binomial distribution $B(n, 1.25/n)$ did not differ significantly. (However, note that whatever the value of n, the mean of the distribution remains unaltered at $\mu = np = 1.25$.) Similarly, in Example 4.2, the actual value of n, at least in the two particular cases examined, was not too critical to the computed probabilities for the binomial distribution $B(n, 3.6/n)$ (with constant mean 3.6).

Now the question arises: can we satisfactorily model the number of monthly soakings for the cyclist, or the number of errors per page of proofs, using a random variable whose distribution depends only on a single parameter, the expected number? In general, can we approximate the two-parameter binomial distribution $B(n, \mu/n)$ by a probability distribution indexed by a single parameter μ?

The answer is that we can, provided $p = \mu/n$ is small (which, in both these examples, it was).

The meaning of the word 'small' will shortly be made more precise.

4.1.2 The Poisson distribution

In both the examples considered so far, the binomial parameter p has been small (and n fairly large): an event which itself is rather unlikely to happen (small p) has had a lot of opportunities (large n) to do so. Overall, the expected number of total occurrences ($\mu = np$) has been non-negligible. It seems, at least in the case of the few test calculations that we have done, that the actual values of n and p are not critical to the values of the binomial probabilities, provided the product np remains the same.

Let us now examine this apparent result. For the binomial distribution $B(n, p)$ or, equivalently, $B(n, \mu/n)$, the probability $p_X(0)$ is given by

$$p_X(0) = (1-p)^n = \left(1 - \frac{\mu}{n}\right)^n.$$

The approximate value of this expression when n is known to be large is $e^{-\mu}$.

So, for n reasonably large, the approximation for $B(n, \mu/n)$ is

$$p_X(0) \simeq e^{-\mu}. \tag{4.1}$$

You might have forgotten the result that for large n,

$$\left(1 + \frac{x}{n}\right)^n \simeq e^x;$$

it is one of the standard results from analysis involving the exponential function. The number e is approximately 2.718 281 8.... Setting x equal to $-\mu$, the particular result given here follows.

You may have seen from your calculations in the preceding exercises that the computation of binomial probabilities can, without the use of a computer, become quite awkward; and time-consuming, for there are many duplicated key-presses. For the simpler calculation of successive terms in the binomial distribution one useful method called *recursive* is to obtain each subsequent term from the previous term by using the relation

$$\frac{p_X(x+1)}{p_X(x)} = \frac{\binom{n}{x+1} p^{x+1} q^{n-x-1}}{\binom{n}{x} p^x q^{n-x}} = \frac{n-x}{x+1} \times \frac{p}{q};$$

$q = 1 - p$

so

$$p_X(x+1) = \frac{n-x}{x+1} \times \frac{p}{q} \times p_X(x).$$

Rewriting p as μ/n and q as $(n-\mu)/n$, this gives the recursion

$$p_X(x+1) = \frac{\mu}{x+1} \times \frac{n-x}{n-\mu} \times p_X(x). \tag{4.2}$$

We have already established in (4.1) an approximation to the binomial probability $p_X(0)$ when the parameter n is large. In the recursion (4.2), when n is large, the second multiplying factor $(n-x)/(n-\mu)$ is approximately equal to one; so we can generate successive terms in the approximating probability distribution by applying the recursive formula

$$p_X(0) = e^{-\mu}, \qquad p_X(x+1) = \frac{\mu}{x+1} p_X(x), \quad x = 0, 1, 2, \ldots. \tag{4.3}$$

Exercise 4.4

Use the recursive scheme defined by (4.3) to write down formulas for the probabilities

(a) $p_X(1)$, (b) $p_X(2)$, (c) $p_X(3)$.

Then

(d) find a general expression for $p_X(x)$.

The definition of the *Poisson distribution* is as follows.

> The random variable X follows a **Poisson distribution with parameter μ** if it has probability mass function
>
> $$p_X(x) = \frac{e^{-\mu}\mu^x}{x!}, \quad x = 0, 1, 2, \ldots. \tag{4.4}$$
>
> This is written $X \sim \text{Poisson}(\mu)$.

The term $x!$ in the denominator denotes the product $1 \times 2 \times \ldots \times x$. The number $0!$ is set equal to 1 by definition.

It is easy to check that the function $p_X(x)$ is indeed a probability function—that is, that the terms sum to one. This vital property has not been lost in the derivation of the function from the binomial probability function. From the polynomial expansion

$$e^{\mu} = \left(1 + \mu + \frac{\mu^2}{2!} + \frac{\mu^3}{3!} + \ldots\right),$$

This is the *Taylor series expansion* for e^{μ}. In some texts it is introduced as the *definition* of the exponential function.

the property follows immediately:

$$\begin{aligned}\sum_{x=0}^{\infty} p_X(x) &= \sum_{x=0}^{\infty} \frac{e^{-\mu}\mu^x}{x!} \\ &= e^{-\mu}\sum_{x=0}^{\infty}\frac{\mu^x}{x!} \\ &= e^{-\mu}\left(1 + \mu + \frac{\mu^2}{2!} + \frac{\mu^3}{3!} + \ldots\right) \\ &= e^{-\mu}e^{\mu} \\ &= 1.\end{aligned}$$

Notice that this is a one-parameter distribution: as will be confirmed later, the parameter μ is (as we might expect) the mean of the Poisson probability distribution. The way it was obtained demonstrates the (equivalent) approximations

$$B(n,p) \approx \text{Poisson}(np) \qquad \text{or} \qquad B\left(n, \frac{\mu}{n}\right) \approx \text{Poisson}(\mu).$$

The symbol '$\approx$' is read as 'has approximately the same distribution as'.

You will recall that this sort of conclusion was precisely the one we sought: we wanted to answer the question 'can we approximate the two-parameter binomial distribution $B(n, \mu/n)$ by a probability distribution indexed by a single parameter μ?'

Shortly, a rough rule will be given for values of the binomial parameters n and p when the Poisson distribution is a useful approximation to the binomial distribution. First, here are a couple of numerical examples.

Example 4.1 continued

Table 4.2 shows the probabilities calculated for the cycling example.

The Poisson approximation gives the following probability distribution of the number of soakings per month:

$$p_X(0) = e^{-1.25} = 0.2865,$$

$$p_X(1) = \frac{e^{-1.25}1.25}{1!} = 0.3581,$$

$$p_X(2) = \frac{e^{-1.25}(1.25)^2}{2!} = 0.2238,$$

and so on.

Table 4.2 Probabilities, monthly downpours while cycling

	$P(X=0)$	$P(X=1)$	$P(X=2)$	$P(X=3)$	$P(X \geq 4)$
$B(50, 1/40)$	0.2820	0.3615	0.2271	0.0932	0.0362
$B(60, 1/48)$	0.2827	0.3610	0.2266	0.0932	0.0365
Poisson(1.25)	0.2865	0.3581	0.2238	0.0933	0.0383

In each case, the value of $P(X \geq 4)$ in the table was obtained by subtraction. (You would obtain from your computer or from tables that $P(X \geq 4)$ when $X \sim B(60, 1/48)$ is 0.0366, to 4 decimal places. This would give

$$P(X=0) + P(X=1) + P(X=2) + P(X=3) + P(X \geq 4)$$
$$= 0.2827 + 0.3610 + 0.2266 + 0.0932 + 0.0366$$
$$= 1.0001$$

(to 4 decimal places), when these probabilities should sum to one. This is the sort of rounding error that *does not matter*.) ■

Exercise 4.5

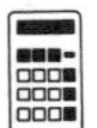

Now complete Table 4.3 for the typographical error model of Example 4.2. (Except for the third row, all the results have been calculated already—in, or just before, Exercise 4.3.) Use your calculator, or your computer if you prefer, to obtain the Poisson probabilities.

Table 4.3 Probabilities, errors in page proofs

	$P(X=0)$	$P(X=1)$	$P(X=2)$	$P(X=3)$	$P(X \geq 4)$
$B(320, 0.01125)$	0.0268	0.0975	0.1769	0.2134	0.4854
$B(360, 0.01)$	0.0268	0.0976	0.1769	0.2133	0.4854
Poisson(3.6)					

Whether or not two probabilities are 'close' is not really an absolute matter: it depends on what you want to use them for, and there are cases where the fourth or fifth decimal place is critical. It is unlikely that you would feel an assessment of the chances of several errors occurring in a page of proofs is one of these critical cases! The question is rather a subjective one. But for a given value of x, any difference in the probabilities listed in Table 4.3 never exceeds 0.001. The approximating Poisson distribution is really very good.

Remember that we seek only an *adequate* model for rare events: in both these examples, the value of the binomial parameter n has been somewhat questionable. The information we started with in each case was a single number—'15 soakings a year on average' (or 1.25 per month) for the cycling example and 'an average of 3.6 errors per page' for the printing example. This single number is all that is needed for a complete specification of the Poisson model.

A rough rule for deciding when the Poisson distribution provides a good approximation to the binomial distribution is as follows.

> Generally, even for quite small values of n, when the event is 'rare enough', occurring, say, with probability $p \leq 0.05$, the approximation $B(n,p) \approx \text{Poisson}(np)$ will still be good; and it will get better as n increases.

4.1.3 Moments of the Poisson distribution

It was mentioned in passing that the indexing parameter μ of the Poisson family of probability distributions is also the mean of the distribution. This can be proved as follows. If $X \sim \text{Poisson}(\mu)$, then

In general, in this course, we shall not be much bothered with formal mathematical proof. But here is an example where just a few lines of algebra constitute the proof of a result, and it is instructive to see this sort of thing occasionally.

$$E(X) = \sum_X x p_X(x) = \sum_{x=0}^{\infty} \frac{x e^{-\mu} \mu^x}{x!}.$$

On expansion, the first term of the summation on the right-hand side is zero; then for each remaining term the x in the numerator cancels out the x in the

$x!$ term in the denominator; so we obtain the result

$$E(X) = 0 + \sum_{x=1}^{\infty} \frac{xe^{-\mu}\mu^x}{x!} = \sum_{x=1}^{\infty} \frac{e^{-\mu}\mu^x}{(x-1)!}.$$

$x! = x \times (x-1)!$

Taking a factor μ and the term $e^{-\mu}$ outside the summation gives

$$\begin{aligned} \mu e^{-\mu} \sum_{x=1}^{\infty} \frac{\mu^{x-1}}{(x-1)!} &= \mu e^{-\mu} \left(1 + \mu + \frac{\mu^2}{2!} + \frac{\mu^3}{3!} + \dots\right) \\ &= \mu e^{-\mu}(e^{\mu}) \\ &= \mu. \end{aligned}$$

Here, we are again making use of the Taylor series expansion for e^{μ}.

Hence we have the required result:

$$E(X) = \mu. \tag{4.5}$$

Calculation of the variance of a Poisson random variable is not quite so straightforward. You need not bother with the details of this mathematical exercise: the variance of X is given by

$$V(X) = \mu. \tag{4.6}$$

So a Poisson random variable has variance equal to its mean μ.

> If the random variable X follows a Poisson distribution with parameter μ, then X has mean and variance
>
> $$E(X) = \mu, \qquad V(X) = \mu.$$

The idea of testing the adequacy of the fit of a proposed probability model to a set of data is one that has been mentioned several times already. One possible test of a proposed Poisson model against a given data set involves calculating the ratio of the *sample variance* to the *sample mean* and seeing how far this ratio deviates from 1. This test is not covered formally in this course, but you will see at the end of this section a data set where the sample variance and sample mean are informally compared.

4.1.4 Calculating Poisson probabilities

Calculation of Poisson probabilities can be a time-consuming and error-prone process even with a calculator. Printed tables of Poisson probabilities are available, at least for selected values of μ. As in the binomial case, there is no convenient formula for the Poisson cumulative distribution function

$$F_X(x) = P(X \leq x) = \sum_{j=0}^{x} \frac{e^{-\mu}\mu^j}{j!};$$

the terms simply do not add conveniently to a closed formula. It is necessary, without a computer, to refer to tables or else explicitly to add successive terms of the probability mass function.

4.1.5 Applications of the Poisson approximation

We have discussed a number of different situations in which an appropriate statistical model is a binomial probability distribution having a small value for the probability p. In such cases the Poisson distribution with matched mean turns out to be a reasonable approximation to the binomial model. There are many other practical situations in which such a model would apply, and hence for which the Poisson distribution could also reasonably be used. Here are a few more examples.

Example 4.3 *Birthdays*

Suppose we have a group of 100 people. Let X be the random variable denoting the number of people in the group whose birthday falls on Christmas Day. Then $p = 1/365$ (making the assumption that births are distributed equally over the year and ignoring Leap Day), $n = 100$, and so the value of μ is $100/365 = 0.274$. The probability, say, that precisely one person in the group has a Christmas birthday is (exact binomial calculation)

Actually, the proportion of births on Christmas Day is well under 1/365: it is a holiday period, and fewer births than usual are induced. The birth rate falls on Sundays, too.

$$100 \times \frac{1}{365} \times \left(\frac{364}{365}\right)^{99} = 0.209,$$

or (Poisson approximation)

$$0.274e^{-0.274} = 0.208.$$

The two probabilities are very close. ∎

Example 4.4 *Defective items*

Suppose that in a process manufacturing identical items there is a small probability p that any item is defective. Making the rather unrealistic assumption of independence from item to item, then the number of defective items in a large batch can be represented approximately by a random variable following a Poisson distribution with the appropriate mean. ∎

Example 4.5 *Colour blindness*

The probability that a man's sight is colour-deficient is about 0.06, and the corresponding figure for women is 0.004. Hence, in a large group of either men or women, the number with colour-deficient sight can be represented as a Poisson variable. ∎

There are various forms of colour-deficiency. This example deals with just one of them. The value $p = 0.06$ in this example exceeds the 0.05 threshold given in our 'rule'—no model will constitute a perfect representation, and with this in mind, the Poisson model may reasonably be applied.

Example 4.6 *Thunderstorms*

Thunderstorms are not common events, even in Britain. There is perhaps some seasonal variation in their frequency, but over the course of a year one might reasonably regard each day as a single 'trial' where something not very likely might nevertheless happen. ∎

Examples 4.3–4.6 are all situations where the Poisson approximation to the binomial model was fairly straightforward to identify. There are many other situations where the derivation is less obvious: we shall now briefly explore one of them.

Example 4.7 Calls at a switchboard

The number of calls received at a telephone switchboard in a fixed time interval of duration t, say, is a realization of some random variable X whose expected value will be proportional to t: call that mean λt. The constant λ is the average rate at which calls come in to the switchboard. To develop a model for this situation, let us imagine the time interval t to be split up into a large number n of very small time intervals of length t/n. Then suppose that the probability that a call is received in one small time interval is p. Provided that the rate of incoming calls remains roughly constant over time, then p may also be assumed to remain constant over the time interval t and to be equal to $\lambda t/n$. Provided n is large enough, so that the intervals t/n are small enough, the probability that more than one call is received in any interval may be regarded as negligible. We can, therefore, think of the random process as a sequence of n independent Bernoulli trials. The total number of calls received will follow a binomial distribution with parameters n and p, where $p = \lambda t/n$, $B(n, \lambda t/n)$. Thus the actual number of calls received in the time interval t may be assumed to follow a Poisson distribution (with mean λt). ■

The letter λ is the Greek lower-case letter 'lambda'.

In this example we have constructed a notional sequence of many trials *only later to ignore them.* The probability distribution of the number of calls received depends only on the time for which they are counted and the average rate at which they are arriving: that is, only on the expected number of arrivals. This is a very simple and intuitive result, to which we shall return in Section 4.4.

Most of our examples have involved processes in one dimension. There are many other practical applications where the Poisson distribution provides a good model: the number of trees in regions of a two-dimensional wood, the number of raisins in slices of cake; the number of live viruses in doses of vaccine. In all cases we have made the assumption that items such as bacteria, raisins, thunderstorms and typographical errors act independently: they have no preferred time of occurrence or preferred location, and they do not associate. If these assumptions do not hold, then the Poisson model will not be suitable. For instance, think about the distribution of individual caterpillars on cabbage plants in a field. They have hatched out from eggs that have been laid in large clusters, so the caterpillars will tend to cluster themselves. In this situation, the Poisson model would break down: the numbers of caterpillars counted on plants would have an observed distribution noticeably different from that suggested by a Poisson model.

The following exercise summarizes the work of this section.

Exercise 4.6

Resistors are very cheap electrical items, and they are easy to make. Assuming they work, they may be assumed to be indestructible. A small proportion (approximately 1 in 20) will not work when they leave the factory. This is a tolerable risk accepted by purchasers: these resistors are not quality tested before being packaged, for to do so would add considerably to their cost.

(a) The resistors are boxed in packages of 50. State an 'exact' model for the number of defective resistors in a box, mentioning any assumptions you make and calculate the probabilities that there are 0, 1, 2, 3, 4 or more than 4 defectives in a box.

(b) Find an approximating distribution for the number of defectives in a box and calculate the same probabilities as you found in (a).

(c) Comment on any differences between your answers in (a) and (b).

This section ends with a final example where the Poisson model was useful.

Example 4.8 *Particle counting*

In 1910 the scientists Rutherford and Geiger reported an experiment in which they counted the number of alpha particles emitted from a radioactive source during intervals of $7\frac{1}{2}$ seconds' duration, for 2612 different intervals. The numbers of particles emitted and the frequencies with which the different counts occurred, are shown in the first two columns of Table 4.4.

Rutherford, E. and Geiger, H. (1910) The probability variations in the distribution of alpha particles. *Philosophical Magazine*, Sixth Series, **20**, 698–704.

The last column of the table, labelled 'Fit', was obtained as follows. There were 10 126 particles counted altogether, implying an average of 3.877 per $7\frac{1}{2}$-second interval. The third column gives the expected frequencies when a Poisson(3.877) model is 'fitted' to the data. You need not bother at this stage with the details of the fitting procedure: the whole question of 'goodness of fit' of a proposed statistical model to a set of data is examined in *Chapter 9*.

Table 4.4 Emissions of alpha particles

Count	Frequency	Fit
0	57	54
1	203	210
2	383	407
3	525	525
4	532	509
5	408	395
6	273	255
7	139	141
8	49	68
9	27	30
10	10	11
11	4	4
12	2	1
> 12	0	1

For the moment, just observe that the Poisson fit to these particular data is very good. Like telephone calls at a switchboard, the physical nature of radioactive particle emission means that such emissions occur entirely at random, independently of one another, but at a rate which remains virtually constant with passing time.

Notice that the sum of the fitted frequencies is 2611 and not 2612: this is the sort of small rounding error which sometimes creeps in with the fitting procedure. It does not matter.

Incidentally, for this data set the sample variance is 3.696, close to the observed sample mean. Remember the property of a Poisson variate, that its mean is equal to its variance. ■

4.2 Drawing random samples

In *Chapter 2*, Section 2.1 following the description of the botanical experiment in which 100 leaves were collected from an ornamental bush and measured, the text continued thus: '... if he [the botanist] had taken a different sample of 100 leaves his calculated average would have been different. This raises the obvious question: if his results vary according to the random choice of leaves in the sample, of what value are the results? Presumably he really hoped to say something about the average length of the population of leaves, and not just of an arbitrarily chosen sample. How can the sample average, which varies from sample to sample, tell us anything about the average length of all the leaves on the bush?'

Here is another situation where the same 'obvious question' is implied.

Example 4.9 A nutritional study

A nutritionist studying the effect of different proportions of protein in the diets of chicks, randomly allocated the chicks to one of four groups and recorded their weights (in grams) after three weeks' growth. The data are given in Table 4.5. Each of the four groups received a different diet.

Table 4.5 Weights of chicks after three weeks' growth (g)

Crowder, M.J. and Hand, D.J. (1990) *Analysis of repeated measures.* Chapman and Hall, London, p. 75.

Normal diet	10% protein replacement	20% protein replacement	40% protein replacement
205	331	256	204
215	167	305	281
202	175	147	200
157	74	341	196
223	265	373	238
157	251	220	205
305	192	178	322
98	233	290	237
124	309	272	264
175	150	321	
205			
96			
266			
142			
157			
117			

Let us focus on only those chicks receiving the normal diet. If x_i is the weight of the ith chick then the total weight of the 16 chicks in the group is

$$w = x_1 + x_2 + \cdots + x_{16} = 205 + 215 + \cdots + 117 = 2844,$$

and the mean weight of the 16 chicks in the group is

$$\overline{x} = \frac{w}{16} = 177.75.$$

With a different allocation of chicks to groups (possibly with different numbers of chicks allocated to the groups), or in a repeated experiment, these results would almost certainly have been different. Each of the recorded weights $x_1, x_2, \ldots, x_{16}$, is in fact a single observation on the random variable X, where X is 'the weight in grams of a randomly selected chick given a normal diet, measured after three weeks' growth'. The observed sample total w is merely a single observation on the random variable

$$W = X_1 + X_2 + \cdots + X_{16},$$

the sum of sixteen independent observations on the random variable X. Finally, the observed sample mean $\overline{x}$ is merely a single observation on the random variable

$$\overline{X} = \frac{X_1 + X_2 + \cdots + X_{16}}{16}.$$

Presumably the nutritionist really wanted to make a meaningful comparison between the effects of the four diets; and in order to do that, it was necessary to be able to say something about the effect of a normal diet before making the comparison. Again, the question suggests itself: if a different allocation had produced different results, as it probably would, what is the worth of the experiment? ■

A different but related question attaching to the whole notion of random sampling is this: if, instead of sampling 100 leaves, the botanist had sampled only ten and used just these ten in the calculation of the sample mean, would his results have been any less useful? The experiment is certainly a simpler one—at any rate, it involves less expenditure of effort. And if the nutritionist simply wanted to compare four diets, why not just choose four chicks (rather than the 45 in the experiment), allocate to each a diet at random, and then compare their weights after three weeks?

We shall now explore the useful consequences of taking large samples in a statistical experiment.

There are two stages in the calculation of a sample mean (once the sample has been collected and the attribute of interest has been measured). These are (i) adding the observations together to obtain the sample total, and (ii) dividing the sample total by the sample size in order to obtain the sample mean. We shall deal with each of these in turn.

In general, if $X_1, X_2, \ldots, X_n$ are random variables with means $\mu_1, \mu_2, \ldots, \mu_n$, respectively, then their sum $X_1 + X_2 + \cdots + X_n$ is also a random variable. This random variable has a probability distribution (which could be quite difficult to work out, depending on the distributions of the components X_i); however, its mean is very easy to work out. The expected value of the sum is, quite simply, the sum of the expected values of the components.

If X_i, $i = 1, 2, \ldots, n$, are random variables with respective means $E(X_i) = \mu_i$, then the mean of their sum $X_1 + X_2 + \cdots + X_n$ is given by

$$E(X_1 + X_2 + \cdots + X_n) = E(X_1) + E(X_2) + \cdots + E(X_n)$$
$$= \mu_1 + \mu_2 + \cdots + \mu_n. \qquad (4.7)$$

This result is stated without proof.

In the particular case of a random sample the components X_i are independent and they are *identically distributed* (being observations on the same attribute of interest) and therefore, in particular, their means are equal. If we write $\mu_i = \mu$ for all i, then we have the particular result that, for a random sample, the sample total has mean

$$E(X_1 + X_2 + \cdots + X_n) = \mu + \mu + \cdots + \mu = n\mu.$$

(Remember that in any experiment, the observed sample total is merely a single observation on a random variable.)

Next, the sample total is divided by the sample size in order to obtain the sample mean, $\overline{X}$ (also a random variable). The sample mean has expected value

$$E(\overline{X}) = E\left(\frac{X_1 + X_2 + \cdots + X_n}{n}\right) = \frac{1}{n}E(X_1 + X_2 + \cdots + X_n).$$

Here, we have used the intuitive result that for a random variable X with constant multiplier a, the expected value of the random variable aX is

$$E(aX) = aE(X).$$

It is possible to prove this result formally, and this is done in the next section, but you might care to reflect as follows. If the random variable X denotes people's height, measured in feet, then the random variable $Y = 12X$ denotes the same thing, measured in inches. Within a population, the mean height in inches, μ_Y, will be twelve times μ_X, the mean height in feet.

Now things become very easy indeed: we have just established that for a random sample the sample total has expected value $n\mu$. It follows that the sample mean has expected value

$$E(\overline{X}) = \frac{1}{n}(n\mu) = \mu. \qquad (4.8)$$

This is a very reassuring result! Having complained that the sample mean is a random variable and questioned the worth of its calculation, we have shown nevertheless that its expected value is the population mean, μ, itself. This confirms our intuitive view that the sample mean is at least in some sense a good indicator or predictor (or, simply, a good *estimator*) for the unknown population mean.

However, notice that the result (4.8) does not depend on the value of n, the sample size. In a sample of size 2, the sample mean has expected value

$$E\left(\frac{X_1 + X_2}{2}\right) = \mu;$$

and the same applies to a sample of size 100. One intuitively feels that a larger sample is 'more accurate' and, in that sense, that its results are 'more useful'. However, the result (4.8) is not concerned with accuracy, and does not imply or even suggest this.

In fact, one's intuition that inferences from larger samples are 'more accurate' is correct, provided a proper interpretation is given to the meaning of the word 'accurate'. The key lies in the precision of the sample mean—to put it another way, its variance.

Here is a further general result that will be necessary. It has been remarked that a sum of random variables is itself a random variable, and at (4.7) we found its mean. Provided one further assumption is made, it is very easy to write down an expression for the variance of the sum.

If X_i, $i = 1, 2, \ldots, n$ are *independent* random variables with respective variances $V(X_i) = \sigma_i^2$, then the variance of their sum $X_1 + X_2 + \cdots + X_n$ is given by

$$\begin{aligned} V(X_1 + X_2 + \cdots + X_n) &= V(X_1) + V(X_2) + \cdots + V(X_n) \\ &= \sigma_1^2 + \sigma_2^2 + \cdots + \sigma_n^2. \qquad (4.9) \end{aligned}$$

This result is stated without proof.

Notice the essential further constraint that the components of the sum should be *independent*. In the case of a random sample this constraint is always assumed to hold, and result (4.9) follows. (In more complicated statistical experiments where individual observations cannot be assumed to be independent of one another, a result corresponding to (4.9) can be obtained for the variance of the sum without great difficulty.)

In the particular case of a random sample, not only are the components independent but they are identically distributed as well, so it follows that their variances are equal. If we write $\sigma_i^2 = \sigma^2$ for all i, then it follows for a random sample that the sample total has variance

$$V(X_1 + X_2 + \cdots + X_n) = \sigma^2 + \sigma^2 + \cdots + \sigma^2 = n\sigma^2.$$

Upon division of the sample total by n, there is a powerful consequence that confirms our intuition that 'larger samples are better', and quantifies the improvement.

The sample mean has variance

$$\begin{aligned} V(\overline{X}) &= V\left(\frac{X_1 + X_2 + \cdots + X_n}{n}\right) \\ &= V\left(\frac{1}{n}(X_1 + X_2 + \cdots + X_n)\right) \\ &= \left(\frac{1}{n}\right)^2 V(X_1 + X_2 + \cdots + X_n) \\ &= \left(\frac{1}{n}\right)^2 (n\sigma^2) \\ &= \frac{\sigma^2}{n}. \end{aligned} \qquad (4.10)$$

Here, we have used the result that if X is a random variable and a is a positive constant, then the variance of the random variable aX is

$$V(aX) = a^2 V(X).$$

The result is not entirely obvious: you might find it easier to use the result that if X is a random variable and a is a positive constant, then the standard deviation of the random variable aX is $SD(aX) = aSD(X)$. These results can be proved formally without much difficulty, and are further discussed in Section 4.3.

Note that in this result (4.10) the sample size n *does* feature: the larger the sample size, the smaller the variance of the sample mean. The sense in which a sample mean obtained from a large sample is a 'more accurate' estimator of the unknown population mean μ than the sample mean obtained from a smaller sample is that it is less likely to be very far away from its expected value—which is, in both cases, the unknown population mean μ.

These two important results can be summarized as follows.

> If X is a random variable with mean μ and variance σ^2, and if $X_1, X_2, \ldots, X_n$ constitutes a random sample from the distribution of X, then the first two moments of the sample mean $\overline{X}$ are
>
> $$E(\overline{X}) = \mu, \qquad V(\overline{X}) = \frac{\sigma^2}{n}.$$

You will notice that nothing has been said about the probability distribution of $\overline{X}$ at this stage, and the main implication behind these results is that large samples are better than smaller ones. In general, exact results are rather difficult to obtain and to apply. However, we shall see in *Chapter 5* that it is quite straightforward to obtain an *approximate* result for the distribution of the sample mean. Using this result approximate probability statements can be made, about the accuracy of $\overline{X}$ as an estimator of μ, about conditions placed on the minimum sample size necessary to achieve a required precision, and so on.

Exercise 4.7

This exercise is a computer simulation exercise designed to illustrate the results of this subsection.

(a) Use your computer to obtain a random sample of size five from the Poisson distribution with mean 8. List the elements of the sample, and find their mean.

(b) Now obtain 100 samples of size five from the Poisson distribution with mean 8. In each case, calculate the sample mean, and store the 100 sample means in a data vector.

(c) Plot a histogram of the data vector of part (b), and find its mean and variance.

(d) Now repeat parts (b) and (c) but with samples of size fifty rather than five, and comment on any differences observed in the data vector of means.

In Section 4.3 we shall look more closely at some general results for random variables.

4.3 Moments of linear functions

A rather general (and not always easy) problem in statistics is illustrated in the following example.

Example 4.10 *Estimating height*

A theodolite is an instrument used by surveyors for measuring angles, and to calculate the heights of tall structures. For instance, in Figure 4.2, the height of the wall (labelled y) is unknown. The theodolite stands a known distance d from the wall on a tripod of known height t. The operator directs the viewing lens of the theodolite at the top of the wall, and records the angle x.

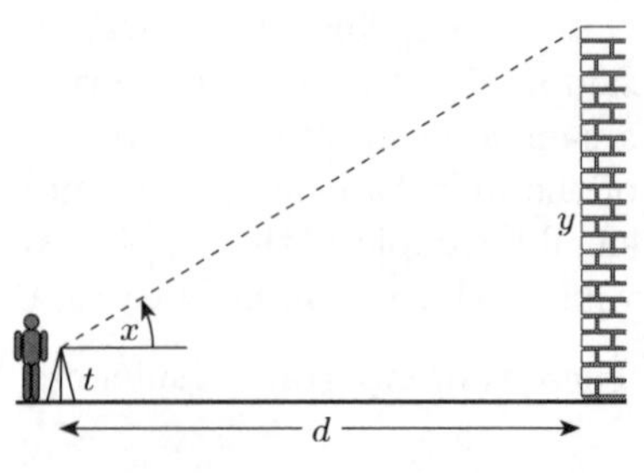

Figure 4.2 Estimating the height of a wall

In terms of the measured angle x, a simple trigonometric argument gives the unknown height of the wall as

$$y = t + d\tan x.$$

In practice, the recorded angle x is an observation on a random variable X: if the experiment were to be repeated, possibly using different operators, one would expect small differences in the recorded angle. This would necessarily imply corresponding differences in the calculated value of y, which should therefore similarly be regarded as due to random variation. In fact, any single calculated value of y is an observation on a random variable Y. The relationship between the random variables X and Y is given by the identity

$$Y = t + d\tan X. \quad \blacksquare$$

In general, any function of a random variable is itself a random variable. This example illustrates a more general problem. Suppose that a random variable X has a probability distribution that is known: what, then, can be said about the probability distribution of an associated random variable Y defined in terms of X through the equation

$$Y = h(X),$$

where $h(\cdot)$ is a known function?

In principle this question can always be answered, and there are some standard results when X is known to follow some standard distribution and where the form of the function $h(\cdot)$ is particularly simple. We shall meet some of these standard results as the course progresses.

However, in this section we will restrict ourselves to a more specific question: given that the random variable X has mean μ and variance σ^2, what can be said about *the mean and variance* of the associated random variable Y defined in terms of X through *the linear function*

$$Y = aX + b,$$

where a and b are constants?

As you can see, this is a much reduced problem. We have already addressed some related questions. Here are some examples where the problem might arise.

Example 4.11 *Chest measurements*

Based on a very large sample of data (see *Chapter 2*, Table 2.17), an adequate probability model for the distribution of chest circumferences (in inches) amongst Scottish soldiery of the last century may be assumed to be given by the normal distribution with mean $\mu = 40$ and standard deviation $\sigma = 2$. It is required to make a comparison with present-day European equivalents, and for this it is required to work in centimetres. There are 2.54 cm to the inch: the question arises 'if the random variable X has mean 40 and standard deviation 2, what are the mean and standard deviation of the associated random variable

$$Y = 2.54X$$

representing a linear function of X with $a = 2.54$ and $b = 0$?' ■

Example 4.12 *Water temperature*

One of the factors in the spread of malaria is the density of the spreading agent (the *Anopheles* mosquito) in the population. An environmental feature that in turn affects this density is the water temperature where the mosquito larvae breed. This will be subject to variation. Suppose that the water temperature is measured in ℃ (degrees Celsius). If measured in ℉ (degrees Fahrenheit) instead, there will still be evidence of variation: the equation that gives the temperature Y (°F) in terms of X (°C) is

$$Y = 1.8X + 32.$$

This represents a linear conversion with $a = 1.8$ and $b = 32$. ■

Example 4.13 *Body weight*

The conversion factor between pounds (lb) and kilograms (kg) is given approximately by 1 lb = 0.455 kg. Americans tend to express body weight measurements in pounds, Europeans (though not usually the inhabitants of Great Britain) in kilograms. For a comparison of the variability in body weight for two cultures, it would be necessary to make a transformation of the style $Y = 0.455X$. ■

Let us now establish the first two moments (i.e. the mean and variance) of the random variable $Y = aX + b$. A formula for the expected value $E[h(X)]$ of a function $h(\cdot)$ of a random variable X is easy to write down. For instance, in the discrete case, this is

$$E[h(X)] = \sum_X h(x)p(x),$$

where the random variable X has probability mass function $p(x)$ and where the summation is taken over the range of X. In the continuous case,

$$E[h(X)] = \int_X h(x)f(x)\,dx,$$

where the random variable X has probability density function $f(x)$ and where the integral is taken over the range of X. So, in the particular case that $h(X)$ takes the form $aX + b$, we have (in the discrete case)

$$\begin{aligned} E(aX + b) &= \textstyle\sum_X (ax + b)p(x) \\ &= \textstyle\sum_X axp(x) + \sum_X bp(x) \\ &= a\textstyle\sum_X xp(x) + b\sum_X p(x) \\ &= aE(X) + b. \end{aligned}$$

The last line follows because $\sum_X xp(x)$ is, by definition, the mean of X, and because $\sum_X p(x)$ is necessarily 1. Similar manipulations show that in the case that X is continuous, then the same formula applies:

This is one of the standard properties of a probability mass function.

$$E(aX + b) = aE(X) + b. \qquad (4.11)$$

Earlier, we used the result $E(aX) = aE(X)$.

Exercise 4.8

(a) In Example 4.11 a random variable X with mean 40 was suggested as a model for the distribution of chest circumferences (measured in inches) amongst Scottish soldiers (based on historical data). What would be the mean circumference if the units of measurement were centimetres?

(b) Suppose that in Example 4.12 the variation in water temperature was modelled by a probability distribution with mean 26 °C. Find the mean temperature in degrees Fahrenheit.

In order to work out the variance of the random variable $aX + b$, we shall start by writing $Y = aX + b$. The variance of Y is, by definition,

$$V(Y) = E[(Y - E(Y))^2]$$

See formula (3.9) in *Chapter 3*.

and the expected value of Y is $E(Y) = aE(X) + b$ from (4.11). So it follows that

$$\begin{aligned} V(Y) &= E[(aX + b - (aE(X) + b))^2] \\ &= E[(aX + b - aE(X) - b)^2] \\ &= E[(aX - aE(X))^2] \\ &= E[a^2(X - E(X))^2] \end{aligned} \tag{4.12}$$

In (4.12) the expression $(X - E(X))^2$ is a random variable; a^2 is a constant and can be taken outside the brackets. Thus

$$V(Y) = a^2E[(X - E(X))^2] = a^2V(X).$$

But Y is just the random variable $aX + b$, so we have obtained the result

$$V(aX + b) = a^2V(X). \tag{4.13}$$

Earlier, we used the result $V(aX) = a^2V(X)$.

It follows from (4.13) (taking square roots of both sides) that

$$SD(aX + b) = \sqrt{a^2V(X)} = |a|SD(X). \tag{4.14}$$

Notice that the constant b does not feature on the right-hand side of (4.13) or (4.14). The modulus sign in (4.14) is important since the constant a could be negative, but a standard deviation is always positive.

Example 4.14 *Sections of a chemical reactor*

Variation in the section temperature across the 1250 sections of a chemical reactor may be assumed to be adequately modelled by a normal distribution with mean 452 ℃ and standard deviation 22 deg C.

These are fictitious data based on a real investigation. See Cox, D.R. and Snell, E.J. (1981) *Applied Statistics.* Chapman and Hall, London, p. 68.

From properties of the normal distribution we know that nearly all the recorded temperatures across the 1250 sections will lie within the range $\mu - 3\sigma$ to $\mu + 3\sigma$, or 386 ℃ to 518 ℃. Converting from ℃ to ℉ leads to a new random variable with mean

$$1.8 \times 452 + 32 = 845.6,$$

and standard deviation

$$1.8 \times 22 = 39.6.$$

As it happens, this new random variable is also normally distributed: none of the preceding results imply this, but it is true nevertheless. You can probably appreciate that altering the scale of measurement will not alter the essential characteristics of the temperature variation: the probability density function reaches a peak at the mean temperature, tailing off symmetrically either side of the mean.

Again, nearly all the temperatures across the sections of the reactor will lie within the range $\mu - 3\sigma$ to $\mu + 3\sigma$: measured in ℉ the lower extreme is

$$845.6 - 3 \times 39.6 = 726.8$$

and the upper extreme is

$$845.6 + 3 \times 39.6 = 964.4.$$

These could have been obtained in a different way by converting 386 ℃ and 518 ℃ (the two likely temperature extremes) to ℉. ■

Exercise 4.9

In the early years of this century, statisticians were very concerned simply with the counting and measurement of random variables, and the recording of those measurements. Anything amenable to measurement and where there was evidence of variation was fair game. Some examples from issues of the journal *Biometrika*, from 1901 to 1902, include the number of sense organs of *Aurelia Aurita*; the number of ridges and furrows in the shells of different species of mollusc; the correlation between the number of stamens and the number of pistils in *Ficaria ranunculoides* during the flowering season; the dimensions of *Hyalopterus trirhodus* (aphids); the dimensions of the human skull; the dimensions of the egg of *Cuculus Canorus* (the cuckoo); the lengths of criminals' left middle fingers (and other dimensions); the number of sepals in *Anemone nemorosa*; the dimensions of the human hand; coat-colour in horses; the epidemiology of smallpox. A very large sample was used in assessing the variation in the lengths of criminals' left middle fingers: the variation may be very adequately modelled by a random variable X with mean 11.55 cm and standard deviation 0.55 cm.

Macdonell, W.R. (1902) On criminal anthropometry and the identification of criminals. *Biometrika*, **1**, 177–227.

It is required for a comparison with Imperial measurements to make a transformation to inches. At 1 in = 2.54 cm, what is the mean and standard deviation in the new scale?

These results for moments of a linear function of a random variable X may be summarized as follows.

> If X is a random variable with mean μ_X and standard deviation σ_X, and if the random variable Y is defined by
>
> $$Y = aX + b,$$
>
> where a and b are constants, then Y has moments
>
> $$\mu_Y = a\mu_X + b;$$
>
> $$\sigma_Y^2 = a^2\sigma_X^2;$$
>
> $$\sigma_Y = |a|\sigma_X.$$

This section ends with a result which provides a very useful alternative formula for the variance of a random variable X.

First, we shall need the incidental result that if $h_1(X)$ and $h_2(X)$ are two functions of a random variable X, then the expected value of their sum is equal to the sum of their expected values:

$$E[h_1(X) + h_2(X)] = E[h_1(X)] + E[h_2(X)]. \tag{4.15}$$

This is easy to check. In the case that X is discrete, for instance, having probability mass function $p(x)$, then

$$\begin{aligned} E[h_1(X) + h_2(X)] &= \textstyle\sum_X (h_1(x) + h_2(x))p(x) \\ &= \textstyle\sum_X (h_1(x)p(x) + h_2(x)p(x)) \\ &= \textstyle\sum_X h_1(x)p(x) + \sum_X h_2(x)p(x) \\ &= E[h_1(x)] + E[h_2(x)]. \end{aligned}$$

Similar manipulations will confirm (4.15) when X is a continuous random variable.

The formula we have been using so far for the variance of a random variable X with expected value μ is

$$V(X) = E[(X - E(X))^2] = E[(X - \mu)^2].$$

Expanding the square on the right-hand side gives

$$V(X) = E[X^2 - 2\mu X + \mu^2].$$

The expression $X^2 - 2\mu X + \mu^2$ can be written in the form $h_1(X) + h_2(X)$ in more than one way, but the most useful way is to set $h_1(X) = X^2$ and $h_2(X) = (-2\mu X + \mu^2)$. Then it follows that

$$V(X) = E(X^2) + E(-2\mu X + \mu^2).$$

Furthermore, we can apply our formula for the mean of a linear function of X, $E(aX + b) = aE(X) + b$, to the last term above. Setting $a = -2\mu$ and $b = \mu^2$ we finally obtain the result

$$\begin{aligned} V(X) &= E(X^2) + (-2\mu E(X) + \mu^2) \\ &= E(X^2) + (-2\mu^2 + \mu^2) \\ &= E(X^2) - \mu^2, \end{aligned}$$

or

$$V(X) = E(X^2) - (E(X))^2 = E(X^2) - \mu_X^2. \tag{4.16}$$

This formulation is sometimes easier to implement than the formula for the variance that we have used up to now.

Example 4.15 A perfect die

In *Chapter 3*, page 110 you saw that the random variable X with probability mass function

$$p(x) = \tfrac{1}{6}, \quad x = 1, 2, \ldots, 6$$

(used as a probability model for the outcome of throws of a perfectly engineered die) has mean

$$E(X) = \mu = 3.5$$

and variance

$$\begin{aligned} V(X) = \sigma^2 &= E[(X-\mu)^2] \\ &= (1-3.5)^2 \times \tfrac{1}{6} + (2-3.5)^2 \times \tfrac{1}{6} + \cdots + (6-3.5)^2 \times \tfrac{1}{6} \\ &= \frac{6.25 + 2.25 + 0.25 + 0.25 + 2.25 + 6.25}{6} \\ &= \frac{17.5}{6} \\ &= 2.92. \end{aligned}$$

Using the alternative formula (4.16) for the variance we first need the value of $E(X^2)$: this is

$$\begin{aligned} E(X^2) &= 1^2 \times \tfrac{1}{6} + 2^2 \times \tfrac{1}{6} + \cdots + 6^2 \times \tfrac{1}{6} \\ &= \tfrac{1}{6} + \tfrac{4}{6} + \tfrac{9}{6} + \tfrac{16}{6} + \tfrac{25}{6} + \tfrac{36}{6} \\ &= \tfrac{91}{6}. \end{aligned}$$

Then

$$V(X) = E(X^2) - \mu^2 = \tfrac{91}{6} - (3.5)^2 = \tfrac{91}{6} - \tfrac{49}{4} = \tfrac{35}{12} = 2.92$$

as before. ■

Exercise 4.10

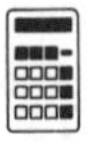

In *Chapter 3* Exercise 3.7, you showed that the variance in the outcome of rolls of a *Double-Five* was $16/6 = 2.7$. (The mean outcome is 4.) Confirm this result for the variance using the formula (4.16).

It was earlier remarked that, in general, exact distributional results for random samples are rather difficult to obtain. There are some exceptions to this rule, four of which we shall now briefly explore.

A Sums of Bernoulli random variables

Suppose that the random variables $X_1, X_2, \ldots, X_n$ are each individually distributed as Bernoulli random variables with the same value of the parameter p. That is, each X_i takes the value 1 with probability p and 0 with probability $q = 1 - p$. Suppose that, as well as being identically distributed, the X_is are independent. Then in *Chapter 2*, Section 2.3, we defined the distribution of the sum $Y = X_1 + X_2 + \cdots + X_n$ to be binomial with parameters n and p, and went on to look at some of the properties of this distribution. In this case the distribution of the sum is not too difficult: it is given by the binomial probability mass function

$$P(Y = y) = \binom{n}{y} p^y q^{n-y}, \quad y = 0, 1, 2, \ldots, n.$$

The intuitive result was also stated without proof in *Chapter 3*, page 102, that the mean of the binomial distribution $B(n,p)$ is the product np. We did not discuss its variance. However, we also found that if X follows a Bernoulli distribution with parameter p, then X has mean and variance

$$E(X) = p, \qquad V(X) = pq.$$

The mean of a Bernoulli distribution was found on page 101 and the variance on page 111.

It follows directly from the results (4.7) and (4.9) of this chapter that

$$E(Y) = E(X_1 + X_2 + \cdots + X_n) = np,$$

$$V(Y) = V(X_1 + X_2 + \cdots + X_n) = npq,$$

thus confirming one result and stating a second for the first time.

> If the random variable Y has a binomial distribution $B(n,p)$, then
>
> $$E(Y) = np, \qquad V(Y) = npq. \qquad (4.17)$$

Exercise 4.11

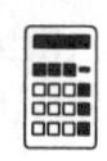

You saw in *Chapter 3*, Example 3.5, that if X is binomial $B(4, 0.4)$ then the individual probabilities for X are as follows.

x	0	1	2	3	4
$p(x)$	0.1296	0.3456	0.3456	0.1536	0.0256

Use the formulas at (4.17) to write down the mean and variance of X and confirm your result for the variance of X using the result

$$V(X) = E(X^2) - (E(X))^2$$

obtained at (4.16).

B Sums of Poisson random variables

> If X_i, $i = 1, 2, \ldots, n$, are independent Poisson variates with respective means μ_i, then their sum $Y = X_1 + X_2 + \cdots + X_n$ is also a Poisson variate, with mean $\mu_1 + \mu_2 + \cdots + \mu_n$: that is,
>
> $$Y \sim \text{Poisson}(\mu_1 + \mu_2 + \cdots + \mu_n). \qquad (4.18)$$

This result is stated without proof.

Notice that, given the distributions of the independent components X_i, you could have written down the mean and the variance of Y immediately using results (4.7) and (4.9). What is not obvious is that the distribution of the sum $Y = X_1 + X_2 + \cdots + X_n$ is also Poisson.

(Incidentally, notice that the result is quite general: there is no requirement here, as there is in the case of a random sample, that the X_is should be identically distributed, only that they should be independent.)

Example 4.5 continued

In the example on colour blindness (Example 4.5) we saw that for one form of colour-deficiency the proportion of male sufferers is about 0.06 and for females it is about 0.004. Although the proportion 0.06 is a little high for our Poisson 'rule of thumb' to hold—the threshold mentioned was $p \leq 0.05$—in a large population the approximation will not be too bad. Suppose that in an assembly of 300 people there are 220 men and 80 women. Then the number of men with colour-deficient sight is approximately Poisson with mean 220×0.06 or 13.2; the number of women with colour-deficient sight is approximately Poisson with mean 80×0.004 or 0.32. The total number of colour-deficient people is, by (4.18), approximately Poisson with mean $13.2 + 0.32$ or 13.52. ■

Example 4.16 *Two species of ant*

Two species of ant, *Messor Wasmanni* and *Cataglyphis bicolor* inhabit northern Greece. In any given region, it might be assumed initially that the number of *Messor* nests follows a Poisson distribution with mean μ_1 proportional to the area of the region under consideration and to the underlying density of *Messor* nests, and that the number of *Cataglyphis* nests follows a Poisson distribution with mean μ_2. In different regions the number of ants' nests will vary according to a Poisson distribution with mean $\mu_1 + \mu_2$. ■

Harkness, R.D. and Isham, V. (1983) A bivariate spatial point pattern of ants' nests. *Applied Statistics*, **32**, 293–303.

There follow two further results that are stated here without proof and further comment. They will be referred to in *Chapter 5* which discusses the normal distribution.

C A linear function of a normal random variable

If the random variable X is normally distributed with mean μ and variance σ^2, then the random variable $aX + b$ is normally distributed with mean $a\mu + b$ and variance $a^2\sigma^2$:

$$aX + b \sim N(a\mu + b, a^2\sigma^2). \qquad (4.19)$$

Again, you could have deduced the mean and the variance of the random variable $aX + b$ from previous results. What is new (although this result was referred to in Example 4.14) is the fact that $aX + b$ is normally distributed.

D Sums of normal random variables

If X_i, $i = 1, 2, \ldots, n$, are independent normally distributed random variables with respective means and variances μ_i and σ_i^2, then their sum $Y = X_1 + X_2 + \cdots + X_n$ is also normally distributed, with mean $\mu_1 + \mu_2 + \cdots + \mu_n$ and variance $\sigma_1^2 + \sigma_2^2 + \cdots + \sigma_n^2$: that is,

$$Y \sim N(\mu_1 + \mu_2 + \cdots + \mu_n, \sigma_1^2 + \sigma_2^2 + \cdots + \sigma_n^2). \tag{4.20}$$

It follows, incidentally, from the results (4.19) and (4.20) that if a random sample of size n is taken from a normal population with mean μ and variance σ^2, then the sample mean $\overline{X}$ is *normally distributed* with mean μ and variance σ^2/n:

$$\overline{X} \sim N\left(\mu, \frac{\sigma^2}{n}\right). \tag{4.21}$$

4.4 The exponential distribution

One of the properties of the Bernoulli process is that the probability of a 'success' at any trial remains the same and equal to the probability of success at all preceding trials, whatever the history of the process up to that trial. So, even a long run of failures (of 0s) does not make the probability of a success (a 1) at the next trial any more likely. There is no question of 'luck being bound to change'. Similarly, a long run of 1s (a 'run of good luck') does not alter the probability of a 1 at subsequent trials. The incidence of 1s may be regarded in a very special sense as 'quite random', in that knowledge of the whole history of the process so far, the outcome of all the trials, does not alter its future development, or the probability laws describing its future development. A statistical experimenter about to embark on trial number 1001 is not helped at all in his assessment of what might or might not happen at the trial by knowing what has happened at any or all of the preceding thousand trials. This property is known as the 'memoryless' property of the Bernoulli process.

In this section we shall meet for the first time (but not examine in great detail) a process rather similar to the Bernoulli process, one which develops, however, in continuous time rather than as a sequence of 0s and 1s at discrete trials. What happens is that with passing time some particular event of interest (which for ease we shall call A) occurs once, and then again, and then again, and then again, ..., and so on. But it does so after intervals (or 'waiting times') that are random in the same sense: knowing how many times A has occurred so far, or how long it is since A last occurred, is of no use or assistance in forecasting the next, or future, occurrences of A. These waiting times correspond roughly to sequences, or runs, of 0s in a realization of a Bernoulli process; when A happens it is rather as though a 1 was recorded. But the waiting times are not integers, and successive occurrences of A are not restricted to discrete opportunities. A more detailed description will be given in *Chapter 12*, where several random processes similar to the Bernoulli process are described.

Here, however, are some examples of the sort of phenomenon to be modelled.

Example 4.17 Random events occurring in continuous time

(a) Even in a machine shop where a regular maintenance programme is carried out, occasional machine breakdowns are an inevitable hazard. However, their occurrence is not forecastable. Sometimes the repair staff are overworked, and sometimes there is nothing for them to do. All that is known from repair records is that the breakdowns occur at an *average* rate of three times a week.

(b) Extreme wind conditions sufficient to ground aircraft are recorded at a small airport on the south coast of England roughly twice a year, but their time of occurrence does not seem to be linked to the time of the year or indeed to anything else. They are always a total surprise.

(c) It is very annoying to break a fingernail. Unfortunately, it happens in a split second and there is nothing one can do about it. Some people are forever breaking nails; others hardly ever do.

(d) Small clicks and buzzes occur sometimes when one is listening to FM broadcasts, but there is nothing one can do about them. They are quite irregular and do not seem traceable to any particular cause.

(e) Police monitoring a stretch of motorway have breakdowns reported to them. There are many more during the day than there are at night, and noticeably more on Friday and Sunday evenings than on Saturday evenings, but over continuous periods of not too long duration the average rate may be regarded as invariant. However, there never seems to be any pattern to the times at which the reported breakdowns occur.

(f) Arrivals at a hospital casualty ward occur without warning, but at average rates which (depending broadly on the time of day and the day of the week) are reasonably stable. ■

Some events, of course, can be forecast with more or less accuracy in the way those in Example 4.17 could not. Here are some examples of forecastable recurrent events.

Example 4.18 Recurrent events showing a non-random pattern

(a) An owner of a battery-operated wristwatch has noticed for the past six years that batteries advertised to last one year have, to within a fortnight or so, lasted a year.

(b) It is in the interest of manufacturers of car tyres to build into their tyres a lifetime with as small a variability as possible (and not too high a mean). It is possible to forecast with some accuracy when a particular set of tyres will have to be replaced.

(c) Patients at a dental surgery arrive by appointment and are seen roughly at the time stated on their appointment card. (The only complication perturbing the appointments system is the occasional unforeseeable emergency arrival!) ■

4.4.1 Random events occurring in continuous time

The discrete-time Bernoulli process is based on the assumption that an event either occurs, or does not occur, at each of a clearly defined sequence of opportunities (called trials). Typical realizations of the Bernoulli process might look like something as shown in Table 4.6. In each case, only the first 25 trials in any realization have been monitored.

Table 4.6 The first 25 trials in three realizations of a Bernoulli process

(i)	0	1	0	0	0	0	0	1	1	0	1	1	0	0	0	1	0	0	0	1	0	1	1	0	0 ...
(ii)	0	0	0	1	0	0	1	1	1	0	0	0	0	0	1	0	1	1	0	0	1	1	0	0	0 ...
(iii)	0	1	0	1	1	0	1	0	0	0	0	1	0	0	0	0	1	1	0	1	0	0	0	0	0 ...

In each of these three examples, the parameter p (the underlying proportion of 1s) happened to be equal to 0.3. We saw in *Chapter 3*, Section 3.3 that the number of trials N up to and including the first success in a Bernoulli process has a geometric distribution with parameter p (mean $1/p$). Here, observed values of N are (i) 2, (ii) 4 and (iii) 2. In fact, by the 'memoryless' property of the Bernoulli process, the number of trials *from one success to the next* also has a geometric distribution with parameter p. In the realizations above there are actually several independent values of N observed. They are: (i) 2, 6, 1, 2, 1, 4, 4, 2, 1—and then a final observation on N where all we know is that it must be at least 3; (ii) 4, 3, 1, 1, 6, 2, 1, 3, 1; (iii) 2, 2, 1, 2, 5, 5, 1, 2.

Events also occur in a random haphazard kind of a way in continuous time —but then there is no notion of a 'trial' or 'opportunity' at which the event of interest might, or might not, occur. A probability model different from the Bernoulli process, one not involving the 'trial' construction, needs to be developed to represent this sort of idea. As well as some of those listed in Example 4.17, examples of random unforecastable events occurring in continuous time include floods, air crashes, earthquakes and other disasters: the point is that by their nature these incidents do not occur according to any sort of schedule. Here, for instance, are some data on serious earthquakes world-wide (see Table 4.7). The table gives the time in days between successive recordings. An earthquake is included if its magnitude was at least 7.5 on the Richter scale, or if over 1000 people were killed. Recording starts on 16 December 1902 (4500 killed in Turkestan) and ends on 4 March 1977 (2000 killed in Vrancea, Romania). There were 63 earthquakes recorded altogether—Table 4.7 shows the 62 'waiting times' between consecutive earthquakes. The numbers should be read across the rows.

Adapted from: The Open University (1981) S237 *The Earth: Structure, Composition and Evolution*, Milton Keynes, The Open University Press.

Table 4.7 Time intervals between major earthquakes (days)

840	157	145	44	33	121	150	280	434	736
584	887	263	1901	695	294	562	721	76	710
46	402	194	759	319	460	40	1336	335	1354
454	36	667	40	556	99	304	375	567	139
780	203	436	30	384	129	9	209	599	83
832	328	246	1617	638	937	735	38	365	92
82	220								

These data can be represented more comprehensibly in a histogram, as shown in Figure 4.3.

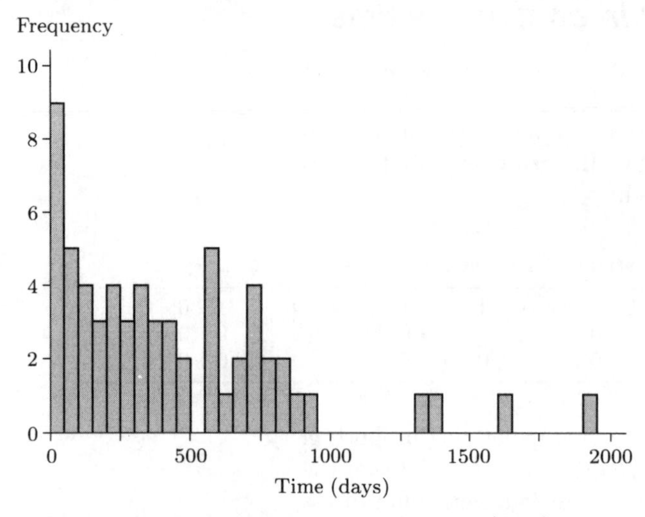

Figure 4.3 Time intervals between major earthquakes (days)

There are no instances of two (or more) earthquakes occurring on the same day, though such a coincidence is not impossible. It might be reasonable at least as a first approximation, therefore, to call any single day an opportunity or trial during which an earthquake might or might not happen. If so, then we can use properties of the geometric model to infer properties of earthquake times: the estimated mean time between earthquakes is

$$\frac{840 + 157 + \ldots + 220}{62} = \frac{27\,107}{62} = 437 \text{ days.}$$

The whole idea of using data for estimating parameters for statistical models is examined in *Chapters 6* and *7*. It is one of the essential applications of statistics.

The estimated probability of a major earthquake occurring somewhere in the world on any one day (tomorrow, say) is $62/27107 = 0.0023$, and the estimated probability (say) of a lull exceeding five years (allowing one leap year—1826 days altogether) is, using the c.d.f. of the geometric distribution,

$$P(N > 1826) = (1 - p)^{1826} = \left(1 - \frac{62}{27\,107}\right)^{1826} = 0.0153,$$

or about 1 in 65. Here, there actually was a lull of 1901 days between the fourteenth and fifteenth serious quakes recorded (one on 3 October 1915 in Nevada, California; and the other in Kansu Shansi in China on 16 December 1920): that is, 1 in 62! This probability is not very great: but then we are looking at the possibility of a lull exceeding 1826 days when the mean lull is only about 437 days: that is, a lull roughly four times longer than average.

Exercise 4.12

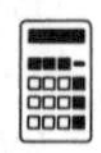

What would the geometric model give for the probability of a time lag between earthquakes exceeding four years? How many cases are there of such a lull being recorded?

What we require to model the waiting times between unforecastable events where there is no natural underlying notion of a trial, is a continuous-time analogue of the geometric distribution. The histogram in Figure 4.3 suggests that a suitable probability model will be one with a highly skewed density function, starting high at 0 and tailing off with longer waiting times (rather as the geometric mass function does). Many density functions could be contrived to possess that general shape: the triangular density is skewed and could provide a useful approximate model. But we also need a further property of the model—one that mimics the 'memoryless' property of the Bernoulli process. It turns out that only one family of density functions satisfies this requirement: it may be parameterized in terms of the average waiting time μ between events as follows.

A model for the waiting time T between consecutive events occurring haphazardly in time is given by the probability distribution with density function

$$f(t) = \frac{1}{\mu}e^{-t/\mu}, \quad t \geq 0, \tag{4.22}$$

where $\mu = E(T)$ is the mean waiting time between events.

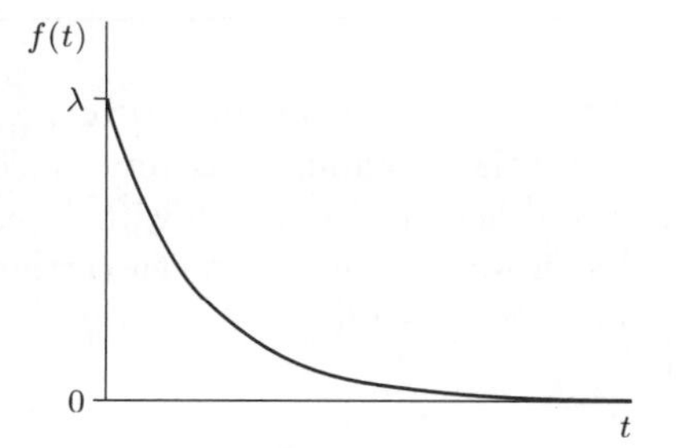

Figure 4.4 The probability density function $f(t) = \lambda e^{-\lambda t}$, $t \geq 0$

Now, if $E(T)$ is the average waiting time between occurrences of the recurrent event A, then the reciprocal of $E(T)$ is the *average rate* at which A occurs. Writing this rate as λ, the probability density function of T can be rewritten

$$f(t) = \lambda e^{-\lambda t}, \quad t \geq 0.$$

Thus we are led to the following definition for the probability distribution of the random variable T, the waiting time between recurrent random events occurring at average rate λ in continuous time.

The random variable T is said to follow an **exponential distribution with parameter $\boldsymbol{\lambda}$** if it has probability density function

$$f(t) = \lambda e^{-\lambda t}, \quad t \geq 0. \tag{4.23}$$

This is written $T \sim M(\lambda)$.

It would be convenient to have parameterized the exponential distribution in terms of its mean μ rather than in terms of the reciprocal of the mean, $\lambda = 1/\mu$. But this would have been to flout convention! The label $M(\cdot)$ comes from the queuing literature and follows the work of the great Russian mathematician Andrei Andreevich Markov (1856–1922). The study of Markov chains—linked sequences of random variables—is a large branch of probability theory. Markov's book, *The Calculus of Probabilities* (1913), is a seminal text. Most of it is rather theoretical, but the last chapters are devoted to the applications of probability theory to insurance problems.

It is important to note that by convention the exponential family is indexed by the parameter $\lambda = 1/\mu$, rather than by the mean μ. The probability density function of the random variable T is sketched in Figure 4.4. Notice its similarity to the shape of the probability function for the discrete geometric random variable N. Both probability distributions are highly skewed.

The cumulative distribution function of the random variable T is given by

$$F(t) = P(T \leq t) = \int_0^t \lambda e^{-\lambda w}\, dw = 1 - e^{-\lambda t}, \quad t \geq 0. \tag{4.24}$$

Result (4.24) is standard and well known. It follows from a straightforward exercise in integration, but you need not worry if this technique is unknown to you.

Just as in cases where the Bernoulli process is likely to be a useful model the indexing parameter p is usually deducible from the description (or estimable from data!), so the mean waiting time $1/\lambda$, or its reciprocal the 'rate' λ, indexes the continuous-time analogue, and may often be estimated from data.

Exercise 4.13

Assume that earthquakes occur world-wide at random but at an estimated average rate of once every 437 days. Assuming a continuous-time model for the phenomenon, find

(a) the probability that no earthquake is experienced world-wide during the first three years of the third millenium (that is, during the years 2000–2002 inclusive. The year 2000 will be a leap year);

(b) the median time between successive earthquakes (that is, the solution x of the equation $F(x) = \frac{1}{2}$);

(c) the proportion of waiting times longer than expected (that is, the proportion of waiting times that exceed 437 days).

You saw from your answer to Exercise 4.13(b) that the median of the exponential distribution is very much less than the mean. Figure 4.4 is reproduced here as Figure 4.5 with λ set equal to 1/437: it shows the median and mean waiting time between earthquakes, and the probability you calculated in Exercise 4.13(c).

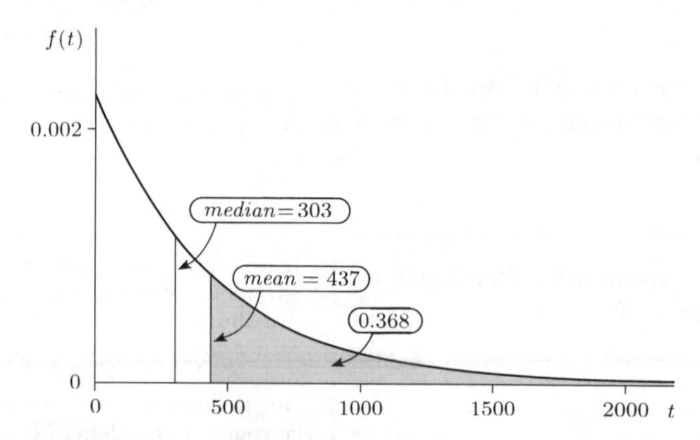

Figure 4.5 A probability model for the waiting time between earthquakes (days)

The consequences of the pronounced skewness of the exponential distribution are very apparent. Notice that Figure 4.5 is a representation of a probability *model* for earthquake waiting times (quite a good one); it is different from the histogram in Figure 4.3, which is a summary of waiting time *data*. When events occur haphazardly 'at random in continuous time' in such a way that they are completely unforecastable, then the model we have just described, a sort of limiting case of the Bernoulli process, is often used. This model is called the *Poisson process*. In *Chapter 12* this whole mathematical construction is investigated further. All that follows now is a stated result that will explain the name.

4.4.2 Counting the events

If events occur at random in continuous time with an average waiting time between consecutive events equal to $\mu = 1/\lambda$, then the average number of events in a time interval of duration t will be λt. (This makes sense. If the number λ is the average rate of occurrence of events, then the expected number in a time interval of duration t is λt.) However, the number of events to occur in such a fixed time interval is a random variable: for instance, for the earthquake data in Table 4.6, it turns out that during the decade from 1 January 1911 to 31 December 1920 there were five earthquakes world-wide; during the next decade there were eight earthquakes; and there were seven during the ten years after that. The numbers 5, 8 and 7 are independent observations on some random variable whose probability distribution is for the moment unknown—but it has mean (allowing two leap years in a decade)

$$\lambda t = \left(\frac{62}{27\,101} \text{ per day}\right) \times (3652 \text{ days}) = 8.35.$$

In fact, the distribution of the number of events in a time interval of duration t is Poisson(λt): roughly, the argument for this result is based on the construction of 'notional trials' that was used in Example 4.7.

The Poisson process

For recurrent events occurring at random at average rate λ in such a way that their occurrence may be modelled as a Poisson process, the number of events to occur during a time interval of duration t is a random variable X, where $X \sim \text{Poisson}(\lambda t)$; the waiting time T between consecutive events follows an exponential distribution $T \sim M(\lambda)$.

Exercise 4.14

Assuming that in a typical decade the expected number of earthquakes world-wide is 8.35, find the probability that there will be

(a) exactly two earthquakes;
(b) at least four earthquakes.

4.4.3 Moments and other properties

It has been stated that the mean of the exponential random variable $T \sim M(\lambda)$ is $1/\lambda$; confirmation of this, and calculation of the standard deviation of the exponential waiting time is a straightforward exercise in integration. Moments of the exponential distribution are given by

$$E(T) = \frac{1}{\lambda}, \qquad V(T) = \frac{1}{\lambda^2}, \qquad SD(T) = \frac{1}{\lambda}. \tag{4.25}$$

Exercise 4.15

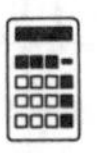

Express the median waiting time between consecutive events in a Poisson process as a fraction of the mean waiting time.

The general result that the median of the exponential distribution is less than 70% of the mean confirms the extreme skewness of the distribution, implied by the sketch of its density function in Figure 4.4.

4.4.4 Simulation of waiting times

It is possible to use a computer to generate a list of typical observations from the exponential distribution $M(\lambda)$. This is a particularly useful activity: suppose you were to generate, say, ten independent observations from $M(\frac{1}{2})$—that is to say, ten observations from an exponential distribution with mean 2. Here (in Table 4.8) is a set typical of what you might have obtained.

Table 4.8 Ten observations from $M(\frac{1}{2})$

1.24	0.07	0.25	4.06	0.43
2.24	0.49	1.37	1.75	3.79

This list of numbers could easily be regarded as merely 'a random sample' of size ten from this particular distribution. However, if you string together a sequence of exponential waiting times, then you obtain a single simulated realization of a Poisson process. For instance, suppose you were told that flood tides occur in a particular Scottish harbour port in a way that is impossible to forecast, but at an average rate of one flood tide every two years. Then a typical sequence of ten flood tides could be simulated by adding together in sequence the ten waiting times listed in Table 4.8. The times at which the simulated flood tides occur would then be as listed in Table 4.9 (times measured in years).

Table 4.9 Simulated times of ten flood tides (years)

Waiting time	Time of flood tide
1.24	1.24
0.07	1.31
0.25	1.56
4.06	5.62
0.43	6.05
2.24	8.29
0.49	8.78
1.37	10.15
1.75	11.90
3.79	15.69

So in this simulation it took over fifteen years for ten flood tides to have been recorded: a little less than the twenty years expected. (There were four very short intervals between floods: 0.07 years is less than a month.) The incidence of flooding can be represented against time as shown in Figure 4.6.

0 2 4 6 8 10 12 14 16

Ten flood tides, time in years

Figure 4.6 Simulated sequence of ten flood tides, times in years

The simulation could be made more elaborate if we knew the probability distribution of flood tide levels. As it happens we do not, but suppose that the simulation showed that the sixth and seventh flood incidents were particularly serious, the tenth less so. Then we might end up with something like the diagram in Figure 4.7.

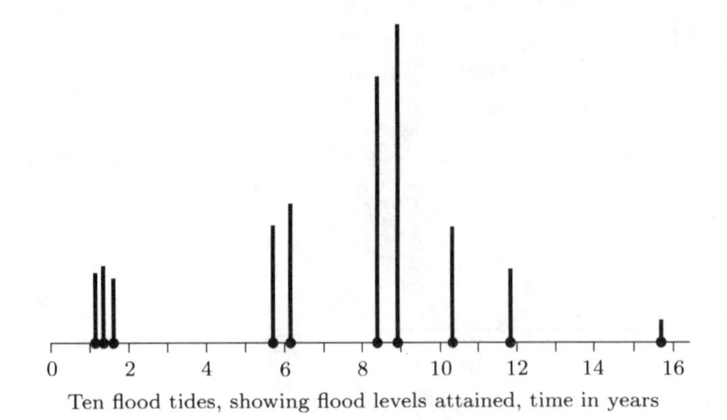

Figure 4.7 Flood tides, showing flood levels attained

The topic of simulation will be examined briefly in Section 4.5. Exercise 4.16 which follows is about the simulation of earthquakes.

Exercise 4.16

(a) Assuming the mean time between earthquakes to be 437 days as suggested by the data in Table 4.6, use your computer to simulate the times of earthquakes world-wide over a twenty-year period. (Ignore leap years: assume 365 days a year.) List the times of occurrence in a table, and on a diagram represent the incidence of earthquakes against time.

(b) How many earthquakes were there in your simulation? How many should you have expected? What is the median number of earthquakes to occur world-wide in a twenty-year period?

Here is a final exercise covering the main points of the section.

Exercise 4.17

Here, our aim is to develop an adequate model for the following data set. The data in Table 4.10 give the time intervals (in seconds) between successive pulses along a nerve fibre. They are extracted from a large data set in which there were 800 pulses recorded, so there were 799 waiting times between pulses. The data in Table 4.10 are the first 200 waiting times.

Cox, D.R. and Lewis, P.A.W. (1966) *The Statistical Analysis of Series of Events.* Chapman and Hall, London, p. 252. Data provided by Dr. P. Fatt and Professor B. Katz, F.R.S., University College London.

Table 4.10 Waiting times between pulses (seconds)

0.21	0.03	0.05	0.11	0.59	0.06	0.18	0.55	0.37	0.09
0.14	0.19	0.02	0.14	0.09	0.05	0.15	0.23	0.15	0.08
0.24	0.16	0.06	0.11	0.15	0.09	0.03	0.21	0.02	0.14
0.24	0.29	0.16	0.07	0.07	0.04	0.02	0.15	0.12	0.26
0.15	0.33	0.06	0.51	0.11	0.28	0.36	0.14	0.55	0.28
0.04	0.01	0.94	0.73	0.05	0.07	0.11	0.38	0.21	0.49
0.38	0.38	0.01	0.06	0.13	0.06	0.01	0.16	0.05	0.10
0.16	0.06	0.06	0.06	0.06	0.11	0.44	0.05	0.09	0.04
0.27	0.50	0.25	0.25	0.08	0.01	0.70	0.04	0.08	0.16
0.38	0.08	0.32	0.39	0.58	0.56	0.74	0.15	0.07	0.26
0.25	0.01	0.17	0.64	0.61	0.15	0.26	0.03	0.05	0.34
0.07	0.10	0.09	0.02	0.30	0.07	0.12	0.01	0.16	0.14
0.49	0.07	0.11	0.35	1.21	0.17	0.01	0.35	0.45	0.07
0.93	0.04	0.96	0.14	1.38	0.15	0.01	0.05	0.23	0.31
0.05	0.05	0.29	0.01	0.74	0.30	0.09	0.02	0.19	0.47
0.01	0.51	0.12	0.12	0.43	0.32	0.09	0.20	0.03	0.05
0.13	0.15	0.05	0.08	0.04	0.09	0.10	0.10	0.26	0.07
0.68	0.15	0.01	0.27	0.05	0.03	0.40	0.04	0.21	0.29
0.24	0.08	0.23	0.10	0.19	0.20	0.26	0.06	0.40	0.51
0.15	1.10	0.16	0.78	0.04	0.27	0.35	0.71	0.15	0.29

(a) Plot a histogram of these data and comment on the shape of your diagram.

(b) Compare

(i) the sample mean with the sample median;

(ii) the sample mean with the sample standard deviation.

(c) Find the lower and upper quartiles for the exponential distribution (expressed in terms of the mean) and compare these with the sample upper and lower quartiles for these data.

(d) Count the number of pulses to have occurred during the first quarter-minute of observation. (Assume the first pulse to have occurred at time zero when the clock was started, so that the first pulse *recorded* occurred at time 0.21, the second at time $0.21 + 0.03 = 0.24$, and so on.) How many occurred during the second quarter-minute?

Your answers to parts (a) to (c) may have enabled you to formulate a model for the incidence of pulses along a nerve fibre.

(e) Assuming your model to be correct, from what probability distribution are the counts you wrote down in part (d) observations?

Here is a final example, one where the model assumptions broke down.

Example 4.19 *Admissions to an intensive care unit*

Data were collected on the arrival times of patients at an intensive care unit—the aim was to identify any systematic variations in arrival rate, in particular, any that might be useful in planning future management of the unit. Table 4.11 gives some of these data. It might initially be supposed that admissions occur in the 'random haphazard' way suggested in the earthquake example. In fact, there are noticeable variations in the data that cannot be ascribed simply to chance variation on the exponential distribution. These are due to variations in the underlying rate of admission both with the time of day and with the day of the week. (The original data give the day and time of admission. Differences have been taken to give the inter-admission waiting times, to the nearest half-hour. The time of observation was from 4 February 1963 to 9 May 1963.)

Cox, D.R. and Snell, E.J. (1981) *Applied Statistics*. Chapman and Hall, London, p. 53. Data collected by Dr. A. Barr, Oxford Regional Hospital Board.

Table 4.11 Admissions to an intensive care unit (hours)

6.0	102.0	59.0	122.0	45.0	7.0	42.0	5.5	28.5	80.0
8.0	62.0	34.0	41.5	98.0	4.5	52.0	144.0	9.5	33.5
77.0	51.0	47.0	2.0	1.0	87.0	0.0	54.0	76.0	0.0
96.0	63.0	6.5	23.5	21.0	96.0	28.5	171.0	116.5	23.0

■

4.5 On simulation

The main activity of this section will be getting your computer to generate large samples of observations from different probability distributions, so that the sample may be regarded as *a genuine data set* drawn from a population where the specified probability distribution holds. Computers can simulate the drawing of a random sample and perform the subsequent statistical analysis of that sample very speedily. You have already seen something of this in *Chapter 2* and at the end of Section 4.2 of this chapter. For instance, one way of investigating the sampling properties of the binomial probability distribution $B(10, \frac{1}{2})$ is to toss ten coins (once) recording, say,

Heads Heads Tails Heads Tails Tails Tails Tails Heads Tails

(thus recording 4 heads altogether) and then repeat the whole experiment another few hundred times. This would be a very time-consuming activity. It would be easier and much more convenient to get the computer to toss the coin—or, rather, get it to generate several hundred observations from $B(10, \frac{1}{2})$

and then list the results—or, perhaps just as usefully, plot them as a bar chart. Here are the results from 1000 tosses of ten coins.

Table 4.12 Ten coins tossed 1000 times by computer: counting the Heads

Heads	0	1	2	3	4	5	6	7	8	9	10
Frequency	0	14	49	130	194	255	188	121	41	8	0

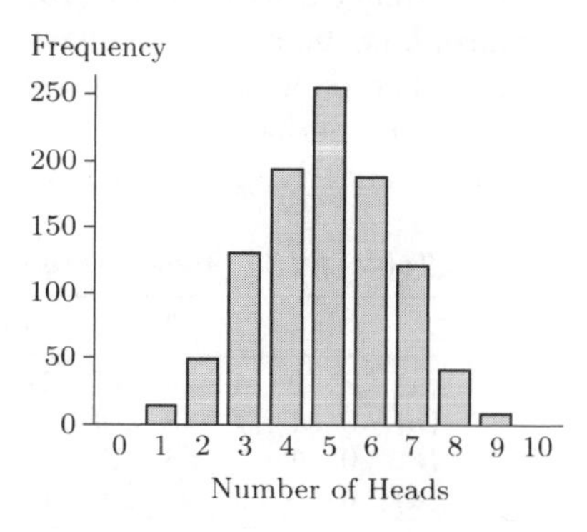

Figure 4.8 1000 observations from $B(10, \frac{1}{2})$

These observations may be plotted on a bar chart as shown in Figure 4.8. This sample of size 1000 seems to be fairly representative of the population from which it is drawn: the symmetric shape of the probability distribution $B(10, \frac{1}{2})$ is evident in the shape of the sample bar chart.

The exercise is called **simulation**, or **computer simulation**. Essentially you can make the computer simulate either realizations of a *random process* like a Bernoulli process (for example, a coin-tossing experiment, sexes of children in a family, sequences of births and deaths in an evolving population, passes and failures in a quality test) or a *random sample*, a list of independent observations on some stated random variable. In the exercises that follow, we shall concentrate mainly on the idea of a random sample. But computer simulation can be used to give an insight into the most complex of random systems, like arrivals and departures at an airport (in theory scheduled, but in practice susceptible to chance perturbations from an infinite array of accidental occurrences), the production line in a factory, or the progress of an epidemic in a school, a country or world-wide.

4.5.1 The uses of simulation

The idea of any simulation of a random process (or sample) that you generate yourself is that it should accurately mimic the probability structure of that process without involving you in what could otherwise be the expensive deployment of experimental equipment; or so that it can save you the time and trouble of observing and recording every detail in a genuine realization of such a process.

It is often the case with a random process that has a simple structure that the answer to every conceivable question about what might or might not happen (and with what probability) in a realization of that process can be deduced directly from theoretical paperwork (that is, from going through an exercise in algebra, or arithmetic). For instance, in a Poisson process model, you *know* the probability distribution of the waiting time between occurrences of the event A; you *know* the probability distribution of the number of occurrences of A in a time interval of given duration. There are, however, four things that a simulation exercise can help you with. These are listed in no particular order of importance.

(a) Simulation will give you some generalized feeling about what realizations of the random process (or samples from the probability distribution) might actually *look like*. This is useful not only so that you can recognize when other realizations or samples are of that pattern, but also so that you can more readily recognize when some other realization or sample does not fall into that pattern. In such circumstances you could therefore deduce that the underlying probability law is not the one you first thought it was. In Table 4.13 and summarized in Figure 4.9, there is listed a sequence of four different random samples obtained from the binomial probability distribution $B(10, \frac{1}{2})$; each sample is of size 100. Did you realize how 'bumpy' a summary bar chart (or histogram) could sometimes be? The point is that a sample of size 100 is really not very large: you need one perhaps ten times the size for some reasonable approximation to the underlying probability distribution to become evident.

Table 4.13 Ten coins tossed 100 times by computer: four samples

	0	1	2	3	4	5	6	7	8	9	10
(a)	0	0	4	14	17	27	23	9	5	1	0
(b)	0	1	4	8	19	27	23	14	4	0	0
(c)	0	1	7	8	17	26	19	14	6	1	1
(d)	0	0	6	20	16	18	20	11	7	2	0

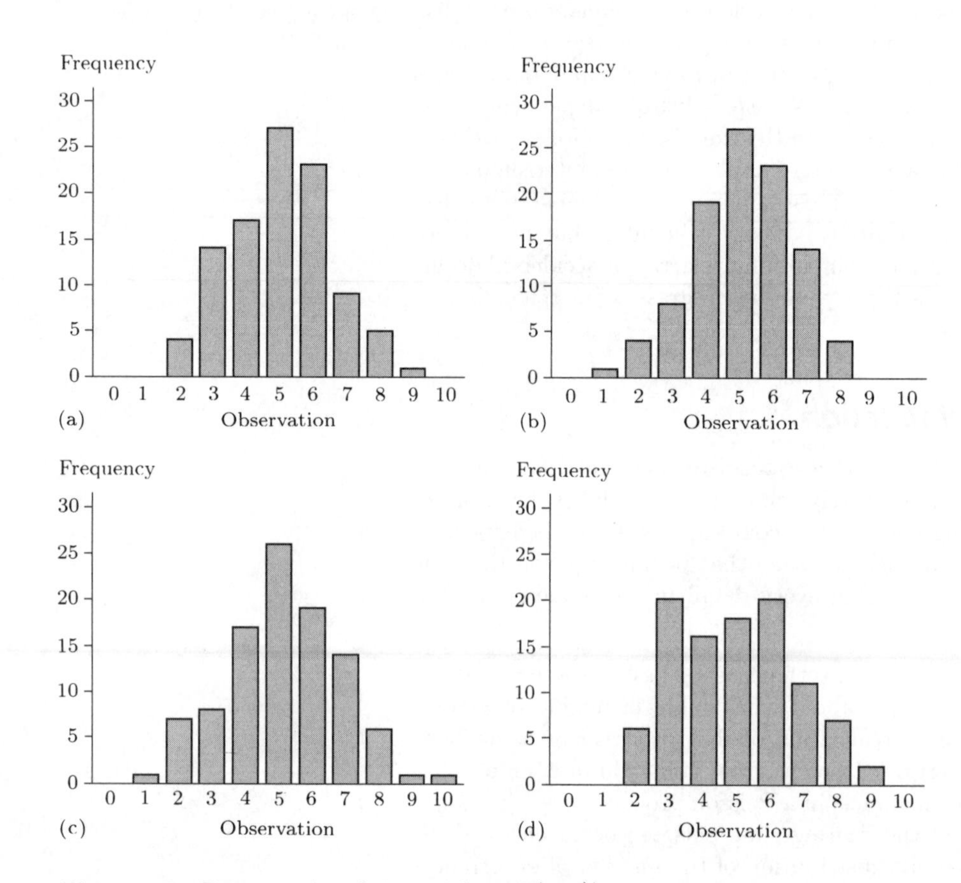

Figure 4.9 Four samples of size 100 from $B(10, \frac{1}{2})$

(b) A simulation can be used to check the conclusions of your own analysis. You might have identified some random variable of interest contained in a random process you are investigating, or perhaps some sampling properties. In deriving the probability distribution of that random variable, or perhaps in identifying some feature of a random sample, you might have made a mistake. Any error would be evident after one or perhaps several simulations of the process, and you can check your work. Alternatively, your conclusions will be confirmed.

(c) Sometimes it is very difficult to obtain the probability distribution of some random variable because the mathematics is too hard. Then you can use simulation to obtain an approximation to it. This will often involve up to several thousands of simulations in order to get a respectable estimate of the proportion of occasions on which a particular random variable equals or exceeds a particular value, but you will have a usable result. For instance, the 1000 observations in Table 4.12 result in a table of relative frequencies approximating the actual binomial probability distribution $B(10, \frac{1}{2})$. The estimated and exact probabilities are shown in Table 4.14—also shown are the rather less accurate estimated probabilities in Table 4.13(b). (The estimates obtained from, say, Table 4.13(d) would have been even worse!)

Table 4.14 Estimated and exact probabilities, $B(10, \frac{1}{2})$

	0	1	2	3	4	5	6	7	8	9	10
Table 4.13(b)	0.00	0.01	0.04	0.08	0.19	0.27	0.23	0.14	0.04	0.00	0.00
Table 4.12	0.000	0.014	0.049	0.130	0.194	0.255	0.188	0.121	0.041	0.008	0.000
Exact calculation	0.001	0.010	0.044	0.117	0.205	0.246	0.205	0.117	0.044	0.010	0.001

(d) Quite often it is only after noting its occurrence in several simulations that some previously unsuspected feature of the random sample, or the random process, will become evident. You can then return to pencil and paper and examine the feature. For instance, the way in which a disease spreads through small closed communities like schools, or through larger communities like towns, or countries or, for that matter, continents, is of great interest to the authorities concerned. Evidently, the element of chance plays a considerable role. However, the mathematics of even quite simple epidemic models which incorporate a chance element turns out to be very intractable. A common procedure then is to look only at the 'average' or 'expected' development of an epidemic. Then certain thresholds can be identified, either side of which the epidemic behaves quite differently. It either 'takes off', with the disease afflicting nearly everybody, or it fades away, affecting almost nobody. These thresholds depend on things like the virulence of the disease, the size of the community in which it first appears, and the contact opportunities between members of the community. The algebra of this 'expected' development can be done on paper. Random models have to be investigated on a computer: and when they are, behaviour of the epidemic either side of the threshold values turns out occasionally (by chance!) to be rather different, and to look rather more like what actually happens in the world.

By the end of the course, some of these uses of simulation will have become more apparent to you. In the exercises that follow, we will concentrate on some simple properties of a random sample from a probability distribution.

4.5.2 Generating random samples

There are no published lists of Poisson random numbers—you would need different lists for different values of the parameter μ. One way of simulating independent observations on a Poisson random variable (that is, of generating a Poisson random sample) is to invert the distribution function appropriately. A second way is to get the computer to generate a Poisson random sample for you. You did this in Exercise 4.7. Actually, the computer's simulation procedures are of course based on a mathematical analysis of the Poisson probability distribution. The computer is not a magic black box—but it does insulate you, the user, from any requirement to understand the mathematical analysis involved. Here is a further exercise in generating Poisson counts.

Exercise 4.18

The aim here is to compare a random sample of observations with the theoretical frequencies for the Poisson distribution.

(a) Generate 20 observations from the Poisson distribution Poisson(3.2) and then tally the data. Repeat the exercise for 50 observations and then 100 observations.

(b) Now generate 1000 observations from the Poisson distribution Poisson(3.2), obtain sample relative frequencies, and compare these with the probability mass function for a Poisson random variable with mean 3.2.

The next exercise demonstrates an example of using a computer in a case where the mathematics would be very difficult indeed. We have seen that in the human population the proportion of males with a particular form of colour-deficient sight is approximately 0.06, while the corresponding proportion for females is approximately 0.004. In a population of n_1 males and n_2 females the number of males with colour-deficient sight is a random variable X_1 following a binomial distribution ($X_1 \sim B(n_1, 0.06)$) and the number of affected females is X_2, where $X_2 \sim B(n_2, 0.004)$. Their sum $X_1 + X_2$ has a complicated distribution that is difficult to write down (and you are not expected to be able to do so). However, using Poisson's approximation for rare events, we can say

$$X_1 \approx \text{Poisson}(0.06n_1), \qquad X_2 \approx \text{Poisson}(0.004n_2),$$

and so the sum $X_1 + X_2$ is approximately Poisson with mean $0.06n_1 + 0.004n_2$. The true binomial model is made more complicated still if the number of males and females is permitted to vary.

Exercise 4.19

(a) Suppose that in an assembly of 100 persons the number of males X is a random variable $B(100, 0.5)$. Simulate the number of males in the assembly and hence deduce the number of females.

(b) Say there are x males. Using an exact binomial model (rather than the Poisson approximation) with $p = 0.06$ simulate the number of colour-deficient males (y_1) and similarly the number of colour-deficient females (y_2) present.

(c) Their sum $w = y_1 + y_2$ is an observation on a random variable W, the total number of colour-deficient people in an assembly of 100. Find w in this case.

The distribution of W is unknown, a rather complex conjunction of binomial variates.

(d) On intuitive grounds alone (that is, without stopping to think too hard!) can you say anything about the expected value of W?

(e) Obtain 1000 independent observations $w_1, w_2, \ldots, w_{1000}$ on the random variable W and store them in a data vector. Calculate the sample mean and variance of this random sample.

Exercise 4.20

The times of occurrence of random unforecastable events such as car accidents or floods may be modelled by assuming that the waiting times between consecutive occurrences come from an exponential distribution with some given mean: that is, that such events occur as a Poisson process. Adding successive waiting times gives the times at which such accidents might typically occur.

(a) Suppose that motor accident claims of a particular kind arrive at an underwriter's office in a way which is not forecastable, but at an average rate of twelve claims a week. (Assume for the sake of this exercise that the office is open for business 24 hours a day, seven days a week.) Calculate the mean time (in hours) between the arrival of successive claims. Simulate the times of arrival of the next 20 claims to arrive after midnight one Sunday night.

(b) Simulate ten weeks of claims. How many claims arrived in the first week? The second week? ... The tenth week? These ten counts are observations on what random variable?

Summary

1. The discrete random variable X is said to follow a Poisson distribution with parameter μ if

$$p_X(x) = \frac{e^{-\mu}\mu^x}{x!}, \quad x = 0, 1, 2, \ldots,$$

and this is written $X \sim \text{Poisson}(\mu)$. The moments of X are $E(X) = \mu$ and $V(X) = \mu$.

2. The distribution of X when $X \sim B(n,p)$ can be well approximated by the Poisson distribution with parameter $\mu = np$, when the parameter p is small (say, less than about 0.05). This may be written

$$B(n,p) \approx \text{Poisson}(np) \qquad \text{or} \qquad B\left(n, \frac{\mu}{n}\right) \approx \text{Poisson}(\mu).$$

Notice that the means of both probability distributions are matched.

3. If X is a random variable with mean μ and variance σ^2 then for constants a and b

$$E(aX+b) = a\mu + b,$$
$$V(aX+b) = a^2\sigma^2,$$

and the standard deviation of $(aX+b)$ is given by

$$SD(aX+b) = |a|\sigma.$$

4. If X_i, $i = 1, 2, \ldots, n$, are random variables with mean μ_i and sum $S = X_1 + \ldots + X_n$, then

$$E(S) = \mu_1 + \mu_2 + \ldots + \mu_n = \sum \mu_i.$$

If X_i, $i = 1, 2, \ldots, n$, are *independent* random variables with variance σ_i^2 and sum $S = X_1 + X_2 + \ldots + X_n$, then

$$V(S) = \sigma_1^2 + \sigma_2^2 + \ldots + \sigma_n^2 = \sum \sigma_i^2.$$

5. In particular, if a random sample of size n is taken from a population with mean μ and variance σ^2, and if the sample mean is written $\overline{X}$, then

$$E(\overline{X}) = \mu, \qquad V(\overline{X}) = \frac{\sigma^2}{n}.$$

6. The sum S of n independent Poisson variates respectively with mean μ_i has a Poisson distribution with mean $\sum \mu_i$: i.e. $S \sim \text{Poisson}(\sum \mu_i)$. The sum S of n independent normal variates respectively with mean μ_i and variance σ_i^2 has a normal distribution with mean $\sum \mu_i$ and variance $\sum \sigma_i^2$: i.e. $S \sim N(\sum \mu_i, \sum \sigma_i^2)$.

7. Recurrent events occurring at a constant average rate λ but otherwise 'at random' in continuous time may be modelled as occurring according to a Poisson process. The waiting time T between consecutive events follows an exponential distribution with parameter λ: this is written $T \sim M(\lambda)$. The random variable T has probability density function

$$f(t) = \lambda e^{-\lambda t}, \quad t \geq 0,$$

and cumulative distribution function

$$F(t) = 1 - e^{-\lambda t}, \quad t \geq 0.$$

The mean of T is $E(T) = 1/\lambda$ and the variance of T is $V(T) = 1/\lambda^2$. (It follows from this that for an exponentially distributed random variable T, $SD(T) = E(T)$.) The distribution of T is highly skewed—for instance, the median of T is only 0.7 times its mean.

8. Observations on $M(\lambda)$ may be simulated by reference to 'exponential random numbers'. By successively adding together exponential waiting times, the times of occurrence of a recurrent event A in a Poisson process can be simulated.

9. In a Poisson process, the number of times the recurrent event A occurs during a time interval of duration t has a Poisson distribution with mean λt, where λ is the average rate of occurrence of A.

Chapter 5
The normal distribution

This chapter deals in detail with one of the most versatile models for variation, the normal distribution or 'bell-shaped' curve. You will learn how to use printed tables to calculate normal probabilities. The normal curve also provides a useful approximation to other probability distributions: this is one of the consequences of the central limit theorem.

In *Chapter 2*, Section 2.4 you were introduced to an important continuous distribution called the normal distribution. It was noted that many real data sets can reasonably be treated as though they were a random sample from the normal distribution and it was remarked that the normal distribution turns out to play a central role in statistical theory as well as in practice. This entire chapter is devoted to the study of the normal distribution.

The chapter begins with a review of all that has been said so far about the normal distribution. The main point to bear in mind is that in many cases a probability model for random variation follows necessarily as a mathematical consequence of certain assumptions: for instance, many random processes can be modelled as sets or sequences of Bernoulli trials, the distribution theory following from the twin assumptions that the trials are independent and that the probability of success from trial to trial remains constant. Quite often, however, data arise from a situation for which no model has been proposed: nevertheless, even when the data sets arise from entirely different sampling contexts, they often seem to acquire a characteristic peaked and symmetric shape that is essentially the same. This shape may often be adequately represented through a normal model. The review is followed by an account of the genesis of the normal distribution.

In Section 5.2, you will discover how to calculate normal probabilities. As for any other continuous probability distribution, probabilities are found by calculating areas under the curve of the probability density function. But for the normal distribution, this is not quite straightforward, because applying the technique of integration does not in this case lead to a formula that is easy to write down. So, in practice, probabilities are found by referring to printed tables, or by using a computer.

The remaining sections of the chapter deal with one of the fundamental theorems in statistics and with some of the consequences of it. It is called the *central limit theorem.* This is a theorem due to Pierre Simon Laplace (1749–1827) that was read before the Academy of Sciences in Paris on 9 April 1810. The theorem is a major mathematical statement: however, we shall be concerned not with the details of its proof, but with its application to statistical problems.

5.1 Some history

5.1.1 Review

The review begins with a set of data collected a long time ago. During the mapping of the state of Massachusetts in America, one hundred readings were taken on the error involved when measuring angles. The error was measured in minutes (a minute is 1/60 of a degree). The data are shown in Table 5.1.

Table 5.1 Errors in angular measurements

Error (in minutes)	*Frequency*
Between +6 and +5	1
Between +5 and +4	2
Between +4 and +3	2
Between +3 and +2	3
Between +2 and +1	13
Between +1 and 0	26
Between 0 and −1	26
Between −1 and −2	17
Between −2 and −3	8
Between −3 and −4	2

United States Coast Survey Report (1854). The error was calculated by subtracting each measurement from 'the most probable' value.

A histogram of this sample is given in Figure 5.1. This graphical representation shows clearly the main characteristics of the data: the histogram is *unimodal* (it possesses just one mode) and it is roughly *symmetric* about that mode.

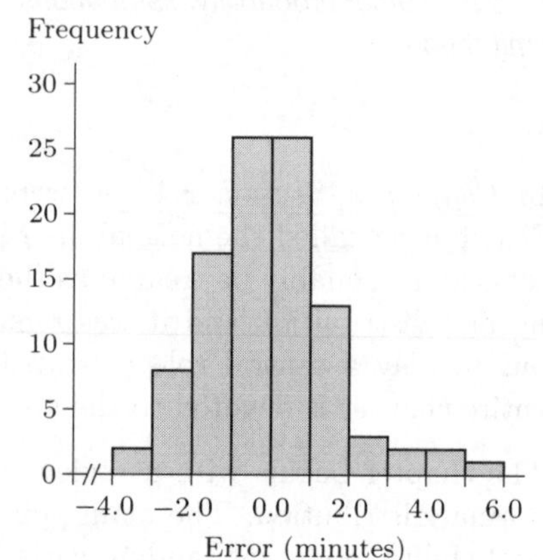

Figure 5.1 Errors in angular measurements (minutes)

Another histogram, which corresponds to a different data set, is shown in Figure 5.2. You have seen these data before.

This is a graphical representation of the sample of Scottish soldiers' chest measurements that you met in *Chapter 2*, Section 2.4. This histogram is also unimodal and roughly symmetric. The common characteristics of the shape of both the histograms in Figures 5.1 and 5.2 are shared with the normal distribution whose p.d.f. is illustrated in Figure 5.3.

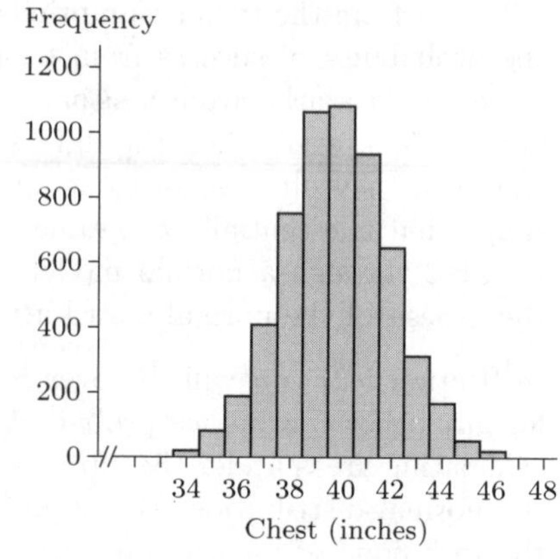

Figure 5.2 Chest measurements of Scottish soldiers (inches)

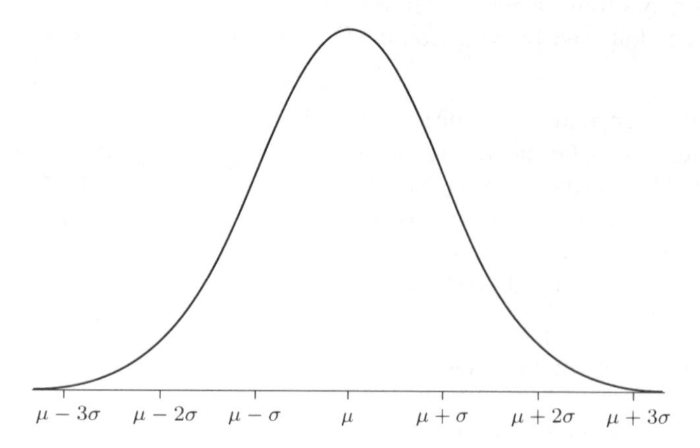

Figure 5.3 The normal p.d.f.

For clarity, the vertical axis has been omitted in this graph of the normal density function.

What is it about Figures 5.1, 5.2 and 5.3 that makes them appear similar? Well, each diagram starts at a low level on the left-hand side, rises steadily until reaching a maximum in the centre and then decreases, at the same rate that it increased, to a low value towards the right-hand side. The diagrams are unimodal and symmetric about their modes (although this symmetry is only approximate for the two data sets). A single descriptive word often used to describe the shape of the normal p.d.f., and likewise histograms of data sets that might be adequately modelled by the normal distribution, is 'bell-shaped'.

Note that there is more than one normal distribution. No single distribution could possibly describe both the data of Figure 5.1, which have their mode around zero and which vary from about -4 minutes of arc to over 5 minutes, and those of Figure 5.2, whose mode is at about 40 inches and which range from approximately 33 inches to 48 inches. In the real world there are many instances of random variation following this kind of pattern: the mode and the range of observed values will alter from random variable to random variable, but the characteristic bell shape of the data will be apparent.

The four probability density functions shown in Figure 5.4 all correspond to different normal distributions.

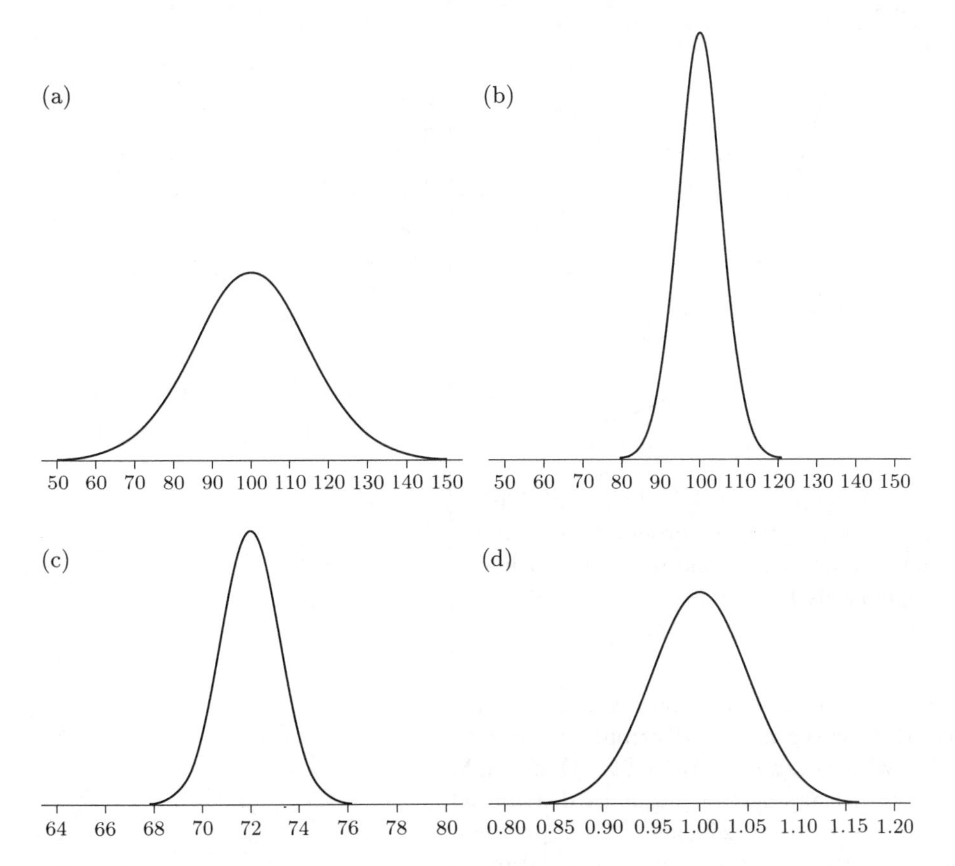

Figure 5.4 Four normal densities

What has been described is another *family* of probability models, just like the binomial family (with two parameters, n and p) and the Poisson family (with one parameter, the mean μ). The normal family has two parameters, one specifying location (the centre of the distribution) and one describing the degree of dispersion. In *Chapter 2* the location parameter was denoted by μ

and the dispersion parameter was denoted by σ; in fact, the parameter μ is the *mean* of the normal distribution and σ is its *standard deviation.*

This information may be summarized as follows. The probability density function for the normal family of random variables is also given.

The normal probability density function

If the continuous random variable X is normally distributed with mean μ and standard deviation σ (variance σ^2) then this may be written

$$X \sim N(\mu, \sigma^2);$$

the probability density function of X is given by

$$f(x) = \frac{1}{\sigma\sqrt{2\pi}} \exp\left[-\frac{1}{2}\left(\frac{x-\mu}{\sigma}\right)^2\right], \quad -\infty < x < \infty. \tag{5.1}$$

As remarked already, you do not need to remember this formula in order to calculate normal probabilities.

A sketch of the p.d.f. of X is as follows.

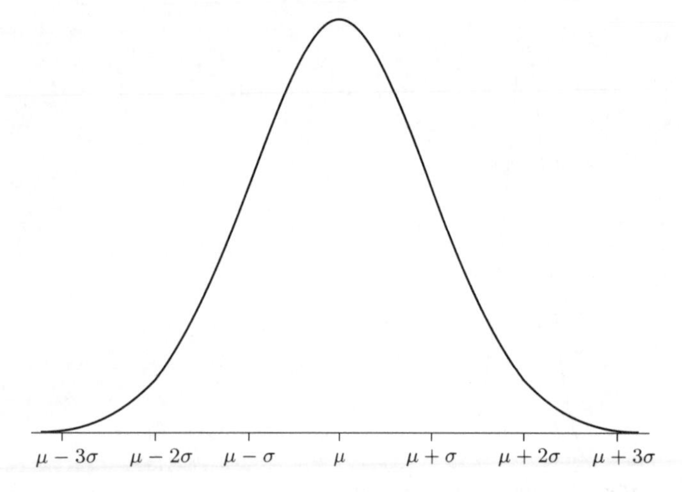

The shape of the density function of X is often called 'bell-shaped'. The p.d.f. of X is positive for all values of x; however, observations more than about three standard deviations away from the mean are rather unlikely. The total area under the curve is 1.

There are very few random variables for which possible observations include all negative and positive numbers. But for the normal distribution, extreme values may be regarded as occurring with negligible probability. One should not say 'the variation in Scottish chest measurements is normally distributed with mean 40 inches and standard deviation about 2 inches' (the implication being that negative observations are possible); rather, say 'the variation in Scottish chest measurements may be adequately modelled by a normal distribution with mean 40 inches and standard deviation 2 inches'.

In the rest of this chapter we shall see many more applications in the real world where different members of the normal family provide good models of variation. But first, we shall explore some of the history of the development of the normal distribution.

Exercise 5.1

Without attempting geometrical calculations, suggest values for the parameters μ and σ for each of the normal probability densities that are shown in Figure 5.4.

5.1.2 An early history

Although the terminology was not standardized until after 1900, the normal distribution itself was certainly known before then (under a variety of different names). The following is a brief account of the history of the normal distribution before the twentieth century.

Credit for the very first appearance of the normal p.d.f. goes to Abraham de Moivre (1667–1754), a Protestant Frenchman who emigrated to London in 1688 to avoid religious persecution and lived there for the rest of his life, becoming an eminent mathematician. Prompted by a desire to compute the probabilities of winning in various games of chance, de Moivre obtained what is now recognized as the normal p.d.f., an approximation to a binomial probability function (these were early days in the history of the binomial distribution). The pamphlet that contains this work was published in 1733. In those days, the binomial distribution was known as a discrete probability distribution in the way we think of discrete distributions today, but it is not generally claimed that de Moivre thought of his normal approximation as defining a continuous probability distribution, although he did note that it defined 'a curve'.

Then, around the end of the first decade of the nineteenth century, two famous figures in the history of science published derivations of the normal distribution. The first, in 1809, was the German Carl Friedrich Gauss (1777–1855); the second, in 1810, was the Frenchman Pierre Simon Laplace (1749–1827). Gauss was a famous astronomer and mathematician. The range of his influence, particularly in mathematical physics, has been enormous: he made strides in celestial mechanics, geometry and geodesy, number theory, optics, electromagnetism, real and complex analysis, theoretical physics and astronomy as well as in statistics. Motivated by problems of measurement in astronomy, Gauss had for a long time recognized the usefulness of the 'principle of least squares', an idea still very frequently used and which you will meet in *Chapter 10*. Allied to this, Gauss had great faith in the use of the mean as the fundamental summary measure of a collection of numbers. Moreover, he wanted to assert that the most probable value of an unknown quantity is the mean of its observed values (that is, in current terminology, that the mean equals the mode). Gauss then, quite rightly, obtained the normal distribution as a probability distribution that would yield these desirable properties: the normal distribution is relevant to the least squares method of estimation and its mode and its mean are one and the same. Having said that, Gauss's argument, or his claims for the consequences of his argument, now look distinctly shaky. He took use of the mean as axiomatic, arguing for its appropriateness in all circumstances, saw that the normal distribution gave the answer he wanted and consequently inferred that the normal distribution should also be the fundamental probability model for variation.

(a)

(b)

Figure 5.5 (a) Gauss and (b) Laplace

The Marquis de Laplace, as he eventually became, lived one of the most influential and successful careers in science. He made major contributions to mathematics and theoretical astronomy as well as to probability and statistics. Laplace must also have been an astute political mover, maintaining a high profile in scientific matters throughout turbulent times in France; he was even Napoleon Bonaparte's Minister of the Interior, if only for six weeks! Laplace's major contribution to the history of the normal distribution was a first version of the *central limit theorem*, a very important idea that you will learn about in Section 5.3. (Laplace's work is actually a major generalization of de Moivre's.) It is the central limit theorem that is largely responsible for the widespread use of the normal distribution in statistics. Laplace, working without knowledge of Gauss's interest in the same subject, presented his theorem early in 1810 as an elegant result in mathematical analysis, but with no hint of the normal curve as a p.d.f. and therefore as a model for random variation. Soon after, Laplace encountered Gauss's work and the enormity of his own achievement hit him. Laplace brought out a sequel to his mathematical memoir in which he showed how the central limit theorem gave a rationale for the choice of the normal curve as a probability distribution, and consequently how the entire development of the principle of least squares fell into place, as Gauss had shown.

This synthesis between the work of Gauss and Laplace provided the basis for all the further interest in and development of statistical methods based on the normal distribution over the ensuing years. Two contemporary derivations of the normal distribution by an Irish-American, Robert Adrain (1775–1843), working in terms of measurement errors, remained in obscurity. It is interesting to note that of all these names in the early history of the normal distribution, it is that of Gauss that is still often appended to the distribution today when, as is often done, the normal distribution is referred to as the *Gaussian distribution.*

The motivating problems behind all this and other early work in mathematical probability were summarized recently by S.M. Stigler thus: 'The problems considered were in a loose sense motivated by other problems, problems in the social sciences, annuities, insurance, meteorology, and medicine; but the paradigm for the mathematical development of the field was the analysis of games of chance'. However, 'Why men of broad vision and wide interests chose such a narrow focus as the dicing table and why the concepts that were developed there were applied to astronomy before they were returned to the fields that originally motivated them, are both interesting questions ...'. Unfortunately, it would be getting too far away from our main focus to discuss them further here.

Stigler, S.M. (1986) *The History of Statistics—The Measurement of Uncertainty before 1900.* The Belknap Press of Harvard University Press.

Such was the progress of the normal distribution in the mid-nineteenth century. The normal distribution was not merely accepted, it was widely advocated as *the* one and only 'law of error'; as, essentially, the only continuous probability distribution that occurred in the world! Much effort, from many people, went into obtaining 'proofs' of the normal law. The idea was to construct a set of assumptions and then to prove that the only continuous distribution satisfying these assumptions was the normal distribution. While some, no doubt, were simply wrong, many of these mathematical derivations were perfectly good characterizations of the normal distribution. That is, the normal distribution followed uniquely from the assumptions. The difficulty lay in the claims for the assumptions themselves. 'Proofs' and arguments about proofs, or at least the assumptions on which they were based, abounded, but it is now known that, as the normal distribution is not universally applicable, all this effort was destined to prove fruitless.

This acceptance of the normal distribution is especially remarkable in light of the fact that other continuous distributions were known at the time. A good example is due to Siméon Denis Poisson (1781–1840) who, as early as 1824, researched the continuous distribution with p.d.f.

Poisson's work on the binomial distribution and the eponymous approximating distribution was described in *Chapter 4*.

$$f(x) = \frac{1}{\pi(1+x^2)}, \quad -\infty < x < \infty,$$

which has very different properties from the normal distribution. An amusing aside is that this distribution now bears the name of Augustin Louis Cauchy (1789–1857) who worked on it twenty years or so later than Poisson did while, on the other hand, Poisson's contribution to the distribution that does bear his name is rather more tenuous compared with those of other researchers (including de Moivre) of earlier times.

What of the role of data in all this? For the most part, arguments were solely mathematical or philosophical, idealized discussions concerning the state of nature. On occasions when data sets were produced, they were ones that tended to support the case for the normal model. Two such samples were illustrated at the beginning of this section. The data on chest measurements of Scottish soldiers were taken from the *Edinburgh Medical and Surgical Journal* of 1817. They are of particular interest because they (or a version of them) were analysed by the Belgian astronomer, meteorologist, sociologist and statistician, Lambert Adolphe Jacques Quetelet (1796–1874) in 1846. Quetelet was a particularly firm believer in, and advocate of, the universal applicability of the normal distribution, and such data sets that do take an approximately normal shape did nothing to challenge that view. Quetelet was also a major figure in first applying theoretical developments to data in the social sciences. The angular data in Table 5.1 are quoted in an 1884 textbook entitled 'A Text-Book on the Method of Least Squares' by Mansfield Merriman, an American author. Again, the book is firmly rooted in the universal appropriateness of the normal distribution.

In a paper written in 1873, the American C.S. Peirce presented analyses of 24 separate tables each containing some 500 experimental observations. Peirce drew smooth densities which, in rather arguable ways, were derived from these data and from which he seemed to infer that his results confirmed (yet again) the practical validity of the normal law. An extensive reanalysis of Peirce's data in 1929 (by E.B. Wilson and M.M. Hilferty) found every one of these sets of data to be incompatible with the normal model in one way or another! These contradictory opinions based on the same observations are presented here more as an interesting anecdote rather than because they actually had any great influence on the history of the normal distribution, but they do nicely reflect the way thinking changed in the late nineteenth century. Peirce's (and Merriman's) contributions were amongst the last from the school of thought that the normal model was the only model necessary to express random variation. By about 1900, so much evidence of non-normal variation had accumulated that the need for alternatives to complement the normal distribution was well appreciated (and by 1929, there would not have been any great consternation at Wilson's and Hilferty's findings). Prime movers in the change of emphasis away from normal models for continuous data were a number of Englishmen including Sir Francis Galton (1822–1911), Francis Ysidro Edgeworth (1845–1926) and Karl Pearson (1857–1936).

But to continue this history of the normal distribution through the times of these important figures and beyond would be to become embroiled in the whole fascinating history of the subject of statistics as it is understood today, so we shall cease our exploration at this point.

There is, however, one interesting gadget to do with the normal distribution developed during the late nineteenth century. It was called the **quincunx**, and was an invention of Francis Galton in 1873 or thereabouts. Figure 5.6 shows a contemporary sketch of Galton's original quincunx; Figure 5.7 is a schematic diagram of the quincunx which more clearly aids the description of its operation. The mathematical sections of good modern science museums often have a working replica of this device, which forms a fascinating exhibit. What does the quincunx do and how does it work? The idea is to obtain in dynamic fashion a physical representation of a binomial distribution. The word 'quincunx' actually means an arrangement of five objects in a square or rectangle with one at each corner and one in the middle; the spots on the '5' face of a die form a good example. Galton's quincunx was made up of lots of these quincunxes. It consists of a glass-enclosed board with several rows of equally spaced pins. Each row of pins is arranged so that each pin in one row is directly beneath the midpoint of the gap between two adjacent pins in the row above; thus each pin is the centre of a quincunx. Metal shot is poured through a funnel directed at the pin in the top row. Each ball of shot can fall left or right of that pin with probability $\frac{1}{2}$.

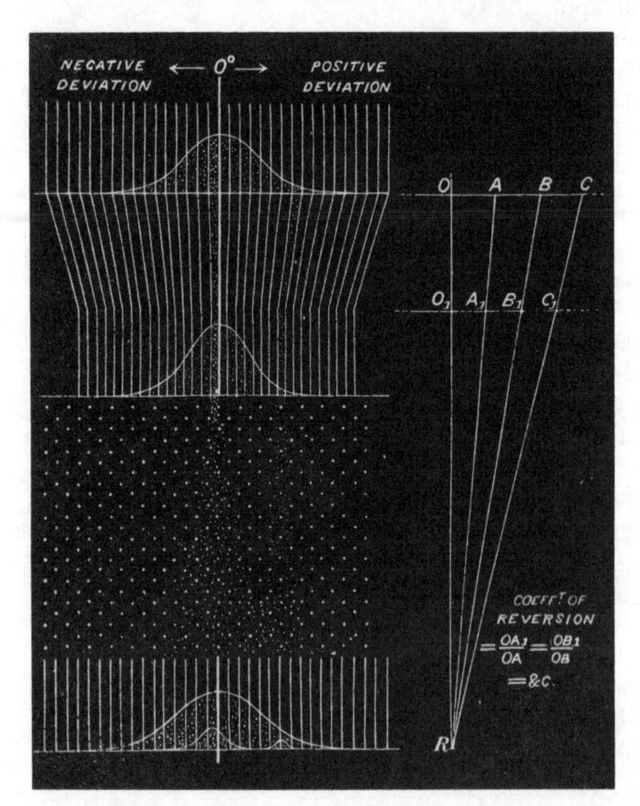

Figure 5.6 Galton's quincunx

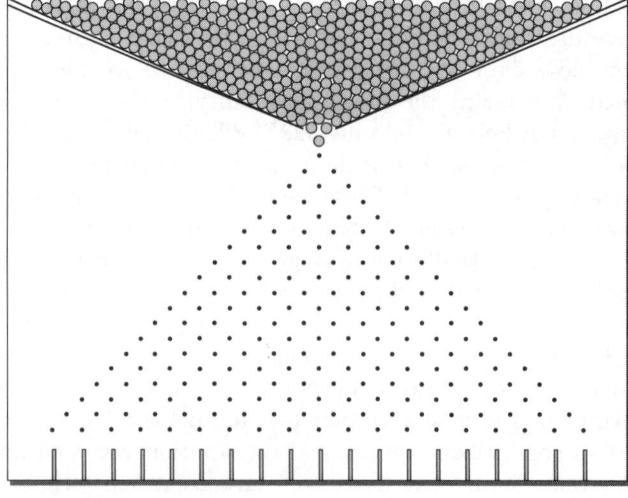

Figure 5.7 Diagram of the quincunx

The same holds for all successive lower pins that the shot hits. Finally, at the bottom, there is a set of vertical columns into which the shot falls, and a kind of histogram or bar chart is formed.

The picture so obtained is roughly that of a unimodal symmetric distribution. In fact, the theoretical distribution corresponding to the histogram formed by the shot is the binomial distribution with parameters $p = \frac{1}{2}$ and n equal to the number of rows of pins, which is 19 in Galton's original device. However, a histogram from the binomial distribution $B(19, \frac{1}{2})$ looks very much the same as a histogram from a normal distribution, so the quincunx also serves as a method of demonstrating normal data. More precisely, the relationship between the binomial distribution $B(19, \frac{1}{2})$ and the normal distribution is a consequence of the central limit theorem. Therefore, Laplace would have understood the reason for us to be equally happy with the quincunx as a device for illustrating the binomial distribution or as a device for illustrating the normal distribution; by the end of this chapter, you will understand why.

5.2 The standard normal distribution

In each of the following examples, a normal distribution has been proposed as an adequate model for the variation observed in the measured attribute.

Example 5.1 *Chest measurements*

After extensive sampling, it was decided to adopt a normal model for the chest measurement in a large population of adult males. Measured in inches, the model parameters were (for the mean) $\mu = 40$ and (for the standard deviation) $\sigma = 2$.

A sketch of this normal density is shown in Figure 5.8. The area under the curve, shown shaded in the diagram, gives (according to the model) the proportion of adult males in the population whose chest measurements are 43 inches or more.

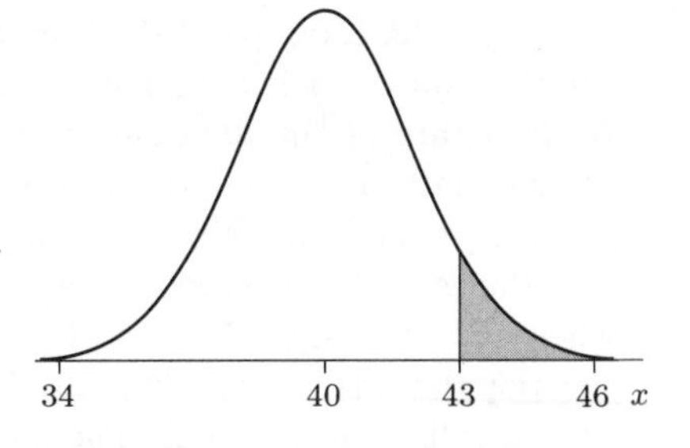

Figure 5.8 A sketch of the normal density $f(x)$, where $X \sim N(40, 4)$

Again, in this diagram the vertical axis has been omitted.

The chest measurement of 43 inches is greater than the average measurement within the population, but it is not very extreme, coming well within 'plus or minus 3 standard deviations'.

The shaded area is given by the integral

$$P(X \geq 43) = \int_{43}^{\infty} \frac{1}{2\sqrt{2\pi}} \exp\left[-\frac{1}{2}\left(\frac{x-40}{2}\right)^2\right] dx$$

(writing $X \sim N(\mu, \sigma^2)$ with $\mu = 40$ and $\sigma = 2$, and using (5.1)). But it is much easier to think of the problem in terms of 'standard deviations away from the mean'. The number 43 is *one and a half* standard deviations above the mean measurement, 40. Our problem is to establish what proportion of the population would be at least as extreme as this. ■

Example 5.2 IQ measurements

There are many different ways of assessing an individual's 'intelligence' (and no single view on exactly what it is that is being assessed, or how best to make the assessment). One test is designed so that in the general population the variability in the scores attained should be normally distributed with mean 100 and standard deviation 15. Denoting this score by the random variable W, then the statistical model is $W \sim N(100, 225)$.

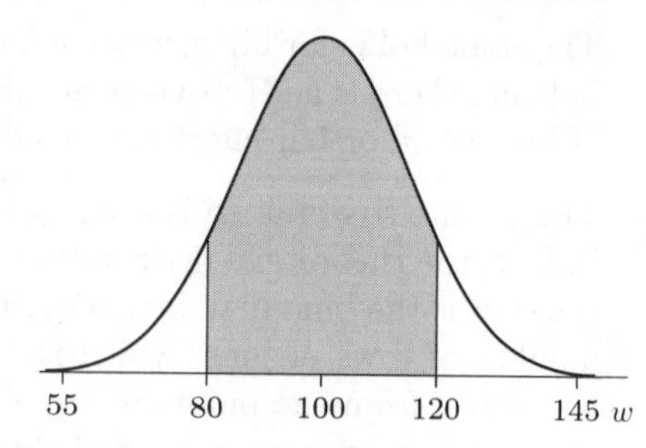

Figure 5.9 A sketch of the normal density $f(w)$, where $W \sim N(100, 225)$

A sketch of the p.d.f. of this normal distribution is given in Figure 5.9. The shaded area in the diagram gives the proportion of individuals who (according to the model) would score between 80 and 120 on the test. The area may be expressed formally as an integral

$$P(80 \leq W \leq 120) = \int_{80}^{120} \frac{1}{15\sqrt{2\pi}} \exp\left[-\frac{1}{2}\left(\frac{w-100}{15}\right)^2\right] dw,$$

but again it is easier to think in terms of a standard measure: how far away from the mean are these two scores? At 20 below the mean, the score of 80 is $\frac{20}{15} = \frac{4}{3} = 1.33$ standard deviations below the mean, and the score of 120 is 1.33 standard deviations above the mean. Our problem reduces to this: what proportion of the population would score within 1.33 standard deviations of the mean (either side)? ■

Example 5.3 Osteoporosis

In *Chapter 2*, Example 2.17 observations were presented on the height of 351 elderly women, taken as part of a study of the bone disease osteoporosis. A histogram of the data suggests that a normal distribution might provide an adequate model for the variation in height of elderly women within the general population. Suppose that the parameters of the proposed model are $\mu = 160$, $\sigma = 6$ (measured in cm; the model may be written $H \sim N(160, 36)$ where H represents the variation in height, in cm, of elderly women within the population). According to this model, the proportion of women over 180 cm tall is rather small. The number 180 is $(180 - 160)/6 = 3.33$ standard deviations above the mean: our problem is to calculate the small area shown in Figure 5.10 or, equivalently, to calculate the integral

$$P(H > 180) = \int_{180}^{\infty} \frac{1}{6\sqrt{2\pi}} \exp\left[-\frac{1}{2}\left(\frac{h-160}{6}\right)^2\right] dh. \quad ■$$

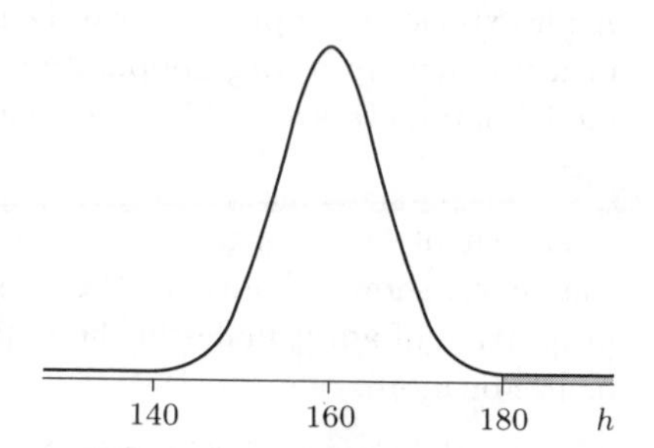

Figure 5.10 A normal model for the variation in height of elderly women

In this diagram the vertical scale has been slightly distorted so that the shaded area is evident: in a diagram drawn to scale it would not show up at all.

5.2.1 The standard normal distribution

In all the foregoing examples, problems about proportions have been expressed in terms of integrals of different normal densities. You have seen that a sketch of the model is a useful aid in clarifying the problem that has been posed. Finally, and almost incidentally, critical values have been *standardized* in terms of deviations from the mean, measured in multiples of the standard deviation.

Is this standardization a useful procedure, or are the original units of measurement essential to the calculation of proportions (that is, probabilities)?

The answer to this question is that it is useful: any normal random variable X with mean μ and standard deviation σ (so that $X \sim N(\mu, \sigma^2)$) may be re-expressed in terms of a standardized normal random variable, usually denoted Z, which has mean 0 and standard deviation 1. Then any probability for observations on X may be calculated in terms of observations on the random variable Z. This result can be proved mathematically; but in this course we shall only be concerned with applying the result. First, the random variable Z will be explicitly defined.

The standard normal distribution

The random variable Z following a normal distribution with mean 0 and standard deviation 1 is said to follow the **standard normal distribution**, written $Z \sim N(0,1)$. The p.d.f. of Z is given by

$$\phi(z) = \frac{1}{\sqrt{2\pi}} e^{-\frac{1}{2}z^2}, \quad -\infty < z < \infty. \tag{5.2}$$

Notice the use of the reserved letter Z for this particular random variable, and of the letter ϕ for the probability density function of Z. This follows the common conventions that you might see elsewhere.

ϕ is the Greek lower-case letter phi, and is pronounced 'fye'.

The graph of the p.d.f. of Z is shown in Figure 5.11. Again, the p.d.f. of Z is positive for any value of z, but observations much less than -3 or greater than $+3$ are unlikely. Integrating this density function gives normal probabilities. (Notice the Greek upper-case letter phi in the following definition. It is conventionally used to denote the c.d.f. of Z.)

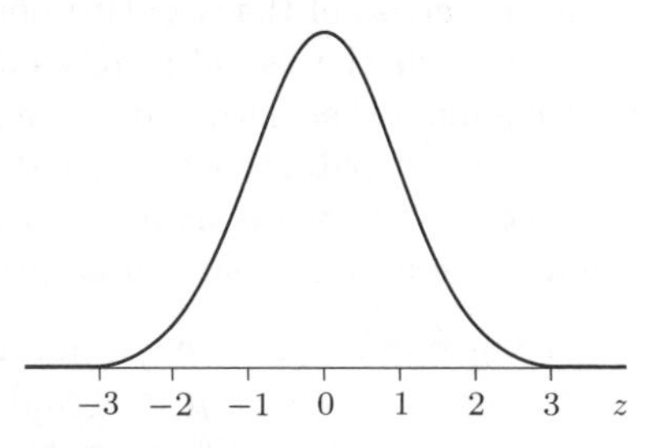

Figure 5.11 The p.d.f. of $Z \sim N(0,1)$, $\phi(z) = \frac{1}{\sqrt{2\pi}} e^{-\frac{1}{2}z^2}$

The c.d.f. of the standard normal variate Z is given by

$$\Phi(z) = P(Z \leq z) = \int_{-\infty}^{z} \frac{1}{\sqrt{2\pi}} e^{-\frac{1}{2}x^2}\,dx. \tag{5.3}$$

It gives the 'area under the curve', shaded in the following diagram of the density of Z (see Figure 5.12).

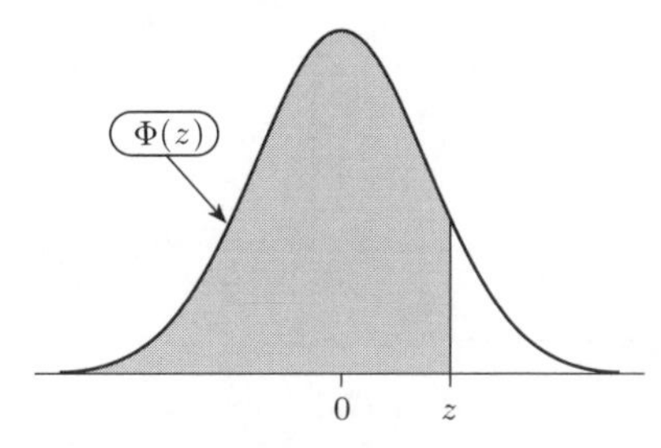

Figure 5.12 The c.d.f. of Z, $\Phi(z) = \int_{-\infty}^{z} \phi(x)\,dx$

Where in other parts of the course the integral notation has been used to describe the area under the curve defined by a probability density function, an explicit formula for the integral has been given, and that formula is used as the starting point in future calculations. In this respect, the normal density is unusual. No explicit formula for $\Phi(z)$ exists, though it is possible to obtain an expression for $\Phi(z)$ in the form of an infinite series of powers of z. So, instead, values of $\Phi(z)$ are obtained from tables or calculated on a computer.

Exercise 5.2

On four rough sketches of the p.d.f. of the standard normal distribution copied from Figure 5.11, shade in the areas corresponding to the following standard normal probabilities.

(a) $P(Z \leq 2)$

(b) $P(Z > 1)$

(c) $P(-1 < Z \leq 1)$

(d) $P(Z \leq -2)$

Before we look at tables which will allow us to attach numerical values to probabilities like those in Exercise 5.2, and before any of the other important properties of the standard normal distribution are discussed, let us pause to establish the essential relationship between the standard normal distribution and other normal distributions. It is this relationship that allows us to calculate probabilities associated with (for example) Victorian soldiers' chest measurements or mapmakers' measurement errors, or any other situation for which the normal distribution provides an adequate model.

Once again, let X follow a normal distribution with arbitrary mean μ and variance σ^2, $X \sim N(\mu, \sigma^2)$, and write Z for the standard normal variate, $Z \sim N(0, 1)$. These two random variables are related as follows.

If $X \sim N(\mu, \sigma^2)$, then the random variable

$$Z = \frac{X - \mu}{\sigma} \sim N(0, 1).$$

Conversely, if $Z \sim N(0, 1)$, then the random variable

$$X = \sigma Z + \mu \sim N(\mu, \sigma^2).$$

The great value of this result is that we can afford to do most of our thinking about normal probabilities in terms of the easier standard normal distribution and then adjust results appropriately, using these simple relationships between X and Z, to answer questions about any given general normal random variable.

Figure 5.13 gives a graphical representation of the idea of standardization.

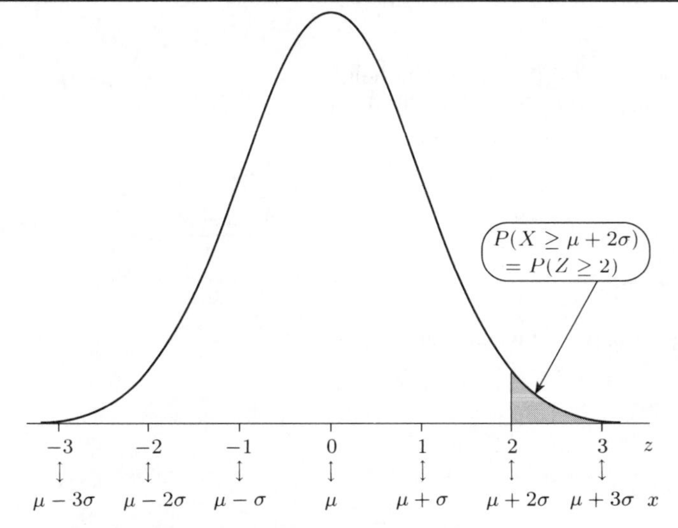

Figure 5.13 Standardization portrayed graphically

The shaded area gives the probability
$$P(X \geq \mu + 2\sigma) = P(Z \geq 2).$$

We can now formalize the procedures of Examples 5.1 to 5.3.

Example 5.1 continued

Our model for chest measurements (in inches) in a population of adult males is normal with mean 40 and standard deviation 2: this was written as $X \sim N(40, 4)$. We can rewrite the required probability $P(X \geq 43)$ as

$$P(X \geq 43) = P\left(Z \geq \frac{43 - 40}{2}\right) = P(Z \geq 1.5).$$

$$z = \frac{x - \mu}{\sigma}$$

This is illustrated in Figure 5.14, which may be compared directly with Figure 5.8. ■

Example 5.2 continued

In this case the random variable of interest is the intelligence score W, where $W \sim N(100, 225)$, and we require the probability $P(80 \leq W \leq 120)$. This may be found by rewriting it as follows:

$$P(80 \leq W \leq 120) = P\left(\frac{80 - 100}{15} \leq Z \leq \frac{120 - 100}{15}\right)$$
$$= P(-1.33 \leq Z \leq 1.33).$$

$$z = \frac{w - \mu}{\sigma}$$

This probability is illustrated by the shaded region in Figure 5.15 (and see also Figure 5.9). ■

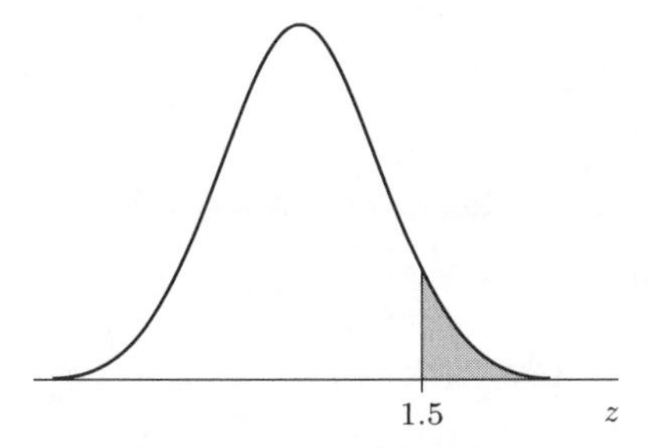

Figure 5.14 The probability $P(Z \geq 1.5)$

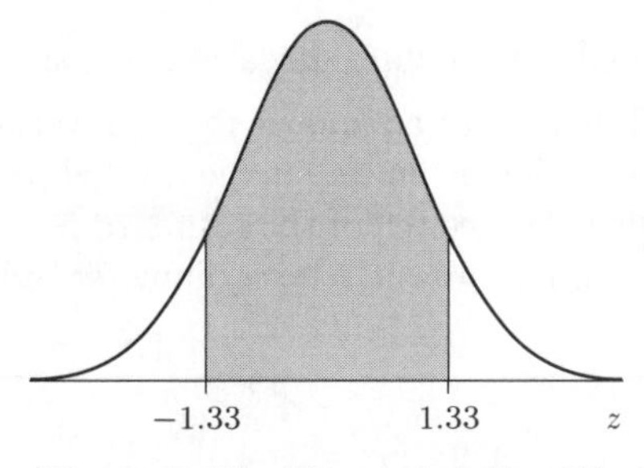

Figure 5.15 The probability $P(-1.33 \leq Z \leq 1.33)$

Example 5.3 continued

In Example 5.3 a normal model $H \sim N(160, 36)$ was proposed for the height distribution of elderly women (measured in cm). We wanted to find the proportion of this population who are over 180 cm tall. This probability $P(H > 180)$ can be rewritten

$$P(H > 180) = P\left(Z > \frac{180 - 160}{6}\right)$$
$$= P(Z > 3.33)$$

$z = \dfrac{h - \mu}{\sigma}$

and is represented by the shaded area in Figure 5.16. The diagram may be compared with that in Figure 5.10.

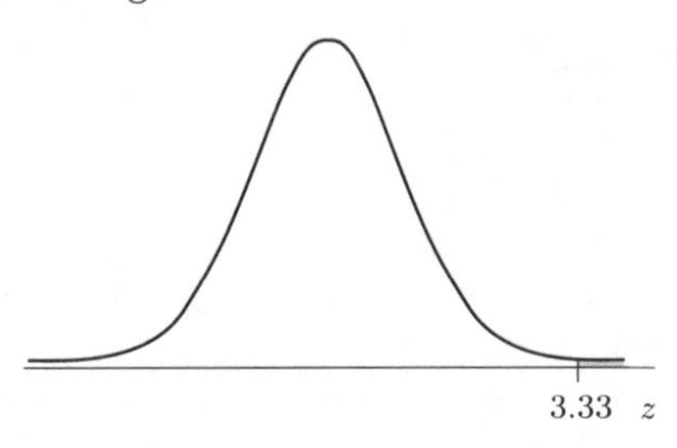

As in Figure 5.10, the vertical scale in this diagram has been slightly distorted.

Figure 5.16 The probability $P(Z > 3.33)$ ■

Exercise 5.3

(a) Measurements were taken on the level of ornithine carbonyltransferase (a liver enzyme) present in individuals suffering from acute viral hepatitis. After a suitable transformation, the corresponding random variable may be assumed to be adequately modelled by a normal distribution with mean 2.60 and standard deviation 0.33. Show on a sketch of the standard normal density the proportion of individuals with this condition, whose measured enzyme level exceeds 3.00.

See *Chapter 2*, Table 2.18.

(b) For individuals suffering from aggressive chronic hepatitis, measurements on the same enzyme are normally distributed with mean 2.65 and standard deviation 0.44. Show on a sketch of the standard normal density the proportion of individuals suffering from aggressive chronic hepatitis with an enzyme level below 1.50.

See *Chapter 2*, Table 2.19.

(c) At a ball-bearing production site, a sample of 10 ball-bearings was taken from the production line and their diameters measured (in mm). The recorded measurements were

1.18 1.42 0.69 0.88 1.62 1.09 1.53 1.02 1.19 1.32.

(i) Find the mean diameter $\overline{x}$ and the standard deviation s for the sample.

(ii) Assuming that a normal model is adequate for the variation in measured diameters, and using $\overline{x}$ as an estimate for the normal parameter μ and s as an estimate for σ, show on a sketch of the standard normal density the proportion of the production output whose diameter is between 0.8 mm and 1.2 mm.

The foregoing approach may be summarized simply as follows.

Calculating normal probabilities

If the random variable X follows a normal distribution with mean μ and variance σ^2, written $X \sim N(\mu, \sigma^2)$, then the probability $P(X \leq x)$ may be written

$$P(X \leq x) = \Phi\left(\frac{x - \mu}{\sigma}\right) \tag{5.4}$$

where $\Phi(\cdot)$ is the c.d.f. of the standard normal distribution.

5.2.2 Tables of the standard normal distribution

We are not yet able to assign numerical values to the probabilities so far represented only as shaded areas under the curve given by the standard normal density function. What is the probability that an IQ score is more than 115? What proportion of Victorian Scottish soldiers had chests measuring 38 inches or less? What is the probability that measurement errors inherent in the process leading to Merriman's data would be less than 2 minutes of arc in absolute value?

The answer to all such questions is found by reference to sets of printed tables, or from a computer. In this subsection you will see how to use the table of standard normal probabilities, Table A2.

You have already seen that any probability statement about the random variable X (when X is $N(\mu, \sigma^2)$) can be re-expressed as a probability statement about Z (the standard normal variate). So only one page of tables is required: we do not need reams of paper to print probabilities for other members of the normal family. To keep things simple, therefore, we shall begin by finding probabilities for values observed on Z, and only later make the simple extension to answering questions about more general normally distributed random variables useful in modelling the real world. The statistics table entitled 'Probabilities for the standard normal distribution' gives the left-hand tail probability

$$P(Z \leq z) = \Phi(z)$$

for values of z from 0 to 4 by steps of 0.01, printed accurate to 4 decimal places. (Other versions of this table might print the probability $P(Z \geq z)$ for a range of values of z; or the probability $P(0 \leq Z \leq z)$; or even $P(-z \leq Z \leq z)$! There are so many variations on possible questions that might be asked, that no one formulation is more convenient than any other.)

Values of z are read off down the leftmost column and across the top row (the top row gives the second decimal place). Thus the probability $P(Z \leq 1.58)$, for example, may be found by reading across the row for $z = 1.5$ until the column headed 8 is found.

Then the entry in the body of the table in the same row and column gives the probability required: in this case, it is 0.9429. (So, only about 6% of a normal population measure in excess of 1.58 standard deviations above the mean.)

As a second example, we can find the probability $P(Z \leq 3.00)$ so frequently mentioned. In the row labelled 3.0 and the column headed 0, the entry in the table is 0.9987, and this is the probability required. It follows that only a proportion 0.0013, about one-tenth of one per cent, will measure in excess of 3 standard deviations above the mean, in a normal population. These probabilities can be illustrated on sketches of the standard normal density, as shown in Figure 5.17.

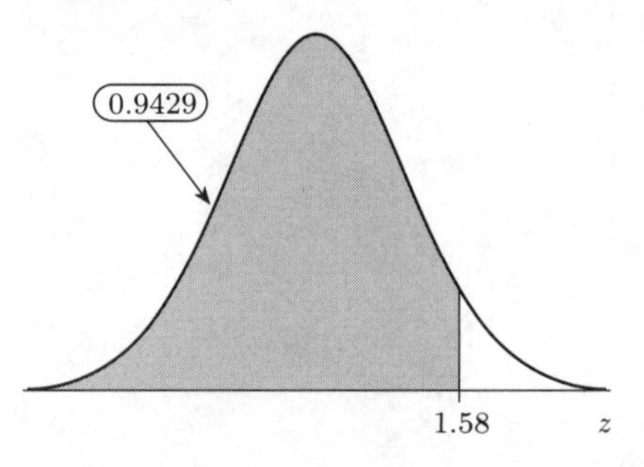

Figure 5.17 (a) $P(Z \leq 1.58)$ (b) $P(Z \leq 3.00)$

Exercise 5.4

Use the table to find the following probabilities.

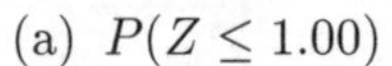

(a) $P(Z \leq 1.00)$
(b) $P(Z \leq 1.96)$
(c) $P(Z \leq 2.25)$

Illustrate these probabilities in sketches of the standard normal density.

Of course, required probabilities will not necessarily always be of the form $P(Z \leq z)$. For instance, we might need to find probabilities such as

$$P(Z \geq 1.50) \text{ or } P(0 \leq Z \leq 1.83) \text{ or } P(-1.33 \leq Z \leq 2.50).$$

In such cases it often helps to draw a rough sketch of what is required and include on the sketch information obtained from tables. The symmetry of the normal distribution will often prove useful; as will the fact that the total area under the standard normal curve is 1. To find $P(Z \geq 1.50)$, for example, we would start with a sketch of the standard normal density, showing the area required, as in Figure 5.18.

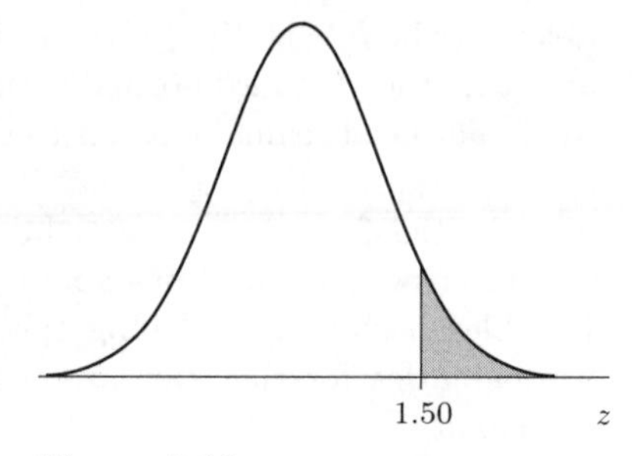

Figure 5.18

From the tables, we find that the probability $P(Z \leq 1.50)$ is 0.9332. By subtraction from 1, it follows that the probability required, the area of the shaded region, is $1 - 0.9332 = 0.0668$. This is illustrated in Figure 5.19.

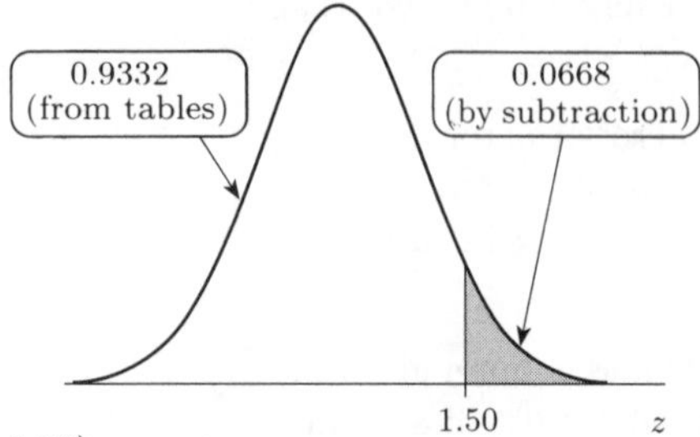

Figure 5.19 $P(Z \geq 1.50)$

Here is a second example. Starting again with a sketch to represent the probability $P(0 \leq Z \leq 1.83)$ (see Figure 5.20), we first obtain from the tables the probability $P(Z \leq 1.83) = 0.9664$. Subtracting 0.5 (the normal density is symmetric about 0), the required probability is found to be 0.4664.

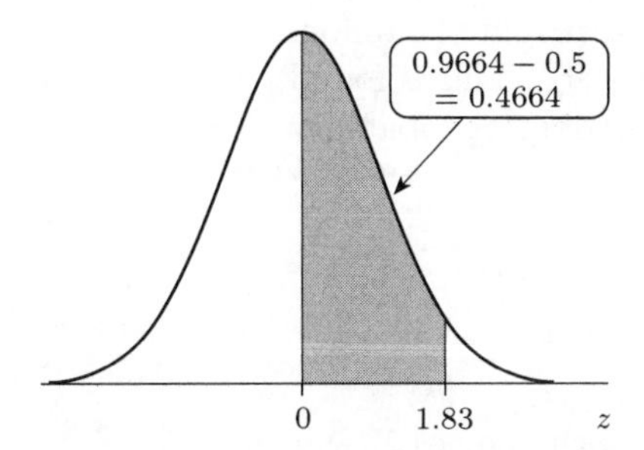

Figure 5.20 $P(0 \leq Z \leq 1.83)$

Exercise 5.5

Using rough sketches of the standard normal density to help you, find the following probabilities.

(a) $P(-1.33 \leq Z \leq 2.50)$

(b) $P(-3.00 \leq Z \leq 3.00)$

(c) $P(0.50 \leq Z \leq 1.50)$

When using the tables to calculate these probabilities, your arithmetic procedures will probably not be identical to those given in the printed solutions; but your final answers should be as printed!

The probability $P(-3.00 \leq Z \leq 3.00)$ in Exercise 5.5(b) could have been written using modulus notation as $P(|Z| \leq 3.00)$. This is a common notation in the context of calculating normal probabilities; it is used in the next exercise.

Exercise 5.6

Find the following probabilities.

(a) $P(|Z| \leq 1.62)$

(b) $P(|Z| \geq 2.45)$

Exercise 5.7

(a) What proportion of a normal population is within one standard deviation of the mean with respect to the attribute being measured?

(b) What proportion is more than two standard deviations away from the mean?

So far the tables have been used only to calculate standard normal probabilities. In real applications the indexing parameters μ and σ for a normal model will usually be other than 0 and 1. However, all that needs to be done in this circumstance is to standardize the original variable X in the usual way, that is, by subtracting its mean μ and dividing by its standard deviation σ:

$$Z = \frac{X - \mu}{\sigma}.$$

***Example 5.4** Calculating normal probabilities after standardization*

According to the particular design of IQ tests which results in scores that are normally distributed with mean 100 and standard deviation 15, what proportion of the population tested will record scores of 120 or more?

The question may be expressed in terms of a normally distributed random variable $X \sim N(100, 225)$ as 'find the probability $P(X \geq 120)$'. This is found by standardizing X, thus transforming the problem into finding a probability involving Z:

$$\begin{aligned} P(X \geq 120) &= P\left(Z \geq \frac{120 - 100}{15}\right) \\ &= P(Z \geq 1.33). \end{aligned}$$

This is found from the tables to be $1 - \Phi(1.33) = 1 - 0.9082 = 0.0918$. Not quite 10% of the population will score 120 or more on tests to this design.

This sort of example demonstrates the importance of the standard deviation in quantifying 'high' scores. Similarly, less than 2.5% of the population will score 130 or more:

$$\begin{aligned} P(X \geq 130) &= P\left(Z \geq \frac{130 - 100}{15}\right) \\ &= P(Z \geq 2.00) \\ &= 1 - 0.9772 \\ &= 0.0228. \quad \blacksquare \end{aligned}$$

Exercise 5.8

A reasonable model for the nineteenth century Scottish soldiers' chest measurements is to take $X \sim N(40, 4)$ (measurements in inches). What proportion of that population would have had chest measurements between 37 inches and 42 inches inclusive?

At this point you might wonder precisely how the actual data—the 5732 Scottish soldiers' chest measurements—enter the calculation. They figure implicitly in the first sentence of the exercise: *a reasonable model* for the distribution of the data is $N(40, 4)$. That the normal distribution provides a reasonable model for the general shape can be seen by looking at the histogram in Figure 5.2. That 40 is a reasonable value to take for μ and 4 for σ^2 can be seen from calculations based on the data—which we shall explore further in *Chapter 6*. Once the data have been used to formulate a reasonable model, then future calculations can be based on that model.

Exercise 5.9

A good model for the angular measurement errors (minutes of arc) mentioned in Section 5.1 is that they be normally distributed with mean 0 and variance 2.75. What is the probability that such an error is positive but less than 2?

Exercise 5.10

Blood plasma nicotine levels in smokers (see *Chapter 2*, Table 2.16) can be modelled as $T \sim N(315, 131^2 = 17\,161)$. (The units are nanograms per millilitre, ng/ml.)

(a) Make a sketch of this distribution marking in $\mu + k\sigma$ for $k = -3, -2, -1, 0, 1, 2, 3$.

(b) What proportion of smokers has nicotine levels lower than 300? Sketch the corresponding area on your graph.

(c) What proportion of smokers has nicotine levels between 300 and 500?

(d) If 20 other smokers are to be tested, what is the probability that at most one has a nicotine level higher than 500?

Here the adequacy of a normal model becomes questionable. Notice that a nicotine level of zero is only $315/131 = 2.40$ standard deviations below the mean. A normal model would thus permit a proportion of $\Phi(-2.40) = 0.008$ negative recordings, though negative recordings are not realizable in practice.

5.2.3 Quantiles

So far questions of this general form have been addressed: if the distribution of the random variable $X \sim N(\mu, \sigma^2)$ is assumed to be an adequate model for the variability observed in some measurable phenomenon, with what probability $P(x_1 \leq X \leq x_2)$ will some future observation lie within stated limits? Given the boundaries illustrated in Figure 5.21, we have used the tables to calculate the shaded area representing the probability $P(x_1 \leq X \leq x_2)$.

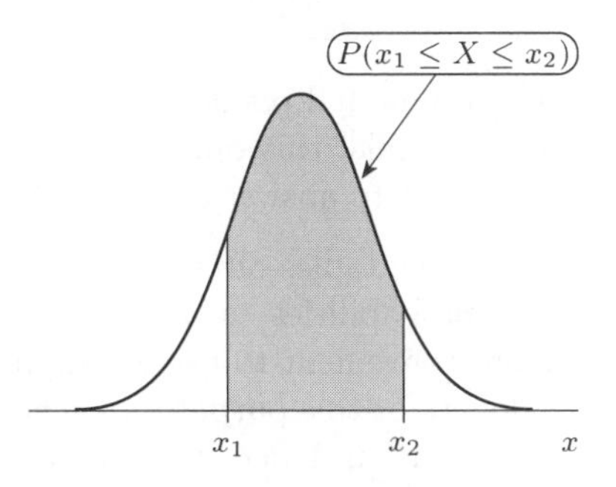

Figure 5.21 The probability $P(x_1 \leq X \leq x_2)$

Conversely, given a probability α we might wish to find x such that $P(X \leq x) = \alpha$. For instance, assuming a good model of IQ scores to be $N(100, 225)$, what score is attained by only the top 2.5% of the population? This problem is illustrated in Figure 5.22. Quantiles were defined in *Chapter 3*, Section 3.5. For a continuous random variable X with c.d.f. $F(x)$, the α-quantile is the value x which is the solution to the equation $F(x) = \alpha$, where $0 < \alpha < 1$. This solution is denoted q_α.

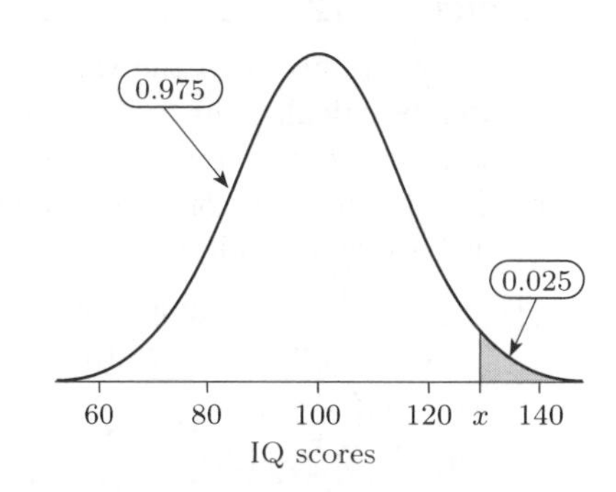

Figure 5.22 The 97.5% point of $N(100, 225)$ (x, unknown)

You may remember these special cases: the lower quartile, $q_{0.25}$ or q_L; the median, $q_{0.5}$ or m; and the upper quartile, $q_{0.75}$ or q_U. These are shown in Figure 5.23 for the standard normal distribution.

The median of Z is clearly 0: this follows from the symmetry of the normal distribution. From the tables, the closest we can get to q_U is to observe that

$$P(Z \leq 0.67) = 0.7486, \quad P(Z \leq 0.68) = 0.7517,$$

so (splitting the difference) perhaps $q_U \simeq 0.675$ or thereabouts. It would be convenient to have available a separate table of standard normal quantiles, and this is provided in Table A3. The table gives values of q_α to 3 decimal places for various values of α from 0.5 to 0.999.

So, for instance, the upper quartile of Z is $q_U = 0.674$; the 97.5% point of Z is $q_{0.975} = 1.96$. If $X \sim N(\mu, \sigma^2)$, then it follows from the relationship

$$X = \sigma Z + \mu$$

that the 97.5% point of X is $1.96\sigma + \mu$. So the unknown IQ score illustrated in Figure 5.22 is

$$x = (1.96)(15) + 100$$
$$= 129.4.$$

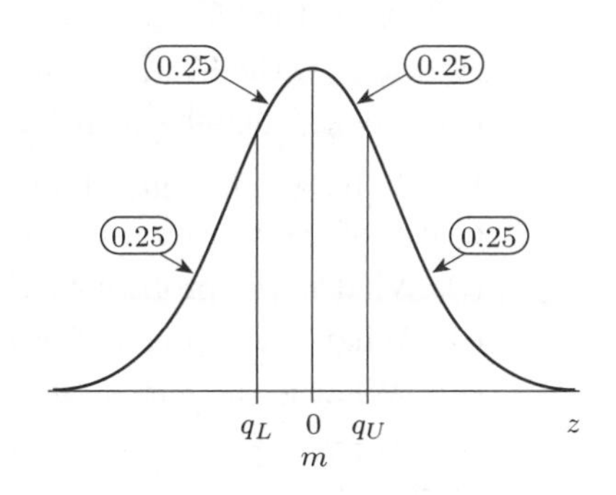

Figure 5.23 q_L, m, q_U for $Z \sim N(0, 1)$

The symmetry of the normal distribution may also be used to find quantiles lower than the median. For instance, the 30% point of Z is

$$q_{0.3} = -q_{0.7} = -0.524.$$

Exercise 5.11

Find $q_{0.2}$, $q_{0.4}$, $q_{0.6}$, $q_{0.8}$ for the distribution of IQ test scores, assuming the normal distribution $N(100, 225)$ to be an adequate model, and illustrate these quantiles in a sketch of the distribution of scores.

There now follows a further exercise summarizing the whole of this section so far. Take this opportunity to investigate the facilities available on your computer to answer this sort of question.

While the tables often provide the quickest and easiest way of obtaining normal probabilities to answer isolated questions, in other circumstances it is more convenient to use a computer, and computer algorithms have been developed for this purpose. In general, too, computers work to a precision much greater than 4 decimal places, and more reliance can be placed on results which, without a computer, involve addition and subtraction of several probabilities read from the tables.

Exercise 5.12

The answers given to the various questions in this exercise are all based on computer calculations. There may be some inconsistencies between these answers and those you would obtain if you were using the tables, with all the implied possibilities of rounding error. However, these inconsistencies should never be very considerable.

(a) The random variable Z has a standard normal distribution $N(0, 1)$. Use your computer to find the following.
  (i) $P(Z \geq 1.7)$
  (ii) $P(Z \geq -1.8)$
  (iii) $P(-1.8 \leq Z \leq 2.5)$
  (iv) $P(1.5 \leq Z \leq 2.8)$
  (v) $q_{0.10}$, the 10% point of the distribution of Z
  (vi) $q_{0.95}$, the 95% point of the distribution of Z
  (vii) $q_{0.975}$, the 97.5% point of the distribution of Z
  (viii) $q_{0.99}$, the 99% point of the distribution of Z

(b) Let X be a randomly chosen individual's score on an IQ test. By the design of the test, it is believed that $X \sim N(100, 225)$.
  (i) What is the probability that X is greater than 125?
  (ii) What is the probability $P(80 \leq X \leq 90)$?
  (iii) What is the median of the distribution of IQ scores?
  (iv) What IQ score is such that only 10% of the population have that score or higher?
  (v) What is the 0.1-quantile of the IQ distribution?

(c) Suppose the heights (in cm) of elderly females follows a normal distribution with mean 160 and standard deviation 6.

(i) What proportion of such females are taller than 166 cm?

(ii) What is the 0.85-quantile of the distribution of females' heights?

(iii) What is the interquartile range of the distribution? (The population interquartile range is the difference between the quartiles.)

(iv) What is the probability that a randomly chosen female has height between 145 and 157 cm?

(d) Nicotine levels in smokers are modelled by a random variable T with a normal distribution $N(315, 17\,161)$.

(i) What is the probability that T is more than 450?

(ii) What is the 0.95-quantile of the nicotine level distribution?

(iii) What is the probability $P(150 \leq T \leq 400)$?

(iv) What is the probability $P(|T - 315| \leq 100)$?

(v) What nicotine level is such that 20% of smokers have a higher level?

(vi) What range of levels is covered by the central 92% of the smoking population?

(vii) What is the probability that a smoker's nicotine level is between 215 and 300 or between 350 and 400?

5.2.4 Other properties of the normal distribution

In *Chapter 4* you looked at some properties of sums and multiples of random variables. In particular, if the random variables $X_1, \ldots, X_n$ are independent with mean μ_i and variance σ_i^2, then their sum $\sum X_i$ has mean and variance

$$E(\sum X_i) = \sum \mu_i, \qquad V(\sum X_i) = \sum \sigma_i^2.$$

You learned the particular result that sums of independent Poisson variates also follow a Poisson distribution.

A corresponding result holds for sums of independent normal random variables: they follow a normal distribution.

If X_i are independent normally distributed random variables with mean μ_i and variance σ_i^2, $i = 1, 2, \ldots, n$, then their sum $\sum X_i$ is also normally distributed, with mean $\sum \mu_i$ and variance $\sum \sigma_i^2$:

$$\sum X_i \sim N(\sum \mu_i, \sum \sigma_i^2).$$

This result is stated without proof.

***Example 5.5** Bags of sugar*

Suppose that the normal distribution provides an adequate model for the weight X of sugar in paper bags of sugar labelled as containing 2 kg. There is some variability, and to avoid penalties the manufacturers overload the bags slightly. Measured in grams, suppose $X \sim N(2003, 1)$.

In fact, items marked with the ℮ symbol next to their weight do weigh 2 kg (or whatever) on average, and that is all that a manufacturer might be required to demonstrate.

It follows that the probability that a bag is underweight is given by

$$\begin{aligned} &P(X < 2000) \\ &= P\left(Z < \frac{2000 - 2003}{1}\right) \\ &= P(Z < -3) \\ &= 0.0013. \end{aligned}$$

So about one bag in a thousand is underweight.

A cook requiring 6 kg of sugar to make marmalade purchases three of the bags. The total amount of sugar purchased is the sum

$$S = X_1 + X_2 + X_3.$$

Assuming independence between the weights of the three bags, their expected total weight is

$$\begin{aligned} E(S) &= \mu_1 + \mu_2 + \mu_3 \\ &= 2003 + 2003 + 2003 \\ &= 6009; \end{aligned}$$

and the variance in the total weight is

$$\begin{aligned} V(S) &= \sigma_1^2 + \sigma_2^2 + \sigma_3^2 \\ &= 3; \end{aligned}$$

that is, $S \sim N(6009, 3)$. The standard deviation in the total weight is $SD(S) = \sqrt{3} = 1.732$ gm.

The probability that altogether the cook has too little sugar for the purpose (less than 6 kg) is given by

$$\begin{aligned} &P(S < 6000) \\ &= P\left(Z < \frac{6000 - 6009}{\sqrt{3}}\right) \\ &= P\left(Z < -\frac{9}{\sqrt{3}}\right) \\ &= P(Z < -5.20). \end{aligned}$$

This probability is negligible. (Your computer, if you are using one, will give you the result 1.02×10^{-7}, about one in ten million!) ■

You also saw in *Chapter 4* that if the random variable X has mean μ and variance σ^2, then for constants a and b, the random variable $aX + b$ has mean and variance

$$E(aX + b) = a\mu + b, \qquad V(aX + b) = a^2\sigma^2.$$

This holds whatever the distribution of X. However, if X is normally distributed, the additional result holds that $aX + b$ is also normally distributed.

If X is normally distributed with mean μ and variance σ^2, written $X \sim N(\mu, \sigma^2)$, and if a and b are constants, then

$$aX + b \sim N(a\mu + b, a^2\sigma^2).$$

5.3 The central limit theorem

In the preceding sections of this chapter and at the first mention of the normal distribution in *Chapter 2*, it has been stressed that the distribution has an important role in statistics as a good approximate model for the variability inherent in measured quantities in all kinds of different contexts.

This section is about one of the fundamental results of statistical theory: it describes particular circumstances where the normal distribution arises not in the real world (chest measurements, enzyme levels, intelligence scores), but at the statistician's desk. The result is stated as a theorem, the **central limit theorem**. It is a theoretical result, and one whose proof involves some deep mathematical analysis: we shall be concerned, however, only with its consequences, which are to ease the procedures involved when seeking to deduce characteristics of a population from characteristics of a sample drawn from that population.

5.3.1 Characteristics of large samples

The central limit theorem is about the distributions of sample means and sample totals. You met these sample quantities in *Chapter 1*. Suppose we have a random sample of size n from a population. The data items in the sample may be listed

$$x_1, x_2, x_3, \ldots, x_n.$$

The *sample total* is simply the sum of all the items in the data set:

$$\begin{aligned} t_n &= x_1 + x_2 + x_3 + \cdots + x_n \\ &= \sum x_i. \end{aligned}$$

The *sample mean* is what is commonly called the 'average', the sample total divided by the sample size:

$$\overline{x}_n = \frac{1}{n}(x_1 + x_2 + \cdots + x_n) = \frac{1}{n}t_n.$$

Notice that in both these labels, t_n and $\overline{x}_n$, the subscript n has been included. This makes explicit the size of the sample from which these statistics have been calculated.

We know that in repeated sampling experiments from the same population and with the same sample size, we would expect to observe variability in the individual data items and also in the summary statistics, the sample total and the sample mean. In any single experiment therefore, the sample total t_n is just one observation on a random variable T_n; and the sample mean $\overline{x}_n$ is just one observation on a random variable $\overline{X}_n$.

You saw in *Chapter 4*, that notwithstanding this variability in the summary statistics, they are useful consequences of the experiment. In particular, assuming the population mean μ and the population variance σ^2 to be unknown, the following important result for the distribution of the mean of samples of size n was obtained:

$$E(\overline{X}_n) = \mu, \qquad V(\overline{X}_n) = \frac{\sigma^2}{n}.$$

That is, if a sample of size n is collected from a large population, and if that sample is averaged to obtain the sample mean, then the number obtained, $\overline{x}_n$, should constitute a reasonable estimate for the unknown population mean μ. Moreover, the larger the sample drawn, the more reliance can be placed on the number obtained, since the larger the value of n, the less deviance that should be observed in $\overline{x}_n$ from its expected value μ.

Chapter 4, page 157

Exercise 5.13

(a) Obtain a sample of size 5 from a Poisson distribution with mean 8, and calculate the sample mean $\overline{x}_5$. Next, obtain 100 observations on the random variable $\overline{X}_5$. How many of the 100 observations (all 'estimating the number 8') are between 6 and 10?

(b) Now obtain a sample of size 20 from a Poisson distribution with mean 8 and calculate the sample mean $\overline{x}_{20}$. Obtain 100 observations altogether on $\overline{X}_{20}$. How many of these are between 6 and 10? How many are between 7 and 9?

(c) Now obtain a sample of size 80 from a Poisson distribution with mean 8, and calculate the sample mean $\overline{x}_{80}$. Obtain 100 observations on $\overline{X}_{80}$, and calculate the number of them that are between 7 and 9.

(d) Summarize in non-technical language any conclusions you feel able to draw from the experiments of parts (a) to (c).

Exercise 5.14

Investigate the sampling properties of means of samples of size 5, 20, 80 from the exponential distribution with mean 8.

In Exercise 5.13, and Exercise 5.14 if you tried it, the same phenomenon should have been evident: that is, variation in the sample mean is reduced as the sample size increases.

But all this is a consequence of a result that you already know, and have known for some time—the point was made in *Chapter 4* that increasing the sample size increases the usefulness of $\overline{x}$ as an estimate for the population mean μ. However, knowledge of the mean $(E(\overline{X}_n) = \mu)$ and variance $(V(\overline{X}_n) = \sigma^2/n)$ of the sample mean does not permit us to make probability statements about likely values of the sample mean, because we still do not know the shape of its probability distribution.

Exercise 5.15

(a) The exponential distribution is very skewed with a long right tail. Figure 5.24 is a sketch of the density for an exponentially distributed random variable with mean 1.

(i) Generate 100 observations on the random variable $\overline{X}_2$ from this distribution; obtain a histogram of these observations.

(ii) Now generate 100 observations on the random variable $\overline{X}_{30}$ from this distribution. Obtain a histogram of these observations.

(iii) Comment on any evident differences in the shape of the two histograms.

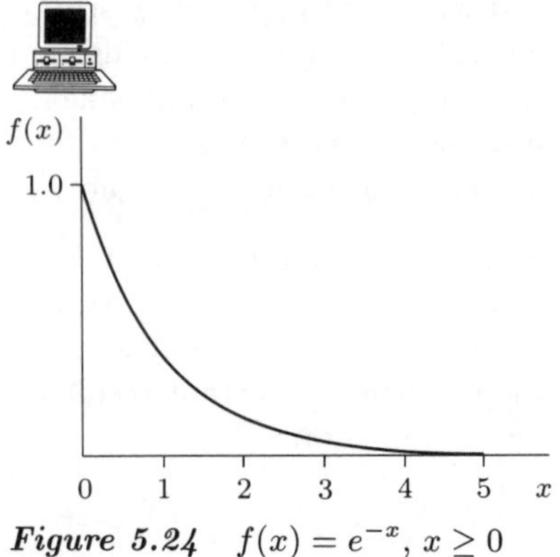

Figure 5.24 $f(x) = e^{-x}, x \geq 0$

(b) The continuous uniform distribution is flat. The density of the uniform distribution $U(0,2)$ (with mean 1) is shown in Figure 5.25.

(i) Generate 100 observations on the random variable $\overline{X}_2$ from this distribution and obtain a histogram of these observations.

(ii) Now generate 100 observations on $\overline{X}_{30}$, and obtain a histogram of the observations.

(iii) Are there differences in the shape of the two histograms?

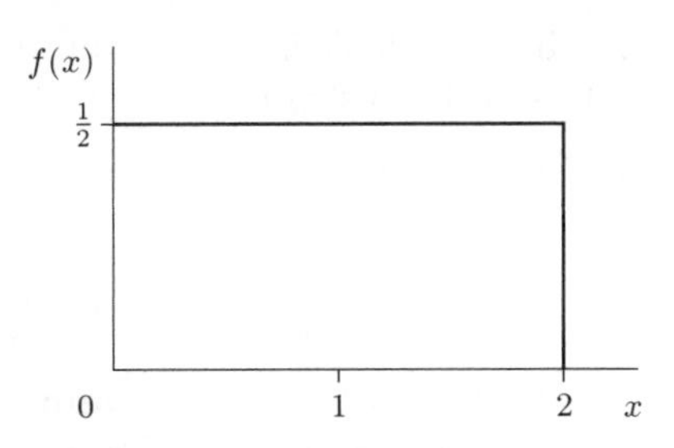

Figure 5.25 The uniform distribution $U(0,2)$

5.3.2 Statement of the theorem

The point illustrated by the solution to Exercise 5.15 is that even for highly non-normal populations, repeated experiments to obtain the sample mean result in observations that peak at the population mean μ, with frequencies tailing off roughly symmetrically above and below the population mean. This is a third phenomenon to add to the two results noted already, giving the following three properties of the sample mean.

(a) In a random sample from a population with unknown mean μ, the sample mean is a good indicator of the unknown number μ $(E(\overline{X}_n) = \mu)$.

(b) The larger the sample, the more reliance can be placed on the sample mean as an estimator for the unknown number μ $(V(\overline{X}) = \sigma^2/n)$.

(c) Notwithstanding any asymmetry in the parent population, and for samples of sufficient size, the sample mean in repeated experiments overestimates or underestimates the population mean μ with roughly equal probability. Specifically, the distribution of the sample mean is approximately 'bell-shaped'.

It is also of interest that this bell-shaped effect happens not just with highly asymmetric parent populations, but also when the parent population is discrete—Figure 5.26 shows the histogram that resulted when 1000 observations were taken on $\overline{X}_{30}$ from a Poisson distribution with mean 2.

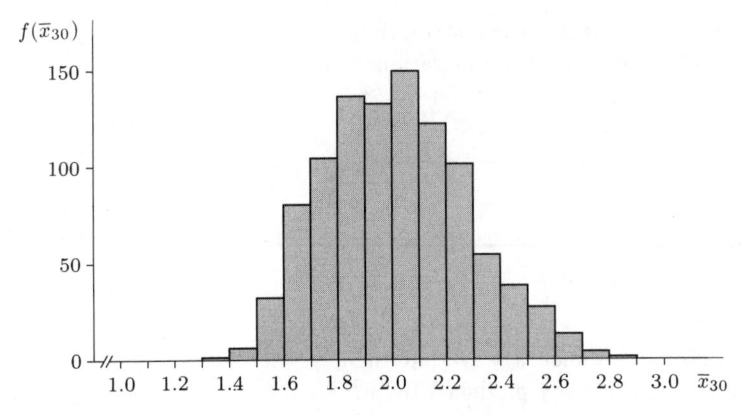

Figure 5.26 1000 observations on $\overline{X}_{30}$ from Poisson(2)

Again, the 'bell-shaped' nature of the distribution of the sample mean is apparent in this case.

Putting these three results together leads us to a statement of the central limit theorem.

The central limit theorem

If $X_1, X_2, \ldots, X_n$ are n independent and identically distributed random observations from a population with mean μ and finite variance σ^2, then for large n the distribution of their mean $\overline{X}_n$ is approximately normal with mean μ and variance σ^2/n: this is written

$$\overline{X}_n \approx N\left(\mu, \frac{\sigma^2}{n}\right). \qquad (5.5)$$

The symbol '$\approx$' is read 'has approximately the same distribution as'.

The theorem is an asymptotic result—that is, the approximation improves as the sample size increases. The quality of the approximation depends on a number of things including the nature of the population from which the n observations are drawn, and one cannot easily formulate a rule such as 'the approximation is good for n at least 30'. There are cases where the approximation is good for n as small as 3; and cases where it is not so good even for very large n. However, certain 'rules of thumb' can be developed for the common applications of this theorem, as you will see. One thing that is certain is that in any sampling context the approximation will get better as the sample size increases.

5.3.3 A corollary to the theorem

We have concentrated so far on the distribution of the mean of a sample of independent identically distributed random variables: this has evident applications to estimation, as we have seen.

As well as the mean $\overline{X}_n$, we might also be interested in the total T_n of n independent identically distributed random variables. This has mean and variance given by

$$E(T_n) = n\mu, \qquad V(T_n) = n\sigma^2.$$

A corollary to the central limit theorem states that for large n the distribution of the sample total T_n is approximately normal, with mean $n\mu$ and variance $n\sigma^2$:

$$T_n \approx N(n\mu, n\sigma^2). \qquad (5.6)$$

***Example 5.6** A traffic census*

In a traffic census, vehicles are passing an observer in such a way that the waiting time between successive vehicles may be adequately modelled by an exponential distribution with mean 15 seconds. As it passes, certain details of each vehicle are recorded on a sheet of paper; each sheet has room to record the details of twenty vehicles.

What, approximately, is the probability that it takes less than six minutes to fill one of the sheets?

If the waiting time T measured in seconds has mean 15, then we know from properties of the exponential distribution that it has standard deviation 15 and variance 225. The time taken to fill a sheet is the sum

See (4.25).

$$W = T_1 + T_2 + \cdots + T_{20}$$

of twenty such waiting times. Assuming the times to be independent, then

$$E(W) = 15 + 15 + \cdots + 15 = 20 \times 15 = 300,$$

and

$$V(W) = 225 + 225 + \cdots + 225 = 20 \times 225 = 4500.$$

Also, by the central limit theorem, W is approximately normally distributed:

$$W \approx N(300, 4500).$$

We need to find the probability that the total time W is less than six minutes: that is, less than 360 seconds, seconds being our unit of measurement. This is given by

$$\begin{aligned} P(W < 360) &\simeq P\left(Z < \frac{360 - 300}{\sqrt{4500}}\right) \\ &= P(Z < 0.89). \end{aligned}$$

From the tables, this probability is 0.8133. (Using a computer directly without introducing incidental approximations yields the answer 0.8145.) ■

Exercise 5.16

A dentist keeps track, over a very long period, of the time T it takes her to attend to individual patients at her surgery. She is able to assess the average duration of a patient's visit, and the variability in duration, as follows:

$$\mu = 20 \text{ minutes}, \qquad \sigma = 15 \text{ minutes}.$$

(In reality, she arrives at these estimates through the sample mean and sample standard deviation of her data collection; but these will suffice as parameter estimates.)

A histogram of her data proves to be extremely jagged, suggestive of none of the families of distributions with which she is familiar. (Although the data set is large, it is not sufficiently large to result in a smooth and informative histogram.)

Her work starts at 9.00 each morning. One day there are 12 patients waiting in the waiting room: her surgery is scheduled to end at noon.

What (approximately) is the probability that she will be able to attend to all 12 patients within the three hours?

Exercise 5.17

Rather than keep an accurate record of individual transactions, the holder of a bank account only records individual deposits into and withdrawals from her account to the nearest pound. Assuming that the error in individual records may be modelled as a continuous uniform random variable $U(-\frac{1}{2}, \frac{1}{2})$, what is the probability that at the end of a year in which there were 400 transactions, her estimate of her bank balance is less than ten pounds in error?

(Remember, if the random variable W is $U(a, b)$, then W has variance $\frac{1}{12}(b-a)^2$.)

5.4 Normal approximations to continuous distributions

The probability density function of the normal distribution is a symmetric bell-shaped curve: many other random variables which are not exactly normally distributed nonetheless have density functions of a qualitatively similar form. So, when, as so often, it is difficult to determine a precise model, using a normal distribution as an approximation and basing our efforts on that is an appealing approach.

In many cases, the central limit theorem is the explanation for the apparently normal nature of a distribution: the random variables we are interested in are really made up of sums or averages of other independent identically distributed random variables, and so the central limit theorem applies to explain the resulting approximate normal distribution. More than that, the central limit theorem tells us the appropriate mean and variance of the approximate normal distribution in terms of the mean and variance of the underlying random variables. So probabilities may be calculated approximately by using the appropriate normal distribution. In Exercises 5.16 and 5.17, you have already done this when given examples of underlying distributions and questions explicitly framed in terms of sums of the associated random variables. But we can also use normal approximations in cases where we know the exact distribution, but where it is not easy to work with the exact result. Examples of this include the binomial distribution—recall from *Chapter 2*, Section 2.3 that binomial random variables are sums of independent identically distributed Bernoulli random variables—and the Poisson distribution (sums of independent Poisson variates are again Poisson variates). Normal approximations to the binomial and Poisson distributions are considered further in Section 5.5.

How large a sample is needed for the central limit theorem to apply? The central limit theorem is a limiting result that, we have seen, we can use as an approximate result for finite sample size: when is that approximation good? Unfortunately, there is no neat single answer to these questions. It all depends on the particular underlying distribution we are concerned with; for the binomial distribution a 'rule of thumb' that has been established over long experience will be given, to provide a reasonable guide. For some distributions, approximate normality can hold for surprisingly small n—like 5 or 10—even when the underlying distribution is very non-normal.

This section deals only with normal approximations to continuous distributions. Normal approximations to discrete distributions will be considered in Section 5.5.

In Exercise 5.13 you used a computer to mimic the repeated drawing of samples from a Poisson distribution; the results of that sampling experiment were illustrated using histograms, and that was your first intimation of the consequences of the central limit theorem. In Exercise 5.14, if you had the time to try it, you would have seen the phenomenon repeated for a continuous model, the highly skewed exponential distribution.

In this section we will also look at the *densities* of means and sums of continuous variables, rather than at histograms, their jagged sampling analogues. That is to say, we shall be considering the exact form of the distribution obtained when continuous random variables are added together.

Now, so far, the only exact result we have used is that the sum of normal random variables is itself normally distributed (remember Example 5.5 where weights of bags of sugar were added together). Even that result was merely stated, and not proved. Otherwise, we have used approximate results based on the central limit theorem. The problem is that, in general, the exact distribution of a sum of continuous random variables is rather difficult to obtain. For the three examples that follow, you are not expected to appreciate all the theoretical detail underlying the results—indeed, not much detail is given. Just try to understand the main message behind the examples. This section is entirely illustrative.

Example 5.7 Summing exponential random variables

The exponential random variable $X \sim M(1)$ has mean 1 and variance 1; its probability density function is given by

$$f(x) = e^{-x}, \quad x \geq 0.$$

The density, sketched in Figure 5.24, illustrates the highly skewed nature of the distribution.

The mean of samples of size 2 from this distribution,

$$\overline{X}_2 = \tfrac{1}{2}(X_1 + X_2),$$

has mean 1 and variance $\frac{1}{2}$. The p.d.f. of $\overline{X}_2$ is not something you need to know, and far less be able to obtain; but the *shape* of the density of $\overline{X}_2$ is given in Figure 5.27.

The variance is σ^2/n, where $\sigma^2 = 1$ and $n = 2$.

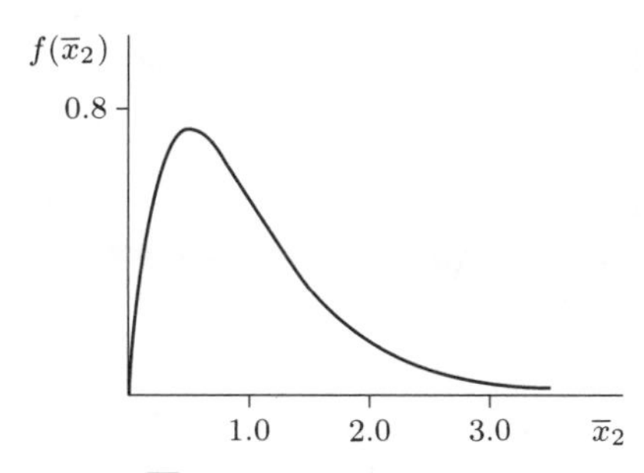

Figure 5.27 The density of $\overline{X}_2$ when $X \sim M(1)$

Already you can see the reduction in the skewness of the density—although far from being symmetric, there is a very apparent peak, and the density tails off either side of this peak.

Figure 5.28 shows the density of $\overline{X}_{10}$, the mean of samples of size 10 from the exponential distribution with mean 1. This random variable has mean 1 and variance $\frac{1}{10}$.

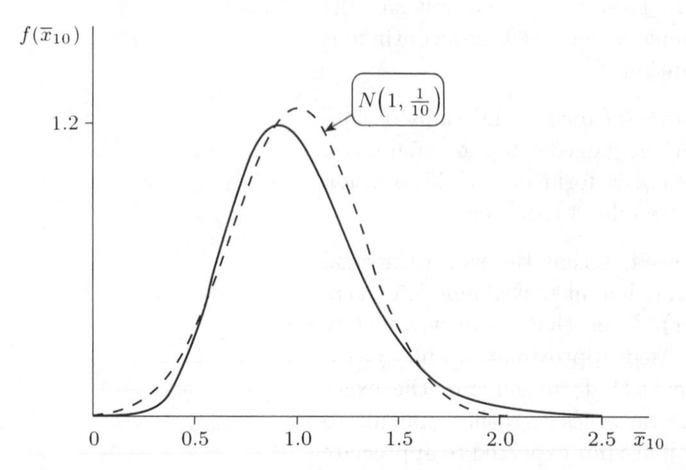

Figure 5.28 The density of $\overline{X}_{10}$ when $X \sim M(1)$

The dashed curve shown in Figure 5.28 is that of the normal density with mean 1 and variance $\frac{1}{10}$. You can see that the two curves are very similar. For modelling purposes, one might as well use the approximating and tractable normal curve—the exact distribution of $\overline{X}_{10}$ is not at all simple. ■

In fact when X is exponential, the distribution of the mean of a random sample from the distribution of X is known to belong to the *gamma family* of distributions. Some computer programs for statistics can supply exact probabilities for this family, eliminating the need for approximate normal probabilities.

Example 5.8 Summing uniform random variables

The uniform random variable $X \sim U(0,1)$ has the density shown in Figure 5.29. The density is flat; the random variable X has a minimum observable value at 0 and a maximum at 1.

The mean of a sample of size 2, $\overline{X}_2 = \frac{1}{2}(X_1 + X_2)$, again has a range extending from 0 to 1. The p.d.f. of $\overline{X}_2$ is symmetric; but now there is a clear mode at the midpoint of the range, $\overline{x}_2 = \frac{1}{2}$. This is shown in Figure 5.30.

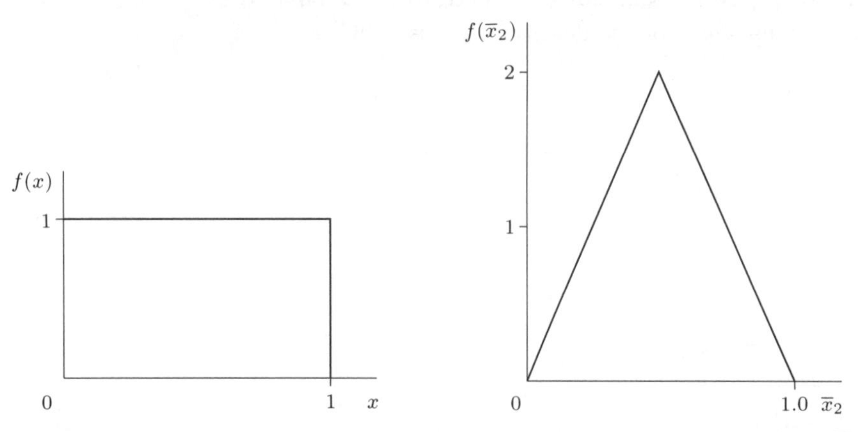

Figure 5.29 The density of X, $X \sim U(0,1)$ ***Figure 5.30*** The density of $\overline{X}_2$, $X \sim U(0,1)$

The mean of a sample of size 3, $\overline{X}_3 = \frac{1}{3}(X_1 + X_2 + X_3)$, again has a density defined only over the range from 0 to 1; the mean of $\overline{X}_3$ is $\frac{1}{2}$; the variance of $\overline{X}_3$ is $\frac{1}{12}/3 = \frac{1}{36}$. Its p.d.f. is drawn in Figure 5.31; the superimposed dotted line is the p.d.f. of the normal distribution with mean $\frac{1}{2}$ and variance $\frac{1}{36}$. You can see that the approximation is already extremely good.

When X is $U(0,1)$, $V(X) = \frac{1}{12}$.

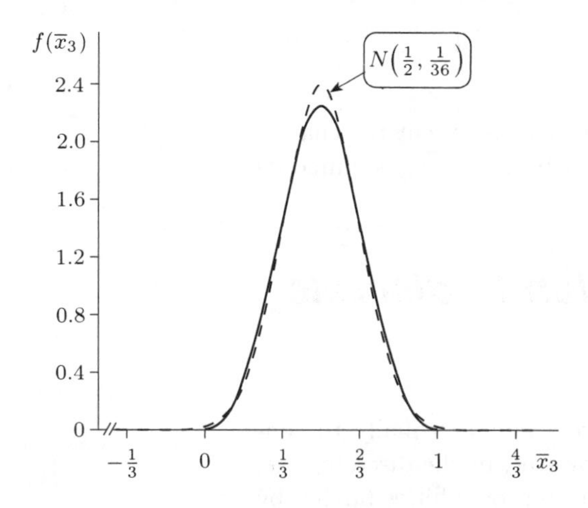

Figure 5.31 The density of $\overline{X}_3$ when $X \sim U(0,1)$ ■

Example 5.9 Summing beta random variables

The random variable X whose p.d.f. is given by

$$f(x) = \frac{x^{-1/6}(1-x)^{-5/6}}{2\pi}, \quad 0 < x < 1,$$

is a member of the *beta family*. (You do not need to know any general properties of the beta family of probability distributions.) It has a highly skewed U-shaped distribution defined over the range $(0, 1)$, as you can see from Figure 5.32. It is not at all easy to obtain the algebraic form of the density of the random variable $\overline{X}_2 = \frac{1}{2}(X_1 + X_2)$, let alone of means of larger samples. Instead, the histograms for 1000 observations on each of the random variables $\overline{X}_2, \overline{X}_{10}, \overline{X}_{20}$ are shown in Figure 5.33.

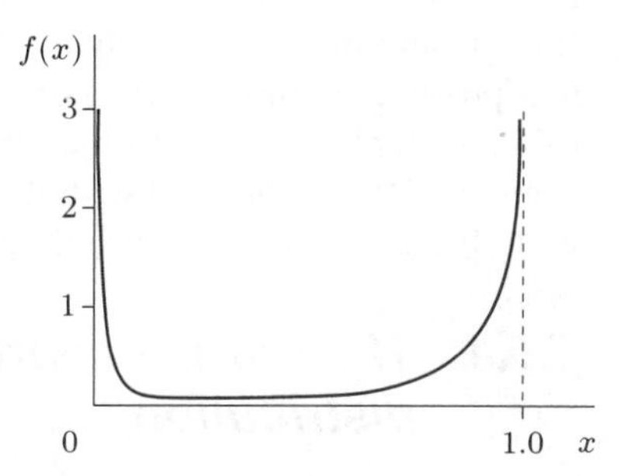

Figure 5.32 $f(x) = x^{-1/6}(1-x)^{-5/6}/2\pi, \quad 0 < x < 1$

In this diagram the scales have been slightly distorted to exaggerate the main features of the U-shaped density.

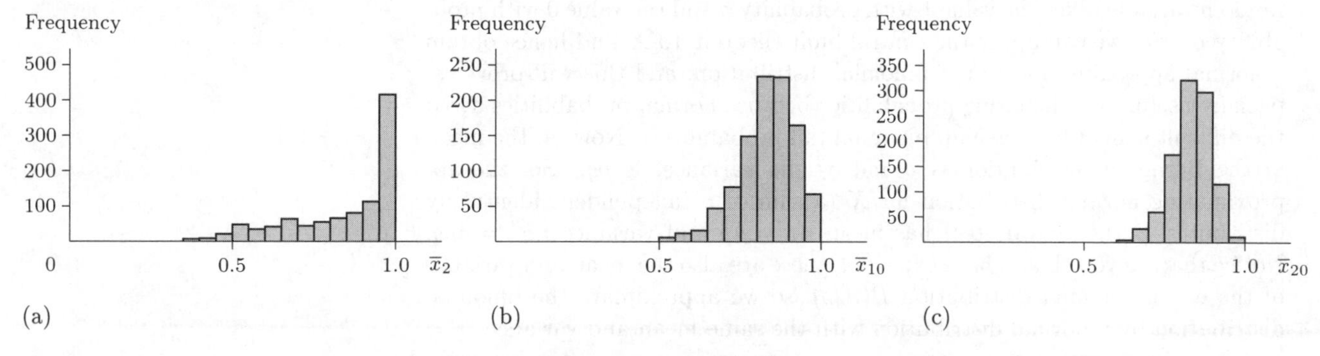

Figure 5.33 (a) 1000 observations on $\overline{X}_2$ (b) 1000 observations on $\overline{X}_{10}$ (c) 1000 observations on $\overline{X}_{20}$ ■

The three histograms are suggestive of the shape of the theoretical density functions for $\overline{X}_2$, $\overline{X}_{10}$ and $\overline{X}_{20}$. You can see that even for samples of size 10 some characteristics of the normal distribution are beginning to become apparent and the distribution of the sample mean for samples of size 20 would appear to be quite usefully approximated by a normal distribution.

The purpose of this section has been to provide graphical support for the statement of the central limit theorem. Irrespective of the shape of the distribution of the random variable X (even when it is flat, or U-shaped) the distribution of the random variable

$$\overline{X}_n = \frac{1}{n}(X_1 + X_2 + \cdots + X_n),$$

the mean of random samples of size n, has shown to some extent the characteristics of the normal distribution: that is, the distribution of $\overline{X}_n$ is unimodal and approximately symmetric.

5.5 The normal approximation to discrete distributions

It has been stressed that the central limit theorem applies equally to continuous and discrete underlying distributions. However, in the discrete case, it is possible to improve the normal approximation to probabilities further by using a simple device called a *continuity correction*. This is developed in the context of the binomial distribution, but the idea is applicable to any discrete underlying distribution, including the Poisson distribution.

5.5.1 The normal approximation to the binomial distribution

The binomial distribution, $B(n,p)$, with parameters n and p is a discrete distribution with probability mass function

$$p_X(x) = \binom{n}{x} p^x q^{n-x}, \qquad x = 0, 1, 2, \ldots, n,$$

where $q = 1 - p$.

You have seen that the random variable $X \sim B(n,p)$ can be thought of as the sum of n independent Bernoulli variates each with parameter p. (A Bernoulli random variable takes the value 1 with probability p and the value 0 with probability q.) So, we can apply the central limit theorem to X and hence obtain a normal approximation to the binomial distribution, and this will prove especially useful for calculating probabilities because normal probabilities avoid the difficult sums that make up binomial tail probabilities. Now, μ, the mean of the Bernoulli distribution is p and σ^2, its variance, is pq. So, the approximating normal distribution for X (a sum of n independent identically distributed Bernoulli variates) has mean $n\mu = np$ and variance $n\sigma^2 = npq$. Notice that, as you should have expected, these are also the mean and variance of the exact binomial distribution $B(n,p)$, so we approximate the binomial distribution by a normal distribution with the same mean and variance.

Example 5.10 *Comparing the distributions $B(16, \frac{1}{2})$ and $N(8,4)$*

As an example of this approximation let us take $n = 16$, $p = \frac{1}{2}$. Then $np = 8$ and $npq = 4$. Graphs of the binomial probability mass function and the approximating normal p.d.f. are shown superimposed in Figure 5.34.

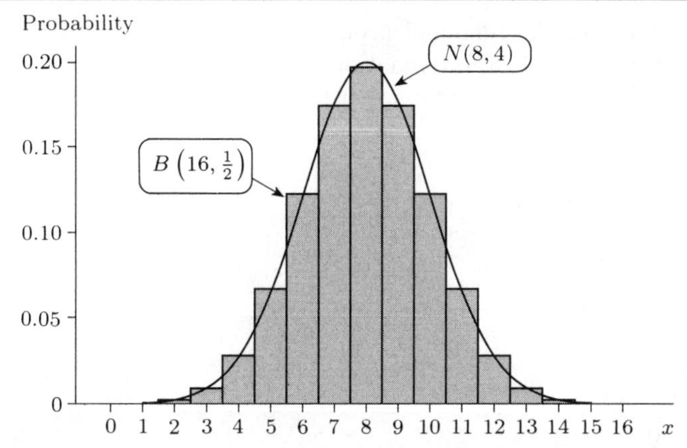

Figure 5.34 The distributions $B\left(16,\frac{1}{2}\right)$ and $N(8,4)$ compared ■

Apart from the obvious differences between a discrete and a continuous distribution, these graphs are really very similar. But what should be our approach if we wish to use the normal approximation to estimate the probability $P(X \leq 6)$, say? In general, we have seen that when approximating the distribution of X (with mean μ and standard deviation σ) by a normal distribution with the same mean and standard deviation, we have used the approximation

$$P(X \leq x) = P\left(Z \leq \frac{x-\mu}{\sigma}\right) = \Phi\left(\frac{x-\mu}{\sigma}\right),$$

where $\Phi(\cdot)$ is the c.d.f. of the standard normal variate Z. In this case, X is binomial with mean $np = 8$ and variance $npq = 4$: so we set μ equal to 8 and σ equal to $\sqrt{4} = 2$. Here, in Table 5.2, are the corresponding binomial and normal c.d.f.s.

Table 5.2 Binomial and normal c.d.f.s

x	Binomial c.d.f.	$\dfrac{x-8}{2}$	$\Phi\left(\dfrac{x-8}{2}\right)$
0	0.0000	−4	0.0000
1	0.0003	−3.5	0.0002
2	0.0021	−3	0.0013
3	0.0106	−2.5	0.0062
4	0.0384	−2	0.0228
5	0.1051	−1.5	0.0668
6	0.2272	−1	0.1587
7	0.4018	−0.5	0.3085
8	0.5982	0	0.5000
9	0.7728	0.5	0.6915
10	0.8949	1	0.8413
11	0.9616	1.5	0.9332
12	0.9894	2	0.9772
13	0.9979	2.5	0.9938
14	0.9997	3	0.9987
15	1.0000	3.5	0.9998
16	1.0000	4	1.0000

You can see that calculated values for the two c.d.f.s are quite close but that, at these particular values of x, the c.d.f. of the normal variate is always smaller than that of the binomial. You can see from Figure 5.35 some indication of why this is happening. This figure shows superimposed on the same diagram the distribution of the binomial random variable $X \sim B(16, \frac{1}{2})$ and the approximating normal random variable $Y \sim N(8,4)$. The shaded area gives the exact binomial probability required, $P(X \leq 6)$, and the hatched area gives the normal probability $P(Y \leq 6)$.

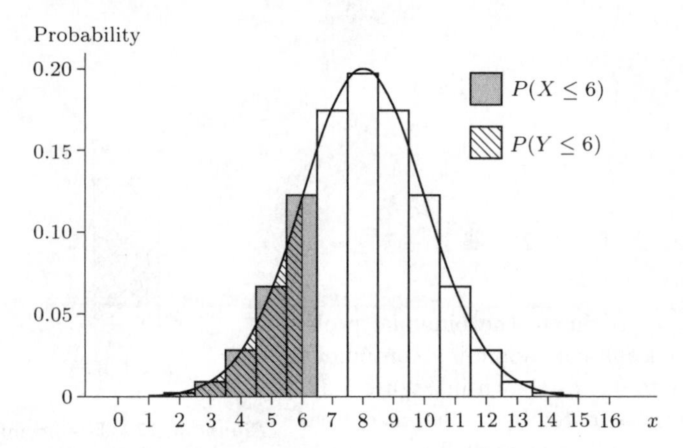

Figure 5.35 The probabilities $P(X \leq 6)$ and $P(Y \leq 6)$ compared

The diagram suggests that a more accurate estimate of the binomial probability $P(X \leq 6)$ would be obtained from the normal approximation $P(Y \leq 6\frac{1}{2})$. This comparison is shown in Figure 5.36.

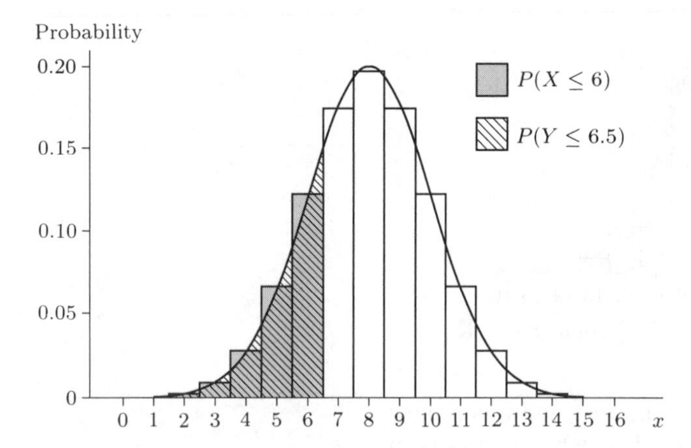

Figure 5.36 The probabilities $P(X \leq 6)$ and $P\left(Y \leq 6\frac{1}{2}\right)$ compared

The normal approximation gives

$$P(Y \leq 6\tfrac{1}{2}) = P\left(Z \leq \frac{6\frac{1}{2} - 8}{2}\right) = P(Z \leq -0.75) = 0.2266$$

and this is indeed very close to the exact value of the binomial probability, 0.2272.

We may also need to use the normal distribution to approximate values of the binomial probability mass function: then the same approach is adopted, as shown in the following example.

Example 5.11 Approximating a binomial probability

Suppose we wish to approximate the binomial probability mass function when $x = 6$. We know that $P(X = 6) = P(X \leq 6) - P(X \leq 5)$ and that this can be approximated by

$$\begin{aligned}
&P(Y \leq 6\tfrac{1}{2}) - P(Y \leq 5\tfrac{1}{2}) \\
&= P\left(Z \leq \frac{6\frac{1}{2} - 8}{2}\right) - P\left(Z \leq \frac{5\frac{1}{2} - 8}{2}\right) \\
&= P(Z \leq -0.75) - P(Z \leq -1.25) \\
&= 0.2266 - 0.1056 \\
&= 0.1210.
\end{aligned}$$

Again, this is reasonably close to the true value of 0.1222. ■

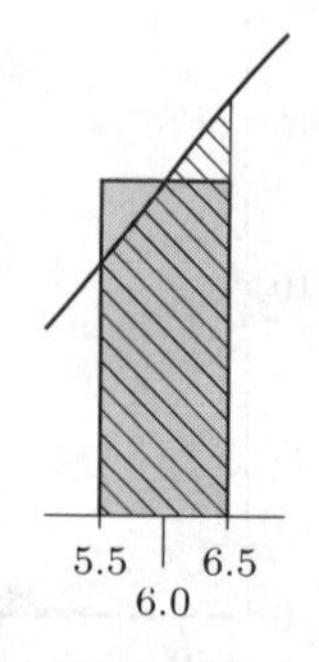

Figure 5.37 The normal approximation to the binomial probability $p_X(6)$

What has been done in this example is to approximate the binomial probability function $p_X(6)$ by the probability that the corresponding continuous random variable lies in the interval from $5\frac{1}{2}$ to $6\frac{1}{2}$, as shown in Figure 5.37. That is, the area of the bar centred on 6 is approximated by the hatched area; you can see that a good approximation ensues. The approach that has been used is an intuitively sensible way to behave, and it is found in general to give extremely satisfactory approximations for quite moderately sized n. This adjustment for discrete distributions to the normal approximation is often called the **continuity correction**.

The continuity correction

When approximating the c.d.f. of an *integer-valued discrete* random variable X using the central limit theorem, write

$$P(X \leq x) \simeq P\left(Y \leq x + \tfrac{1}{2}\right), \qquad (5.7)$$

where Y is normally distributed with the same mean and variance as X: that is, $Y \sim N(\mu_X, \sigma_X^2)$.

In the case where X is a binomial random variable with mean np and variance npq, the expression (5.7) becomes

$$P(X \leq x) \simeq P\left(Y \leq x + \tfrac{1}{2}\right) = P\left(Z \leq \frac{x + \frac{1}{2} - np}{\sqrt{npq}}\right) = \Phi\left(\frac{x + \frac{1}{2} - np}{\sqrt{npq}}\right).$$

Exercise 5.18

If $X \sim B\left(16, \frac{1}{2}\right)$ use the normal distribution to approximate the probability $P(12 \leq X \leq 15)$. Compare your answer with the true value of the probability.

So far we have looked at approximations to the binomial distribution when $p = \frac{1}{2}$ and found them to be reasonably good. We should expect the approximation to work best in this case because when $p = \frac{1}{2}$ the probability mass function of the binomial distribution is symmetric like the normal p.d.f. This is, of course, precisely what was driving Galton's exhibition of approximate normality using the quincunx, as was described in Section 5.1. However, the approximation is also useful when p is not equal to $\frac{1}{2}$. Here is a diagram of the $B(20, 0.3)$ probability mass function and the approximating $N(6, 4.2)$ p.d.f. to help convince you of this.

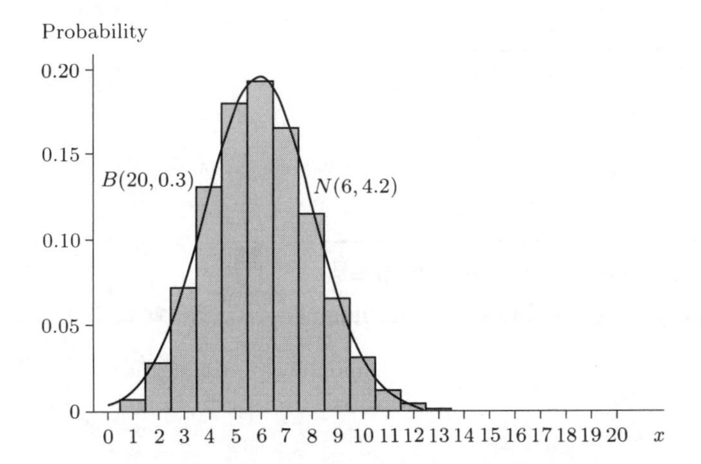

Figure 5.38 The probability distributions $B(20, 0.3)$ and $N(6, 4.2)$ compared

Example 5.12 An asymmetric binomial distribution

Suppose the random variable X has a binomial distribution $B(20, 0.3)$. Then the mean of X is $\mu = np = 6$, and its variance is $\sigma^2 = npq = 4.2$. Then, for example,

$$P(X \leq 5) = 0.4164$$

while (writing $Y \sim N(6, 4.2)$ and employing a continuity correction)

$$P(Y \leq 5\tfrac{1}{2}) = P\left(Z \leq \frac{5\frac{1}{2} - 6}{\sqrt{4.2}}\right) = P(Z \leq -0.24)$$

and from the tables this is 0.4052. Similarly,

$$P(X = 4) = 0.1304.$$

This probability may be approximated by

$$P(3\tfrac{1}{2} \leq Y \leq 4\tfrac{1}{2}) = P\left(\frac{3\frac{1}{2} - 6}{\sqrt{4.2}} \leq Z \leq \frac{4\frac{1}{2} - 6}{\sqrt{4.2}}\right) = P(-1.22 \leq Z \leq -0.73),$$

shown in the hatched area in Figure 5.39. From the tables this probability is $0.2327 - 0.1112 = 0.1215$. The normal approximation to the exact binomial probability required is not as good in this example as it was in Example 5.11, but you can see that any differences are not serious.

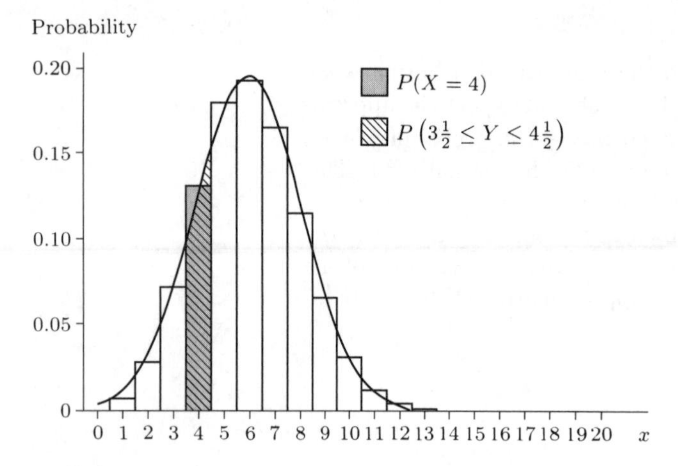

Figure 5.39 ■

Exercise 5.19

Suppose that the binomial random variable X has parameters $n = 25$, $p = \frac{1}{4}$.

(a) Use your computer to obtain the following probabilities to 6 decimal places.

(i) $P(X = 5)$

(ii) $P(X = 6)$

(iii) $P(X = 7)$

(iv) $P(X = 8)$

(b) Write down the probability $P(6 \leq X \leq 8)$.

(c) Give the parameters of the normal approximation to this binomial distribution indicated by the central limit theorem.

(d) Use the normal approximation to find the following probabilities, by rewriting these probabilities in terms of the normal random variable Y approximating the distribution of X.

(i) $P(X = 6)$

(ii) $P(X = 7)$

(iii) $P(X = 8)$

(iv) $P(6 \leq X \leq 8)$

You have seen that the central limit theorem provides us with a very useful approximation to the binomial distribution provided n is fairly large; it is also important that the binomial distribution is not too far from being symmetric. Asymmetry is most evident when p is close to either 0 or 1. Here is a rough rule for deciding when it may be appropriate to use a normal approximation for a binomial distribution.

> The c.d.f. of the normal distribution with a continuity correction provides a usable approximation to the c.d.f. of the binomial distribution $B(n, p)$ when both $np \geq 5$ and $nq \geq 5$, where $q = 1 - p$.

This rule is of course fairly arbitrary and whether or not it works depends on how close an approximation is required, but it provides a reasonable basis for deciding when the approximation may be used.

5.5.2 Normal approximations to other discrete distributions

This chapter ends by looking at two more discrete distributions, one already familiar to you, and their normal approximations. The method used is the same as for the binomial distribution; the c.d.f. of our discrete random variable is approximated by that of a continuous random variable with a normal distribution, by invoking the central limit theorem, and a continuity correction is included to improve the approximation to calculated probabilities.

The familiar case is the Poisson distribution. It was mentioned in *Chapter 4*, Section 4.3 that if $X_1, X_2, \ldots, X_n$ are independent Poisson variates with means all equal to μ, then their total also follows a Poisson distribution:

$$\sum_{i=1}^{n} X_i \sim \text{Poisson}(n\mu).$$

So, if we wish to approximate by a normal distribution a Poisson distribution with large mean μ, then we may *think of* the Poisson distribution as arising as a sum of a large number n of independent Poisson variates, each with mean μ/n. By the central limit theorem, their sum will be approximately normally distributed.

In other words, the central limit theorem tells us that a Poisson distribution with mean μ may be approximated by a normal distribution with the same mean and variance. In this case, these are both equal to μ.

For instance, if X is Poisson(16), then we might try the approximation

$$P(X \leq x) \simeq P\left(Y \leq x + \tfrac{1}{2}\right),$$

where $Y \sim N(16, 16)$, and

$$P(X = x) \simeq P\left(x - \tfrac{1}{2} \leq Y \leq x + \tfrac{1}{2}\right).$$

Values based on this last approximation are given in Table 5.3.

Table 5.3 Exact and approximate Poisson probabilities

x	$p_X(x)$	Approximation
0	0.0000	0.0000
1	0.0000	0.0001
2	0.0000	0.0002
3	0.0001	0.0005
4	0.0003	0.0011
5	0.0010	0.0023
6	0.0026	0.0044
7	0.0060	0.0080
8	0.0120	0.0136
9	0.0213	0.0217
10	0.0341	0.0325
11	0.0496	0.0457
12	0.0661	0.0605
13	0.0814	0.0752
14	0.0930	0.0878
15	0.0992	0.0964
16	0.0992	0.0995
17	0.0934	0.0964
18	0.0830	0.0878
19	0.0699	0.0752
20	0.0559	0.0605
21	0.0426	0.0457
22	0.0310	0.0325
23	0.0216	0.0217
24	0.0144	0.0136
25	0.0092	0.0080
26	0.0057	0.0044
27	0.0034	0.0023
28	0.0019	0.0011
29	0.0011	0.0005
30	0.0006	0.0002
> 30	0.0006	0.0001

Although the approximating probabilities are not *very* close, they are always within about 0.006 of the true probability, and so usually would be sufficiently accurate.

Exercise 5.20

If the random variable X has a Poisson distribution with mean 40, use your computer to find as accurately as you can the probability $P(30 \leq X \leq 45)$ and then find an approximation to this probability using the central limit theorem.

The point at which the normal approximation $N(\mu, \mu)$ to the Poisson distribution Poisson(μ) becomes a useful approximation depends essentially on the purpose to which the resulting calculations will be put; but a rough rule which many practitioners use is that μ should be at least 30. As μ becomes larger than 30 the approximation gets better and better.

The next example is about sums of discrete uniform distributions, a context that so far we have not considered.

Example 5.13 *Rolling three dice*

Finally, let us try rolling some dice! The probability distribution of the total score when three fair dice are rolled is actually quite difficult to find: it is the sum of three independent identically distributed discrete uniform scores on $1, 2, 3, \ldots, 6$. Let us try using the central limit theorem to obtain an approximation to the probability that the total score exceeds 15.

We know that if X is uniformly distributed on the integers $1, 2, \ldots, n$, then

$$E(X) = \mu = \frac{n+1}{2}, \qquad V(X) = \sigma^2 = \frac{n^2 - 1}{12}.$$

These results were stated in *Chapter 3*.

Setting n equal to 6 gives

$$\mu = \frac{6+1}{2} = 3.5, \qquad \sigma^2 = \frac{35}{12} = 2.9167.$$

The sum $S = X_1 + X_2 + X_3$ of three independent scores has mean $3\mu = 10.5$ and variance $3\sigma^2 = 8.75$. Then the probability required (that the total score S exceeds 15) may be approximated by writing

$$P(S > 15) \simeq P(Y \geq 15.5),$$

where $Y \sim N(10.5, 8.75)$. This is

$$P(Y \geq 15.5) = P\left(Z \geq \frac{15.5 - 10.5}{\sqrt{8.75}}\right) = 0.0455.$$

In fact, the distribution of S is given in the following table.

Table 5.4 The distribution of the total score S, when three dice are rolled

s	3	4	5	6	7	8	9	10	11	12	13	14	15	16	17	18
$P(S = s)$	$\frac{1}{216}$	$\frac{3}{216}$	$\frac{6}{216}$	$\frac{10}{216}$	$\frac{15}{216}$	$\frac{21}{216}$	$\frac{25}{216}$	$\frac{27}{216}$	$\frac{27}{216}$	$\frac{25}{216}$	$\frac{21}{216}$	$\frac{15}{216}$	$\frac{10}{216}$	$\frac{6}{216}$	$\frac{3}{216}$	$\frac{1}{216}$

The probability that the total score on a throw of three dice exceeds 15 is therefore

$$\tfrac{6}{216} + \tfrac{3}{216} + \tfrac{1}{216} = \tfrac{10}{216} = 0.0463;$$

and so the normal approximation to this probability is not bad. ■

Summary

1. Many manifestations of variability in the real world may be adequately modelled by the two-parameter normal distribution $N(\mu, \sigma^2)$ with mean μ and variance σ^2 (standard deviation σ).
2. The p.d.f. of the normal random variable $X \sim N(\mu, \sigma^2)$ is given by

$$f(x) = \frac{1}{\sigma\sqrt{2\pi}} \exp\left[-\frac{1}{2}\left(\frac{x-\mu}{\sigma}\right)^2\right], \quad -\infty < x < \infty.$$

The p.d.f. of the standard normal random variable $Z \sim N(0,1)$ is given by

$$\phi(z) = \frac{1}{\sqrt{2\pi}} e^{-\frac{1}{2}z^2}, \quad -\infty < z < \infty.$$

The c.d.f. of Z is given by

$$\Phi(z) = \int_{-\infty}^{z} \phi(x)\, dx.$$

3. In the usual way, probabilities are found by integrating under the normal curve, but this approach does not yield a closed formula. Instead, reference is made to tables of standard normal probabilities and the relationship between $X \sim N(\mu, \sigma^2)$ and the standard normal variable $Z \sim N(0,1)$:

$$\frac{X - \mu}{\sigma} = Z$$

and so

$$P(X \leq x) = P\left(Z \leq \frac{x-\mu}{\sigma}\right) = \Phi\left(\frac{x-\mu}{\sigma}\right).$$

4. Standard normal quantiles are defined by

$$q_\alpha = P(Z \leq \alpha).$$

To find the quantiles of $X \sim N(\mu, \sigma^2)$ it is necessary to use the relation

$$X = \sigma Z + \mu.$$

5. If the random variables $X_i, i = 1, \ldots, n$, are independent normally distributed random variables with mean μ_i and variance σ_i^2, then their sum is also normally distributed:

$$X_1 + X_2 + \cdots + X_n \sim N(\textstyle\sum \mu_i, \sum \sigma_i^2);$$

also, if X is normally distributed with mean μ and variance σ^2, then for constants a and b,

$$aX + b \sim N(a\mu + b, a^2\sigma^2).$$

6. The central limit theorem states that if $X_1, X_2, \ldots, X_n$ are independent identically distributed random variables with mean μ and variance σ^2, then their mean $\overline{X}_n$ has an approximate normal distribution

$$\overline{X}_n = \frac{X_1 + X_2 + \cdots + X_n}{n} \approx N\left(\mu, \frac{\sigma^2}{n}\right).$$

Equivalently, their sum T_n has an approximate normal distribution

$$T_n = X_1 + X_2 + \cdots + X_n \approx N(n\mu, n\sigma^2).$$

The symbol '$\approx$' is read 'has approximately the same distribution as'.

7. By the central limit theorem, if X is binomial $B(n, p)$, then the distribution of X may be approximated by a normal model with corresponding mean $\mu = np$ and corresponding variance $\sigma^2 = npq$, where $q = 1 - p$. The approximation will be useful if both np and nq are at least 5.
8. By the central limit theorem, if X is Poisson(μ), then the distribution of X may be approximated by a normal model with mean μ and variance μ. The approximation will be useful if μ is at least 30.
9. When approximating the distribution of a discrete integer-valued random variable X with mean μ and variance σ^2 by a normally distributed random variable $Y \sim N(\mu, \sigma^2)$, it is appropriate to use a continuity correction:

$$P(X \leq x) \simeq P\left(Y \leq x + \tfrac{1}{2}\right)$$

and

$$P(X = x) \simeq P\left(x - \tfrac{1}{2} \leq Y \leq x + \tfrac{1}{2}\right).$$

Chapter 6
Point estimation

This chapter deals with one of the central problems of statistics, that of using a sample of data to estimate the parameters for a hypothesized population model from which the data are assumed to arise. There are many approaches to estimation, and several of these are mentioned; but one method has very wide acceptance, the method of maximum likelihood.

So far in the course, a major distinction has been drawn between *sample* quantities on the one hand—values calculated from data, such as the sample mean and the sample standard deviation—and corresponding *population* quantities on the other. The latter arise when a statistical model is assumed to be an adequate representation of the underlying variation in the population. Usually the model family is specified (binomial, Poisson, normal, ...) but the indexing parameters (the binomial probability p, the Poisson mean μ, the normal variance σ^2, ...) might be unknown—indeed, they usually will be unknown. Often one of the main reasons for collecting data is to *estimate*, from a sample, the value of the model parameter or parameters.

Here are two examples.

Example 6.1 Counts of females in queues

In a paper about graphical methods for testing model quality, an experiment is described to count the number of females in each of 100 queues, all of length ten, at a London underground train station.

Jinkinson, R.A. and Slater, M. (1981) Critical discussion of a graphical method for identifying discrete distributions. *The Statistician*, **30**, 239–248.

The number of females in each queue could theoretically take any value between 0 and 10 inclusive (though, in fact, no queue from the 100 observed was constituted entirely of females). Table 6.1 shows the frequencies for the different counts.

Table 6.1 Numbers of females in 100 queues of length ten

Count	0	1	2	3	4	5	6	7	8	9	10
Frequency	1	3	4	23	25	19	18	5	1	1	0

A possible model for the observed variation is that the probability distribution of the number of females in a queue is binomial $B(10, p)$, where the parameter p (not known) is the underlying proportion of female passengers using the London underground transport system during the time of the experiment.

The parameter p may be estimated from this sample in an intuitive way by calculating

$$\frac{\text{total number of females}}{\text{total number of passengers}}$$
$$= \frac{1 \times 0 + 3 \times 1 + 4 \times 2 + \cdots + 1 \times 9 + 0 \times 10}{10 \times 100} = \frac{435}{1000} = 0.435,$$

or just under $\frac{1}{2}$.

Parameter estimates are not always 'obvious' or 'intuitive', as we shall see. Then you have to use mathematics as a guide; otherwise, mathematics can be used to confirm your intuition.

In this case, as it happens, the binomial model fits the data very well; but it could have turned out that the binomial model provided a less than adequate representation of the situation. One of the assumptions underlying the model is that of independence. In this case that means that the gender of any person in a queue does not affect, and is not affected by, the gender of others. If, for instance, there had been too many Male–Female couples standing together, then the observed frequency of 4s, 5s and 6s in the counts would have been too high. ■

Example 6.2 Counts of the leech Helobdella

An experiment is described in Jeffers (1978) in which 103 water samples were collected and the number of specimens of the leech *Helobdella* contained in each sample was counted. More than half of the samples collected (58 of them) were free of this contamination, but all the other samples contained at least one leech—three contained five or more. Table 6.2 gives the frequencies of the different counts for the 103 samples.

Jeffers, J.N.R. (1978) *An introduction to systems analysis with ecological applications.* Edward Arnold, London.

Table 6.2 Counts of the leech *Helobdella* in 103 water samples

Count	0	1	2	3	4	5	6	7	8	≥9
Frequency	58	25	13	2	2	1	1	0	1	0

One model that might be thought to be at least reasonable for the observed variation in the counts is that they follow a Poisson distribution. The Poisson distribution is indexed by one parameter, its mean μ. You saw in *Chapter 4*, Section 4.2 that the sample mean $\overline{X}$ has some 'good' properties as an estimator for a population mean μ. (For instance, it has expectation μ: this was shown in (4.8) in *Chapter 4*.) In this case, the sample mean is

$$\overline{x} = \frac{0 \times 58 + 1 \times 25 + 2 \times 13 + \cdots + 8 \times 1}{58 + 25 + 13 + \cdots + 1} = \frac{84}{103} = 0.816.$$

This constitutes an estimate, based on this sample, for the unknown Poisson mean μ.

(In fact, it turns out that the observed variation here is not very well expressed by a Poisson model. The reason is that the leeches tend to cluster together within the water mass from which they were drawn: that is, they are not independently located. Model testing is examined in *Chapter 9* of the course.) ■

There are many questions that might be asked about model parameters, and in this course *Chapters 6* to *8* are devoted to answering such questions. One question is of the general form: is it reasonable to believe that the value of a particular parameter is ...? (zero, or one-half, or positive, or whatever). Appropriate sampling procedures, followed by an analysis of the data collected, permit informative tests of such hypotheses. For example, is it reasonable to believe the proposition that at least half the passengers using London underground trains are female, given this moderate evidence to the contrary? The subject of hypothesis testing is examined in *Chapter 8*.

Another, related, matter of interest is this: on the basis of these data, and assuming an underlying model for the variation observed, what single value, or what range of values, is plausible for the indexing parameter?

Here, there are two questions posed. The first asks for the single best guess, or *estimate*, of the value of the parameter. To find this, we shall apply an estimating formula, or *estimator*, to the data available. This is a question about **point estimation** and it is the subject of the present chapter. The second question asks for a range of credible values for an unknown model parameter, based on a sample of data. Such a range is called a *confidence interval*: the idea of interval estimation is discussed in *Chapter 7*.

In Sections 6.1 and 6.2 we look at a number of different situations in which data have been collected on a random variable, and where there is a clear problem of parameter estimation. A number of possible approaches to the problem is described (there are many others).

Often, a somewhat involved mathematical procedure leads to an estimator which after all is the obvious common-sense estimator to choose, though this is not always the case. In Section 6.3 we explore one particular approach which has common acceptance, leading to estimators with good properties, and that does not usually conflict badly with the dictates of common sense. This is known as the *method of maximum likelihood*.

In Section 6.4 we examine a number of particular examples. As well as revisiting earlier examples, we briefly explore estimation problems where one method or another fails (for whatever reason) but where the method of maximum likelihood proves to be useful and informative. (Nevertheless, similar examples could be found to embarrass almost any estimating procedure, including maximum likelihood.) Most of the exercises in this section require the use of a computer for the numerical procedures involved.

Section 6.5 is very reliant on computer simulation. A new continuous probability distribution, the *chi-squared distribution*, is introduced. The distribution is relevant to the problem of estimating the variance of a normal distribution.

6.1 Principles of point estimation

Point estimation is the process of using the data available to estimate the unknown value of a parameter, when some representative statistical model has been proposed for the variation observed in some chance phenomenon. The *point estimate* obtained from the data will be a single number. Here are some examples. All the examples illustrate important features of point estimation. These essential features are summarized after Example 6.5 and restated at the end of the section.

Example 6.2 continued

The *Helobdella* experiment resulted in a data list comprising 103 observations on a discrete random variable (58 0s, 25 1s, and so on). For the purposes of drawing useful inferences from the experiment, these data were assumed to be *independent* observations on the same random variable, a Poisson variate with unknown mean μ. One obvious estimate for the parameter μ is the sample mean

$$\overline{x} = \frac{0 \times 58 + 1 \times 25 + 2 \times 13 + \cdots + 8 \times 1}{58 + 25 + 13 + \cdots + 1} = \frac{84}{103} = 0.816.$$

In this case our data set comprised a particular collection of 103 independent observations $X_1, X_2, \ldots, X_{103}$ on the random variable $X \sim \text{Poisson}(\mu)$; our estimate of μ was the corresponding particular observation on the sample mean, the random variable

$$\overline{X}_{(1)} = \frac{X_1 + X_2 + \cdots + X_{103}}{103}.$$

Suppose a similar study was carried out under similar circumstances, but on that occasion only 48 water samples were collected; as before, the number of leeches in each sample was counted. Assuming the same Poisson model, then the new estimate of the unknown parameter μ would be

$$\overline{x}_{(2)} = \frac{x_1 + x_2 + \cdots + x_{48}}{48}.$$

This is an observation on the random variable

$$\overline{X}_{(2)} = \frac{X_1 + X_2 + \cdots + X_{48}}{48}. \quad \blacksquare$$

Exercise 6.1

The first experiment (the one actually carried out) resulted in an estimate for μ of $\overline{x}_{(1)} = 0.816$, an observation on the random variable $\overline{X}_{(1)}$. The second (notional) experiment resulted in an estimate $\overline{x}_{(2)}$ which was an observation on the random variable $\overline{X}_{(2)}$.

Write down in terms of the unknown parameter μ the mean and variance of the random variable $\overline{X}_{(1)}$ and of the random variable $\overline{X}_{(2)}$.

In this example we had a *procedure* or *estimating formula* which may be expressed thus: collect a total of n water samples and count the number of leeches $X_1, X_2, \ldots, X_n$ in each sample; find the total number of leeches $X_1 + X_2 + \cdots + X_n$ and divide this number by n to obtain the average number of leeches in a sample of water. In other words, the random variable

$$\overline{X} = \frac{X_1 + X_2 + \cdots + X_n}{n}$$

is the estimating formula for μ. It is called an **estimator** for μ. With different data sets, different values of the estimator will be obtained (for example, 84/103 or perhaps 30/48). These are **estimates** for μ.

Example 6.3 *Alveolar-bronchiolar adenomas in mice*

In a research experiment into the incidence of alveolar-bronchiolar adenomas in mice, several groups of mice were examined. One of the groups contained 54 mice: after examination, six of the 54 mice were found to have adenomas.

Tamura, R.N. and Young, S.S. (1987) A stabilized moment estimator for the beta-binomial distribution. *Biometrics*, **43**, 813–824. An adenoma is a benign tumour originating in a gland.

Assuming independence from subject to subject, the experiment consists of observing an outcome r on a binomial random variable $R \sim B(n,p)$. In this case the number observed was $r = 6$ and the sample size was $n = 54$; the obvious estimate of the proportion p is $\frac{6}{54} = \frac{1}{9}$, or about 11%.

A different experiment might have involved a different number, n, of subjects; assuming the experimental design to be the same, and making the same assumptions, then the number of affected mice is a binomial random variable $R \sim B(n,p)$, and the experiment will result in the estimate r/n. Here, our procedure or estimating formula is: observe the value of the random variable R and divide this observed value by n. So R/n is the estimating formula, or *estimator*, for p. In different experiments, different values of the estimator will be obtained. These are *estimates* $r_1/n_1, r_2/n_2, \ldots$.

Indeed, the experiment involved altogether 23 groups of mice. Examination of the other 22 groups resulted in several different estimates for the proportion of affected mice in the wider population, from as low as $\frac{0}{20}$ through $\frac{4}{47}$ to estimates as high as $\frac{4}{20} = \frac{1}{5}$. ■

Example 6.4 Sand flies

An experiment was performed in which sand flies were caught in two different light traps; then the numbers of male and female flies were counted in each trap.

Christiensen, H.A., Herrer, A. and Telford, S.R. (1972) Enzootic cutaneous leishmaniasis in Eastern Panama. II: Entomological investigations. *Annals of Tropical Medicine and Parasitology*, **66**, 55–66.

The first trap was set three feet above the ground: when the traps were lifted it contained 173 male sand flies and 150 females—an observed proportion of 173/323 males, or about 54% (just over one-half). The second trap was set 35 feet above the ground: on inspection it was found to contain 125 males and 73 females.

Thus we have two rather different estimates of the proportion of male sand flies in the population: whether or not the difference is a 'real' difference (fewer females venturing far above the ground) or due simply to random variation is the sort of question that is the subject of *Chapter 8*. At first sight, we have here one estimate $r_1/n_1 = 173/323 = 0.54$ for the proportion p_1 of males in the sand flies to be found 3 feet above ground level, and another estimate $r_2/n_2 = 125/198 = 0.63$ for the proportion p_2 of males in the sand fly population at 35 feet above ground. ■

Example 6.5 Epileptic seizures

A number of patients with intractable epilepsy controlled by anticonvulsant drugs was observed for times between three months and five years, and information about the number of daily seizures suffered by each patient was recorded. One of the patients was observed for 422 consecutive days. Table 6.3 gives the frequencies for the daily seizure counts.

Albert, P.S. (1991) A two-state Markov mixture model for a time series of epileptic seizure counts. *Biometrics*, **47**, 1371–1381.

Table 6.3 Counts of epileptic seizures, daily over 422 days

Count	0	1	2	3	4	5	6
Frequency	263	90	32	23	9	3	2

Assuming a Poisson model for the variation in the daily counts, a reasonable estimator for the indexing parameter (the mean μ) is given by the sample mean $\overline{X}$. In this case, for this particular data set, the corresponding estimate is

$$\overline{x} = \frac{0 \times 263 + 1 \times 90 + 2 \times 32 + 3 \times 23 + 4 \times 9 + 5 \times 3 + 6 \times 2}{263 + 90 + 32 + 23 + 9 + 3 + 2}$$
$$= \frac{286}{422}$$
$$= 0.68. \quad \blacksquare$$

The preceding examples illustrate the following essential features of a point estimation problem:

(a) some *data*—if we have no data then we cannot make an estimate;

(b) a *probability model* for the way the data were generated—in the above examples, $X \sim \text{Poisson}(\mu)$ or $R \sim B(n,p)$;

(c) the model involves a parameter whose value is unknown—this is the value we wish to estimate;

(d) an estimating formula or *estimator* for the parameter, irrespective of any particular data values obtained, depending only on the model;

(e) the value of the estimator given by the data, that is, the *estimate* for the parameter.

The probability model adopted may or may not be quite adequate to explain the variation observed: but this is a problem of goodness of fit, not primarily one of estimation, and it is addressed later in the course.

Notation

It is useful to introduce some notation here. In statistical work, it is often convenient to denote an estimate of a parameter by adding a circumflex or 'hat' symbol $\widehat{\ }$ thus: we might write that the water experiment resulted in the estimated mean incidence of leeches $\widehat{\mu} = 0.816$. Similarly, the estimated proportion of males in the sand fly population at 35 feet above ground level was $\widehat{p}_2 = 0.63$. Most statisticians (in speaking, or reading to themselves) conventionally use the phrase 'μ hat' or 'p hat' for simplicity.

The same notation is also used for an *estimator*—the estimator of p in the above examples is $\widehat{p} = R/n$; similarly, the estimator of μ is $\widehat{\mu} = \overline{X} = (X_1 + X_2 + \cdots + X_n)/n$. Notice the essential difference here. The estimate $\widehat{p}$ is a number obtained from data while the estimator $\widehat{p}$ is a random variable expressing an estimating formula. The estimate $\widehat{\mu} = \overline{x}$ is a number obtained from a data sample by adding together all the items in the sample and dividing by the sample size. The estimator $\widehat{\mu} = \overline{X}$ is a random variable. As a random variable, an estimator will itself follow some probability distribution, and this is called the **sampling distribution of the estimator**.

It would be unwieldy to develop a separate notation to distinguish estimates and estimators: you will find in practice that there is little scope for confusion.

By looking in particular at summary measures of the sampling distribution of an estimator, particularly its mean and variance, we can get a good idea of how well the estimator in question can be expected to perform (that is, how 'accurate', or inaccurate, the estimates obtained from data might turn out to be). For instance, it would be useful if the estimator turns out to have expected value equal to the parameter we are interested in estimating. (In the case of a Poisson mean, for instance, we have used the fact that $E(\overline{X}) = \mu$ as the basis for our notion of a 'good' estimator. Also, if the random variable R

has a binomial distribution $B(n, p)$, then $E(R) = np$ and so $E(R/n) = p$.) It will be useful too if the estimator has a small variance: then we can expect estimates resulting from statistical experiments to be close to the parameter we are trying to estimate. In any given estimation problem, there is not always one clear estimator to use: there may be several possible formulas that could be applied. The question that naturally arises is: which formula is likely to lead to 'better' estimates? The next example illustrates this.

Example 6.6 Divorces in England and Wales, 1975–1980

The next data set is about the annual numbers of divorces in England and Wales for the six years between 1975 and 1980. The data are listed in Table 6.4. They appear to show some sort of linear upward trend, though there is not an exactly constant increase (and perhaps one would be rather surprised if there was). The aim of this example is to demonstrate that there are at least three sensible estimates that could be proposed for the underlying trend. The idea of estimating slopes through scattered data points is one that will be discussed in detail in *Chapter 10*. All you need to do for the moment is appreciate that in any context there may be more than one way of obtaining from data a 'sensible' estimate for an unknown parameter, and that there are simple methods for choosing from these estimates one that might be better than the others.

Data from *Marriage and Divorce Statistics*, Office of Population Censuses and Surveys, HMSO. For the purposes of this example, it is important that the data exhibit some sort of trend. More recent data than these (1985–1990) do not: there appears to be a levelling off.

Table 6.4 Annual numbers of divorces in England and Wales, 1975–1980

Year	1975	1976	1977	1978	1979	1980
Number of divorces (thousands)	120.5	126.7	129.1	143.7	138.7	148.3

Let us denote the number of divorces (thousands) in any particular year by y, and the year, reckoned from 1900 for convenience, by x. We can then plot the data as six points (x_i, y_i) $(i = 1, 2, \ldots, 6)$ on a scatter diagram as in Figure 6.1; so, for example, $x_1 = 75$, $y_1 = 120.5$; $x_6 = 80$, $y_6 = 148.3$. The six points are labelled P_1, P_2, P_3, P_4, P_5, P_6 going from left to right, i.e. P_1 is (x_1, y_1), and so on.

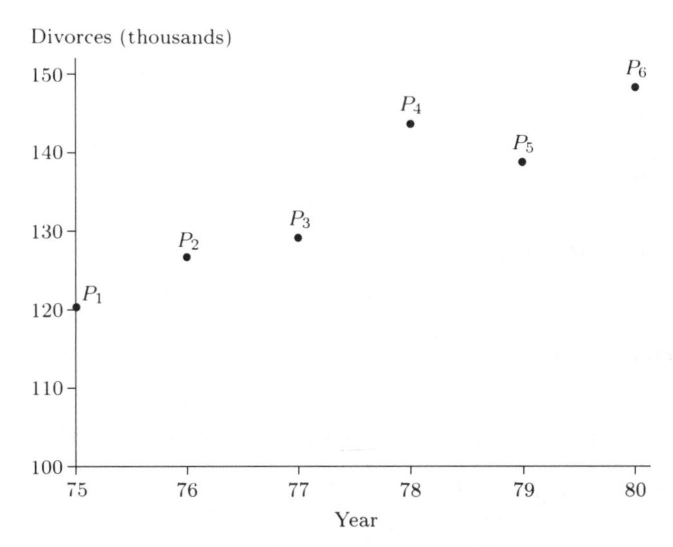

Figure 6.1 Divorces in England and Wales, 1975–1980 (thousands)

The pattern of the six points appears roughly linear, with positive slope; in other words, there appears to have been, over the six years covered by the data, an underlying upwards trend in the annual number of divorces. Our assumption will be that there is a trend line which truly underlies the six points; it has a slope β, say, which can be interpreted as the *annual rate of increase of divorces.* The number β is a *parameter* whose value we do not know; but we wish to *estimate* it.

In previous examples there was a unique 'obvious' way of estimating μ, namely $\widehat{\mu} = \overline{X}$, and of estimating p, namely $\widehat{p} = R/n$. Here, this is not the case; one can think of several apparently sensible ways of estimating β. Three estimators of β we might consider are:

(1) $\widehat{\beta}_1$ = the slope of the line joining the first and last points P_1 and P_6;

(2) $\widehat{\beta}_2$ = the slope of the line joining the midpoint of P_1P_2 to the midpoint of P_5P_6;

(3) $\widehat{\beta}_3$ = the slope of the line joining the 'centre of gravity' of the first three points P_1, P_2, P_3 to the centre of gravity of the last three points P_4, P_5, P_6.

The technical term for 'centre of gravity' is *centroid.* The centroid of the points P_1, P_2, P_3 has coordinates $\frac{1}{3}(P_1 + P_2 + P_3)$.

Other estimators are possible (you can probably think of some) and different ones are, in fact, usually preferred; but these three will suffice to make the important points. The three lines with slopes $\widehat{\beta}_1$, $\widehat{\beta}_2$ and $\widehat{\beta}_3$ are shown in Figure 6.2; these are labelled $\widehat{l}_1$, $\widehat{l}_2$, $\widehat{l}_3$.

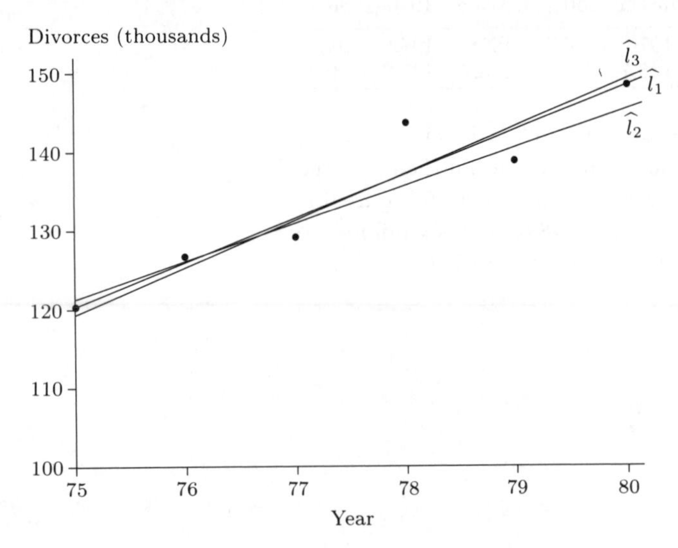

Figure 6.2 Divorces in England and Wales, 1975–1980: trend lines

With this data set, these three estimating procedures give the following three *estimates* (each to one decimal place):

$$\widehat{\beta}_1 = \frac{148.3 - 120.5}{80 - 75} = \frac{27.8}{5} = 5.6,$$

$$\widehat{\beta}_2 = \frac{(138.7 + 148.3)/2 - (120.5 + 126.7)/2}{79.5 - 75.5} = \frac{19.9}{4} = 5.0,$$

$$\widehat{\beta}_3 = \frac{(143.7 + 138.7 + 148.3)/3 - (120.5 + 126.7 + 129.1)/3}{79 - 76} = \frac{18.13}{3} = 6.0.$$

Which one of these three estimates should we use? Is one of them better than the others? For that matter, what does 'better' mean? The word 'better' does not refer to a specific value, like 5.6 or 5.0, but to the *estimating formula* which produces such a value. So we shall need to compare the properties of the various estimators: that is, we shall need to compare the properties of the sampling distributions of the estimators.

In order to make these comparisons we must first decide on a sensible probability model for the data: in other words, we must decide where randomness enters the situation, and how to model that randomness.

This is not quite obvious: there are certainly deviations in the data to be observed from a straight line, so the observed points (x_i, y_i) must be scattered above and below the trend line, wherever it may be. On the other hand, there is no real sense in which the data observed constitute part of a statistical experiment, an experiment that could be *repeated* on a similar occasion when different results would be recorded. There was only one year 1975, and during that year there were 120.5 thousands of divorces in England and Wales; and in a sense that is all that need, or can, be said.

Nevertheless, it is very common that deviations about a perceived trend are observed, and equally common to model those deviations as though they were evidence of random variation. This is what we shall do here.

We are assuming that there is an underlying linear trend in the number of divorces y year by year (that is, with increasing x). In other words, our assumption is that the underlying trend may be modelled as a straight line with equation

$$y = \alpha + \beta x,$$

where the parameter β is the slope we are trying to estimate.

Now, take any particular year, $x_1 = 75$, say. The observed number of divorces (in thousands) is $y_1 = 120.5$. On the other hand, the trend model, if it was accurate, predicts a total of $\alpha + \beta x_1 = \alpha + 75\beta$ thousands of divorces in 1975. The difference

$$y_1 - (\alpha + \beta x_1) = 120.5 - (\alpha + 75\beta)$$

between the observed value and that predicted by the model is a single observation on a random variable W, say. The scatter of the observed points $P_i = (x_i, y_i)$ above and below the trend line is measured by the fluctuations or deviations

$$\begin{aligned}
w_1 &= 120.5 - (\alpha + 75\beta)\\
w_2 &= 126.7 - (\alpha + 76\beta)\\
w_3 &= 129.1 - (\alpha + 77\beta)\\
w_4 &= 143.7 - (\alpha + 78\beta)\\
w_5 &= 138.7 - (\alpha + 79\beta)\\
w_6 &= 148.3 - (\alpha + 80\beta);
\end{aligned}$$

these six observations on the random variable W are illustrated in Figure 6.3. (For the purposes of illustration, a choice was made in this diagram for sensible values of α and β—but you need to realize that no calculations have in fact yet been performed.)

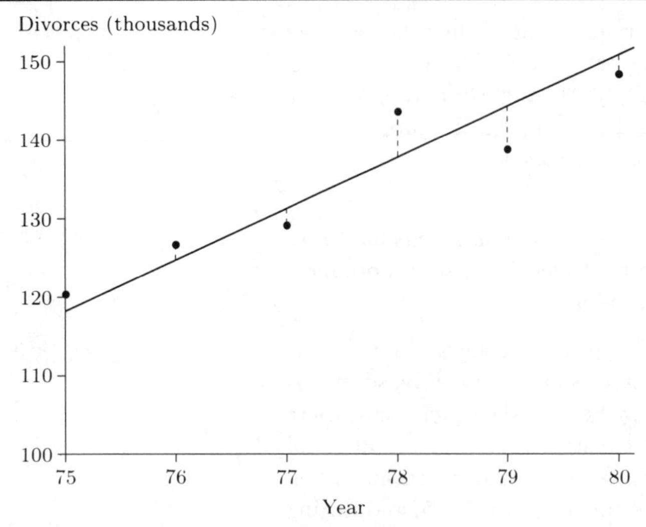

Figure 6.3 Six observations on the random variable W

We shall assume that the observations $w_1, w_2, \ldots, w_6$ may be treated as though they were *independent* observations on the *same* random variable W; further, that the random variable W has mean 0 (observations occur above and below the trend line model); finally, that the random variable W has non-zero variance σ^2 (the problem is that there is evidence of deviation from the trend line).

So our model for the data set in Table 6.4 is

$$Y_i = \alpha + \beta x_i + W_i, \quad i = 1, 2, \ldots, 6.$$

There are three parameters for the model, α, β and σ^2, the values of which are all unknown.

Now, having constructed this model for the data, we can discuss relevant aspects of the sampling distributions of each of $\widehat{\beta}_1$, $\widehat{\beta}_2$ and $\widehat{\beta}_3$, treating the estimate

$$\widehat{\beta}_1 = \frac{y_6 - y_1}{x_6 - x_1},$$

say, exactly as though it was an observation on the random variable (the estimator)

$$\widehat{\beta}_1 = \frac{Y_6 - Y_1}{x_6 - x_1}.$$

Notice that here is a good example of the situation where we use the notation $\widehat{\beta}_1$ to refer not just to the *estimate* of β_1, a number calculated from data, but also to the estimating formula, which we call the *estimator.*

We can work out the mean and the variance of the estimator $\widehat{\beta}_1$: this will give us some idea of its usefulness as an estimator for the trend slope β (the only one of the three unknown model parameters in which, for the present, we are interested). To do this, we need first to find out the mean and variance of the random variable $Y_i = \alpha + \beta x_i + W_i$. ■

Exercise 6.2

Remembering that α, β and x_i are just constants, and that W_i has mean 0 and variance σ^2 for all $i = 1, 2, \ldots, 6$, calculate the mean and variance of the random variable Y_i.

Example 6.6 continued

From the solution to Exercise 6.2, we have the results

$$E(Y_i) = \alpha + \beta x_i, \qquad V(Y_i) = \sigma^2. \tag{6.1}$$

The expected value of the estimator $\widehat{\beta}_1$ is given by

$$E(\widehat{\beta}_1) = E\left(\frac{Y_6 - Y_1}{x_6 - x_1}\right) = \frac{1}{x_6 - x_1}E(Y_6 - Y_1).$$

(Here, the result that has been used is that for any random variable N and for any constant c, the expectation $E(cN)$ is equal to $cE(N)$). Using (6.1) this reduces to

Chapter 4, (4.11).

$$\begin{aligned} E(\widehat{\beta}_1) &= \frac{1}{x_6 - x_1}(E(Y_6) - E(Y_1)) \\ &= \frac{1}{x_6 - x_1}((\alpha + \beta x_6) - (\alpha + \beta x_1)) \\ &= \frac{1}{x_6 - x_1}(\beta x_6 - \beta x_1) \\ &= \beta. \end{aligned}$$

In other words, the expected value of the estimator $\widehat{\beta}_1$ is simply the unknown parameter β. This is encouraging, for our purpose in constructing the estimator $\widehat{\beta}_1$, was to provide a useful estimate of β. ■

Exercise 6.3

(a) Write down the estimators $\widehat{\beta}_2$ and $\widehat{\beta}_3$ in terms of $Y_1, Y_2, \ldots, Y_6$ and $x_1, x_2, \ldots, x_6$.

Hint Look at how their numerical values, the estimates $\widehat{\beta}_2$ and $\widehat{\beta}_3$, were obtained.

(b) Find the expected values of the estimators $\widehat{\beta}_2$ and $\widehat{\beta}_3$.

So all of the three suggested estimators, $\widehat{\beta}_1$, $\widehat{\beta}_2$ and $\widehat{\beta}_3$, have the desirable property that they have expectation β. Each of the estimators is said to be **unbiased** for β.

> An estimator $\widehat{\theta}$ for an unknown model parameter θ is said to be **unbiased for $\boldsymbol{\theta}$** (or, simply, **unbiased**) if it has expectation
>
> $$E(\widehat{\theta}) = \theta.$$

Example 6.6 continued

There is therefore, so far, nothing to choose between the three estimators $\widehat{\beta}_1$, $\widehat{\beta}_2$ and $\widehat{\beta}_3$, and in that sense there is no clue which is the 'better' estimator, and therefore no indication about how much reliance could be placed on the three estimates

$$\widehat{\beta}_1 = 5.6, \qquad \widehat{\beta}_2 = 5.0, \qquad \widehat{\beta}_3 = 6.0$$

for the slope of the underlying trend line.

What about the variances of the three estimators? As they all have the same mean, it seems particularly sensible to choose as the best of these estimators the one with the least amount of variability about that mean, as measured by the variance. In *Chapter 4*, you were shown how to calculate the variance of a constant multiple of a sum of independent random variables: if Y_1 and Y_2 are independent, then

$$V(c(Y_1 + Y_2)) = c^2 V(Y_1 + Y_2) = c^2(V(Y_1) + V(Y_2)).$$

Since the random variables $W_1, W_2, \ldots, W_6$ are independent by assumption, it follows that the random variables $Y_1, Y_2, \ldots, Y_6$ which just add constants (albeit unknown ones) to these random variables must be independent as well. Therefore, we can apply the variance formula to work out the variances of the estimators. Here is the first one:

$$\begin{aligned} V(\widehat{\beta}_1) &= V\left(\frac{Y_6 - Y_1}{x_6 - x_1}\right) \\ &= \frac{1}{(x_6 - x_1)^2}(V(Y_6) + V(-Y_1)) \\ &= \frac{1}{(x_6 - x_1)^2}(V(Y_6) + V(Y_1)) \\ &= \frac{1}{(x_6 - x_1)^2}(\sigma^2 + \sigma^2) \\ &= \frac{2\sigma^2}{(x_6 - x_1)^2}. \end{aligned}$$

Notice that here we have used the result
$$V(-Y_1) = (-1)^2 V(Y_1) = V(Y_1).$$

Using the particular values for x_1 and x_6 given in Table 6.4, we have the final result

$$V(\widehat{\beta}_1) = \frac{2\sigma^2}{5^2} = 0.0800\sigma^2. \quad \blacksquare$$

Exercise 6.4

Find in terms of σ^2 the variances $V(\widehat{\beta}_2)$ and $V(\widehat{\beta}_3)$ for the other two estimators of slope that have been suggested.

Example 6.6 continued

From the solution to Exercise 6.4, it follows that the variances of the three estimators are ordered from least to greatest as

$$V(\widehat{\beta}_2) < V(\widehat{\beta}_3) < V(\widehat{\beta}_1);$$

so by this reckoning the estimator $\widehat{\beta}_2$, having the smallest variance, is the best of the three estimators of β. $\blacksquare$

Let us briefly summarize the main points made so far.

In a scenario involving point estimation we have

- *data*;
- a *model* for the data;
- an unknown *parameter* (or more than one) in the model, which we wish to *estimate*;
- irrespective of the particular values in our data set, an estimating formula or *estimator*;
- an *estimate* of the parameter; this will be different for different data sets: thus, an estimator is a random variable, an estimate is just a number;
- the *sampling distribution* of the estimator which will tell us how useful our estimator, and hence our estimate, is.

Moreover, we can say what the word 'useful' means here. It is useful if an estimator is *unbiased* and it is also useful if it has a *small variance*.

6.2 Methods of estimation

Where do we get our estimators from? In all the preceding examples, we have simply used a bit of common sense. But in Example 6.6, we devised three different estimators for the model parameter β; without much difficulty, others could have been invented. When estimating the binomial parameter p, the estimator $\widehat{p} = R/n$ is a very natural one to use, but what about the estimator $(R+1)/(n+2)$, for instance, which possesses some attractive properties? Or $\left(R+\frac{1}{2}\right)/(n+1)$?

One quite desirable property of both these rival estimators for the binomial parameter p is that if in a sequence of n Bernoulli trials the number of successes observed is either $r = 0$ or $r = n$, it does not follow that the estimate $\widehat{p}$ is respectively 0 or 1, suggesting in a rather absolute way 'impossibility' or 'certainty'.

Given several competing estimators for some parameter, it has been indicated that we might be able to choose between them on the basis of properties of their respective sampling distributions. But we may well be able to continue inventing estimators *ad infinitum*. Or, in more complicated modelling situations than these basic ones, we might be lost for ideas for obtaining any reasonable estimators at all. What is needed is a more systematic method for deriving sensible estimators in the first place. Just as an estimate is the numerical value resulting from applying an estimator to a particular set of data, so it would be helpful if an estimator were to be the result of applying some general estimation technique to the problem at hand.

Once again, however, there is no single estimation technique which is universally 'best', or even always appropriate! What follows in this section are very brief outlines of two of the more popular estimation techniques—there are many others—after which we shall concentrate on just one for the rest of the chapter. This third technique is known as the method of maximum likelihood, an approach to the problem which has many useful properties.

While it might be discouraging that there are several different estimation techniques that could be applied in any given sampling context, however, it often turns out that many estimation techniques, relying on different principles, result in exactly the same estimating formula for an unknown parameter. This is an encouraging finding.

For simplicity, we shall let θ, say, be the single parameter of interest here.

6.2.1 The method of least squares

Suppose $X_1, X_2, \ldots, X_n$ is our random sample, and that the parameter θ we wish to estimate is the mean of the distribution from which the random sample was drawn.

Now, each X_i should, loosely speaking, be 'somewhere near' θ (although there must be some variability about θ: quite how much there is depends on the amount of dispersion in the population). Then each difference $X_i - \theta$ should be 'fairly small'. A reasonable estimate of θ might then be chosen to try to make all the differences as small as possible simultaneously or, at least, to make the sum of the squares of all these differences,

$$S = \sum_{i=1}^{n} (X_i - \theta)^2,$$

as small as we can. Squaring the differences makes them all positive so that (possibly large) positive and negative differences do not cancel each other out.

The approach exemplified here, minimizing sums of squared differences between observed data and what might have been expected, is known as the **principle of least squares**. An alternative approach might be to choose our estimate of θ to minimize the sum of the absolute differences,

In *Chapter 10*, this principle is adopted when fitting straight lines to data, just as we tried to do in Example 6.6.

$$S = \sum_{i=1}^{n} |X_i - \theta|.$$

In general this approach leads to intractable algebra, and of the two, the least squares approach is not only easier, but results in estimators whose general properties are more easily discerned.

For arithmetic ease, the data in the next example are artificial.

Example 6.7 An artificial data set—the method of least squares

Suppose that observations $x_1 = 3$, $x_2 = 4$, $x_3 = 8$ were collected in a random sample of size 3 from a population with unknown mean, θ. The observed sum of squared differences is given by

$$\begin{aligned}
&\sum_{i=1}^{3} (x_i - \theta)^2 \\
&= (x_1 - \theta)^2 + (x_2 - \theta)^2 + (x_3 - \theta)^2 \\
&= (3 - \theta)^2 + (4 - \theta)^2 + (8 - \theta)^2 \\
&= \left(9 - 6\theta + \theta^2\right) + \left(16 - 8\theta + \theta^2\right) + \left(64 - 16\theta + \theta^2\right) \\
&= 89 - 30\theta + 3\theta^2.
\end{aligned}$$

This expression is a function of θ, and takes different values as θ takes different values as shown in the following table: we want to identify the value of θ for which this sum of squares is minimized.

θ	0	1	2	3	4	5	6	7	8	...
$89 - 30\theta + 3\theta^2$	89	62	41	26	17	14	17	26	41	...

From this table, it looks as though the sum of squares is high for low values of θ around 0, 1 and 2, attains a minimum at or around $\theta = 5$, and then climbs again. This makes sense: based on observations $x_1 = 3$, $x_2 = 4$ and $x_3 = 8$, a guess at an underlying mean of, say, $\theta = 1$ is not a very sensible guess, nor is a guess of $\theta = 8$.

You can see from the graph of the function $89 - 30\theta + 3\theta^2$ in Figure 6.4 that a minimum is indeed attained at the point $\theta = 5$. You may also have observed already that in this case the sample mean is $\overline{x} = 5$, the 'common sense' estimate, based on this sample, of the unknown population mean.

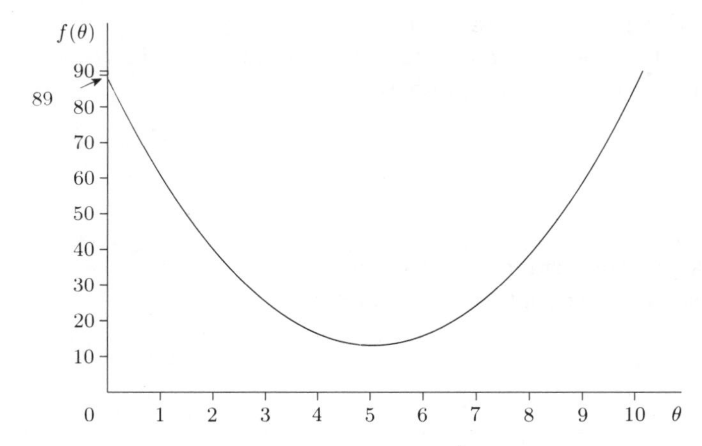

Figure 6.4 Graph of the function $89 - 30\theta + 3\theta^2$

In fact, for a general random sample of size n from a population with unknown mean θ, the expression for the sum of squared differences may be written

$$\sum_{i=1}^{n}(X_i - \theta)^2 = \sum_{i=1}^{n}\left(X_i^2 - 2\theta X_i + \theta^2\right) = \sum_{i=1}^{n} X_i^2 - 2\theta\sum_{i=1}^{n} X_i + n\theta^2.$$

This is a quadratic function of θ: the expression is minimized at the point

$$\theta = \frac{\sum_{i=1}^{n} X_i}{n} = \overline{X},$$

the sample mean.

This result may be obtained either by completing the square or by applying the calculus technique of differentiation. You need not worry about the algebraic details here. You will see in Sections 6.3 and 6.4 that there are 'standard estimators' for common models which do not need to be derived from first principles every time they are used.

That is, the sample mean $\overline{X}$ of a random sample $X_1, X_2, \ldots, X_n$ from a population with unknown mean θ is the **least squares estimator** $\widehat{\theta}$ of θ. From the sampling properties of $\overline{X}$, conclusions may be drawn about the usefulness of $\overline{X}$ as an estimator of θ. In this case, applying a formal principle has resulted in the same estimator as that suggested by common sense. ■

Let us now move on to a discussion of a second approach to estimation, the method of moments.

6.2.2 The method of moments

The **method of moments** is a very intuitive approach to the problem of parameter estimation: the argument is as follows.

On the one hand, we have seen defined in the course summary *population moments* as expressions of the variability in a population; on the other hand, for random samples drawn from such populations, we have calculated *sample moments* to summarize the data. Population moments involve unknown parameters; sample moments are numbers. The method of moments matches analogous moments to obtain estimates for the unknown parameters.

For instance, in a sample from the normal distribution $N(\mu, \sigma^2)$ with unknown mean and variance, this approach would involve matching the unknown parameters μ and σ^2 precisely to their sample analogues $\overline{x}$ and s^2; so the parameter estimates are

$$\widehat{\mu} = \overline{x}, \qquad \widehat{\sigma}^2 = s^2.$$

Alternatively, for the geometric distribution with unknown parameter p, we know that the distribution mean is $\mu = 1/p$. Matching the first population moment with the first sample moment to obtain an estimator for p, we would have $\overline{x} = 1/\widehat{p}$; so

See (3.19) in *Chapter 3*.

$$\widehat{p} = \frac{1}{\overline{x}}.$$

The method of moments raises a number of questions, of which perhaps the most immediate is: which moment do we match? For instance, the variance of the geometric distribution is $(1-p)/p^2$; matching the population variance with the sample variance, we would obtain

$$s^2 = \frac{1-\widehat{p}}{\widehat{p}^2}$$

which may be written as a quadratic equation for $\widehat{p}$ as

$$s^2\widehat{p}^2 + \widehat{p} - 1 = 0;$$

this has solution

$$\widehat{p} = \frac{\sqrt{1+4s^2} - 1}{2s^2}.$$

This is also a moment estimator for p, but for any particular random sample it will not usually be equal to our first guess, $\widehat{p} = 1/\overline{x}$. Which estimator may be better to use depends on a comparison of their sampling distributions, properties of which are not necessarily obvious.

For the Poisson distribution with parameter μ, both the population mean and the population variance are equal to μ—should we use the sample mean $\overline{x}$ or the sample variance s^2 as an estimator $\widehat{\mu}$ for μ?

Similarly, a population median is not strictly a 'moment' in the sense that the word has been used in this course—it is a quantile—but it conveys a useful numerical summary measure of a probability distribution. The median of the normal distribution is also μ—would the median of a normal sample be a 'better' estimator for μ than the sample mean?

Answers to these questions can always be obtained by reference to the sampling distribution of the proposed estimator. In general, a rule which works well in practice is: use as many sample moments as there are parameters requiring estimation, starting with the sample mean (first), the sample variance (second) and the sample skewness (third), and so on. (In fact, in this course we shall not be dealing with three-parameter distributions, and so we shall never need to use the sample skewness in a point estimation problem.)

Exercise 6.5

Using this rule, write down moment estimators for

(a) the Poisson parameter μ, given a random sample $X_1, X_2, \ldots, X_n$ from a Poisson distribution;

(b) the geometric parameter p, given a random sample $X_1, X_2, \ldots, X_n$ from a geometric distribution;

(c) the normal parameters μ and σ^2, given a random sample $X_1, X_2, \ldots, X_n$ from a normal distribution with unknown mean and variance;

(d) the exponential parameter λ, given a random sample $X_1, X_2, \ldots, X_n$ from an exponential distribution;

(e) the binomial parameter p, given a random sample $X_1, X_2, \ldots, X_n$ from a binomial distribution $B(m, p)$. (Notice the binomial parameter m here; the number n refers to the sample size. Assume m is known.)

Of course, practical estimates corresponding to these estimators but based on data samples, will not usually be *equal* to the population parameter they purport to estimate. For instance, here is a random sample of size 8 generated by computer from a Poisson distribution with mean 3:

5 5 2 3 4 6 4 1.

The moment estimate for the Poisson mean μ is the sample mean $\overline{x}$:

$$\widehat{\mu} = \overline{x} = \frac{5+5+2+3+4+6+4+1}{8} = \frac{30}{8} = 3.75.$$

A second random sample of the same size from the same distribution resulted in the estimate

$$\widehat{\mu} = \overline{x} = \frac{4+2+5+2+4+1+1+1}{8} = \frac{20}{8} = 2.50.$$

The first overestimates the true (though usually unknown) population parameter; the second underestimates it. However, we do know that for any random sample of size n drawn from a population with mean μ and standard deviation σ, the sample mean has mean and standard deviation given by

$$E(\overline{X}) = \mu, \qquad SD(\overline{X}) = \sqrt{\frac{\sigma^2}{n}}.$$

So, samples of size 8 drawn from a Poisson distribution with mean 3 (and, therefore, variance 3) have mean and standard deviation

$$E(\widehat{\mu}) = 3, \qquad SD(\widehat{\mu}) = \sqrt{\frac{3}{8}} = 0.6,$$

and the variation observed is not very surprising.

In the geometric case (Exercise 6.5(b)) you obtained the parameter estimator

$$\widehat{p} = \frac{1}{\overline{X}};$$

and for the exponential case (Exercise 6.5(d)) the parameter estimator was

$$\widehat{\lambda} = \frac{1}{\overline{X}}.$$

Each estimator is the reciprocal of the sample mean. Now, we know for samples from a population with mean μ that the sample mean $\overline{X}$ is unbiased for $\mu : E(\overline{X}) = \mu$. Unfortunately, it does not follow from this that the reciprocal of the sample mean has expectation $1/\mu$—if this were true, we would have the useful and desirable result in the geometric case that $E(\widehat{p})$ was $1/\mu = 1/(1/p) = p$. In fact, the estimator $\widehat{p} = 1/\overline{X}$ is not unbiased for p: the expectation $E(\widehat{p})$ is only *approximately* equal to p; however, for large samples the approximation is good.

See *Chapter 4*, (4.8).

The exact value of $E(\widehat{p})$ turns out to be very complicated indeed; it would not be useful to write it down here.

Similarly, for a random sample of size n from a population where the variation is assumed to follow an exponential distribution, it can be shown that the moment estimator $\widehat{\lambda} = 1/\overline{X}$ for λ has expectation

$$E(\widehat{\lambda}) = \lambda\left(1 + \frac{1}{n-1}\right)$$

which is not equal to λ: the estimator is **biased**. For large samples (that is, large n) the bias becomes negligible, but for small samples it is considerable. For example, for samples X_1, X_2 of size 2 from an exponential distribution, the moment estimator of λ is

$$\widehat{\lambda} = \frac{1}{\overline{X}} = \frac{2}{X_1 + X_2}.$$

The estimator has expectation

$$E(\widehat{\lambda}) = \lambda\left(1 + \frac{1}{n-1}\right) = 2\lambda;$$

in this case the bias is very considerable since $\widehat{\lambda}$ has expectation twice the value of the parameter for which it is the estimating formula.

Exercise 6.6

(a) Use a computer to simulate 1000 random samples of size 2 from an exponential distribution with parameter $\lambda = 5$ (that is, with mean $\frac{1}{\lambda} = 0.2$), and hence obtain 1000 independent estimates of the (usually unknown) parameter λ.

(b) What is the mean of your 1000 estimates?

Exercise 6.7

(a) What would be the moment estimator $\widehat{\mu}$ for μ from a random sample $X_1, X_2, \ldots, X_n$ from an exponential distribution with unknown mean μ: that is, with p.d.f.

$$f(x) = \frac{1}{\mu}e^{-x/\mu}, \quad x \geq 0?$$

(b) What is the expected value of $\widehat{\mu}$?

(c) Use a computer to simulate 1000 random samples of size 2 from an exponential distribution with mean $\mu = 0.2$, and hence obtain 1000 independent estimates of the (usually unknown) mean μ.

(d) What is the mean of the 1000 estimates?

The following two exercises summarize the work of this section. You will find that apart from keying in the individual data points to your calculator, or to your computer, very little extra work is required to obtain the estimates requested.

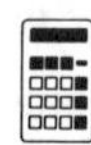

Exercise 6.8

The ecologist E.C. Pielou was interested in the pattern of healthy and diseased trees—the disease that was the subject of her research was '*Armillaria* root rot'—in a plantation of Douglas firs. Several thin lines of trees through the plantation (called 'transects') were examined. The lengths of unbroken runs of healthy and diseased trees were recorded. The observations made on a total of 109 runs of diseased trees are given in Table 6.5.

Pielou, E.C. (1963) Runs of healthy and diseased trees in transects through an infected forest. *Biometrics*, **19**, 603–614.

Table 6.5 Run lengths of diseased trees in an infected plantation

Run length	1	2	3	4	5	6
Number of runs	71	28	5	2	2	1

Pielou proposed that the geometric distribution might be a good model for these data, and showed that this was so. The geometric distribution has probability mass function

$$p(x) = (1-p)^{x-1}p, \quad x = 1, 2, \ldots$$

where the parameter p is, in this context, unknown. (Here, p is the proportion of healthy trees in the plantation.) Figure 6.5(a) shows a bar chart of the data in Table 6.5. Figure 6.5(b) gives a sketch of the probability mass function of a particular choice of geometric distribution (that is, one with a particular choice of p) to confirm that a geometric fit is a reasonable modelling assumption.

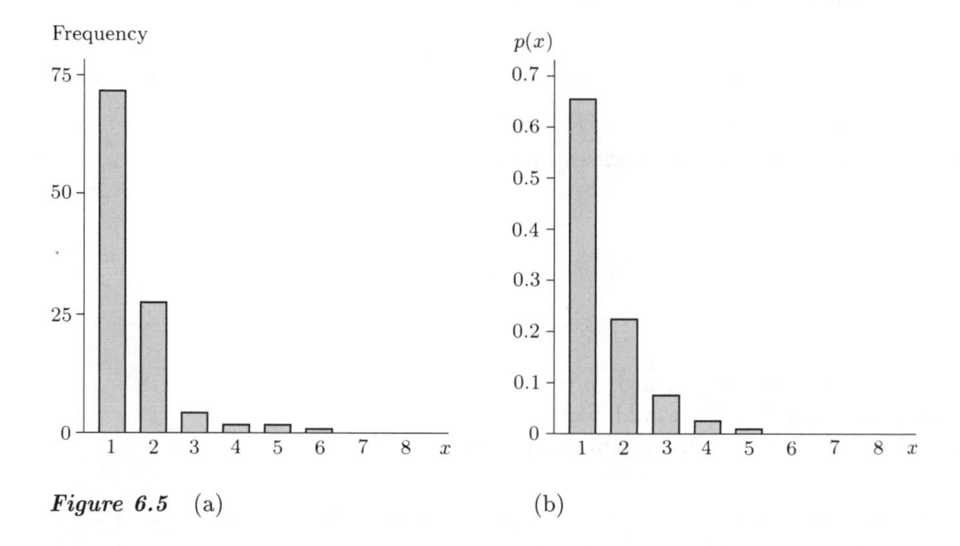

Figure 6.5 (a) (b)

Using these data obtain a moment estimate $\widehat{p}$ for the geometric parameter p.

Exercise 6.9

Chapter 4, Table 4.7 lists the 62 time intervals (in days) between successive earthquakes world-wide. There is a histogram of the data in Figure 4.3 and in Figure 4.5 the results of fitting an exponential model to the data are shown. (The exponential fit looks very good.)

(a) Using the data obtain a moment estimate $\widehat{\lambda}$ for the exponential parameter λ, and quantify any bias there may be in the estimate. What are the units of the estimate $\widehat{\lambda}$?

(b) Using the data obtain a moment estimate $\widehat{\mu}$ for the underlying exponential mean μ, and say whether or not the corresponding estimator is biased. What are the units of the estimate $\widehat{\mu}$?

As you can see, the method of moments has many attractive properties as a principle on which to base an approach to estimation. It is an intuitive and straightforward approach to follow. However, there are some occasions (that is, particular sampling contexts) where the method fails and others where it cannot be applied at all.

6.3 The method of maximum likelihood

The following data set illustrates a situation that would not occur very often in practice: the data are artificial and are designed to illustrate a point.

Example 6.8 *An artificial data set—the method of moments*

Three observations were collected on a continuous uniform random variable $X \sim U(0, \theta)$, where the parameter θ is unknown—the aim of the sample was to estimate θ. The data recorded were

$$x_1 = 3.2, \quad x_2 = 2.9, \quad x_3 = 13.1.$$

The sample mean is $\overline{x} = \frac{1}{3}(3.2 + 2.9 + 13.1) = 6.4$. The mean of the uniform distribution $U(0, \theta)$ is $\frac{1}{2}\theta$. Matching the two to obtain a moment estimate for θ gives

$$\overline{x} = \tfrac{1}{2}\widehat{\theta}$$

or

$$\widehat{\theta} = 2\overline{x} = 12.8.$$

The trouble with this estimate is that it is so obviously wrong—a probability model defined to take values only over the range $[0, 12.8]$ would not permit an observation as high as 13.1, and yet that was the third value obtained in the sample. ■

Example 6.9 *Vehicle occupancy*

In this course we shall consider quite a number of statistical models for variation: new models are being developed all the time as researchers attempt to refine the quality of the fit of models to data. One context where modelling is very important for forecasting purposes is in traffic research: data are collected on traffic flow (private and public transport, and freight volumes) and vehicle occupancy, amongst other things. Tripathi and Gupta (1985) were interested in fitting a particular two-parameter model to data on the numbers of passengers (including the driver) in private cars. The model they attempted to fit is a discrete probability distribution defined over the positive integers $1, 2, \ldots$. (It is a versatile model, fitting even very skewed data moderately well.) The data Tripathi and Gupta had to work on is given in Table 6.6. It gives the numbers of occupants of 1469 cars (including the driver).

Tripathi, R.C. and Gupta, R.C. (1985) A generalization of the log-series distribution. *Communications in Statistics (Theory and Methods)*, **14**, 1779-1799. The log-series distribution is quite a sophisticated discrete probability distribution, and details of it (or of generalizations of it) are not included here.

Table 6.6 Counts of occupants of 1469 private cars

Count	1	2	3	4	5	≥ 6
Frequency	902	403	106	38	16	4

The problem here is that the data are **censored**: for whatever reason, the actual number of occupants of any of the cars that contained six persons or more was not precisely recorded: all that is known is that in the sample there were four vehicles as full as that (or fuller).

A consequence of this is that the method of moments cannot be applied to estimate the two parameters of the log-series distribution, for the sample mean and the sample standard deviation are unknown. ■

Example 6.10 *Soil organisms*

This is a second example of the same phenomenon: censored data. An area of soil was divided into 240 regions of equal area (called 'quadrats') and in each quadrat the number of colonies of bacteria found was counted. The data are given in Table 6.7.

Jones, P.C.T., Mollison, J.E. and Quenouille, M.H. (1948) A technique for the quantitative estimation of soil micro-organisms. *J. Gen. Microbiology*, **2**, 54–69.

Table 6.7 Colonies of bacteria in 240 quadrats

Count	0	1	2	3	4	5	≥ 6
Frequency	11	37	64	55	37	24	12

Again, for whatever reason, precise records for those quadrats supporting more than five colonies were not kept. It is not possible to calculate sample moments for this data set. ■

In cases where a statistical method fails, it does not necessarily follow that the method is 'wrong'—we have seen in Section 6.2 that the method of moments can provide extremely useful estimators for population parameters with very good properties. It is simply that the method is not entirely reliable. In other cases, there may be insufficient information to apply it. In fact, it can be shown in the case of uniform parameter estimation described in Example 6.8 that the alternative estimator

$$\widehat{\theta} = \left(1 + \frac{1}{n}\right) X_{\max},$$

(where the random variable $X_{\max}$ is the largest observation in a sample of size n) has good variance properties and is unbiased for θ. Furthermore, it is an estimator that will not lead to a logical contradiction! Indeed, it makes rather good sense: it says 'take the largest observation, and add a bit'.

For the two examples cited of censored data, it would not be very difficult to obtain really very good estimates of the sample moments, and these could be used in the estimation procedure for the model parameters.

Exercise 6.10

For the data in Example 6.8, write down the estimate $\widehat{\theta}$ for the unknown uniform parameter θ, based on this alternative estimator.

The point here is that it is possible in the context of any estimation procedure to invent sampling scenarios that result in estimates which are nonsensical, or which do not permit the estimation procedure to be followed. There is one method, the method of maximum likelihood, which has pleasing intuitive properties (like the method of moments) but which is also applicable in many sampling contexts.

6.3.1 Discrete probability models

For ease, let us consider in general a case where data are collected on a discrete random variable X. Suppose that the variation in observations on X is to be modelled by a probability distribution indexed by a single unknown parameter, θ. The random variable X therefore has a probability mass function which may be written

$$P(X = x) = p(x;\theta), \quad x = 0, 1, 2, \ldots .$$

Notice that here it is emphasized that there is a parameter θ involved in the probability mass function, by explicitly including it in the expression $p(x;\theta)$ on the right-hand side.

Suppose that, for the purposes of estimating the value of θ, a random sample of size n is collected.

The probability that X_1 takes the value x_1, say, is $p(x_1;\theta)$; the probability that X_2 equals x_2 is $p(x_2;\theta)$; and so on. The random variables X_1 and X_2 are independent (by implication of the phrase 'random sample') so it follows that the probability that $X_1 = x_1$ *and* $X_2 = x_2$ is

$$P(X_1 = x_1, X_2 = x_2) = p(x_1;\theta) \times p(x_2;\theta).$$

Indeed, the whole sample arises as a realization of the collection of n mutually independent random variables $X_1, X_2, \ldots, X_n$, and hence the probability of obtaining the observed collection $x_1, x_2, \ldots, x_n$ of sample values is

$$p(x_1;\theta) \times p(x_2;\theta) \times \ldots \times p(x_n;\theta). \tag{6.2}$$

Now this expression tells us the probability that our actual sample arose, given the true, but unknown, value of θ. As we do not know θ, we cannot be sure what the true value of (6.2) is. What we can try to do, however, is to work out the probability given by (6.2) for various guessed values of θ, and see what it turns out to be. Taking this to the limit of considering all possible values of θ, we can think of (6.2) as a function of θ. What this function tells us is how likely we are to obtain our particular sample for each particular value of θ. So, it seems reasonable to estimate θ to be the value that gives maximum probability to the sample that actually arose: that is, we should choose $\widehat{\theta}$ to *maximize* (6.2).

The product (6.2) is called the **likelihood of $\boldsymbol{\theta}$ for the sample $\boldsymbol{x_1, x_2, \ldots, x_n}$**—or, usually, simply the **likelihood of $\boldsymbol{\theta}$**. An approach that asks 'What value of θ maximizes the chance of observing the random sample that was, in fact, obtained?' is intuitively a very appealing one. This approach is known as the **method of maximum likelihood**.

> **The method of maximum likelihood**
>
> If several independent observations $X_1, X_2, \ldots, X_n$ are collected on the discrete random variable X with probability mass function
>
> $$p_X(x) = p(x; \theta),$$
>
> then the product
>
> $$p(X_1, X_2, \ldots, X_n; \theta) = p(X_1; \theta) \times p(X_2; \theta) \times \ldots \times p(X_n; \theta)$$
>
> is known as the **likelihood of θ for the random sample $X_1, X_2, \ldots, X_n$.**
>
> The value $\widehat{\theta}$ of θ at which the likelihood is maximized is known as the **maximum likelihood estimator for θ.**

The estimator $\widehat{\theta}$ is itself a random variable and has a sampling distribution: the mean and variance of this distribution yield useful information about the precision of the estimating procedure.

Here is an example.

Example 6.11 *Estimating the parameter of a geometric distribution*

For ease, let us consider the very small artificial data set first introduced in Example 6.7. Here, there were three observations $x_1 = 3$, $x_2 = 4$ and $x_3 = 8$. Let us suppose that these are observations from a geometric distribution with unknown parameter θ. The moment estimator for the geometric parameter θ is $\widehat{\theta} = 1/\overline{X}$, so for these data the corresponding moment estimate is $\widehat{\theta} = 1/\overline{x} = 1/5 = 0.2$.

For consistency with the current development, we shall temporarily refer to the parameter indexing the geometric distribution as θ (conventionally, we would refer to it as p).

The probability mass function for the geometric distribution (using the more developed notation) is

$$p(x; \theta) = (1-\theta)^{x-1}\theta, \quad x = 1, 2, 3, \ldots .$$

It follows that the likelihood for this particular random sample of size 3 is given by

$$\begin{aligned} &p(x_1, x_2, x_3; \theta) \\ &= p(x_1; \theta) \times p(x_2; \theta) \times p(x_3; \theta) \\ &= (1-\theta)^{x_1-1}\theta \times (1-\theta)^{x_2-1}\theta \times (1-\theta)^{x_3-1}\theta \\ &= (1-\theta)^{3-1}\theta \times (1-\theta)^{4-1}\theta \times (1-\theta)^{8-1}\theta \\ &= (1-\theta)^{2+3+7}\theta^3 \\ &= (1-\theta)^{12}\theta^3. \end{aligned}$$

We now have, for the particular sample observed, a function of the unknown parameter θ. For different values of θ, the function will itself take different values: what we need to find is the value of θ at which the function is maximized.

We could start by drawing up a table of values as we did when estimating the mean of the sample using the method of least squares (see Example 6.7). In this case, we would obtain the following.

θ	0	0.2	0.4	0.6	0.8	1
$(1-\theta)^{12}\theta^3$	0	0.00054976	0.00013931	0.00000362	0.00000000	0

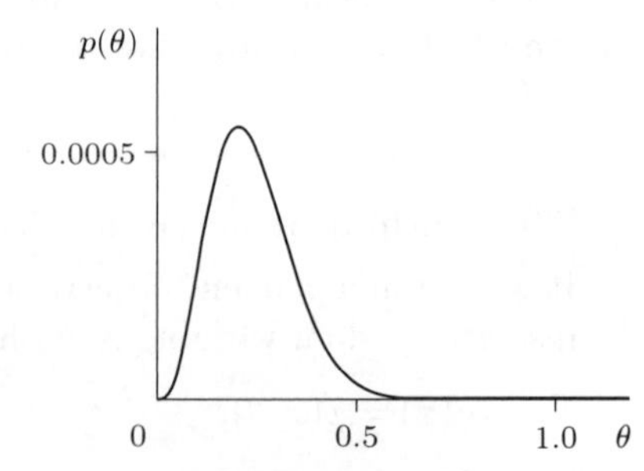

Figure 6.6 The likelihood $(1-\theta)^{12}\theta^3$, for $0 \leq \theta \leq 1$

These calculations suggest that the likelihood is maximized somewhere between $\theta = 0$ and $\theta = 0.4$—possibly at $\theta = 0.2$ itself. A graph of the likelihood is shown in Figure 6.6. As you can see from the graph, the likelihood is maximized at the value $\theta = 0.2$. In this case, the maximum likelihood estimate of θ is the same as the moment estimate. ■

A note about finding maximum likelihood estimators

In order to obtain from first principles estimating formulas (that is, estimators) for model parameters using the method of maximum likelihood, it is (generally speaking) essential to have a working knowledge of the algebraic technique of differentiation.

For instance, suppose a random sample $x_1, x_2, \ldots, x_n$ of size n was collected from a Poisson population with unknown mean θ. Then the likelihood of θ for the sample is given by

$$\begin{aligned}
&p(x_1;\theta)p(x_2;\theta)\ldots p(x_n;\theta) \\
&= \frac{e^{-\theta}\theta^{x_1}}{x_1!} \times \frac{e^{-\theta}\theta^{x_2}}{x_2!} \times \cdots \times \frac{e^{-\theta}\theta^{x_n}}{x_n!} \\
&= \frac{(e^{-\theta} \times e^{-\theta} \times \cdots \times e^{-\theta}) \times (\theta^{x_1} \times \theta^{x_2} \times \cdots \times \theta^{x_n})}{x_1!x_2!\ldots x_n!} \\
&= \frac{e^{-n\theta}\theta^{\Sigma x_i}}{x_1!x_2!\ldots x_n!} \\
&= \text{constant} \times e^{n\theta}\theta^{\Sigma x_i}.
\end{aligned}$$

In this case the constant term does not involve the parameter θ, and so the value of θ that maximizes the likelihood is simply the value of θ that maximizes the expression

$$e^{-n\theta}\theta^{\Sigma x_i}.$$

Here, the algebra of differentiation can be employed to deduce that the maximum likelihood estimate $\widehat{\theta}$ of θ, for a Poisson sample $x_1, x_2, \ldots, x_n$ is, in fact, the sample mean

$$\widehat{\theta} = \overline{x} = \frac{\sum x_i}{n}.$$

In general, this is an observation on a random variable: the maximum likelihood estimator of a Poisson mean, based on a random sample $X_1, X_2, \ldots, X_n$, is the sample mean $\overline{X}$.

(So, in this case, the maximum likelihood estimator is the same as the minimum least squares estimator and the moment estimator, and they all appeal to common sense.)

However, when obtaining maximum likelihood estimates (that is, numbers based on a sample of data, as in Example 6.11) differentiation is not required, for there is an armoury of numerical, computational or graphical approaches available. In this course, maximum likelihood *estimators* for the parameters of the more common probability models will be stated as standard results (although you should understand the principle underlying the way in which they are obtained) and it will be assumed in the data-oriented exercises in the rest of the chapter that you have software capable of the numerical procedures necessary to obtain maximum likelihood *estimates*.

Here is another example.

Example 6.3 continued

In the example about adenomas in mice, there was one group of 54 mice, six of which had adenomas. Assuming an underlying proportion θ (unknown) of afflicted mice in the wider population, and independence within the group, the probability of this event is

$$p(6;\theta) = \binom{54}{6}\theta^6(1-\theta)^{48} = 25\,827\,165\,\theta^6(1-\theta)^{48}.$$

This is the likelihood of θ for the data observed; and it is maximized at $\widehat{\theta} = \frac{6}{54} = \frac{1}{9}$. ■

The following example is of a rather different type from the previous two, in that the probability model being applied (and which is indexed by a single unknown parameter) is not one of the standard families. However, the principle is exactly the same: we seek to find the value of the parameter that maximizes the likelihood for the sample that was actually observed.

Example 6.12 *The leaves of Indian creeper plants*

In *Chapter 3*, Example 3.12 a genetics experiment was referred to in which the leaf characteristics of Indian creeper plants *Pharbitis nil* were observed. Altogether four different combinations of leaf-type were possible. According to one theory, the different combinations should have been observed according to the relative frequencies

$$\tfrac{9}{16} : \tfrac{3}{16} : \tfrac{3}{16} : \tfrac{1}{16};$$

in one experiment the observed frequencies were

$$187 : 35 : 37 : 31,$$

and on the basis of these data that theory was rejected.

In *Chapter 9* you will read about a procedure for comparing observed frequencies with those expected if some theory were true.

An alternative theory allows for the phenomenon known as *genetic linkage* and assumes that the observations might have arisen from a probability distribution indexed by a parameter θ as follows:

$$\tfrac{9}{16} + \theta : \tfrac{3}{16} - \theta : \tfrac{3}{16} - \theta : \tfrac{1}{16} + \theta.$$

So for the Indian creeper data we know that leaf characteristics assigned a probability of $\frac{9}{16} + \theta$ were observed 187 times; characteristics assigned a probability of $\frac{3}{16} - \theta$ were observed a total of $35 + 37 = 72$ times; and characteristics assigned a probability of $\frac{1}{16} + \theta$ were observed 31 times. The likelihood of θ for the sample observed is therefore given by writing down the probability of this conjunction of events: it is

$$\left(\tfrac{9}{16} + \theta\right)^{187} \left(\tfrac{3}{16} - \theta\right)^{72} \left(\tfrac{1}{16} + \theta\right)^{31}.$$

Again, if you are familiar with the technique of differentiation you will be able to locate the value $\widehat{\theta}$ of θ at which the likelihood is maximized. Without it, graphical or other computational methods can be implemented—it turns out that the likelihood is maximized at $\theta = 0.0584$. So the new estimated probabilities for the four leaf characteristics, on the basis of this experiment, are

$$0.6209 : 0.1291 : 0.1291 : 0.1209.$$

You can see that this model would appear to provide a much better fit to the observed data. ■

So far the three estimation procedures we have explored necessarily involve some quite high-level algebraic (or, at the very least, numerical) skills. Most of the time, however, the resulting estimators have an intuitive basis supported by common sense.

In Section 6.4 you will be encouraged to explore the facilities available on your computer for the calculation of maximum likelihood estimates when you do not always have an estimating formula. Table 6.8 shows a list of standard results for maximum likelihood estimators for the parameters of the more well-known discrete probability models. These estimators assume a random sample $X_1, X_2, \ldots, X_n$ with sample mean $\overline{X}$ (and, for the uniform distribution, sample maximum $X_{\max}$).

Table 6.8 Standard results for discrete probability models: maximum likelihood estimators

Probability distribution	Estimator	Properties
Poisson(μ)	$\widehat{\mu} = \overline{X}$	$E(\widehat{\mu}) = \mu$
Bernoulli(p)	$\widehat{p} = \overline{X}$	$E(\widehat{p}) = p$
$B(m,p)$	$\widehat{p} = \overline{X}/m$	$E(\widehat{p}) = p$
$G(p)$	$\widehat{p} = 1/\overline{X}$	$\widehat{p}$ is biased
Uniform($1, 2, \ldots, m$)	$\widehat{m} = X_{\max}$	$\widehat{m}$ is biased

In both cases where the estimator is biased ($\widehat{p}$ for the geometric distribution $G(p)$, and $\widehat{m}$ for the uniform model) the exact value of the mean of the estimator is known; but it would not be particularly useful or helpful to write it down.

The table raises the following question: if different estimating procedures result in the same estimator, and if the estimator is (by and large) the common-sense estimator that one might have guessed without any supporting theory at all, what is the point of a formal theoretical development?

The answer is that maximum likelihood estimators (as distinct from other estimators based on different procedures) have particular statistical properties. If these estimators happen to be the same as those obtained by other means (such as the method of moments; or just guessing) then so much the better.

But it is only for maximum likelihood estimators that these properties have a sound theoretical basis. A statement of these important properties follows (their derivation is not important here).

6.3.2 Properties of maximum likelihood estimators

With one exception (the estimator of the normal variance, which you found in Exercise 6.5 and to which we shall turn again at the start of Section 6.5), every estimator obtained in Exercise 6.5 by the method of moments is the same as that which would have been obtained using the method of maximum likelihood. Statistical theory (quite difficult theory, mathematically, and we shall not go into the details of it) tells us that maximum likelihood estimators possess 'good' properties, including the following.

(a) Maximum likelihood estimators are often *unbiased* (i.e. $E(\widehat{\theta}) = \theta$); if not, then they are *asymptotically unbiased*. That is,

$$E(\widehat{\theta}) \to \theta \text{ as } n \to \infty,$$

Read the symbol '$\to$' as 'tends to'.

where n is the sample size. From this we may deduce that maximum likelihood estimators are approximately unbiased for large sample sizes.

(b) Maximum likelihood estimators are *consistent*. Roughly speaking (things get rather technical here) this means that their variance $V(\widehat{\theta})$ tends to 0 with increasing sample size:

$$V(\widehat{\theta}) \to 0 \text{ as } n \to \infty.$$

So maximum likelihood estimators (that is, estimators obtained by applying the method of maximum likelihood) possess the sort of useful properties we identified in Section 6.1: for instance, if they are not unbiased for θ, (and they often are) then for large samples they are at least approximately unbiased.

(It is possible to state useful conclusions not just about the *mean* and *variance* of maximum likelihood estimators, but their *sampling distribution* as well. Maximum likelihood estimators are asymptotically normally distributed. A reasonable assumption is that, for large samples, they are approximately normally distributed. However, this kind of result requires a certain amount of supporting theory before it can be confidently applied, and it will not be pursued further in this course.)

6.4 More about maximum likelihood estimation

For a large part of this section, you will be encouraged to obtain maximum likelihood estimates for particular sampling scenarios using your computer. However, we have so far restricted attention to discrete probability models; before embarking on the exercises, first it is necessary to develop the maximum likelihood approach for continuous random variables.

6.4.1 Continuous probability models

In the following example the Pareto probability model is introduced.

Example 6.13 Annual wages (USA)

The data in Table 6.9 give the annual wages (in multiples of 100 US dollars) of a random sample of 30 production line workers in a large American industrial firm.

Dyer, D. (1981) Structural probability bounds for the strong Pareto law. *Canadian Journal of Statistics*, **9**, p. 71.

Table 6.9 Annual wages (hundreds of US dollars)

112	154	119	108	112	156	123	103	115	107
125	119	128	132	107	151	103	104	116	140
108	105	158	104	119	111	101	157	112	115

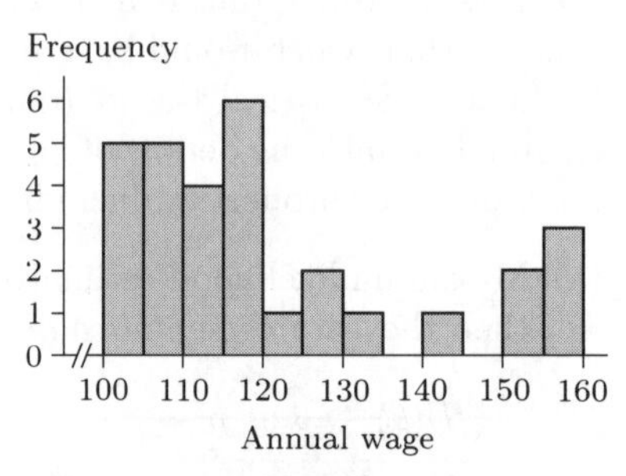

Figure 6.7 Annual wage data (hundreds of US dollars)

A histogram of these data is shown in Figure 6.7. A fairly standard probability model for variation in income is the **Pareto probability distribution**. This is a two-parameter continuous probability distribution with probability density function given by

$$f(x) = \frac{\theta K^\theta}{x^{\theta+1}}, \quad x \geq K,$$

where the two parameters are $\theta > 1$ and $K > 0$. The parameter K represents some minimum value that the random variable X can take, and in any given context the value of K is usually known (hence the range $x \geq K$). The parameter θ is not usually known, and needs to be estimated from a sample of data. The densities shown in Figure 6.8 all have K set equal to 1, and show Pareto densities for $\theta = 2, 3$ and 10. Notice that the smaller the value of θ, the more dispersed the distribution.

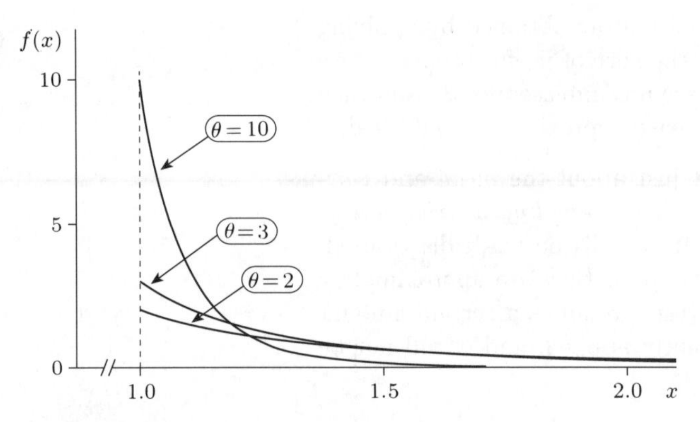

Figure 6.8 Three different Pareto densities

The cumulative distribution function for the Pareto random variable is obtained using integration:

$$F(x) = P(X \leq x) = \int_K^x f(v)\,dv = 1 - \left(\frac{K}{x}\right)^\theta, \quad x \geq K.$$

It is represented by the area of the shaded region shown in Figure 6.9. The mean of the Pareto distribution is given by

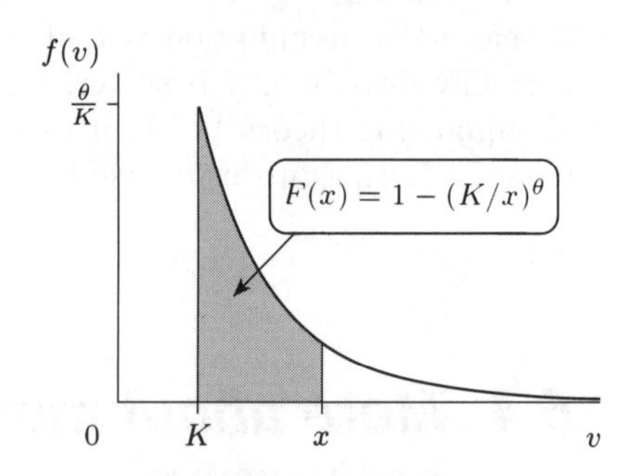

Figure 6.9 The probability $1 - (K/x)^\theta$

$$\mu = \int_K^\infty x f(x)\,dx = \frac{K\theta}{\theta - 1}.$$

This description may be summarized as follows.

Again, if you are unfamiliar with integration, do not worry: these results for $F(\cdot)$ and μ are standard, and will not generally require proof.

The Pareto distribution

The continuous random variable X with probability density function

$$f(x) = \frac{\theta K^\theta}{x^{\theta+1}}, \quad x \geq K,$$

where $\theta > 1$ and $K > 0$, is said to follow **a Pareto distribution with parameters K and θ**; this may be written

$$X \sim \text{Pareto}(K, \theta).$$

The c.d.f. of X is given by

$$F(x) = 1 - \left(\frac{K}{x}\right)^\theta, \quad x \geq K,$$

and the mean of X is

$$\mu = \frac{K\theta}{\theta - 1}.$$

An appropriate model for the variation in the US annual wage data is the Pareto probability distribution Pareto(100, θ) where $K = 100$ (i.e. US\$10 000) is an assumed minimum annual wage. That leaves θ as the single parameter requiring estimation. ■

However, notice the peak in the histogram for values above 150. We shall return to this data set in *Chapter 9*.

Exercise 6.11

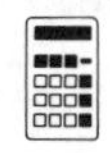

(a) Write down the mean of the Pareto probability distribution with parameters $K = 100$, θ unknown.

(b) Use the method of moments and the data of Table 6.9 to estimate the parameter θ for an assumed Pareto model.

You should have found for the Pareto model that a moment estimator for the parameter θ when K is known is

$$\widehat{\theta} = \frac{\overline{X}}{\overline{X} - K}.$$

What can we say about properties of this estimator? Since these depend on the sampling distribution of the random variable $\widehat{\theta} = \overline{X}/(\overline{X} - K)$, the answer is 'rather little'. At a glance, it is not even obvious whether or not the estimator is unbiased.

One possible option is to perform a simulation exercise: use the computer to generate a very large number of random samples from known Pareto densities, and try to deduce what we can from the observed sampling variation in $\widehat{\theta}$.

An alternative is to obtain the maximum likelihood estimator of θ. We know that this estimator is unbiased, or approximately unbiased for large samples, and that it is consistent. If it turns out to be the same as our moment estimator, then we can deduce those properties for the moment estimator too.

We have not so far tried to determine the likelihood of the unknown parameter θ for a random sample $x_1, x_2, \ldots, x_n$ from a *continuous* distribution. In the discrete case we were able to say

$$\begin{aligned} &P(X_1 = x_1, X_2 = x_2, \ldots, X_n = x_n) \\ &= P(X_1 = x_1)P(X_2 = x_2) \ldots P(X_n = x_n) \\ &= p(x_1; \theta)p(x_2; \theta) \ldots p(x_n; \theta) \end{aligned}$$

and work from there.

Remember the notation $p(x; \theta)$ for the probability mass function, which emphasizes that it also features the unknown parameter θ.

From a simple limiting argument, it turns out that the likelihood of an unknown parameter θ for a random sample $x_1, x_2, \ldots, x_n$ from a continuous distribution may be written as the product

$$f(x_1; \theta)f(x_2; \theta) \ldots f(x_n; \theta),$$

where the function $f(\cdot)$ is the probability density function of the random variable X, and the notation again expresses the dependence of the p.d.f. on the parameter θ. The method of maximum likelihood involves finding the value of θ that maximizes this product.

Example 6.14 Estimating the exponential parameter

For a random sample $x_1, x_2, \ldots, x_n$ from an exponential distribution with unknown parameter θ, the corresponding probability density function is

$$f(x; \theta) = \theta e^{-\theta x}, \quad x \geq 0,$$

and so the likelihood of θ for the sample is

$$\begin{aligned} &f(x_1; \theta)f(x_2; \theta) \ldots f(x_n; \theta) \\ &= \theta e^{-\theta x_1} \theta e^{-\theta x_2} \ldots \theta e^{-\theta x_n} \\ &= \theta^n e^{-\theta(x_1 + x_2 + \cdots + x_n)} \\ &= \theta^n e^{-\theta \Sigma x_i}. \end{aligned}$$

Viewed as a function of θ, the likelihood attains a maximum at the value

$$\theta = \frac{n}{\sum x_i} = \frac{1}{\overline{x}}.$$

The point θ at which the likelihood attains its maximum may be found using differentiation. It is not important that you should be able to confirm this result yourself.

In other words, the maximum likelihood estimator $\widehat{\theta}$ of the exponential parameter θ based on a random sample, is the reciprocal of the sample mean:

$$\widehat{\theta} = \frac{1}{\overline{X}}.$$

This is the same as the moment estimator. ■

Example 6.13 continued

For a random sample $x_1, x_2, \ldots, x_n$ from a Pareto distribution, the likelihood of θ is given by

$$\begin{aligned} &f(x_1, \theta)f(x_2; \theta) \ldots f(x_n; \theta) \\ &= \frac{\theta K^\theta}{x_1^{\theta+1}} \cdot \frac{\theta K^\theta}{x_2^{\theta+1}} \cdots \frac{\theta K^\theta}{x_n^{\theta+1}} \\ &= \frac{\theta^n K^{n\theta}}{(x_1 x_2 \ldots x_n)^{\theta+1}}. \end{aligned}$$

This is maximized at the point

$$\theta = \frac{n}{\sum_{i=1}^{n} \log(x_i/K)}.$$

The point θ at which the likelihood attains its maximum may be found using differentiation.

So the maximum likelihood estimator of θ for a random sample of size n from a Pareto(K, θ) distribution, assuming K is known, is

$$\widehat{\theta}_{ML} = \frac{n}{\sum_{i=1}^{n} \log(X_i/K)}.$$

Notice the convenient use of subscripts to distinguish the two estimators: $\widehat{\theta}_{ML}$ for the maximum likelihood estimator, and $\widehat{\theta}_{MM}$ for that obtained using the method of moments.

This is quite different from the moment estimator

$$\widehat{\theta}_{MM} = \frac{\overline{X}}{\overline{X} - K}$$

that we found earlier.

So unfortunately, in this case, we are no nearer deducing sampling properties of the moment estimator. However, with a computer, the maximum likelihood estimate of θ in any given sampling context is not more difficult to obtain than the moment estimator, and is the one to use in order to find numerical estimates, since its properties are known. ■

Exercise 6.12

Compare the moment estimate $\widehat{\theta}_{MM}$ and the maximum likelihood estimate $\widehat{\theta}_{ML}$ for the US annual wage data in Table 6.9.

Standard results for continuous probability models

Table 6.10 lists some standard results for maximum likelihood estimators of the parameters of the more common continuous probability models.

Table 6.10 Standard results for continuous probability models: maximum likelihood estimators

Probability distribution	Estimator	Properties
$M(\lambda)$	$\widehat{\lambda} = 1/\overline{X}$	$\widehat{\lambda}$ is biased
$N(\mu, \sigma^2)$	$\widehat{\mu} = \overline{X}$	$E(\widehat{\mu}) = \mu$
Uniform$(0, \theta)$	$\widehat{\theta} = X_{\max}$	$\widehat{\theta}$ is biased
Pareto(K, θ)	$\widehat{\theta} = \dfrac{n}{\sum \log(X_i/K)}$	$\widehat{\theta}$ is biased

It is possible to prove that for the Pareto distribution the maximum likelihood estimator is biased: in fact, $E(1/\widehat{\theta}) = 1/\theta$.

6.4.2 Computer exercises

Generally speaking, in a problem involving maximum likelihood estimation, you should use standard results whenever they can be applied. Otherwise, you will need to explore the facilities available on your computer. Many statistical packages include as standard functions probability mass functions, density functions and cumulative distribution functions. To identify maximum likelihood estimates, you might need to use your machine's plotting facilities; but it may be equipped with procedures for identifying maximum and minimum points on curves. Finally, a small amount of programming might be necessary.

All the exercises are about finding estimates (numbers based on samples of data) rather than deriving estimators (formulas), which requires different skills not developed in this course.

The first exercise is about estimating the underlying proportion in a sampled population, which possesses some attribute of interest.

Exercise 6.13

It is algebraically quite involved to prove that in independent experiments to estimate the binomial parameter p, where the jth experiment resulted in r_j successes from m_j trials ($j = 1, 2, \ldots, n$), the maximum likelihood estimate of p is given by

$$\widehat{p} = \frac{r_1 + r_2 + \cdots + r_n}{m_1 + m_2 + \cdots + m_n};$$

but this intuitive result (where p is estimated by the total number of successes divided by the total number of trials performed) is true and you can take it on trust.

(a) For the 23 different groups of mice referred to in Example 6.3, the results for all the groups were as follows.

0/12	0/20	1/20	1/15	3/20	4/47	0/12	0/20
1/19	2/25	3/20	6/54	0/10	0/19	1/19	2/22
3/18	8/49	1/10	0/17	1/17	2/20	4/20	

Use these data to calculate a maximum likelihood estimate for the underlying proportion of mice afflicted with alveolar-bronchiolar adenomas.

(b) In a famous experiment (though it took place a long time ago) the number of normal *Drosophila melanogaster* and the number of vestigial *Drosophila melanogaster* were counted in each of eleven bottles. The numbers (normal : vestigial) observed were as follows.

$25:1 \quad 80:15 \quad 38:12 \quad 52:8 \quad 9:0 \quad 21:7 \quad 33:6 \quad 24:2 \quad 30:7$

$51:7 \quad 56:3$

Use these data to estimate the proportion of normal *Drosophila* in the population from which the bottles were drawn.

The data are cited in Haldane, J.B.S. (1955) The rapid calculation of χ^2 as a test of homogeneity from a $2 \times n$ table. *Biometrika*, **42**, 519–520.

Exercise 6.14 is about the results of a genetics experiment.

Exercise 6.14

In 1918, T. Bregger crossed a pure-breeding variety of maize, having coloured starchy seeds, with another, having colourless waxy seeds. All the resulting first-generation seeds were coloured and starchy. Plants grown from these seeds were crossed with colourless waxy pure-breeders. The resulting seeds were counted. There were 147 coloured starchy seeds (CS), 65 coloured waxy seeds (CW), 58 colourless starchy seeds (NS) and 133 colourless waxy seeds (NW). According to genetic theory, the seeds should have been produced in the ratios

$$\text{CS} : \text{CW} : \text{NS} : \text{NW} = \tfrac{1}{2}(1-r) : \tfrac{1}{2}r : \tfrac{1}{2}r : \tfrac{1}{2}(1-r),$$

where the number r is called the recombinant fraction.

Find the maximum likelihood estimate $\widehat{r}$ for these data.

Bregger, T. (1918) Linkage in maize: the C-aleurone factor and wax endosperm. *American Naturalist*, **52**, 57–61.

Exercises 6.15 and 6.16 are about censored data.

Exercise 6.15

Although there are probability models available that would provide a better fit to the vehicle occupancy data of Example 6.9 than the geometric distribution, the geometric model is not altogether valueless. Find the maximum likelihood estimate $\widehat{p}$ for p, assuming a geometric fit.

Hint If X is $G(p)$, then

$$P(X \geq x) = (1-p)^{x-1}, \quad x = 1, 2, \ldots .$$

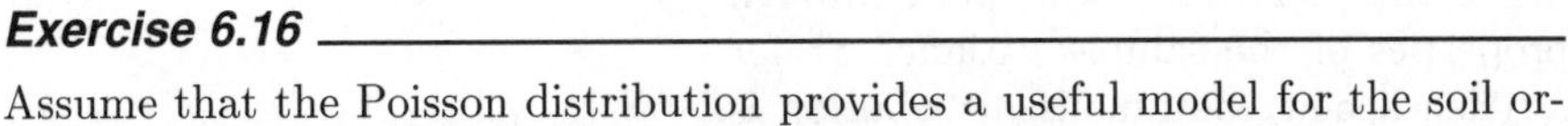

Exercise 6.16

Assume that the Poisson distribution provides a useful model for the soil organism data of Example 6.10 (that is, that the Poisson distribution provides a good model for the variation observed in the numbers of colonies per quadrat). Obtain the maximum likelihood estimate of the Poisson parameter μ.

Note Remember that for the Poisson distribution there is no convenient formula for the tail probability $P(X \geq x)$. You will only be able to answer this question if you are very competent at algebra, or if your computer knows about Poisson tail probabilities.

Exercises 6.17 and 6.18 only require you to use the tables of standard results for maximum likelihood estimators, Tables 6.8 and 6.10.

Exercise 6.17

In *Chapter 4*, Table 4.10 data are given for 200 waiting times between consecutive nerve pulses. Use these data to estimate the pulse rate (per second).

Exercise 6.18

Assuming a normal model for the variation in height, obtain a maximum likelihood estimate for the mean height of the population of women from which the sample in *Chapter 2*, Table 2.15 was drawn.

Since discovering in Exercise 6.5(c) the moment estimator for the variance parameter σ^2 of a normal distribution, nothing more has been said about this. The moment estimator is

$$S^2 = \frac{1}{n-1}\sum(X_i - \overline{X})^2,$$

Recall that we write the estimator as S^2, using an upper-case S, to emphasize that the estimator is a random variable.

based on a random sample $X_1, X_2, \ldots, X_n$ from a normal distribution with unknown mean μ and unknown variance σ^2.

(A certain amount of statistical theory exists about how to estimate the parameter σ^2 when the value of μ is known. However, the sampling context in which one normal parameter is known but the other is not is very rare, and in this course only the case where neither parameter is known will be described.)

So far, nothing has been said about the sampling distribution of the estimator S^2. We do not even know whether the estimator is unbiased, and we know nothing of its sampling variation. Nor do we know the maximum likelihood estimator of the variance of a normal distribution.

6.5 Estimating a normal variance

The main points of this section are illustrated using computer simulation, so you should try to find the time to do the exercises as you work through it. First, we shall investigate the properties of the sample variance S^2 for samples from a normal population. Then a new continuous distribution is introduced and explored. This probability distribution enables us to complete our description of the sampling distribution of S^2.

6.5.1 The sample variance for a normal distribution

This subsection begins with a statement of the maximum likelihood estimator for the normal variance. This result may be deduced algebraically, but here it is presented without proof.

The maximum likelihood estimator for the normal variance σ^2, based on a random sample $X_1, X_2, \ldots, X_n$, is given by

$$\widehat{\sigma}^2 = \frac{1}{n}\sum(X_i - \overline{X})^2 = \frac{n-1}{n}S^2 = \left(1 - \frac{1}{n}\right)S^2.$$

You can see that there is a small difference here between the maximum likelihood estimator $\widehat{\sigma}^2$ and the moment estimator S^2. (In fact, for large samples, there is very little to choose between the two estimators.) Each estimator has different properties (and hence the provision on many statistical calculators of two different buttons for calculating the sample standard deviation according to two different definitions). The main distinction, as we shall see, is that the moment estimator S^2 is unbiased for the normal variance; the maximum likelihood estimator possesses a small amount of bias. In some elementary approaches to statistics, the estimator S^2 is not mentioned at all.

The exercises in this section generally require the use of a computer.

Exercise 6.19

(a) Obtain 3 observations x_1, x_2, x_3 on the random variable $X \sim N(100, 25)$, and calculate the sample variance s^2.

Your computed value of s^2 in part (a) might or might not have provided a useful estimate for the normal variance, known in this case to be 25. On the basis of one observation, it is not very sensible to try to comment on the usefulness of S^2 as an estimator of σ^2.

(b) Now obtain 100 random samples all of size three from the normal distribution $N(100, 25)$, and for each of the samples calculate the sample variance s^2. This means that you now have 100 observations on the random variable S^2.

(i) Calculate the mean of your sample of 100.

(ii) Plot a histogram of your sample of 100.

(iii) Find the variance of your sample.

You should have found in your experiment in Exercise 6.19 that the distribution of the estimator S^2 is highly skewed and that there was considerable variation in the different values of s^2 yielded by your different samples. Nevertheless, the mean of your sample of 100 should not have been too far from 25, the value of σ^2 for which S^2 is our suggested estimator.

It would be interesting to see whether larger samples lead to a more accurate estimation procedure. Try the next exercise.

Exercise 6.20

Obtain 100 random samples of size 10 from the normal distribution $N(100, 25)$, and for each of the samples calculate the sample variance s^2. This means that you now have 100 observations on the random variable S^2.

(a) Calculate the mean of your sample of 100.

(b) Plot a histogram of your sample of 100.

(c) Calculate the variance of your sample of 100.

In Exercise 6.20 you should have noticed that in a very important sense S^2 has become a better estimator as a consequence of increasing from three to ten the size of the sample drawn. For the estimator has become considerably less dispersed around the central value of about 25.

Did you also notice from your histogram that the distribution of S^2 appears to be very much less skewed than it was before?

It appears as though the sample variance S^2 as an estimator of the normal variance σ^2 possesses two useful properties. First, the estimator may be unbiased; and second, it has a variance that is reduced if larger samples are collected. In fact, this is true: the estimator is unbiased and consistent. One can go further and state the *distribution* of the sample variance S^2, from which it is possible to quantify its usefulness as an estimator for σ^2.

Before doing so, it is necessary to introduce another continuous probability distribution, the **chi-squared distribution**. Its properties will now be explored.

The word chi is pronounced 'kye', referring to a Greek letter to be introduced shortly.

6.5.2 The chi-squared distribution

Next, we investigate what happens when observations on the standard normal random variable Z are squared, and added.

Exercise 6.21

(a) If Z is $N(0,1)$, use your computer to obtain the probabilities

(i) $P(-1 \leq Z \leq 1)$;

(ii) $P(-\sqrt{2} \leq Z \leq \sqrt{2})$;

(iii) $P(-\sqrt{3} \leq Z \leq \sqrt{3})$;

(iv) $P(-2 \leq Z \leq 2)$.

(b) Now define the random variable

$$W = Z^2.$$

That is, to obtain an observation on the random variable W, first obtain an observation on Z, and then square it. Since it is the result of a squaring operation, W cannot be negative, whatever the sign of Z. Use your answers to part (a) to write down the probabilities

(i) $P(W \leq 1)$;

(ii) $P(W \leq 2)$;

(iii) $P(W \leq 3)$;

(iv) $P(W \leq 4)$.

(c) Now use your computer to obtain 1000 independent observations on Z, and then square them, so that you have 1000 independent observations on W. Find the proportion of

(i) observations less than 1;

(ii) observations less than 2;

(iii) observations less than 3;

(iv) observations less than 4.

(d) Obtain a histogram of your sample of observations.

(e) The random variable W has mean and variance

$$E(W) = \mu_W, \qquad V(W) = \sigma_W^2.$$

The values of these two moments are at this stage unknown, although they could be deduced algebraically. Instead, use your sample of 1000 observations on W to obtain moment estimates for μ_W and σ_W^2.

Your solution to parts (c), (d) and (e) of Exercise 6.21 will have been different to that obtained in the printed solutions, because your computer will have drawn a different random sample of observations on Z initially. However, you should have noticed that your histogram is very skewed; and you may have obtained summary sample moments $\overline{w}$ not too far from 1, and s_W^2 not too far from 2. As you saw, when the exercise was run to obtain a solution for printing, the summary statistics

$$\overline{w} = \widehat{\mu}_W = 1.032, \qquad s_W^2 = 2.203$$

were obtained.

Actually, the theoretical mean μ_W is quite easy to obtain. We know W is defined to be Z^2, so

$$\mu_W = E(W) = E(Z^2).$$

Since, by definition, $V(Z) = E(Z^2) - (E(Z))^2$, it follows that

$$\mu_W = V(Z) + (E(Z))^2 = \sigma_Z^2 + \mu_Z^2.$$

But Z is $N(0,1)$, so $\mu_Z = 0$ and $\sigma_Z^2 = 1$; therefore

$$\mu_W = 0 + 1^2 = 1.$$

It is not quite so easy to obtain the variance σ_W^2 of W and you are spared the details. In fact,

$$\sigma_W^2 = 2.$$

Exercise 6.22

Write down the mean and variance of the random variable

$$W = Z_1^2 + Z_2^2 + \cdots + Z_r^2,$$

where $Z_i, i = 1, 2, \ldots, r$, are independent observations on the standard normal variate Z.

By the theory of sums of independent random variables, the random variable $W = Z_1^2 + Z_2^2 + \cdots + Z_r^2$ has mean r and variance $2r$.

The chi-squared distribution

The continuous random variable W given by

$$W = Z_1^2 + Z_2^2 + \cdots + Z_r^2,$$

obtained as the sum of r independent squared observations on the standard normal variate Z, is said to follow a *chi-squared distribution with parameter* r, written

$$W \sim \chi^2(r).$$

The mean of W is $\mu_W = r$ and the variance of W is $\sigma_W^2 = 2r$; and W is strictly positive. For historical reasons, the parameter r is usually given the label 'degrees of freedom'. So W is said to follow a **chi-squared distribution with r degrees of freedom.**

The Greek letter χ is pronounced 'kye'.

You will notice that the p.d.f. of the random variable $W \sim \chi^2(r)$ has not been given; this is because it is not, for present purposes, very useful. For instance, the density does not integrate tidily to give a very convenient formula for calculating probabilities of the form $F_W(w) = P(W \leq w)$. These probabilities generally have to be deduced from tables, or need to be computed. (The same was true for tail probabilities for the standard normal variate Z.) In Figure 6.10 examples of chi-squared densities are shown for several degrees of freedom. Notice that for small values of the parameter the distribution is very skewed; for larger values the distribution looks quite different, appearing almost bell-shaped. This makes sense: for large values of r, W may be regarded as the sum of a large number of independent identically distributed random variables, and so the central limit theorem is playing its part in the shape of the distribution of the sum.

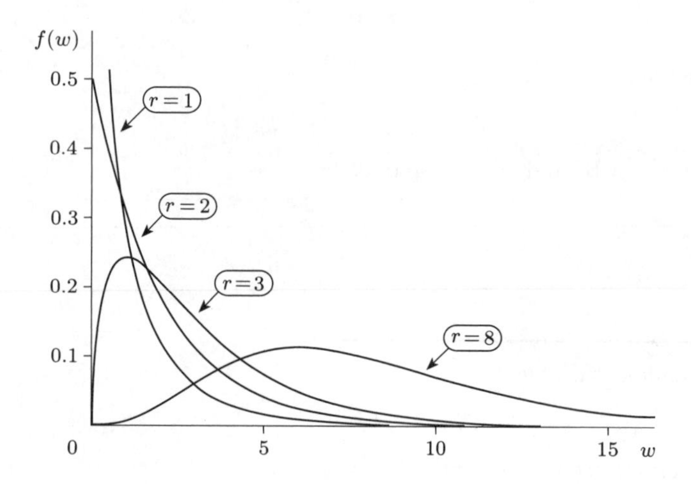

Figure 6.10 Chi-squared densities

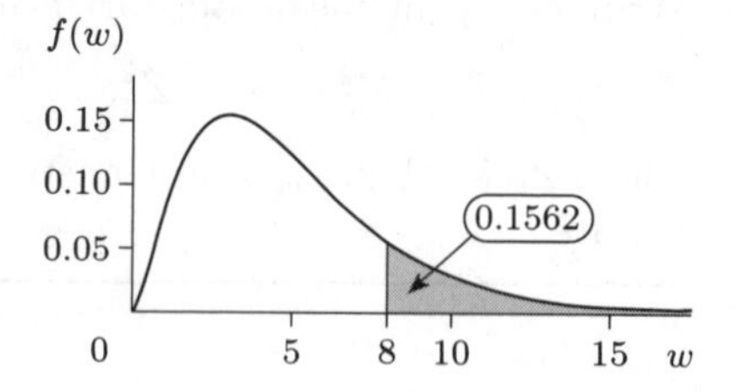

Figure 6.11 $P(W \geq 8)$ when $W \sim \chi^2(5)$

Most statistical computer programs contain the appropriate commands to calculate chi-squared probabilities. For instance, if $W \sim \chi^2(5)$, then the area of the region shown in Figure 6.11 gives the probability

$$P(W \geq 8) = 0.1562.$$

Exercise 6.23

Use your computer to calculate the following probabilities.

(a) $P(W \geq 4.8)$ when $W \sim \chi^2(6)$

(b) $P(W \geq 12.5)$ when $W \sim \chi^2(8)$

(c) $P(W \leq 10.0)$ when $W \sim \chi^2(19)$

(d) $P(W \geq 30.0)$ when $W \sim \chi^2(19)$

Similarly, you should be able to obtain quantiles for chi-squared densities from your computer. For instance, if $W \sim \chi^2(5)$ then W has lower and upper quartiles, and median given by

$$q_{0.25} = 2.67, \qquad q_{0.50} = 4.35, \qquad q_{0.75} = 6.63.$$

These points are shown in Figure 6.12.

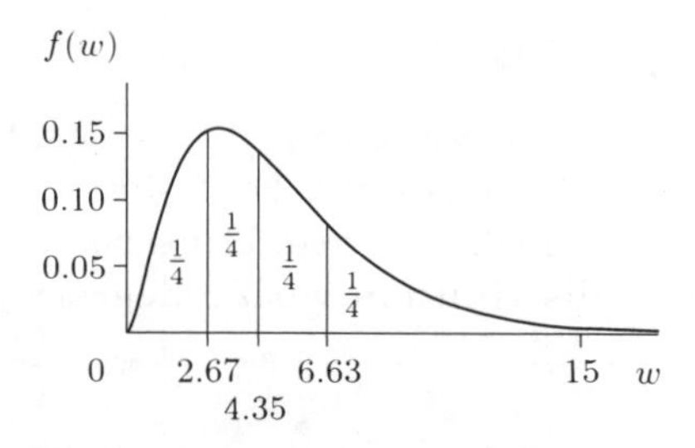

Figure 6.12 Quartiles of the $\chi^2(5)$ distribution

Exercise 6.24

Use your computer to calculate the following quantiles.

(a) $q_{0.2}$ when $W \sim \chi^2(8)$

(b) $q_{0.95}$ when $W \sim \chi^2(8)$

(c) $q_{0.1}$ when $W \sim \chi^2(19)$

(d) $q_{0.5}$ when $W \sim \chi^2(19)$

(e) $q_{0.99}$ when $W \sim \chi^2(19)$

Illustrate your findings in (c) to (e) in a sketch.

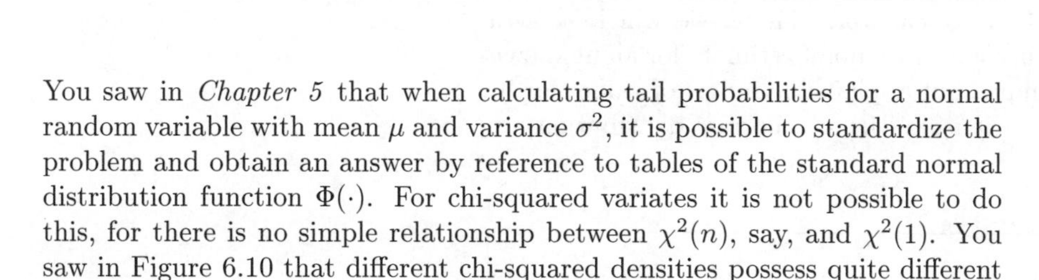

You saw in *Chapter 5* that when calculating tail probabilities for a normal random variable with mean μ and variance σ^2, it is possible to standardize the problem and obtain an answer by reference to tables of the standard normal distribution function $\Phi(\cdot)$. For chi-squared variates it is not possible to do this, for there is no simple relationship between $\chi^2(n)$, say, and $\chi^2(1)$. You saw in Figure 6.10 that different chi-squared densities possess quite different shapes.

Many pages of tables would therefore be needed to print values of the distribution function $F_W(w)$ for different values of w and for different degrees of freedom! In general, the need for them is not sufficient to warrant the publishing effort.

However, it is quite easy to print selected quantiles of the chi-squared distribution, and this is done in Table A6. Different rows of the table correspond to different values of the degrees of freedom, r; different columns correspond to different probabilities, α; the entry in the body of the table gives the α-quantile, q_α, for $\chi^2(r)$.

Exercise 6.25

Use the printed table to write down the following quantiles.

(a) $q_{0.05}$ when $W \sim \chi^2(23)$

(b) $q_{0.10}$ when $W \sim \chi^2(9)$

(c) $q_{0.50}$ when $W \sim \chi^2(12)$

(d) $q_{0.90}$ when $W \sim \chi^2(17)$

(e) $q_{0.95}$ when $W \sim \chi^2(1)$

That concludes our theoretical introduction to the chi-squared distribution. We now return to the estimator S^2 of a normal variance σ^2, where the reason for its introduction is finally explained. The result given in Exercise 6.26 for the sampling distribution of S^2 is not proved, but it is one with which you should become familiar.

Exercise 6.26

Using only the information that the sample variance S^2 in samples of size n from a normal distribution with variance σ^2, has the sampling distribution

$$\frac{(n-1)S^2}{\sigma^2} \sim \chi^2(n-1),$$

establish the two results

$$E(S^2) = \sigma^2, \qquad V(S^2) = \frac{2\sigma^4}{n-1}.$$

These results will be useful in the next chapter. There, we will look at a procedure that enables us to produce not just a point estimate for an unknown parameter based on a random sample, but a 'plausible range' of values for it. The main results for the sample variance S^2 from a normal distribution are summarized in the following box.

> The sample variance S^2 obtained for a random sample of size n from a normal distribution with unknown mean μ and unknown variance σ^2 has mean
>
> $$E(S^2) = \sigma^2$$
>
> and variance
>
> $$V(S^2) = \frac{2\sigma^4}{n-1}.$$
>
> The distribution of S^2 is given by
>
> $$\frac{(n-1)S^2}{\sigma^2} \sim \chi^2(n-1).$$

Summary

1. There are many ways of obtaining estimating formulas (that is, estimators) for unknown model parameters; when these formulas are applied in a data context, the resulting number provides an estimate for the unknown parameter. The quality of different estimates can be assessed by discovering properties of the sampling distribution of the corresponding estimator.

2. Not all estimating procedures are applicable in all data contexts, and not all estimating procedures are guaranteed always to give sensible estimates. Most require numerical or algebraic computations of a high order. Two of the most important methods are the method of moments and the method of maximum likelihood. In a given context the two methods often lead to the same estimator (though not always); maximum likelihood estimators are known to possess useful properties.

3. If X is a discrete random variable with probability mass function $p(x;\theta)$, where θ is an unknown parameter, then the likelihood of θ for the random sample $X_1, X_2, \ldots, X_n$ is given by the product

$$p(X_1, X_2, \ldots, X_n; \theta) = p(X_1;\theta)p(X_2;\theta)\ldots p(X_n;\theta).$$

If X is a continuous random variable with probability density function $f(x;\theta)$, where θ is an unknown parameter, then the likelihood of θ for the random sample $X_1, X_2, \ldots, X_n$ is given by the product

$$p(X_1, X_2, \ldots, X_n; \theta) = f(X_1;\theta)f(X_2;\theta)\ldots f(X_n;\theta).$$

4. A maximum likelihood estimate $\widehat{\theta}$ of θ may be obtained in either case by finding the value of θ which maximizes the likelihood of θ. This estimate will be an observation on the corresponding estimator, a random variable. Maximum likelihood estimators possess two useful properties.

(i) The maximum likelihood estimator $\widehat{\theta}$ for θ is asymptotically unbiased:

$$E(\widehat{\theta}) \to \theta \text{ as } n \to \infty,$$

where n is the sample size.

(ii) The maximum likelihood estimator $\widehat{\theta}$ for θ is consistent:

$$V(\widehat{\theta}) \to 0 \text{ as } n \to \infty.$$

5. The continuous random variable $W = Z_1^2 + Z_2^2 + \cdots + Z_r^2$, the sum of r independent squared observations on the standard normal variate Z, is said to follow a chi-squared distribution with r degrees of freedom. The random variable W has mean r and variance $2r$; the distribution is written $W \sim \chi^2(r)$.

6. The moment estimator S^2 for the variance σ^2 of a normal distribution, based on a sample of size n, is unbiased with variance $2\sigma^4/(n-1)$. The sampling distribution of the estimator S^2 is given by

$$\frac{(n-1)S^2}{\sigma^2} \sim \chi^2(n-1).$$

7. The maximum likelihood estimator $\widehat{\sigma}^2$ for the variance of a normal distribution based on a sample of size n is given by

$$\widehat{\sigma}^2 = \frac{n-1}{n}S^2.$$

Chapter 7
Estimation with confidence

In this chapter the idea of using samples of data to provide estimates for unknown model parameter values is further developed. However, instead of stating a single number (a point estimate), the aim is to provide a range of plausible values for the parameters.

In this chapter the ideas of estimation which were introduced and described in *Chapter 6* are extended. In *Chapter 6*, Example 6.2 an experiment in which 103 samples of water were collected from the same source was described. The number of specimens of the leech *Helobdella* in each sample was counted. A Poisson model was suggested for the variation observed in the counts. The data collected during the experiment yielded an estimate (the sample mean) $\overline{x} = \widehat{\mu} = 0.816$ for the Poisson parameter μ, that is, for the underlying average number of leeches per water sample.

It was pointed out that had a second experiment been performed using the same water source, then the numbers of leeches collected would probably have been different, and the sample mean used to estimate the underlying mean μ would probably have been different. It has already been remarked several times that in a random sample the observed sample mean $\overline{x}$ is just one observation on a random variable $\overline{X}$. Quite conceivably, in this experiment, $\overline{x}$ might have been as low as 0.7 or even lower; it might have been as high as 1, or perhaps 2, or 3. In the initial experiment (the one actually performed), there were a lot of 0s observed (samples containing no leeches at all), quite a few 1s and 2s, and just a few 3s and 4s. None of the 103 samples contained more than eight leeches. It would therefore be very surprising to find that the underlying mean incidence of leeches was as high as 6, say, and it is almost incredible (though, of course, it is not impossible) that the underlying mean could be as high as 8, or higher still.

This chapter is about using the results of statistical experiments to obtain some idea of a plausible *range* of values for some unknown population characteristic (maybe an average, or a rate, or a proportion). Assuming a reasonable statistical model, this characteristic will always be expressible in terms of the parameters of the model. The minimum and maximum values of this range are called *confidence limits*, and the range of plausible or 'believable' values is called a *confidence interval.*

Section 7.1 of the chapter deals with the situation where only a single observation has been collected in order to shed light on the area of investigation: our random sample is of size one. You will see how confidence intervals for the indexing parameters of some of the standard probability models with which you are already familiar are calculated. You already know from *Chapter 4* how important it is in a sampling context, generally speaking, to collect as much data as possible, and in this chapter the influence of sample size on the usefulness of the conclusions drawn will become very apparent. However, circumstances do occur where a sample is necessarily small, and this approach (drawing a sample of size one) provides an easy introduction to the topic.

See (4.10).

In Section 7.2 you will see the consequences of increasing the sample size. It is very useful to be able to say with some high degree of confidence that the value of some unknown parameter is between certain limits. It is more useful still if those limits are not very far apart. Specifically, the confidence limits obtained through a sampling experiment have a tendency to move closer and closer together as the sample size increases: the resulting confidence interval becomes narrower and more 'precise'. All the calculations in this section require a computer. (The problem is simple enough to state, but the arithmetic involved in calculating confidence limits can become surprisingly difficult.)

In Sections 7.1 and 7.2 we look at confidence intervals for descriptive parameters such as a Poisson mean μ, a Bernoulli probability p, an exponential mean, and so on. Sections 7.3 and 7.4 are based on normal theory. In Section 7.3, it is assumed that interest centres around a population where the variation observed may be adequately modelled by a normal distribution. We have seen many contexts where this is a reasonable assumption. Methods for obtaining confidence intervals for the two parameters of the normal distribution, the mean μ and the standard deviation σ, are developed. You will see that in order to write down confidence limits for μ it becomes necessary to make use of a statistical distribution introduced for the first time in this chapter. This is Student's t-distribution, named after W. S. Gosset who published under the pseudonym 'Student' at the turn of the century. In this case the calculations are fairly simple and only require reference to statistical tables and a calculator.

See *Chapter 2*, p. 60.

In Section 7.4, it is assumed that samples are large enough for the central limit theorem to be used. This is very commonly the case in practice, and again calculations are based on the normal distribution. Even when (as in Section 7.2) the underlying model is Poisson or binomial, for instance, approximate confidence limits can be calculated using normal tables and a calculator.

In the first four sections of the chapter, it is always assumed that an adequate underlying statistical model (normal, Poisson, binomial, and so on) for the variation observed in a particular context has been identified. Section 7.5 discusses what to do when all you have is a long list of numbers, and not the remotest idea about what might constitute a good model for the underlying variation in the population from which they have been drawn. You will see that, provided your sample is large enough, normal distribution theory can be used to find approximate confidence limits for the underlying population mean.

7.1 Samples of size one

It is not at all common for statistical inferences to be based on just one *datum* (the singular form of data, 'the thing given'). More often some sort of replication in a statistical experiment is possible, at not too great a cost, and with a consequent enhancement in the precision of the conclusions that may be drawn. However, it will be useful to begin this description of what is involved in setting up a confidence interval by taking a moderately simplified approach in which only one observation has been taken on the random variable of interest.

7.1.1 Some examples

***Example 7.1** Accident counts*

In an investigation into accident proneness in children, numbers of injuries were counted for 621 children over the eight-year period between the ages of 4 and 11. The early history of each child (aged 4 to 7) was compared with their later history (aged 8 to 11). One child experienced a total of 3 injuries between the ages of 4 and 7. (We shall return to the full data set in Section 7.4.)

Mellinger, C.D., Gaffey, W.R., Sylwester, D.L. and Manheimer, D.I. (1965) A mathematical model with applications to a study of accident repeatedness among children. *J. American Statistical Association*, **60**, 1046–1059. In the context of this investigation, an 'accident' was one requiring professional medical attention.

For our purposes a Poisson model may be assumed to describe adequately the variation in the number of accidents experienced by children over a four-year period (though, actually, it was an aim of the research exercise to show that a more elaborate model is required in this context). Then the number 3 represents a single observation on the random variable N, say, where $N \sim \text{Poisson}(\mu)$. The sample mean $\widehat{\mu} = 3$ is a maximum likelihood estimate of μ.

This is useful information yielded by the investigation, and this single data point tells us something about the underlying childhood accident rate. It suggests that accidents happen ($\mu > 0$), but not every day, or even every month—perhaps an average of once a year. It would be useful to go further and state, with some confidence, a *range* of credible values for μ. Can we devise a procedure that enables us to make a **confidence statement** along the lines: with 90% confidence, and on the basis of this observation, the value of μ lies between 1.5 and 4.8? (Or whatever the confidence limits might turn out to be.) ■

***Example 7.2** Coal-mining disasters*

Data were collected on the time intervals in days between disasters in coal mines in Britain from 15 March 1851 to 22 March 1962 inclusive. In this context, an industrial accident is called a 'disaster' if ten or more men were killed. There were altogether 191 such accidents. The data set is a famous one, attracting much interest from analysts, not least because the original published set contained several errors. This kind of delving and reanalysis is a common feature of statistical activity.

Jarrett, R.G. (1979) A note on the intervals between coal-mining disasters. *Biometrika*, **66**, 191–193. The original data are to be found in Maguire, B.A., Pearson, E.S. and Wynn, A.H.A. (1952) The time intervals between industrial accidents. *Biometrika*, **39**, 168–80. The errors are corrected and the information is substantially updated in the 1979 paper. We shall return to the full data set in Section 7.4.

The first accident to occur after 15 March 1851 took place on 19 August that year. If data collection had ceased at that point, then the single observation of 157 days would have been collected on a random variable T which may, for our purposes, be supposed to follow an exponential distribution with unknown mean μ. This number provides an estimate of the mean time interval (in days) between consecutive accidents. Can the number also (and, arguably, more usefully) be used to furnish a confidence interval for the average time between consecutive accidents? ■

Example 7.3 Absenteeism

The absences of each of 113 students from a lecture course were recorded over the 24 lectures for which the course ran. There were eleven lectures during the first term and thirteen in the second. Assuming a binomial model for the number of missed lectures in each term, an estimate for the binomial parameter p (the proportion of lectures missed overall) is provided by the information that during the first term (11 lectures) one student missed four of them. The estimate of p is $\widehat{p} = 4/11 = 0.364$. This offers a guide to the underlying value of p, a measure of students' propensity to miss a lecture. Can the information be developed to provide confidence limits for p?

Ishii, G. and Hayakawa, R. (1960) On the compound binomial distribution. *Annals Institute of Statistical Mathematics, Tokyo*, **12**, 69–80.

An analysis of the extended data set shows that not every student has the same probability for missing a lecture—some students are simply more committed than others—and a binomial model does not provide a good fit. Permitting the absence probability to vary from student to student according to some other statistical distribution could lead to a more realistic, and a better, model. The authors of the paper in which these data were published discuss at some length the fitting of such different models. At the moment, however, the binomial model is as refined a model as we have available to us. For many purposes, it is quite adequate. ■

Example 7.4 Diseased trees

In *Chapter 6*, Table 6.5 data are given on the lengths of runs of diseased trees in an infected plantation of Douglas firs. The disease was *Armillaria* root rot, and interest is centred on assessing how infectious the disease was. The geometric distribution turns out to be a good model for the variation in run length. There was one run of 4 infected trees. What can this tell us about the indexing parameter p for the geometric distribution in this context?

Pielou, E.C. (1963) Runs of healthy and diseased trees in transects through an infected forest. *Biometrics*, **19**, 603–614.
A run of infected trees is a line of trees with the disease, bounded at each end by a healthy tree.

The maximum likelihood estimate of p, based on this single observation, is $\widehat{p} = 1/\overline{x} = 1/4 = 0.25$. It would be useful to develop this into a confidence statement such as: a 95% confidence interval for p, based on the single observation 4, is given by the confidence limits $p = 0.1$ to $p = 0.5$ (or whatever the limits might be). ■

These four examples all have three features in common. The first is *data* (3 injuries; 157 days; 4 out of 11; 4 trees). The second is that, in every case, some sort of *probability model* has been suggested (Poisson, exponential, binomial, geometric). The third feature is that in each case, too, a descriptive *model parameter* has been identified and related to the data. Let us now pursue the first of these examples further.

Example 7.1 continued

In Example 7.1, a single observation of 3 was recorded on a random variable assumed to follow a Poisson distribution with unknown mean μ. Although the data set is extremely small, it provides an estimate for μ: it is the sample mean, 3. The purpose of this chapter is to answer two questions. The first is: how large could μ actually be before an observation as low as 3 starts to seem unlikely? This upper limit for μ is called an **upper confidence limit** for μ. The second question is similar: how small could μ actually be before an observation as high as 3 starts to seem unlikely? The answer to this question provides a **lower confidence limit** for μ. The range of values spanned by the lower and upper confidence limits is called a **confidence interval** for μ.

To proceed further, we need to decide what in this context is meant by the word 'unlikely'. For instance, suppose the underlying rate for accidents in early childhood (that is, over the four years from age 4 to age 7) was, in fact, $\mu = 5$. (This is higher than 3, but scarcely renders the observation 3 an incredible one.) In fact, the probability of recording a value as low as that observed (i.e. 3 or lower) is given by the probability $P(X \leq 3)$ where $X \sim \text{Poisson}(5)$. This probability is

$$\begin{aligned} P(X \leq 3) &= P(X = 0) + P(X = 1) + P(X = 2) + P(X = 3) \\ &= e^{-5}\left(\frac{5^0}{0!} + \frac{5^1}{1!} + \frac{5^2}{2!} + \frac{5^3}{3!}\right) \\ &= 0.265, \end{aligned}$$

which is not particularly small. Extending this argument, if μ were actually as high as 8, then the probability of observing a count of 3 or lower would be

$$\begin{aligned} P(X \leq 3) &= P(X = 0) + P(X = 1) + P(X = 2) + P(X = 3) \\ &= e^{-8}\left(\frac{8^0}{0!} + \frac{8^1}{1!} + \frac{8^2}{2!} + \frac{8^3}{3!}\right) \\ &= 0.042. \end{aligned}$$

This is much smaller: there is a probability of only 4% of observing a value as low as 3. If the childhood accident rate μ were actually as high as 12, then the probability of a child suffering fewer than four accidents would turn out to be just 0.002, which is very small indeed. Inverting that last remark, the observation $x = 3$ suggests that to put the underlying childhood accident rate as high as 12 would be seriously to overestimate it.

Of course, it is perfectly possible that the value of μ is really as high as 12, and an event with a very small probability has happened. This possibility must never be forgotten when constructing confidence intervals.

What about low values of μ? In this case we need to assess the value of μ at which an observation as high as 3 (i.e. 3 or higher) starts to look implausible. If we try setting μ equal to 1, say, then

$$\begin{aligned} P(X \geq 3) &= 1 - P(X \leq 2) \\ &= 1 - (P(X = 0) + P(X = 1) + P(X = 2)) \\ &= 1 - e^{-1}\left(\frac{1^0}{0!} + \frac{1^1}{1!} + \frac{1^2}{2!}\right) \\ &= 1 - 0.920 \\ &= 0.080. \end{aligned}$$

This is beginning to look a little unlikely. If we try setting μ equal to 0.5, then

$$\begin{aligned} P(X \geq 3) &= 1 - P(X \leq 2) \\ &= 1 - (P(X = 0) + P(X = 1) + P(X = 2)) \\ &= 1 - e^{-0.5}\left(\frac{0.5^0}{0!} + \frac{0.5^1}{1!} + \frac{0.5^2}{2!}\right) \\ &= 1 - 0.986 \\ &= 0.014. \end{aligned}$$

This is even less likely: a value for μ as low as 0.5 is beginning to look implausible, based on the single observation $x = 3$. ■

What we have done here is to propose a particular value for μ and then to assess, conditional on that value, the chances of observing a count as extreme as the one that actually occurred. If the chances are small, making that observation an unlikely one under the proposed conditions, then we have low confidence in the original proposed value for μ. A statement about random variables and probability is interpreted in terms of parameters and confidence.

It is a very common practice to express confidence statements in terms of percentages, as in: a 95% confidence interval for μ, based on the observation $x = 3$, is the interval from ... to .... There is nothing special about the **confidence level** of 95%, and you will often see levels of 90% and 99% mentioned. What this means is that an event that might have occurred with probability 0.95 is thought to be 'reasonable', whereas something occurring with probability merely 0.05 will (for the purposes of confident estimation) be considered 'unlikely'.

Now, an event might be considered unlikely because the observed count x is surprisingly high, or because it is surprisingly low. Let us make these events 'equally' surprising. This suggests that we reserve $\frac{1}{2} \times 0.05 = 0.025$ or 2.5% for probability statements about high values of x, and 2.5% for probability statements about low values of x. A 95% confidence interval can be obtained in this case by first solving for μ the equation

$$P(X \leq 3) = e^{-\mu}\left(1 + \mu + \frac{\mu^2}{2!} + \frac{\mu^3}{3!}\right) = 0.025 \tag{7.1}$$

(giving the upper confidence limit); and second, solving for μ the equation

$$P(X \geq 3) = 1 - P(X \leq 2) = 1 - e^{-\mu}\left(1 + \mu + \frac{\mu^2}{2!}\right) = 0.025 \tag{7.2}$$

giving the lower confidence limit.

The phrase 'solving an equation for μ' means rewriting the equation with μ as the subject. In other words, the equation has to be rearranged so that μ is set equal to an expression that just involves numbers. It is not possible to do this with either (7.1) or (7.2), and other numerical methods need to be followed. As you will discover later in Section 7.2, the first equation has solution $\mu = 8.8$, while the second equation has solution $\mu = 0.62$. We can complete our calculations with a confidence statement for the unknown Poisson mean μ, as follows, demonstrating a common notation for lower and upper confidence limits, and the notation that will be used in this course.

See Exercise 7.8(a)(ii).

A 95% confidence interval for the Poisson mean μ, based on the single observation 3, is

$$\mu_- = 0.62 \quad \text{to} \quad \mu_+ = 8.8.$$

Here, both the upper and lower confidence limits have been given to two significant figures (rather than, say, two decimal places).

The confidence interval may be written

$$(\mu_-, \mu_+) = (0.62, 8.8).$$

You may have observed that the confidence interval is very wide, with the upper confidence limit nearly fifteen times the lower confidence limit! This is an unavoidable consequence of the sparsity of the data. We shall see that larger samples typically reduce the width of confidence intervals very usefully.

The approach may be summarized as follows. A single observation x is collected on the random variable X, where X follows a specified probability model with unknown parameter θ. Then a 95% confidence interval for θ is provided by solving separately for θ the equations

$$P(X \leq x) = 0.025 \quad \text{and} \quad P(X \geq x) = 0.025.$$

It has already been remarked that there is nothing special about the confidence level of 95%, although it is a common one to choose. Depending on the purpose for which confidence statements are formulated, selected levels might be as low as 90% or as high as 99% or 99.9%. It is common to write the level generally as $100(1-\alpha)\%$, and to solve for θ the two equations

$$P(X \leq x) = \tfrac{1}{2}\alpha \quad \text{and} \quad P(X \geq x) = \tfrac{1}{2}\alpha.$$

There are other ways of setting up the two equations. One is to write them in the form

$$P(X \leq x) = p\alpha \text{ and } P(X \geq x) = q\alpha,$$

where $p + q = 1$, and choose p and q to minimize the difference $(\theta_+ - \theta_-)$ or the ratio (θ_+/θ_-). In this course we shall adopt the convention of setting $p = q = \frac{1}{2}$.

In general, the arithmetic of such computations is difficult, and numerical techniques beyond the scope of this course have to be adopted. Alternatively, many statistical computer packages make such techniques unnecessary, returning confidence limits at the press of a key. The exercises of Section 7.2 assume that you have access to such a computer package.

Confidence intervals

Suppose that a single observation x is collected on a random variable X, following a specified probability distribution with unknown parameter θ. Then a confidence interval (θ_-, θ_+) for θ with level $100(1-\alpha)\%$ is provided by separately solving for θ the two equations

$$P(X \leq x) = \tfrac{1}{2}\alpha \quad \text{and} \quad P(X \geq x) = \tfrac{1}{2}\alpha.$$

In some cases the arithmetic is straightforward, as in the following example.

Example 7.2 continued

In the example on coal-mining disasters, the single number 157 (days) was obtained as an observation on an exponentially distributed random variable T with mean μ. Suppose that what is now required is a 90% confidence interval for μ, based on the single observation $t = 157$. The procedure is as follows.

The required confidence level is $100(1-\alpha)\% = 90\%$, so $\frac{1}{2}\alpha = 0.05$. For an exponentially distributed random variable T with mean μ, the c.d.f. is given by

$$P(T \leq t) = 1 - e^{-t/\mu}, \quad t \geq 0.$$

The equation $P(T \leq t) = 0.05$, which yields an upper confidence level for μ in the exponential case, may be solved as follows.

$$P(T \leq t) = 1 - e^{-t/\mu} = 0.05$$

gives

$$e^{-t/\mu} = 0.95$$

so

$$\frac{t}{\mu} = -\log(0.95)$$

or

$$\mu = \frac{t}{-\log(0.95)} = \frac{157}{0.051293} = 3060.8$$

or, say,

$$\mu_{+} = 3060 \text{ days.} \quad \blacksquare$$

Here, the upper confidence limit has been given to three significant figures.

Exercise 7.1

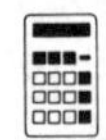

Calculate the corresponding lower confidence limit μ_{-} for μ, the average number of days between disasters, assuming an exponential model and based on the single observation 157 days. Write out explicitly a confidence statement about the value of μ.

Notice that in Exercise 7.1 it is again evident that calculation of confidence limits requires three things: *data*, a *model* and a statement of the *parameter* of interest.

Another feature you might have noticed is the extreme width of the confidence interval that results! (This was true for the Poisson mean as well.) The time interval observed was about five months; the confidence limits for the mean time between disasters are from 'rather less than two months' to 'more than eight years'. This confidence interval is not very informative, due partly to the highly skewed shape of the exponential distribution, but, again, mainly to the dearth of data.

Here is another example where the arithmetic is straightforward.

Example 7.5 A crooked die

A gambler, who believes that a die has been loaded in such a way that rolling a 6 is less likely than it ought to be (i.e. $p < 1/6$), discovers in a single experiment that it takes 13 rolls to obtain a 6. If the die is fair, then the number of throws necessary to achieve the first 6 is a random variable N following a geometric distribution with parameter $p = 1/6$. The expected value of N is $\mu_N = 1/p = 6$. Certainly, an observation as high as 13 suggests that p is smaller than $1/6$; indeed, the maximum likelihood estimate of p is

$$\widehat{p} = \tfrac{1}{13} = 0.077.$$

But it is quite possible for an observation as high as 13 to be observed even if the die is fair. Here, in Table 7.1 are some simulated observations on the random variable $N \sim G(p)$, where $p = 1/6$.

Table 7.1 20 simulated observations on $N \sim G(p)$, $p = 1/6$

2	1	11	3	6	1	4	12	7	1
9	2	11	6	2	7	1	1	1	4

In none of these twenty simulations was a count as high as 13 observed; but there were two 11s and one 12. For a random variable N following a geometric distribution with parameter p, the c.d.f. of N is given by

$$P(N \leq n) = 1 - (1-p)^n, \quad n = 1, 2, 3, \ldots.$$

See (3.17) in *Chapter 3*.

If $p = 1/6$, an observation as extreme as 13 (i.e. 13 or more) has probability

$$\begin{aligned} P(N \geq 13) &= 1 - P(N \leq 12) \\ &= 1 - \left(1 - \left(1 - \tfrac{1}{6}\right)^{12}\right) \\ &= \left(\tfrac{5}{6}\right)^{12} \\ &= 0.112, \end{aligned}$$

so an observation as large as 13 is not extraordinary. What we could do to try to clarify matters for the gambler is to calculate confidence limits for p based on the single datum 13. Let us set the confidence level required at 95%. Then $100(1-\alpha)\% = 95\%$ so $\frac{1}{2}\alpha = 0.025$. First, we solve for p the equation

$$P(N \leq 13) = 1 - (1-p)^{13} = 0.025.$$

The solution is

$$p = 1 - (1 - 0.025)^{1/13} = 1 - 0.998 = 0.002,$$

so $p_- = 0.002$.

(Notice that in this case, solving the equation $P(N \leq n) = \frac{1}{2}\alpha$ has yielded a lower confidence limit for the parameter: for here, it is low values of p that render low values of N unlikely.) ■

Exercise 7.2

Assuming a geometric model, find the corresponding upper confidence limit p_+ for p, and complete your confidence statement for p, based on the single observation $n = 13$.

Exercise 7.3

Find a 90% confidence interval for p, and compare it with your 95% confidence interval.

So the 95% confidence interval for p, based on these data and assuming an underlying geometric model, is given by

$$(p_-, p_+) = (0.002, 0.265),$$

and the 90% confidence interval is

$$(p_-, p_+) = (0.004, 0.221).$$

Notice particularly that the number $p = 1/6 = 0.167$ is included both in the 90% and 95% confidence intervals, which are extremely wide: the case for

'collecting more data' is seriously pressing. There seems no particular reason to suppose that the die is unfairly loaded away from 6s, even though the experiment in Example 7.5 took 13 rolls of the die to achieve a 6. The idea of performing statistical experiments to explore propositions is a very important one, and one that is pursued in *Chapter 8*.

Exercise 7.4

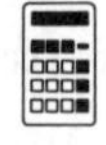

In Example 7.4, one run of diseased trees was of length 4. Assuming a geometric model for the distribution of run length, use this observation to construct a 99% confidence interval for the mean length μ of runs of diseased trees.

Note Here, you can establish confidence limits for p and then simply make use of the relationship for a geometric random variable that $\mu = 1/p$ to rewrite the upper and lower confidence limits. Or you could reparameterize the geometric distribution in terms of μ rather than p by saying

$$P(N \leq n) = 1 - (1-p)^n = 1 - \left(1 - \tfrac{1}{\mu}\right)^n,$$

and then work directly in terms of the parameter μ.

Example 7.3 continued

The only one of the four examples with which this section was introduced and which has not yet been followed through is Example 7.3, in which 4 absences were observed in a total of 11 lectures. The maximum likelihood (and common-sense) estimate for p, the underlying probability of absence, is $\widehat{p} = 4/11 = 0.364$. Again, it would be useful to develop this estimate into a confidence interval for p. Assuming independence from lecture to lecture (which is probably rather a strong assumption in this context) then we have a single observation $x = 4$ on a binomially distributed random variable $X \sim B(11, p)$. We need therefore to solve the two equations

$$P(X \leq 4) = \tfrac{1}{2}\alpha \quad \text{and} \quad P(X \geq 4) = \tfrac{1}{2}\alpha$$

for p. The first reduces to

$$\begin{aligned} P(X \leq 4) &= P(X=0) + P(X=1) + P(X=2) + P(X=3) + P(X=4) \\ &= \binom{11}{0}p^0q^{11} + \binom{11}{1}p^1q^{10} + \binom{11}{2}p^2q^9 + \binom{11}{3}p^3q^8 + \binom{11}{4}p^4q^7 \\ &= (1-p)^{11} + 11p(1-p)^{10} + 55p^2(1-p)^9 + 165p^3(1-p)^8 + 330p^4(1-p)^7 \\ &= \tfrac{1}{2}\alpha. \end{aligned}$$

The algebra here is not at all convenient (as it has been in the case of exponential and geometric probability calculations). The equation reduces to

$$(1-p)^7(210p^4 + 84p^3 + 28p^2 + 7p + 1) = \tfrac{1}{2}\alpha,$$

which is not soluble by usual methods. Actually, the 90% upper confidence limit ($\tfrac{1}{2}\alpha = 0.05$) is given by $p_+ = 0.6502$. This kind of calculation can be executed with a single command if you have a computer with the appropriate software installed. The second equation $P(X \geq 4) = \frac{1}{2}\alpha$ reduces to

$$1 - (1-p)^8(120p^3 + 36p^2 + 8p + 1) = \tfrac{1}{2}\alpha.$$

This equation has the solution (when $\frac{1}{2}\alpha = 0.05$) $p_- = 0.1351$. So we might conclude: a 90% confidence interval for p, based on observing 4 successes in 11 trials and assuming independence from trial to trial, is given by $(p_-, p_+) = (0.1351, 0.6502)$. ■

7.1.2 Interpreting a confidence interval

You have seen that a requirement to construct a confidence interval for an unknown parameter θ is the collection of data. In different random samples, the data will vary. It follows (rather as in the case of point estimation) that the resulting confidence limits θ_- and θ_+ are themselves random variables. The resulting confidence interval (θ_-, θ_+) is called a **random interval**. What are its relevant properties?

This question will be explored in one very simple case. The principle is the same in other sampling contexts, but the algebra can become rather involved.

Suppose that a single observation X has been collected from a population assumed to follow an exponential distribution with mean μ. The 2.5% point of this distribution is the solution $q_{0.025}$ of the equation $1 - e^{-q/\mu} = 0.025$, or $q_{0.025} = -\mu \log(0.975) = 0.025\mu$. Similarly, the 97.5% point is $q_{0.975} = 3.69\mu$. So we can write

$$P(0.025\mu \leq X \leq 3.69\mu) = 0.95.$$

Then the double inequality on the left-hand side can be rewritten with μ as the subject:

$$P\left(\frac{X}{3.69} \leq \mu \leq \frac{X}{0.025}\right) = P(0.27X \leq \mu \leq 39.5X) = 0.95.$$

This is a statement about the random interval $0.27X$ to $39.5X$: it is a very wide interval, stretching from about one-quarter of the single observed data point to forty times it. The point is that with probability 0.95, the random interval contains the unknown number μ.

That is to say: if the statistical experiment were repeated many times, and if the **random interval**

$$(\mu_-, \mu_+) = (0.27X, 39.5X)$$

were computed each time, then approximately 95% of these intervals would contain the unknown number μ. It is a common error to complete the statistical investigation with a 'confidence' statement such as: with probability 0.95, the value of μ is between μ_- and μ_+. But the number μ is not a random variable. It is important to remember that a confidence statement follows from a probability statement about a random interval, which might, or might not, contain the constant μ.

7.1.3 Some further exercises

Exercise 7.5

In *Chapter 2*, Example 2.8, a triangular model $T \sim \text{Triangular}(20)$ was suggested for the waiting time (in seconds) between consecutive vehicles using a particular grade of road.

(a) Find in terms of α and t a defining formula for the $100(1-\alpha)\%$ confidence interval for the parameter θ of a triangular density, based on the single observation t.

(b) Use your formula to give a 95% confidence interval for the parameter θ of a triangular distribution, based on the single observation $t = 5$.

Exercise 7.6

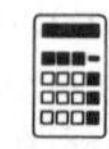

Firefly racing dinghies carry sail numbers consisting of a capital letter F followed by one or more digits. The dinghies are numbered consecutively as they are manufactured, starting at F1.

The total number of *Firefly* dinghies manufactured is not known. However, one dinghy has been sighted. It bore the sail number F3433. On the basis of this one observation, and stating your model for the population from which this one observation was drawn, calculate a 90% confidence interval for the total number of *Firefly* dinghies manufactured to date.

7.2 Small samples

In Section 7.1 only very small data sets (samples of size one) were considered. In all cases where you had to calculate a confidence interval for some unknown model parameter, one recurring feature was the extreme width of the intervals found. In this section and in Section 7.4 you will see the beneficial consequences of using larger samples.

As in Section 7.1, the derivation of a confidence interval for a population characteristic requires *data*, and specification of *a model* for the inherent variation. The characteristic of interest will be expressible in terms of the model parameter.

In most of the cases that we shall look at in this section, the underlying statistical theory is based on the sample total; for the sort of models used in this course much is known about the distribution of sums of random variables. In Section 7.4 the samples will be assumed to be sufficiently large for the central limit theorem to be implemented: the sample total may be assumed to be approximately normally distributed. However, in this section, the samples are small and the exact distribution of the sample total is used. Brief explanations are given of the principles on which interval estimation in this section is based, but the details are attended to by the computer.

However, the sample total is not always the appropriate statistic to use, and at the end of this section you will see an example where this is the case.

You have already seen that even when the sample properties are well known and easy to write down, the arithmetic involved in calculating confidence limits is not necessarily straightforward. We have so far been unable directly to deduce confidence limits for a binomial probability p or a Poisson mean μ, for instance, although you have seen examples of them.

The exercises in this section require the use of a computer to cope with the arithmetic. The problems are stated: you need to explore and use the facilities of your computer in order to solve them.

7.2.1 Confidence intervals for the Bernoulli parameter

A researcher interested in the proportion p of members of a population who possess a particular attribute is not likely to base an assessment of the value of p on a single observation X, taking the value 0 (attribute not possessed) or 1 (attribute possessed). It is more likely that a larger sample $X_1, X_2, \ldots, X_n$ ($n > 1$) will be collected. Assuming independence between sample responses, then the sample total $X_1 + X_2 + \cdots + X_n$ follows a binomial distribution $B(n,p)$ and an estimate of p is provided by the sample mean

$$\widehat{p} = \overline{X} = \frac{X_1 + X_2 + \cdots + X_n}{n}.$$

We have already seen in principle how to extract a $100(1-\alpha)\%$ confidence interval for p, based on a statistical experiment in which x successes were observed in a total of n trials.

Exercise 7.7

Use your computer to calculate confidence intervals for the Bernoulli parameter p in each of the following situations.

(a) A total of 4 successes is observed in 11 trials (so $\widehat{p} = 4/11 = 0.364$). Find (i) a 90% confidence interval for p and (ii) a 95% confidence interval for p based on these data.

(b) In a similar follow-up experiment, the sample size is doubled to 22; and, as it happens, a total of 8 successes is observed. So $\widehat{p} = 0.364$ as before. Find (i) a 90% confidence interval for p and (ii) a 95% confidence interval for p based on these data. Can you explain why using a larger sample provides narrower confidence intervals?

(c) Find a 99% confidence interval for a Bernoulli parameter p based on observing 4 successes in 5 trials.

(d) *Simulation* Use your computer to generate 10 observations on the binomial distribution $B(20, 0.3)$. (You will obtain a sequence of success counts—3 out of 20, 8 out of 20, 2 out of 20, ..., and so on.) For each of these 10 experimental results, calculate the corresponding 90% confidence interval for p. How many of your intervals contain the value 0.3? How many would you have expected to contain the value 0.3?

See the solution to this exercise for a note about confidence intervals when the underlying probability distribution is discrete.

7.2.2 Confidence intervals for a Poisson mean

You have seen the sort of calculation involved in obtaining confidence limits for a Poisson parameter μ when a single data point is available. What happens, when more than one data point is sampled, is quite straightforward: the calculations are almost the same. If n independent observations $X_1, X_2, \ldots, X_n$ are taken from a Poisson distribution with mean μ, then their sum $T = X_1 + X_2 + \cdots + X_n$ follows a Poisson distribution with mean $\mu_T = n\mu$. A confidence interval is found for μ_T based on the *single* observation t. Then the confidence limits for μ_T can be converted to confidence limits for μ simply by dividing by n, the sample size.

Exercise 7.8

Use your computer to calculate confidence intervals for the Poisson parameter μ in each of the following situations.

(a) A child is observed for one year, during which period he suffers 3 minor accidents. Use this information to calculate (i) a 90% confidence interval and (ii) a 95% confidence interval for the underlying annual accident rate for boys of his age.

(b) Seven boys, all of the same age, are observed for a year. They suffer 4, 4, 3, 0, 5, 3, 2 minor accidents respectively. Use this information (i) to estimate the underlying annual accident rate μ for boys of their age and (ii) to calculate 90% and 95% confidence intervals for μ.

(c) Six girls, all of the same age, are observed for a year. Between them they suffer a total of 20 minor accidents. Use this information (i) to estimate the underlying annual accident rate μ for girls of their age and (ii) to calculate 90% and 95% confidence intervals for μ.

Notice that in part (b) of this exercise all the information was given; in part (c) only summary data were provided.

7.2.3 Confidence intervals for an exponential mean

If n independent observations are taken from an exponential distribution with mean μ, then their sum follows a probability distribution which is a member of the two-parameter **gamma** family. In this case the values of the two parameters are n and μ, but the details are, in this context, unimportant. Enough is known about relevant properties of the gamma distribution to enable computer routines to be written for the evaluation of confidence limits for exponential parameters based on more than one observation.

Exercise 7.9

(a) The data in *Chapter 4*, Table 4.7 give 62 time intervals in days between consecutive serious earthquakes world-wide. Use these data to construct a 90% confidence interval for the mean time interval between earthquakes, stating any probability model you assume.

(b) The data listed in *Chapter 2*, Table 2.11 are the time intervals (in seconds) between successive vehicles using the Kwinana Freeway one morning in Perth, Western Australia.

(i) Use these data to estimate the *rate* (vehicles per minute) at which vehicles pass the observation point. (ii) Calculate a 90% confidence interval for the rate.

The last part of this exercise involves a computer simulation.

(c) Generate 20 observations from the exponential distribution $M(1)$, and use your data to estimate the (usually unknown) population mean; then use the data to calculate a 90% confidence interval for the population mean. Repeat the process 10 times. In each case, examine whether the resulting confidence interval contains the number 1.

7.2.4 Confidence intervals for a geometric parameter

If n independent observations are taken from a geometric distribution with parameter p, then their sum follows a probability distribution which is a member of the two-parameter **negative binomial** family. The values of the two parameters are n and p, but again the details in this context are unimportant. Enough is known about relevant properties of the negative binomial distribution to enable the development of computer routines for the evaluation of confidence limits for geometric parameters based on more than one observation.

Exercise 7.10

In this exercise you will need to use your computer to count the number of rolls N necessary to achieve a 6, when a loaded die with $P(6) = 1/10$ is rolled.

(a) Obtain 10 observations on N, and use these observations to obtain a 90% confidence interval for the (usually unknown) probability $p = P(6)$. Does your interval contain the value $p = 1/10$? Does it contain the value $p = 1/6$?

(b) Now obtain 100 observations on N, and use them to obtain a 90% confidence interval for p. How does the width of this interval compare with the width of the interval you obtained in part (a)? Does your new interval contain the value $p = 1/10$? Does it contain the value $p = 1/6$?

7.2.5 Postscript

So far in this section we have seen the useful consequences of drawing a larger sample of data in order to calculate confidence intervals for unknown model parameters: the larger the sample, the narrower (that is, the more precise) the resulting interval tends to be.

Your computer will have (or should have) insulated you from the algebraic detail of the computations involved, but for all four of the standard cases considered so far, inference was based on the sample total. We have not considered samples from the triangular or Pareto distributions, or from the uniform distribution, all of which you have met in the course so far. It is sometimes far from easy to obtain the distribution of the sum $X_1 + X_2 + \cdots + X_n$ of a random sample from some stated distribution. In such a case the only thing to do (if you cannot calculate exact results numerically) is to make sure that the sample drawn is so large that the sample total may be assumed to be approximately normally distributed (by the central limit theorem). We shall look at this in Section 7.4.

To round off this section, we shall look at a situation where the sample total is not an appropriate statistic to use when calculating a confidence interval, but where an exact computation is nevertheless possible.

The uniform distribution is rather unusual. Recall the *Firefly* example in Exercise 7.6. If, say, five *Firefly* dinghies had been sighted (rather than just one), then would that have helped us further? Suppose the numbers sighted had been F3433 (as before) and then F1326, F378, F1826, F1314. It might seem as though the numbers 1326, 378, 1826, and 1314 really provide no further information at all about the total number of *Firefly* dinghies manufactured altogether: that number is evidently at least 3433, as was apparent from the first sighting. In fact, a useful inference can be based not on the sample total but on the *sample maximum.* The maximum $X_{\max}$ of a random sample of size n from a discrete uniform distribution $U(1,2,\ldots,\theta)$ has c.d.f.

$$P(X_{\max} \leq x) = \left(\frac{x}{\theta}\right)^n, \quad x = 1,2,\ldots,\theta.$$

The corresponding $100(1-\alpha)\%$ confidence interval for θ based on n observations with maximum value x is found by solving the two equations

$$P(X_{\max} \leq x) = \left(\frac{x}{\theta}\right)^n = \tfrac{1}{2}\alpha$$

and

$$P(X_{\max} \geq x) = 1 - P(X_{\max} \leq x - 1) = 1 - \left(\frac{x-1}{\theta}\right)^n = \tfrac{1}{2}\alpha.$$

So, for instance, the 90% confidence interval for θ based on the five observations with maximum 3433 is found by solving the two equations

$$P(X_{\max} \leq 3433) = \left(\frac{3433}{\theta}\right)^5 = 0.05,$$

which has solution $\theta_+ = 3433/0.05^{1/5} = 6250$; and

$$P(X_{\max} \geq 3433) = 1 - P(X_{\max} \leq 3432) = 1 - \left(\frac{3432}{\theta}\right)^5 = 0.05,$$

with solution $\theta_- = 3432/0.95^{1/5} = 3467$. The corresponding confidence interval for θ is given by $(\theta_-, \theta_+) = (3467, 6250)$. If you compare this with your result in Exercise 7.6 you can see that, again, a larger sample has resulted in a more useful confidence interval (because it is narrower).

7.3 Confidence intervals for the parameters of a normal distribution

In the previous sections, we did not consider the fundamental problem of constructing confidence intervals for the two parameters of a normal population, based on a random sample drawn from that population. This should now be quite straightforward: all that is required is to write down probability statements about statistics derived from the random sample (such as the sample mean and sample standard deviation) and invert those statements so that the unknown parameters μ and σ become the subjects of confidence statements.

7.3.1 Confidence intervals for the normal mean

Here is a typical example of the sort of statistical experiment which arises, where a normal model may reasonably be assumed, but where nothing is known either about the indexing mean μ or about the standard deviation σ.

Example 7.6 *A mechanical kitchen timer*

A kitchen timer is a small alarm clock that, by turning a dial, can be set to ring after any length of time between one minute and an hour. It is useful as a reminder to somebody working in a kitchen that some critical stage has been reached. The usefulness of such timers is not restricted to the kitchen, of course.

An enthusiastic cook was interested in the accuracy of his own kitchen timer, and on ten different occasions set it to ring after a five-minute delay (300 seconds). The ten different time intervals recorded on a stop-watch are shown in Table 7.2.

Data provided by B.J.R. Bailey, University of Southampton.

Table 7.2 Ten time delays (seconds)

293.7	296.2	296.4	294.0	297.3
293.7	294.3	291.3	295.1	296.1

Assuming that the stop-watch itself was an accurate measuring instrument, then the only variability from the 300 seconds' delay intended in the times recorded is due to difficulties in actually setting the time (that is, positioning the dial) and to mechanical malfunction in the operation of the timer. Assuming a normal model $N(\mu, \sigma^2)$ for the variation in the actual times recorded when a time of five minutes is set, then the data yield the parameter estimates

$$\overline{x} = 294.81, \quad s^2 = 3.1232.$$

That is, an estimate of the normal mean μ is the sample mean $\widehat{\mu} = \overline{x} = 294.81$ seconds (about 4m 55s—five seconds short of the five-minute interval set); an estimate of the normal standard deviation σ is the sample standard deviation $s = \sqrt{3.1232} = 1.77$ seconds. ■

Can we say more about the average time delay other than that it seems to be about 4m 55s? Perhaps an actual mean of five minutes (the delay intended) is plausible ... but, given these data, how confident could one be in putting this proposition forward?

If $\overline{X}$ is the mean of a random sample of size n from a normal distribution with mean μ and standard deviation σ, then the distribution of $\overline{X}$ is given by

$$\overline{X} \sim N\left(\mu, \frac{\sigma^2}{n}\right);$$

on standardizing, this is equivalent to

$$Z = \left(\frac{\overline{X} - \mu}{\sigma/\sqrt{n}}\right) \sim N(0, 1). \tag{7.3}$$

This result enables us to write down probability statements of the general form

$$P\left(-z \leq \frac{\overline{X} - \mu}{\sigma/\sqrt{n}} \leq z\right) = 1 - \alpha.$$

Based on what we know about normal probabilities, we might write, say,

$$P\left(-1.96 \leq \frac{\overline{X}-\mu}{\sigma/\sqrt{n}} \leq 1.96\right) = 0.95.$$

This follows since 95% of observations on a normal population are within 1.96 standard deviations of the mean. If the population mean μ was the only unknown quantity in this probability statement we could rewrite the double inequality with μ as the subject, as follows:

$$\begin{aligned} 0.95 &= P\left(-1.96 \leq \frac{\overline{X}-\mu}{\sigma/\sqrt{n}} \leq 1.96\right) \\ &= P\left(-1.96\frac{\sigma}{\sqrt{n}} \leq \overline{X}-\mu \leq 1.96\frac{\sigma}{\sqrt{n}}\right) \\ &= P\left(-\overline{X}-1.96\frac{\sigma}{\sqrt{n}} \leq -\mu \leq -\overline{X}+1.96\frac{\sigma}{\sqrt{n}}\right) \\ &= P\left(\overline{X}-1.96\frac{\sigma}{\sqrt{n}} \leq \mu \leq \overline{X}+1.96\frac{\sigma}{\sqrt{n}}\right). \end{aligned}$$

Thus we obtain a random interval, centred on the sample mean $\overline{X}$, which with probability 0.95 contains the unknown population mean μ. So, apparently, a 95% confidence interval for μ based on a sample $x_1, x_2, \ldots, x_n$ with sample mean $\overline{x}$, is given by

$$(\mu_-, \mu_+) = \left(\overline{x}-1.96\frac{\sigma}{\sqrt{n}}, \overline{x}+1.96\frac{\sigma}{\sqrt{n}}\right).$$

This result is useful if the normal standard deviation σ is known. Usually, however, in any context where the mean is unknown then the standard deviation will be unknown as well. So this approach cannot be used for making inferences about the unknown parameter μ, because of the presence of the unknown term σ. What is to be done? The obvious solution is to replace σ by its estimator S, the sample standard deviation. Remember that S is the square root of the sample variance

$$S^2 = \frac{1}{n-1}\sum_{i=1}^{n}(X_i - \overline{X})^2$$

and $E(S^2) = \sigma^2$, so the substitution seems quite sensible.

Let us now define a new random variable T (as at (7.3)) by

$$T = \frac{\overline{X}-\mu}{S/\sqrt{n}}. \tag{7.4}$$

In this expression the right-hand side has the property that the unknown parameter σ does not feature. We can then go on to make probability statements of the kind

$$P\left(-t \leq \frac{\overline{X}-\mu}{S/\sqrt{n}} \leq t\right) = 1-\alpha$$

(exactly as before), but first we have to find the appropriate values of t. That is, before we can use the random variable T to make inferences about μ, we need to know its properties, and in particular its probability distribution. Since we have obtained T from Z by replacing a constant, σ, by a random variable, S, you will not be surprised to be told that T is not normally distributed.

In fact there is not just one distribution for T, but a whole *family* of distributions, one for each value of the sample size n, $(n = 2, 3, \ldots)$.

The distribution is often known as **Student's *t*-distribution** since it was first derived, in 1908, by W.S. Gosset who wrote under the pseudonym of 'Student'. The family is indexed by means of a parameter called the degrees of freedom (just as in the case of a χ^2 variate—see *Chapter 6*). Thus we write $T \sim t(\nu)$ to denote that the random variable T has a t-distribution with ν degrees of freedom.

'Student' (1908) The probable error of a mean. *Biometrika*, **6**, 1–25. See also p. 60.

Like the standard normal variate Z each member of this family of distributions, for $\nu = 1, 2, 3, \ldots$, is symmetric about 0: the numerator of T, the difference $\overline{X} - \mu$, is as likely to be negative as it is to be positive. But in view of its dependence on two sample statistics (the sample standard deviation S as well as the sample mean $\overline{X}$) the random variable T is, in a sense, more variable than Z, and its probability density function has 'longer tails' than that of Z. All this is clearly seen in Figure 7.1 which shows the probability density function of Z together with those of $t(1)$, $t(3)$ and $t(7)$—that is, the t-distributions with 1, 3 and 7 degrees of freedom.

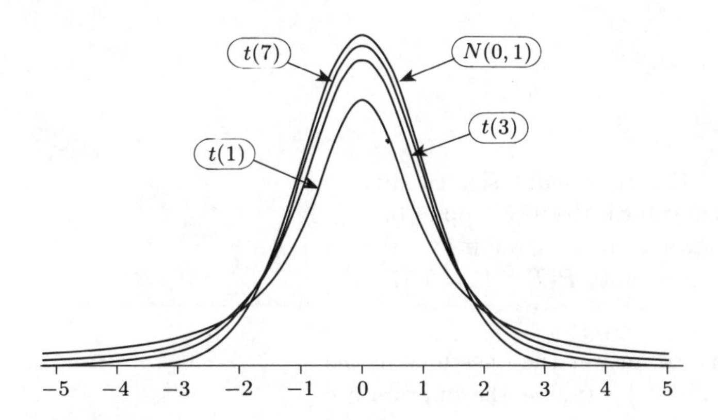

Figure 7.1 The densities of $t(1), t(3), t(7)$ and $Z \sim N(0, 1)$

The larger the value of ν, the closer is the distribution of $t(\nu)$ to a standard normal distribution. This is illustrated in the comparative sketches of the corresponding probability density functions in Figure 7.1. This makes sense: for a large sample size, S should be an accurate estimator for σ, and the distribution of the random variable T will be close to the distribution of the standard normal variate Z.

In order to make inferences about the unknown parameter μ using Student's t-distribution, it is necessary to obtain critical values from statistical tables.

A table of critical values of $t(\nu)$ for different values of ν is given as Table A5 in the Appendix. As in the case of the normal distribution, a sketch always helps. Here, in Figure 7.2, are some examples showing different critical values for Student's t-distribution, for different degrees of freedom ν. You should check that you can find these critical values using the table.

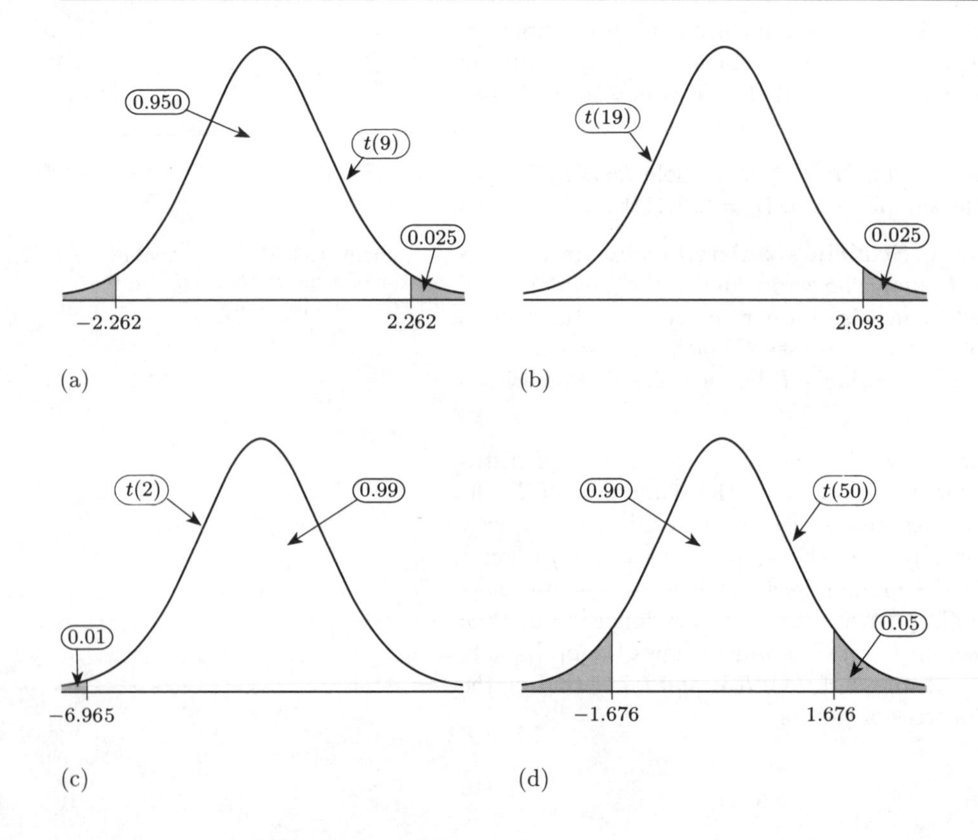

Figure 7.2 Critical values of $t(\nu)$

For example, in Figure 7.2(a), the size of the right-hand shaded area is $\frac{1}{2}(1 - 0.95) = 0.025$: the corresponding critical value is the 97.5% point ($q_{0.975}$) of $t(9)$. Looking along the ninth row of the table (corresponding to $\nu = 9$) and under the column corresponding to the probability $P(T \leq t) = 0.975$, the value of t is 2.262.

You should check that you can use the table to answer the questions in the exercise that follows, and confirm your findings by typing the appropriate command at your computer.

Exercise 7.11

(a) If $\nu = 29$, determine t such that $P(T \leq t) = 0.95$.

(b) If $\nu = 30$, determine t such that $P(T \geq t) = 0.05$.

(c) If $\nu = 5$, determine t such that $P(T \leq t) = 0.01$.

(d) If $\nu = 9$, determine t such that $P(|T| \leq t) = 0.95$.

You have just used tables of the t-distribution and your computer to determine the value of stated percentage points; that is, the value of the critical point t which (for given values of ν and α) will permit probability statements of the form

$$P(-t \leq T \leq t) = 1 - \alpha,$$

where $T \sim t(\nu)$. Now, suppose in a sampling context that $\overline{X}$ is the mean of a random sample from a normal distribution with mean μ, and that S is the sample standard deviation. In this context, the random variable T defined at (7.4) has Student's t-distribution with degrees of freedom $\nu = n - 1$.

In a random sample of size n from a normal distribution with mean μ, the random variable

$$T = \frac{\overline{X} - \mu}{S/\sqrt{n}}$$

(where $\overline{X}$ is the sample mean and S is the sample standard deviation) follows **Student's t-distribution** (or simply a **t-distribution**) with $n-1$ degrees of freedom. This is written

$$T = \frac{\overline{X} - \mu}{S/\sqrt{n}} \sim t(n-1).$$

This result is stated without proof.

An important consequence of this result is that it is possible to make probability statements of the form

$$P\left(-t \leq \frac{\overline{X} - \mu}{S/\sqrt{n}} \leq t\right) = 1 - \alpha,$$

where t is the $100(1 - \frac{1}{2}\alpha)\%$ point of $t(n-1)$. This probability statement may be rewritten

$$P\left(\overline{X} - \frac{tS}{\sqrt{n}} \leq \mu \leq \overline{X} + \frac{tS}{\sqrt{n}}\right) = 1 - \alpha,$$

which is a probability statement about a random interval, centred on the sample mean $\overline{X}$, which with probability $1 - \alpha$ contains the unknown population mean μ. The corresponding confidence interval for μ, based on a random sample $x_1, x_2, \ldots, x_n$, is

$$(\mu_-, \mu_+) = \left(\overline{x} - \frac{ts}{\sqrt{n}}, \overline{x} + \frac{ts}{\sqrt{n}}\right).$$

Here, $\overline{x}$ is the sample mean, s is the sample standard deviation, n is the sample size and t is a critical value obtained from tables of Student's t-distribution (or from your computer).

Example 7.6 continued

If it is assumed that the time until the kitchen timer alarm bell sounds is normally distributed, but with unknown mean and variance, then we are now in a position to construct a confidence interval for the unknown mean waiting time μ. We know that the sample statistics are $\overline{x} = 294.81$ and $s = \sqrt{3.1232} = 1.77$. For a 90% confidence interval, say, with $\nu = n - 1 = 9$

degrees of freedom, the critical t-value is 1.833 (using $\frac{1}{2}\alpha = 0.05$). Hence the confidence interval we obtain for μ, based on these data, is

$$\begin{aligned}(\mu_-, \mu_+) &= \left(\overline{x} - \frac{ts}{\sqrt{n}}, \overline{x} + \frac{ts}{\sqrt{n}}\right)\\ &= \left(294.81 - \frac{1.833 \times 1.77}{\sqrt{10}}, 294.81 + \frac{1.833 \times 1.77}{\sqrt{10}}\right)\\ &= (294.81 - 1.03, 294.81 + 1.03)\\ &= (293.8, 295.8).\end{aligned}$$

Notice that the number 300 (seconds, a five-minute delay) is not contained within this confidence interval! (Also, the confidence interval is usefully *narrow.*) So there is some statistical evidence here that the timer is consistently going off early. We shall return to this very important point in *Chapter 8*. ■

Exercise 7.12

In 1928 the London and North-Eastern Railway ran the locomotive *Lemberg* with an experimental boiler pressure of 220 lb, in five trial runs. Several random variables were measured, one being the coal consumption, in pounds per draw-bar horse-power hour; the resulting observations were

3.27 3.17 3.24 2.92 2.99.

Regarding these data as a random sample from a normal distribution, construct a 95% confidence interval for the population mean μ (that is, the mean coal consumption at a boiler pressure of 220 lb).

Exercise 7.13

The following data (see Table 7.3) are from 'Student's' 1908 paper. Several illustrative data sets were used in the paper; this one is taken from a table by A.R. Cushny and A.R. Peebles in the *Journal of Physiology* (1904) showing the effects of the optical isomers of hyoscyamine hydrobromide in producing sleep.

The sleep of ten patients was measured without hypnotic and after treatment (a) with D-hyoscyamine hydrobromide and (b) with L-hyoscyamine hydrobromide. The average number of hours' sleep gained by the use of the drug was tabulated. Here, D- denotes *dextro* and L- denotes *laevo*, a classification system for stereoisomers.

The table is *not* exactly as printed in *Biometrika*. It is a pity that in this fundamental paper there should have been a typographical error: this has been corrected in Table 7.3 (and it was not difficult to isolate). A minus sign indicates a net sleep loss.

For the purposes of this exercise, only consider the ten numbers in the column headed L−D. This summarizes any differences between the two treatments. (Looking at the values it seems as though treatment with L-hyoscyamine hydrobromide was consistently the more effective.)

Table 7.3 Sleep gain after drug treatment

Patient	D	L	L − D
1	+0.7	+1.9	+1.2
2	−1.6	+0.8	+2.4
3	−0.2	+1.1	+1.3
4	−1.2	+0.1	+1.3
5	−0.1	−0.1	0
6	+3.4	+4.4	+1.0
7	+3.7	+5.5	+1.8
8	+0.8	+1.6	+0.8
9	0	+4.6	+4.6
10	+2.0	+3.4	+1.4

(a) Calculate the sample mean $\overline{x}$ and the sample standard deviation s for these data, and hence construct a 95% confidence interval for the unknown mean difference μ, assuming the differences to be normally distributed.

(b) Say whether the number 0 is contained in the confidence interval you found in part (a). Can you deduce anything from your study?

In general, for larger samples, the calculation of the statistics $\overline{x}$ and s may be inconvenient. Not only can a computer perform this arithmetic task, but it can then go on to 'look up critical percentage points' against the appropriate t-distribution and hence calculate confidence intervals. You do not have to do anything at all, other than key in the appropriate command. Your computer probably carries out this exercise quite easily.

Exercise 7.14

Confirm your answers to Exercises 7.12 and 7.13 by obtaining the confidence intervals sought directly from your computer.

7.3.2 Confidence intervals for the normal variance

The normal distribution is indexed by two parameters: the second is its variance, denoted σ^2. In contexts where this quantity is unknown, it may often be useful to draw a random sample of observations in order to obtain a confidence interval. Let us start with an example.

Example 7.7 *Breaking strengths*

In the production of synthetic fibres, it is important that the fibres produced are consistent in quality. One aspect of this is that the tensile strength of the fibres should not vary too much. A sample of eight pieces of fibre produced by a new process is taken, and the tensile strength (in kg) of each fibre was tested. The sample mean was $\overline{x} = 150.72\,\text{kg}$ and the sample variance was $s^2 = 37.75\,\text{kg}^2$. A confidence interval for the variance is required. ■

Assuming a normal model, we might wish to find, say, a 90% confidence interval for the population variance σ^2. The theoretical distribution underlying statistical inferences about a normal variance is the χ^2-distribution, which as you saw in *Chapter 6* is actually a whole family of distributions whose indexing parameter (the degrees of freedom) is, in this particular context, dependent upon the sample size. Here, again, is the sampling distribution of the variance of normal random samples of size n, when the parent population is $N(\mu, \sigma^2)$:

$$\frac{(n-1)S^2}{\sigma^2} \sim \chi^2(n-1). \qquad (7.5)$$

This result was first used in Exercise 6.26.

In Figure 7.3 you can see how the chi-squared density changes shape with increasing degrees of freedom.

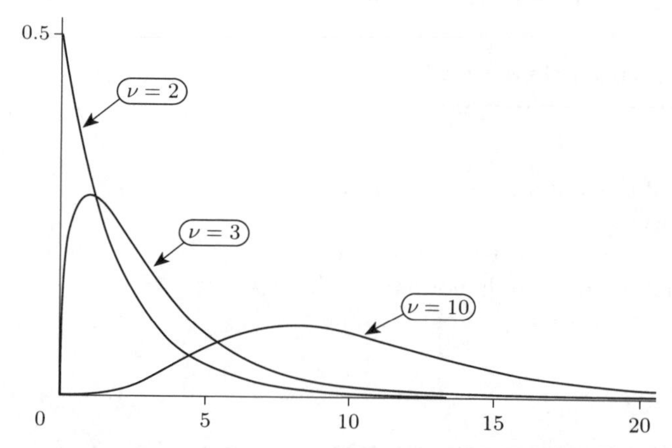

Figure 7.3 The density of $\chi^2(\nu)$, $\nu = 2, 3, 10$

For small values of the parameter ν the χ^2 density is very highly skewed; for increasing values of the parameter the distribution becomes less skewed.

Corresponding critical points for the chi-squared distribution can be obtained by reference to statistical tables (or by keying in the appropriate command on your computer). For instance, the 5% point of $\chi^2(4)$ is 0.711; the 97.5% point is 11.143. This information is shown graphically in Figure 7.4. You should check that you can obtain these values from statistical tables, and from your computer.

See also Exercises 6.23 and 6.24.

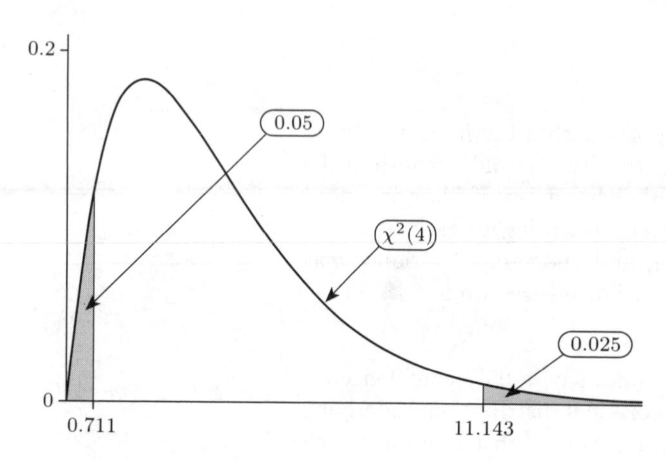

Figure 7.4 Critical points of $\chi^2(4)$

A confidence interval for an unknown variance σ^2 based on a random sample $x_1, x_2, \ldots, x_n$ from a normal distribution may be found as follows. As usual, we begin with a probability statement based on the distribution of the estimator (in this case, S^2). Using (7.5) we can write

$$P\left(c_L \leq \frac{(n-1)S^2}{\sigma^2} \leq c_U\right) = 1 - \alpha, \tag{7.6}$$

where $c_L = q_{\frac{1}{2}\alpha}$ is the 'left-hand' critical point of the $\chi^2(n-1)$ distribution; and $c_U = q_{1-\frac{1}{2}\alpha}$ is the 'right-hand' critical point.

The double inequality on the left-hand side of (7.6) can be written with the variance σ^2 as the subject:

$$P\left(\frac{(n-1)S^2}{c_U} \leq \sigma^2 \leq \frac{(n-1)S^2}{c_L}\right) = 1 - \alpha. \tag{7.7}$$

This is a probability statement about a random interval which, with probability $1 - \alpha$, will contain the unknown variance σ^2. The corresponding confidence interval for σ^2, based on a random sample with standard deviation s, is given by

$$(\sigma_-^2, \sigma_+^2) = \left(\frac{(n-1)s^2}{c_U}, \frac{(n-1)s^2}{c_L}\right). \tag{7.8}$$

Example 7.8 Finding a confidence interval for a normal variance

For inferences from normal samples of size 10, say, we would refer to tables of the $\chi^2(9)$ distribution: the 5% point, for example, is 3.325, the 95% point is 16.919. Different samples of size 10 from $N(\mu, \sigma^2)$ will give rise to different observations on the sample variance S^2—it is a random variable—and we can write down the probability statement

$$P\left(3.325 \leq \frac{9S^2}{\sigma^2} \leq 16.919\right) = 0.90.$$

The corresponding 90% confidence interval for σ^2 is

$$\begin{aligned}(\sigma_-^2, \sigma_+^2) &= \left(\frac{(n-1)s^2}{c_U}, \frac{(n-1)s^2}{c_L}\right) \\ &= \left(\frac{9s^2}{16.919}, \frac{9s^2}{3.325}\right) = (0.53s^2, 2.71s^2).\end{aligned}$$

Notice that in the whole of this example, knowledge of the sample mean is unnecessary. ■

Example 7.6 continued

For the kitchen timer data, we have $(n-1)s^2 = 9 \times 3.1232 = 28.109$; it follows that a 90% confidence interval for σ^2, using $c_L = 3.325$, $c_U = 16.919$, is given by

$$(\sigma_-^2, \sigma_+^2) = \left(\frac{28.109}{16.919}, \frac{28.109}{3.325}\right) = (1.66, 8.45).$$

(This may be compared with the point estimate $s^2 = 3.12$.) ■

Exercise 7.15

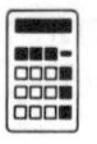

For the breaking strength data of Example 7.7, find (a) a 90% confidence interval and (b) a 95% confidence interval for σ^2.

7.3.3 Confidence intervals for the normal standard deviation

Returning once more to the probability statement (7.7),

$$P\left(\frac{(n-1)S^2}{c_U} \leq \sigma^2 \leq \frac{(n-1)S^2}{c_L}\right) = 1 - \alpha,$$

and then taking square roots, we obtain

$$P\left(S\sqrt{\frac{n-1}{c_U}} \leq \sigma \leq S\sqrt{\frac{n-1}{c_L}}\right) = 1 - \alpha.$$

It follows that a $100(1-\alpha)\%$ confidence interval for σ, based on a random sample with standard deviation s, is given by

$$(\sigma_-, \sigma_+) = \left(s\sqrt{\frac{n-1}{c_U}}, s\sqrt{\frac{n-1}{c_L}}\right).$$

In other words, the confidence limits for σ are just the square roots of the respective confidence limits for σ^2.

Exercise 7.16

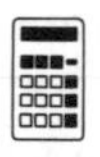

For the breaking strength data of Example 7.7, (a) estimate σ; and (b) find a 90% confidence interval for σ. (Use your answers from Exercise 7.15.)

7.4 Larger samples

We have seen that confidence intervals for the parameters of a normal distribution have been fairly easy to develop by following these rules:

(i) write down a probability statement about a sample statistic whose distribution involves the parameter for which the confidence interval is required;

(ii) turn this probability statement into a statement about a random interval which might or might not contain the unknown parameter;

(iii) collect a random sample;

(iv) calculate numerical values for the extremes of the random interval;

(v) call your single observation on that random interval a confidence interval.

In this section we return to non-normal models such as the exponential and Bernoulli distributions, and develop techniques for finding confidence intervals for unknown parameters with the assumption that the random samples we have drawn, in order to perform such inferences, are large enough for the central limit theorem to apply. (Where this assumption cannot be made, we need to use the exact methods as discussed in Section 7.2.)

Essentially, the idea is this: if $X_1, X_2, \ldots, X_n$ are observations on a random variable X, then the central limit theorem says that

$$\sum_{i=1}^{n} X_i = X_1 + X_2 + X_3 + \cdots + X_n \approx N\left(n\mu_X, n\sigma_X^2\right),$$

where the moments μ_X and σ_X^2 are respectively the mean and variance of the random variable X. That is, the sample total is approximately normally distributed. The moments μ_X and σ_X^2 will usually feature the unknown parameter (or parameters, if there is more than one) that we wish to estimate. So this result can be used to write down approximate probability statements about the sample total which can be turned into confidence statements about the unknown parameters in the usual way.

7.4.1 The exponential model

One of the simplest applications of this idea is to data for which an exponential distribution is assumed to provide an adequate model.

Example 7.9 Coal-mining disasters

Here is the full data set from the investigation of Example 7.2. Table 7.4 gives the times (in days) between consecutive disasters in coal mines in Britain between 15 March 1851 and 22 March 1962. (The data are to be read across rows.) This time interval covers 40 550 days. There were 191 explosions altogether, including one on each of the first and last days of the investigation. So the data involve 190 numbers whose sum is 40 549. The 0 occurs because there were two disasters on 6 December 1875.

Table 7.4 Times (in days) between disasters

157	123	2	124	12	4	10	216	80	12
33	66	232	826	40	12	29	190	97	65
186	23	92	197	431	16	154	95	25	19
78	202	36	110	276	16	88	225	53	17
538	187	34	101	41	139	42	1	250	80
3	324	56	31	96	70	41	93	24	91
143	16	27	144	45	6	208	29	112	43
193	134	420	95	125	34	127	218	2	0
378	36	15	31	215	11	137	4	15	72
96	124	50	120	203	176	55	93	59	315
59	61	1	13	189	345	20	81	286	114
108	188	233	28	22	61	78	99	326	275
54	217	113	32	388	151	361	312	354	307
275	78	17	1205	644	467	871	48	123	456
498	49	131	182	255	194	224	566	462	228
806	517	1643	54	326	1312	348	745	217	120
275	20	66	292	4	368	307	336	19	329
330	312	536	145	75	364	37	19	156	47
129	1630	29	217	7	18	1358	2366	952	632

Suppose that these data represent a random sample from an exponential distribution and that we are interested in the mean time interval between disasters. So in this example we have data, a specified model, and we have isolated a parameter of interest.

Suppose we denote the exponential mean by μ. Then denoting by X_i the ith element of the sample, we have $\mu_X = \mu$, $\sigma_X^2 = \mu^2$, and it follows that

The variance of an exponential distribution is equal to the square of its mean: see (4.25).

$$E\left(\sum_{i=1}^{n} X_i\right) = n\mu \quad \text{and} \quad V\left(\sum_{i=1}^{n} X_i\right) = n\mu^2.$$

Applying the central limit theorem,

$$\frac{\sum_{i=1}^{n} X_i - n\mu}{\mu\sqrt{n}} = \frac{\overline{X} - \mu}{\mu/\sqrt{n}} \approx N(0,1). \tag{7.9}$$

Notice the '$\approx$' symbol: the normal distribution is only an approximate model for the sample total when the sample size is large, and we must remember this when making confidence statements. There is no point in giving confidence limits to six significant figures.

Now let us proceed as before to try to find a 95% confidence interval for the exponential mean in the case of the coal-mining data. Notice that the only sample statistic in the expression at (7.9) is the sample mean $\overline{X}$; it is not necessary to calculate the sample standard deviation. Using (7.9), we can write the probability statement

$$P\left(-1.96 \leq \frac{\overline{X}-\mu}{\mu/\sqrt{n}} \leq 1.96\right) \simeq 0.95.$$

This is a statement about the random variable $\overline{X}$. We render it into a statement about a random interval by rearranging the double inequality so that μ is the subject. By rewriting the inequality in this way, we obtain

$$P\left(\frac{\overline{X}}{1+1.96/\sqrt{n}} \leq \mu \leq \frac{\overline{X}}{1-1.96/\sqrt{n}}\right) \simeq 0.95. \tag{7.10}$$

This says that the random interval defined by transforming the sample mean in the manner described will, with high probability, contain the unknown population mean μ: it provides a 95% confidence interval for μ.

For the coal-mining disaster data, the mean of the $n = 190$ waiting times is $\overline{x} = 213.416$ (days). Based on these data, the corresponding observation on the random interval defined at (7.10) is

$$\begin{aligned}(\mu_-, \mu_+) &\simeq \left(\frac{\overline{x}}{1+1.96/\sqrt{n}}, \frac{\overline{x}}{1-1.96/\sqrt{n}}\right)\\ &= \left(\frac{213.416}{1+1.96/\sqrt{190}}, \frac{213.416}{1-1.96/\sqrt{190}}\right)\\ &= \left(\frac{213.416}{1+0.14219}, \frac{213.416}{1-0.141219}\right)\\ &= (186.8, 248.8). \quad \blacksquare\end{aligned}$$

Since for exponential random samples the maximum likelihood estimate $\widehat{\mu}$ of μ is the sample mean $\overline{x}$, the preceding discussion can be summarized as follows.

Approximate large-sample confidence interval for an exponential mean

An approximate $100(1-\alpha)\%$ confidence interval for the mean μ of an exponential random variable, based on a random sample $x_1, x_2, \ldots, x_n$, is given by

$$(\mu_-, \mu_+) \simeq \left(\frac{\widehat{\mu}}{1+z/\sqrt{n}}, \frac{\widehat{\mu}}{1-z/\sqrt{n}}\right),$$

where z is the $100(1-\frac{1}{2}\alpha)\%$ point of the standard normal distribution and $\widehat{\mu} = \overline{x}$ is the maximum likelihood estimate of μ.

Notice that the fact that these are *approximate* limits is demonstrated by the presence of the term, $1 - z/\sqrt{n}$, which if the sample is too small or if the confidence level sought is too high, will be negative. This does not make sense.

Exercise 7.17

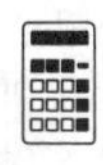

Use the earthquake data of *Chapter 4*, Table 4.7 to establish a 95% confidence interval for the mean time between serious earthquakes world-wide. State any modelling assumptions you make. (Use your answers to Exercise 7.9.)

Exercise 7.18

(a) Establish a 90% confidence interval for the mean traffic rate (vehicles per minute) for the Kwinana Freeway, based on the traffic data of Table 2.11, stating any assumptions you make during the calculation.

(b) Comment on any differences between the confidence interval you have found in part (a) and the one you calculated in Exercise 7.9.

A confidence interval which is wide is less useful than one which is narrow. Large samples give narrower and hence more useful confidence limits than small samples. It is possible to use the general expression for the confidence limits to determine how large the sample should be before the experiment is performed. This is demonstrated in the following example.

***Example 7.10** Determination of the sample size*

Suppose that a sample of observations from an exponential distribution is required to produce a 90% confidence interval for the mean, with both upper and lower confidence limits within 5% of the estimated mean. This implies that both the following inequalities must hold:

$$\frac{\widehat{\mu}}{1 + z/\sqrt{n}} \geq 0.95\widehat{\mu} \qquad \text{and} \qquad \frac{\widehat{\mu}}{1 - z/\sqrt{n}} \leq 1.05\widehat{\mu},$$

where $z = 1.645$. The first inequality gives $\sqrt{n} \geq 19z$, or $n \geq 976.9$; the second inequality gives $\sqrt{n} \geq 21z$, or $n \geq 1193.4$. So a sample size of at least 1194 (say 1200, perhaps) will ensure a sufficiently narrow confidence interval at the specified 90% level. ■

Exercise 7.19

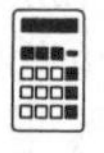

A sample of observations from an exponential distribution is required. It must be sufficiently large to ensure that a 95% confidence interval for the mean has upper and lower confidence limits both within 3% of the estimated mean. What minimum sample size will achieve this?

7.4.2 The Poisson model

Assume that we have a random sample of size n from a Poisson distribution with unknown mean μ. Then, by the central limit theorem, the sample total

$\sum_{i=1}^{n} X_i$ is approximately normally distributed with mean $n\mu$ and variance $n\mu$. For instance, we can make the probability statement

$$P\left(-1.645 \leq \frac{\sum_{i=1}^{n} X_i - n\mu}{\sqrt{n\mu}} \leq 1.645\right) \simeq 0.90.$$

For a Poisson distribution, the variance is equal to the mean. See *Chapter 4*, Subsection 4.1.3.

In order to calculate a confidence interval for μ, we need to solve the double inequality inside $P(\cdot)$ and make μ the subject. To generalize what follows to confidence levels other than 90%, we shall first replace the number 1.645 by z. Then, squaring both sides, we obtain (after a little algebra) the following quadratic inequality in μ:

$$n\mu^2 - (2n\overline{X} + z^2)\mu + n\overline{X}^2 \leq 0.$$

The inequality is satisfied for μ lying between the two solutions of the quadratic, which are

$$\frac{2n\overline{X} + z^2 \pm \sqrt{4n\overline{X}z^2 + z^4}}{2n}.$$

The notation '$\pm$' is a convenient abbreviation for the lower and upper confidence limits. For example, $(a \pm b)$ means $(a - b, a + b)$.

This is a very complicated expression that could nevertheless be used to define upper and lower confidence limits for μ. However, it is based on an approximation (that induced by applying the central limit theorem and using approximate normal theory) and, in the interests of simplifying things a little, a second approximation is now introduced which should not seriously perturb matters further. In the numerator of the expression above, terms involving n will dominate, since n is assumed to be large. Including only these dominant terms, the two solutions are approximately

$$\frac{2n\overline{X} \pm \sqrt{4n\overline{X}z^2}}{2n} = \overline{X} \pm z\sqrt{\frac{\overline{X}}{n}}.$$

Of course, in the Poisson case, the sample mean $\overline{X}$ is simply the maximum likelihood estimator $\widehat{\mu}$. So, writing these solutions in their most general form, we have the following result.

Approximate large-sample confidence interval for a Poisson mean

If a random sample $x_1, x_2, \ldots, x_n$ is obtained from a population where the Poisson distribution is assumed to provide an adequate model for variation, then an approximate $100(1-\alpha)\%$ confidence interval for the population mean μ is given by

$$\left(\widehat{\mu} - z\sqrt{\frac{\widehat{\mu}}{n}}, \widehat{\mu} + z\sqrt{\frac{\widehat{\mu}}{n}}\right),$$

where $\widehat{\mu}$ is equal to $\overline{x}$, the sample mean, and z is the $100(1 - \frac{1}{2}\alpha)\%$ point of the standard normal distribution.

Exercise 7.20

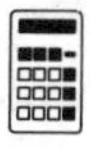

The full childhood accident data initially described in Example 7.1 are given in Table 7.5.

Table 7.5 621 childhood accident counts, aged 4 to 7 and 8 to 11

		Injuries aged 4 to 7								
		0	1	2	3	4	5	6	7	
	0	101	76	35	15	7	3	3	0	240
	1	67	61	32	14	12	4	1	1	192
Injuries	2	24	36	22	15	6	1	2	1	107
aged 8	3	10	19	10	5	2	4	0	2	52
to 11	4	1	7	3	4	2	0	0	0	17
	5	2	1	4	2	0	0	0	0	9
	6	1	1	1	1	0	0	0	0	4
		206	201	107	56	29	12	6	4	621

The columns give injury counts between the ages of 4 and 7; the rows give the counts between the ages of 8 and 11. For instance, by reference to the column labelled '2' and the row labelled '1', you can see that 32 of the 621 children sampled experienced two accidents between the ages of 4 and 7, followed by just one accident between the ages of 8 and 11.

Summing across diagonals gives the frequencies for the total number of accidents sustained over the complete eight-year period. For example, the total number of children who had two accidents is $24 + 61 + 35 = 120$. The data are summarized in Table 7.6.

Table 7.6 621 childhood accident counts, aged 4 to 11

0	1	2	3	4	5	6	7	8	9	10
101	143	120	93	63	49	23	13	12	2	2

Use the data in Table 7.6 to calculate a 95% confidence interval for the average number of accidents sustained by children between the ages of 4 and 11. State any modelling assumptions you make.

A reasonable rule of thumb for applying the central limit theorem to Poisson data in order to calculate approximate confidence intervals for the mean μ, is that the sample total should be at least 30. (So, in this case, notice that you need not worry about the sample size.)

7.4.3 The geometric model

Approximate confidence intervals for the parameter p of a geometric distribution, and for the mean $\mu = 1/p$, are given overleaf.

You are spared the details of the algebra here. The result is important. Notice that the confidence limits for the mean μ in this case are *not* the reciprocals of those for the parameter p. This is because the limits were obtained in each case through an approximation, and what gets 'cut' differs in the two cases.

Approximate large-sample confidence intervals for the parameter of a geometric distribution

An approximate $100(1-\alpha)\%$ confidence interval for the parameter p of a geometric distribution, based on a random sample $x_1, x_2, \ldots, x_n$ is given by

$$(p_-, p_+) = \left(\widehat{p} - z\widehat{p}\sqrt{\frac{1-\widehat{p}}{n}}, \widehat{p} + z\widehat{p}\sqrt{\frac{1-\widehat{p}}{n}}\right),$$

where $\widehat{p} = 1/\overline{x}$ is the reciprocal of the sample mean and z is the $100(1-\frac{1}{2}\alpha)\%$ point of the standard normal distribution. Equivalently, an approximate $100(1-\alpha)\%$ confidence interval for the population mean μ is given by

$$(\mu_-, \mu_+) = \left(\widehat{\mu} - z\sqrt{\frac{\widehat{\mu}(\widehat{\mu}-1)}{n}}, \widehat{\mu} + z\sqrt{\frac{\widehat{\mu}(\widehat{\mu}-1)}{n}}\right),$$

where $\widehat{\mu}$ is the sample mean $\overline{x}$.

Exercise 7.21

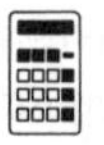

Chapter 6, Table 6.5 gives the lengths of runs of Douglas firs infected with *Armillaria* root rot. Use the data to find a 99% confidence interval for the mean length of a run, and state any modelling assumptions you make.

7.4.4 The Bernoulli model

Approximate confidence intervals for the Bernoulli parameter, the probability p, may be obtained as follows.

Approximate large-sample confidence interval for the Bernoulli probability p

An approximate $100(1-\alpha)\%$ confidence interval for the Bernoulli probability p, based on observing x successes in a sequence of n independent Bernoulli trials, is given by

$$(p_-, p_+) = \left(\widehat{p} - z\sqrt{\frac{\widehat{p}(1-\widehat{p})}{n}}, \widehat{p} + z\sqrt{\frac{\widehat{p}(1-\widehat{p})}{n}}\right),$$

where $\widehat{p} = x/n$ is the maximum likelihood estimate of p, and where z is the $100(1-\frac{1}{2}\alpha)\%$ point of the standard normal distribution.

Example 7.11 *Smokers*

Of a random sample of 7383 adults aged 18 and over chosen from the electoral register in England, Wales and Scotland in part of a study into low blood pressure, a proportion of 32.8% were found to be smokers.

Wessely, S., Nickson, J. and Cox, B. (1990) Symptoms of low blood pressure: a population study. *British Medical Journal*, **301**, 362–365.

The reference does not give the actual number of smokers in the sample of 7383. From other information supplied, it is possible only to deduce that the number is between 2419 and 2421. But the number is not necessary to calculate the confidence interval. Based on these data, a 95% confidence interval for the unknown proportion p of smokers in the population is

$$\begin{aligned}(p_-, p_+) &= \left(\widehat{p} - z\sqrt{\frac{\widehat{p}(1-\widehat{p})}{n}}, \widehat{p} + z\sqrt{\frac{\widehat{p}(1-\widehat{p})}{n}}\right) \\ &= \left(0.328 - 1.96\sqrt{\frac{0.328 \times 0.672}{7383}}, 0.328 + 1.96\sqrt{\frac{0.328 \times 0.672}{7383}}\right) \\ &= (0.328 - 0.011, 0.328 + 0.011) \\ &= (0.317, 0.339). \quad \blacksquare\end{aligned}$$

Notice that in the Bernoulli case the width of the confidence interval is $2z\sqrt{\widehat{p}(1-\widehat{p})/n}$. Now, it is fairly easy to see that for $0 \leq \widehat{p} \leq 1$ the maximum value that the product $\widehat{p}(1-\widehat{p})$ can take is $\frac{1}{4}$, so the maximum possible width of the confidence interval is $z/\sqrt{n}$. If, a 98% confidence interval for p (so $z = 2.326$) not wider than 0.05 is required, then in the worst case the sample size will have to be at least 2165.

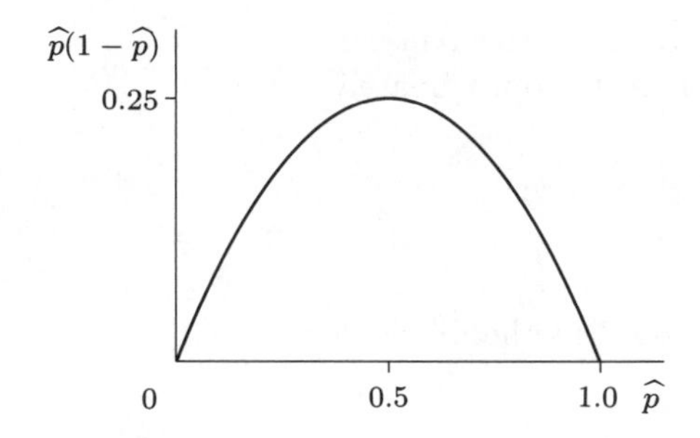

Figure 7.5 gives the graph of the function $\widehat{p}(1-\widehat{p})$ for $0 \leq \widehat{p} \leq 1$. The graph is symmetric about $\widehat{p} = \frac{1}{2}$, where it attains its maximum value of $\frac{1}{2} \times \frac{1}{2} = \frac{1}{4}$.

Figure 7.5 The graph of $\widehat{p}(1-\widehat{p})$, $0 \leq \widehat{p} \leq 1$

As long as the number of 1s (Yeses) and the number of 0s (Noes) in your sample are both more than about five or six, then approximate confidence intervals calculated after applying the central limit theorem and using normal critical values should be reasonably accurate.

Exercise 7.22

An experiment was undertaken to examine the association between eye and hair colour in samples of Scottish school children.

(a) Of 5387 children sampled in Caithness, 286 had hair classified as 'red'. Use these data to establish a 90% confidence interval for the proportion of Scottish children who have red hair. State any modelling and sampling assumptions you make.

(b) Of 5789 fair-haired children sampled in Aberdeen, 1368 had blue eyes. Use this information to determine a 95% confidence interval for the proportion of fair-haired Scottish children who have blue eyes. State any assumptions you make.

These are the famous 'Scottish school children' data of the 1940s. See Fisher, R.A. (1940) The precision of discriminant functions. *Annals of Eugenics*, **10**, 422–429. See also Maung, K. (1941) Measurement of association in a contingency table with special reference to the pigmentation of hair and eye colour of Scottish school children. *Annals of Eugenics*, **11**, 189–205.

7.5 Inference without a model

In all the cases we have met so far some sort of model for the variation observed has been postulated. This raises the question of how to begin a statistical analysis when no model for the data seems apparent either from physical considerations or from some underlying structure in the 'shape' of the data.

What should your approach be if you are simply presented with a large mass of data, with the request (i) to estimate the mean μ of the population from which the data were drawn, and (ii) to construct confidence intervals for the population mean?

In this section we shall only be concerned with confidence intervals for the population *mean*, and not with other population characteristics.

Probably the first thing you need to assume is that the data do indeed constitute a random sample from the population. (This does not necessarily hold, however, and you should not take this assumption lightly.)

Next, it is probably reasonable to suppose that the best estimate for the unknown population mean μ is the sample mean $\overline{x}$. (This will not necessarily be the maximum likelihood estimate for μ; but the sample mean $\overline{x}$ will be unbiased for μ.)

Then, if the sample size n is reasonably large, the central limit theorem can be applied, and you can assume that the sample mean $\overline{X}$ is approximately normally distributed. That is,

$$\overline{X} \approx N\left(\mu, \frac{\sigma^2}{n}\right),$$

where σ^2 is the unknown variance in the population. From this there follows the approximate probability statement

$$P\left(\mu - z\frac{\sigma}{\sqrt{n}} \leq \overline{X} \leq \mu + z\frac{\sigma}{\sqrt{n}}\right) \simeq 1 - \alpha,$$

where z is the $100(1 - \frac{1}{2}\alpha)\%$ point of the standard normal distribution. This can be rewritten in terms of a probability statement about a random interval:

$$P\left(\overline{X} - z\frac{\sigma}{\sqrt{n}} \leq \mu \leq \overline{X} + z\frac{\sigma}{\sqrt{n}}\right) = 1 - \alpha. \qquad (7.11)$$

The standard deviation σ is of course unknown. If it was known that the underlying variation in the population was normal, then you would substitute the sample standard deviation s for σ, and the appropriate critical t-value for z. Here, the obvious thing to do first is to replace the unknown standard deviation σ by its estimate s. (There are arguments that σ might just as well be replaced by the maximum likelihood estimate $\widehat{\sigma}$ with the n-divisor as by s; but these are not necessarily very good arguments.)

There is, however, no good reason for replacing z in (7.11) by a critical t-value, even if the sample size is not great: the development of Student's t-distribution was as an accurate mathematical 'formula' for making probability statements about the distribution of the sample mean $\overline{X}$ when the parent population is known to be normal, which is not the case here. So, in the end, what happens is quite simple: we just substitute s for σ and leave z as it is. This leads to the following rule.

For a random sample $x_1, x_2, \ldots, x_n$ drawn from a population where no distributional assumptions can be made, an approximate $100(1-\alpha)\%$ confidence interval for the population mean μ is given by

$$(\mu_-, \mu_+) = \left(\overline{x} - z\frac{s}{\sqrt{n}}, \overline{x} + z\frac{s}{\sqrt{n}}\right),$$

where n is the sample size, $\overline{x}$ is the sample mean, s is the sample standard deviation and z is the $100(1-\frac{1}{2}\alpha)\%$ point of the standard normal distribution.

Exercise 7.23

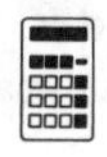

Chapter 2, Table 2.9 gives data on library book usage: the number of times that each of 122 books was borrowed during the course of a year was counted.

Without making any distributional assumptions about a statistical model for the variation in the number of annual withdrawals, use the data in Table 2.9 to calculate a 90% confidence interval for the mean number of loans in a year.

Exercise 7.24

An experiment was conducted into the effects of environmental pollutants upon animals. For 65 Anacapa pelican eggs, the concentration, in parts per million of PCB (polychlorinated biphenyl, an industrial pollutant), was measured, along with the thickness of the shell in millimetres.

Risebrough, R.W. (1972) Effects of environmental pollutants upon animals other than man. In *Proceedings of the 6th Berkeley Symposium on Mathematics and Statistics*, VI. University of California Press, 443–463.

The data are summarized in Table 7.7.

Table 7.7 PCB concentration (ppm) and eggshell thickness (mm) for 65 pelican eggs

ppm	mm	ppm	mm	ppm	mm	ppm	mm
452	0.14	184	0.19	115	0.20	315	0.20
139	0.21	177	0.22	214	0.22	356	0.22
166	0.23	246	0.23	177	0.23	289	0.23
175	0.24	296	0.25	205	0.25	324	0.26
260	0.26	188	0.26	208	0.26	109	0.27
204	0.28	89	0.28	320	0.28	265	0.29
138	0.29	198	0.29	191	0.29	193	0.29
316	0.29	122	0.30	305	0.30	203	0.30
396	0.30	250	0.30	230	0.30	214	0.30
46	0.31	256	0.31	204	0.32	150	0.34
218	0.34	261	0.34	143	0.35	229	0.35
173	0.36	132	0.36	175	0.36	236	0.37
220	0.37	212	0.37	119	0.39	144	0.39
147	0.39	171	0.40	216	0.41	232	0.41
216	0.42	164	0.42	185	0.42	87	0.44
216	0.46	199	0.46	236	0.47	237	0.49
206	0.49						

We shall discover methods for exploring the association between shell thickness (or thinness) and degree of contamination later in the course. For the moment, use the data on shell thickness to determine a 95% confidence interval for the mean thickness of Anacapa pelican eggs.

Summary

1. If a single observation x is taken on a random variable X with a stated probability distribution indexed by a parameter θ whose value is unknown, then a $100(1-\alpha)\%$ confidence interval (θ_-, θ_+) may be found by solving separately for θ the two equations

$$P(X \leq x) = \tfrac{1}{2}\alpha \quad \text{and} \quad P(X \geq x) = \tfrac{1}{2}\alpha.$$

The two solutions θ_- and θ_+ are called respectively the lower and upper confidence limits for the parameter θ.

2. The interpretation of a confidence interval is as follows: in independent repeated experiments each resulting in the statement of a $100(1-\alpha)\%$ confidence interval for an unknown parameter θ, the expected proportion of intervals that actually contain the number θ is $1-\alpha$.

3. For small samples from populations where the standard models are assumed, exact confidence limits may be calculated by reference to a statistical computer package; occasionally exact arithmetic computations are possible.

4. For a random sample $X_1, X_2, \ldots, X_n$ of size n from the normal distribution with unknown mean μ and unknown standard deviation σ, the sampling distributions of the sample mean $\overline{X}$ and the sample variance S^2 are as follows.

The random variable

$$T = \frac{\overline{X} - \mu}{S/\sqrt{n}}$$

is said to follow Student's t-distribution with $n-1$ degrees of freedom. This is written $T \sim t(n-1)$. Also,

$$\frac{(n-1)S^2}{\sigma^2} \sim \chi^2(n-1).$$

These distributions may be used to calculate confidence intervals for the parameters μ and σ^2 (or σ).

5. For larger samples from standard distributions, the central limit theorem may be applied to construct approximate confidence intervals for unknown parameters. (Occasionally, further approximations are also introduced.)

An approximate large-sample confidence interval for an exponential mean μ is

$$(\mu_-, \mu_+) = \left(\frac{\widehat{\mu}}{1 + z/\sqrt{n}}, \frac{\widehat{\mu}}{1 - z/\sqrt{n}}\right).$$

An approximate large-sample confidence interval for a Poisson mean μ is

$$\left(\widehat{\mu} - z\sqrt{\frac{\widehat{\mu}}{n}}, \widehat{\mu} + z\sqrt{\frac{\widehat{\mu}}{n}}\right).$$

An approximate large-sample confidence interval for the parameter p in a geometric distribution is given by

$$(p_-, p_+) = \left(\widehat{p} - z\widehat{p}\sqrt{\frac{1-\widehat{p}}{n}}, \widehat{p} + z\widehat{p}\sqrt{\frac{1-\widehat{p}}{n}}\right).$$

An approximate large-sample confidence interval for the mean μ of a geometric distribution is given by

$$(\mu_-, \mu_+) = \left(\widehat{\mu} - z\sqrt{\frac{\widehat{\mu}(\widehat{\mu}-1)}{n}}, \widehat{\mu} + z\sqrt{\frac{\widehat{\mu}(\widehat{\mu}-1)}{n}}\right).$$

An approximate large-sample confidence interval for the Bernoulli probability p is given by

$$(p_-, p_+) = \left(\widehat{p} - z\sqrt{\frac{\widehat{p}(1-\widehat{p})}{n}}, \widehat{p} + z\sqrt{\frac{\widehat{p}(1-\widehat{p})}{n}}\right).$$

6. An approximate $100(1-\alpha)\%$ confidence interval for the mean μ of a population where no distributional assumptions have been made may be obtained by using

$$(\mu_-, \mu_+) = \left(\overline{x} - z\frac{s}{\sqrt{n}}, \overline{x} + z\frac{s}{\sqrt{n}}\right),$$

where n is the sample size, $\overline{x}$ is the sample mean, s is the sample standard deviation and z is the $100(1-\frac{1}{2}\alpha)\%$ point of the standard normal distribution; the sample taken is assumed to be sufficiently large that the central limit theorem may be applied.

Chapter 8
Testing hypotheses

This chapter is about using statistics to test the validity of claims, propositions or hypotheses. There are several ways of doing this, and the area is one which continues to excite some controversy among philosophers of statistics. Essentially, however, most approaches to testing reduce to the same question: does the sample we have collected support the claim made about characteristics of the population, or is there evidence that the claim may be false?

In *Chapters 6* and *7* samples of data were collected in order to explore properties of the population from which those samples were drawn. Point and interval estimates were made for population parameters. The point has repeatedly been made that statistics derived from a random sample (for example, the sample mean and variance, and estimators generally) are themselves random variables. In this respect, point estimates only 'suggest' corresponding values for an unknown parameter. Arguably more usefully, confidence intervals derived from data can be used to suggest a range of plausible values for the parameter. In this context, the variation is evident in the upper and lower confidence limits, which are themselves random variables.

This chapter completes the investigation by providing you with methods for investigating claims. Examples of the sorts of claims advanced that are amenable to statistical examination include the following.

> Eight out of ten dogs prefer Pupkins to any other dog food.
>
> Women enjoy watching soccer as much as men do.
>
> Drug A is better at bringing pain relief than drug B is.

At the time of writing, there is no dog food on the market called Pupkins, and no allusion to any other trade name is intended here.

Before considering how such a claim might be tested, it is important to be absolutely clear about what it means, for only then can a suitable testing procedure be devised.

Let us consider the first claim, possibly advanced by the manufacturer of the dog food in the course of an advertising campaign: 'Eight out of ten dogs prefer Pupkins to any other dog food.' This is not too problematical: it appears to mean that in the relevant population of dogs (perhaps all those in Britain), 80% of dogs presented with a choice of all available dog foods would select Pupkins; the other 20% would make a selection from the remainder.

One way to test this would be to take a sample of dogs from the population of dogs in Britain, offer them the full array of available foods, and keep a record of which of them preferred Pupkins to all others. (There are some problems of definition here: does the phrase 'any other dog food' mean 'any other dog food you can buy in the shops' or could it mean 'anything at all one might reasonably offer a dog to eat'? This second interpretation, if it is not too extreme an interpretation, raises real problems for the designer of the testing strategy.)

This could be an expensive test in the cost of materials, and not necessarily an easy one to apply: it is not entirely clear how you persuade a dog to abstain from eating for as long as necessary for it to be made aware of the choice available. Maybe some alternative design, in which different dogs are offered a limited choice in different combinations, but where relevant conclusions about preferences could still be drawn, could be contrived: it is this sort of problem that it is one of the tasks of a statistician to address.

Notice that there is implicit in the claim the idea that '*at least* 80% of dogs prefer Pupkins'. One would not enter into dispute with the manufacturer if the evidence suggested that, say, 85% or 90% of dogs preferred Pupkins. (One might advise a manufacturer in these circumstances that their claim was, if anything, rather modest.) One would contest the claim only if there was evidence that their claim was inflated, and that in fact the underlying proportion was less than 80%. This is an example of what is known as a *one-sided test.*

If in a small sample of 20 dogs, only 15 dogs showed the claimed preference for Pupkins, an observed sample proportion of 75%, only an unreasonable person would challenge the claim of an underlying 80%, for there is the usual perturbing feature of random variation to understand and make allowance for. Perhaps only if as few as 11 or 12 dogs, say, demonstrated the claimed preference (only just over half of the sample) would one seriously begin to doubt the manufacturer's figure. Or perhaps fewer still, if one wanted really strong evidence against it. (You might usefully pause for a moment here, and ask yourself at what point you would start to entertain doubts about the claimed preference level.) What constitutes sufficient evidence to reject a claim or hypothesis, or at least to cast doubt upon it, is what this chapter is all about.

The second claim was 'Women enjoy watching soccer as much as men do.' Again, it is useful to spend a short time being specific about what this means. It could refer to a kind of measure of enjoyment to be taken on people indulging in various activities (eating, watching television, listening to music, doing housework, and so on). Perhaps the statement means that in the population of women, the distribution of this measure for watching soccer is the same as the distribution of the measure among men. In principle this claim could be tested, though it begs the question of how the measurement might be taken.

Alternatively, the intention behind the claim could be that in the population of those who enjoy watching soccer, the number of women is not substantially different from the number of men. This is a more easily comprehended notion, probably a reasonable interpretation of what was intended, and certainly one that is more easily tested. One might sample the spectators at one or more soccer matches, and count the men and women sampled; then devise some procedure for assessing whether or not the proportion of women is substantially different from $\frac{1}{2}$—that is, whether the sampled proportion is very much less than $\frac{1}{2}$ or very much greater than $\frac{1}{2}$. In either case, the claim would be rejected in the face of evidence to the contrary. This is an example of a *two-sided test.*

(But it is still not entirely obvious that we have selected quite the right test procedure here. Is $\frac{1}{2}$ the appropriate fraction to test? Were 'men' and 'women' intended to include boys and girls? Some of them enjoy watching soccer. Did the word 'watching' mean 'at a football ground' or was the intention to cover television broadcasts of games? Finally, was the real intention behind the claim that 'women enjoy watching soccer *at least* as much as men do (contrary to what you might think)'? In which case the appropriate test would be one-sided.)

As regards the final example, 'Drug A is better at bringing pain relief than drug B is', similar questions are raised. What does the word 'better' mean in this context? Faster? More efficacious at relieving intense pain? Or merely more cost-effective? Once that is clarified, a test procedure may then be designed. Here, it is worth noting that no numbers are included in the claim (like the explicit 80% in the first example). We are not (for the purposes of exploring the hypothesis) concerned with the absolute performance of drug A or of drug B, but with the difference in their performance. Notice, incidentally, the claim that drug A is better than drug B, not merely different from drug B: an appropriate hypothesis test in this case will be one-sided.

Usually matters do not require this sort of microscopic and pedantic interpretation, but it is important to be very clear about what is being claimed, and what is being tested, and how. The British physicist, mathematician, biologist and geneticist, Ronald Aylmer Fisher (1890–1962) of whom more later, and whose contributions to the discipline of statistics are immense, viewed hypothesis testing as an art, not reducible to a procedural task list. He wrote: 'Constructive imagination, together with much knowledge based on experience of data of the same kind, must be exercised before deciding on what hypotheses are worth testing, and in what respects. Only when this fundamental thinking has been accomplished can the problem be given a mathematical form.'

Fisher himself claims not to have been a mathematician, though most biographers classify him as one!

Fisher, R.A. (1939) On "Student". *Annals of Eugenics*, **9**, 1–9.

In Section 8.1, a straightforward approach to testing is taken as follows. First, the claim is reinterpreted as a statement about the value of some unknown population parameter. Then a random sample is taken from the population in such a way that light is shed on the value of the parameter: in particular, so that a confidence interval can be developed for it. This means that we will need a *probability model* for the variation observed. If the confidence interval contains the hypothesized parameter value, then the conclusion is reached that the sample provides no evidence to dismiss the claim. However, if the interval does not contain the hypothesized value, then the conclusion is reached that there is evidence to doubt the claim.

In this way, a decision rule has been developed for deciding whether or not to reject a hypothesis. Notice that the decision rule will depend on the level of confidence adopted. As a method for illustrating the approach, tests will be described in this section for hypotheses about the value of a Bernoulli parameter p and for a Poisson mean μ.

Notice, incidentally, the wording used here: whatever the results of the test, there is no implication that the claim is accepted as 'true'. There is a principle of falsifiability rather than verifiability in operation here. A hypothesis is either rejected in the light of the evidence, or not rejected (because there is insufficient evidence to reject it). In other words, statistics may be used not to prove the truth of something, merely to provide more or less evidence for its falsity.

In fact, in stating the conclusion of a test one might speak loosely of 'acceptance' of the null hypothesis, or of the null hypothesis being 'true', if for no other reason than that the language is sometimes less awkward. But you should be aware of the problems involved in attempting to use statistics to 'prove' something.

In Section 8.2, an approach called *fixed-level testing* is taken. The approach is similar to, but not quite the same as, the approach outlined in Section 8.1. Tests for the value of a normal mean μ are described, and in particular you will learn about *Student's t-test for differences.* This latter test is appropriate where the same individual has had two measurements taken under different circumstances (for example, pulse rate before and after light exercise, or reaction times one hour after ingesting alcohol and after a day's abstinence) and the main question is whether perceived average differences are 'real', or merely manifestations of random variation.

In Section 8.3, an approach called *significance testing* is described; this is the approach that is followed in the rest of the chapter. The main feature of a significance test is that rather than provide a decision rule for the rejection of a hypothesis (though it can be used to provide a decision in a very straightforward way), it provides an assessment of the extent to which the data support the null hypothesis. Again, a number of specific tests appropriate to particular circumstances are described. Some of the tests involve rather awkward arithmetic calculations, the details of which you are spared, but as in other parts of the course it is assumed that you have access to statistical software which would enable you to perform these tests yourself. For larger samples, the central limit theorem applies and approximate normal distribution theory can be used.

In practice the three approaches involve some assumptions and procedures that are common to all: the differences reside mainly in the form of words in which the conclusions of tests are stated. In most usual situations, the results of a significance test can always be called upon to provide a decision, while the first two approaches do not allow the casual assessment of data that a significance test permits.

In all of these first three sections, we shall be dealing with one-sample tests. These are tests where a random sample has been drawn from a single population in order to test some claim about the characteristics of that population. This provides an easy context in which to introduce the testing approaches.

However, it is more commonly the case in practice that what is being tested is a claim that two populations are similar in some respect. To proceed with such a test, samples are drawn from both populations and any differences between the two samples are scrutinized for their importance. Are they 'real' differences, or merely evidence of random variation? These questions are addressed in Sections 8.4 and 8.5. In Section 8.4 we shall study the *two-sample t-test*, one of the most important comparative tests available to the statistician. In Section 8.5 some exact tests for small samples are described—again, the arithmetic can become very awkward, and it is assumed that you have a computer available to you when you need to perform these tests.

8.1 An approach using confidence intervals

8.1.1 Exact tests

In this approach, direct use is made of confidence intervals developed according to the methods of *Chapter 7*. The aim of this approach is to determine from data whether or not a hypothesized parameter value is plausible, by listing those values that do seem plausible and seeing whether the hypothesized value features on the list. This approach can be illustrated most easily by means of an example.

Example 8.1 *Testing a hypothesis about a proportion*

This example is about testing the random generator within a computer program. As you saw in *Chapter 3*, Example 3.17, where a faulty program generated a perfectly regular sequence of the ten digits $0, 1, 2, \ldots, 9$

0 1 2 ... 9 0 1 2 ...,

there is more to a strict test in this context than a simplistic assessment of digit frequencies; but, for the moment, let us keep things simple.

More sophisticated tests for sequences are developed in *Chapter 12*.

The problem in this case is to use a computer to simulate successive rolls of a perfect die, and to count the proportion of 6s. If the observed proportion is sufficiently close to $\frac{1}{6}$, then the program will be deemed satisfactory. If not (i.e. too many 6s, or too few), then the program will be called unsatisfactory and some alternative way found to simulate the rolls of the die.

The results of 100 rolls are shown in Table 8.1. In what is intended to be a sequence of Bernoulli trials (or, strictly speaking, since the computer algorithm is really rather complicated, in what is intended to be indistinguishable from a sequence of Bernoulli trials) a 0 indicates that a $1, 2, 3, 4$ or 5 was rolled; a 1 indicates a 6.

Table 8.1 Throwing a 6: computer simulation

0	0	0	0	0	0	0	0	0	0	0	0	0	0	0	0	0	0	0	0
0	0	0	0	0	0	0	0	1	0	0	0	0	0	1	1	1	0	0	0
0	0	0	0	0	1	0	0	0	0	0	0	0	0	0	0	0	0	0	0
0	0	0	0	0	1	0	0	1	0	0	0	0	0	0	0	0	0	0	0
0	1	0	1	0	0	0	0	0	0	0	0	0	0	0	1	0	0	0	0

In this example there were ten 6s thrown in a total of 100 rolls of the supposedly fair die, rather fewer than expected. The question arises: does this experiment provide any substantial evidence that the die is biased (that is, that the program generating the throws is flawed)?

An exact 90% confidence interval for p, the underlying proportion of 6s, is provided by the methods of *Chapter 7*; assuming a binomial model $B(100, p)$ for the number of 6s to occur in 100 rolls of the die, it is

$$(p_-, p_+) = (0.0553, 0.1637).$$

A confidence interval can be used to provide a decision rule for a test of a hypothesis about the value of a model parameter. The most noticeable feature of this confidence interval is that it does not contain the theoretical (or assumed) underlying value $p = \frac{1}{6} = 0.1667$. The conclusions of this simple test may be stated as follows.

In a test of the hypothesis that $p = \frac{1}{6}$, there was evidence at the 10% level of significance to reject the hypothesis in favour of the alternative that $p \neq \frac{1}{6}$; in fact, having performed the test, there is some indication from the results of the test that $p < \frac{1}{6}$. ■

There are a number of features of the testing procedure to notice here. Most obviously, the raw material for the statistical testing of a hypothesis is the same as that for the construction of a confidence interval: we require data; we need an underlying probability model; and we need to have identified a model parameter relevant to the question we are interested in answering.

What has altered is the form of the final statement: rather than listing a range of credible values for an unknown parameter, at some level of confidence, a statement is made about whether or not a **hypothesis** about a particular parameter value is tenable, at some assigned **level of significance**. Notice that in this example the significance level has been expressed as a percentage, and is equal to 100 minus the confidence level used to perform the test. (This is just the way the conventional language has developed.)

Example 8.1 continued

An exact 95% confidence interval for the binomial parameter p, based on a count of 10 successes in 100 trials, is

$$(p_-, p_+) = (0.0490, 0.1762).$$

In this case the hypothesized value $p = \frac{1}{6}$ is contained in the confidence interval: in other words, at this level of confidence, and based on these data, it seems a plausible value. The conclusions of the corresponding test may be stated as follows.

In a test of the hypothesis that $p = \frac{1}{6}$, there was insufficient evidence at the 0.05 level of significance to reject the hypothesis in favour of the alternative that $p \neq \frac{1}{6}$. ■

Here the significance level has been expressed as a number (0.05) between 0 and 1, rather than as a percentage (5%). Either formulation is common.

It is worth noticing that in our statement of the hypothesis $p = \frac{1}{6}$, and in the context of the problem, there has been the implication that either a sample proportion too high (suggesting $p > \frac{1}{6}$) or a sample proportion too low (suggesting $p < \frac{1}{6}$) would both offer evidence to reject the hypothesis. In the jargon, we have performed a **two-sided test** (sometimes called a **two-tailed test**). In other contexts, the implication will be that only extreme values in one direction would offer serious evidence against a proposition. The dog food example illustrates this: to refute the manufacturer's claim that the preference rate was 80%, only evidence suggesting it was lower than this would normally be of interest. But, in general, you need to have a very clear understanding of what is being suggested, that is, of the implications of a hypothesis, when you are setting up an appropriate test.

The approach may be summarized as follows. Characteristics of a population may be expressed in terms of a parameter θ in such a way that the hypothesis under test takes the form $\theta = \theta_0$, where θ_0 is some specified value. This is called the **null hypothesis** and is written

As in *Chapters 6* and *7*, the convention of using θ as the symbol for a general parameter is adopted here.

$$H_0 : \theta = \theta_0.$$

The **alternative hypothesis** is that the claim is false: this is written

$$H_1 : \theta \neq \theta_0.$$

In other testing scenarios, the alternative hypothesis might be $H_1 : \theta > \theta_0$ or $H_1 : \theta < \theta_0$. We shall look at one-sided tests involving hypotheses like these later.

We then collect data on an appropriate random variable where the variation observed may be expressed through a probability model indexed by θ. Using the data, we construct a $100(1 - \alpha)\%$ confidence interval for θ in the form

$$(\theta_-, \theta_+).$$

Our decision rule for rejecting the null hypothesis H_0 depends on whether the hypothesized value θ_0 of θ is, or is not, contained in this list of plausible values.

Use this approach when attempting the following exercise. Be explicit in your statement of the null hypothesis H_0, of the alternative hypothesis H_1 and of the underlying probability model on which your confidence interval, and therefore your decision rule, is based.

Exercise 8.1

A different computer, and a different statistical package, was used to generate the results of a sequence of Bernoulli trials, where the intention was that the probability of success at any trial should be $p = \frac{2}{3}$. The results of a sequence of 25 trials are shown in Table 8.2.

Table 8.2 Computer simulation: $p = \frac{2}{3}$

0	0	0	1	0	0	1	0	0	1	0	1	0
1	0	1	1	0	0	1	1	0	0	0	1	

Ignoring the fact that a simple test of the observed sample proportion against the hypothesized value $p = \frac{2}{3}$ does not itself constitute a test that the results were generated independently, perform a test as follows.

(a) Find a 95% confidence interval for p based on the observed sequence of trials.

(b) Hence perform a test at significance level 0.05 of the hypothesis that the underlying proportion of successes is $p = \frac{2}{3}$, against the alternative hypothesis that the underlying proportion is different from $\frac{2}{3}$.

(c) In a similar way, perform a test at significance level 0.01.

Example 8.2 illustrates testing a statement about a Poisson mean.

Example 8.2 Cycle usage

In *Chapter 4*, Example 4.1 it was stated that on average, a typical cyclist will get caught in the rain on about 15 occasions a year. A cyclist interested in checking this claim kept a diary for a year of meteorological occurrences (among other things) during cycle rides. During the year she got wet 8 times. Is there sufficient evidence here to challenge the claim?

Ballantine, R. (1975) *Richard's Bicycle Book.* Pan, Great Britain.

Before we can proceed to answer this question, several matters need to be clarified. First, we require a probability model on which to base a hypothesis test. (We have the data; but so far, no explicit statement of two other requirements for a test—a model and a parameter.) Previously, a Poisson model for the incidence of rain has been assumed: let us continue with this assumption. Second, we need to state a parameter whose value is the subject of the test. In this case an obvious choice is the Poisson mean μ. Third, we need to be clear whether this is a two-sided test. There is no indication that before the data were collected, the experimenter suspected Richard's estimate of 15 occasions a year to be an overestimate—this is only apparent after the data were collected. So it is reasonable to suppose that the claim under test is the claim $H_0 : \mu = 15$; the alternative hypothesis is $H_1 : \mu \neq 15$.

If the Poisson mean μ is the obvious choice of parameter to test, nevertheless it is not the only choice. Another interpretation, for example, is that the median number of occasions when it rains is 15.

Finally, there is one rather subtle matter to be sorted out. Richard's estimate of 15 times a year was for a 'typical' cyclist. It may be that the person performing the test (that is, the person who kept the diary yielding the data) is atypical in some way: perhaps she never goes out if the weather forecaster mentions the word 'rain', or perhaps she goes out only on Sundays, or perhaps she is a city delivery courier whose job consists largely of riding around on a bicycle. Any one of these considerations dooms the test of Richard's claim. Let us assume (the whole matter is rather imprecise) that the test is a fair one.

No significance level has been specified for the test. An exact 90% confidence interval for a Poisson mean μ, based on the single observation 8, is given by

See *Chapter 7*, Subsection 7.2.2.

$$(\mu_-, \mu_+) = (3.98, 14.43).$$

This confidence interval does not contain the hypothesized value $\mu = 15$ and your conclusion should therefore be as follows.

Based on the single observation 8, and assuming a Poisson model for the variation in the incidence of wet journeys for a typical cyclist, the null hypothesis $H_0 : \mu = 15$ is rejected at the 10% level of significance (or at level of significance 0.10), in favour of the alternative hypothesis that μ is somewhat different from this, $H_1 : \mu \neq 15$. (In fact, there is some evidence from the data that the hypothesized value is an overestimate.)

But notice how close things are: the value $\mu = 15$ is only just outside the 90% confidence interval obtained from the data. A 95% interval for the Poisson mean μ is furnished by

$$(\mu_-, \mu_+) = (3.45, 15.76),$$

and this wider confidence interval does contain the hypothesized value. In this case one would reach the following conclusion.

Based on the single observation 8 per year and assuming a Poisson model, there is insufficient evidence at the 5% level of significance to reject the hypothesis that the annual average is 15.

Notice that the sparsity of data has resulted in very wide confidence intervals: two years' data, or more, would provide a much more informative test. Perhaps the observation 8 is unusually low, and the mean really is about 15; perhaps Richard's figure really does overestimate things, and the observation 8 is much more representative of what one might expect. ■

The next exercise asks you to test a hypothesis about traffic rate, based on the Kwinana Freeway traffic data in *Chapter 2*, Table 2.11. The wording of the exercise is very explicit about the probability model you should assume for the observed variation in the waiting times. Again, be clear about the parameter you are testing and write down explicit statements of the null and alternative hypotheses.

Exercise 8.2

The Kwinana Freeway data list the waiting times (in seconds) between successive vehicles in free-flowing traffic. Assuming an exponential model for the variation in waiting times, test the hypothesis that the mean traffic flow rate is 10 vehicles per minute. Perform the test at levels of significance 0.10, 0.05 and 0.01.

In all the examples of tests used in this section, the hypothesis under test has taken the form $H_0 : \theta = \theta_0$, where θ_0 is some specified value (e.g. $\mu = 15$ or $p = \frac{1}{6}$). The alternative hypothesis has been the converse of this, that is, of the form $H_1 : \theta \neq \theta_0$. In other words, all the tests that have been performed have been two-sided.

There is a very simple reason for this: in the whole of *Chapter 7*, only two-sided confidence intervals of the form (θ_-, θ_+) were constructed. So-called 'one-sided' confidence intervals of the form

$$(0, \theta_+) \qquad \text{or} \qquad (\theta_-, \infty)$$

or perhaps $(-\infty, \theta_+)$ or $(\theta_-, 1)$ are an almost immediate extension, and may be used to test one-sided hypotheses. In Section 8.2 a testing approach is introduced that will permit one-sided tests (as well as two-sided tests) to be performed in a very direct and evident way.

One-sided confidence intervals include one limit drawn from data; the other limit follows from consideration of the nature of the parameter. For instance, any probability is at least 0, and at most 1; a Poisson mean cannot be negative. This course does not deal explicitly with one-sided confidence intervals.

8.1.2 Large-sample tests

Just as in the case of constructing confidence intervals from large collections of data (see *Chapter 7*, Section 7.4), the central limit theorem can be cited in order that tests may be based on normal distribution theory. (If nothing else, this reduces your reliance on computer software.)

A normal approximation to a confidence interval is used in Example 8.3.

Example 8.3 Yeast cells on a microscope slide

Some of 'Student's' original experiments involved counting the numbers of yeast cells found on a microscope slide. The results of one experiment are given in *Chapter 2*, Table 2.5. This table gives the numbers of yeast cells observed in each of 400 small squares on a slide. The number varies between 0 and 5. It is required to use these data to test the null hypothesis

$$H_0 : \mu = 0.6$$

that the mean number of cells per slide is 0.6, against the alternative hypothesis

$$H_1 : \mu \neq 0.6.$$

Suppose that a Poisson model is assumed for the variation in counts. So for the purposes of our test we have identified an appropriate probability model with an indexing parameter directly relevant to the hypothesis of interest; also, we have data. The observed sample mean is

$$\overline{x} = \frac{0 \times 213 + 1 \times 128 + \cdots + 5 \times 1}{213 + 128 + \cdots + 1} = \frac{273}{400} = 0.6825.$$

No significance level has been stipulated for the test: let us use $\alpha = 0.05$. Then an approximate 95% confidence interval for the unknown Poisson mean μ is given by

Use of the symbol α in a testing context matches its use in the specification of confidence levels. See *Chapter 7*, Section 7.4 for the derivation of large-sample confidence intervals.

$$\begin{aligned}
(\mu_-, \mu_+) &= \left(\widehat{\mu} - z\sqrt{\frac{\widehat{\mu}}{n}}, \widehat{\mu} + z\sqrt{\frac{\widehat{\mu}}{n}}\right) \\
&= \left(\overline{x} - 1.96\sqrt{\frac{\overline{x}}{n}}, \overline{x} + 1.96\sqrt{\frac{\overline{x}}{n}}\right) \\
&= \left(0.6825 - 1.96\sqrt{\frac{0.6825}{400}}, 0.6825 + 1.96\sqrt{\frac{0.6825}{400}}\right) \\
&= (0.6015, 0.7635).
\end{aligned}$$

This confidence interval does not contain the hypothesized mean $\mu_0 = 0.6$, and so the null hypothesis is rejected in favour of the alternative hypothesis at level of significance 0.05. ■

In the following exercise, you should use the appropriate large-sample confidence interval to perform your test (that is, a confidence interval based on the normal distribution).

Exercise 8.3

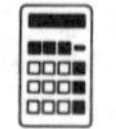 

This exercise is based on data published by Gregor Mendel, an early explorer of the science of genetics. Several of his experiments reduce to counting the number of successes in a sequence of Bernoulli trials in order to test a hypothesis about the value of the Bernoulli parameter p, the underlying probability of success.

In this exercise, you are invited to follow Mendel's footsteps in an analysis of his experimental data.

(a) Test the null hypothesis $H_0 : p = \frac{3}{4}$ against the alternative $H_1 : p \neq \frac{3}{4}$, based on an observed 787 successes in 1064 trials. (These are the results of Mendel's seventh experiment in what became known as the 'first series'. In this case he was counting the number of yellow peas in a collection of first-generation pea plants. The 'failures' were green peas.) Use a 10% significance level for your test.

(b) Test the null hypothesis $H_0 : p = \frac{2}{3}$ given 60 successes in 100 trials (fifth experiment, second series). Perform your test at level of significance 0.05.

(In 1936, R.A. Fisher concluded on the basis of tests that Mendel had falsified his data. The results Mendel quoted from his own experiments seemed 'too good to be true' in that they varied too little from the results that would have been expected, if Mendel's genetic theories had been correct. In experiments like this, Fisher argued that random variation would make it unlikely that the estimate $\widehat{p}$ of p would turn out to be as close to the hypothesized value of p as Mendel had reported.

Fisher, R.A. (1936) Has Mendel's work been rediscovered? *Annals of Science*, **1**, 115–137.

Fisher's conclusion was refuted in 1984 and in subsequent papers by the researcher, Ira Pilgrim, and the case has been much discussed in the literature. Altogether 14 of Mendel's experiments have been analysed, seven from each of the first and second series. In 1985, Monaghan and Corcos wrote of the controversy: 'There seems to be no satisfactory solution to this problem at present, at least not in the statistics'. Pilgrim's 1986 paper ends with the words 'I can conclude from the above that there is no reason whatever to question Mendel's honesty'.)

Pilgrim, I. (1984) The too-good-to-be-true paradox and Gregor Mendel. *J. Heredity*, **75**, 501–502.

Monaghan, F. and Corcos, A. (1985) Chi-square and Mendel's experiments: where's the bias? *J. Heredity*, **76**, 307–309 and Pilgrim, I. (1986) A solution to the too-good-to-be-true paradox and Gregor Mendel. *J. Heredity*, **77**, 218–220.

In Section 8.2 an alternative approach to testing is described. In this approach, it is not the hypothesized parameter value that comes under direct scrutiny: instead, the data are examined to see whether they are consistent with the hypothesis or whether they depart from what might have been expected and in a manner forecast by the alternative hypothesis.

8.2 Fixed-level testing

Although there is no particularly compelling reason for this, you have seen that it is very common to utter confidence statements at predetermined levels of 90%, 95% and 99%. Similarly, it is very common to perform tests of hypotheses at predetermined significance levels. This is what is meant by a **fixed-level test**. It is common to choose levels such as 10%, 5% and 1%. In this section an interpretation of a significance level as a probability is provided. Also, you will see how to perform one-sided tests.

The aim of this approach is, as before, to develop a decision rule for rejection of a null hypothesis in favour of a stated alternative, at some predetermined level of significance. It will become obvious to you that the method of hypothesis testing advanced in this section can be applied in any context where a clear probability model has been specified. It is therefore not the intention here to offer examples and illustrations of all possible contexts, but only of a small selection of typical testing scenarios, so that the procedure is clear (even if the method is not exemplified in every possible case you might come across).

8.2.1 Performing a fixed-level test

We have seen that in order to perform a hypothesis test, the first requirement is for a clear statement of what that hypothesis is: this is usually expressed in terms of the parameter of a probability model. This implies, just as for the construction of confidence intervals, that we need to have decided on a usable probability model; that this model should be indexed by a parameter which really does encapsulate the intention behind the statement of the hypothesis; and, of course, we need data on which to base our test.

Further, we need to decide on an alternative hypothesis which will indicate the sort of departure from expectation that would be of interest—that is, we need to determine whether our test is to be one-sided or two-sided.

It will be useful if at this stage a little more terminology is introduced. You have seen that the hypothesis to be tested is usually called the *null hypothesis*, and the symbol H_0 is used to denote the null hypothesis. The hypothesis to be regarded as an alternative to this is called the *alternative hypothesis* and is denoted by the symbol H_1. In general, the choice of the alternative hypothesis depends on our purpose in performing the hypothesis test, and should reflect this purpose.

Their general statement might take the form

$$H_0 : \theta = \theta_0, \qquad H_1 : \theta \neq \theta_0.$$

Possible variations on this (depending on the intention behind the claim under investigation) are

$$H_0 : \theta = \theta_0, \qquad H_1 : \theta > \theta_0;$$

or

$$H_0 : \theta = \theta_0, \qquad H_1 : \theta < \theta_0.$$

Of course, these are only helpful statements if the role of the parameter θ_0 is clear: this means that data relevant to the hypothesis under test need to be collected, and a random variable indicative for the test needs to have been identified. The statistic used to test a hypothesis (often the sample mean or the sample total; possibly the sample maximum, the sample median or some other quantity calculated from the data) is called the **test statistic**. The distribution of the test statistic if the null hypothesis H_0 were true is called the **distribution of the test statistic under the null hypothesis** or, more conveniently, the **null distribution** of the test statistic.

8.2.2 Testing a hypothesis about a normal mean

As before, let us begin with an example. This will serve to illustrate the main features of a fixed-level test.

Example 8.4 Pretzels

This is an example about a claim made on packaged goods about what the package contains—a common situation. A company producing snack foods used a machine to package pretzels in bags with a labelled weight of 454 grams. Every so often, the product was monitored by taking a selection of bags from the production line and weighing them. In one experiment 50 bags were weighed. The results of the experiment are shown in Table 8.3.

Table 8.3 Weights of 50 bags of pretzels (grams)

464	450	450	456	452	433	446	446	450	447
442	438	452	447	460	450	453	456	446	433
448	450	439	452	459	454	456	454	452	449
463	449	447	466	446	447	450	449	457	464
468	447	433	464	469	457	454	451	453	443

Weiss, N.A. and Hassett, M.J. (1991) *Introductory Statistics*, 3rd edition, Addison-Wesley, Massachusetts.

The purpose in collecting these data is to determine whether the machine is 'working correctly'. That phrase itself is open to more than one interpretation (for example, is the machine sealing the bags of pretzels adequately?) but let us take it to mean that the average weight of bags produced on the line is indeed 454 grams. In other words, we wish to test the null hypothesis

$$H_0 : \mu = 454.$$

'Working incorrectly' could mean simply that the bags are either underweight or overweight; and that is what we shall take it to mean here. But it is worth remarking that the consequences of selling underweight bags (possible legal action under the trade laws) are quite different from the consequences of selling overweight bags (additional manufacturing costs and reduced profits), and it may be that the original purpose of the test was to explore only whether one of these was occurring. That would imply a one-sided test; however, we have decided to set up a two-sided test and therefore write the alternative hypothesis as

$$H_1 : \mu \neq 454.$$

Our next requirement is to set up a test statistic and the corresponding null distribution. We shall need a probability model for the variation observed. Figure 8.1 shows a histogram of the data.

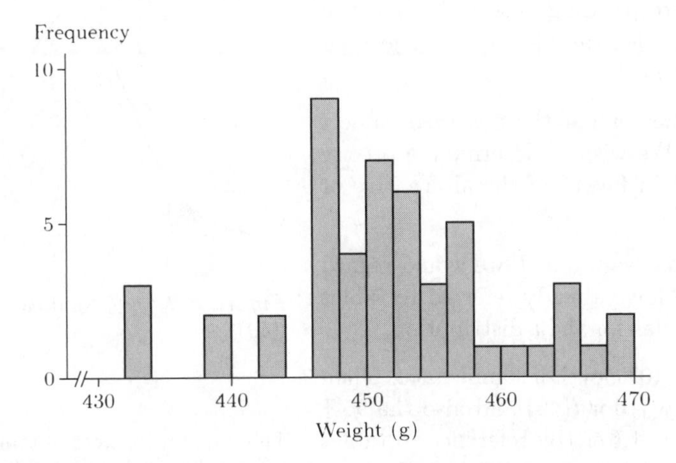

Figure 8.1 Weights of 50 bags of pretzels (grams)

The sample size of 50 is not very large, and this is reflected in the relative jaggedness of the histogram in the figure; but it does seem that a normal model might be adequate for our purposes. This achieved, we need to identify a test statistic, based on the sample, whose distribution involves the unknown parameter μ, but no others.

Another good reason for choosing a normal model here is that no better one springs very easily to mind.

This sort of problem is familiar from *Chapter 7*. Denoting by X the actual weights of bags of pretzels, we adopt the model

$$X \sim N(\mu, \sigma^2)$$

to reflect the variation in weights. For samples of size 50 from this distribution, either the sample total or the sample mean has a reasonably simple distribution, involving the unknown parameter μ:

$$\sum_{i=1}^{50} X_i \sim N(50\mu, 50\sigma^2) \qquad \text{and} \qquad \overline{X} \sim N\left(\mu, \frac{\sigma^2}{50}\right).$$

Unfortunately, both these statistics also have distributions that involve the parameter σ^2, a **nuisance parameter** in that it is unknown and anyway irrelevant to the matter in which we are most interested. However, we know one other useful test statistic for samples from a normal population, whose distribution involves the population mean μ but not the population variance σ^2.

We know that for a sample from a normal distribution with unknown mean μ and unknown variance σ^2, the distribution of the sample mean $\overline{X}$ is usefully given in terms of Student's t-distribution as

$$T = \frac{\overline{X} - \mu}{S/\sqrt{n}} \sim t(n-1), \qquad (8.1)$$

See *Chapter 7*, page 285.

where n is the sample size and S is the sample standard deviation.

If we use this as our test statistic, then the null distribution of T (that is, the distribution of T if the null distribution $H_0 : \mu = 454$ were true) for samples of size 50 is given by

$$T = \frac{\overline{X} - 454}{S/\sqrt{50}} \sim t(49). \qquad (8.2)$$

You can see from the form of the test statistic T in (8.2) that if the observed sample mean $\overline{x}$ is less than 454 then the corresponding observed value t of the test statistic T will be negative; if the observed sample mean $\overline{x}$ is greater than 454 then the observed value of T will be positive.

The idea of fixed-level testing is to see whether or not the observed value t of T is consistent with this null distribution. We wish to determine a precise rule for whether we reject the null hypothesis in favour of the alternative or whether, in fact, we accept the null hypothesis.

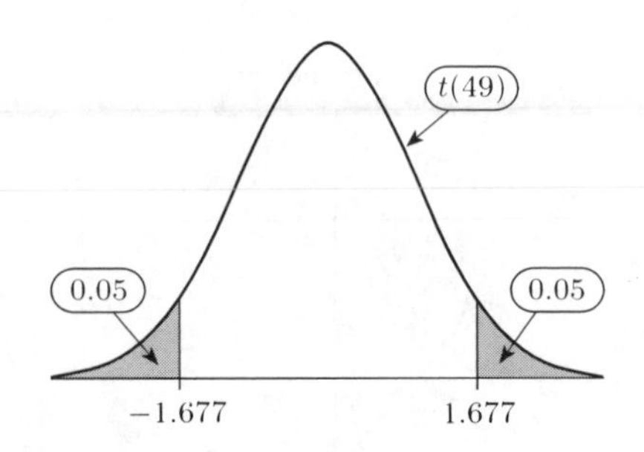

Figure 8.2 $q_{0.05}$ and $q_{0.95}$ for $t(49)$

We achieve this by determining in advance what values of T we would regard as extreme. This is fairly straightforward: we have already referred to tables of Student's t-distribution to determine quantiles for the t-distribution.

Let us suppose a test is to be performed at the 10% level of significance. Then our procedure is to identify the 5% quantile $q_{0.05}$ for $t(49)$ and also the 95% quantile $q_{0.95}$. Here, $q_{0.05} = -1.677$ and $q_{0.95} = 1.677$ (by reference to tables or a computer and using the symmetry of the t-distribution). These values are shown in Figure 8.2. Our decision rule for the test is: if the observed value t of the test statistic T is so small (i.e. less than -1.677) or so large (i.e. greater than 1.677) that a significant departure from expectation is indicated, then we will reject the null hypothesis in favour of the alternative hypothesis H_1 at level 0.10: for there is evidence inconsistent with H_0. The shaded region in Figure 8.3 illustrates this decision criterion: it is called the **rejection region**.

These quantiles were obtained from a computer and not from Table A5, which contains no row corresponding to 49 degrees of freedom for the t-distribution. However extensive your tables are, they cannot cover every eventuality.

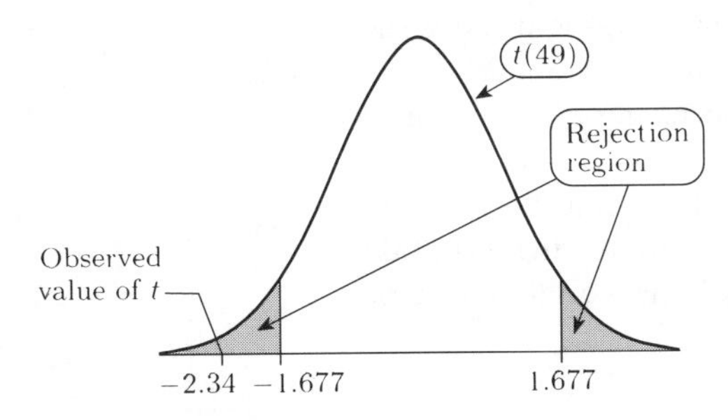

Figure 8.3 The observed value $t = -2.34$ of the test statistic T

At the penultimate stage in the test, we need to calculate the test statistic for the observed sample. For the pretzels data in Table 8.3, we have the summary statistics

$$\overline{x} = 451.22, \qquad s = 8.40,$$

and, therefore, the corresponding value of t is

$$t = \frac{\overline{x} - \mu}{s/\sqrt{n}} = \frac{451.22 - 454}{8.40/\sqrt{50}} = -2.34.$$

Thus, in this example, the observed value of the test statistic lies in the **rejection region** (see Figure 8.3). Finally, we need to state the conclusions of the test.

Based on the sample collected, there is evidence at the 10% level of significance that the mean weight of bags of pretzels from the production line is not equal to the hypothesized value of 454 grams. (In fact, the data suggest that the bags are underweight.) ■

Sometimes the phrase 'the underlying mean is *significantly different* from the hypothesized mean', is used.

The strategy for a fixed-level test may be summarized as follows.

Fixed-level testing

In a fixed-level test:

1 determine the null hypothesis H_0 and the alternative hypothesis H_1 appropriately (for a one- or two-sided test);

2 decide what data to collect that will be informative for the test;

3 determine a suitable test statistic and the null distribution of the test statistic (that is, the distribution of the test statistic when H_0 is true);

4 use the stated level of the test and the form of H_1 to determine the rejection region for the test; for this, you will need to calculate quantiles of the null distribution;

5 collect your data and evaluate the observed value of the test statistic for the sample;

6 by determining whether or not the observed value of the test statistic lies in the rejection region, decide whether or not to reject the null hypothesis in favour of the alternative;

7 state your conclusions clearly.

In many testing contexts, the appropriate test statistic to use is usually the sample mean or the sample total. Often the null distribution is tractable (that is, reasonably easy to work with) and well-known (for example, the sum of independent observations on a Poisson variate itself has a Poisson distribution; in samples drawn from a normal population the sample mean is itself normally distributed). However, there is more to it than this: in any testing situation where data have been collected, all sorts of features of the sample could be used to test the null hypothesis, such as

the sample mean;
the sample total;
the sample median;
the sample maximum;

to name just a few. All of these may be useful indicators of the truth or otherwise of a hypothesis, but some are more useful—leading to more *powerful* tests—than others. In a technical sense, one test of a particular set of hypotheses is more powerful than another if it leads to a higher probability of rejecting the null hypothesis when the null hypothesis is false. Identification of powerful tests involves subtle (and often mathematically quite difficult) considerations—we shall not enter into them, but at the end of Section 8.3 some famous names from the fundamental development of tests for hypotheses are mentioned.

The test used in Example 8.4 for the mean of a normal distribution is called the ***t*-test** or **Student's *t*-test** (though, actually, R.A. Fisher had a lot to do with its development). This is one of the most commonly used tests in statistics, for as we have seen the normal distribution is a useful model for variation with many different applications.

Exercise 8.4

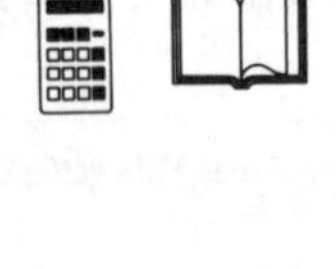

Most individuals, if required to draw a rectangle (for example, when composing a picture) would produce something not too 'square' and not too 'oblong'. A typical rectangle is shown in Figure 8.4.

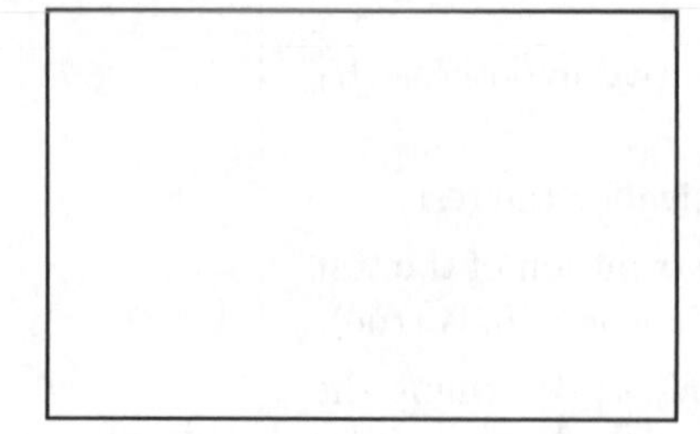

Figure 8.4 A typical rectangle

The Greeks called a rectangle 'golden' if the ratio of its width to its length was $\frac{1}{2}(\sqrt{5}-1) = 0.618$. The Shoshoni Indians used beaded rectangles to decorate their leather goods. The data in Table 8.4 are the width-to-length ratios for twenty rectangles, analysed as part of a study in experimental aesthetics.

DuBois, C. (1960) *Lowie's Selected Papers in Anthropology*, University of California Press, pp. 137–142.

Table 8.4 Width-to-length ratios, Shoshoni rectangles

0.693	0.662	0.690	0.606	0.570	0.749	0.672	0.628	0.609	0.844
0.654	0.615	0.668	0.601	0.576	0.670	0.606	0.611	0.553	0.933

Assuming a normal model for the variation in observed ratios, test the null hypothesis

$$H_0 : \mu = 0.618$$

against the alternative hypothesis

$$H_1 : \mu \neq 0.618$$

at the 5% level of significance.

In Example 8.4 and in Exercise 8.4, the aim was to test for a particular specified mean μ_0. There is one important further application of the t-test where the data take the form of differences, and the aim is to test the hypothesis

$$H_0 : \mu = 0$$

that the mean difference is zero.

8.2.3 Student's t-test for zero mean difference

In Example 8.4 a test was performed of the hypothesis $H_0 : \mu = 454$ assuming a normal model with unknown variance, and involving a test statistic following Student's t-distribution. This is an important test with many applications.

One particular application is where the observations are the differences in *matched pairs of observations.* An example of this was given in *Chapter 7*, Table 7.3, where Student's data on the effects of two different hypnotics on sleep duration are listed.

The aim of the test was to determine whether there was a significant difference in sleep gain between the hypnotics L-hyoscyamine hydrobromide and D-hyoscyamine hydrobromide. An interesting preliminary test, however, is whether the hypnotics themselves have an effect.

The sleep gain (measured in hours) for the ten individuals who were prescribed L-hyoscyamine hydrobromide are reproduced here in Table 8.5.

Table 8.5 Sleep gain (hours)

Patient	Gain
1	1.9
2	0.8
3	1.1
4	0.1
5	−0.1
6	4.4
7	5.5
8	1.6
9	4.6
10	3.4

These data are individual *differences* between the length of time asleep after taking L-hyoscyamine hydrobromide and the length of time asleep after taking no drug. The differences are all positive except the fifth, and this remark alone suggests that the hypnotic L-hyoscyamine hydrobromide is effective at prolonging sleep. However, a formal test of the hypothesis that the prescription in fact makes no difference to the duration of sleep might take the form

$$H_0 : \mu = 0,$$

where the parameter μ is the mean underlying sleep gain. For a test of this hypothesis, it is necessary to provide a model for the variability in observed gain. This data set is very small and a histogram is not likely to display much in the way of persuasive evidence for or against a normal model; however, we do require a continuous model for variation that will permit negative as well as positive observations, and in this regard the normal model is essentially the only one available to us.

Here, the random variable has been written D to represent the 'difference'. There is the small possibility of confusion with the effects of D-hyoscyamine hydrobromide, and the variable could have been written X or even G (for 'gain'); but the notation D is fairly standard and for that reason has been adopted here.

Again, the obvious test statistic to use in this context is based around the sample mean $\overline{D}$, but we need to take account of the fact that the variance σ^2

in sleep gain under the hypnotic is unknown. Assuming that the observed differences d_i, $i = 1, 2, \ldots, 10$, are independent observations on a normal random variable

$$D \sim N(\mu, \sigma^2),$$

then under the null hypothesis $H_0 : \mu = 0$, the sample mean $\overline{D}$ has a t-distribution

$$\frac{\overline{D}}{S/\sqrt{n}} \sim t(n-1). \tag{8.3}$$

In this case it will be interesting to pursue a one-sided test to reflect one's suspicion (or even one's aim in administering the dose) that the hypnotic is an effective prolonger of sleep. Let us therefore write the alternative hypothesis as

$$H_1 : \mu > 0.$$

The null distribution of the test statistic is given at (8.3). This may be used to define the rejection region for the test. Now, the implication of the way the test has been designed is that the null hypothesis will be rejected in favour of the alternative hypothesis if there is sufficient evidence of positive gain; that is, if the test statistic (notice the numerator $\overline{D}$) is sufficiently large—and positive. We shall need to compare the observed value of the test statistic with the appropriate quantile of the t-distribution with $n - 1 = 9$ degrees of freedom, i.e. $t(9)$. For a test at level, say, 0.10, and assuming the one-sided alternative hypothesis $H_1 : \mu > 0$, the relevant quantile is

$$q_{0.90} = 1.383.$$

This is shown in Figure 8.5; the shaded area gives the rejection region for the test.

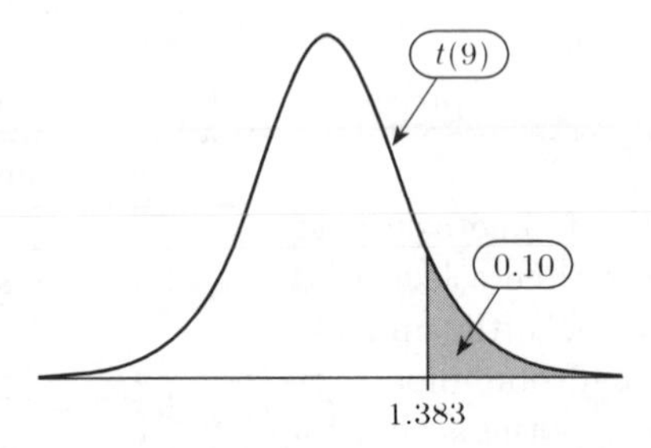

Figure 8.5 The rejection region for a one-sided test at level 0.10

Now we proceed to calculation of the observed value of the test statistic. In this case the sample standard deviation of the observed gains is $s = 2.00$ and the sample size is $n = 10$. The sample mean is $\overline{d} = 2.33$. The observed value of the test statistic is therefore

$$\frac{\overline{d}}{s/\sqrt{10}} = \frac{2.33}{2.00/\sqrt{10}} = 3.68.$$

You can see that the observed value of the test statistic $t = 3.68$ is well inside the rejection region: the effect of the hypnotic L-hyoscyamine hydrobromide is a very pronounced one—this confirms formally our earlier observation that an effect is probable since all the differences bar one are positive.

Exercise 8.5

Use the data from *Chapter 7*, Table 7.3 in a one-sided test at level 0.05 of the hypothesis that the hypnotic D-hyoscyamine hydrobromide has no effect, against the alternative hypothesis that it leads to a net sleep gain. Start with an explicit statement of the hypotheses H_0 and H_1, and follow the pattern of Example 8.4 as you go through the various stages of the test.

Exercise 8.6 relates to the results of one of Darwin's experiments.

Exercise 8.6

Darwin measured differences in height for 15 pairs of plants of the species *Zea mays*. (Each plant had parents grown from the same seed—one plant in each pair was the progeny of a cross-fertilization, the other of a self-fertilization. Darwin's measurements were the differences in height between cross-fertilized and self-fertilized progeny.) The data are given in Table 8.6. The units of measurement are eighths of an inch.

The data are quoted in Fisher, R.A. (1942) *The Design of Experiments*, 3rd edition, Oliver and Boyd, London, p. 27.

(a) Supposing that the observed differences d_i, $i = 1, 2, 3, \ldots, 15$, are independent observations on a normally distributed random variable D with mean μ and variance σ^2, state appropriate null and alternative hypotheses for a two-sided test of the hypothesis that there is no difference between the heights of progeny of cross-fertilized and self-fertilized plants, and state the null distribution of an appropriate test statistic.

(b) Obtain the form of the rejection region for the test you defined in part (a), assuming a 10% significance level.

(c) Calculate the value of the test statistic for this data set, and state the conclusions of your test.

Table 8.6 Differences in plant height ($\frac{1}{8}$ inch)

Pair	Difference
1	49
2	−67
3	8
4	16
5	6
6	23
7	28
8	41
9	14
10	29
11	56
12	24
13	75
14	60
15	−48

8.2.4 Fixed-level testing for discrete distributions

For discrete distributions the fixed-level testing approach is almost identical, although there may be minor difficulties in determining the rejection region. This is merely because, as you saw in *Chapter 3*, Section 3.5, it is not a straightforward matter to identify quantiles of discrete distributions. An example will illustrate the problem.

Example 8.5 *Anopheles farauti mosquitoes*

Researchers needed to evaluate the effectiveness of an insecticide (dieldrin) in killing *Anopheles farauti* mosquitoes. The theory was that resistance to dieldrin was due to a single dominant gene, and that in an appropriately selected sample of the mosquitoes, there should be 50% susceptibility to the insecticide. To test this hypothesis

The results of one such experiment are reported by Osborn, J.F. (1979) *Statistical Exercises in Medical Research*, Blackwell, Oxford. In a sample of 465 mosquitoes, 264 died.

$$H_0 : p = \tfrac{1}{2}$$

against the alternative hypothesis

$$H_1 : p \neq \tfrac{1}{2}$$

it was decided to test the insecticide on a small sample of 30 mosquitoes at level of significance $\alpha = 0.05$. The number of mosquitoes R for which the insecticide proved lethal would be counted.

Under the null hypothesis the test statistic R has a binomial distribution $B(30, \frac{1}{2})$. Figure 8.6 shows the null distribution of R.

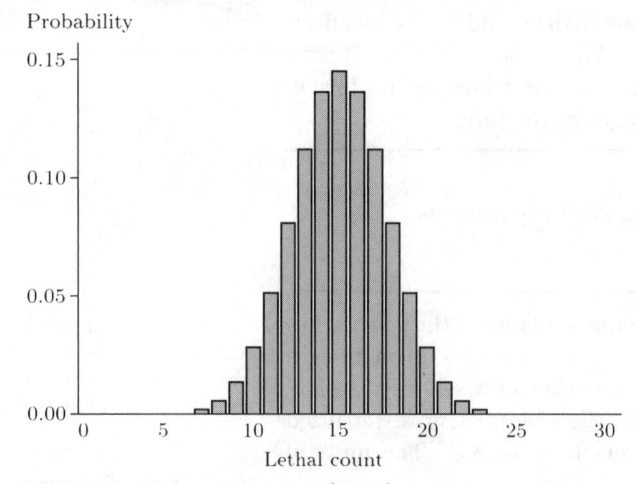

Figure 8.6 The null distribution $R \sim B(30, \frac{1}{2})$

The rejection region will include very low or very high observed values of R, indicating respectively a lethal count lower or higher than expected. Suppose that the intended size of the rejection region is 0.05. Therefore it is required to find the value of r satisfying the probability statement

$$P(R \leq r) = P(R \geq 30 - r) = 0.025. \tag{8.4}$$

Now, some useful probabilities for the binomial distribution $B(30, \frac{1}{2})$ are

$$P(R \leq 9) = P(R \geq 21) = 0.0214;$$

$$P(R \leq 10) = P(R \geq 20) = 0.0494.$$

One of these is just below the required value 0.025; the other is somewhat above it. (So, no value of r exactly satisfies the requirement given at (8.4).)

The closest one can get to the required significance level is to define the rejection region as shown in either one of the two diagrams in Figure 8.7. In one case the significance level is under 0.05 (since $2 \times 0.0214 = 0.0428$); in the second case it substantially exceeds 0.05 ($2 \times 0.0494 \simeq 0.1$).

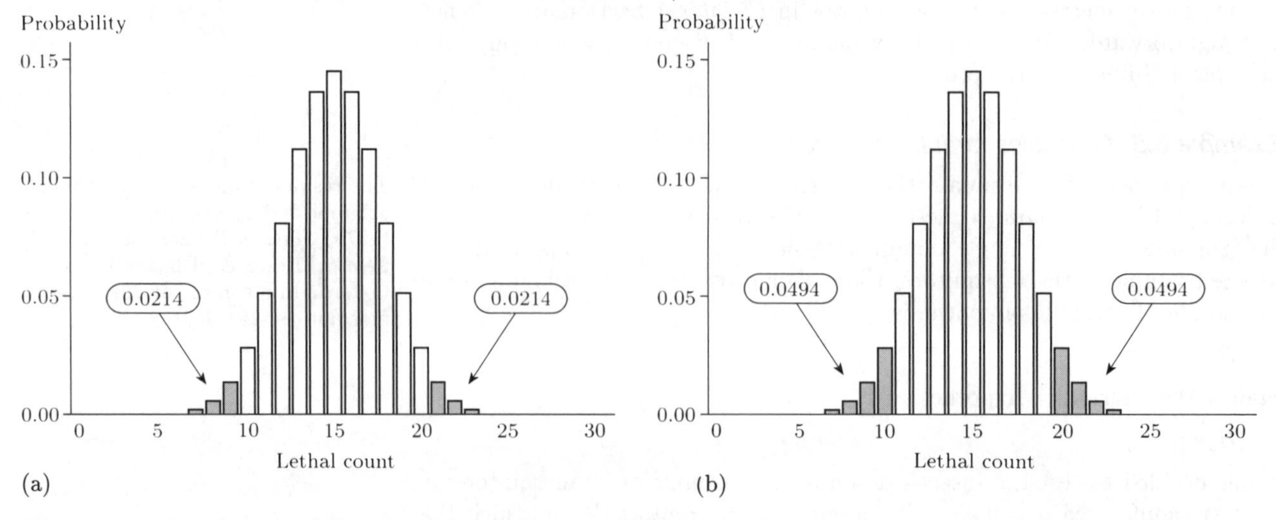

Figure 8.7 Possible rejection regions of approximate size 0.05

In this case a decision to reject the null hypothesis if the observed value r of R is less than 10 ($r \leq 9$) or greater than 20 ($r \geq 21$) gives a significance level of about 4%, close to the 5% intended.

This sort of problem will also arise with one-sided tests: it is a consequence of the nature of the probability mass function for a discrete random variable. However, it is not a problem you should become too concerned about: the fixed-level approach to testing has the remarkable and essentially unreasonable preoccupation with 'tidy' significance levels like 10%, 5% and 1%. If it turns out to be necessary to fix the level at 12% or 4% or 0.08% then (as long as there is a clear statement of what has occurred) the test simply proceeds according to the approximate level set. ■

Exercise 8.7

Determine a rejection region for a two-sided test of the null hypothesis $H_0 : \mu = 3.0$ for a Poisson mean, so that the level of the test is as close as possible to $\alpha = 0.10$. Assume

(a) a sample of size 1;

(b) a sample of size 5;

(c) a sample of size 10

is drawn, and in each case be clear about your test statistic and its distribution under the null hypothesis.

8.2.5 A few comments

Interpreting the significance level

You have seen that the rejection region for a fixed-level hypothesis test is defined by identifying those values of the test statistic that under the null hypothesis would be most extreme (according to whether the test was two-sided or one-sided). It constitutes a summary of those results that would appear to be so inconsistent with the null hypothesis that it is rejected. But, of course, from the very definition of the rejection region, you can see that it is calculated from the null distribution which assumes the null hypothesis H_0 to be true; so what we have is

$$\text{The significance level} = \alpha = P(\text{rejecting } H_0 \text{ when } H_0 \text{ is true}).$$

The act of rejecting H_0 when H_0 is true is called a **Type I error**, and it is conventional to take acceptable values for its probability as 1%, 5%, or 10%, and not usually more. A Type I error is an error which, in the nature of things, the designer of the test will not know has been committed: but notice that its probability is entirely within the designer's control.

The power of a test

Of course, there is another sort of error, and this is where the designer of the test accepts the null hypothesis H_0 even though it is false. This is an equally unfortunate outcome of the testing scenario: it is called a **Type II error**. Moreover, having fixed the level of the test and therefore having defined the

Notice the use of the word 'accepts' here, rather than 'fails to reject'. It simply makes for less awkward language.

rejection region, it is one over which the user has no direct control. Indeed, the smaller the rejection region, the less scope there is for a Type I error, but the greater the likelihood that a Type II error could be made.

What most aptly measures the usefulness of a hypothesis test is the probability

$$P(\text{rejecting } H_0 \text{ when } H_0 \text{ is false}).$$

This is simply one minus the probability of a Type II error (in other words, the probability of avoiding that error) and it is called the **power** of the test. Earlier on it was mentioned that in order to make a test as powerful as possible it was important to select an appropriate and informative test statistic. The mathematics of power, or its arithmetic at least, can become rather involved and, without some idea of the manner in which departures from H_0 might be manifested, it is difficult to make useful remarks about it.

However, as we have repeatedly discovered in other analytic pursuits, one way of improving the power of a test and simultaneously constraining the probability of a Type I error (that is, keeping down the significance level) is to increase the sample size.

Composite hypotheses

The statement of the null hypothesis in the form

$$H_0 : \theta = \theta_0$$

is not always easy or natural. Sometimes one's intentions would be better expressed by proposing a list or *range* of values for a parameter. The dog food example at the beginning of this chapter is a good illustration of this: if it appeared that more than the claimed 80% of dogs preferred Pupkins to any other dog food, one would not wish to dispute the manufacturer's claim which, in its essentials, states simply that a lot of dogs like Pupkins. An alternative statement of null and alternative hypotheses which more accurately reflects the true state of affairs under test is

$$H_0 : p \geq 0.8, \qquad H_1 : p < 0.8.$$

A hypothesis of the form $\theta = \theta_0$, isolating a particular value in the set of possible parameter values, is called a **simple hypothesis**. A hypothesis of the form $\theta \neq \theta_0$ or $\theta < \theta_0$ or $\theta > \theta_0$ is called a **composite hypothesis**. Here, a list or range of parameter values is hypothesized.

A typical representation of composite null and alternative hypotheses could therefore be stated as follows:

$$H_0 : \theta \geq \theta_0, \qquad H_1 : \theta < \theta_0.$$

Again, the theoretical consequences of a composite null hypothesis for the power of a test, the identification of a suitable test statistic and even the meaning of the phrase 'null distribution' are not immediately obvious, and mathematically things are far from simple.

However, it turns out in practice that the probability of making a Type I error is at its greatest when the actual value of θ is at the boundary between H_0 and H_1 (that is, when θ is equal to θ_0, in the example quoted). If a test is designed according to this worst-case scenario, then it cannot be criticized on the grounds that it appears a better test than it is. If it makes sense to write a null hypothesis H_0 as a composite hypothesis, then this should be done: calculations for the rejection region should be based as before on a null distribution whose parameter is located at the boundary between H_0 and H_1.

8.3 Significance testing

In this section we shall look at a third approach to the problem of testing a claim: it is called **significance testing** and it has become a common method for assessing a hypothesis. This section starts with a brief explanation of the approach, and a description of what the technique involves; but shortly we shall pause for a while and discuss why there should be so many approaches to what seems a straightforward problem to describe.

Both the approaches described so far have involved the setting up of a null distribution, the probability distribution of an appropriate test statistic if the null hypothesis H_0 were true. For example, in testing a hypothesized Bernoulli probability θ ($H_0 : \theta = \theta_0$) we might set up a sequence of n Bernoulli trials and count the total number r of successes. The test statistic is the random variable R; the null distribution of R is

$$R \sim B(n, \theta_0).$$

In the confidence level approach based on the techniques of *Chapter 7*, we use the observed value r of R to construct a $100(1-\alpha)\%$ confidence interval for θ; then, depending on whether or not that interval (θ_-, θ_+) contains the hypothesized value θ_0, we either reject the null hypothesis $H_0 : \theta = \theta_0$ (or not) at significance level α (or $100\alpha\%$) in favour of the alternative hypothesis $H_1 : \theta \neq \theta_0$.

In a fixed-level test, we identify lower and upper quantiles $q_{\alpha/2}$ and $q_{1-\alpha/2}$ of the null distribution $B(n, \theta_0)$; then, depending on whether or not the observed value r of R is in the defined rejection region, the null hypothesis is or is not rejected at significance level α.

The test might be one-sided, in which case the rejection region will be determined by either the lower quantile q_α or the upper quantile $q_{1-\alpha}$ of the null distribution, depending on the direction of the test. In the case of a discrete null distribution, the significance level α may be only approximately attained.

You should notice that each of these two tests permits a decision rule for the user of the test to follow.

Although in some respects the two approaches are quite different, you should also note that for a given significance level α and assuming that the same data are used in both cases, either strategy will always lead to the same decision being taken in respect of a particular hypothesis under test. This is an easy finding to illustrate, though slightly less easy to prove—and in the case of discrete data, one needs to be clear about the value of α—and we shall not spend more time on this. However, it is an important equivalence of which you should be aware.

The third approach that is described in this course also requires statement of a null hypothesis and calculation of a test statistic, and the collection of data in order to test that hypothesis. What happens next is what makes this approach different from the first two. The test results not in a stated decision (for example, 'reject H_0') but in a *number* called the **significance probability**, denoted SP. Broadly speaking, this number describes the extent to which the data support the null hypothesis: if the statistical experiment were to be repeated on many subsequent occasions (collect some data and evaluate the test statistic), and if the null hypothesis were true, the SP represents the proportion of future experiments that would offer less support for the null hypothesis than the experiment that was, in fact, performed. The higher the significance probability, therefore, the more the data support the null hypothesis. Subsection 8.3.1 shows an example of the test in practice.

8.3.1 Performing a significance test: testing a Bernoulli probability

Many examples in genetics involve testing the value of a Bernoulli probability p, for it is a field where there is much interest in the fraction of a population displaying a particular attribute. Often sample sizes are relatively small, and exact distribution theory is appropriate.

***Example 8.6** The colour of seed cotyledons in the edible pea*

Mendel observed that seed cotyledons in the edible pea may be either yellow or green and that the peas themselves appear either yellow or green. (These were the subject of his second experiment in the first series: he observed 6022 yellow peas and 2001 green peas in a harvest of 8023 peas bred in particular circumstances, offering support for his theory that on genetic principles the proportion of yellow peas under such circumstances should be $\frac{3}{4}$.)

In a smaller experiment, 12 yellow peas were found in a harvest of 20 peas. It was required to use these data in a **significance test** of the hypothesis

$$H_0 : p = \tfrac{3}{4}.$$

The obvious test statistic to use in this context is the number of yellow peas (N, say), which in repeated experiments of the same size would follow a binomial distribution $N \sim B(20, p)$. Under the hypothesis $H_0 : p = \frac{3}{4}$, the null distribution of N is binomial $B(20, \frac{3}{4})$. The number observed was $n = 12$.

A diagram of the null distribution is shown in Figure 8.8; the shaded regions in the diagram show the possible counts which are themselves no more likely than the count observed (that is, all those observations on the random variable N such that $p_N(n) \leq p_N(12)$.

Perhaps a phrase more wieldy—though less precise—than that used is to refer to those counts 'at least as extreme as' the observed count.

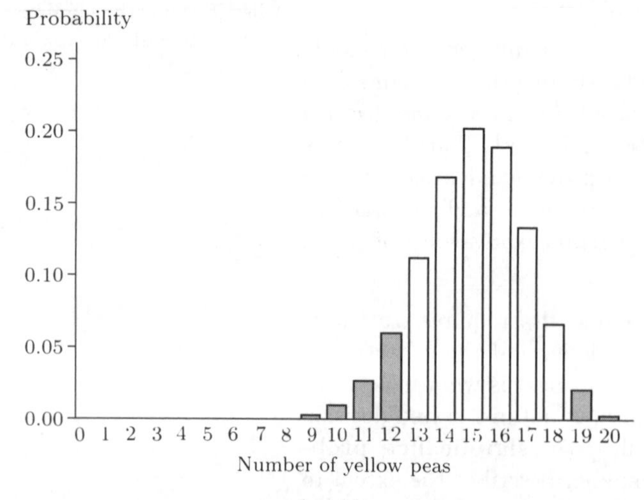

Figure 8.8 The null distribution $B(20, \frac{3}{4})$ and counts no more likely than that observed, $n = 12$

Not shown on the diagram but given in Table 8.7 are the corresponding probabilities (to four decimal places) for the binomial distribution $B(20, \frac{3}{4})$; you can see from the table that all the counts $n = 0, 1, 2, \ldots, 11$ and $n = 19, 20$ are less likely than the observed count $n = 12$, which is also included in the shaded region. This table was used in order to draw the diagram in Figure 8.8.

Table 8.7 The binomial probability distribution $B(20, \frac{3}{4})$

x	$p(x)$
0	0.0000
1	0.0000
2	0.0000
3	0.0000
4	0.0000
5	0.0000
6	0.0000
7	0.0002
8	0.0008
9	0.0030
10	0.0099
11	0.0271
12	0.0609
13	0.1124
14	0.1686
15	0.2023
16	0.1897
17	0.1339
18	0.0669
19	0.0211
20	0.0032

The significance probability for the test is given by the sum of the two shaded tail areas, and is (again, accurate to four decimal places)

$$SP = P(N \leq 12) + P(N \geq 19) = 0.1018 + 0.0243 = 0.1261.$$

Of these extremes, small values of n (including the observed value $n = 12$) would suggest that the underlying value of p is in fact less than the hypothesized $\frac{3}{4}$; those at the other extreme of the null distribution would suggest that the underlying value of p exceeds $\frac{3}{4}$. It is common to conclude a significance test with a statement such as

$$SP(\text{obtained direction}) = 0.1018$$

$$SP(\text{opposite direction}) = 0.0243$$

$$SP(\text{total}) = 0.1261,$$

and an interpretation of the significance probability. In this case there is little evidence that the underlying value of p is different from the hypothesized value $p = \frac{3}{4}$.

This completes the significance test. ■

The procedure for a significance test may be summarized as follows.

Significance testing

In a significance test:

1 determine the null hypothesis H_0;

2 decide what data to collect that will be informative for the test;

3 determine a suitable test statistic and the null distribution of the test statistic (that is, the distribution of the test statistic when H_0 is true);

4 collect your data and evaluate the observed value of the test statistic for the sample;

5 identify all other values of the test statistic that under the null hypothesis are no more likely than the value that was observed;

6 these 'extreme values' will usually fall into two classes, each suggesting some departure from the null hypothesis. The class containing the observed experimental outcome contributes to the SP in the obtained direction (that is, suggestive of one type of departure from the null hypothesis). The other class contributes to the SP in the opposite direction;

7 interpret the SP.

When the null distribution of the test statistic is multimodal, the approach described here falters at Step 6, because there may be more than two groups of 'unlikely' values. There is still no universal agreement about the 'best' approach to test a hypothesis (see Subsection 8.3.4); this approach will usually be adequate for our purposes.

Figure 8.9 illustrates these 'extreme values'.

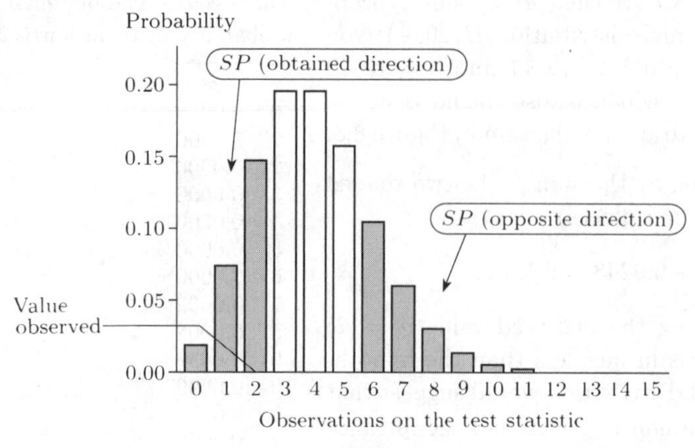

Figure 8.9 Evaluating a significance probability

Notice that there is no requirement to conclude a significance test with a decision on whether or not to reject the null hypothesis. This would only make sense if some alternative hypothesis had been identified. The significance probability is a measure of the extent to which the data support the null hypothesis, and the test ends here. However, it is easy to extend the test to incorporate a decision rule based on whether the SP exceeds or does not exceed some predetermined value. The 'obtained direction' offers a clue to an appropriate alternative.

Exercise 8.8

The coat colour of grey rabbits depends on genetic characteristics inherited from generation to generation. There are five possible colour combinations: normal grey, chinchilla (a kind of silver grey), light grey, Himalayan (white with black extremities) and albino. In one large population under study, genetic theory forecast that different coloured grey rabbits should occur in the respective relative frequencies

$$\frac{3}{4} : \frac{1}{16} : \frac{1}{8} : \frac{21}{400} : \frac{1}{100}.$$

An efficient test of the theory would involve matching the observed frequencies of all colour combinations in a large sample with the expected frequencies if the theory were a valid one (and this topic is covered in *Chapter 9*). In one test a small random sample of 18 adult rabbits from the population was taken, and the number of light grey rabbits was counted.

(a) What is the probability distribution of the number of light grey rabbits in the sample, on the assumption that the forecast frequencies are correct?

(b) In fact there were four light-greys. What is the evidence that the theory is faulty?

8.3.2 Testing a Poisson mean

When testing hypotheses about the value of a Poisson mean, the same sort of distributional considerations apply here as became evident in *Chapter 7* when calculating confidence intervals for a Poisson mean. If in testing the null hypothesis $H_0 : \mu = \mu_0$ a sample of size n is collected, then the null distribution of the sample total T is Poisson with mean $n\mu_0$. The significance probability SP can be calculated as though the observed sample total t was a single observation on the random variable $T \sim \text{Poisson}(n\mu_0)$.

***Example 8.7** Breakdowns*

In large organizations, central facilities such as printers and photocopiers are often conveniently located in order to provide access to large numbers of personnel. The breakdown incidence is usually monitored and some sort of record is kept of machines' reliability. One printer had an average breakdown rate of 3 times a week. It was made less accessible by being moved up one floor in the building in which it was located. Over the next six weeks the numbers of breakdowns recorded weekly were $3, 4, 2, 1, 1, 2$.

We want to perform a significance test of the hypothesis that the breakdown rate has remained unchanged.

Assuming a Poisson model for the number of breakdowns per week, and writing

$$H_0 : \mu = 3$$

and using as our test statistic the total number T of breakdowns over the six-week period after the move, then the null distribution of the random variable T is $T \sim \text{Poisson}(18)$. The observed value t of T is

$$t = 3 + 4 + 2 + 1 + 1 + 2 = 13.$$

For the Poisson distribution with mean 18, key parts of the probability mass function are as follows.

x	0	1	2	...	12	13	14	...	21	22	23	24	...
$p(x)$	0.000	0.000	0.000	...	0.037	0.051	0.065	...	0.068	0.056	0.044	0.033	...

The calculated SP is

You will need your computer to calculate these probabilities.

$$SP(\text{obtained direction}) = P(T \leq 13) = 0.143$$

$$SP(\text{opposite direction}) = P(T \geq 23) = 0.145$$

$$SP(\text{total}) = 0.288.$$

Again, the SP is not remarkably small. In particular, there is scant evidence that the move has reduced the breakdown rate, as might have been intended. ■

You should be aware that some statisticians, and some statistical software, calculate the SP for an exact test such as this in a slightly different way: the difference lies in the way that the SP in the opposite direction is dealt with. The difference hardly ever has practical importance.

You can see that the procedure can become quite intricate, for it involves scanning the Poisson probability distribution to identify those counts less likely than the count observed. Although this is an easy task to describe, it can take a little time. Explore the facilities available on your computer in attempting Exercise 8.9.

Exercise 8.9

A total of 33 insect traps were set out across sand dunes and the numbers of different insects caught in a fixed time were counted. Table 8.8 gives the number of traps containing various numbers of insects of the taxa *Staphylinoidea*.

Gilchrist, W. (1984) *Statistical Modelling*, Wiley, Chichester, p. 132. The original purpose of the experiment was to test the quality of the fit of a Poisson model to the data: in this exercise, the Poisson model is assumed to be adequate.

Table 8.8 *Staphylinoidea* in 33 traps

Count	0	1	2	3	4	5	6	≥ 7
Frequency	10	9	5	5	1	2	1	0

Assuming a Poisson model for the variation in the counts of trapped insects, perform a significance test of the hypothesis $H_0 : \mu = 1$, and state your conclusions. How different from 1 is the sample mean catch?

8.3.3 Large-sample approximations

For significance tests on the Bernoulli parameter p, the Poisson mean μ and the exponential mean μ, the central limit theorem can be applied if the sample size is large, and approximate normal distribution theory may be used.

Significance tests for the Bernoulli parameter

In a significance test of the null hypothesis $H_0 : p = p_0$ for a Bernoulli parameter p, suppose that n trials are performed and the number of successes, X, is counted. The exact null distribution of X is binomial $B(n, p_0)$. A significance test of H_0 may be based on the approximating normal distribution

$$X \approx N(np_0, np_0q_0)$$

where $q_0 = 1 - p_0$, for n 'large enough' (say, so that $np_0 > 5$ and $nq_0 > 5$).

Significance tests for the Poisson mean

In a significance test of the null hypothesis $H_0 : \mu = \mu_0$ for a Poisson mean, suppose that a sample of size n is collected. The null distribution of T, the sample total, is Poisson$(n\mu_0)$. A significance test of H_0 may be based on the approximating normal distribution

$$T \approx N(n\mu_0, n\mu_0)$$

as long as $n\mu_0$ is greater than about 30.

Significance tests for the exponential mean

In a significance test of the null hypothesis $H_0 : \mu = \mu_0$ for an exponential mean, the sample collected is of size n. The null distribution of T, the sample total, is a member of the gamma family: the gamma distribution is not particularly tractable. As long as the sample size is large enough (say, more than about 30), a significance test of H_0 may be based on the approximating normal distribution

$$T \approx N\left(n\mu_0, n\mu_0^2\right).$$

8.3.4 Neyman, Pearson and Fisher

The twentieth century has been enlivened by a number of philosophical disputes among statistical practitioners. One of the more hotly argued is that between R.A. Fisher and the duo made up of Egon Pearson (1895–1980) and Jerzy Neyman (1894–1981).

Figure 8.10 R.A. Fisher

Fisher has already been mentioned during the course of this chapter. At the age of 22 on graduation from Cambridge University, Fisher worked for three years as a statistician in London and then until 1919 as a schoolteacher (and not a good one, according to contemporary sources). From 1919 until 1933 he worked at Rothamsted Experimental Station, the agricultural research establishment near Harpenden in Hertfordshire, England; in 1925 he published the famous text entitled *Statistical Methods for Research Workers.* After leaving Rothamsted he was Professor of Eugenics at University College London until 1943, after which he was appointed Professor of Genetics at Cambridge. His papers on theoretical statistics form the foundation of much of modern statistics. Many of his methods are used world-wide to this day, including the analysis of variance that will be mentioned briefly in Subsection 8.4.2.

Figure 8.11 Egon Pearson

Egon Pearson was the son of Karl Pearson (1857–1937), arguably the founder of modern statistics. Egon worked in the Department of Applied Statistics at University College London (headed by his father) from 1921. In 1933, on Karl's retirement, he took over the chair of the department, which he headed until his own retirement in 1960. During the key period of the dispute between Fisher and Neyman and Pearson, their departments occupied different floors of the same building at University College.

Jerzy Neyman was born in Bendery near the border between Russia and Romania. He was educated at the University of Kharkov in the Ukraine and lectured there until going to live in Poland in 1921. He was a lecturer at the University of Warsaw when in 1925 he visited London and met Egon Pearson. The pair, much of an age, struck up an immediate and close personal and professional relationship.

Figure 8.12 Jerzy Neyman

In 1933 Neyman and Pearson published a paper 'On the Problem of the Most Efficient Tests of Statistical Hypotheses' in the *Philosophical Transactions of The Royal Society, Series A*, **231**, 289–337. Theirs is basically the fixed-level approach of Section 8.2. Essentially their work was generated by concern that there should be some criterion other than intuition to provide a guide to what test statistic to utilize in performing a hypothesis test, and this in turn implied the strict requirement for an alternative hypothesis.

In many cases, a statistical test is used more or less to assess the data, and not (necessarily) to reach any firm conclusion. This is the idea behind a significance test, and seems to have been the attitude of Fisher, who was thinking of research situations and not of cases where the background of the problem requires a clear decision. Fisher's approach corresponds in most respects to the approach described in this section; however, he would not have agreed with everything you have read here.

Fisher's approach requires three components: a null distribution for the test statistic, an ordering of all possible observations of the test statistic according to their degree of support for the null hypothesis and, finally, a measure of deviation from the null hypothesis as the chance that anything even more extreme was observed.

Within repeated experiments, the idea of outcomes more discordant with a null hypothesis than others is fairly clear. However, with different experiments or when using different test statistics, it is not at all clear whether a significance probability as an absolute measure of accord, one that can be compared across experiments, is a useful notion. This is an important criticism of the approach.

The approach of Neyman and Pearson offers an alternative, but some key concepts were always rejected by Fisher. Among other things, he considered the use of a pre-specified alternative hypothesis to be inappropriate for scientific investigations. He maintained that the fixed-level approach was that of mere mathematicians, without experience in the natural sciences. As well as subtle and irreconcilable philosophical and theoretical incompatibilities between the two approaches, there is no doubt that the controversy was fuelled by personal antipathies as well. Peters (1987) writes: 'Fisher was a fighter rather than a modest and charitable academic.'

Peters, W.S. (1987) *Counting for Something—Statistical Principles and Personalities*, Springer-Verlag, New York.

There is an interesting postscript to all this, described by the statistician Florence David, who visited University College as a tutor: 'Most of the time I was babysitting for Neyman, explaining to the students what the hell he was up to ... I saw the lot of them. Went flyfishing with Gosset. A nice man. Went to Fisher's seminars with Cochran and that gang. Endured Karl Pearson. Spent three years with Neyman. ... They were all jealous of one another, afraid someone would get ahead. Gosset didn't have a jealous bone in his body.' Gosset's modesty and diffidence were renowned. In a letter to Fisher he spoke of his difficulties with calculating the tables of quantiles for his t-distribution. He (Gosset) had left an updated version of the table with Karl Pearson, editor of the journal *Biometrika*, in which an earlier version had been published: '... when I came back on my way to Dublin I found that he [Pearson] agreed with me and that the *new* table was wrong. On further investigation both tables were found to be perfectly rotten. All 0.1 and 0.2 wrong in the fourth place, mostly it is true by 0.0001 only ... The fact is that I was even more ignorant when I made the first table than I am now ... Anyhow the old man is just about fed up with me as a computer and wouldn't even let me correct my own table. I don't blame him either. ... Whether he will have anything to do with our table I don't know ... It has been rather a miserable fortnight finding out what an ass I made of myself and from the point of view of the new table, wholly wasted. However, I begin work again tomorrow.'

Reid, C. (1982) *Neyman from Life*, Springer-Verlag, New York, page 53.

Fisher Box, J. (1981) Gosset, Fisher and the t-distribution. *American Statistician*, **35**, 61–66. Joan Fisher Box is one of R.A. Fisher's daughters. In 1978 she published a biography of her distinguished father, entitled *R.A. Fisher: The Life of a Scientist.* John Wiley & Sons, New York.

8.4 Comparing the means of two normal populations

Up to now we have considered tests of hypotheses that a model parameter takes a specified value. Such hypotheses are not uncommon in genetics and in manufacturing contexts, and in quality control. However, a more common testing situation is where independent samples are drawn from two different populations in order to test some hypothesis about differences in population characteristics for some measured attribute. This section and Section 8.5 are devoted to this topic.

Figure 8.13 W.S. Gosset ('Student')

There have been many examples of this sort of sampling context in the course. In *Chapter 1*, Example 1.3, the birth weights (see Table 1.4) in kilograms of 50 infants displaying the symptoms of severe idiopathic respiratory distress syndrome were listed. In more than half the cases, the child unfortunately died. It seems possible that there were significant differences in birth weight between those children who died and those who survived. If this is so, then birth weight could be used as an indicator for children needing very special care and attention. A formal test might suggest confirmation of this apparent difference, in which case preparations to offer that care and attention could be made.

In *Chapter 2*, Example 2.7, it seems possible (again, without performing any sort of formal test) that, given some appropriate stimulus, unpleasant memories (see Figure 2.10) are more difficult to recall than pleasant ones—or, at least, their retrieval seems to take longer. To psychologists interested in memory retention and retrieval, this finding (if it is true) is a significant one.

Example 2.20 was about measurements on a liver enzyme (ornithine carbonyltransferase) for two different sets of individuals, 57 patients suffering from acute viral hepatitis and 40 from aggressive chronic hepatitis. Again, there seem to be differences (see Figure 2.23). The purpose of the investigation was to determine whether it was possible to distinguish between patient groups on the basis of this measurement: this would be a very useful discriminant aid. A formal test will help to decide whether such an approach is feasible.

In Exercise 2.4 an experiment to do with gender differences was described, and the question was raised whether the observed proportion of 71 out of 100 was significantly different from the observed proportion of 89 out of 105. Again, a formal test can help here.

The most general question that could be asked is this: is the pattern of variation in the measured attribute the same in one population as it is in the other? In other words (denoting the respective distribution functions $F_1(\cdot)$ and $F_2(\cdot)$), we might suggest the hypothesis

$$H_0 : F_1(x) = F_2(x) \quad \text{for all } x.$$

However, it may be that our main interest resides in the average measure for the two populations and therefore in testing the hypotheses (writing μ_1 and μ_2 for the two population means and m_1 and m_2 for the population medians)

$$H_0 : \mu_1 = \mu_2,$$

or perhaps

$$H_0 : m_1 = m_2.$$

Other tests may be designed to compare other population moments or other population quantiles. In this section a test for comparing the means of two normal populations is described.

8.4.1 The two-sample t-test

The **two-sample *t*-test** is one of the most useful tests available to you. Under certain assumptions, it permits a test of the null hypothesis

$$H_0 : \mu_1 = \mu_2$$

for the means μ_1 and μ_2 of two distinct populations.

These assumptions are that the variation in the first population may be modelled adequately by a normal distribution with mean μ_1 and variance σ^2, and that the variation in the second population may be modelled by a normal distribution with mean μ_2 and variance σ^2. That is,

$$X_1 \sim N(\mu_1, \sigma^2), \qquad X_2 \sim N(\mu_2, \sigma^2).$$

The assumption of normality is one that, as you have seen, is often well approximated in practice for many different measured attributes in many different contexts—in any case, it is an assumption that is easy to check in an informal way using a histogram.

More formal tests for normality are described in *Chapter 9*.

However, notice the second assumption that the variance in both populations is the same. It will almost invariably be the case that the sample variances s_1^2 and s_2^2 for the two samples will differ, and thus the question is raised of how pronounced this difference might be before it suggests that the assumption of equal variances $\sigma_1^2 = \sigma_2^2$ is a faulty assumption. To put it another way, it appears that a t-test for the equality of two normal means ought itself always to be prefaced by a formal test for the equality of the two variances!

This is an approach that is sometimes followed. In this course we will always informally check the variances before embarking on a t-test. However (depending on the sample size) if one sample variance is larger than the other by a factor of less than about 3, it will be assumed that the assumption of equal variances for the t-test is not adrift.

Current practice suggests that a factor rather higher than this is not badly damaging to the conclusions of the t-test, particularly when the two sample sizes are not too different.

So, assuming that the twin assumptions of normality and equal variances are satisfied, the two-sample t-test proceeds as follows.

Having decided on a background probability model, the next thing to determine is a test statistic relevant to the null hypothesis $H_0 : \mu_1 = \mu_2$. The respective estimators for the two population means are the sample means $\overline{X}_1$ and $\overline{X}_2$: a useful statistic indicative of the difference $\mu_1 - \mu_2$ is surely the difference between the sample means, $\overline{X}_1 - \overline{X}_2$. In fact, this difference is not only useful, but *powerful* in a technical sense: it can be shown that this is the best statistic to choose when the assumptions of normality and equal variances are satisfied.

Denoting by n_1 and n_2 the two sample sizes, then (as a consequence of normal distribution theory) we have the results

$$\overline{X}_1 \sim N\left(\mu_1, \frac{\sigma^2}{n_1}\right), \qquad \overline{X}_2 \sim N\left(\mu_2, \frac{\sigma^2}{n_2}\right). \tag{8.5}$$

See (4.8) and (4.10).

Assuming the two samples to be independent of one another, then it follows from this that the difference $\overline{X}_1 - \overline{X}_2$ has a normal distribution

$$\overline{X}_1 - \overline{X}_2 \sim N\left(\mu_1 - \mu_2, \frac{\sigma^2}{n_1} + \frac{\sigma^2}{n_2}\right). \tag{8.6}$$

See (4.9).

Now we need to eliminate the nuisance parameter σ^2 from our test. Lacking information about the value of the common variance σ^2, what happened previously in this sort of situation was that the parameter σ^2 was replaced by its estimator. Since both the first sample variance S_1^2 and the second sample variance S_2^2 are candidates, the question is: which estimator do we use? Is there some combination of them that would be a better estimator than either alone?

We already know (see *Chapter 6*, Exercise 6.26) that the sample variance from a normal random sample has first two moments

$$E(S_1^2) = \sigma^2, \qquad V(S_1^2) = \frac{2\sigma^4}{n_1 - 1};$$

$$E(S_2^2) = \sigma^2, \qquad V(S_2^2) = \frac{2\sigma^4}{n_2 - 1}.$$

Both estimators are unbiased for σ^2; of the two, the better one would be the one with the smaller variance, that is, the one based on the larger sample. But it would make sense, intuitively, to use information from both samples, and it turns out that an unbiased estimator for the unknown parameter σ^2 with the smallest possible variance is the estimator

The estimator S_p^2 is unbiased for σ^2, and has variance

$$V(S_p^2) = \frac{2\sigma^4}{n_1 + n_2 - 2}.$$

$$S_p^2 = \frac{(n_1 - 1)S_1^2 + (n_2 - 1)S_2^2}{n_1 + n_2 - 2}, \qquad (8.7)$$

and this is the estimator we shall use. Because it involves a combination of the two estimators S_1^2 and S_2^2 it is called the **pooled estimator for the common variance**. It follows from (8.6) (not quite directly, but the details are not important) that the quantity

$$\frac{(\overline{X}_1 - \overline{X}_2) - (\mu_1 - \mu_2)}{S_p\sqrt{\dfrac{1}{n_1} + \dfrac{1}{n_2}}}$$

has a t-distribution with $n_1 + n_2 - 2$ degrees of freedom. Under the null hypothesis $H_0 : \mu_1 = \mu_2$ the difference $\mu_1 - \mu_2$ in the numerator vanishes; an appropriate test statistic for the null hypothesis that two normal populations have the same mean has the null distribution

$$T = \frac{\overline{X}_1 - \overline{X}_2}{S_p\sqrt{\dfrac{1}{n_1} + \dfrac{1}{n_2}}} \sim t(n_1 + n_2 - 2). \qquad (8.8)$$

This may be used as the basis for a significance test. (Of course, it may also be used to develop confidence intervals for the difference between two population means or as the basis for a fixed-level test. It is not the intention of this chapter to take every possible approach to every exploration of a null hypothesis.)

Example 8.8 Infants with SIRDS

In *Chapter 1*, Example 1.3 a sample was described of infants all displaying severe idiopathic respiratory distress syndrome: the infants had been weighed at birth and their birth weights (in kg) recorded. It was also noted that 27 infants died (while 23 survived). A preliminary comparative boxplot (see Figure 1.23) suggests that there may be a significant difference in birth weights between those who survived and those who did not.

It is possible to explore this suggestion using the two-sample t-test. Neither sample is very large, but neither appears very skewed, and in both cases a histogram (see Figure 8.14) suggests that a normal model for the variation observed might be adequate.

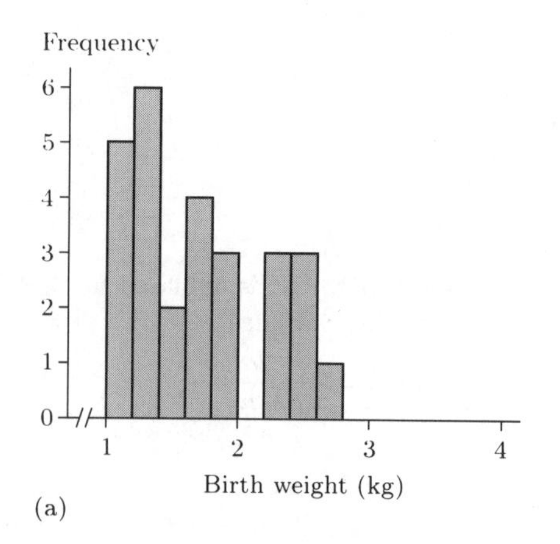

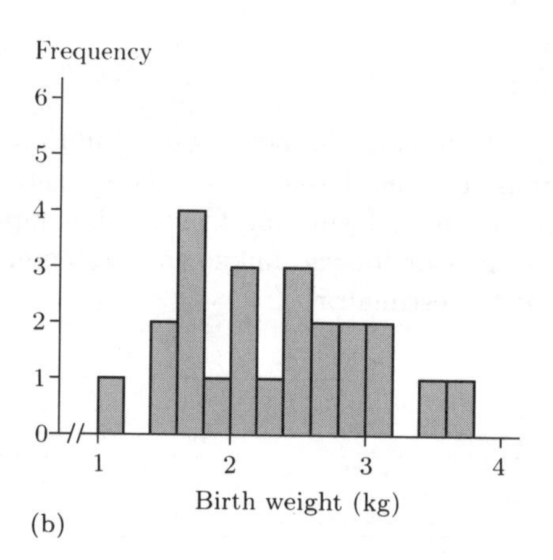

Figure 8.14 Histograms for (a) 27 infants who died, and (b) 23 infants who survived

Before formally embarking on the test, we ought to check the sample variances. Let us take as the first sample the birth weights of the 27 children who died, and as the second the birth weights of the 23 who survived. Then

$$s_1^2 = 0.268, \qquad s_2^2 = 0.442,$$

(to three decimal places). Without any formal criteria on which to base an assessment, it is difficult to say whether or not these estimates suggest different underlying variances; in fact, the larger of the two sample variances is less than twice the smaller; according to the rough guide that a ratio of up to about 3 is acceptable, this suggests that it is reasonable to embark on the t-test.

We also require the summary statistics

$$n_1 = 27, \qquad \overline{x}_1 = 1.692;$$
$$n_2 = 23, \qquad \overline{x}_2 = 2.307;$$

and it follows from this and (8.7) that the pooled estimate for the unknown variance σ^2 is

$$s_p^2 = \frac{(n_1-1)s_1^2 + (n_2-1)s_2^2}{n_1+n_2-2} = \frac{26 \times 0.268 + 22 \times 0.442}{27+23-2} = 0.348.$$

In these calculations, intermediate and final results are all shown accurate to three decimal places.

Finally, using (8.8), the observed value t of the test statistic T is

$$t = \frac{\overline{x}_1 - \overline{x}_2}{s_p\sqrt{\frac{1}{n_1}+\frac{1}{n_2}}} = \frac{1.692 - 2.307}{\sqrt{0.348} \times \sqrt{\frac{1}{27}+\frac{1}{23}}} = -3.68.$$

This needs to be compared against Student's t-distribution with $27 + 23 - 2 = 48$ degrees of freedom, that is, $t(48)$. Figure 8.15 illustrates the corresponding significance probabilities (obtained from a computer) which may be stated as

$$SP(\text{obtained direction}) = 0.0003$$
$$SP(\text{opposite direction}) = 0.0003$$
$$SP(\text{total}) = 0.0006.$$

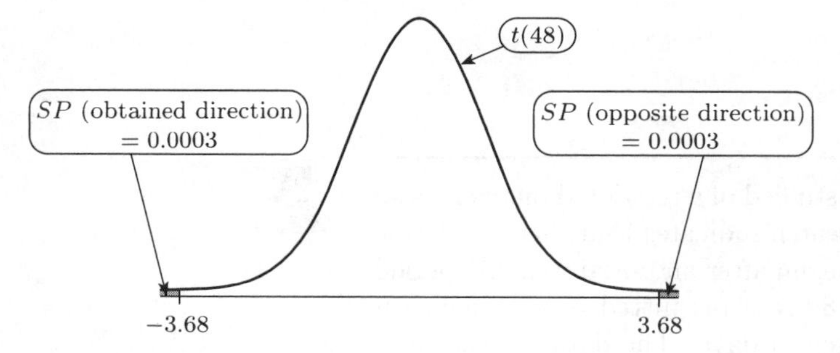

Figure 8.15 Calculating a SP

In fact, in this context, there already was the suspicion that the birth weights of the children who died were significantly lower than those of the children who survived. In a test of the null hypothesis of zero difference against a one-sided alternative, the obtained SP is 0.0003. This is very low; there is considerable evidence to reject the null hypothesis of zero difference in favour of the suggested alternative. ■

Example 8.9 Memories

In *Chapter 2*, Table 2.10 there are listed memory recall times (in seconds) for twenty pleasant memories and twenty unpleasant memories. A comparative boxplot was drawn to summarize the data in Figure 2.10. The data are very skewed. If the two sample variances are calculated they are found to be

$$s_1^2 = 2.0695, \qquad s_2^2 = 9.3833.$$

The ratio of larger to smaller is about 5. The opportunity to use a two-sample t-test seems doomed on both counts: neither population looks remotely normal, and the variances do not look similar. ■

Exercise 8.10

In *Chapter 3*, Example 3.1 data were considered on the maximum breadths (in mm) of 84 Etruscan skulls and 70 modern Italian skulls (see Table 3.1). The question of interest was whether there was a significant difference between the two distributions. If this may be reinterpreted as a difference between the mean maximum breadths for the two populations, then perhaps the two-sample t-test may be advanced to provide an answer.

(a) A comparative boxplot for the two samples is shown in Figure 3.1. Histograms for both data sets are drawn in Figure 3.2. Use these graphical representations of the data to comment on the assumption of normality underlying the two-sample t-test.

(b) Calculate the two sample variances and comment informally on the assumption of a common variance.

(c) If you consider it appropriate to do so, carry out a t-test of the null hypothesis that there is no difference between the mean maximum head breadth for Etruscans and for modern Italian males, against a two-sided alternative. Give your answer as a significance probability, and comment on your findings.

Here is a further exercise.

Exercise 8.11

The effect on the total lifespan of rats was studied of a restricted diet versus an *ad libitum* diet (that is, free eating). Research indicates that diet restriction might affect longevity. Treatments were begun after an initial weaning period on 106 rats given a restricted diet, and 89 rats permitted to eat whenever they wished to do so. Lifespan is measured in days. The data are shown in Table 8.9.

Table 8.9 Lifespans of rats (days)

Berger, R.L., Boos, D.D. and Guess, F.M. (1988) Tests and confidence sets for comparing two mean residual life functions. *Biometrics*, **44**, 103–115.

106 rats given the restricted diet									
105	193	211	236	302	363	389	390	391	403
530	604	605	630	716	718	727	731	749	769
770	789	804	810	811	833	868	871	875	893
897	901	906	907	919	923	931	940	957	958
961	962	974	979	982	1001	1008	1010	1011	1012
1014	1017	1032	1039	1045	1046	1047	1057	1063	1070
1073	1076	1085	1090	1094	1099	1107	1119	1120	1128
1129	1131	1133	1136	1138	1144	1149	1160	1166	1170
1173	1181	1183	1188	1190	1203	1206	1209	1218	1220
1221	1228	1230	1231	1233	1239	1244	1258	1268	1294
1316	1327	1328	1369	1393	1435				
89 rats given the *ad libitum* diet									
89	104	387	465	479	494	496	514	532	536
545	547	548	582	606	609	619	620	621	630
635	639	648	652	653	654	660	665	667	668
670	675	677	678	678	681	684	688	694	695
697	698	702	704	710	711	712	715	716	717
720	721	730	731	732	733	735	736	738	739
741	743	746	749	751	753	764	765	768	770
773	777	779	780	788	791	794	796	799	801
806	807	815	836	838	850	859	894	963	

Take the opportunity to explore the facilities of your computer in testing whether there are differences in the mean lifespan for the two dietary regimes.

8.4.2 Postscript: comparing more than two means

Frequently in statistics there are three populations or more from which samples have been drawn. Then the question arises of how to make comparisons between the samples, and how to draw valid conclusions about differences (if any) between the populations. In *Chapter 1*, Table 1.16 gives the amounts of nitrogen-bound bovine serum albumen (BSA) used to treat three groups of diabetic mice. A comparative boxplot provides an informal test for differences between the three groups; but how do we conduct a formal test?

In *Chapter 4* a nutritional study was described (Example 4.9) in which 45 chicks were randomly allocated to four groups given different diets; after three weeks the chicks were weighed in order to assess the different effects (if any) of the diets. The data are given in Table 4.5. It would be interesting to know how to analyse these data in order to exhibit any significant differences between the groups.

Here is another example.

Example 8.10 Silver content of Byzantine coins

The silver content (% Ag) of a number of Byzantine coins discovered in Cyprus was determined. Nine of the coins came from the first coinage of the reign of King Manuel I, Comnenus (1143–80); there were seven from the second coinage minted several years later and four from the third coinage (later still); another seven were from the fourth coinage. The question is: were there differences in the silver content of coins minted early and late in Manuel's reign? The data are given in Table 8.10.

Hendy, M.F. and Charles, J.A. (1970) The production techniques, silver content and circulation history of the twelfth-century Byzantine Trachy. *Archaeometry*, **12**, 13–21.

Table 8.10 Silver content (% Ag) of coins

1st	2nd	3rd	4th
5.9	6.9	4.9	5.3
6.8	9.0	5.5	5.6
6.4	6.6	4.6	5.5
7.0	8.1	4.5	5.1
6.6	9.3		6.2
7.7	9.2		5.8
7.2	8.6		5.8
6.9			
6.2			

What is of most interest here is whether there is any significant difference in the silver content of the coins with passing time. (There is a suspicion that the content was steadily reduced: this could be tested according to some appropriate one-sided test.) ■

It is important to remember that in this context one should not run individual t-tests on each of all the possible pairs of groups selected from a data collection—the tests would not be independent. The appropriate methodology to adopt is called *analysis of variance*. This is a technique invented by R.A. Fisher. The technique is mentioned again, briefly, in *Chapter 14*.

8.5 Other comparisons

8.5.1 Comparing two binomial probabilities

One testing context that arises frequently in practice is where the proportion of individuals possessing an attribute is observed in samples from two different populations. In most cases the two observed proportions will be different. Assuming the underlying proportion in the first population to be p_1 and in the second to be p_2, then the hypothesis to be tested is

$$H_0 : p_1 = p_2.$$

Notice here that what is *not* under test is the actual value of the underlying proportion in either population: only that the proportion is the same in both populations. This fact is important later.

Suppose the sample drawn from the first population is of size n_1 and the sample drawn from the second population is of size n_2. In each case the number in the sample possessing the attribute of interest is a random variable; assuming independence within each sample, then (denoting the numbers in respective samples possessing the attribute by R_1 and R_2)

$$R_1 \sim B(n_1, p_1) \qquad \text{and} \qquad R_2 \sim B(n_2, p_2).$$

Under the null hypothesis $H_0 : p_1 = p_2$, the null distributions of R_1 and R_2 become

$$R_1 \sim B(n_1, p) \qquad \text{and} \qquad R_2 \sim B(n_2, p),$$

where the parameter p is unknown and irrelevant to the hypothesis that is being tested: in that sense it is a nuisance parameter. While it is there, it perturbs any further analysis based on the observed fractions r_1/n_1 and r_2/n_2.

There is a test which resolves this difficulty by so composing the random variables R_1 and R_2 and the numbers n_1 and n_2 that the parameter p vanishes in the algebra. It is known as **Fisher's exact test for the equality of two proportions**. The details of its development are somewhat complicated, and in any case the resulting arithmetic (the enumeration of possible cases and the assessment of which of them are more or less 'extreme' than that observed) is very drawn out—you really only need to know that the test exists and what it is called.

There is more on Fisher's exact test in *Chapter 11*.

Many statistics software packages include routines for running Fisher's exact test, and it is assumed that you have access to such a package.

Example 8.11 The sand fly data

In *Chapter 6*, Example 6.4, data were given on the proportions of male sand flies to be found in traps at two different altitudes. At low altitude there were 173 males observed in a total of 323 flies caught in a trap, an observed proportion of 53.6% males; at higher altitude there were 125 males observed in a total of 198 flies caught: 63.1%. This second proportion is higher: is it 'significantly' higher?

This can be set up as a significance test. It is very important to realize that the required inputs for Fisher's exact test are the four numbers r_1, n_1, r_2 and n_2 rather than simply the two observed proportions r_1/n_1 and r_2/n_2. Fisher's test gives the significance probabilities for these data:

$$SP(\text{obtained direction}) = 0.020$$
$$SP(\text{opposite direction}) = 0.016$$
$$SP(\text{total}) = 0.036.$$

In this context the SP in the obtained direction consists of all those 'extreme' events tending to support the finding that the second proportion is higher than the first. It may be that the researcher had a suspicion that in high-flying sand flies there is a higher proportion of males than in those flying at low altitude: the SP for such a one-sided test is 0.020. This is very small: there is considerable evidence that the suspicion is an accurate one. ■

Exercise 8.12

(a) In *Chapter 2*, Exercise 2.4 an experiment designed to test people's willingness to help others was described. The question was whether the sex of the person requiring help was an important feature. In the experiment described, 71 male students out of 100 requiring help were given it; 89 female students out of 105 were helped. Use Fisher's exact test to explore any differences between the two proportions 71/100 and 89/105.

(b) When people suffer brain damage (for instance, following major accidents or other traumas), it is important to make an assessment of the degree of mental facility that remains. Many tests have been devised for this. One test involves the completion of logical syllogisms. Here are two examples.

All dogs are animals.
All animals are black.
All dogs are black.

All women are humans.
No humans have wings.
No women have wings.

Notice that the sentences that constitute a syllogism do not themselves have to be true: the conclusion merely has to follow from the premisses as a valid argument.

In one test an individual who had suffered brain damage to some degree was able to provide an accurate conclusion in 8 out of 20 cases. A second person classified as 'normal' (that is, who was not known to have suffered any damage to the brain) provided 11 correct conclusions to 20 sets of premisses. Use Fisher's exact test to quantify any differences in levels of attainment. (Provide a clear statement of the hypothesis you are testing.)

Data were provided by Dr S.L. Channon, Middlesex Hospital, University College London.

8.5.2 Comparing two Poisson means

In this case we shall assume that the null hypothesis under test is given by

$$H_0 : \mu_1 = \mu_2$$

and that the variation in both populations is adequately modelled by a Poisson distribution. In this case too, it is necessary to find a method to eliminate a nuisance parameter. Suppose that in order to test the null hypothesis H_0 the data take the following form. A random sample of n_1 observations $X_1, X_2, \ldots, X_{n_1}$ is drawn from the first population; each X_i is Poisson(μ_1) and therefore the sample total is Poisson($n_1\mu_1$):

$$T_1 = X_1 + X_2 + \ldots + X_{n_1} = \sum_{i=1}^{n_1} X_i \sim \text{Poisson}(n_1\mu_1).$$

Similarly, if the second sample (that is, a sample drawn from a second population) consists of n_2 observations each independently following a Poisson distribution with mean μ_2, then the distribution of the sample total T_2 is Poisson($n_2\mu_2$):

$$T_2 \sim \text{Poisson}(n_2\mu_2).$$

Under the null hypothesis $H_0 : \mu_1 = \mu_2$ the two sample totals are independent with Poisson distributions

$$T_1 \sim \text{Poisson}(n_1\mu), \qquad T_2 \sim \text{Poisson}(n_2\mu).$$

Again, the unspecified parameter μ (the common mean) is a nuisance parameter: it is unknown and irrelevant to the null hypothesis; but without some further algebra it is undeniably there. A test for the equality of two

Poisson means takes advantage of certain convenient properties of the Poisson distribution to produce a test statistic appropriate for testing H_0.

The algebra is somewhat involved, and a detailed development of the test would inevitably include some new notation; however, you are spared these details. Briefly, the idea is as follows. Suppose you knew neither T_1 nor T_2, but you did know their total was equal to t, say. Then T_1 could be any of $0, 1, 2, \ldots, t$. Moreover, the larger the sample size n_1 relative to n_2, the larger the expected value of T_1 relative to T_2. Under the null hypothesis $H_0 : \mu_1 = \mu_2$, the null distribution of T_1 conditional on knowing the total $T_1 + T_2 = t$, turns out to be binomial. We denote this by T_1^*:

$$T_1^* \sim B\left(t, \frac{n_1}{n_1 + n_2}\right). \tag{8.9}$$

Those of you who are familiar with the standard notation for conditional probabilities will recognize that the notation used here is slightly non-standard. All that is required is to emphasize that the random variable T_1^*, constrained to be between 0 and t, is binomial (not Poisson, like T_1).

The test is based on this conditional distribution. Here is an example where the test is applied.

Example 8.12 Comparing accident rates

A local authority wished to investigate the consequences for the traffic accident rate of painting designs on traffic roundabouts which, it was hoped, would attract the attention of drivers as they approached, rendering them more aware of the imminent hazard. For three months before experimenting at a particular roundabout known to be an accident black spot, a record was kept of all minor incidents. Monthly counts were 3, 1 and 1. Then the roundabout was painted with chevrons in a high-intensity yellow shade. For the next four months, the accident counts were 1, 0, 2 and 0 respectively. These data were to be used to investigate whether the mean monthly accident rate had changed—in particular, whether it had decreased.

In the absence of indications to the contrary, a Poisson model may be used for the variation in monthly accident counts. Then the hypothesis under test is $H_0 : \mu_1 = \mu_2$, where μ_1 is the mean monthly accident rate before the roundabout was painted, and μ_2 is the mean monthly rate after the painting was carried out. In the notation already developed,

$$n_1 = 3, \quad n_2 = 4, \quad t = 3 + 1 + 1 + 1 + 0 + 2 + 0 = 8;$$

and the observed value of T_1^* is $t_1^* = 3 + 1 + 1 = 5$.

The hypothesis test reduces to considering the observation $t_1^* = 5$ on the binomial random variable $T_1^* \sim B(t, n_1/(n_1 + n_2))$, that is, $T_1^* \sim B(8, \frac{3}{7})$. The probability distribution of T_1^* is given in Table 8.11.

Table 8.11 The probability distribution of T_1^*

t_1^*	0	1	2	3	4	5	6	7	8
$p(t_1^*)$	0.011	0.068	0.179	0.269	0.252	0.151	0.057	0.012	0.001

From this, the SP is given by

$$\begin{aligned} SP(\text{obtained direction}) &= P(T_1^* \geq 5) \\ &= 0.151 + 0.057 + 0.012 + 0.001 \\ &= 0.221. \end{aligned}$$

Since there was particular interest in whether the accident rate has decreased, this is all that needs to be calculated. The SP in the obtained direction (suggestive of a decrease) is relatively high: in fact, there is little evidence to suppose that the underlying mean accident rate has changed. ■

Now try Exercise 8.13.

Exercise 8.13

(a) Sahai and Misra (1992) describe an experiment in which a biologist counted diatoms in water from two different sources. (A diatom is a member of a class of microscopic algae with flinty shells in two halves.)

Sahai, H. and Misra, S.C. (1992) Comparing means of two Poisson distributions. *Math. Scientist*, **17**, 60–67.

In a basic preliminary experiment, 3 diatoms were found in a small amount of water from one source and 6 in an identical amount of water from a second source. Assuming a Poisson model for the variation in counts, explore whether there is a difference in the underlying mean density of diatoms in the two water sources.

(b) Snedecor and Cochran (1989) describe an experiment in which poppy plants were counted in regions of equal area where two different plant treatments had been used. Four regions received Treatment 1, eight received Treatment 2. The plant counts are given in Table 8.12.

Snedecor, G.W. and Cochran, W.G. (1989) *Statistical Methods*, 9th edition, Iowa State University Press.

Table 8.12 Counts of poppy plants

Treatment 1	77	61	157	52				
Treatment 2	17	31	87	16	18	26	77	20

In this case one scarcely needs statistics to deduce that a difference in means exists between the two treatments. However, use the test procedure to explore whether there is a significant difference in means and, if you think it appropriate, incorporate a normal approximation in your analysis.

Summary

In this chapter three methods have been described for testing hypotheses. The three approaches have certain features in common: the first two, if applied to the same set of data, would yield the same conclusion. The third approach permits a quantitative assessment of the extent to which a set of data supports a hypothesis.

1. A simple approach to testing the null hypothesis $H_0 : \theta = \theta_0$ against the two-sided alternative hypothesis $H_1 : \theta \neq \theta_0$ is to use data to obtain a confidence interval (θ_-, θ_+) for θ. For a test at level α, a $100(1-\alpha)\%$ confidence interval should be used. Depending on whether or not the interval (θ_-, θ_+) contains the hypothesized value θ_0, the null hypothesis is 'accepted', or rejected in favour of the alternative.

2. A second approach is called fixed-level testing. A test statistic, whose distribution under the null hypothesis is known, is calculated for a set of data. If its value falls in the tails of the null distribution (or in one of the tails of the null distribution, in a one-sided test) then it is said to have fallen in the rejection region and the null hypothesis is rejected. The size of the rejection region is determined by the predetermined level α at which the test is performed. The strategy for fixed-level testing is summarized in the box on page 317.

3. When sampling from a discrete population, the required level of the test may be only approximately attained. This is because it is not always possible to obtain exact quantiles for discrete distributions.

4. For tests about a normal mean, Student's t-distribution is required. In particular, the data may take the form of differences. A test of the null hypothesis $H_0 : \mu = 0$ is commonly known as Student's t-test for zero mean difference.

5. The conclusions of a hypothesis test may be in error. The act of rejecting H_0 when H_0 is true is called a Type I error, and has probability α. Alternatively, the null hypothesis might be 'accepted' when it is false, and this is called a Type II error. The probability of avoiding a Type II error (that is, the probability of rejecting H_0 when H_0 is false) is called the power of the test. The mathematics of power are rather complicated: they depend among other things on the selection of the test statistic and on the size of the sample taken.

6. The third approach to hypothesis testing described in this chapter is called significance testing. It requires statement of a null hypothesis and of a statistic to be used for testing that hypothesis, but not, strictly, statement of an alternative hypothesis. The approach results in a number called the significance probability (SP) which quantifies the extent to which the data support the null hypothesis. The approach is summarized in the box on page 327.

7. Usually the SP comprises two components—that in the obtained direction, and that in the opposite direction. These may be used as part of a decision procedure to reject the null hypothesis in favour of stated one- or two-sided alternatives.

8. A test known as Student's two-sample t-test may be used to compare the means of two populations. The assumptions of the test are that the variation in either population may be modelled by a normal distribution with the same variance in each population. The test includes calculation of the pooled estimator for this common variance

$$S_p^2 = \frac{(n_1 - 1)S_1^2 + (n_2 - 1)S_2^2}{n_1 + n_2 - 2},$$

and of the value of the test statistic t with distribution

$$T = \frac{\overline{X}_1 - \overline{X}_2}{S_p\sqrt{\frac{1}{n_1} + \frac{1}{n_2}}} \sim t(n_1 + n_2 - 2)$$

under the null hypothesis, $H_0 : \mu_1 = \mu_2$.

9. There are very many other occasions where it may be necessary to compare two populations. In this course two such further comparisons are described. The first is Fisher's exact test for the equality of two proportions. The test is algebraically and arithmetically not quite straightforward, and it is assumed that you would have access to the appropriate computer software were you to use the test.

10. A test for the comparison of two Poisson means which reduces to assessing the value of a conditional test statistic against a binomial distribution is described.

Chapter 9
Examining the assumptions

In this chapter two methods are described for examining the quality of the fit of a hypothesized probability model to a set of data. The validity of some test procedures may depend on the model being a good one, and if it seems evident that this may not be the case, some approaches are suggested for further exploration of the data.

By now you have met many of the most fundamental ideas of statistics. You have seen that variability in a population can be represented by probability models. With the aid of a few common-sense assumptions, you are able to produce probability distributions for both discrete and continuous random variables. In *Chapters 6* to *8* you used those modelling distributions to provide methods of using data to answer important, practical questions.

You began by using the data to provide numerical estimates of unknown model parameters and, in *Chapter 6*, you met the powerful and logical method of maximum likelihood. This method relies upon the chosen probability model being plausibly correct, so you need to be fairly sure that your modelling is good. The same is true when it comes to calculating confidence intervals. You will recall, from *Chapter 7*, that it is vital to know the underlying distribution in order to perform the calculations. *Chapter 8* dealt with the testing of hypotheses and, once again, in order to be able to calculate significance probabilities, it was necessary to assume an underlying probability model. Of course, the central limit theorem can sometimes be applied: for large enough sample sizes, the normal distribution can be used and you can proceed without having to specify the actual population distribution; but the majority of data sets you have met so far have not usually been all that large.

Clearly, the time has come to examine and test the plausibility of the models we use. Now that we have seen the methods in action, we can re-examine these models and ask ourselves what could possibly go wrong. We shall attempt to answer five main questions, and we shall also see how these questions lead to good practical procedures by revisiting some previous analyses.

Are the distributional assumptions plausible?

This is the obvious question, yet it is far from easy to answer. Up to now you have looked at histograms of the data and, because they have shown the right sort of shape, it has been assumed that all is well. This is, at best, an imprecise procedure, and we need more accurate ways of checking the usefulness—the 'fit'—of a model. Ideally we need a quick, graphical method of checking which can be done easily on a computer, and which is fairly precise. The method needs to be able to show discrepancy between the data and a proposed model in such a way that we can identify the causes and be guided to the appropriate action.

Can we quantify any discrepancy between the data and the model?

We need a formal method for numerically quantifying discrepancy between the data and the model. This would give us the ability to test the hypothesis that a proposed model is adequate.

If the distributional assumptions are not plausible, what can we do?

We could 'modify the model': this is the first possibility that comes to mind. Alternatively, we might try to fit a model to transformed data—either we shall be able to find a mathematical transformation of the data and the transformed data can be modelled well, or no such useful transformation can be found. In the former case, you will see in this chapter how to decide which transformation to use.

Is a modelling distribution necessary for answering the questions we are asking of the data?

You have already seen how the central limit theorem can be used with large sample sizes to side-step the issue of requiring a suitable model for the variation in the population. We shall see that, when the sample sizes are small and all else fails, we can proceed by adopting a distribution-free approach to estimation and hypothesis testing.

Are the data typical or do they contain some 'unlucky' values?

There is always a possibility that a data value in your sample is far from being typical: the very clever child, the person who lives to be 110, the multimillionaire, the athlete with the very low heart rate, the 335-day pregnancy. One of these atypical values, in a small sample, can have a disproportionately large effect upon a statistical investigation. How can such aberrations be detected and how should you allow for them in your calculations? Can you use a method which is not sensitive to such aberrations?

In this chapter the emphasis is on making sure that the methods you have practised and used are valid for the data in question, and on what steps to take when the assumptions you need do not appear to be justified. Occasionally, new data sets will be explored; but we shall also look again at many of those we have already met. Let us begin by questioning some of the assumptions of normality made in *Chapter 8*.

9.1 Are the distributional assumptions plausible?

Let us reconsider two data sets in which an assumption of normality would be a necessary first stage for answering the question posed.

The first data set comprises 20 width-to-length ratios for beaded rectangles used by the Shoshoni American Indians in decorating their leather goods. The question of interest was whether these ratios approximated to the 'golden ratio' of the Greeks, $\frac{1}{2}(\sqrt{5}-1)$ or about 0.618. In order to answer this question, a one-sample t-test was performed. This test requires an assumption of normality. How can you check the validity of this assumption? With a data set as small as this, you cannot sensibly use a histogram to support the assumption because there are insufficient data points to suggest any structure in the variation observed.

See *Chapter 8*, Table 8.4

See Exercise 8.4

A data set was also described in which the silver content of coins from four mintings of the twelfth-century Byzantine Trachy was measured. The question was: did the silver content alter with successive mintings? You were advised that separate comparisons of different mintings using two-sample t-tests would not yield independent results: however, a single comparison of, say, the fourth minting with the first would be a valid statistical procedure; and the single two-sample t-test would also require an assumption of normality. The data are reproduced in Table 9.1.

See *Chapter 8*, Table 8.10

The two-sample t-test also requires that the two populations from which the samples are drawn have equal variances. In this case the ratio of the larger sample variance to the smaller is about 2.25 (less than 3), so there is little evidence to refute this assumption.

Table 9.1 Silver content of coins: first and fourth coinage (% Ag)

First coinage	5.9	6.8	6.4	7.0	6.6	7.7	7.2	6.9	6.2
Fourth coinage	5.3	5.6	5.5	5.1	6.2	5.8	5.8		

With a data set as small as this, it is difficult to see how the assumption of normality might be tested: again, there are insufficient data points for a histogram to yield a useful message. What is needed is a graphical method for checking distributional assumptions about the data which can be used for small data sets. In this section an exploratory method known as **probability plotting** is introduced. This is excellent for performing a quick check and is also ideally suited to computer analyses.

9.1.1 Probability plotting for normal distributions

The idea behind probability plotting is a simple one. Suppose that you have n observations $y_1, y_2, \ldots, y_n$, and that you wish to know whether they may plausibly have arisen from a normal distribution. First, re-arrange them into ascending order. Denote the ith ordered observation by $y_{(i)}$, so that

$$y_{(1)} \leq y_{(2)} \leq y_{(3)} \leq \cdots \leq y_{(n)}.$$

Next, plot the ordered data points $y_{(i)}$ against the corresponding standard normal quantiles x_i given by solving the equation

$$\Phi(x_i) = \frac{i}{n+1}, \tag{9.1}$$

for all $i = 1, 2, \ldots, n$, where $\Phi(\cdot)$ is the c.d.f. of the standard normal distribution. If the data are normal, then the n points $(x_i, y_{(i)})$ should lie on a

The process of re-arranging data points into ascending order was introduced in *Chapter 1* as the first stage in calculating sample quartiles. The reason for denoting the data points y_i (rather than, say, x_i) will become apparent when the probability plot is constructed.

straight line. Of course, the inherent random variation means that a perfect straight line will not generally be produced, but if what you see approximates to a straight line, then the data may plausibly be assumed to have arisen from a normal distribution.

The values x_i, $i = 1, 2, \ldots, n$, which are the solutions to the family of equations (9.1) are known as **normal scores**, and most statistical computer packages will produce a vector of normal scores corresponding to an input data vector, easing the task of producing the resulting plot.

In some texts the phrase 'normal score' refers not to the solution of the equation $\Phi(x_i) = i/(n+1)$, but to the expected value x_i of the ith data point in a random sample of size n from the standard normal distribution, arranged in ascending order. Differences between the two definitions are in this context negligible.

Here is an example where probability plotting is used to test an assumption of normality.

Example 9.1 A normal probability plot for silver content

Let us construct a normal probability plot for the first coinage in Table 9.1. As the sample size is $n = 9$, we determine the points x_i for which $\Phi(x_i) = i/10$. For instance, $x_5 = q_{0.50} = 0$ is the median, the 50% point of the standard normal distribution; $x_7 = q_{0.70} = 0.524$ is the 70% point of the standard normal distribution. These points are listed in Table 9.2. The points $y_{(i)}$ are the nine data points, listed in ascending order.

Table 9.2 Silver content: first coinage (% Ag)

i	$y_{(i)}$	$i/10$	x_i
1	5.9	0.1	−1.282
2	6.2	0.2	−0.842
3	6.4	0.3	−0.524
4	6.6	0.4	−0.253
5	6.8	0.5	0.000
6	6.9	0.6	0.253
7	7.0	0.7	0.524
8	7.2	0.8	0.842
9	7.7	0.9	1.282

Notice that the lower half of the normal scores are the same as the upper half of the normal scores apart from the sign, with 0 for the middle value since n is odd.

Now $y_{(i)}$ is plotted against x_i, for $i = 1, 2, \ldots, 9$.

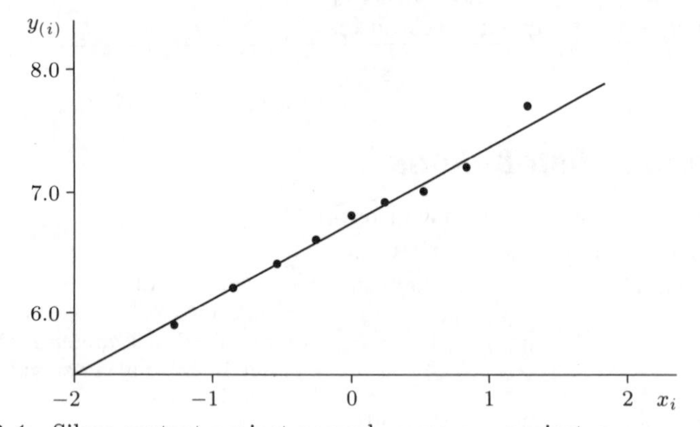

Figure 9.1 Silver content against normal scores, $y_{(i)}$ against x_i

In this diagram no complicated arithmetic has been used to draw a straight line through the nine points. The line drawn simply looks 'reasonable'.

You can see that a straight line provides a reasonable fit to the points $(x_i, y_{(i)})$, and that there is no obvious curve which would fit better. We can conclude that the normal distribution provides a reasonable model for the variation observed.

Incidentally, the intercept of the fitted straight line (that is, the value of y when $x = 0$) provides an estimate of μ, the mean silver content, and the slope provides an estimate of σ, its standard deviation. In this context we are not particularly interested in these estimates: we merely wish to check our assumption of normality for a subsequent t-test. In any case, it is not clear what criteria might be applied to construct the 'best' straight line through the data points, and therefore it is not clear how 'best' estimates might be achieved. But the notion of using probability plotting as a method of parameter estimation is a potentially useful one. ■

Here is a second example.

Example 9.2 A normal probability plot for the Shoshoni data

A probability plot for the Shoshoni rectangles data in *Chapter 8*, Table 8.4 is shown in Figure 9.2. It does not suggest an acceptable straight line through the points $(x_i, y_{(i)})$, and we should be much less sanguine about employing a t-test. This calls into question our conclusions following the t-test that was in fact performed in Exercise 8.4. We shall return to these data in Section 9.4 where we shall see how the problem of non-normality is tackled.

See Exercise 9.14.

The data for the probability plot in Figure 9.2 are given in Table 9.3. For example, x_7 is the solution of the equation

$$\Phi(x_7) = \tfrac{7}{21} = \tfrac{1}{3};$$

so $x_7 = -0.431$.

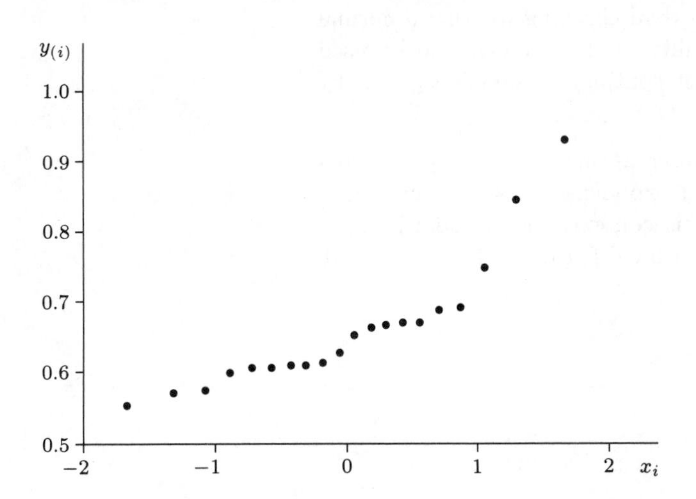

Figure 9.2 Width-to-length ratios of Shoshoni rectangles against normal scores ■

Table 9.3 Normal scores for the Shoshoni rectangles data

i	$y_{(i)}$	$i/21$	x_i
1	0.553	1/21	−1.668
2	0.570	2/21	−1.309
3	0.576	3/21	−1.068
4	0.601	4/21	−0.876
5	0.606	5/21	−0.712
6	0.606	6/21	−0.566
7	0.609	7/21	−0.431
8	0.611	8/21	−0.303
9	0.615	9/21	−0.180
10	0.628	10/21	−0.060
11	0.654	11/21	0.060
12	0.662	12/21	0.180
13	0.668	13/21	0.303
14	0.670	14/21	0.431
15	0.672	15/21	0.566
16	0.690	16/21	0.712
17	0.693	17/21	0.876
18	0.749	18/21	1.068
19	0.844	19/21	1.309
20	0.933	20/21	1.668

The usefulness of the method just described is that it is ideal for performing a quick check for normality when you are using a computer. Most statistical packages either provide normal scores for any data vector or have a direct normal probability plotting routine. Thus any data vector can be checked for normality with a few keystrokes. There are computer exercises later in the section. First of all you should make sure you thoroughly understand the idea by constructing normal probability plots the long way, using your tables.

Exercise 9.1

Construct a normal probability plot for the fourth coinage data in Table 9.1. Are you satisfied that the normality assumption is a reasonable one?

Exercise 9.2

Table 9.4 gives the corneal thicknesses (in microns) of 8 people, each of whom had one eye affected by glaucoma.

Table 9.4 Corneal thickness in patients with glaucoma (microns)

Patient	1	2	3	4	5	6	7	8
Glaucomatous eye	488	478	480	426	440	410	458	460
Normal eye	484	478	492	444	436	398	464	476

A t-test for zero mean difference relies on the assumption that the variation in observed differences may be modelled by a normal distribution. Construct a normal probability plot for the differences. Is the normality assumption reasonable?

9.1.2 Probability plotting for other continuous distributions

Probability plotting has a wider application than checking whether a normal distribution provides a satisfactory probability model. It can also be used to investigate the quality of the fit of other continuous models to data by proceeding in a similar way.

Suppose, for example, that independent observations $y_1, y_2, \ldots, y_n$ are believed to arise from a population where an exponential distribution might provide a useful model for variation. In this context, the 'standard' exponential distribution is that having mean 1, with c.d.f. $F(x) = 1 - e^{-x}$, $x \geq 0$. The solution of the equation

$$F(x_i) = 1 - e^{-x_i} = \frac{i}{n+1}$$

is

$$x_i = -\log\left(\frac{n+1-i}{n+1}\right) \tag{9.2}$$

for all $i = 1, 2, \ldots, n$. If an exponential model is appropriate, the points $(x_i, y_{(i)})$ will lie approximately on a straight line: in this case the straight line must pass through the origin and has slope μ, the population mean. Thus the parameter μ may again be estimated from the plot, provided you are not too rigorous over what constitutes a 'good' estimate.

Here is an example where the quality of the fit of an exponential model for waiting times is explored.

Example 9.3 *Memory recall times*

A data set on memory recall times (in seconds) of pleasant and unpleasant memories was described in *Chapter 2* (see Table 2.10). Boxplots of the recall times for each kind of memory were skewed (see Figure 2.10). Could the data be modelled by separate exponential distributions? Figure 9.3 shows an exponential probability plot for the recall times of pleasant memories. Also shown in the figure is a straight line through the origin, where an attempt has been made to fit the line to the points. You can see that these data are not well modelled by an exponential distribution.

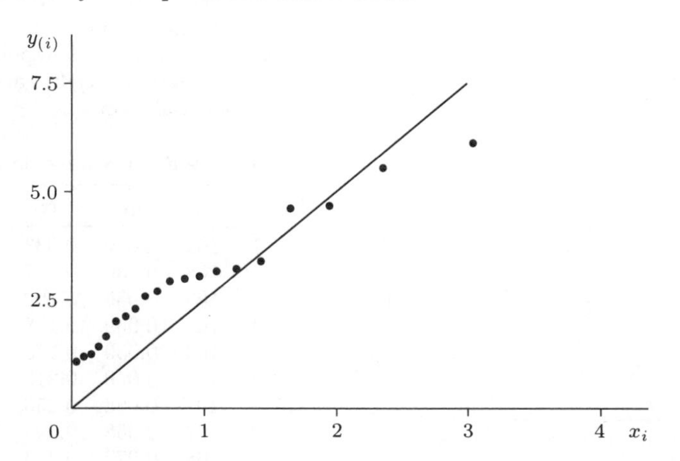

Figure 9.3 Recall times of pleasant memories against exponential scores ■

The data for the exponential probability plot in Figure 9.3 are given in Table 9.5. For example,

$$x_{13} = -\log\left(\frac{20+1-13}{20+1}\right) = -\log \tfrac{8}{21} = 0.965.$$

Table 9.5 Exponential scores for pleasant memories

i	$y_{(i)}$	x_i
1	1.07	0.049
2	1.17	0.100
3	1.22	0.154
4	1.42	0.211
5	1.63	0.272
6	1.98	0.336
7	2.12	0.405
8	2.32	0.480
9	2.56	0.560
10	2.70	0.647
11	2.93	0.742
12	2.97	0.847
13	3.03	0.965
14	3.15	1.099
15	3.22	1.253
16	3.42	1.435
17	4.63	1.658
18	4.70	1.946
19	5.55	2.351
20	6.17	3.045

Exercise 9.3

Construct an exponential probability plot for the recall times of unpleasant memories. Is an exponential distribution a reasonable model here?

Probability plotting is a useful technique for checking the fit of many continuous models. However, you need to bear in mind that not all distributions will produce a straight line directly: an example is the Pareto distribution. As you saw in *Chapter 6*, Section 6.4, its c.d.f. is given by

$$F(w) = 1 - \left(\frac{K}{w}\right)^{\theta}, \quad x \geq K, \ \theta > 1.$$

The parameter K (usually known) provides a lower limit on the range of the random variable W; the parameter θ is usually unknown. We saw in *Chapter 6* how to derive estimates for θ from data, assuming a Pareto model to be appropriate. Unfortunately there is no 'standard' Pareto distribution free of the parameter θ. (The reasons for this are rather complicated, and you need not concern yourself with them.) However, it turns out that the transformed variable

$$Y = \log\left(\frac{W}{K}\right)$$

has an exponential distribution with parameter θ (i.e. mean $1/\theta$). Plotting the transformed data points $y_{(i)}$ against exponential scores will itself constitute a graphical test of the assumption that the original data arise from a Pareto distribution. (Also, if the assumption seems a reasonable one, the fitted line will have slope approximately equal to the reciprocal of the unknown parameter θ.)

Figure 9.4 shows a probability plot of the data on annual wages in the USA (see Table 6.9 in *Chapter 6*).

The data for the plot shown in Figure 9.4 are given in Table 9.6. The original data values are denoted $w_1, w_2, \ldots, w_{30}$. The transformed variable is denoted by $y_i = \log(w_i/100)$. The points $x_i = -\log((31-i)/31)$ are exponential scores.

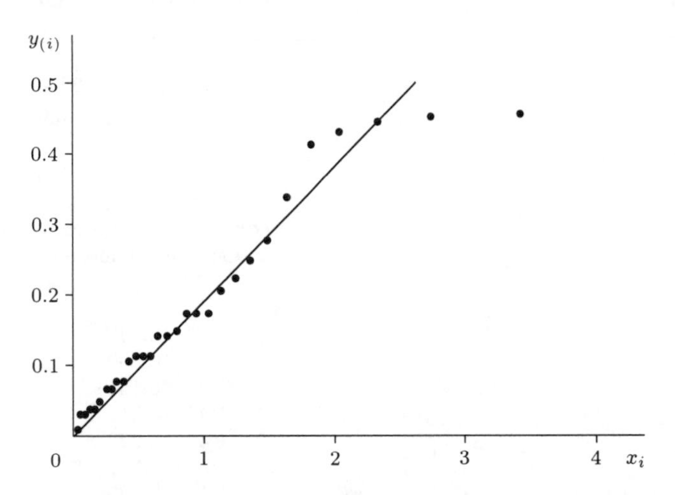

Figure 9.4 Transformed wage data against exponential scores ($K = 100$)

Table 9.6 US wage data

i	$w_{(i)}$	$y_{(i)}$	x_i
1	101	0.010	0.033
2	103	0.030	0.067
3	103	0.030	0.102
4	104	0.039	0.138
5	104	0.039	0.176
6	105	0.049	0.215
7	107	0.068	0.256
8	107	0.068	0.298
9	108	0.077	0.343
10	108	0.077	0.389
11	111	0.104	0.438
12	112	0.113	0.490
13	112	0.113	0.544
14	112	0.113	0.601
15	115	0.140	0.661
16	115	0.140	0.726
17	116	0.148	0.795
18	119	0.174	0.869
19	119	0.174	0.949
20	119	0.174	1.036
21	123	0.207	1.131
22	125	0.223	1.237
23	128	0.247	1.355
24	132	0.278	1.488
25	140	0.336	1.642
26	151	0.412	1.825
27	154	0.432	2.048
28	156	0.445	2.335
29	157	0.451	2.741
30	158	0.457	3.434

You can see from the plot that a straight line would fit reasonably well, except where recorded wages are high. So the Pareto model does not adequately represent the variation inherent in the high-income levels. (This feature of the data was also apparent from the histogram in Figure 6.7.)

Exercises 9.4 to 9.6 allow you to explore the facilities for probability plotting available on your computer.

Exercise 9.4

In order to compare skull dimensions, observations on the maximum head breadth (measured in mm) were taken on the skulls of 84 Etruscan males and those of 70 modern Italian males. The data are given in *Chapter 3*, Table 3.1. A comparative boxplot is given in Figure 3.1 (suggesting significant differences) and in Exercise 8.10 of *Chapter 8* a formal t-test was undertaken. This test assumes normal variation in both populations (see the histograms in Figure 3.2). Use the technique of probability plotting to explore the validity of these assumptions.

Exercise 9.5

(a) Data are given in *Chapter 4*, Table 4.7 on the waiting time (in days) between successive serious earthquakes world-wide. A modelling assumption was made that the variation in these data may be plausibly represented by an exponential distribution. Use an appropriate probability plot to explore this assumption.

(b) Investigate whether the coal-mining disaster data of *Chapter 7*, Table 7.4, may be plausibly modelled by an exponential distribution. (This model was assumed, for instance, in the calculation of confidence intervals.)

Exercise 9.6

Use your computer to simulate 30 observations on the Pareto distribution with parameters $K = 8$, $\theta = 4$. Assuming K is known, use an appropriate probability plotting technique to assess the usefulness of your computer's Pareto simulation routine.

9.2 Can we quantify discrepancy between the data and the model?

The method of probability plotting is a useful technique for exploring whether or not a particular probability model fits observed data, but it has certain disadvantages. If the points on a plot all lie on a straight line, then the model fits the data very well; but random variability in the data makes this an unlikely eventuality. Even if the model fits well, the points will not lie *exactly* on the line. How far from the line can they be before you should conclude that the model does not fit?

Probability plotting is an exploratory technique used more as a general indication of the quality of the fit of a hypothesized model to a set of data. It has the virtues of speed and convenience, but it does not provide the answer to inferential questions about the quality of the fit of the model to the data.

In this section, a hypothesis test is introduced that can be used with discrete data to decide how reasonable it is to assume that a particular probability model fits a particular data set. It provides a test of the null hypothesis that that particular model did indeed generate the data. The method is, therefore, an *inferential* rather than an exploratory one. You will also see that the method may be adapted to make inferences about continuous distributions.

9.2.1 Goodness-of-fit of discrete distributions

The method is demonstrated through an example.

Example 9.4 *Testing a Poisson fit*

In *Chapter 4*, Table 4.4, data on the emissions of α-particles were presented and the observation was made that the Poisson distribution provided a very good fit.

There were 2612 equal time intervals during which emissions occurred: 57 of them contained zero emissions, 203 contained one emission, and so on. The expected frequencies may be calculated by assuming a Poisson distribution with mean 3.877 so that the expected number of intervals containing i emissions, E_i, is calculated from

$$E_i = 2612 \times \frac{3.877^i e^{-3.877}}{i!}.$$

For example, the expected frequency for two emissions is

$$E_2 = 2612 \times \frac{3.877^2 e^{-3.877}}{2!} = 406.61.$$

In Table 4.4 the expected frequencies (in the column headed 'Fit') were given to the nearest whole number. In Table 9.7 the expected frequencies are listed to two decimal places.

Table 9.7 Emissions of α-particles: observed and expected frequencies, assuming a Poisson model

Count	Observed frequency	Expected frequency
0	57	54.10
1	203	209.75
2	383	406.61
3	525	525.47
4	532	509.31
5	408	394.92
6	273	255.19
7	139	141.34
8	49	68.50
9	27	29.51
10	10	11.44
11	4	4.03
12	2	1.30
> 12	0	0.53

One need only compare the second and third columns (that is, the observed and expected frequencies) to see that the fit looks very good: but can we judge *how* good? Obviously, if the differences between the observed frequencies and expected frequencies are large, then the Poisson model is not a good one, and we should use those differences in a way which allows an overall assessment. Let us call the observed frequencies O_i and the expected frequencies E_i, so that each difference can be written in the form $(O_i - E_i)$. These differences are included in Table 9.8.

Table 9.8 Differences between observed and expected frequencies

Count	O_i	E_i	$(O_i - E_i)$
0	57	54.10	2.90
1	203	209.75	−6.75
2	383	406.61	−23.61
3	525	525.47	−0.47
4	532	509.31	22.69
5	408	394.92	13.08
6	273	255.19	17.81
7	139	141.34	−2.34
8	49	68.50	−19.50
9	27	29.51	−2.51
10	10	11.44	−1.44
11	4	4.03	−0.03
12	2	1.30	0.70
> 12	0	0.53	−0.53

Since we are not interested in whether the differences are positive or negative but only in their magnitudes, it makes sense to use *squared* differences in an overall assessment of the quality of fit. However, even though the model looks quite good and most of the differences are small, one or two of them will turn out to be inordinately large when squared. For example, a difference of 2.90 in an expected frequency of 54.10 (Count = 0) is virtually the same

An alternative approach is to use the modulus of the differences, $|O_i - E_i|$; however, using the squared differences $(O_i - E_i)^2$ leads to a more tractable test statistic.

percentage difference as the difference of −23.61 in an expected frequency of 406.61 (Count = 2), and yet once the differences are squared, one is much larger than the other, misrepresenting the discrepancy between the data and the model.

The solution to this problem is to *scale* the squared differences by dividing by the expected frequency. But why does dividing by E_i give the right scaling? Why not E_i^2 or some other function of E_i? Alas, there is no easy intuitive explanation of why we must divide by E_i, but there are sound mathematical reasons. It is possible to show that this particular choice of scaling results in a test statistic with a familiar distribution; but it requires some careful mathematics, and we need not go into the details.

Thus we use the scaled squared differences $(O_i - E_i)^2/E_i$, and add them together to give the sum

$$\sum \frac{(O_i - E_i)^2}{E_i}$$

as an overall measure of the fit of the model. This is the statistic used in the **chi-squared goodness-of-fit test** devised in 1900 by Karl Pearson. The test is based upon the approximate distribution of this statistic, which is that of a chi-squared random variable, and may be described generally as follows.

The influence of Karl Pearson on the development of the science of statistics has already been mentioned in *Chapter 8*.

The chi-squared goodness-of-fit test

Suppose that in a random sample of size n, each of the observations can be classified into one of k distinct classes or categories and that the number of observations out of a total of n falling into category i is denoted by O_i. Corresponding to a statistical hypothesis, a model is set up which defines the respective probabilities $\theta_1, \theta_2, \ldots, \theta_k$ of an observation falling into one of the k categories. The expected number of observations falling into category i is therefore $E_i = n\theta_i$ and, for large n, the distribution of the quantity

$$\chi^2 = \sum_{i=1}^{k} \frac{(O_i - E_i)^2}{E_i},$$

providing a measure of the quality of the fit of the model to the data, is approximately that of a chi-squared random variable with $(k-1)$ degrees of freedom. This is written

$$\chi^2 = \sum_{i=1}^{k} \frac{(O_i - E_i)^2}{E_i} \approx \chi^2(k-1).$$

It may be that the fitted model includes a parameter or parameters whose values are estimated from the original data. If setting up the statistical model requires using the data to estimate p parameters, then

$$\chi^2 = \sum_{i=1}^{k} \frac{(O_i - E_i)^2}{E_i} \approx \chi^2(k-p-1).$$

Here, $n = 2612$.

Here, $k = 14$.

In this case, we have probabilities $\theta_0, \theta_1, \ldots, \theta_{12}, \theta_{13}$ with

$$\theta_i = \frac{3.877^i e^{-3.877}}{i!}$$

for $i = 0, 1, \ldots, 12$, and

$$\theta_{13} = P(X > 12)$$

when $X \sim \text{Poisson}(3.877)$.

In this case the data were used to estimate the Poisson mean μ: $\widehat{\mu} = \overline{x}$, where $\overline{x}$ is the sample mean. So, here, $p = 1$.

This result is presented without proof, and you should merely observe that it is a consequence of the central limit theorem. This does pose the problem of the adequacy of the approximation and here is a simple rule of thumb for you to follow.

> The chi-squared approximation is adequate if no expected frequency is less than 5; otherwise it may not be good enough.

Here, the last three expected frequencies (4.03, 1.30 and 0.53) in Table 9.8 are all less than 5.

There is a simple remedy when categories have expected frequencies less than 5—just pool two or more categories to make one larger category. Look at Table 9.8 again. You can see that the last three categories have expected frequencies below 5, so they are pooled to produce the categories given in Table 9.9.

Table 9.9 Emissions of α-particles: calculating the value of χ^2

Count	O_i	E_i	$(O_i - E_i)$	$(O_i - E_i)^2/E_i$
0	57	54.10	2.90	0.155
1	203	209.75	−6.75	0.217
2	383	406.61	−23.61	1.371
3	525	525.47	−0.47	0.000
4	532	509.31	22.69	1.011
5	408	394.92	13.08	0.433
6	273	255.19	17.81	1.243
7	139	141.34	−2.34	0.039
8	49	68.50	−19.50	5.551
9	27	29.51	−2.51	0.213
10	10	11.44	−1.44	0.181
>10	6	5.86	0.14	0.003

There are now 12 categories, and the value of the chi-squared test statistic is

$$\chi^2 = \sum \frac{(O_i - E_i)^2}{E_i} = 0.155 + 0.217 + \cdots + 0.003 = 10.4.$$

One parameter has been estimated, namely the mean μ of the Poisson distribution, so $p = 1$, and the chi-squared distribution against which the statistic is to be tested has $(12 - 1 - 1) = 10$ degrees of freedom.

Remember that the χ^2 test statistic measures the extent to which observed frequencies differ from those expected under the hypothesized model: the higher the value of χ^2, the greater the discrepancy between the data and the model. Thus only high values of χ^2 contribute to the significance probability.

It is possible to argue that low values of χ^2 suggest a fit so good that the data are suspect, showing less variation than might be expected. This is the sort of approach taken when scrutinizing data that are thought to be 'too good to be true', as in *Chapter 8*, Exercise 8.3.

The upper tail area of $\chi^2(10)$ cut off at 10.4 is 0.41 (that is, the SP of the test is 0.41), and thus we have no evidence in favour of rejecting the null hypothesis that the data may be fitted by a Poisson distribution. ■

Here is a second illustration of the method.

Example 9.5 Testing a binomial fit

Chapter 2, Example 2.15 introduced data from an experiment on visual perception which was argued to be a result of a sequence of Bernoulli trials each with probability of success equal to 0.29. Table 9.10 gives observed and expected frequencies calculated from the hypothesized binomial probability distribution $B(16, 0.29)$, with sample size $n = 1000$.

For instance,

$$\begin{aligned} E_2 &= 1000\theta_2 \\ &= 1000 \binom{16}{2} (0.29)^2 (0.71)^{14} \\ &= 83.48. \end{aligned}$$

Table 9.10 Counts on a random screen pattern

Count	O_i	E_i
0	2	4.17
1	28	27.25
2	93	83.48
3	159	159.13
4	184	211.23
5	195	207.07
6	171	155.06
7	92	90.48
8	45	41.58
9	24	15.09
10	6	4.32
11	1	0.96
12	0	0.16
13	0	0.02
14	0	0.00
15	0	0.00
16	0	0.00

The first two and the last seven expected frequencies need to be pooled to give categories containing not less than 5; this is done in Table 9.11. Then there are 10 categories with

$$\chi^2 = \sum \frac{(O_i - E_i)^2}{E_i} = 0.064 + 1.086 + \cdots + 0.434 = 13.0.$$

Table 9.11 Counts on a random screen pattern: calculating the value of χ^2

Count	O_i	E_i	$(O_i - E_i)$	$(O_i - E_i)^2/E_i$
0 or 1	30	31.42	−1.42	0.064
2	93	83.48	9.52	1.086
3	159	159.13	−0.13	0.000
4	184	211.23	−27.23	3.510
5	195	207.07	−12.07	0.704
6	171	155.06	15.94	1.639
7	92	90.48	1.52	0.026
8	45	41.58	3.42	0.281
9	24	15.09	8.91	5.261
>9	7	5.46	1.54	0.434

No parameter has been estimated (the hypothesized binomial model was fully specified, including the value $p = 0.29$) so that the chi-squared distribution against which χ^2 is tested has $(10 - 1) = 9$ degrees of freedom and the SP of the test is therefore 0.163. We therefore conclude that there is little evidence to reject the null hypothesis that the observations are from a binomial distribution $B(16, 0.29)$. ■

Exercise 9.7

A genetics experiment was described in *Chapter 3*, Example 3.12, in which the leaves of Indian creeper plants *Pharbitis nil* were characterized by type. Of 290 offspring plants observed, four types of leaf occurred with frequencies 187: 35: 37: 31.

(a) According to one simplistic theory, the four types should have occurred in the ratios

$$\tfrac{9}{16} : \tfrac{3}{16} : \tfrac{3}{16} : \tfrac{1}{16}.$$

Use a chi-squared test of goodness-of-fit to show that the data offer considerable evidence that the theory is unfounded.

(b) In *Chapter 6*, Example 6.12, a more developed theory allowing for genetic linkage is proposed: in this case, after estimation of one parameter, the hypothesized proportions are $0.6209 : 0.1291 : 0.1291 : 0.1209$ respectively. Test this model against the data, using an appropriate goodness-of-fit test.

Exercise 9.8

In *Chapter 6*, Exercise 6.8, Pielou's data on *Armillaria* root rot in a plantation of Douglas firs were quoted. Several transects through the plantation were examined and the lengths of runs of healthy and diseased trees were recorded.

Chapter 6, Figure 6.5(a) showed a bar chart of the data together with a geometric probability distribution with estimated parameter $p = 109/166 = 0.657$ in order to persuade you that Pielou's assumption of a geometric modelling distribution was not unreasonable. Confirm this model by carrying out a chi-squared test of goodness-of-fit.

9.2.2 Goodness-of-fit for continuous models

Up to now, only discrete distributions and other models where the categories are clear-cut, have been tested for goodness-of-fit. With discrete data, it is easy to define the outcomes and to count their frequencies of occurrence, O_i. The only complication arises when expected frequencies turn out to be less than 5, and when this occurs we pool frequencies in order to cope.

When data arise from continuous distributions it is necessary, in order to perform the chi-squared test of goodness-of-fit, to classify the data into different groups, and to count the observed frequencies in each group. As you saw when drawing different histograms of the same data set in *Chapter 1* (Figures 1.10 and 1.11), the group classifications are within your control. So to use the chi-squared test, a decision has to be made about where to draw the group borderlines.

An example will make this clear.

Example 9.6 *Testing a normal fit*

Enzyme measurements for 57 patients suffering from acute viral hepatitis are recorded in *Chapter 2*, Table 2.18. A histogram of these data is given in Figure 2.22(a). It was suggested that a normal model might be adequate for the variation in the data. In order to perform a chi-squared test for a hypothesized normal model, it is necessary first to group the data into categories. Suppose the data are grouped as in Table 9.12.

In Table 9.12 data points on boundaries have been allocated to the higher of the two possible groups.

The data are now in a form we can use for a chi-squared test of goodness-of-fit, but we shall need to calculate the expected frequencies for each of the thirteen groups identified in Table 9.12. The first step, therefore, is to find estimates of the two parameters of the normal distribution, namely the mean, μ, and the standard deviation, σ. In fact, you have already seen how to estimate the mean and variance of a normal distribution in *Chapter 6*, where you obtained $\overline{x} = \sum x_i/n$ and $s^2 = \sum(x_i - \overline{x})^2/(n-1)$ as respective estimates. For the data in Table 9.12, the estimates are $\overline{x} = 2.587$ and $s^2 = 0.107$. Next, we use these estimates to calculate probabilities and, in turn, expected frequencies.

Table 9.12 Grouped enzyme measurements

Grouping	Frequency
< 1.6	0
1.6–1.8	1
1.8–2.0	2
2.0–2.2	3
2.2–2.4	8
2.4–2.6	18
2.6–2.8	11
2.8–3.0	9
3.0–3.2	3
3.2–3.4	1
3.4–3.6	0
3.6–3.8	1
≥ 3.8	0

There were 57 patients: so, for example, the expected frequency for the second group is $57P(1.6 \leq X < 1.8)$ where $X \sim N(2.587, 0.107)$. The probabilities may be calculated directly on your computer or obtained from tables of the normal distribution. In the latter case you need to establish the normal z-value corresponding to the group boundaries. For example, for the boundary 1.6, the corresponding z-value is

$$\frac{1.6 - \overline{x}}{s} = \frac{1.6 - 2.587}{\sqrt{0.107}} = -3.017.$$

For the boundary 1.8, the z-value is

$$\frac{1.8 - \overline{x}}{s} = \frac{1.8 - 2.587}{\sqrt{0.107}} = -2.406.$$

In fact, the maximum likelihood estimate for a normal variance, based on a random sample $x_1, x_2, \ldots, x_n$, is $\sum(x_i - \overline{x})^2/n$. In what follows the unbiased estimate $s^2 = \sum(x_i - \overline{x})^2/(n-1)$ for σ^2 is used.

Now,

$$P(-3.017 \leq Z < -2.406) = 0.0068,$$

so the expected frequency in the group from 1.6 to 1.8 is $57 \times 0.0068 = 0.39$ (to two decimal places).

Expected frequencies are now incorporated in Table 9.13.

Table 9.13 Grouped enzyme measurements

Grouping	O_i	E_i
< 1.6	0	0.07
1.6–1.8	1	0.39
1.8–2.0	2	1.61
2.0–2.2	3	4.68
2.2–2.4	8	9.43
2.4–2.6	18	13.23
2.6–2.8	11	12.92
2.8–3.0	9	8.78
3.0–3.2	3	4.16
3.2–3.4	1	1.37
3.4–3.6	0	0.31
3.6–3.8	1	0.05
≥ 3.8	0	0.01

Table 9.14 is the result of pooling groups with expected frequencies less than 5 and performing the chi-squared calculations.

Table 9.14 Grouped enzyme measurements

Grouping	O_i	E_i	$O_i - E_i$	$(O_i - E_i)^2/E_i$
< 2.2	6	6.75	-0.75	0.083
2.2–2.4	8	9.43	-1.43	0.217
2.4–2.6	18	13.23	4.77	1.720
2.6–2.8	11	12.92	-1.92	0.285
2.8–3.0	9	8.78	0.22	0.006
≥ 3.0	5	5.90	-0.90	0.137

Now, as usual, we compute the χ^2 test statistic: this is

$$\chi^2 = \sum \frac{(O_i - E_i)^2}{E_i} = 0.083 + 0.217 + \cdots + 0.137 = 2.45.$$

There are six categories; but two parameters have been estimated from the data, so we test the observed value of χ^2 against $\chi^2(6-2-1)$, that is, against $\chi^2(3)$. The SP is 0.48 and there is no reason to disbelieve the assumption of normality made earlier. ■

Exercise 9.9

In *Chapter 2*, Example 2.18, a supposed normal data set was presented comprising blood plasma nicotine levels for 55 smokers.

Classify the data into groups with borderlines 150, 250, 350, 450, and carry out a chi-squared test of goodness-of-fit for a normal distribution.

Exercise 9.10

Chapter 2, Table 2.8 lists annual maxima for daily rainfall over a period of 47 years.

Classify the data into groups with borderlines 600, 1000, 1400, 1800, 2200, and carry out a chi-squared test of goodness-of-fit for a normal distribution.

9.3 If the distributional assumptions are not plausible, what can we do?

In Sections 9.1 and 9.2 you have seen how the goodness-of-fit of a proposed probability model may be assessed and how doubts can arise about the suitability of models which, at first, seem plausible. In Example 9.3 and Exercise 9.3, for example, you saw that the memory recall times which appeared first in *Chapter 2* were not well-fitted by an exponential distribution. Perhaps a normal modelling distribution would be better. Figure 9.5 gives a normal probability plot for the recall times for pleasant memories.

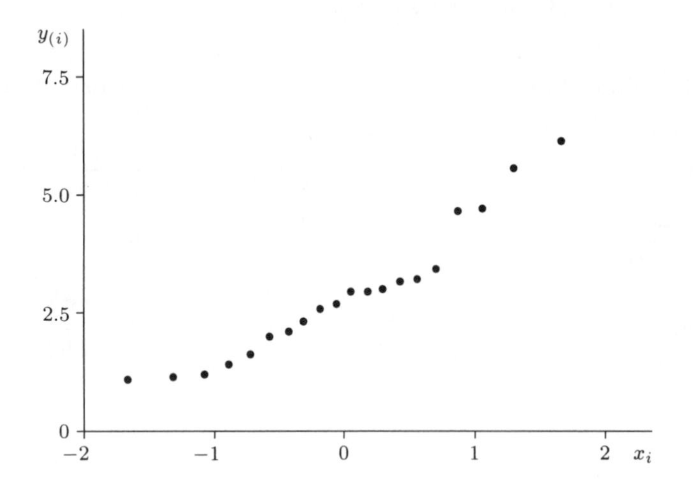

Figure 9.5 Recall times for pleasant memories against normal scores

You can see from the figure that the fit is moderately good; however, should you try it for yourself, you will find that the normal probability plot for unpleasant memories is rather worse. We therefore have a problem, because we would like to perform a test of the hypothesis that the mean recall times for pleasant memories and unpleasant memories are no different; but how can this be done without a plausible modelling distribution for memory recall times?

Table 9.4 lists data relating to people with one glaucomatous eye. The problem was to determine whether the mean corneal thickness of unaffected eyes is the same as that of glaucomatous eyes. A possible approach is to perform a one-sample t-test for zero difference on the individual differences, but this test assumes a normal model for the differences: in Exercise 9.2 you constructed a probability plot that cast doubt upon the assumption. The t-test concludes that there is no significant difference ($SP = 0.33$), but now that you have seen a normal probability plot for these data, would you trust this result?

In this section you will see that all may not be lost and that, even when initial distributional assumptions appear to be unjustified, it is often possible to *transform* the data to a form which allows suitable modelling.

9.3.1 Transformations

Four histograms of data sets each comprising 300 data points are shown in Figure 9.6.

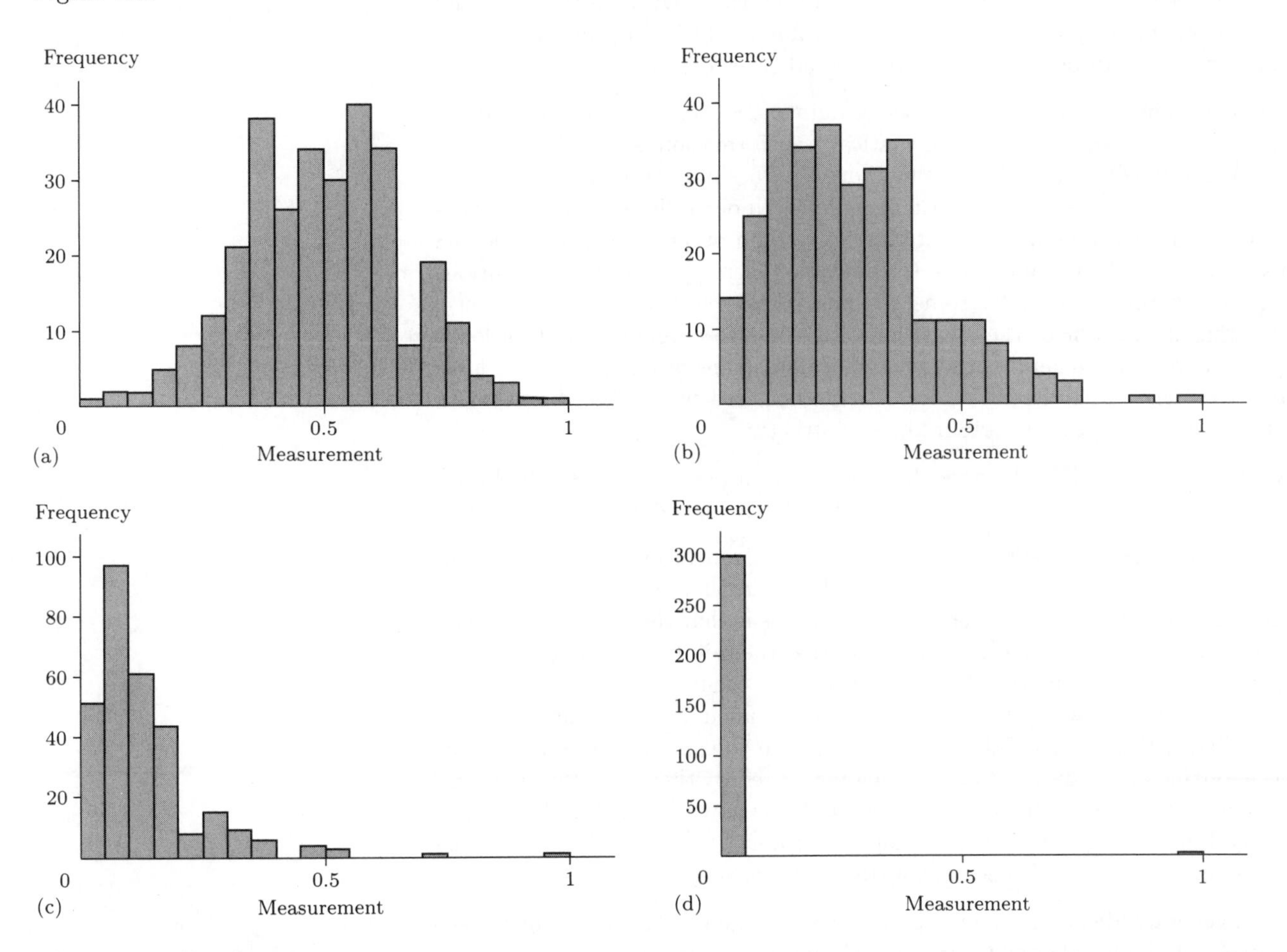

Figure 9.6 Four histograms

Histogram (a) looks as if the data could be normally distributed, but (b), (c) and (d) are progressively more skewed and appear to be anything but normal. Their appearances are confirmed by the respective measures of skewness, which are -0.02, 0.89, 2.62 and 17.21. It may surprise you to learn that all four histograms actually come from the *same* normally distributed data set! Histogram (a) was produced by generating a normal sample of size 300 on a computer. You can also obtain each data point used for histogram (a) by taking the square root of each of the data points in histogram (b), or by taking one plus half the logarithm of the data points in histogram (c), or by taking the reciprocal of 263 times those in histogram (d). You can do this because the data for (b), (c) and (d) were generated from the data for (a) in the first place, and the transformations suggested are the inverses of the ones used for generating (b), (c) and (d).

Since the data in histogram (a) are plausibly normally distributed, there should be nothing wrong in using them to carry out a t-test. For example, you could test the null hypothesis that the mean of the population from which they were drawn is 0.5. However, it would not be legitimate to carry out t-tests on the data in (b), (c) or (d) because the variation is far from normal. Now, suppose that data resembling those in (b), (c) or (d) were to arise in practice. It would clearly be in order to transform them with an appropriate transformation and then to carry out t-tests based on the assumption of normality. If the data looked like those in histogram (b), then the correct procedure would be to take square roots. Taking the logarithm would be appropriate for (c) and taking a reciprocal transformation would cater for (d).

The main aim in transforming a set of data $x_1, x_2, \ldots, x_n$ to a different set $y_1, y_2, \ldots, y_n$ by means of a mathematical transformation is usually to render the transformed data more plausibly normal. Possibilities include $y_i = x_i^2$ or $y_i = 1/x_i$ or $y_i = \log(x_i)$: the list is endless. Under different circumstances some transformations will be better or worse than others at finally achieving a satisfactorily 'bell-shaped' look to the data. (Clearly, some transformations will be inappropriate under some circumstances—one cannot, for instance, take the square root or the logarithm of a negative number.) Quite a lot is known about the general consequences of data transformations of the form $y_i = (\alpha + x_i)^\beta$ for different values of α and β, and when the original data (that is, the x_is) possess certain kinds of structure.

Above all, one needs to be sure that the question posed of the original data can be answered, or approximately answered, following a data transformation.

Sometimes experimentalists object to such data transformations: their argument is that transformations distort the data and that such massaging is an unjustifiable 'fiddle'. A counter-argument to this is that the mechanism which produced the data has inadvertently distorted them in the first place, so that a statistical transformation is merely restoring them to their proper form. Another objection which is frequently heard is that in a given context the required statistical approach should be decided on theoretical grounds without even looking at the data, let alone being influenced by them, or altering them. Such a view was popular some years ago, but it is now regarded as discredited by modern statisticians. There is nothing against the idea of looking at the shape of your data while choosing a statistical technique.

Some general indications of an approach can be made: if the original data are positive and highly skewed with many relatively small values and fewer higher values (with possibly some very high values indeed) then data transformations of the form

$$y_i = \sqrt{x_i} = x_i^{1/2}, \quad y_i = x_i^{1/3}, \quad y_i = \log(x_i)$$

will all tend to have a greater reducing effect on higher x-values than on lower values: the overall effect will be to reduce the skewness in the data, possibly quite substantially, and potentially very usefully, if the transformed data values become more symmetric. (But notice that the log transformation would have a very considerable stretching effect on values of x between 0 and 1.) The **ladder of powers** lists transformations of the form

$$\ldots, x^{-2}, x^{-1}, x^{-1/2}, \log x, x^{1/2}, x^{1}, x^{2}, \ldots.$$

The transformation x^1 leaves the value of x as it is. Provided $x > 1$, powers above 1 on the ladder expand the high values relative to the low values, and powers below 1 have the reverse effect. The further up or down the ladder from x^1, the greater the effect. Notice where the transformation $\log x$ (not, in fact, a power transformation) is located on the ladder.

Examples 9.7 and 9.8 which follow show that it is useful to transform data prior to analysis. The area is a difficult one, requiring experience and inspiration; just read the examples for what you can get out of them, and do not worry too much about the arithmetic detail.

9.3.2 Making use of transformations

The area of selecting suitable data transformations is one in which the computer is an invaluable tool. You can move along the ladder of powers trying different transformations until sample skewness is either largely removed or at least reduced as much as possible. Of course, you should keep the transformations as simple as possible and use only powers such as -1 or $\frac{1}{2}$: you should not attempt to interpolate by investigating such powers as, say, 0.41. By keeping things simple you are roughly satisfying the assumptions while retaining a straightforward interpretation, and that is what you should aim for. In the case of a single sample, a normal probability plot can then be used as a check and, if it shows a passable straight line, a test may be carried out with confidence in the statistical technique chosen.

When testing two samples, the data are transformed, the respective transformed means are subtracted from each group and the residuals thus obtained are checked with a probability plot. Let us try a couple of examples, to enable you to become familiar with the method before you try some exercises for yourself. Example 9.7 is an example with paired data.

Example 9.7 Error gravity scores

Table 9.15 gives error gravity scores for a sample of native teachers of English and a sample of Greek teachers of English for each of 32 English sentences analysed by students.

Woods, A., Fletcher, P. and Hughes, A. (1986) *Statistics in language studies*, Cambridge University Press, p. 201. An error gravity score quantifies the seriousness of an error as perceived by a teacher—here, teachers of English judged written material.

Table 9.15 Error gravity scores of English and Greek teachers

Sentence	1	2	3	4	5	6	7	8	9	10	11	12	13	14	15	16
English	22	16	42	25	31	36	29	24	29	18	23	22	31	21	27	32
Greek	36	9	29	35	34	23	25	31	35	21	33	13	22	29	25	25
Sentence	17	18	19	20	21	22	23	24	25	26	27	28	29	30	31	32
English	23	18	30	31	20	21	29	22	26	20	29	18	23	25	27	11
Greek	39	19	28	41	25	17	26	37	34	28	33	24	37	33	39	20

In order to test the hypothesis that the mean scores given by English and Greek teachers are the same, the scores for each of the sentences are differenced and the differences obtained are tested for a mean of zero. Assuming that the differences are normally distributed, a t-test would be appropriate here. The sample skewness for the 32 differences is 0.405, which is rather high.

Using a power transformation to reduce skewness needs care here because some of the data points are negative. Since the minimum difference is -16, we first add 16 to each of them so that all differences are now non-negative. If we use a transformation x^{β}, testing for a mean difference of 0 is now the same as testing the transformed data for a mean of 16^{β}. We are trying to remove positive skewness, so let us try a power of $\beta = 1/2$. Our transformed variable is now $(\text{Difference} + 16)^{1/2}$, and has sample skewness -0.420.

Strictly, some bias is introduced here: even if the mean difference is 0, the mean of the transformed variable $(\text{Difference} + 16)^{\beta}$ is not quite 16^{β}. Despite this bias, the procedure is a useful one.

Obviously we have gone too far, so we could try using a power of $\beta = 3/4$. This final transformation yields a variable which is $(\text{Difference} + 16)^{3/4}$: the sample skewness is 0.094.

We now test the transformed data for a mean of $16^{3/4}$ (that is, a mean of 8) and obtain a SP of 0.014. The SP obtained by testing the untransformed data for a mean of zero is 0.035. The transformed data now need to be checked for normality. Normal probability plots for both the untransformed and the transformed data are shown in Figure 9.7.

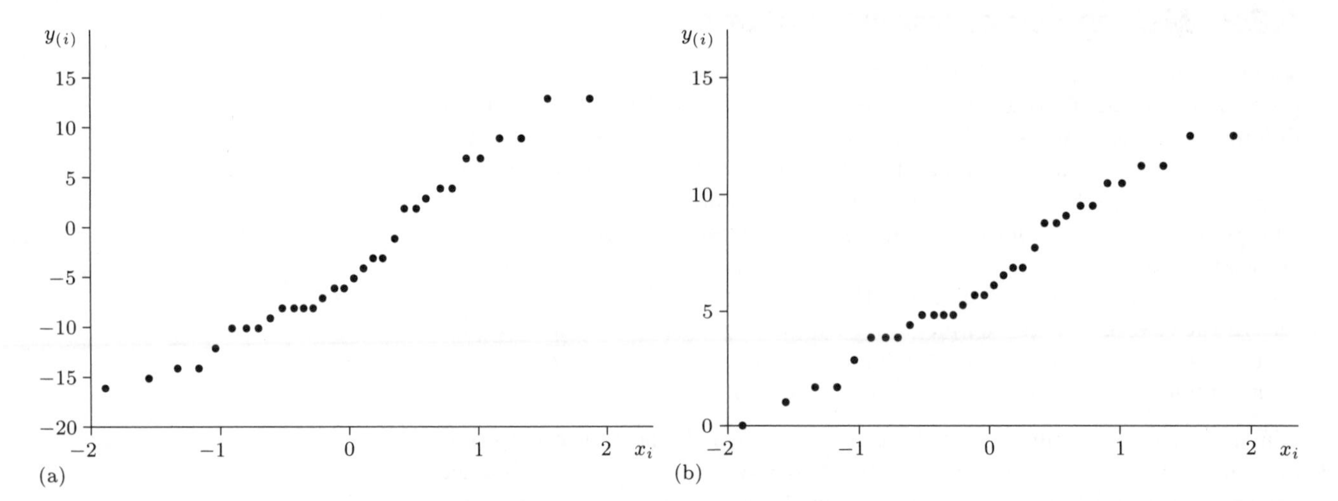

Figure 9.7 Normal probability plots for (a) untransformed and (b) transformed error gravity scores

The difference is very slight here. The plot for the untransformed data, Figure 9.7(a), does show a slight curvature, but this is acceptable. Figure 9.7(b) shows straightening of the curvature and an improvement in the fit. After transformation we can happily reject the null hypothesis of no difference with a SP of 0.014. ■

Example 9.8 Ontario Protestant mothers

In *Chapter 1* you saw some data on family sizes of Ontario Protestant mothers. One group comprised mothers with six years of formal education or less, the other was for mothers who had been educated for seven years or more. These

See Table 1.15.

data were, in fact, part of a larger data set. The mothers in question were married at age 15–19, whereas the complete data set also includes those mothers married at age 20–24. The data for the mothers married at age 20–24 are given in Table 9.16.

Table 9.16 Sizes of Protestant families in Ontario—mothers aged 20–24

Mothers educated for 6 years or less													
Family size	0	1	2	3	4	5	6	7	8	9	10		
Frequency	2	3	2	11	9	1	9	2	2	0	2		
Mothers educated for 7 years or more													
Family size	0	1	2	3	4	5	6	7	8	9	10	11	12
Frequency	8	17	21	26	17	13	7	6	1	3	1	0	1

For these two data sets, the sample skewnesses are 0.434 and 0.967 respectively.

There are no negative data points. A little trial and error shows that the transformation $y = x^{3/4}$ works well, producing a slight negative skewness in one group and a positive skewness in the other: the values are -0.076 and 0.389 respectively.

This transformation is probably as good as any on these data, so now a t-test of the hypothesis that there is no difference between the mean family size for either of the two groups can be performed. A t-statistic of 2.218 is obtained which, on 162 degrees of freedom, yields a SP of 0.028. This suggests a significant difference between the two groups.

How tenable is the assumption that the transformed data are plausibly normal? This can be checked by obtaining residuals and plotting them against normal scores. For the first group the mean, in this case 2.886, is subtracted from the transformed data. The mean of the second group, 2.382, is also subtracted from the transformed data, and the two sets of residuals thus obtained are pooled together into one batch. The combined residuals for the two groups are then plotted against normal scores in the usual way to obtain the probability plot shown in Figure 9.8.

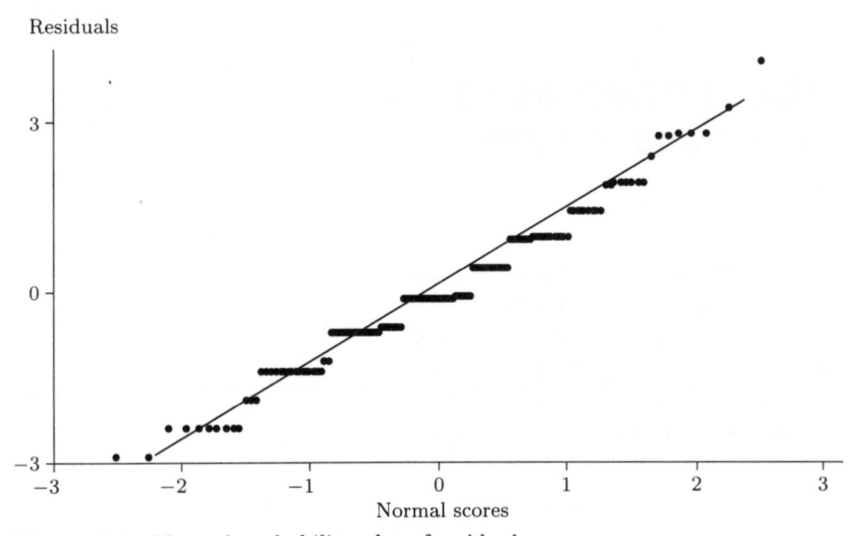

Figure 9.8 Normal probability plot of residuals

With so many points on it, the graph is a little blurred, but it does show an acceptable straight line and therefore we can be confident about rejecting the null hypothesis that there is no difference between the groups: in fact, the data suggest that mothers with more years' education tend to have fewer children. ■

Exercise 9.11

In *Chapter 1* you were presented with a data set on birth weights of 50 children suffering from severe idiopathic respiratory distress syndrome (see *Chapter 1*, Table 1.4).

Decide whether a test of equal mean birth weights for the two groups should first involve a transformation of the data. If you decide to transform them, find a suitable transformation.

Exercise 9.12

Using your conclusions from Exercise 9.11, carry out a test of the null hypothesis that the mean birth weights of the two groups of children are the same. Check the validity of your assumptions.

Exercise 9.13

In Example 9.8 you looked at data on family sizes of mothers married when aged 20–24. These data were part of a data set which also includes family sizes of mothers married when aged 15–19. This latter group appeared in *Chapter 1*, Table 1.15.

Repeat the analysis of Example 9.8 for the 15–19 age group and discuss the validity of the assumptions you make.

9.4 Is a modelling distribution necessary for answering the questions we are asking of the data?

The normal probability plot you produced for Exercise 9.13 may have left you in some doubt about the validity of assuming normality even for the transformed data. While the idea of transforming the data to plausible normality is an attractive one, it is clear that many data sets are not amenable to this technique. If application of the ladder of powers does not produce a satisfactory model, we shall be compelled to fall back upon a technique which does not require a model. In other words, we need a technique which we can use when all else fails.

9.4.1 Early ideas—the sign test

Distribution-free statistical tests can be traced back at least as far as 1710, when John Arbuthnot produced the first recorded instance of such a technique. The fundamental principle behind his test is simple. For the 82-year period from 1629 to 1710 (inclusive) he observed from City of London records that the annual number of births of boys always exceeded the annual number of births of girls. If he were to believe births of either sex to be equally likely, the probability of such an outcome would be $\left(\frac{1}{2}\right)^{82}$, which is a very tiny probability indeed. He therefore refused to believe it (in statistical parlance he rejected the null hypothesis of boys and girls being equally likely) and concluded that the probability of a boy was greater than that of a girl. He further concluded that the observation constituted clear evidence for divine providence since, with wars and diseases resulting in a higher death-rate for males, God had compensated by arranging for more males to be produced, and hence arranged 'for every woman her proper husband'.

Arbuthnot, J. (1710) An Argument for Divine Providence, taken from the constant Regularity observ'd in the Births of both Sexes. *Philosophical Transactions*, **27**, 186–190. Arbuthnot's researches were described in *Chapter 3*, when a binomial model for the distribution of the sexes in families was considered. Presumably the data for 1710 did not cover the whole year.

Notice that Arbuthnot's test makes no assumptions about the distribution of births for either sex. Nowadays his test is called the **sign test**, under which name it appears in most elementary statistics textbooks. In spite of its frequent appearance in texts, it is hardly ever used in practice. The principle remains the same as Arbuthnot's; if you have paired data, calculate the differences and count the number of + signs obtained. If the distribution of differences is symmetric about zero, you can expect roughly as many + signs as − signs and you can obtain a significance probability by using a binomial distribution $B\left(n, \frac{1}{2}\right)$ (where n is the sample size). Arbuthnot subtracted the number of girls recorded from the number of boys for each of 82 years and obtained 82 + signs.

Example 9.9 *Corneal thickness—the sign test*

In Exercise 9.2 you constructed a normal probability plot for the data on corneal thickness in patients with glaucoma. These are paired data so you calculated the differences between pairs, thereby creating a single sample of differences. The null hypothesis that there is no difference between the corneal thickness of a glaucomatous eye and a normal eye can be explored by testing the individual differences for a mean of zero using a t-test. On page 363 the remark was made that this analysis is suspect in the light of the normal probability plot of the data.

In the sign test, zeros are ignored. Looking at the differences listed in Table S9.1, there are three + signs and four − signs or, to put it another way, three + signs out of seven. Assuming equal probability for individual + and − signs, the probability of observing up to 3 + signs out of 7 is

There is a similar approach to the sign test in which zeros are incorporated into the analysis.

$$\sum_{x=0}^{3} \binom{7}{x} \left(\tfrac{1}{2}\right)^7 = 0.5;$$

the total SP for the test is 1, and we can conclude that there is no evidence of a difference in typical corneal thickness. ■

The sign test has a degree of historical interest, but it is rarely used in practice. The reason for this is that it throws away too much valuable information and, as a result, is not very powerful: it is prone to failing to reject a null hypothesis in all but the most obvious cases. The information thrown away is information on the *size* of the difference. Among the differences for glaucomatous patients, patient 4 with a difference of -18 microns is given exactly the same importance as patient 1 with a difference of 4 microns. This seems unsatisfactory, and it fell to Frank Wilcoxon in 1945 to propose a method of testing which takes the size of the differences into account.

Wilcoxon, F. (1945) Individual comparisons by ranking methods. *Biometrics Bulletin*, **1**, 80–83.

9.4.2 The Wilcoxon signed rank test

Wilcoxon's idea was to replace individual differences by **ranks**. These are allocated to the absolute values of the differences, the smallest being given a rank of 1, the next smallest a rank of 2, and so on. The differences then have their signs restored to them and the ranks are added up separately for each of the positive sign and negative sign groups. If the total for one group is very small (which means that the total for the other group must be very large since they add up to a fixed total), the null hypothesis that there is no difference will be rejected. A simple example should make this method clear.

***Example 9.10** Corneal thickness—the Wilcoxon signed rank test*

We shall use the paired data on corneal thickness in patients with glaucoma from Example 9.9. Table 9.17 separates the differences from their associated signs.

Table 9.17 Corneal thickness in patients with glaucoma

Patient	1	2	3	4	5	6	7	8
Glaucomatous eye	488	478	480	426	440	410	458	460
Normal eye	484	478	492	444	436	398	464	476
Sign of difference	+		−	−	+	+	−	−
|Difference|	4	0	12	18	4	12	6	16
Rank	$1\frac{1}{2}$		$4\frac{1}{2}$	7	$1\frac{1}{2}$	$4\frac{1}{2}$	3	6

There are two important things to notice in Table 9.17. The first is that the difference of zero for patient 2 has not been included in the reckoning; it is ignored and the sample size is regarded as being 7, not 8. The second is that where two differences have the same absolute value, an *average* rank is given. In the table, the two lowest absolute values are tied on 4. Since the two lowest ranks are equal to 1 and 2, each is allocated rank $\frac{1}{2}(1+2) = 1\frac{1}{2}$. The same has happened where two absolute differences are tied on 12. They have ranks 4 and 5, so each is given a rank of $\frac{1}{2}(4+5) = 4\frac{1}{2}$. Now we reintroduce the signs. The sum of the positive ranks is

$$w_+ = 1\tfrac{1}{2} + 1\tfrac{1}{2} + 4\tfrac{1}{2} = 7\tfrac{1}{2};$$

the sum of the negative ranks is

$$w_- = 4\tfrac{1}{2} + 7 + 3 + 6 = 20\tfrac{1}{2}.$$

The sum $w_+ + w_-$ is always equal to $1 + 2 + \cdots + n = \frac{1}{2}n(n+1)$, where n is the sample size. (Here, $w_+ + w_- = 28$; the sample size is $n = 7$.) The test statistic for the Wilcoxon signed rank test is w_+: under the null hypothesis of zero difference, values of w_+ that are extremely small or extremely large will lead to rejection of the null hypothesis. The null distribution of the test statistic w_+ is rather complicated, relying upon a complete enumeration of cases: to obtain the SP for a Wilcoxon signed rank test, reference is usually

In some approaches, the test statistic is taken to be the minimum of w_+ and w_-, that is, $\min(w_+, w_-)$. This has some advantages, but information is lost about the precise nature of the differences that make up the sample.

made to a computer. In this case, with $n = 7$ and $w_+ = 7\frac{1}{2}$, the test results are

$$SP(\text{obtained direction}) = SP(\text{opposite direction}) = 0.172;$$
$$SP(\text{total}) = 0.344.$$

The null distribution of w_+ is symmetric.

There appears to be no significant difference in corneal thickness. ■

As for the sign test, there is an alternative procedure in which zeros are incorporated into the analysis. After all, they offer contributory evidence for a hypothesis of mean zero difference. Except for very small samples, the alternative approach does not usually lead to substantially different conclusions. Here, incorporating the zero in Table 9.17, the signed rank test leads to a total SP of 0.391.

The procedure for conducting the Wilcoxon signed rank test for zero difference is summarized as follows.

The Wilcoxon signed rank test for zero difference

1. Obtain a data vector $d_1, d_2, \ldots, d_n$ of differences with 0s deleted.
2. Without regard to sign, order the differences from least to greatest, and allocate rank i to the ith absolute difference. In the event of ties, allocate the average rank to the tied absolute differences.
3. Now reintroduce the signs of the differences. Denote by w_+ the sum of the positive ranks. This is the Wilcoxon signed rank test statistic.
4. Obtain the SP for your test (usually by reference to a computer running appropriate software).
5. State your conclusions.

Now try the following exercise.

Exercise 9.14

Doubt has already been cast on the assumption of normality for the Shoshoni rectangles data (see Figure 9.2). A test for the Greek standard which does not make this assumption might be more appropriate. The data are given in *Chapter 8*, Table 8.4.

(a) Obtain a table of differences by subtracting the Greek standard from each of the ratios in Table 8.4 and allocate signed ranks to each difference.

(b) Use the signed ranks to calculate a value of the Wilcoxon signed rank test statistic w_+ and use it to test the null hypothesis that the Shoshoni rectangles conform to the Greek standard.

(c) Compare your result with that obtained under normal assumptions.

In fact, there is a sense in which the central limit theorem operates with the Wilcoxon signed rank statistic. For a sample of size n, it can be shown that under the null hypothesis of zero difference,

$$E(W_+) = \frac{n(n+1)}{4} \quad \text{and} \quad V(W_+) = \frac{n(n+1)(2n+1)}{24};$$

and it can further be shown that

$$Z = \frac{W_+ - E(W_+)}{SD(W_+)}$$

has a distribution which is approximately standard normal. The approximation is quite good, but should not be used for sample sizes that are very small.

***Example 9.11** Corneal thickness — normal approximation for the Wilcoxon signed rank test*

In Example 9.10 we looked at differences in corneal thickness of eyes of patients with one normal and one glaucomatous eye. There were seven such differences (excluding one with zero difference), so that

$$E(W_+) = \frac{7 \times 8}{4} = 14, \quad V(W_+) = \frac{7 \times 8 \times 15}{24} = 35.$$

In this case the sum of positive ranks is $w_+ = 7\frac{1}{2}$ so the corresponding observed value of Z is

$$z = \frac{w_+ - 14}{\sqrt{35}} = \frac{7\frac{1}{2} - 14}{\sqrt{35}} = -1.099;$$

the number $z = -1.099$ is at the 13.6% point of the standard normal distribution. We then have

$$SP(\text{obtained direction}) = SP(\text{opposite direction}) = 0.136;$$
$$SP(\text{total}) = 0.272.$$

Here, the sample size is only 7; the approximate SP is noticeably different from that given in Example 9.10. ■

Exercise 9.15

Use the value of the Wilcoxon signed rank statistic obtained in Exercise 9.14 and a normal approximation to test the null hypothesis that the Shoshoni rectangles conform to the Greek standard. Compare your SP with that obtained in Exercise 9.14.

9.4.3 The Mann–Whitney–Wilcoxon test

The idea of using ranks instead of the data values is a logical and appealing one. Furthermore, it has an obvious extension to testing two groups of data when a two-sample t-test may not be applicable because of lack of normality. The test itself was first proposed by H.B. Mann and D.R. Whitney in 1947, and modified by Wilcoxon; it turns out to be very nearly as powerful as the two-sample t-test. Strictly speaking, a test based on ranks does not test the same null hypothesis as the t-test, which tests for equal means. However, it is nevertheless a test of equality of locations of the two groups and using it as an alternative to the two-sample t-test is an approximation often made

Mann, H.B. and Whitney, D.R. (1947) On a test of whether one of two random variables is stochastically larger than the other. *Annals of Mathematical Statistics*, **18**, 50–60.

in practice. It may not be theoretically valid to regard the Mann–Whitney–Wilcoxon test as equivalent to a t-test, but using it is a valid alternative as a test of location.

Suppose we have two independent samples, A and B: the Mann–Whitney–Wilcoxon test may be used to test the hypothesis that the samples arise from the same population. The procedure is as follows.

Calculating the Mann–Whitney–Wilcoxon test statistic

1. Pool the two samples and then sort the combined data into ascending order (but do not lose sight of which data value belongs to which sample).
2. Allocate a rank to each data value, the smallest being given rank 1. As usual, if two or more data values are equal, allocate the average of the ranks to each.
3. Add up the ranks for each sample, writing

   $$u_A = \text{the sum of the ranks for sample A}$$
   $$u_B = \text{the sum of the ranks for sample B}$$

   Notice that if sample A is of size n_A, and if sample B is of size n_B, then the sum of u_A and u_B is

   $$\begin{aligned} u_A + u_B &= 1 + 2 + \cdots + (n_A + n_B) \\ &= \tfrac{1}{2}(n_A + n_B)(n_A + n_B + 1), \end{aligned}$$

   providing a useful check on your arithmetic.
4. The Mann–Whitney–Wilcoxon test statistic is u_A: very small or very large observed values imply rejection of the null hypothesis, suggesting respectively that A-values are 'too frequently' smaller than or larger than B-values.

The observed value u_A of U_A may then be compared with the null distribution of U_A to yield a SP for the test. Again, the null distribution of U_A is complicated: its derivation relies on combinatorial arguments, and this calculation would normally require the use of a computer program.

Alternatively, the null distribution of U_A may be approximated by a normal distribution:

$$U_A \approx N\left(\frac{n_A(n_A + n_B + 1)}{2}, \frac{n_A n_B(n_A + n_B + 1)}{12}\right).$$

The normal approximation is valid as long as the number of tied values (that is, values that are the same) in the pooled data set is not too great.

Thus approximation can be used for quite modest values of n_A and n_B; say, each of size 8 or more.

Example 9.12 Dopamine activity

In a study into the causes of schizophrenia, 25 hospitalized schizophrenic patients were treated with antipsychotic medication, and after a period of time were classified as psychotic or non-psychotic by hospital staff. Samples of cerebro-spinal fluid were taken from each patient and assayed for dopamine b-hydroxylase enzyme activity. The data are given in Table 9.18: the units are nmol/(ml)(hr)/mg of protein.

Sternberg, D.E., Van Kammen, D.P. and Bunney, W.E. (1982) Schizophrenia: dopamine b-hydroxylase activity and treatment response. *Science*, **216**, 1423–1425.

Table 9.18 Dopamine b-hydroxylase activity (nmol/(ml)(hr)/mg)

(A) Judged non-psychotic

0.0104	0.0105	0.0112	0.0116	0.0130	0.0145	0.0154	0.0156
0.0170	0.0180	0.0200	0.0200	0.0210	0.0230	0.0252	

(B) Judged psychotic

0.0150	0.0204	0.0208	0.0222	0.0226	0.0245	0.0270	0.0275
0.0306	0.0320						

The data may be pooled and ranked as shown in Table 9.19.

Table 9.19 Pooled and ranked data

	0.0104	0.0105	0.0112	0.0116	0.0130	0.0145	0.0150
Sample	A	A	A	A	A	A	B
Rank	1	2	3	4	5	6	7
	0.0154	0.0156	0.0170	0.0180	0.0200	0.0200	0.0204
Sample	A	A	A	A	A	A	B
Rank	8	9	10	11	$12\frac{1}{2}$	$12\frac{1}{2}$	14
	0.0208	0.0210	0.0222	0.0226	0.0230	0.0245	0.0252
Sample	B	A	B	B	A	B	A
Rank	15	16	17	18	19	20	21
	0.0270	0.0275	0.0306	0.0320			
Sample	B	B	B	B			
Rank	22	23	24	25			

The sample sizes are $n_A = 15$ and $n_B = 10$; so $n_A + n_B = 25$. Totalling the ranks gives

$$u_A = 140, \quad u_B = 185;$$

their sum is $140 + 185 = 325$. Also,

$$\tfrac{1}{2}(n_A + n_B)(n_A + n_B + 1) = \tfrac{1}{2}(25)(26) = 325$$

(a useful check on your progress so far, if you are not using a computer).

The expected value of U_A under the null hypothesis that the two samples are from identical populations is

$$\frac{n_A(n_A + n_B + 1)}{2} = \frac{15(15 + 10 + 1)}{2} = 195.$$

The observed value $u_A = 140$ is substantially smaller than this, but is it significantly smaller? When there are ties in the data (as there are here) the null distribution of U_A will be asymmetric; and for small samples, the distribution can be fiercely multimodal. Computation, and interpretation, of significance probabilities in such a context becomes quite difficult. Exact computation (feasible only with the aid of a machine) gives

$$SP(\text{total}) = 0.0015.$$

Alternatively, the variance of U_A under the null hypothesis is

$$\frac{n_A n_B(n_A + n_B + 1)}{12} = \frac{15 \times 10 \times 26}{12} = 325.$$

The observed value u_A has a corresponding z-value

$$z = \frac{140 - 195}{\sqrt{325}} = -\frac{55}{\sqrt{325}} = -3.051;$$

so the SP based on an appropriate normal distribution is given by

$$SP(\text{total}) = 2 \times \Phi(-3.051) = 0.002,$$

Clearly the dopamine activity for the two groups is very different. ■

Now try the following exercise.

Exercise 9.16

In Section 9.3 you saw how poor was the fit of a normal distribution to the data on recall of pleasant and unpleasant memories. Carry out a distribution-free test of the null hypothesis that there is no difference in recall time between pleasant and unpleasant memories. The data are given in *Chapter 2*, Table 2.10.

In this section you have seen that, by the simple expedient of replacing data values by ranks, it is possible to carry out tests of statistical hypotheses without making distributional assumptions.

9.5 Are the data typical or do they contain some 'unlucky' values?

From the moment that boxplots were introduced in *Chapter 1* as a graphical representation of variability in a set of data, you have been aware of the notion of statistical outliers. Similarly, in probability plots, occasionally there may be evidence of an unusual outlying point, or set of points, suggesting that apart from a few exceptional observations, the hypothesized model is adequate to describe the variability observed. Of course, outliers are more disconcerting if they are found in very small data sets.

The study of outliers and how to treat them is somewhat complex: it is only possible to give a little general guidance in this course. Broadly speaking, their treatment can be separated into two basic approaches which depend upon how many outliers appear in the data and how far you are prepared to go in believing that you have been unlucky enough to obtain a few atypical values—rather than believing that the distributional assumptions are not viable.

9.5.1 Data containing very few apparent outliers

Up to now, it has usually been suggested that in a data set where relatively few data points appear to be atypical, then it would be wise to exclude them from any statistical analysis. (The alternative is to include them, and accept that possibly they might have very considerable influence on the general conclusions.)

Example 9.13

In Example 9.8 we looked at the numbers of children born to mothers married at age 20–24, and found a transformation $y = x^{3/4}$, after which we treated the data as normal and carried out a t-test. It is instructive to look at a comparative boxplot of the transformed data.

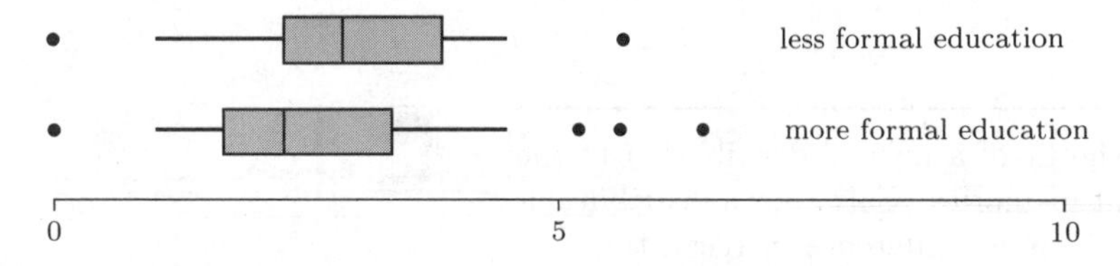

Figure 9.9 Comparative boxplot of transformed data

There is a case for regarding all of the transformed data values outside the range of the whiskers as possible outliers. If they are omitted, and a two-sample t-test is carried out on the remaining data values, a t-statistic of 2.643 on 145 degrees of freedom is obtained with a total SP of 0.009. This may be contrasted with the result, obtained in Example 9.8, of 2.218 on 162 degrees of freedom for a total SP of 0.028. Regarding the values within the range of the whiskers as somehow more 'typical' produces a more 'convincing' conclusion. ■

But throwing away 17 outliers is a rather drastic step and you would need to be very sure of your ground. Merely 'lying outside the whiskers' is hardly a sufficient reason for trimming the data. You should certainly suspect such data points, and look at them more closely. Perhaps there is additional evidence that they are non-typical: for example, they might have an unusual medical history.

9.5.2 Data containing too many apparent outliers

Are the outliers in Figure 9.9 really outliers, or should we adopt a modelling distribution with fat tails and a high probability of values far from the mean? How many data values is it reasonable to treat as outliers? The previous example must be pushing the limits; out of a total of 164 data points, we have assumed no fewer than 17 to be outliers.

There is no hard and fast rule which decides how much the data may be trimmed while assumptions of normality are retained. In practice, it is best to remove no more than one or two values and, if in doubt, to keep all the values and to revert to a distribution-free method. Using ranks instead of data values loses information about how far apart the values actually are but, on the other hand, it removes sensitivity to abnormally large or abnormally small values. If decisions about which method to use seem unduly vague, you should remember that there is no definitively right or wrong way of performing any kind of statistical analysis. All you can do is use your common sense to the best of your ability.

Exercise 9.17

The following data are a set of radio-carbon age determinations, in years, from the Lake Lamoka site: eight samples were used.

Long, A. and Rippeteau, B. (1974) Testing contemporaneity and averaging radio-carbon dates. *American Antiquity*, **39,** 205–215.

Table 9.20 Radio-carbon dating

Sample number	Radio-carbon age determination
C-288	2419
M-26	2485
C-367	3433
M-195	2575
M-911	2521
M-912	2451
Y-1279	2550
Y-1280	2540

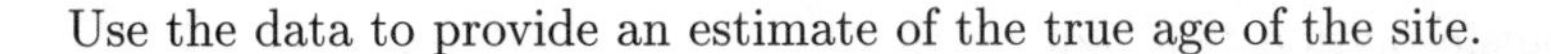

Use the data to provide an estimate of the true age of the site.

Summary

1. When sample sizes are small, the technique of probability plotting can be used to test informally the hypothesis that a set of data $y_1, y_2, \ldots, y_n$ arises from some stated family of probability distributions. To test for normality, the points $y_{(1)}, y_{(2)}, \ldots, y_{(n)}$ in ascending order are plotted against the points $x_1, x_2, \ldots, x_n$, where

$$\Phi(x_i) = \frac{i}{n+1}, \quad i = 1, 2, \ldots, n.$$

If a straight line fits the scatter plot of points, then the normal model is plausible. The parameters μ and σ can be roughly estimated by the intercept and slope of the line respectively.

2. To test a hypothesized exponential model, the points $y_{(1)}, y_{(2)}, \ldots, y_{(n)}$ are plotted against the points

$$x_i = -\log\left(\frac{n+1-i}{n+1}\right), \quad i = 1, 2, \ldots, n.$$

If the exponential model is plausible a straight line through the origin, with slope equal to the mean of the exponential distribution, will fit the scatter plot of points.

3. Tests for other probability models may require a transformation of the data before the idea of probability plotting can be applied.

4. The chi-squared test for goodness of fit relies on a comparison of frequencies observed and frequencies expected under a hypothesized model. The test statistic is

$$\chi^2 = \sum_{i=1}^{k} \frac{(O_i - E_i)^2}{E_i},$$

where the data are allocated to k categories in such a way that the expected frequencies in each category are at least 5. Under the null hypothesis that the data arise from the hypothesized model, the test statistic χ^2 follows a chi-squared distribution with $k - p - 1$ degrees of freedom, where p is the number of parameters that required estimation in order to calculate the expected frequencies. The SP of the chi-squared test is given by the upper tail probability of $\chi^2(k - p - 1)$ for values exceeding χ^2.

5. Sometimes if data are heavily skewed a transformation that reduces the skewness will also render the data more 'bell-shaped' and in that sense more plausibly normal. Experimentation with the ladder of powers (including the log transformation) often leads to a helpful power transformation.

6. A test for zero difference analogous to Student's t-test for zero mean difference, but not requiring the assumption of normality, is provided by the Wilcoxon signed rank test. Significance probabilities for the test statistic w_+ (the sum of positive ranks when any zeros in the data are deleted and the data are ranked) may be calculated exactly if appropriate software is available, or approximately using the result

$$W_+ \approx N\left(\frac{n(n+1)}{4}, \frac{n(n+1)(2n+1)}{24}\right),$$

where n is the sample size.

7. The Mann–Whitney–Wilcoxon test may be used to test the hypothesis that two samples of data (A and B) arise from the same population, where no normality assumptions are made. Significance probabilities for the test statistic U_A (the sum of the A-ranks after pooling of the data) may be calculated exactly (using a computer), or approximately using

$$U_A \approx N\left(\frac{n_A(n_A + n_B + 1)}{2}, \frac{n_A n_B(n_A + n_B + 1)}{12}\right),$$

where n_A and n_B are the respective sample sizes.

Chapter 10
Explanatory relations

In this chapter observations are collected on a random variable whose mean alters with changing circumstances (for instance, with passing time). Techniques are developed in an attempt to model this dependence and to make predictions.

So far in this course models for variation have been developed that are sufficient for the entire population under study. For instance, in *Chapter 2*, Example 2.17, data were collected on the heights of elderly women: a sample of size 351 from the community was used. The data collection formed part of a study into the disease osteoporosis, and in this context the population of elderly women was the only one of interest. There was evident variation in the sample (see Figure 2.19) and it was suggested that a normal distribution with appropriately fitted parameters μ and σ might be adequate to model the variation in the population. We did not seek to model the heights of females in general (because in this context the variation in height of teenage girls, for instance, was of no interest). But in the population in general, we know that height depends on age at least up to about 15 or 16 years old. Clothing manufacturers, for instance, would need to be aware of the extent of this variation, and how it depended on age. Figure 10.1 illustrates the variation in height of schoolboys aged between 6 and 10 years old, taken from an early study in America. For each of the five age groups, a histogram for the variation in heights was drawn. Then each histogram was composed into the single diagram in the way shown.

This early study explored the relationship between the age and heights of 24 500 Boston schoolboys, conducted for the Massachusetts Board of Health in 1877 by H.P. Bowditch. The figure is adapted from Peters, W.S. (1987) *Counting for Something—Statistical Principles and Personalities.* Springer-Verlag, New York, p. 90.

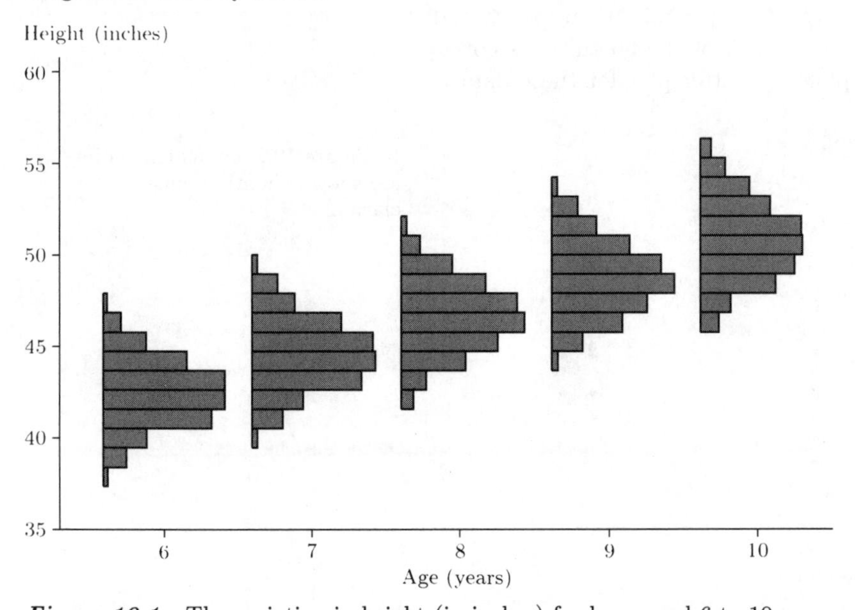

Figure 10.1 The variation in height (in inches) for boys aged 6 to 10

It is clear that the distribution in heights alters with the increasing age of the boys. In fact, the figure suggests very clearly for each age group that a normal distribution would provide a very adequate model for the variation in heights; that the mean increases roughly linearly with age (at least between the ages 6 and 10; the trend would not continue into middle age, of course); it also seems that the variance in height does not alter much with age, if at all. So, perhaps a model where the parameter μ altered linearly with age but where σ did not alter at all, would provide a very good model for the variation in height of this developing population.

This kind of dependence is the subject of the present chapter. Here are two more examples.

Example 10.1 Finger-tapping

An experiment was carried out to investigate the effect of the stimulant caffeine on performance on a simple physical task. Thirty male college students were trained in finger-tapping. They were then randomly divided up into three groups of 10, and the students in each group received different doses of caffeine (0, 100 and 200 mg). Two hours after treatment, each subject was required to do finger-tapping, and the number of taps achieved per minute was recorded. The question of interest was: does caffeine affect performance on this task?

Draper, N.R. and Smith, H. (1981) *Applied regression analysis*, 2nd edn. John Wiley and Sons, New York, p. 425. Finger-tapping is a fairly standard psychological task performed by subjects to assess alertness through manual (or, strictly, digital) dexterity. Of course, a more informative experiment might have been to collect 'Before' and 'After' tapping frequency data for each of the 30 subjects, and then to explore the difference achieved for each subject, and how that difference related to the caffeine dose. However, we must deduce what we can from the data supplied.

The recorded figures are given for each of the 30 subjects in Table 10.1.

Table 10.1 Finger-tapping data

Caffeine dose (mg)	Finger-taps per minute									
0	242	245	244	248	247	248	242	244	246	242
100	248	246	245	247	248	250	247	246	243	244
200	246	248	250	252	248	250	246	248	245	250

There are not enough data here to construct meaningful or informative histograms for each of the three dose levels. The appropriate graphical representation to use is the scatter plot, where the caffeine dose is measured along the horizontal axis, and recorded finger-taps per minute are measured up the vertical axis. The result recorded for each of the 30 subjects corresponds to a single point on the scatter plot. A scatter plot for these data is given in Figure 10.2.

In Figure 10.2, coincident points are shown slightly displaced for clarity.

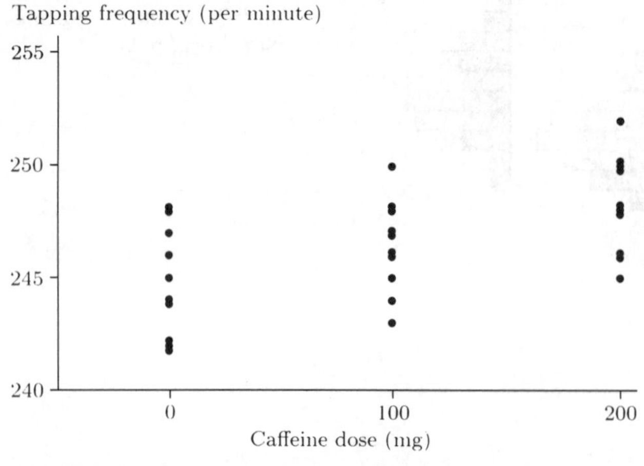

Tapping frequency (per minute)
255
250
245
240
0
100
200
Caffeine dose (mg)

Figure 10.2 Tapping frequency against caffeine dose

In this example, it is not possible to deduce very much about the shape of the variation in tapping frequencies at each dose level, but the range is easy to determine (a range of about 8 beats per minute, from somewhere in the mid-240s to somewhere in the mid-250s). More significantly, there is some evidence of a possible upwards trend from the graph in Figure 10.2—that is, there is some evidence that increasing the caffeine dose leads to an increase in tapping frequency. The questions that might be asked are: is there a genuine underlying trend or is its manifestation here merely an accident of the data? If the trend is there, how can we model it? How can we account for the variation between responses at the same dose level? One might then use the model to predict finger-tapping frequencies at higher (or just different) dose levels. Without more knowledge of the effect of caffeine on the human physiology, one should be wary of extending the trend and predicting finger-tapping performance at very much higher doses than 200 mg. ■

This is a typical problem in the statistical exercise known as **regression**. Regression analysis involves the development and use of statistical techniques designed to reflect the way in which variation in an observed random variable changes with changing circumstances.

Example 10.2 Forbes' data

In the 1840s and 1850s the Scottish physicist James Forbes was interested in developing a method for estimating altitude on a hillside from measurement of the boiling point of water there. The temperature at which water boils is affected by atmospheric pressure, which in turn is affected by altitude. Forbes concluded that it would be possible for climbers to estimate their height from the temperature at which water boiled. Carrying altimeters (that is, barometers) up and down hills is a tricky business: boiling a pan of water and measuring the temperature at which boiling point occurs is less troublesome.

Forbes, J.D. (1857) Further experiments and remarks on the measurement of heights by the boiling point of water. *Transactions of the Royal Society of Edinburgh*, **21**, 135–143.

It would be useful to have data available for the way boiling point varies with altitude: in fact, the data in Table 10.2 give the boiling point (℉) and atmospheric pressure (inches Hg—that is, inches of mercury) at 17 locations in the Alps and in Scotland.

Table 10.2 Forbes' data

Boiling point (℉)	Pressure (inches Hg)
194.5	20.79
194.3	20.79
197.9	22.40
198.4	22.67
199.4	23.15
199.9	23.35
200.9	23.89
201.1	23.99
201.4	24.02
201.3	24.01
203.6	25.14
204.6	26.57
209.5	28.49
208.6	27.76
210.7	29.04
211.9	29.88
212.2	30.06

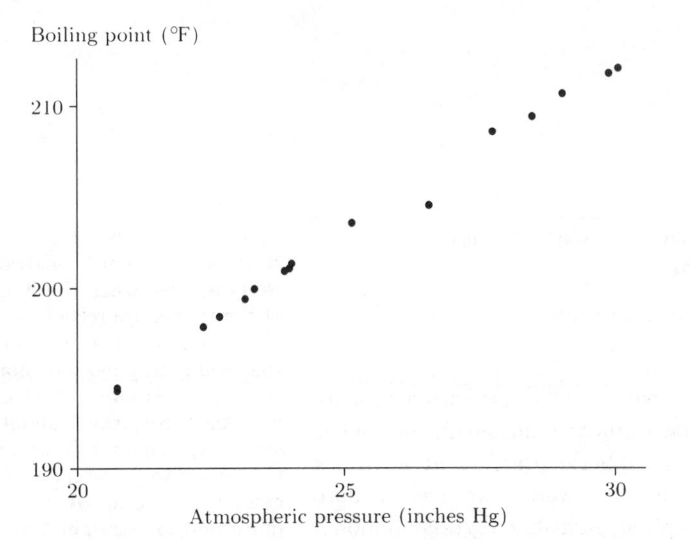

Figure 10.3 Forbes' data: boiling point of water against atmospheric pressure

It is the different atmospheric pressure prevailing at each location that 'causes' differences in the observed boiling point of water. You may know that the higher the altitude, then the lower the pressure, and the lower the boiling point of water. In Figure 10.3 pressure is measured along the horizontal axis, and temperature is measured up the vertical axis. There is a very strong suggestion of a straight-line relationship between the boiling point of water and atmospheric pressure. It is a feature of science that many relationships can be formulated (sometimes approximately) in very simple terms, in which case they are sometimes called *natural laws* or *laws of nature*—or just *laws*. For example, Hooke's Law states that the extension produced in a spring is proportional to the external force applied to the spring. In fact, most such 'laws' merely articulate experimental results perceived to be true under certain circumstances. ■

However, one should not jump too rapidly to easy and convenient conclusions. Often, it is found that a list of data pairs appears to suggest a linear law (a straight-line relationship) over the range investigated, but when further measurements are taken beyond that domain, it becomes clear that a non-linear law may be required overall. The case of atmospheric pressure and altitude provides an example of this. The scatter plot in Figure 10.4 shows atmospheric pressure (as a percentage of pressure at sea level) plotted against altitude (in metres, at various points on the earth's surface). You can see that pressure decreases with increasing altitude, and that the relationship appears to be linear, at least over the range explored, which was from sea level up to 1000 m.

Figures 10.4 and 10.5 are from The Open University (1992) MS284 *An Introduction to Calculus*, Unit 7, *Numbers from Nature*, Milton Keynes, The Open University.

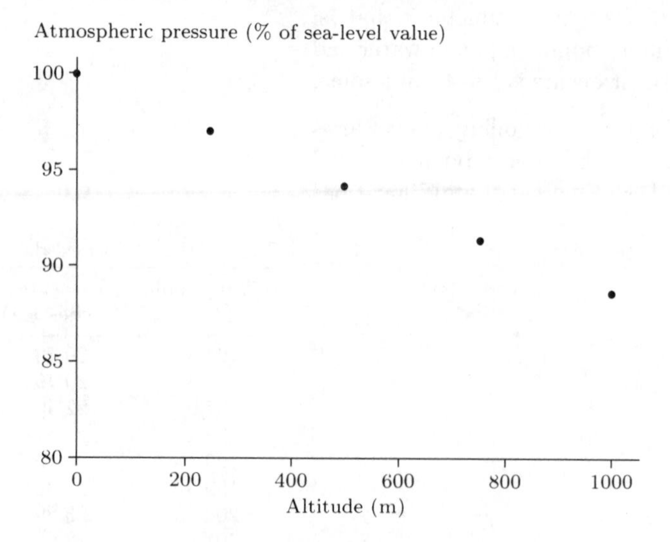

Figure 10.4 Pressure at different altitudes up to 1000 metres

You should be aware that statisticians often fit a straight line to data, even when the limitations of the model are recognized and when there are reasons to know that something more elaborate is really appropriate. Those of you who know something about Taylor series expansions will understand that some very complicated curves can be approximated, and often quite well, by straight lines, over limited domains.

If, however, you were to explore what happens when further measurements are taken outside this range, you would find that a linear relationship no longer holds, as the plot of measured values of atmospheric pressure at altitudes up to 30 000 metres shows in Figure 10.5. In other words, we need a more sophisticated mathematical model than a simple straight-line regression model to represent the dependence.

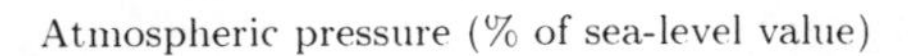

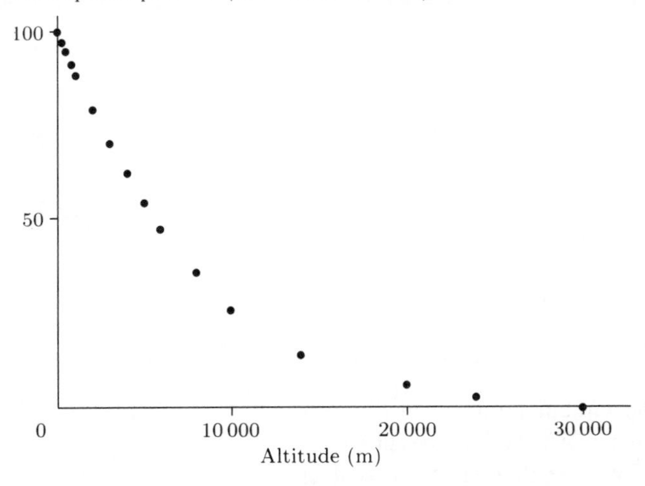

Figure 10.5 Pressure at different altitudes up to 30 000 metres

In order to plot a scatter diagram for this sort of 'paired' data, we need to establish a criterion for which of the two measured variables to plot along the horizontal axis, and which variable to plot up the vertical axis.

Often, the data list will have arisen as the result of an experiment specifically designed to investigate the effect that changes in one variable have on the other. For example, the effect on the extended length of a stretched rubber band when different masses are suspended from it; or the change in recorded atmospheric pressure following a change in altitude; or how the time taken for a planet to orbit the Sun depends on its distance from the Sun.

In yet another context, researchers might be interested in the way car engine size affects urban fuel consumption, or in how driving the same car at different speeds alters its fuel consumption.

In all these cases one would naturally speak of changes in one variable having an effect on, or causing differences in, the resulting measurements of another. For example, 'driving this car at a constant speed of 70 m.p.h. leads to fuel consumption rates roughly double those when the same car is driven at 40 m.p.h.'; or 'climbing from sea level to a height of 1000 metres caused a drop of about ten per cent in recorded atmospheric pressure'. It would not be at all natural or even sensible to speak of reduced fuel consumption 'causing' a change in speed; or of altered atmospheric pressure 'affecting' a climber's height; or of one car engine being larger than another 'because' it uses more fuel.

The case is not always clear-cut: but usually there is a sense in which controllable or intentional differences in one variable (often the basis of the design of the experiment) lead to changes in another.

The first variable is called the **explanatory variable** and is conventionally plotted horizontally, the second is called the **response variable** and is plotted vertically.

There are many different terms for the two variables. Others include the *independent variable* and the *dependent variable*. The explanatory variable may also be called the *predictor variable* or the *regressor*.

One can think of a scatter diagram as displaying some sort of functional relationship between the response and explanatory variables (without drawing too strong an analogy because of the variability involved). It is our aim in this chapter to discover something about the function which best expresses that relationship.

10.1 Some examples and a brief history

There are many contexts in which regression data arise, and in this section several examples are given. In each case, questions posed are those relating to what the data can tell us about the regression relationship that exists between the response and the explanatory variables.

10.1.1 Some examples

Example 10.3 revisits a data set from earlier in the course.

Example 10.3 Divorce data

In *Chapter 6*, Table 6.4, data are given on the annual numbers of divorces recorded in England and Wales between the years 1975 and 1980. The data suggest a roughly linear upward trend, but of course there is some scatter about the trend due to random variation.

While it would not be entirely sensible to speak of the passage of time as 'causing' an increase in divorce, the problem here can be posed as one in regression, because the divorce rate seems to vary with time, and in a way that is interesting enough to be worth investigating.

When these data were introduced in *Chapter 6*, a straight line model of the form

$$Y_i = \alpha + \beta x_i + W_i, \quad i = 1, 2, \ldots, 6$$

was suggested, where x_i, $i = 1, 2, \ldots, 6$, denotes the year (reckoned from 1900 for convenience), Y_i denotes the annual number of divorces (in thousands) and where W_i is a random term to account for the scatter. The underlying straight-line trend is given by the term $\alpha + \beta x_i$; and in Figure 10.6 three trend lines $\widehat{l}_1$, $\widehat{l}_2$ and $\widehat{l}_3$, with slopes $\widehat{\beta}_1$, $\widehat{\beta}_2$ and $\widehat{\beta}_3$ respectively, are suggested.

Figure 10.6 is a copy of Figure 6.2 in *Chapter 6*, included here for convenience.

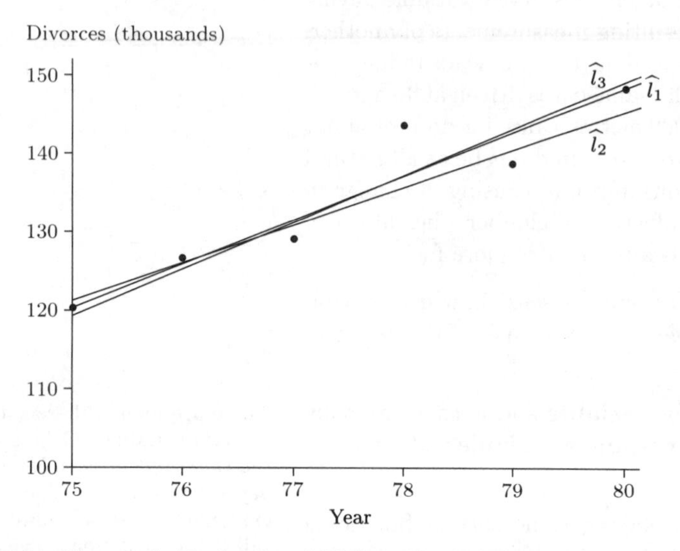

Figure 10.6 Divorces in England and Wales, 1975–1980: trend lines

An essential fact to remember is that the estimator of the slope is a random variable. Each of the three lines shown in Figure 10.6 is an estimate of the true underlying trend line, resulting from applying three different estimating formulas to the collected data. The question was: which of the three lines in Figure 10.6 was in some sense the best expression for the underlying increase in the divorce rate? To answer this question, the accuracy of the estimator for the slope of each line was assessed. It was established that in fact all three lines had expected slope β—and in that sense, there is nothing to choose between them. However, the slope estimator that had the smallest variance was $\widehat{\beta}_2$, and in that sense the second of the three lines, $\widehat{l}_2$, is the best expression of the underlying trend.

The slope of the line $\widehat{l}_1$ was written $\widehat{\beta}_1$, and so on. We found

$$V(\widehat{\beta}_1) = 0.0800\sigma^2,$$
$$V(\widehat{\beta}_2) = 0.0625\sigma^2,$$
$$V(\widehat{\beta}_3) = 0.0741\sigma^2.$$

The line could be used, for instance, to forecast future divorce rates on the assumption that the upward trend would continue into the 1980s. In fact, that would have been a false assumption, for it turns out that while the number of divorces continued its annual rise, there was a noticeable reduction in the rate of increase. Extrapolation can be dangerous! ■

Example 10.4 describes the results of an investigation into the strength of wooden beams.

Example 10.4 The strength of beams

The data set lists the results of an investigation into specific gravity and moisture as *predictors* of the strength of timber beams. Ten beams were used in the experiment. The data are given in Table 10.3. The response variable (strength) is given with the two explanatory variables (specific gravity and moisture content).

Draper, N.R. and Stoneman, D.M. (1966) Testing for inclusion of variables in linear regression by a randomization technique. *Technometrics*, **8**, 695–699.

Table 10.3 The strength of beams

Strength	Specific gravity	Moisture content
11.14	0.499	11.1
12.74	0.558	8.9
13.13	0.604	8.8
11.51	0.441	8.9
12.38	0.550	8.8
12.60	0.528	9.9
11.13	0.418	10.7
11.70	0.480	10.5
11.02	0.406	10.5
11.42	0.467	10.7

A plot of strength against specific gravity (see Figure 10.7(a)) strongly suggests some sort of linear relationship between response and explanatory variable—though possibly there is an outlier at (0.499, 11.14). On the other hand, the scatter plot of strength against moisture content (see Figure 10.7(b)), while suggestive of a downward trend, is not so convincing.

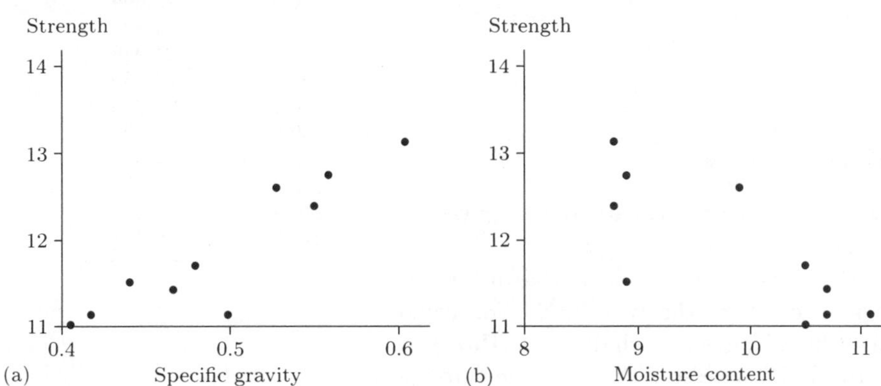

Figure 10.7 (a) Strength against specific gravity (b) Strength against moisture content

It need not necessarily be the case that the suggestion of a functional relationship is as strong as that suggested by Figure 10.7(a). We shall see examples where there is rather little evidence of it, but where regression analysis has nevertheless been usefully employed. ■

Example 10.5 extends Forbes' researches over a wider range of investigation. Forbes considered variation in boiling point of water for atmospheric pressures ranging from about 21 to 30 inches Hg. In this example we explore the results of experiments taken over the range from 15 inches Hg to about 30 inches Hg.

Example 10.5 Hooker's data

As well as his own results Forbes quotes Dr Joseph Hooker's researches into the relationship between altitude and the boiling point of water. The data in Table 10.4 are the results of 31 measurements taken in the Himalayas.

A scatter plot for these data is shown in Figure 10.8. Again, there is the strong suggestion of a linear relationship between the response variable and the explanatory variable—that is, of some physical 'law' connecting the two, which may usefully be expressed as a linear law. Just as for the divorce data, a reasonable model might again be written in the form

$$Y_i = \alpha + \beta x_i + W_i,$$

at least for values of x between about 15 and about 30. We have not explored the usefulness of the model outside this range. It is interesting that Hooker was apparently able to measure atmospheric pressure to an accuracy of three decimal places (whereas Forbes' data are given to two decimal places).

Table 10.4 Hooker's data

Boiling point (°F)	Pressure (inches Hg)
180.6	15.376
181.0	15.919
181.9	15.928
181.9	16.106
182.4	16.235
183.2	16.385
184.1	16.817
184.6	16.881
184.1	16.959
185.6	17.062
186.0	17.221
185.7	17.267
188.8	18.356
188.5	18.507
189.5	18.869
190.6	19.386
191.1	19.490
191.4	19.758
193.6	20.212
193.4	20.480
195.6	21.605
196.3	21.654
197.0	21.892
196.4	21.928
199.5	23.030
200.1	23.369
200.6	23.726
202.5	24.697
208.4	27.972
210.2	28.559
210.8	29.211

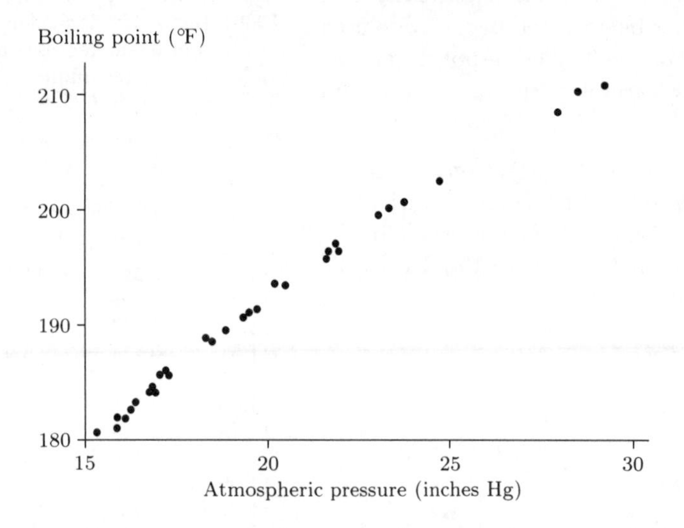

Figure 10.8 Hooker's data: boiling point of water against atmospheric pressure

As a comparison of the two sets of results, it would be interesting to fit straight lines to each set, and then to compare the two lines. You could explore this informally, fitting straight lines by eye to each data set. However, in Section 10.2 we shall establish criteria for assessing the best line through scattered data. If the results (that is, the best lines) turn out to be different for the two sets of data, it will be necessary (but also interesting) to speculate on whether that difference is a real difference, or just a manifestation of random variation, and if so, on the causes of the difference. ■

Example 10.6 is about comparing distances between two points as suggested by a map, and as the corresponding road distances turn out to be in practice.

Example 10.6 Map (straight-line) and road distances compared

Maps can sometimes be deceiving in the impression they give of distances between two locations. The data in Table 10.5 give the map distance in miles (that is, the straight-line distance) and the distance by road between twenty different pairs of locations in Sheffield, England. The data raise the questions: what is the relationship between the two variables? How well can the road distance be *predicted* from the map distance?

Gilchrist, W. (1984) *Statistical modelling.* John Wiley and Sons, Chichester, p. 5.

Table 10.5 Map distances and actual road distances in Sheffield, England

Road distance (miles)	Straight-line distance (miles)
10.7	9.5
11.7	9.8
6.5	5.0
25.6	19.0
29.4	23.0
16.3	14.6
17.2	15.2
9.5	8.3
18.4	11.4
28.8	21.6
19.7	11.8
31.2	26.5
16.6	12.1
6.5	4.8
29.0	22.0
25.7	21.7
40.5	28.2
26.5	18.0
14.2	12.1
33.1	28.0

In every case the road distance exceeds the map distance. One might have expected this: roads tend to have bends, adding to the distance between two points. A scatter plot of road distance against map distance is given in Figure 10.9. Again, the plot suggests a roughly linear relationship between the two measures. (That is, the plot suggests that some sort of linear relationship between the two measures might provide an adequate model.)

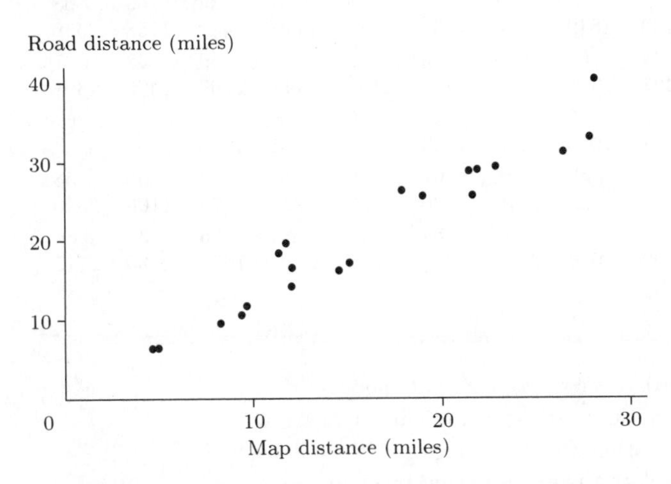

Figure 10.9 Road distance against map distance between pairs of locations in Sheffield (miles)

Can you see that here the appropriate model might be a little different to that considered in previous examples? If the map distance between two points is zero (that is, if the two points are the same) then the road distance will also be zero. The straight line fitted to the data should go through the origin. Our model relating Y_i (road distances) to x_i (map distances) might now reasonably be expressed in the form

$$Y_i = \gamma x_i + W_i,$$

where the parameter γ represents the factor by which map distances need to be multiplied to give an estimate of the actual road distance (which may or may not be a useful estimate). Again, the random term W_i accounts for the scatter identified in the data. ■

The letter γ is the Greek lower-case letter gamma. It is useful to distinguish in this way between the slopes of the two straight line models $Y_i = \alpha + \beta x_i + W_i$ and $Y_i = \gamma x_i + W_i$.

In Example 10.7 there is very little pattern to be identified in the scatter of data points.

Example 10.7 Examination scores

It is not always the case that there is a clear relationship between two variables: sometimes a scatter plot suggests that there is no relationship at all, despite what one might have suspected, or expected. Table 10.6 gives examination scores (out of 75) for 134 candidates, and the number of seconds that each candidate needed to complete the paper.

Basak, I., Balch, W.R. and Basak, P. (1992) Skewness: asymptotic critical values for a test related to Pearson's measure. *J. Applied Statistics*, **19**, 479–487.

Table 10.6 Scores (out of 75) and duration (seconds)

Score	49	49	70	55	52	55	61	65	57	71	49	48	49	69
Time	2860	2063	2013	2000	1420	1934	1519	2735	2329	1590	1699	1816	1824	1899
Score	44	53	49	52	53	36	61	68	67	53	33	64	57	56
Time	1714	1741	1968	1721	2120	1435	1909	1707	1431	2024	1725	1634	1949	1278
Score	41	40	42	40	51	53	62	61	49	54	57	71	45	70
Time	1677	1945	1754	1200	1307	1895	1798	1375	2665	1743	1722	2562	2277	1579
Score	58	62	28	72	37	67	51	55	68	58	61	43	60	53
Time	1785	1068	1411	1162	1646	1489	1769	1550	1313	2472	2036	1914	1910	2730
Score	51	51	60	64	66	52	45	48	51	73	63	32	59	68
Time	2235	1993	1613	1532	2339	2109	1649	2238	1733	1981	1440	1482	1758	2540
Score	35	64	62	51	52	44	64	65	56	52	59	66	42	67
Time	1637	1779	1069	1929	2605	1491	1321	1326	1797	1158	1595	2105	1496	1301
Score	48	56	47	68	58	59	45	31	47	56	38	47	65	61
Time	2467	1265	3813	1216	1167	1767	1683	1648	1144	1162	1460	1726	1862	3284
Score	45	63	66	44	57	56	56	54	61	58	46	62	68	58
Time	1683	1654	2725	1992	1332	1840	1704	1510	3000	1758	1604	1475	1106	2040
Score	47	66	61	58	45	55	54	54	54	41	65	66	38	51
Time	1594	1215	1418	1828	2305	1902	2013	2026	1875	2227	2325	1674	2435	2715
Score	49	49	51	42	61	69	42	53						
Time	1773	1656	2320	1908	1853	1302	2161	1715						

A scatter plot of these data is given in Figure 10.10. You can see that there is simply a cloud of points, with no particular functional relationship (linear or otherwise) suggested between the variables. In other words, it does not look as though the time it takes a candidate to complete the examination provides any good prediction for what the candidate's score might be.

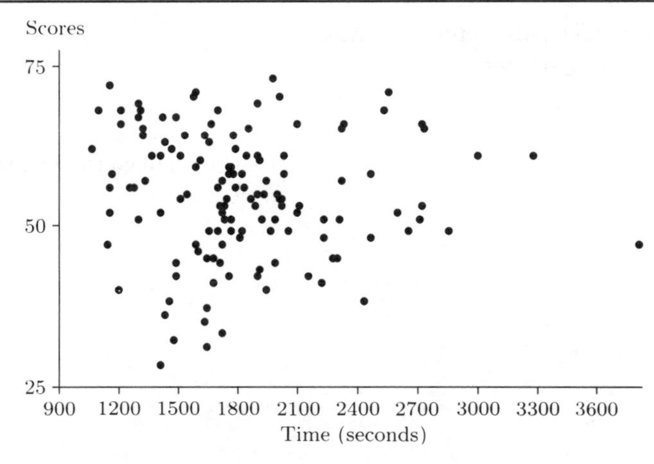

Figure 10.10 Examination scores against duration

Often, an apparent functional relationship between explanatory variable and response variable prompts one to speculate about its cause. Here, one might usefully speculate about the causes for the *lack* of any identifiable link between examination time and score. These data make it all the more interesting that, in some tests for assessing intelligence, the time taken for candidates to complete the paper, as well as their score on the paper, is used to provide an overall assessment. ■

Example 10.8 demonstrates that a functional relationship identified between explanatory variable and response variable is not necessarily a linear one.

Example 10.8 d'Arcy Thompson's duckweed data

In his book *On Growth and Form* d'Arcy Thompson recounts an experiment into the growth of duckweed. Growth was monitored by counting duckweed fronds at weekly intervals for eight weeks, starting one week after the introduction of a single duckweed plantlet into a growth medium (in this case, pure water). The data are given in Table 10.7.

d'Arcy Thompson refers to work summarized in Bottomley, W.B. (1914) Some accessory factors in plant growth and nutrition. *Proceedings of the Royal Society, Series B*, **88**, 237–247. Two growth media are compared in Bottomley's paper: pure water (the data given here), and water to which a small amount of peat fertilizer ('auxitone') was added. The fronds were counted using an instrument known as an *auxiometer*.

Table 10.7 Duckweed growth, monitored weekly

Week	1	2	3	4	5	6	7	8
Fronds	30	52	77	135	211	326	550	1052

It is known that initially (week 0) there were 20 fronds on the plantlet that was placed in the water. A scatter plot of these data is given in Figure 10.11. You can see the very strong suggestion of a functional relationship between duckweed growth and passing time: but the relationship is not a linear one, and it would not be useful in this case to go through the procedures for fitting a straight line.

There are two possible approaches in a case such as this. The first is to try to fit a curve to the data, rather than a straight line. What is needed is some sort of power or exponential law to model the growth observed. Here, an obvious candidate might be a formula expressing exponential growth, say,

$$Y_i = 20e^{\lambda x_i} + W_i,$$

where time x is measured in weeks. Here, the model parameter requiring estimation is λ. The random term W_i accounts for any scatter.

We might wish to constrain any fitted curve to go through the point $(0, 20)$.

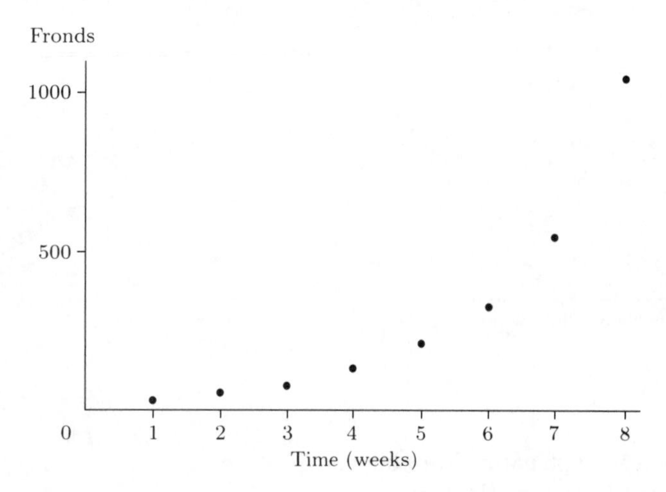

Figure 10.11 Duckweed growth

Alternatively, it might be possible to transform the frond count in some way that the resulting variable (maybe log(fronds)) varies approximately linearly with time. We shall briefly explore regression models where the relationship is not linear in Section 10.5. ∎

Here is a further example in which, once again, a clear pattern is evident relating response and explanatory variables, but the relationship is not linear.

Example 10.9 Paper strength data

The tensile strength (p.s.i.) of Kraft paper was measured against the percentage of hardwood in the batch of pulp from which the paper was produced. There were 19 observations recorded. The data are given in Table 10.8.

Table 10.8 Paper strength against hardwood content

Strength (p.s.i)	Hardwood content (%)
6.3	1.0
11.1	1.5
20.0	2.0
24.0	3.0
26.1	4.0
30.0	4.5
33.8	5.0
34.0	5.5
38.1	6.0
39.9	6.5
42.0	7.0
46.1	8.0
53.1	9.0
52.0	10.0
52.5	11.0
48.0	12.0
42.8	13.0
27.8	14.0
21.9	15.0

Joglekar, G., Schuenemeyer, J.H. and LaRiccia, V. (1989) Lack-of-fit testing when replicates are not available. *American Statistician*, **43**, 135–143. The word 'Kraft' actually refers to a method of paper production. The paper is of a thick brown type used for wrapping.

If tensile strength is plotted against hardwood content as in Figure 10.12, you can see once more a very evident functional relationship, but it is not a linear one.

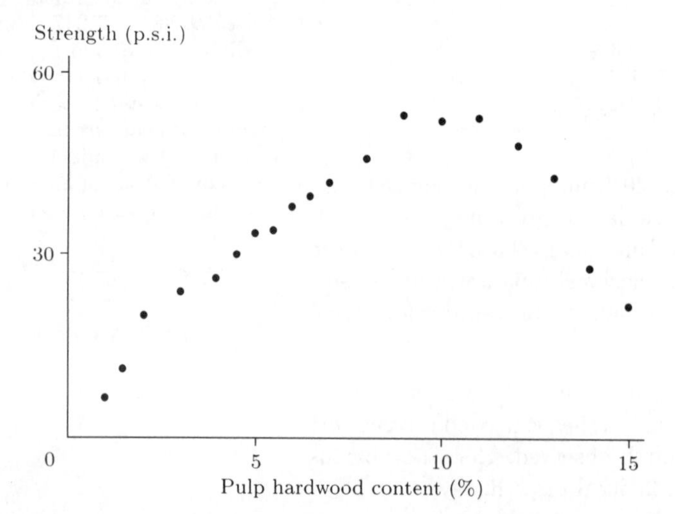

Figure 10.12 Paper strength against pulp hardwood content

It appears (from this experiment, anyway), that Kraft paper is at its strongest for some intermediate level of pulp hardwood content (about 10%). It would be the task of a statistician, given the data, to provide as good an estimate as possible of the optimal hardwood content. ■

The word 'optimal' used here suggests that strength is the only consideration for the ideal pulp hardwood content. There may in fact be other constraints, such as production costs.

In Example 10.10, there is a clear downward trend. What is unusual in this case is that the scatter increases with increasing x. This, if it is a genuine increase, conflicts with the regression model

$$Y_i = \alpha + \beta x_i + W_i,$$

where, as we shall see, it is a model assumption that W_i has mean 0 and *constant* variance σ^2.

Peixoto, J.L. (1990) A property of well-formulated polynomial regression models. *American Statistician*, **44**, 26–30. The authors report a study in which a linear relationship is assumed between temperature and latitude; then, after adjusting for latitude, a cubic polynomial in longitude accurately predicts temperature. This is an example of *multiple regression*—more than one explanatory variable—and you need not worry about the details.

Example 10.10 *Temperature differences on the Earth's surface*

The data listed in Table 10.9 give the normal average January minimum temperature (℉) with latitude (°N) and longitude (°W) for 56 cities in the United States. (Average minimum temperature for January is found by adding together the daily minimum temperatures and dividing by 31. For this table the January average minima for the years 1931 to 1960 were averaged over the 30 years.)

Table 10.9 Temperature (℉) against latitude (°N) and longitude (°W)

City	Temperature (℉)	Latitude (°N)	Longitude (°W)	City	Temperature (℉)	Latitude (°N)	Longitude (°W)
Mobile, AL	44	31.2	88.5	Omaha, NB	13	41.9	96.1
Montgomery, AL	38	32.9	86.8	Concord, NH	11	43.5	71.9
Phoenix, AZ	35	33.6	112.5	Atlantic City, NJ	27	39.8	75.3
Little Rock, AR	31	35.4	92.8	Albuquerque, NM	24	35.1	106.7
Los Angeles, CA	47	34.3	118.7	Albany, NY	14	42.6	73.7
San Francisco, CA	42	38.4	123.0	New York, NY	27	40.8	74.6
Denver, CO	15	40.7	105.3	Charlotte, NC	34	35.9	81.5
New Haven, CT	22	41.7	73.4	Raleigh, NC	31	36.4	78.9
Wilmington, DE	26	40.5	76.3	Bismarck, ND	0	47.1	101.0
Washington, DC	30	39.7	77.5	Cincinnati, OH	26	39.2	85.0
Jacksonville, FL	45	31.0	82.3	Cleveland, OH	21	42.3	82.5
Key West, FL	65	25.0	82.0	Oklahoma City, OK	28	35.9	97.5
Miami, FL	58	26.3	80.7	Portland, OR	33	45.6	123.2
Atlanta, GA	37	33.9	85.0	Harrisburg, PA	24	40.9	77.8
Boise, ID	22	43.7	117.1	Philadelphia, PA	24	40.9	75.5
Chicago, IL	19	42.3	88.0	Charleston, SC	38	33.3	80.8
Indianapolis, IN	21	39.8	86.9	Nashville, TN	31	36.7	87.6
Des Moines, IA	11	41.8	93.6	Amarillo, TX	24	35.6	101.9
Wichita, KS	22	38.1	97.6	Galveston, TX	49	29.4	95.5
Louisville, KY	27	39.0	86.5	Houston, TX	44	30.1	95.9
New Orleans, LA	45	30.8	90.2	Salt Lake City, UT	18	41.1	112.3
Portland, ME	12	44.2	70.5	Burlington, VT	7	45.0	73.9
Baltimore, MD	25	39.7	77.3	Norfolk, VA	32	37.0	76.6
Boston, MA	23	42.7	71.4	Seattle, WA	33	48.1	122.5
Detroit, MI	21	43.1	83.9	Spokane, WA	19	48.1	117.9
Minneapolis, MN	2	45.9	93.9	Madison, WI	9	43.4	90.2
St Louis, MO	24	39.3	90.5	Milwaukee, WI	13	43.3	88.1
Helena, MT	8	47.1	112.4	Cheyenne, WY	14	41.2	104.9

If you were to plot temperature against latitude you would observe a rather interesting phenomenon. (There is really little to be seen from a plot of temperature against longitude.)

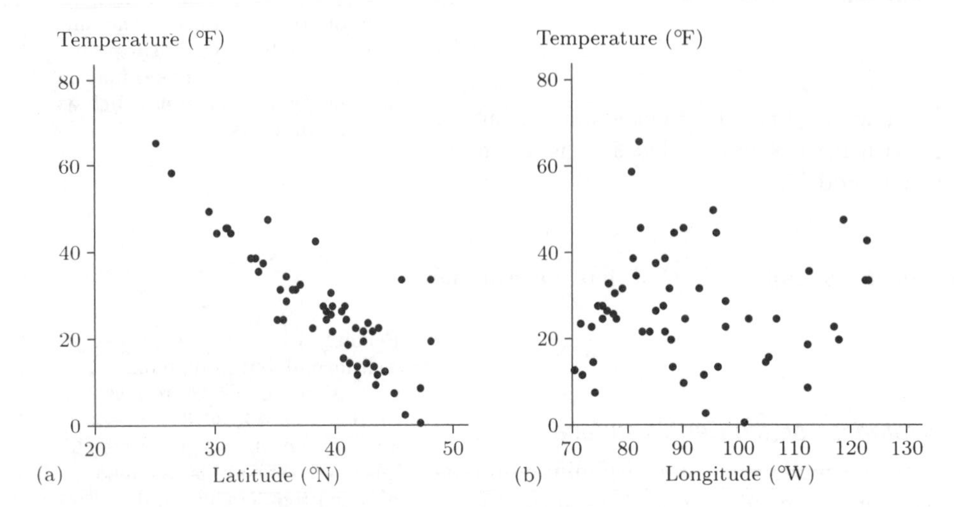

Figure 10.13 (a) Temperature against latitude (b) Temperature against longitude

It is evident from Figure 10.13(a) that while a straight line might be adequately fitted to these data the assumption of constant variance in the error term appears to be broken. A transformation of the data that reduced the scatter in the temperature for the higher (more northerly) latitudes might be worth considering. ■

Before moving on to Section 10.2, let us briefly trace some of the history of the important subject of regression.

10.1.2 A brief history of regression

Linear regression is currently one of the most used and useful statistical tools. But a little over a hundred years ago, the term and its essential ideas were unknown. It was thanks to the work of Sir Francis Galton in the 1880s that regression became recognized, first as a statistical phenomenon, and then as a tool of great potential. The details of how the concept of regression evolved in Galton's work are fascinating (but intricate): it is worthwhile to put regression in its historical context, and to outline the kind of practical problems and mysteries that led to its development.

Figure 10.14 Sir Francis Galton (1822–1911)

An experiment in 1875 on sweet peas led Galton a long way towards the idea of regression. The experiment, conducted by Galton himself with the help of some friends, yielded the weights of the seeds of the progeny of seven groups of sweet peas, the groups themselves classified according to increasing parental seed weight. In the framework of this chapter, Galton wished to *regress* offspring seed weight (the response variable, Y) on parental seed weight (the explanatory variable, x). What Galton observed—without the aid of scatter plots, incidentally—was that the variation in offspring seed weights was much the same, irrespective of their classification.

What had concerned Galton was whether or not heavy parental seeds led to heavier offspring seeds (a major question of heredity); what he found was that to some extent there was a parental weight effect, but smaller than might have been expected. Offspring of parents with heavy seeds had seeds heavier than usual, but the difference was not as marked for the offspring as for the parents. Similarly, offspring of parents with light seeds had seeds lighter than usual, but the differences were not so marked. Galton initially called this phenomenon *reversion*; he wrote that the seeds' mean weight 'reverted, or regressed, toward mediocrity'.

In the 1880s, Galton continued his development of regression ideas in the context of heredity by changing attention to human populations. His analysis of the sweet pea data in 1877 had left several questions unanswered; many of these problems were solved in 1885 by an analysis of data on the heights of human parents and their offspring, which again exhibited the phenomenon of regression. The key probability model in this analysis was the *bivariate normal distribution* which you will meet briefly in Section 11.5 of *Chapter 11*.

In 1889 Galton published *Natural Inheritance*, a seminal text which stimulated a great deal of subsequent research. In the introduction to the book, Galton wrote of regression: 'I have a great subject to write upon. It is full of interest of its own. It familiarizes us with the measurement of variability, and with curious laws of chance that apply to a vast diversity of social subjects.' In modern terms, Galton's work was just a start: the development of the techniques of regression analysis was continued by (amongst others) Francis Ysidro Edgeworth (1845–1926), W.F.E. Weldon (1860–1906) and Karl Pearson (1857–1936). It is still an important area for research.

In Section 10.2 we turn our attention to the problem of fitting the regression model suggested by the data.

10.2 Fitting the chosen regression model

In most of the examples in Section 10.1, a scatter plot of the response variable against an explanatory variable suggests at least the possibility of some kind of functional relationship between the two variables. (The exception is the examination score data of Example 10.7, where there seems to be no clear correspondence between a candidate's score and the length of time taken by the candidate to complete the paper.) Sometimes the relationship appears to be linear; on other occasions there is an evident curve underlying the disposition of points in the scatter plot.

Whatever the underlying relationship may or may not be, evident in all the plots has been the scatter induced by random variation. In order to make some kind of estimate of the underlying functional relationship between the response variable and the explanatory variable, this random scatter will need to be explicit in our probability model.

Writing the response variable Y (upper case to indicate that it is a random variable) and the explanatory variable x, then the most general regression model can be written

$$Y_i = h(x_i) + W_i,$$

where $h(\cdot)$ represents some function (perhaps polynomial or exponential or trigonometric). Our aim is to identify, within such a family of functions, the function that best fits the scattered points (x_i, y_i). In this course, attention is restricted to straight-line functions. However, the need for a more developed modelling approach than this is evident, and in Section 10.5 other approaches are explored.

In fact, it is possible (and not very difficult), given the full list of points (x_i, y_i), to construct a formula for the curve that will pass through all the points, so that all the w_is are zero. However, the resulting formula will contain as many terms as there are fitted points, and in this regard will not be a particularly useful or helpful representation.

10.2.1 Fitting the least squares straight line through the origin

In Example 10.6 actual road distances between locations in Sheffield were compared with direct distances taken from a map. It was decided to fit the model

$$Y_i = \gamma x_i + W_i, \quad i = 1, 2, \ldots, 20$$

(a straight line passing through the origin) to the data.

It is one of the assumptions of the regression model that the random terms W_i are *independent*, with mean 0 and *constant* variance σ^2: these terms explain the scatter around the underlying regression line $y = \gamma x$ that is evident from the scatter plot. The observed differences

$$w_i = y_i - \gamma x_i, \quad i = 1, 2, \ldots, 20,$$

which are independent observations on the random variable W, are shown in Figure 10.15. These differences are called **residuals**; they represent the deviations between observed data and a fitted model. Notice that in Figure 10.15 the line $y = \gamma x$ has been drawn for illustrative purposes only: we do not yet know the value of the slope, γ, that corresponds to the best straight line through the data.

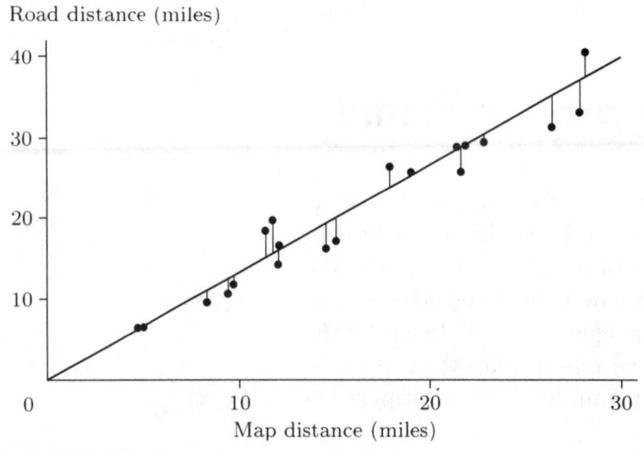

Figure 10.15 Residuals $w_i = y_i - \gamma x_i$

The residuals are positive for all the observations y_i above the fitted line ($w_i = y_i - \gamma x_i > 0$) and negative for observations below the fitted line. It is useful to remember the relationship

RESIDUAL = DATA − FIT

in all modelling exercises: it is *the* essential statement of departures from the model evident in the data.

The line drawn in Figure 10.15 seems to provide a good fit to the data—it was deliberately designed to look reasonable: how do we determine the equation of the line which in some sense is better than any other?

The answer to this question is found in the **principle of least squares**. The residuals w_i are a quantitative measure of the quality of the fit of the line to the data at the point x_i. If the absolute value of w_i is small then, at the point x_i at least, the line 'fits the data' well. Large negative residuals are as indicative of a poor fit as large positive residuals. An overall measure of the quality (actually, of the 'badness') of the fit is encapsulated by the number

$$\sum_{i=1}^{n} w_i^2 = \sum_{i=1}^{n} (y_i - \gamma x_i)^2; \tag{10.1}$$

that is, by the *sum of squared residuals* or, as it is usually known, the **residual sum of squares**. There is one unknown quantity in (10.1): it is the slope γ. The value of γ that minimizes the residual sum of squares is called the **least squares estimator** of the slope of the regression line, and is written $\widehat{\gamma}$.

This we shall take as our criterion for the best line through the data.

There are several ways to find the value $\widehat{\gamma}$ of γ that will achieve this minimum. If you know about the technique of differentiation, then you will know that this can be used in a very straightforward way to locate minimum points. Alternatively, the expression (10.1) can be written as a quadratic expression in γ as follows.

$$\begin{aligned}\sum w_i^2 &= \sum (y_i - \gamma x_i)^2 \\ &= \sum (x_i^2\gamma^2 - 2x_i y_i \gamma + y_i^2) \\ &= \left(\sum x_i^2\right)\gamma^2 - \left(2\sum x_i y_i\right)\gamma + \left(\sum y_i^2\right)\end{aligned}$$

Notice that here the limits $i = 1$ and $i = n$ have been dropped from the summation symbols. For the rest of this chapter, and where the limits are obvious as they are here, they will not usually be included, in the interests of clear presentation.

This expression can be rewritten

$$\sum w_i^2 = \left(\sum x_i^2\right)\left(\gamma - \frac{\sum x_i y_i}{\sum x_i^2}\right)^2 + \left(\sum y_i^2 - \frac{(\sum x_i y_i)^2}{\sum x_i^2}\right); \tag{10.2}$$

The algebraic technique of 'completing the square' was used in (10.2). You do not need to understand the techniques or to follow all the details of this derivation of the estimator $\widehat{\gamma}$.

and in (10.2) only the term

$$\left(\gamma - \frac{\sum x_i y_i}{\sum x_i^2}\right)^2$$

involves γ. The residual sum of squares is minimized when this term is zero, that is, when

$$\gamma = \frac{\sum x_i y_i}{\sum x_i^2}; \tag{10.3}$$

this value is written $\widehat{\gamma}$.

Thus according to the least squares criterion, this is the slope of the best straight line (through the origin) through the scattered points. The equation of the line can be written

$$y = \widehat{\gamma} x = \frac{\sum x_i y_i}{\sum x_i^2} x.$$

Example 10.6 continued

In the case of the Sheffield map data,

$$\begin{aligned}\sum x_i y_i &= (9.5 \times 10.7) + (9.8 \times 11.7) + \cdots + (28.0 \times 33.1)\\ &= 101.65 + 114.66 + \cdots + 926.80\\ &= 8026.25,\end{aligned}$$

and

$$\begin{aligned}\sum x_i^2 &= 9.5^2 + 9.8^2 + \cdots + 28.0^2\\ &= 90.25 + 96.04 + \cdots + 784.00\\ &= 6226.38.\end{aligned}$$

So the least squares estimate of the slope γ is

$$\widehat{\gamma} = \frac{\sum x_i y_i}{\sum x_i^2} = \frac{8026.25}{6226.38} = 1.289.$$

So the fitted straight line through the scattered data points has equation

$$y = 1.289x,$$

or, perhaps more intelligibly,

Road distance $= 1.289 \times$ Map distance.

The fitted straight line is shown in Figure 10.16. You can see that the fit is really quite good; the residuals are not large.

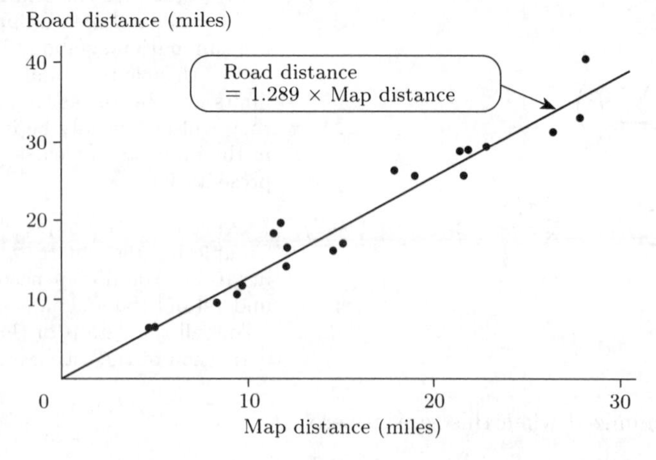

Figure 10.16 Straight-line fit superimposed on the Sheffield scatter plot

It is useful to denote the fitted value of y corresponding to the value x_i by $\widehat{y}_i$. Here, $\widehat{y}_i = \widehat{\gamma} x_i$. In fact, the residual sum of squares, the quantity we wished to minimize, is given by

$$\begin{aligned}\sum (y_i - \widehat{y}_i)^2 = \sum (y_i - \widehat{\gamma} x_i)^2 &= \sum y_i^2 - \frac{(\sum x_i y_i)^2}{\sum x_i^2}\\ &= \sum y_i^2 - \widehat{\gamma}^2 \sum x_i^2\\ &= 107.30. \qquad (10.4)\end{aligned}$$

■

This result will be useful later. Do not worry about the algebraic details which lead to the final expression for the residual sum of squares $\sum(y_i - \widehat{y}_i)^2$. They are included purely for illustrative purposes. There are many different ways that this could be written down. The easiest and most common way of calculating the residual sum of squares is as part of a regression analysis pursued on a computer. This kind of dense arithmetic calculation is nowadays very rarely performed by hand!

You can estimate the corresponding road distance y_0 for any particular map distance x_0 by finding

$$y_0 = \widehat{\gamma}x_0.$$

For instance, the predicted road distance for a map distance of $x_0 = 18$ miles is

$$y_0 = \widehat{\gamma}x_0 = 1.289 \times 18 = 23.2 \text{ miles}.$$

Notice that the map distance $x_0 = 18$ is well within the range covered by the original data set (which had map distances from about 5 miles to about 30 miles). Also, it must not be forgotten that the experiment covers the correspondence between map and road distances within the city of Sheffield, and would not extend as a useful estimator of the distance between Sheffield and Birmingham (map distance 67 miles; according to one of the road associations the shortest practicable road distance is 77 miles). There is no particular reason why this estimated model should be useful within, say, Milton Keynes, where local topography is quite different to that to be found in Sheffield.

However, even for the particular journey identified in Sheffield, it is not at all likely that the road distance y_0 will turn out to be exactly 23.2 miles—this is just what the model *predicts*. The doubt surrounding the actual distance y_0 can be quantified, and we shall see in Section 10.4 principles for the construction of a *prediction interval* for a response, given a particular value for the explanatory variable.

Try the following exercise which is about estimating the size of a population. Incidentally, the exercise has been set up as a calculator exercise in order to encourage you to practise keying in paired data in the form $(x_1, y_1), (x_2, y_2), \ldots, (x_n, y_n)$ to your calculator, and to become familiar with the key routines necessary to access such quantities as $\sum x_i^2$ and $\sum x_i y_i$. Most machines, particularly if they supply the standard statistical measures such as sample mean and sample standard deviation, will do this. However, in general, particularly for data sets containing as many as 25 data pairs or more, a computer is usually used for the least squares analysis.

Exercise 10.1

In a botanical experiment a researcher wanted to estimate the number of individuals of a particular species of beetle (*Diaperus maculatus*) within fruiting bodies (brackets) of the birch bracket fungus *Polyporus betulinus*. (This is a shelf fungus that grows on the trunks of dead birch trees.) When the brackets are stored in the laboratory, the beetle larvae within them mature over several weeks—the adults then emerge and can be removed and counted.

Pielou, E.C. (1974) *Population and Community Ecology—Principles and Methods.* Gordon and Breach, New York, pp. 117–121.

A sample of 25 brackets was collected. Their weights (in grams) and the number of beetles they were shown eventually to contain are given in Table 10.10.

Table 10.10 Number of beetles against weight of bracket (grams)

Number of beetles	Weight of birch bracket (grams)
16	62
50	226
26	175
73	255
95	226
40	99
40	150
3	25
49	200
0	77
56	178
91	283
48	192
25	125
61	177
64	307
0	42
10	99
1	122
15	88
98	201
60	183
0	63
98	296
15	162

(a) The research problem of interest is to investigate whether the number of beetles a shelf fungus contains might reasonably be predicted from the weight of the fungus. With this in mind, sketch a scatter plot of these data.

(b) Comment on the usefulness of fitting a straight line through the origin to these data, obtain the least squares estimate of the slope of the line, and include the least squares line on your plot.

(c) Predict the number of beetles that would eventually emerge from a fungus found to weigh 240 g at collection.

(d) Calculate the residual sum of squares $\sum(y_i - \widehat{y})^2$ for these data.

The work of this subsection may be summarized as follows.

The least squares straight line through the origin

If a scatter plot of data points $(x_i, y_i), i = 1, 2, \ldots, n$, suggests that a regression model of the form

$$Y_i = \gamma x_i + W_i, \quad i = 1, 2, \ldots, n,$$

might be an appropriate model, where the random terms W_i are independent with mean 0 and variance σ^2, then the least squares estimate $\widehat{\gamma}$ of the regression slope γ is given by

$$\widehat{\gamma} = \frac{\sum x_i y_i}{\sum x_i^2}.$$

There is very much more that might be said here: for instance, it might be useful for future work first to use the data points (x_i, y_i) to estimate the second of the two model parameters, σ^2. Secondly, the estimate $\widehat{\gamma} = \sum x_i y_i / \sum x_i^2$ is just one observation on the random variable $\widehat{\gamma} = \sum x_i Y_i / \sum x_i^2$. We have not formally determined that $\widehat{\gamma}$ is unbiased for γ (in fact, it is, and this is not difficult to show, as we shall see in Subsection 10.3.1). Thirdly, it would be very useful to assess the quality of the fit of the model to the data, and obtain some quantitative assessment on whether or not the straight-line fit is useful. Finally, in other contexts, the fitted straight line might be constrained not through the origin, but through some other identifiable point (x_0, y_0) known to the researcher to be a keypoint.

Some of this sort of detail is covered in Subsection 10.2.2, where the most general unconstrained straight line is fitted to a set of data pairs (x_i, y_i). For this course, it is sufficient that you understand the principles on which the straight line through the origin was obtained, and that you recognize some of the features of the model.

10.2.2 Fitting the unconstrained least squares straight line

We saw, in Figures 10.1, 10.2, 10.3, 10.6, 10.7(a) and 10.8, cases where it seemed a straight-line model would fit the scattered data points (x_i, y_i) moderately well (in some cases, very well). An important point to realize is that it is not necessary to formulate any reason why this should be so, based on any knowledge of principles governing the relationship between explanatory and response variables. For the statistical model

$$Y_i = \alpha + \beta x_i + W_i, \quad i = 1, 2, \ldots, n,$$

the observed residuals may be written

$$w_i = y_i - (\alpha + \beta x_i), \quad i = 1, 2, \ldots, n.$$

The residuals for Forbes' temperature data are shown in Figure 10.17. (Here, again, the fitted line has no optimal properties, for we do not know at this stage what the best estimates of α and β are. The line was fitted by eye.)

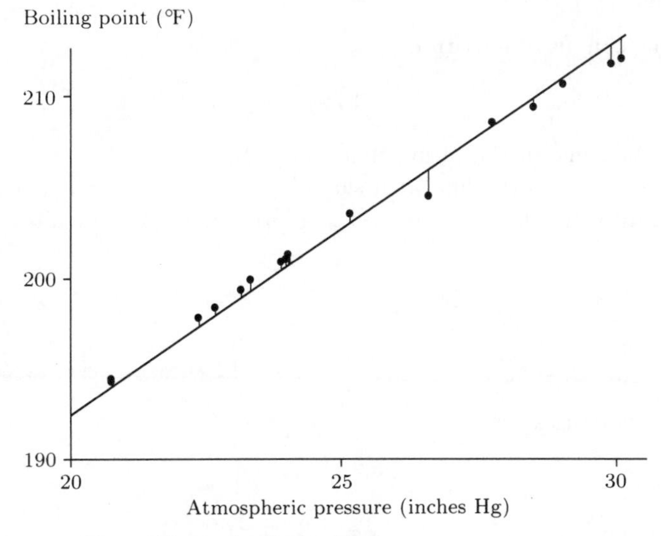

Figure 10.17 The residuals for Forbes' data

In this case the residual sum of squares is given by

$$\sum w_i^2 = \sum (y_i - (\alpha + \beta x_i))^2, \tag{10.5}$$

and our aim in fitting the best straight line through the data points is to contrive the line to minimize the residual sum of squares. (You should always remember that this is not the only criterion to yield a useful fitted line—you saw this in Figure 10.6—but it is one of the most practicable. We shall see in Subsection 10.3.2 that least squares estimators can possess optimal properties.)

It is possible to rewrite (10.5): first, as a quadratic expression in α; secondly, as a quadratic expression in β; and finally to identify the minimum points of each quadratic. This is not a difficult exercise (you saw the same idea used when fitting the least squares straight line constrained through the origin) although it is not a particularly elegant one, for the algebra gets quite untidy. Alternatively, the technique of partial differentiation will yield the required values $\widehat{\alpha}$ of α and $\widehat{\beta}$ of β where (10.5) is minimized.

However, the results are quite standard, and for present purposes it will be sufficient simply to write the estimates down. The least squares estimate of the slope β of the straight line through the points $(x_1, y_1), (x_2, y_2), \ldots, (x_n, y_n)$ is given by

$$\widehat{\beta} = \frac{n\sum x_i y_i - \sum x_i \sum y_i}{n\sum x_i^2 - (\sum x_i)^2}. \tag{10.6}$$

Again, for simplicity, the limits $i = 1$ and $i = n$ on the summation symbols have not been included here.

A similar expression can be written down for $\widehat{\alpha}$, the least squares estimate of α, the constant term in the regression line. However, it is easier to use the value of $\widehat{\beta}$ so that $\widehat{\alpha}$ can be written

$$\widehat{\alpha} = \frac{1}{n}\left(\sum y_i - \widehat{\beta}\sum x_i\right) = \overline{y} - \widehat{\beta}\overline{x}. \tag{10.7}$$

Then, using (10.6) and (10.7), the equation of the least squares regression line through the data points $(x_1, y_1), (x_2, y_2), \ldots, (x_n, y_n)$ can be written

$$y = \widehat{\alpha} + \widehat{\beta}x = (\overline{y} - \widehat{\beta}\overline{x}) + \widehat{\beta}x. \tag{10.8}$$

Incidentally, notice that the equation of the line can also be written

$$(y - \overline{y}) = \widehat{\beta}(x - \overline{x}). \tag{10.9}$$

One reason for writing the least squares regression line in the form (10.9) is that it exhibits an easily-stated property of the line: it is the line with slope $\widehat{\beta}$, passing through the point $(\overline{x}, \overline{y})$, the **centroid** of the data.

Example 10.2 continued

For Forbes' data of Example 10.2 the summary statistics are

$$n = 17, \qquad \sum x_i = 426, \qquad \sum y_i = 3450.2,$$

$$\sum x_i^2 = 10\,820.9966, \qquad \sum x_i y_i = 86\,735.495.$$

So the slope estimate is given by

$$\widehat{\beta} = \frac{n\sum x_i y_i - \sum x_i \sum y_i}{n\sum x_i^2 - (\sum x_i)^2}$$
$$= \frac{17 \times 86\,735.495 - 426 \times 3450.2}{17 \times 10\,820.9966 - 426^2}$$
$$= \frac{4718.215}{2480.9422} = 1.9018;$$

and

$$\widehat{\alpha} = \overline{y} - \widehat{\beta}\,\overline{x}$$
$$= \tfrac{1}{17}(3450.2 - 1.9018 \times 426) = 155.296.$$

So the fitted line has equation

$$y = 155.3 + 1.9x.$$

Notice that at this point the number of decimal places has been reduced: the idea of a 'model' is to be helpful, not to provide six-figure predictions.

More clearly, the regression relationship between explanatory variable and response can be written

Boiling point = 155.3 + 1.90 × Atmospheric pressure,

where temperature is measured in °F and pressure in inches Hg. The least squares line through Forbes' data is shown in Figure 10.18.

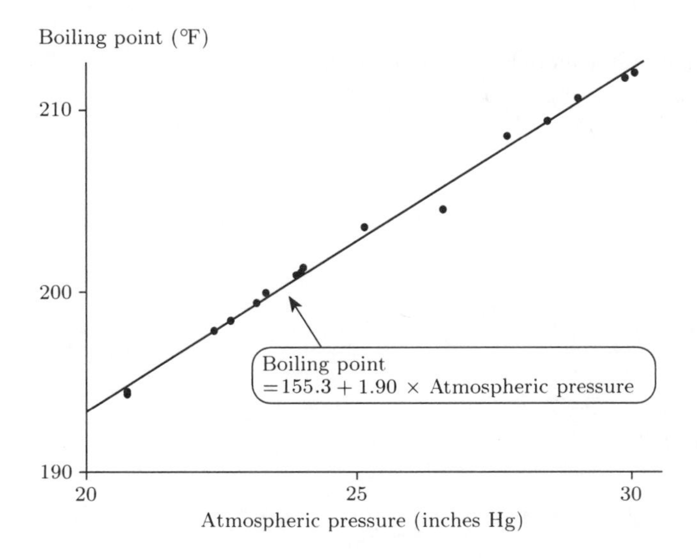

Figure 10.18 Forbes' data with the fitted least squares line

This line could be used for prediction, and that is a common purpose of regression analysis. But remember that Forbes' results formed just part of a series of experiments to predict altitude from the temperature of boiling water, rather than the temperature of boiling water from atmospheric pressure! ■

These results may be summarized as follows.

The least squares regression line through scattered data

If a scatter plot of data points (x_i, y_i), $i = 1, 2, \ldots, n$, suggests that a regression model of the form

$$Y_i = \alpha + \beta x_i + W_i, \quad i = 1, 2, \ldots, n,$$

might be an appropriate statistical model, where the random terms W_i have mean 0 and variance σ^2, then the least squares estimate of the slope of the regression line is

$$\widehat{\beta} = \frac{n\sum x_i y_i - \sum x_i \sum y_i}{n\sum x_i^2 - (\sum x_i)^2},$$

and the least squares estimate of the constant term α is given by

$$\widehat{\alpha} = \overline{y} - \widehat{\beta}\overline{x}.$$

The line passes through the centroid of the data, the point $(\overline{x}, \overline{y})$: it is sometimes convenient to write the equation of the line as

$$y - \overline{y} = \widehat{\beta}(x - \overline{x}).$$

This is also a useful formulation for *remembering* the equation of the fitted line.

Let us return to the divorce data in *Chapter 6* and consider this fourth option for a fitted line ($\widehat{l_4}$, say) through the data.

Example 10.3 continued

We already have three suggested straight-line trend models for the divorce data in *Chapter 6* summarized in Figure 10.6. Now we have a fourth model: that is, the least squares regression line. This has slope

$$\begin{aligned}\widehat{\beta} &= \frac{n\sum x_i y_i - \sum x_i \sum y_i}{n\sum x_i^2 - (\sum x_i)^2} \\ &= \frac{6 \times 62\,637.3 - 465 \times 807}{6 \times 36\,055 - 465^2} \\ &= \frac{568.8}{105} = 5.42,\end{aligned}$$

and constant term

$$\widehat{\alpha} = \overline{y} - \widehat{\beta}\overline{x} = \tfrac{1}{6}(807 - \widehat{\beta} \times 465) = -285.33.$$

So the fitted line has equation

$$y = -285.33 + 5.42x.$$

Alternatively, the regression relationship between explanatory variable (passing time, in this case) and response variable can be written

$$\text{Annual divorces} = -285.33 + 5.42 \times \text{Year},$$

where divorces are counted in thousands, and the year is reckoned from 1900. ■

Is the fourth slope estimate better than any of the preceding three? To answer this question we need to look at the estimating formula (in other words, the estimator) from which the estimates were obtained. Sampling properties of the estimators $\widehat{\alpha}$ and $\widehat{\beta}$ are addressed in Section 10.3.

After fitting the regression line, the residual sum of squares is

$$\sum(y_i - \widehat{y}_i)^2 = \sum(y_i - (\widehat{\alpha} + \widehat{\beta}x_i))^2$$
$$= \sum(y_i - \overline{y})^2 - \widehat{\beta}^2 \sum(x_i - \overline{x})^2. \qquad (10.10)$$

Again, there are many ways in which this sum can be written. This is the most convenient for computational purposes. The most convenient computational method of all (in this context) is to use regression software. However, note particularly that you can get $\sum(y_i - \overline{y})^2$ and $\sum(x_i - \overline{x})^2$ very easily from a calculator by thinking of the y_is and x_is as random samples: the sums are each $(n-1)$ times the sample variance.

The result given at (10.10) is an important one because of its use in assessing the model, calculating confidence intervals and testing hypotheses. We shall see examples of this in Section 10.3. There now follow several exercises. You are encouraged to use both your calculator and computer (as directed) when answering the questions in Exercises 10.2 to 10.7. Many calculators will provide the regression estimators immediately (after keying in the data points) without the need for intermediate calculations of summary statistics such as $\sum x_i$, $\sum x_i y_i$, and so on.

Exercise 10.2

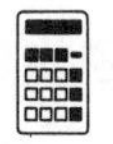

The scatter plot in Figure 10.7(a) (see Example 10.4) suggests that a straight line model might be useful to express the relationship between the strength of wooden beams and their specific gravity. Find the equation of the least squares regression line for these data.

Exercise 10.3

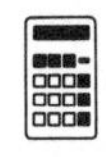

Find the least squares regression line for the data on the finger-tapping frequencies in Example 10.2.

Exercise 10.3 raises two interesting points. The first is that Figure 10.2 makes explicit the variation in tapping frequency at each dose level, and it would be useful to obtain an estimate of this variation. So far, we do not have the necessary methods for this. Secondly, it is possible that the trend line perceived in Figure 10.2 is of no significance, and that ingestion of caffeine has, in fact, no effect on finger-tapping performance. This proposition could be explored with a formal test of the hypothesis

$$H_0 : \beta = 0.$$

But in order to test this hypothesis, we need a statistical model for the variation in the slope estimator $\widehat{\beta}$. This problem is addressed in Subsection 10.3.2.

Try Exercise 10.4 using your computer.

Exercise 10.4

Forbes' data giving the boiling point of water (°F) at different atmospheric pressures (inches Hg) are given in Table 10.2. Hooker's data are listed in Table 10.4. Respective scatter plots are given in Figures 10.3 and 10.8. In both cases a straight line is a reasonable regression model to attempt to fit.

(a) Obtain least squares estimates $\widehat{\alpha}$ and $\widehat{\beta}$ for the parameters α and β when the model $Y_i = \alpha + \beta x_i + W_i$ is fitted to Forbes' data.

(b) Obtain least squares estimates for α and β for Hooker's data.

You know the answers to part (a) from the work on page 401: the purpose of this exercise is to check that you can use your computer.

The question was posed earlier: do the data collected by Forbes and Hooker provide significantly different fitted lines expressing the relationship between boiling point and atmospheric pressure? It is possible to formally compare two regression slopes and, indeed, two regression lines, in order to provide an answer to this question and others like it. However, the details are somewhat intricate and beyond the scope of this course. What one can do in a case such as this is to plot the two sets of data, and the two fitted lines, on the same axes. This is shown in Figure 10.19.

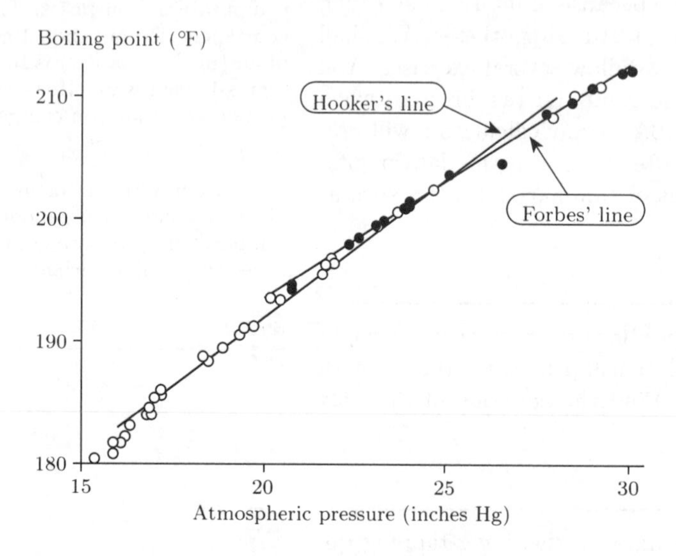

Figure 10.19 Forbes' data (•) and Hooker's data (∘)

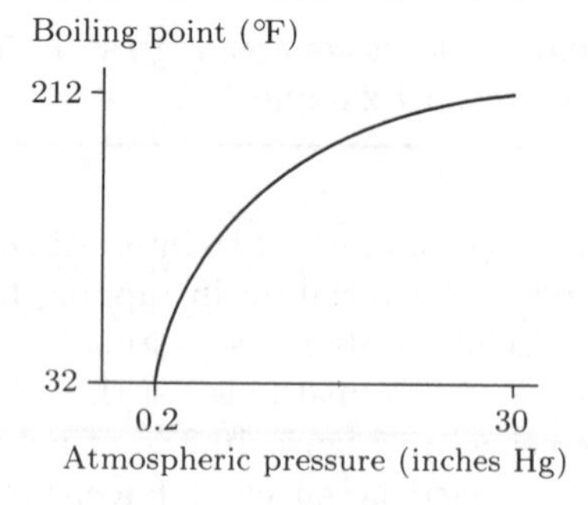

Figure 10.20 The effect of varying atmospheric pressure on the boiling point of water

Can you see the suggestion of a curve now? (Or, perhaps, the drawn lines are merely deceiving the eye and there is no real curve there ....) In fact, under ideal conditions (but, almost certainly, not those to be found on mountain sides in Scotland, the Alps and the Himalayas) the effect of changing atmospheric pressure on the boiling point of water can be carefully monitored, and is illustrated in Figure 10.20 (which is not drawn to scale).

Statistical researches do not always answer questions, or only answer questions: often they generate further questions and different avenues of research.

Apart from the labour of keying in the data, you should find Exercises 10.5 to 10.7 very straightforward, assuming you have access to the appropriate software.

Exercise 10.5

In 1975, the British government set up a Resources Allocation Working Party to 'review the arrangement for distributing National Health Service capital and revenue'. It was decided to base regional resource allocation on death rate within regions, (or, more precisely, on a 'standardized mortality rate'). But NHS resources need to reflect regional variations in 'chronic sickness' long-standing health problems that require medical treatment. The question then, which was a controversial one at the time, is: are death rates a good predictor of sickness rates? The data shown in Table 10.11 address this question: they

Morbidity rates are a measure of sickness. You do not need to understand the technical details of the calculation of mortality and morbidity rates to answer the questions posed in this exercise.

show standardized mortality rates per 10 000 and standardized morbidity rates per 1000 for ten regions of England and Wales, for the period 1972–1973.

Forster, J. (1977) Mortality, morbidity and resource allocation. *The Lancet*, **1**, 997–998. See also Marsh, C. (1988) *Exploring Data—An introduction to data analysis for social scientists.* Polity Press, Cambridge.

Table 10.11 Standardized mortality and morbidity rates, 1972–73

Region	Mortality rate (per 10 000)	Morbidity rate (per 1000)
North	132.7	228.2
Yorkshire	126.8	235.2
North West	132.8	218.6
East Midlands	119.2	222.0
West Midlands	124.8	210.5
East Anglia	108.2	205.0
Greater London	116.3	202.6
South East	109.5	189.6
South West	112.2	186.6
Wales	128.6	249.9

(a) Bearing in mind the research question of interest, plot the data points on an appropriate scatter plot.

(b) Commenting on the appropriateness of a straight line model, calculate the least squares estimates for α and β based on these data.

Exercise 10.6

The following data list fuel consumption (in miles per gallon) against kerb weight (in kg) for a sample of 42 British diesel motor cars.

These data were extracted from the February 1992 edition of *Diesel Car* published by Merricks Publishing Ltd.

Table 10.12 Weights and mileage per gallon for British diesel cars

Kerb weight (kg)	1090	1300	720	1035	990	1380	870	1040	1130
Miles per gallon	48.3	40.5	60.9	49.1	48.4	41.9	51.0	46.1	48.9
Kerb weight(kg)	875	1010	1085	1120	1120	1325	1130	1320	1370
Miles per gallon	54.6	52.9	52.9	43.7	43.7	35.6	41.5	38.9	38.2
Kerb weight (kg)	880	950	1080	1080	1080	1430	1007	1040	1095
Miles per gallon	57.8	48.2	47.0	45.6	36.8	41.9	46.1	46.3	51.4
Kerb weight (kg)	1145	1160	1470	950	1015	1105	1175	1040	855
Miles per gallon	48.8	47.6	42.7	42.3	41.9	49.1	46.7	56.1	54.6
Kerb weight (kg)	1005	961	1353	985	1190	1436			
Miles per qallon	49.0	47.9	41.7	47.4	53.3	36.5			

It is required to use these data to explore the relationship between kerb weight and fuel consumption.

(a) Plot the data on an appropriate scatter plot, explaining how you decided to label the axes.

(b) Commenting informally on the quality of the fit, estimate model parameters for a straight line fit to the data.

Notice that for these data, the fitted linear relationship (were it to be extrapolated to the right) suggests that eventually cars which are very heavy indeed will return zero or negative fuel consumption figures. This anomaly would matter more if such heavy cars (about 3000 kg, or 3 tons) were common; but they are not. (Some heads of state, for instance, like to use armour-plated vehicles which are rather heavy. This data set is a good example of the dangers of over-interpretation of the regression model: one needs to be wary of extrapolating the line too far from the domain of the experiment.)

Exercise 10.7

Crickets make their chirping sound by sliding one wing cover back and forth over the other, very rapidly. Naturalists have long recognized a linear relationship between frequency of chirping and temperature, and interest centres on seeing if one can measure temperature approximately by calibrating the chirping of crickets. The precise nature of the relationship varies from species to species. Table 10.13 lists 15 frequency-temperature observations for the striped ground cricket, *Nemobius fasciatus fasciatus*.

Pierce, G.W. (1949) *The Songs of Insects.* Harvard University Press, USA.

Table 10.13 Temperature and chirping frequency

Chirps per second	20.0	16.0	19.8	18.4	17.1	15.5	14.7	17.1
Temperature (℉)	88.6	71.6	93.3	84.3	80.6	75.2	69.7	82.0
Chirps per second	15.4	16.2	15.0	17.2	16.0	17.0	14.4	
Temperature (℉)	69.4	83.3	79.6	82.6	80.6	83.5	76.3	

This is an interesting research question because without too much thought one might have supposed the obvious way to treat these data is to take chirping frequency as the response variable, and temperature as the explanatory variable. In fact, the problem is posed the other way round, as a prediction problem for temperature, given chirping frequency. Plot the data accordingly, find the equation of the least squares straight line fit to the data and estimate the temperature based on a chirping frequency of 18 chirps per second. (This question anticipates the material of Section 10.4, which is all about prediction.)

We now turn our attention to an assessment of the accuracy and precision (in other words, of the usefulness) of the regression estimators.

10.3 Sampling properties of the model

In this section, sampling properties of the estimators are explored and summarized. Once these are known (or assumed) then they can be used for the construction of confidence intervals and for hypothesis testing. It is not the intention in this section to provide an exhaustive list of results, or to offer illustrations of every sort of question that might be put to a statistician by a researcher in a regression context. However, some results are useful and important, and these are stated (usually without proof); and some questions are interesting and occur very commonly in this context. These are dealt with in the following subsections.

10.3.1 Sampling distributions of the estimators

So far we have assumed that the random terms W_i, responsible for the scatter usually evident in regression problems, are independent of one another with mean 0 and constant variance σ^2.

The straight line through the origin

The least squares estimator of the slope of the best straight line through the origin is

$$\widehat{\gamma} = \frac{\sum x_i Y_i}{\sum x_i^2}.$$

Remember, the x_is are not random variables.

Then, it follows that

$$\begin{aligned} E(\widehat{\gamma}) = E\left(\frac{\sum x_i Y_i}{\sum x_i^2}\right) &= \frac{1}{\sum x_i^2} E\left(\sum x_i Y_i\right) \\ &= \frac{1}{\sum x_i^2} \sum x_i E(Y_i) \\ &= \frac{1}{\sum x_i^2} \sum x_i \gamma x_i = \gamma; \end{aligned}$$

The limits $i = 1$ and $i = n$ have again been omitted here, in the interests of clarity.

so the estimator $\widehat{\gamma}$ for the slope of the constrained line is unbiased. In fact, the variance of the estimator is

$$V(\widehat{\gamma}) = \frac{\sigma^2}{\sum x_i^2}.$$

This has useful consequences. One of the aims of the design of any statistical experiment must be to improve, where you can, the precision of the results. In this case we see that the slope estimator is unbiased regardless of the details of the experimental design. However, notice that the variance is reduced not only if further observations are taken, but also if the values taken on the explanatory variable are further away from the origin. This makes sense—you know that the line must go through the origin: if measurements are taken close to the origin, small errors will have a large effect on the slope of the fitted line. Further away from the origin, even quite large errors will not have much effect on the fit of the line.

These results are useful when you can control the explanatory variable. For instance, the designer of a controlled trial for a new pain-killer can alter the dose and observe the different responses. The social scientist interested in changing divorce rates with time cannot strictly control the calendar.

Now suppose that the random terms are normally distributed. Then the distribution of the estimator $\widehat{\gamma}$ is also normal.

The distribution of the slope estimator

If the random terms W_i in the model

$$Y_i = \gamma x_i + W_i, \quad i = 1, 2, \ldots, n,$$

are assumed to be independent and normally distributed with mean 0 and variance σ^2, then the distribution of the slope estimator is normal:

$$\widehat{\gamma} \sim N\left(\gamma, \frac{\sigma^2}{\sum x_i^2}\right). \tag{10.11}$$

Of course, this is not always a useful result for inferential purposes because the value of the parameter σ^2 is not usually known. In this case the parameter is replaced by its estimator S^2, where

$$S^2 = \frac{\sum (Y_i - \widehat{Y}_i)^2}{n-1}.$$

Notice the numerator in the estimator of σ^2: it is the familiar residual sum of squares, after fitting the least squares line.

This estimator is unbiased for σ^2: the maximum likelihood estimator for σ^2 has denominator n. As usual in this course (when estimating variance) we use the unbiased estimator. The probability distribution of S^2 is given by

$$\frac{(n-1)S^2}{\sigma^2} \sim \chi^2(n-1). \tag{10.12}$$

These results are stated without proof. The parameter σ^2 is a nuisance parameter: generally, its value is not known. Finally (and this is an important result describing the sampling distribution of the least squares estimator $\widehat{\gamma}$ but not involving σ^2): (10.11) and (10.12) together yield the distribution of the estimator $\widehat{\gamma}$. It is given by

$$\frac{\widehat{\gamma}-\gamma}{S/\sqrt{\sum x_i^2}} \sim t(n-1). \tag{10.13}$$

Example 10.11 illustrates one application of this result.

Example 10.11 *A confidence interval for the slope*

For the Sheffield map and road distance data, the least squares estimator for the slope γ was $\widehat{\gamma} = 1.289$. The estimate of the underlying variance σ^2 is

$$s^2 = \frac{\sum(y_i - \widehat{y}_i)^2}{n-1},$$

where the numerator, the residual sum of squares, is given by

$$\sum(y_i - \widehat{y}_i)^2 = 107.30.$$

See (10.4).

So $s^2 = 107.30/19 = 5.647$. It follows from (10.13) that a 90% confidence interval for γ is given by

$$\left(\widehat{\gamma} - q\frac{s}{\sqrt{\sum x_i^2}}, \widehat{\gamma} + q\frac{s}{\sqrt{\sum x_i^2}}\right)$$
$$= \left(1.289 - 1.729\frac{\sqrt{5.647}}{\sqrt{6226.38}}, 1.289 + 1.729\frac{\sqrt{5.647}}{\sqrt{6226.38}}\right)$$
$$= (1.289 - 0.052, 1.289 + 0.052) = (1.237, 1.341),$$

using the fact that the 95% quantile for $t(19)$ is given by $q = q_{0.95} = 1.729$. ■

The unconstrained straight line

In this case, assuming only that the random terms W_i are independent, with mean 0 and variance σ^2, it can be shown that

$$E(\widehat{\alpha}) = \alpha, \qquad V(\widehat{\alpha}) = \frac{\sigma^2}{n}; \tag{10.14}$$

and

$$E(\widehat{\beta}) = \beta, \qquad V(\widehat{\beta}) = \frac{\sigma^2}{\sum(x_i - \overline{x})^2}. \tag{10.15}$$

The calculations involved in (10.14) and (10.15) are not entirely straightforward. It is important to note that both estimators are unbiased, and to recognize the useful consequences of a large sample of data (large n). Also, in this case, the variance in the slope estimator is reduced if the x-values are widely dispersed. (If you imagine that the numbers $x_1, x_2, \ldots, x_n$ constitute a random sample, then the number $\sum(x_i - \overline{x})^2$ is proportional to the sample variance.)

Now, it is also true that the estimator

$$S^2 = \frac{\sum(Y_i - \widehat{Y}_i)^2}{n-2}$$

Notice the denominator $(n-2)$ in the two-parameter case.

is unbiased for σ^2. If, in addition, it is assumed that the random terms W_i are normally distributed, then

$$\widehat{\beta} \sim N\left(\beta, \frac{\sigma^2}{\sum(x_i - \overline{x})^2}\right), \qquad \frac{(n-2)S^2}{\sigma^2} \sim \chi^2(n-2). \tag{10.16}$$

These results are stated without proof.

Eliminating the nuisance parameter σ^2 in the usual way, then the distribution of the slope estimator $\widehat{\beta}$ is given by

$$\frac{\widehat{\beta} - \beta}{S/\sqrt{\sum(x_i - \overline{x})^2}} \sim t(n-2). \tag{10.17}$$

We shall see in Subsection 10.3.3 the useful consequences of this result: it can be used for calculating confidence intervals for β and for testing hypotheses about β.

Example 10.3 continued

In Example 10.3 the least squares estimator $\widehat{\beta}$ for the slope of the underlying trend was introduced, and we speculated whether or not this estimator was better than some or any of the preceding three slope estimators. They are all unbiased. In fact, from (10.15), the least squares estimator has variance

See pages 230 and 385.

$$\frac{\sigma^2}{\sum(x_i - \overline{x})^2} = \frac{\sigma^2}{17.5} = 0.057\sigma^2.$$

This is smaller than any of the variances of the three alternative estimators. In other words, this *estimating procedure* possesses better properties than the other three. This does not offer a guarantee that the slope estimate 5.42 ($\widehat{\beta}_4$, say) is any more accurate than any of the other three estimates
($\widehat{\beta}_1 = 5.6$, $\widehat{\beta}_2 = 5.0$, $\widehat{\beta}_3 = 6.0$). ■

10.3.2 Maximum likelihood estimation

This brief subsection contains an argument that you should try to follow. The conclusions are important: it illuminates the optimal properties of the least squares estimators $\widehat{\alpha}$ and $\widehat{\beta}$.

The estimators $\widehat{\alpha}$ and $\widehat{\beta}$ have been obtained through minimizing the sum of squared residuals. The principle of least squares is a very reasonable criterion (though, as was pointed out in the context of the data set on divorces, not the only reasonable one).

When the assumption of normality is made for the random error terms W_i, $i = 1, 2, \ldots, n$, then

$$W_i \sim N(0, \sigma^2), \quad i = 1, 2, \ldots, n;$$

equivalently, since $Y_i = \alpha + \beta x_i + W_i$,

$$Y_i \sim N(\alpha + \beta x_i, \sigma^2), \quad i = 1, 2, \ldots, n.$$

Maximizing the likelihood of α and β for the random sample $y_1, y_2, \ldots, y_n$ amounts to locating the maximum of the product

$$f(y_1) \times f(y_2) \times \ldots \times f(y_n)$$

$$= \left(\frac{1}{\sigma\sqrt{2\pi}}\right)^n \exp\left[-\frac{1}{2}\sum_{i=1}^{n}\left(\frac{y_i - (\alpha + \beta x_i)}{\sigma}\right)^2\right].$$

$f(y_i)$ is the p.d.f. of Y_i when $Y_i \sim N(\alpha + \beta x_i, \sigma^2)$.

This in turn reduces to locating the minimum of the sum

$$\sum_{i=1}^{n}\left(\frac{y_i - (\alpha + \beta x_i)}{\sigma}\right)^2,$$

or, simply, locating the minimum of the sum

$$\sum_{i=1}^{n}(y_i - (\alpha + \beta x_i))^2.$$

This is precisely the least squares criterion: it follows that under the assumption of normality, the least squares estimators $\widehat{\alpha}$ and $\widehat{\beta}$ of α and β are also the maximum likelihood estimators (and so possess all the optimal properties of maximum likelihood estimators).

10.3.3 Is the slope of the regression line 0?

We are now in a position to answer the question posed more than once before: is the slope 0?

We know that the sampling distribution of $\widehat{\beta}$ is given by (10.17). This can be used to provide a confidence interval for $\widehat{\beta}$ (which may or may not contain the value 0, and in that way constitutes a test of the hypothesis $H_0 : \beta = 0$); alternatively, we can obtain a *SP* for the null hypothesis $H_0 : \beta = 0$.

Example 10.1 continued

For the finger-tapping example the least squares regression line is given by

$$y = 244.75 + 0.0175x,$$

See Exercise 10.3.

or

$$\text{Tapping frequency} = 244.75 + 0.0175 \times \text{Caffeine dose},$$

where taps are counted per minute, and the caffeine dose is measured in mg.

The question was: does caffeine have any effect on tapping frequency? To answer this question, one approach is to test the hypothesis

$$H_0 : \beta = 0.$$

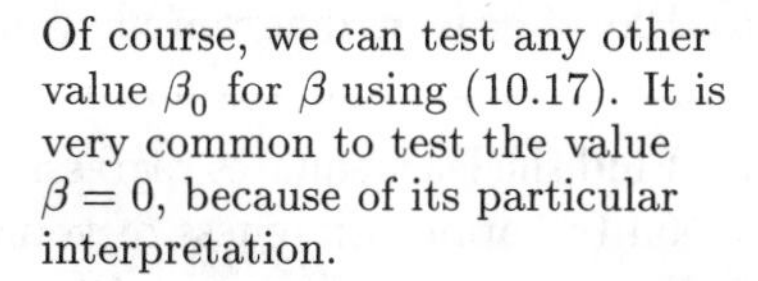

Of course, we can test any other value β_0 for β using (10.17). It is very common to test the value $\beta = 0$, because of its particular interpretation.

Certainly the estimated value $\widehat{\beta} = 0.0175$ seems quite a small number in absolute terms, but it needs to be assessed in the context of the overall variation. For this example the residual sum of squares is given by

$$\sum(y_i - \widehat{y}_i)^2 = 134.25,$$

You can use (10.10) here, or, more easily, a computer with regression software.

and so our estimate of σ^2 is given by $s^2 = 134.25/28 = 4.795$. Under the null hypothesis $H_0 : \beta = 0$ we need to compare the value

$$\frac{\widehat{\beta} - 0}{s/\sqrt{\sum(x_i - \overline{x})^2}} = \frac{0.0175 - 0}{\sqrt{4.795}/\sqrt{200000}} = 3.574$$

against $t(n-2)$, that is, $t(28)$. This gives us the results

$$SP(\text{obtained direction}) = SP(\text{opposite direction}) = 0.00065;$$
$$SP(\text{total}) = 0.0013.$$

This SP is extremely small: the null hypothesis of zero effect is rejected. ■

Now try the following exercises.

Exercise 10.8

The examination scores data in Table 10.6 suggest that the amount of time a candidate takes to complete the paper is of no value as an indicator of the final score. Pose this formally as a hypothesis in a regression context, and test the hypothesis.

Exercise 10.9

The data in Table 10.14 give the average level of aflatoxin (parts per billion) and the percentage of non-contaminated peanuts in 34 batches of peanuts in a sample of 120 pounds.

Draper, N.R. and Smith, H. (1981) *Applied regression analysis*, 2nd edn. John Wiley and Sons, New York, p. 63.

Table 10.14 Percentage of non-contaminated against aflatoxin

Percentage not contaminated	Aflatoxin (parts per billion)	Percentage not contaminated	Aflatoxin (parts per billion)
99.971	3.0	99.788	71.1
99.942	18.8	99.956	12.3
99.863	46.8	99.858	25.8
99.979	4.7	99.821	71.3
99.932	18.9	99.972	12.5
99.811	46.8	99.987	30.6
99.982	8.3	99.830	83.2
99.908	21.7	99.889	12.6
99.877	58.1	99.958	36.2
99.971	9.3	99.718	83.6
99.970	21.9	99.961	15.9
99.798	62.3	99.909	39.8
99.957	9.9	99.642	99.5
99.985	22.8	99.982	16.7
99.855	70.6	99.859	44.3
99.961	11.0	99.658	111.2
99.933	24.2	99.975	18.8

The aim is to investigate the relationship between the two variables, and to predict the percentage of non-contaminated peanuts from toxin levels.

(a) Plot a scatter diagram showing the percentage of peanuts not contaminated against aflatoxin level.

(b) Find the least squares regression line through the data points, commenting on the appropriateness of your model.

(c) Test the proposition that aflatoxin level is not a useful indicator of the percentage of non-contaminated peanuts in a batch.

10.4 The prediction problem

A major use of regression is to predict future values of the response variable given new values of the explanatory variable. Suppose x_0 is the value of the explanatory variable associated with an individual whose response Y_0 is not known. Then the obvious *predictor* of Y_0 is

$$\widehat{y}_0 = \widehat{\alpha} + \widehat{\beta}x_0.$$

Example 10.12 *The consequences of higher doses of caffeine*

According to the model fitted to the finger-tapping data in Exercise 10.3, the predicted finger-tapping frequency for a dose of 400 mg of caffeine is

$$\widehat{\alpha} + \widehat{\beta}x_0 = 244.75 + 0.0175 \times 400 = 251.75 \text{ taps per minute.}$$

It is possible but unlikely that this exact frequency will be attained. Even if the trend continues in the way suggested by the data in Table 10.1 (and it is possible that larger doses of caffeine will have quite different effects) the estimate takes no account of the variation in tapping frequency at a given dose (which, as we know, is considerable). ■

In general, *any* such prediction will be wrong! After all, in this case, we have merely identified the point on the fitted line at the value $x_0 = 400$, shown in Figure 10.21. And we should not really expect responses to fall exactly on that line.

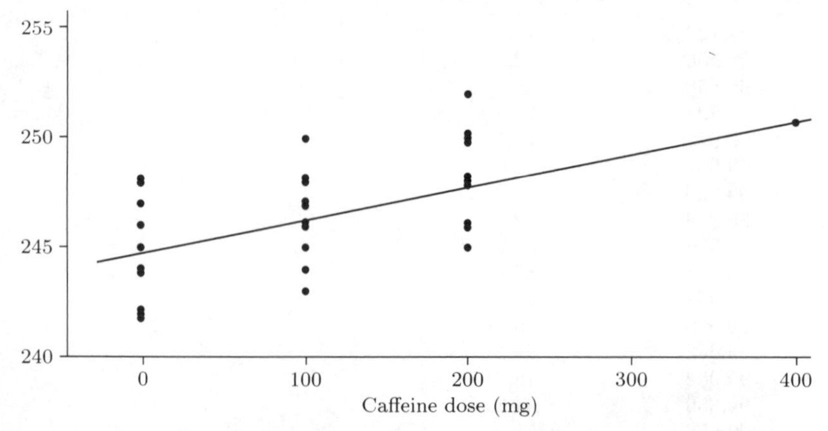

Figure 10.21 Estimating tapping frequency attained after a dose of 400 mg of caffeine

In evaluating the usefulness of predictions based on regression analyses, it is important to recognize that there are two sources of error in a prediction. First, we know (for this has been our model all along) that for a given value x_0 of the explanatory variable, the response is a random variable

$$Y_0 = \alpha + \beta x_0 + W_0$$

with mean $\alpha + \beta x_0$ and variance σ^2. Our prediction $\widehat{\alpha} + \widehat{\beta} x_0$ estimates the mean of the distribution of Y_0. This estimate will itself be subject to error, for the numbers α and β are unknown and have been estimated by $\widehat{\alpha}$ and $\widehat{\beta}$. Second, the prediction takes no account of the error term W_0.

10.4.1 A confidence interval for the mean

Assuming only that $W_i, i = 1, 2, \ldots, n$, are independent with mean 0 and variance σ^2, then the random variable

$$\widehat{\alpha} + \widehat{\beta} x_0$$

has mean

$$E(\widehat{\alpha} + \widehat{\beta} x_0) = \alpha + \beta x_0$$

and variance

$$V(\widehat{\alpha} + \widehat{\beta} x_0) = \left(\frac{(x_0 - \overline{x})^2}{\sum (x_i - \overline{x})^2} + \frac{1}{n} \right) \sigma^2.$$

In this result, the quantity

$$\overline{x} = \frac{x_1 + x_2 + \ldots + x_n}{n}$$

is the mean of the original list of x-values, not including the value x_0.

The algebra is a little awkward here, because the estimators $\widehat{\alpha}$ and $\widehat{\beta}$ are not independent; this result is included without proof.

If it is further assumed that the random terms W_i are normally distributed, then it turns out (eliminating the nuisance parameter σ in the usual way) that

$$\frac{(\widehat{\alpha} + \widehat{\beta} x_0) - (\alpha + \beta x_0)}{S \sqrt{\dfrac{(x_0 - \overline{x})^2}{\sum (x_i - \overline{x})^2} + \dfrac{1}{n}}} \sim t(n - 2), \tag{10.18}$$

where $S^2 = \sum (y_i - \widehat{y}_i)^2 / (n - 2)$ is, as usual, the unbiased estimator for σ^2.

The result (10.18) can be used to provide a confidence interval for the expected value $\alpha + \beta x_0$ of Y_0.

Example 10.12 continued

For the finger-tapping data the estimated mean tapping frequency in response to a dose of $x_0 = 400$ mg of caffeine is 251.75 taps per minute. We know that

$$n = 30, \qquad \overline{x} = 100, \qquad \sum (x_i - \overline{x})^2 = 200\,000.$$

Also, we have found

$$s^2 = \frac{\sum (y_i - \widehat{y}_i)^2}{n - 2} = \frac{134.25}{28} = 4.795.$$

To find a 90% confidence interval for the mean tapping frequency for a dose of 400 mg, we require the 95% quantile of $t(28)$: this is $q_{0.95} = 1.701$. So a 90% confidence interval is given by

$$\left(\widehat{\alpha} + \widehat{\beta}x_0 \pm q_{0.95}s\sqrt{\frac{(x_0 - \overline{x})^2}{\sum(x_i - \overline{x})^2} + \frac{1}{n}}\right)$$

$$= \left(251.75 \pm 1.701\sqrt{4.795}\sqrt{\frac{300^2}{200\,000} + \frac{1}{30}}\right)$$

$$= (251.75 \pm 2.590) = (249.2, 254.3).$$

The notation '±' is a convenient abbreviation for the lower and upper confidence limits. For example, $(a \pm b)$ means $(a - b, a + b)$.

Notice that this is not a confidence interval for the finger-tapping frequency that the next student given a dose of 400 mg of caffeine might attain: it is simply a confidence interval for the unknown parameter $\alpha + 400\beta$. ■

Now try Exercise 10.10.

Exercise 10.10

Table 10.15 gives the measures of resistance to breathing, to be regressed against the heights of a sample of 24 children suffering from cystic fibrosis.

Table 10.15 Breathing resistance and height (cm) for 24 children suffering from cystic fibrosis

Resistance	13.8	8.2	9.0	12.5	21.1	6.8	17.0	11.0	8.2
Height	89	93	92	101	95	89	97	97	111
Resistance	12.7	8.5	10.0	11.6	9.5	15.0	13.5	11.0	11.0
Height	102	103	108	103	105	109	93	98	103
Resistance	8.8	9.5	9.2	15.0	7.0	6.3			
Height	108	106	109	111	111	116			

Cogswell, J.J. (1973) Forced oscillation technique for determination of resistance to breathing in children. *Archives of Diseases in Children*, **48**, 259–266.

Find a 95% confidence interval for the mean breathing resistance for sufferers who are 100 cm tall.

10.4.2 A prediction interval for the response

It is possible to develop a **prediction interval** for Y_0. This is similar to finding a confidence interval for an unknown parameter (but, of course, Y_0 is a random variable, not an unknown constant).

Writing

$$\widehat{Y}_0 = \widehat{\alpha} + \widehat{\beta}x_0 + W_0,$$

it follows that

$$\begin{aligned} E(\widehat{Y}_0) &= E(\widehat{\alpha} + \widehat{\beta}x_0 + W_0) \\ &= E(\widehat{\alpha}) + E(\widehat{\beta}x_0) + E(W_0) \\ &= \alpha + \beta x_0 + 0 \\ &= \alpha + \beta x_0. \end{aligned}$$

Also,

$$V(\widehat{Y}_0) = V(\widehat{\alpha} + \widehat{\beta}x_0 + W_0)$$
$$= V(\widehat{\alpha} + \widehat{\beta}x_0) + V(W_0)$$

where the two bracketed terms are independent. Since $V(W_0) = \sigma^2$, this expression reduces to

$$V(\widehat{Y}_0) = \left(\frac{(x_0 - \overline{x})^2}{\sum(x_i - \overline{x})^2} + \frac{1}{n} + 1\right)\sigma^2.$$

The distribution theory here is again slightly awkward. A prediction interval for an observation on a random variable is not the same as a confidence interval for a parameter: however, each is calculated in a similar way. A 'guess' is made; allowing for uncertainty, lower and upper limits are calculated, bracketing the guess. A $100(1-\alpha)\%$ *prediction interval* for Y_0 is given by

$$\left(\widehat{\alpha} + \widehat{\beta}x_0 - qs\sqrt{\frac{(x_0 - \overline{x})^2}{\sum(x_i - \overline{x})^2} + \frac{1}{n} + 1}, \widehat{\alpha} + \widehat{\beta}x_0 + qs\sqrt{\frac{(x_0 - \overline{x})^2}{\sum(x_i - \overline{x})^2} + \frac{1}{n} + 1}\right),$$

where s is the unbiased estimate for σ, and q is the $100(1-\alpha/2)\%$ quantile of $t(n-2)$.

The result looks worse than it is: the following example shows how a prediction interval is obtained. The work involved is scarcely more than that involved in finding a confidence interval.

Example 10.12 continued

A 90% prediction interval for the finger-tapping frequency attained *by an individual* after a 400 mg dose of caffeine (assuming the extrapolation to be reasonable, which on physiological grounds it might not be) is

$$\left(\widehat{\alpha} + \widehat{\beta}x_0 \pm q_{0.95}s\sqrt{\frac{(x_0 - \overline{x})^2}{\sum(x_i - \overline{x})^2} + \frac{1}{n} + 1}\right)$$
$$= \left(251.75 \pm 1.701\sqrt{4.795}\sqrt{\frac{300^2}{200\,000} + \frac{1}{30} + 1}\right)$$
$$= (251.75 \pm 4.536) = (247.2, 256.3)$$

which is a very much wider interval than the confidence interval for the mean response. The reason is that it is the prediction interval for the response of an individual, and has to allow for the variation between individuals at all dose levels. ■

Exercises 10.11 to 10.13 are on this topic. The calculations can get rather involved: you are recommended to use your computer here.

Exercise 10.11

Cogswell also gave data on resistance to breathing regressed against height for a further sample of 42 children suffering from asthma. These data are given in Table 10.16.

Table 10.16 Breathing resistance and height (cm) for asthmatic children

Resistance	25.6	15.1	9.8	7.5	10.1	12.5	9.1	17.0	5.5
Height	90	97	97	104	119	106	113	116	118
Resistance	15.7	6.4	8.8	10.1	5.0	5.8	12.8	10.0	8.0
Height	119	117	122	122	124	129	130	135	127
Resistance	23.8	7.5	12.1	6.2	8.3	3.5	10.2	16.1	10.1
Height	120	121	126	123	125	118	122	120	133
Resistance	12.1	11.2	9.4	15.6	7.9	18.7	8.3	7.5	8.5
Height	125	123	141	121	128	125	140	140	140
Resistance	8.9	7.9	5.7	9.5	6.5	15.0			
Height	148	145	138	148	132	134			

Obtain a 95% prediction interval for the breathing resistance of a child 100 cm tall.

Exercise 10.12

Prediction of temperature from cricket chirping rate was a prime motivation for collecting the data given in Table 10.13. Find a 99% prediction interval for the temperature when the cricket chirping level is 18.

Exercise 10.13

Find a 95% prediction interval for the fuel consumption of a diesel car weighing 1080 kg (see Table 10.12).

10.5 The assumptions of the regression model

In this final section we shall consider some of the assumptions of the regression model, and what could go wrong. Competence at this kind of data exploration comes with practice and experience. You should read the commentaries and understand the motivation behind them, but do not worry if at this stage the approach adopted is not always entirely obvious.

The first assumption made is that the random error term W_i in the model

$$Y_i = \alpha + \beta x_i + W_i$$

has constant variance σ^2. We saw in Figure 10.13(a) a case where this assumption was clearly broken: the scatter away from an apparent trend line was becoming more pronounced with increasing x.

One way of dealing with this is to assume that the model is

$$Y_i = \alpha + \beta x_i + W_{0i}$$

say, where W_{0i} has mean 0 (as before) and variance $\sigma^2 x_i^2$ (that is, standard deviation σx_i, proportional to x_i. Dividing through by x_i gives the model

$$\frac{Y_i}{x_i} = \beta + \frac{\alpha}{x_i} + \frac{W_{0i}}{x_i}.$$

Can you see that this is simply a linear regression model with new explanatory variable $1/x_i$ and new response variable Y_i/x_i? The constant term in the new model is β; the slope is α; and the random term W_{0i}/x_i has mean 0 and constant variance σ^2, as required. After making this transformation, you can regress the response variable Y_i/x_i against the explanatory variable $1/x_i$, estimating parameters in the usual way.

Most approaches in an assessment of the model assumptions are not as mathematical as this! But it *is* a very neat transformation of the model.

Sometimes there is a very obvious but non-linear shape to the data (as with the duckweed data of Example 10.8 and the paper strength data of Example 10.9). For example, one might fit to the paper data the model

$$Y_i = \alpha + \beta x_i + \gamma x_i^2 + W_i,$$

choosing the parameters α, β and γ to minimize the sum of squared residuals

$$\sum_{i=1}^{n}(y_i - (\alpha + \beta x_i + \gamma x_i^2))^2.$$

This approach extends to higher powers of x_i. Other regression functions might be appropriate: you need to be rather careful that the assumption of constant variance holds, here. For instance, variation in the duckweed counts is likely to increase as fast as the counts themselves, with passing time. A model of the form

The approach is called polynomial regression.

See Example 10.8.

$$Y_i = 20e^{\lambda x_i} + W_i,$$

Remember, there were 20 duckweed fronds at week zero, and the experiment continued for eight weeks.

where W_i has mean 0 and constant variance σ^2, is probably not a useful one, since the assumptions are so badly broken. On the other hand, if you transform the duckweed count by taking logarithms, you obtain a scatter diagram suggestive of a linear fit, and the assumption of constant variance is probably a much more reasonable one. The new scatter plot is shown in Figure 10.22.

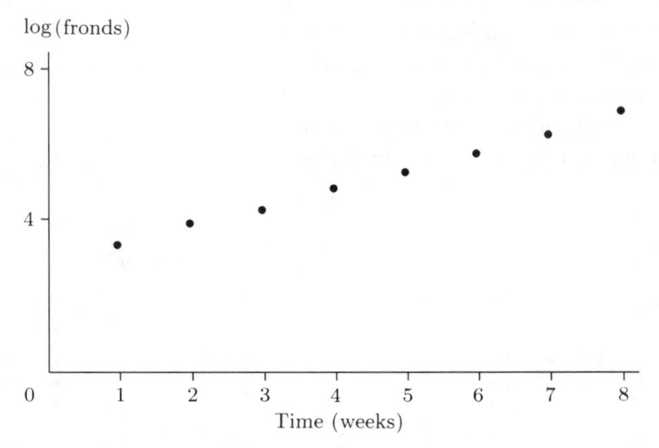

Figure 10.22 Duckweed data transformed: log(fronds) against time (weeks)

Example 10.13 is another example in which various transformations were useful: however, they were performed not on the response variable but on the explanatory variable.

Example 10.13 Tearing factor for paper

The data given in Table 10.17 come from an experiment in which five different manufacturing pressures were each applied to four different sheets of paper, in order to investigate the effect of pressure on the 'tearing factor' of the paper.

The data are reported in Williams, E.J. (1959) *Regression Analysis.* John Wiley and Sons, New York.

Table 10.17 Tearing factor data

Pressure	Tearing factors			
35.0	112	119	117	113
49.5	108	99	112	118
70.0	120	106	102	109
99.0	110	101	99	104
140.0	100	102	96	101

A scatter plot for these data is given in Figure 10.23.

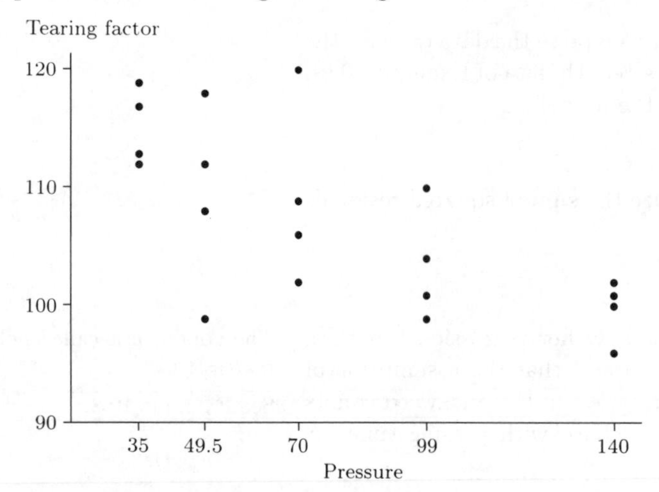

Figure 10.23 Tearing factor versus pressure

It does not seem unreasonable to fit a straight line to the raw data in this example. Nevertheless, there does appear to be some evidence of curvature. A good place to start on the ladder of powers is with the logarithmic transformation.

The next powers either way on the ladder of powers also give reasonable, but not quite so good, roughly linear scatter plots (see Figure 10.24). More to the point, an additional factor that makes the logarithmic transformation seem appropriate in this particular case is that the original x-values appear to have been chosen by the experimenter to be equally spaced on a logarithmic scale in Figure 10.24; thus it seems that the experimenter expected to use the explanatory variable log(pressure) on the basis, one imagines, of previous experience and knowledge.

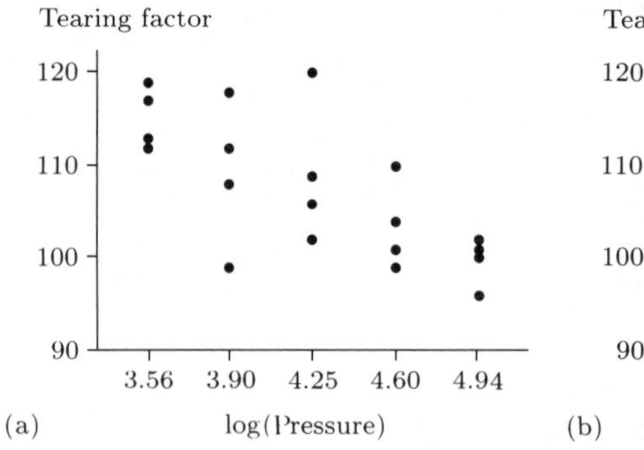

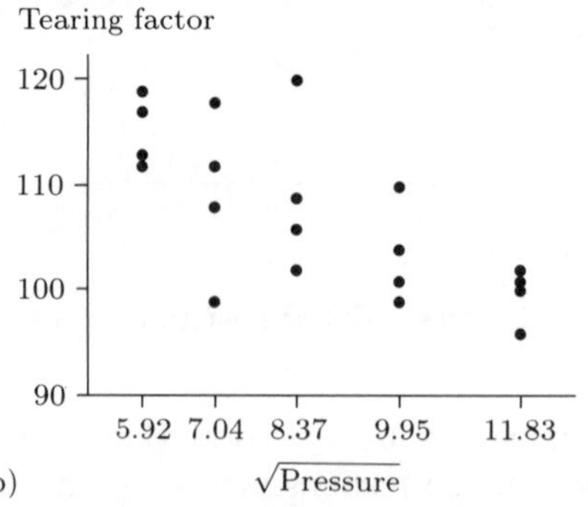

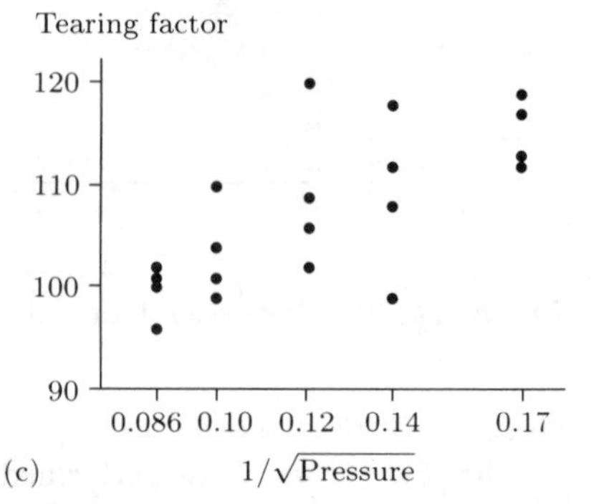

Figure 10.24 Tearing factor versus (a) log(Pressure) (b) $\sqrt{\text{Pressure}}$ (c) $1/\sqrt{\text{Pressure}}$

You can see from this example that there are no fixed 'rules' about appropriate transformations to use.

What makes this example artificial is that we are being forced to act as though we had not been consulted about the design of the experiment: ideally, in practice the statistician would have been involved from the beginning, and would know the designer's intentions. The logarithmic scale is suggested because each of the numbers 49.5, 70.0, 99.0 and 140.0 is very close to $\sqrt{2}$ times the previous number. ■

The problem of *outliers* is an important one in regression analysis, as in any other statistical analysis of data. We saw in Figure 10.7(a) the suggestion of a possible outlier. The difficulty here is that in a regression problem a single point that seems far away from some others may not be aberrant, but in fact it may be a very useful and informative point. We saw, for instance, in the case of a straight line constrained through the origin, that it is very useful to select values x_i of the response variable as far from the origin as possible. Then the point is not an outlier but an *influential point*.

Roughly, an outlier is a point 'far away' in the y-direction, while an influential point is 'far away' in the x-direction.

Example 10.14 shows that checking for outliers is not always clear cut.

Example 10.14 Lung cancer and smoking

As part of the developments in the 1950s that established the link between cigarette smoking and lung cancer, Richard Doll published data from 11 western countries relating to the male death rate from lung cancer in 1950 and per capita consumption of cigarettes in 1930. The data are given in Table 10.18.

Doll, R. (1955) Etiology of lung cancer. *Advances in Cancer Research*, **3**.

A scatter plot (showing the least squares regression line, and with country labels attached) is given in Figure 10.25.

Table 10.18 Lung cancer death rates and cigarette consumption

Country	Cigarettes	Cancer death rate
Iceland	220	58
Norway	250	90
Sweden	310	115
Canada	510	150
Denmark	380	165
Australia	455	170
United States	1280	190
Holland	460	245
Switzerland	530	250
Finland	1115	350
Great Britain	1145	465

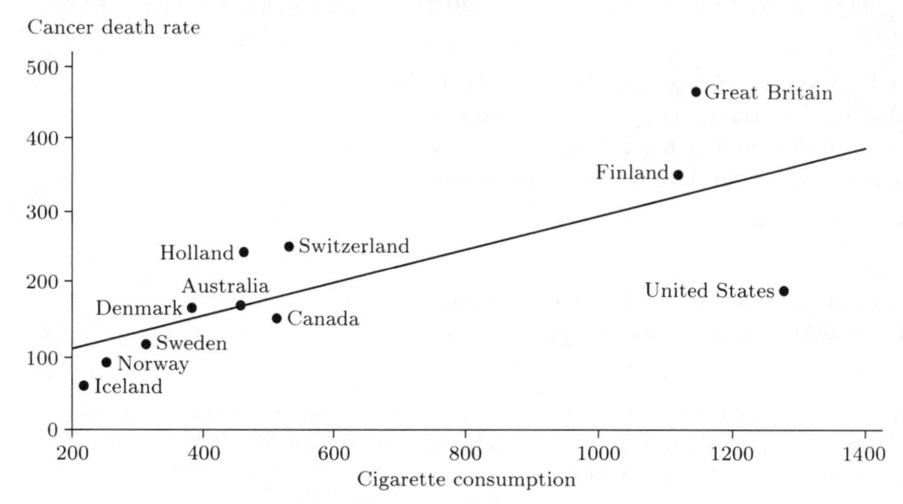

Figure 10.25 Scatter plot of lung cancer death rate against cigarette consumption

In Figure 10.25, the US seems a definite outlier from the linear trend and possibly Great Britain is aberrant. Are both the US and Britain out of step with the rest of the data? The problem is that the point corresponding to the US has exerted a big downward tension on the fitted regression line, making it rather shallow, and dragging it well away from the point corresponding to Britain (and so making that point look like an outlying point). If the point for the US is ignored in fitting the line, a noticeable change occurs, as you can see from Figure 10.26. It appears that Great Britain is not out of step at all with the other nine countries.

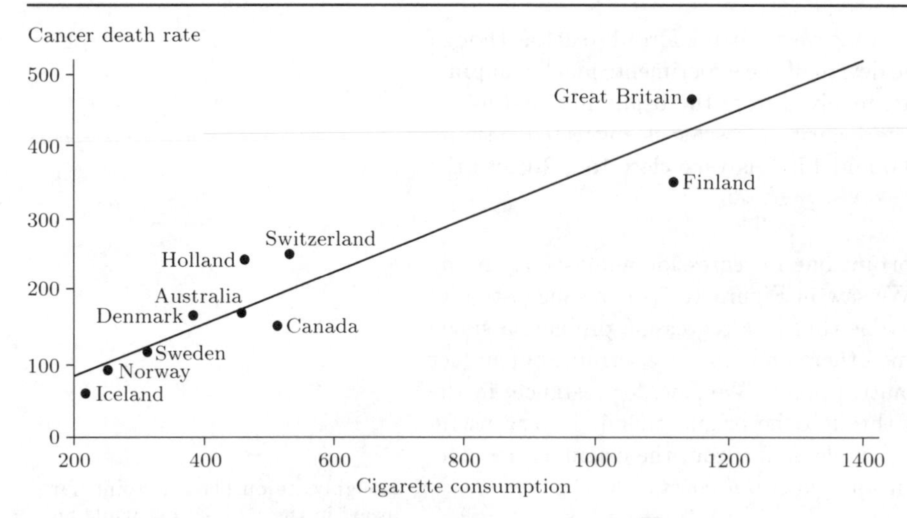

Figure 10.26 Scatter plot with the US point removed ■

Regression diagnostics is a huge field of study, and there is no limit to the ways in which one might adapt the models, or tweak and transform the data. We have already performed some informal regression diagnostics: we have always done the sensible thing and looked at scatter plots of the data first, rather than blindly 'applying the technique' which may in some cases be entirely inappropriate.

One final and very important additional diagnostic is the use of a **residual plot**. This is a way of graphically exploring the residuals $w_i = y_i - \widehat{y}_i$ after the fitting has taken place.

Remember,
RESIDUAL = DATA − FIT.

Discrepancies from the model are not always as easy to spot from the basic scatter plot as from residual plots. Any distinct relationship in the data tends to obscure patterns in the remaining variation about that relationship. A way around this is to subtract the fitted relationship from the data and concentrate on the residuals

$$w_i = y_i - \widehat{y}_i.$$

If all of our modelling assumptions are correct, each w_i will be an observation from a normal distribution with mean zero and an unknown but constant variance.

The **residual plot** is the scatter plot of residuals w_i against the x_i. The residuals have a mean of zero, and the residual plot should show the residuals fluctuating about zero in a random, unpatterned fashion. We look for a pattern in the residual plot as evidence that an assumption may be violated. The residual plot has a variety of rather small advantages over the scatter plot of responses against the explanatory variable.

Sometimes residuals are plotted against the fitted values: that is to say, a scatter plot of w_i against $\widehat{y}_i$ is scrutinized.

Figure 10.27 illustrates four typical residual plots. Figure 10.27(a) is a residual plot with no apparent pattern of any kind in the residuals: this is the type of plot that accords with our assumptions. Figure 10.27(b) shows a definite pattern. As we move from left to right, from smaller to larger xs, the residuals are first negative, then positive, then negative again. Such a residual plot is (usually) associated with an explanatory relation that is other than linear ($Y_i = f(x_i) + W_i$ for some function $f(\cdot)$ other than a straight line): the straight-line model has systematically over-estimated the level of the response

Possibly $f(\cdot)$ is quadratic in this case.

in some regions and under-estimated it in others. In Figure 10.27(c) the pattern of residuals is indicative of a variance σ^2 that is not constant. As the explanatory variable gets larger, so too does the variability of the residuals (and hence of the responses). Figure 10.27(d) shows a residual plot which is similar to that in Figure 10.27(a) in most respects except for a single observation which produces a residual considerably larger in magnitude than any of the others. The plotted point may correspond to an outlier, an observation so disparate in size as to suggest that it was not generated by the same process as were the other data points.

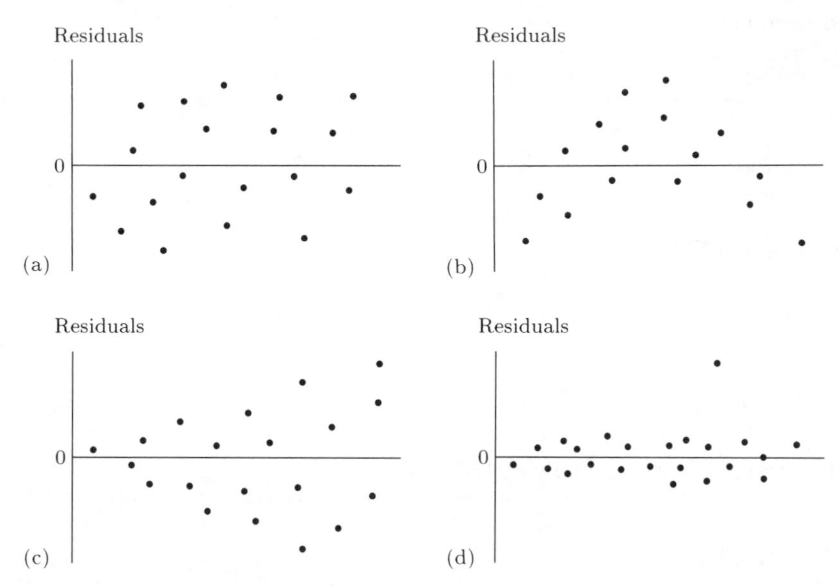

Figure 10.27 Residual patterns: (a) unpatterned (assumptions tenable) (b) a systematic discrepancy (c) variance not constant (d) an outlier

Finally, the assumption of a normal distribution for the ws can be checked through a standard graphical check for normality using the residuals. Given the ideas in *Chapter 9*, Section 9.1, a natural approach is to construct probability plots for the residuals.

In this subsection we have seen some of the informal checks and considerations that occur in a regression analysis. In this chapter we have scratched the surface of an important and potentially very informative technique, one of the most important available to you. As well as being aware of the technical aspects of a regression analysis, you should also have observed the essential requirement to begin your analysis with a graphical representation of the data, to see what messages are evident from that.

Summary

1. When the distribution of a random variable Y depends on the value taken by some associated variable x, then this can be represented by a regression model, with x the explanatory variable, and Y the response variable. Less generally, the mean of Y might alter with x.

2. A common regression model takes the form

$$Y_i = \gamma x_i + W_i,$$

where the random terms W_i are independent with mean 0 and constant variance σ^2; then

$$E(Y_i) = \gamma x_i, \qquad V(Y_i) = \sigma^2.$$

3. The parameter γ may be estimated through the principle of least squares, by minimizing the sum of squared residuals

$$\sum w_i^2 = \sum (y_i - \gamma x_i)^2;$$

then the appropriate estimator is

$$\widehat{\gamma} = \frac{\sum x_i Y_i}{\sum x_i^2};$$

the estimator $\widehat{\gamma}$ is unbiased and $V(\widehat{\gamma}) = \sigma^2 / \sum x_i^2$. Also

$$S^2 = \frac{\sum (Y_i - \widehat{Y}_i)^2}{n-1}$$

is an unbiased estimator for σ^2.

4. If it is also assumed that the random terms W_i are normally distributed, then

$$\widehat{\gamma} \sim N\left(\gamma, \frac{\sigma^2}{\sum x_i^2}\right), \qquad \frac{(n-1)S^2}{\sigma^2} \sim \chi^2(n-1),$$

and, in particular,

$$\frac{\widehat{\gamma} - \gamma}{S/\sqrt{\sum x_i^2}} \sim t(n-1).$$

5. The most common straight-line model for regression is given by

$$Y_i = \alpha + \beta x_i + W_i,$$

where the random terms W_i are independent with mean 0 and constant variance σ^2; then

$$E(Y_i) = \alpha + \beta x_i, \qquad V(Y_i) = \sigma^2.$$

6. The parameters α and β have least squares estimators obtained by minimizing the sum of squared residuals

$$\sum w_i^2 = \sum (y_i - (\alpha + \beta x_i))^2.$$

These estimators are

$$\widehat{\alpha} = \overline{Y} - \widehat{\beta}\overline{x}, \qquad \widehat{\beta} = \frac{n \sum x_i Y_i - \sum x_i \sum Y_i}{n \sum x_i^2 - (\sum x_i)^2}.$$

The least squares regression line

$$(y - \overline{y}) = \widehat{\beta}(x - \overline{x})$$

passes through the centroid of the data, the point $(\overline{x}, \overline{y})$. Both estimators are unbiased; also

$$V(\widehat{\alpha}) = \frac{\sigma^2}{n}, \qquad V(\widehat{\beta}) = \frac{\sigma^2}{\sum(x_i - \overline{x})^2}.$$

An unbiased estimator for σ^2 is given by

$$S^2 = \frac{\sum(Y_i - \widehat{Y}_i)^2}{n-2};$$

a useful way of calculating the sum of squared residuals $\sum(y_i - \widehat{y})^2$ for a particular data set is by writing

$$\sum(y_i - \widehat{y}_i)^2 = \sum(y_i - \overline{y})^2 - \widehat{\beta}^2 \sum(x_i - \overline{x})^2.$$

7. If the random terms W_i are normally distributed then the least squares estimators $\widehat{\alpha}$ and $\widehat{\beta}$ are also the maximum likelihood estimators for α and β; the slope estimator $\widehat{\beta}$ has distribution

$$\frac{\widehat{\beta} - \beta}{S/\sqrt{\sum(x_i - \overline{x})^2}} \sim t(n-2).$$

A $100(1-\alpha)\%$ confidence interval for the mean response at x_0, $\alpha + \beta x_0$, is given by

$$\left(\widehat{\alpha} + \widehat{\beta} x_0 \pm qs\sqrt{\frac{(x_0 - \overline{x})^2}{\sum(x_i - \overline{x})^2} + \frac{1}{n}}\right),$$

where q is the $100(1-\alpha/2)\%$ quantile of $t(n-2)$. A $100(1-\alpha)\%$ prediction interval for the response of an individual at x_0 is given by

$$\left(\widehat{\alpha} + \widehat{\beta} x_0 \pm qs\sqrt{\frac{(x_0 - \overline{x})^2}{\sum(x_i - \overline{x})^2} + \frac{1}{n} + 1}\right),$$

where q is the $100(1-\alpha/2)\%$ quantile of $t(n-2)$.

Chapter 11
Related variables

So far in the course, just one attribute (age, temperature, weight, and so on) has normally been measured on a random sample from some population. In this chapter we explore situations where more than one attribute is measured, and interest centres on how the attributes vary together (for example, height and weight). We learn how to quantify any perceived association between variables, and how to test a hypothesis that there is, in fact, no association between them. A new probability model, the bivariate normal distribution, is introduced.

Like the previous chapter, *Chapter 11* is concerned with ideas and techniques for data consisting of pairs of variables; that is, with data in the form $(X_1, Y_1), (X_2, Y_2), \ldots, (X_n, Y_n)$. *Chapter 10* concentrated on regression analysis, and a key idea there was that one of the variables involved was treated as the explanatory variable, and the other as the response variable. In this chapter, the two variables are not distinguished in that way. We shall not be concerned with trying to explain how measurements on the variable Y change in response to changes in the variable X, but instead we shall treat the two variables on an equal footing.

For instance, Figure 11.1 is a scatter plot of data on the heights (in cm) and weights (in kg) of 30 eleven-year-old girls. Rather than asking questions about how a girl's weight depends on her height, in this chapter we shall ask how the weights and heights of girls vary together. What do we mean when we say that the two variables, height and weight, are related? How can we measure how closely related they are?

Data provided by A.T. Graham, The Open University.

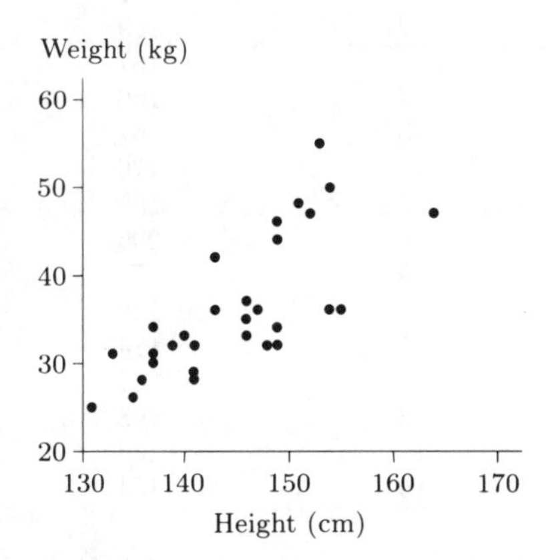

Figure 11.1 Scatter plot of the heights and weights of 30 eleven-year-old girls in a Bradford school

In situations of this sort, it is often useful to treat each pair of observations on the two random variables as *one* observation on a kind of two-dimensional random variable—a *bivariate random variable.* The resulting data are often called *bivariate data.* Section 11.1 describes in more detail what is meant by bivariate data, as well as exploring the idea of what it means for two variables to be related. Section 11.2 develops ways of attaching a numerical measure to the strength of the relationship between two random variables, and for using such a measure in a hypothesis test of whether the variables really are related at all. Sections 11.3 and 11.4 are concerned with data on pairs of variables, each of which is discrete and can take only a small number of values. An example to which we shall return comes from a study of risk factors for heart disease. A number of individuals were given a score on a four-point scale to indicate the amount that they snored at night. The second variable involved took only two values; it was a Yes–No measure of whether each individual had heart disease. The question of interest was whether heart disease and snoring frequency are related. Such data are very often presented in a table called a *contingency table*. Methods for testing for relationships between variables of this sort are rather different from those covered in Section 11.2 for continuous data and other forms of discrete data; one of the tests turns out to be a form of the chi-squared goodness-of-fit test that you met in *Chapter 9*. Finally, Section 11.5 returns to continuous data and presents a probability model for bivariate data—the *bivariate normal distribution.*

11.1 Bivariate data

11.1.1 Scatter plots and relationships

Let us begin with some examples of bivariate data.

Example 11.1 Systolic and diastolic blood pressures

The data in Table 11.1 come from a study of the effect of a drug, captopril, on blood pressure in human patients who had moderate essential hypertension (moderately raised blood pressure). The pressure of the blood inside the body varies as the heart beats, and a blood pressure measurement generally produces two values: the *systolic* pressure, which is the maximum pressure as the heart contracts, and the *diastolic* pressure, which is the minimum pressure. The data in Table 11.1 are readings taken on the fifteen patients before they were given the drug, captopril.

Table 11.1 Blood pressure measurements for 15 patients before treatment with captopril (mm Hg)

Patient number	Systolic blood pressure (mm Hg)	Diastolic blood pressure (mm Hg)
1	210	130
2	169	122
3	187	124
4	160	104
5	167	112
6	176	101
7	185	121
8	206	124
9	173	115
10	146	102
11	174	98
12	201	119
13	198	106
14	148	107
15	154	100

MacGregor, G.A., Markandu, N.D., Roulston, J.E. and Jones, J.C. (1979) Essential hypertension: effect of an oral inhibitor of angiotensin-converting enzyme. *British Medical Journal*, **2**, 1106–1109.

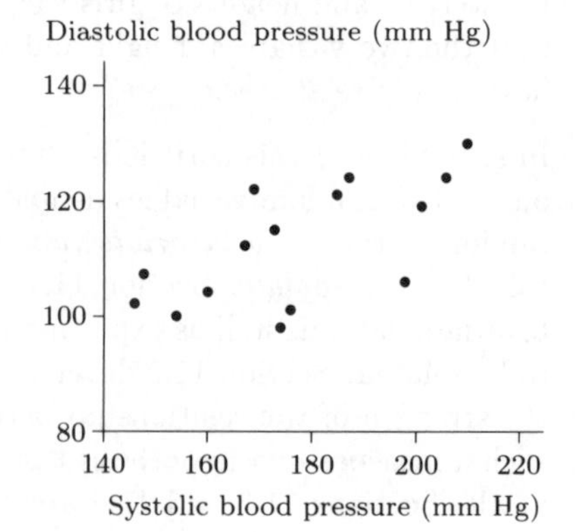

Figure 11.2 Systolic and diastolic blood pressures for 15 patients

In Figure 11.2 the data are plotted on a scatter plot. The question of interest is: how do these two measurements vary together in patients with this condition? From the scatter plot, it appears that there is something of a tendency for patients who have high systolic blood pressures to have high diastolic pressures as well. The pattern of points on the scatter plot slopes upwards from left to right, as it did in several of the scatter plots you met in *Chapter 10*. In fact, it would be *possible* to analyse the data using the regression methods discussed in *Chapter 10*. But there is a problem here.

Which of the two variables would you choose to be the explanatory variable, and which the response variable? There seems to be no clear answer. We are neither investigating how diastolic blood pressure changes in response to changes in systolic blood pressure, nor the other way round. In Figure 11.2, systolic pressure is plotted on the horizontal axis, but in the context of this investigation this was an arbitrary choice. It would have been perfectly feasible to plot the data the other way round. We cannot use the regression methods of *Chapter 10*, because those methods do not treat the two variables on an equal footing. ■

The data in Example 11.1 provide evidence that, for people with moderate essential hypertension at least, people with relatively high systolic blood pressure tend to have relatively high diastolic pressure as well. People with low systolic pressure tend to have low diastolic pressure. And, importantly, these statements work the other way round too. People with high diastolic pressure tend to have high systolic pressure. Another way to think of this is as follows. Suppose you choose at random a patient with moderate essential hypertension. On the basis of the data in Table 11.1, you could say something about what his or her diastolic blood pressure might be. Without performing any calculations, you would probably find it surprising if their diastolic blood pressure fell a long way outside the range from about 95 to about 130 mm Hg.

However, suppose you were now told that this person's systolic blood pressure was 200 mm Hg. On the range of values of systolic pressure represented in Figure 11.2, this is a high value. One might expect, then, that this person's diastolic pressure would be relatively high too. You might think it quite unlikely, for instance, to find that their diastolic pressure was as low as 100. In other words, knowing their systolic pressure has provided information about their diastolic pressure. Similarly, knowing their diastolic pressure would tell you something about their systolic pressure, if you did not know it already.

In intuitive terms, this is what it means for two random variables to be *related*. Knowing the value of one of the variables tells you something about the value of the other variable.

When two variables are related, it is often possible to describe in simple terms the manner of the relationship. In Example 11.1, one might say that the variables are *positively* related, because the two variables tend to be both high at the same time or both low at the same time. The pattern of points on the scatter plot slopes upwards from left to right.

We are not doing regression analysis here; but if we were, the regression line would have a positive slope.

Example 11.2 Socio-economic data on US states

Figure 11.3 shows a scatter plot of data on 47 states of the USA, taken from a study of crime rates and their determinants. The variable on the horizontal axis is a measure of the educational level of residents of each state: it is the mean number of years' schooling of the population aged 25 and over. The variable on the vertical axis is a measure of the inequality of income distribution in the state: it is the percentage of families who earned below one-half of the overall median income.

Vandaele, W. (1978) Participation in illegitimate activities: Erlich revisited. In Blumstein, A., Cohen, J. and Nagin, D., (eds) *Deterrence and incapacitation: estimating the effects of criminal sanctions on crime rates.* National Academy of Sciences, Washington, DC, pp. 270–335. The data relate to the calendar year 1960.

Again, the aim here is not to describe or investigate how one of the variables changes in response to changes in the other. It is to describe how the variables vary together, or in other words how they are related. The variables clearly

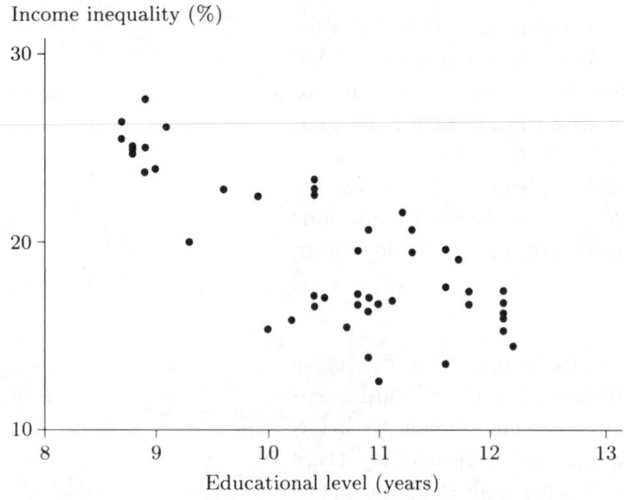

Figure 11.3 Educational level and income inequality

are related. States with relatively high average educational level tend to have relatively low income inequality, on the measure used here; and states with low educational level have high income inequality. Therefore, knowing something about the educational level of a state tells you something about its income inequality, and vice versa. The relationship is different from that in Example 11.1 though. In Example 11.1, a high value of one variable was associated with a high value of the other and a low value of one variable was associated with a low value of the other. In this example the association works the other way round: low values of one variable go with high values of the other. The pattern of points on the scatter plot slopes downwards from left to right. Figure 11.3 shows a *negative* relationship between the variables. ■

Variables can be related in other ways. Figure 11.4 is a scatter plot of two economic variables, the percentage of the UK workforce that is unemployed and the percentage change in wage rates, for the years 1861 to 1913.

Phillips, A.W. (1958) The relationship between unemployment and the rate of change of money wage rates in the United Kingdom, 1861–1957. *Economica*, **25**, 283–299.

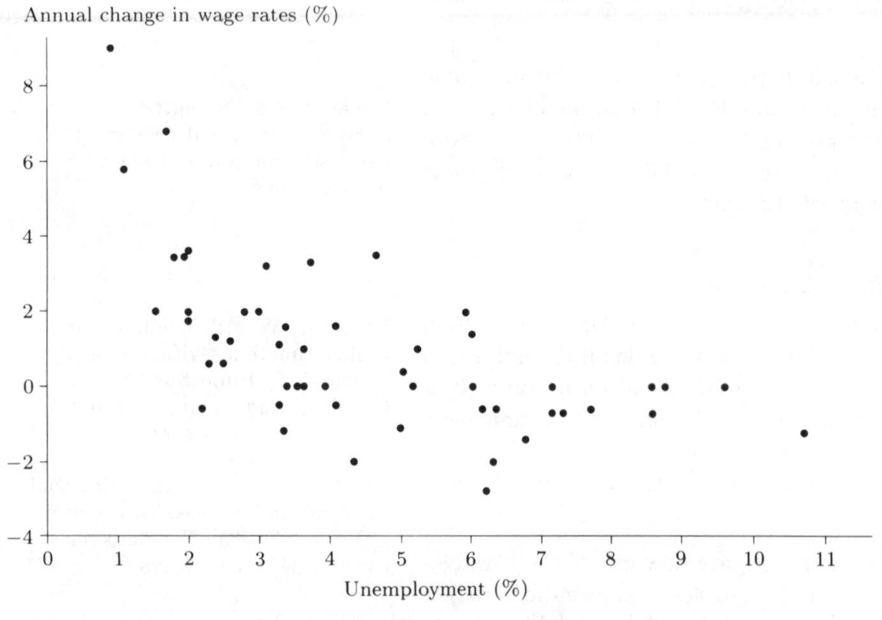

Figure 11.4 Change in wage rate (%) against unemployment (%)

Here, the two variables are negatively related because there is a tendency for high values of one of them to go with low values of the other. The pattern of points on the scatter plot slopes downwards from left to right. The difference between this scatter plot and Figure 11.3 is that, in Figure 11.3, the points slope downwards in a more or less linear (straight-line) way, but in Figure 11.4 the points show a clearly curved pattern. However, they are still negatively related.

Figure 11.5 shows a rather different pattern.

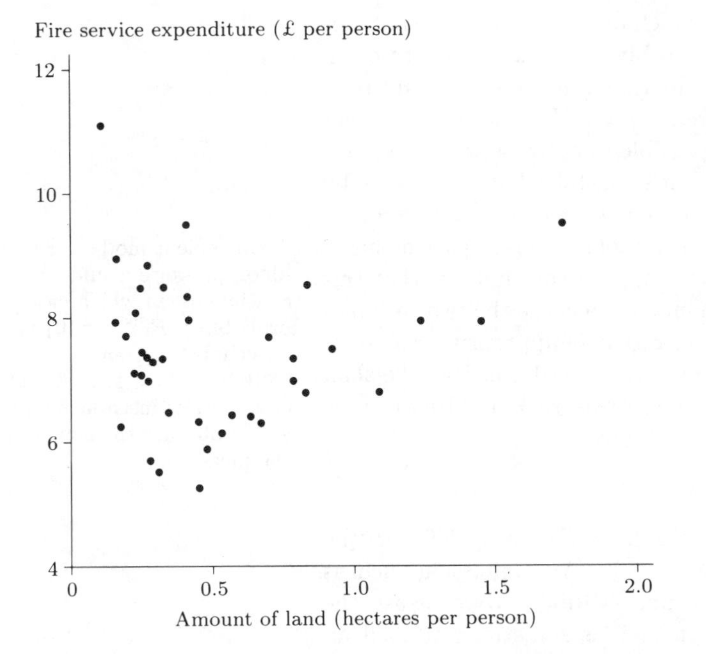

Figure 11.5 Population density and fire service expenditure

The Open University (1983) MDST242 *Statistics in Society.* Unit A3: *Relationships*, Milton Keynes, The Open University.

Here, each data point corresponds to a non-metropolitan county of England. The two variables are the density of population (hectares per person) and the expenditure (in £) per head of population on the fire service. The scatter plot shows a fairly clear pattern, but it suggests a curve. Counties with low population density and high population density both spend relatively large amounts on the fire service, while points with medium population density spend relatively small amounts. Therefore, knowing the population density of a county tells you something about its expenditure on the fire service and vice versa. The two variables are related, but in this case it is not very appropriate to describe the relationship either as positive or as negative.

Exercise 11.1

For each of the scatter plots in Figure 11.6, state whether the variables involved are related, and if they are, say whether the relationship is positive, negative or neither of these.

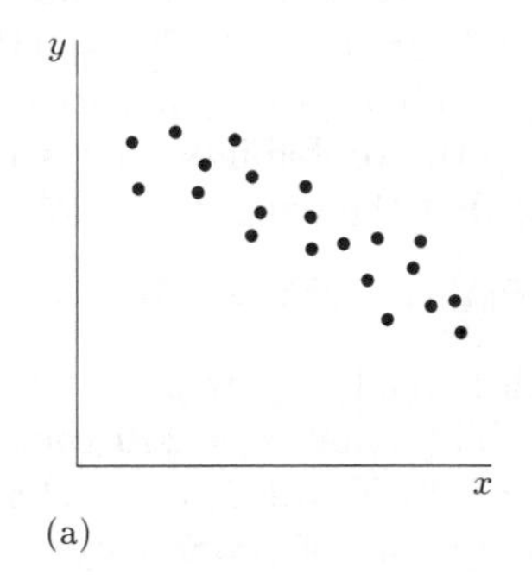

(a)

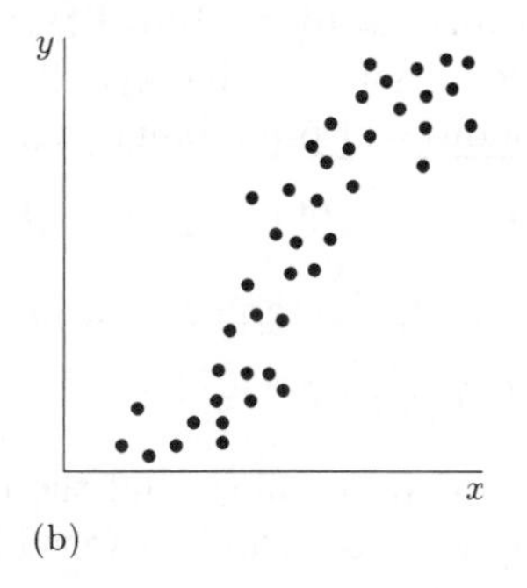

(b)

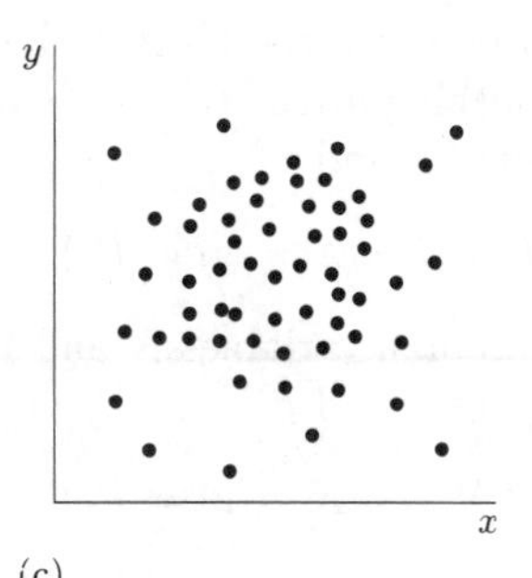

(c)

Figure 11.6 Three scatter plots

An important feature of all the examples we have looked at so far is that in each case both the variables involved are, or can be thought of as, random variables. Bivariate data are data giving values of pairs of random variables—hence the name. There is a contrast here with the data used in regression. In many regression situations, the explanatory variable is not random, though the response variable is. You may recall the data in Section 10.1 on duckweed. There, the explanatory variable is the week number and there is nothing random about that. The response, the number of duckweed fronds, is a random variable. Since only one of the variables is random, it would be inappropriate to ask the question: how do week number and number of duckweed fronds vary together? In this section and Section 11.2, all the data that we shall look at are bivariate data.

Depending on the question of interest, it may well be appropriate to use regression methods on bivariate data as well as on data where one of the variables is not random.

See Example 10.8.

It is possible to be more formal (though for much of this chapter we shall not need to be) about the idea of a relationship in bivariate data. Denote by X and Y the two random variables involved. In Example 11.1, X could be a randomly chosen patient's systolic blood pressure and Y would be his or her diastolic blood pressure. Then it would be possible to describe what is known about the value of a randomly chosen patient's diastolic blood pressure by giving the probability distribution of Y. If you knew this, you could find, for instance, $P(Y = 110)$, the probability that a randomly chosen patient has a diastolic blood pressure of 110 mm Hg. Now, suppose you find out that this patient has a systolic blood pressure of 200 mm Hg. Because the two variables X and Y are related, $P(Y = 110)$ will no longer give an appropriate value for the probability that the patient's diastolic pressure is 110 mm Hg. We shall denote the probability that a patient's diastolic pressure Y is 110 when we know that the patient's systolic pressure X is 200 by

A convenient model for variation in blood pressure would be continuous, in which case the probability $P(Y = 110)$ might strictly be written $P(109.5 \leq Y < 110.5)$. But in this case a strict insistence on notation would obscure the simple nature of the message.

$$P(Y = 110|X = 200),$$

which is read as 'the probability that $Y = 110$ *given that* $X = 200$', or 'the probability that $Y = 110$ *conditional on* $X = 200$'. An expression such as $P(Y = 110|X = 200)$ is called a **conditional probability**. (By contrast, the ordinary sort of probability without a | sign in it is sometimes termed an **unconditional probability**.) Since X and Y are related, it is the case that the conditional probability that $Y = 110$ given $X = 200$ is different from the unconditional probability that $Y = 110$; that is,

$$P(Y = 110|X = 200) \neq P(Y = 110).$$

In general, if X and Y are any two random variables, we define

$$P(Y = y|X = x)$$

to be the probability that the random variable Y takes the value y when it is known that the random variable X takes the value x.

The conditional probability $P(Y = y|X = x)$ is read as 'the probability that $Y = y$ *given that* $X = x$', or 'the probability that $Y = y$ *conditional on* $X = x$'.

The random variables X and Y are *not* related if knowing the value of X tells you nothing about the value of Y. That is, X and Y are not related if, for *all* values of x and y,

$$P(Y = y|X = x) = P(Y = y).$$

The random variables X and Y are related if, for at least some of the possible values of x and y,

$$P(Y = y|X = x) \neq P(Y = y), \tag{11.1}$$

that is, if knowing the value of X can tell you something about the distribution of Y that you did not already know.

Exercise 11.2

(a) Suppose that X and Y are two random variables which take values on the integers $0, 1, 2, \ldots$, and so on. It is known that $P(Y = 10) = 0.3$, and $P(Y = 10|X = 4) = 0.4$. Are the random variables X and Y related?

(b) Suppose that W and Z are two more random variables taking values on the integers $0, 1, 2, \ldots$. It is known that $P(Z = 5) = 0.4$, and also that $P(Z = 5|W = 4) = 0.4$. Are the random variables W and Z related?

This probability definition of what it means for two random variables to be related may seem a little unsatisfactory to you for two reasons. First, nothing has been said about how the conditional probabilities that have been defined can be estimated from data. If we do not know what the value of the probability $P(Y = 110|X = 200)$ actually is, how do we decide whether or not it is different from $P(Y = 110)$, in order to conclude whether Y and X are related? In order to estimate these probabilities, usually we must use a probability model for the random variables involved. One way to model bivariate data is discussed in Section 11.5 and further discussion will be deferred until then. Furthermore, methods of testing whether two random variables are related generally do not involve calculating probabilities of this sort directly.

Second, it has been emphasized that the idea of related variables involved treating the two variables on an equal footing; but here we looked at the conditional probability $P(Y = y|X = x)$, and this expression treats X and Y differently. In fact, we could just as well have defined X and Y to be related if

$$P(X = x|Y = y) \neq P(X = x), \tag{11.2}$$

for some values of x and y. It can be shown that the definitions given by (11.1) and (11.2) always agree.

You might suspect that this idea of a relationship between two random variables has something to do with the idea of *independence* of random variables that you met first in *Chapter 3*. You would be right. It can be shown that two random variables are related in the sense we have just discussed if they are not independent in the sense defined in *Chapter 3*. If two variables are independent, they are not related.

See, say, Exercise 3.8.

Section 11.2 of this chapter is concerned with ways of measuring how strong a relationship is in data of the sort we have been looking at. But Sections 11.3 and 11.4 are concerned with relationships between variables of a rather different sort, and we now turn briefly to introduce this kind of data.

11.1.2 Relationships in discrete data

Again let us start with an example.

Example 11.3 *Snoring frequency and heart disease*

The data in Table 11.2 come from a study which investigated whether snoring was related to various diseases. A large number of individuals were surveyed and classified according to the amount they snored, on the basis of reports from their spouses. A four-fold classification of the amount they snored was used. In addition, the researchers recorded whether or not each person had

Norton, P.G. and Dunn, E.V. (1985) Snoring as a risk factor for disease: an epidemiological survey. *British Medical Journal*, **291**, 630–632.

certain diseases. These particular data relate to the presence or absence of heart disease. The table gives counts of the number of people who fell into various categories. The top left-hand number shows, for instance, that 24 out of the 2484 people involved were non-snorers who had heart disease. The numbers in the right-hand column are the row totals and show, for instance, that 2374 of the 2484 people involved did not have heart disease. The bottom row gives the column totals: for instance, 213 people snored nearly every night.

Table 11.2 Snoring frequency and heart disease

Heart disease	Non-snorers	Occasional snorers	Snore nearly every night	Snore every night	Total
Yes	24	35	21	30	110
No	1355	603	192	224	2374
Total	1379	638	213	254	2484

Though these data look very different in form from those we have looked at so far, they are similar in several respects. Each individual surveyed provided an observation on two discrete random variables. The first, X, can take two values: Yes or No, depending on whether the individual has heart disease. The second, Y, can take four values: Non-snorer, Occasional snorer, Snore nearly every night and Snore every night, depending on how often he or she snores. Thus the data set consists of 2484 values of the pair of random variables X and Y. In 24 of the pairs, X takes the value Yes and Y takes the value Non-snorer. In 192 of them, X takes the value No and Y takes the value Snore nearly every night. Thus these are bivariate data. We can therefore ask the question: are X and Y independent? Formal methods for answering this question are developed in Section 11.4; but Exercise 11.3 will give an informal answer. ■

Up until now in the course, it has always been insisted that random variables should be real-valued (that is, their values should be numbers, and not words like 'never', 'occasional', 'often', 'always'). In the case of Bernoulli trials, we have been careful to identify with one outcome the number 1, and with the other the number 0. This identification is essential when calculating means and variances, and the strict definition of a random variable requires that it should be real-valued. Here, however, it would be merely cumbersome to attach numbers (0, 1, 2, 3, say) to outcomes and since no modelling is taking place here, let us not bother to do so. Data of this kind are often called **categorical data**.

Exercise 11.3

(a) On the basis of the data in Table 11.2, what would you estimate to be the (unconditional) probability that the random variable X takes the value Yes?

(b) What would you estimate to be the conditional probability that X takes the value Yes given that Y takes the value Snore every night?

Hint This is a probability conditional on Y taking the value Snore every night, so it is only the people who snore every night who provide direct information about it.

Do you think X and Y are related?

This does not entirely answer the question of whether X and Y are related: we have not taken into account the possibility that the result is a 'fluke' resulting from sampling variability. But this example shows that the notions of bivariate data, and of related random variables, crop up in discrete data of this kind as well as in data of the sort that can be plotted in scatter plots.

This point will be discussed in Section 11.4.

We shall now look at formal ways of measuring the strength of a relationship between variables.

11.2 Measures of association

This section will develop ways of measuring the strength of a relationship between two random variables, or the strength of the **association** or **correlation** between them as it is sometimes termed. The methods apply to continuous bivariate data, and can also be applied to discrete bivariate data of certain kinds, as long as it makes sense to plot the data on a scatter plot.

11.2.1 The Pearson correlation coefficient

In Section 11.1, you saw that bivariate data on two random variables might indicate that the two variables are related. Variables can be related positively or negatively (or in some other way); and in some cases the relationship can be reasonably represented by a straight line, whereas in others it cannot. There are other aspects to relationships between variables. Compare the three scatter plots in Figure 11.7, which again give data for 47 US states in 1959–60.

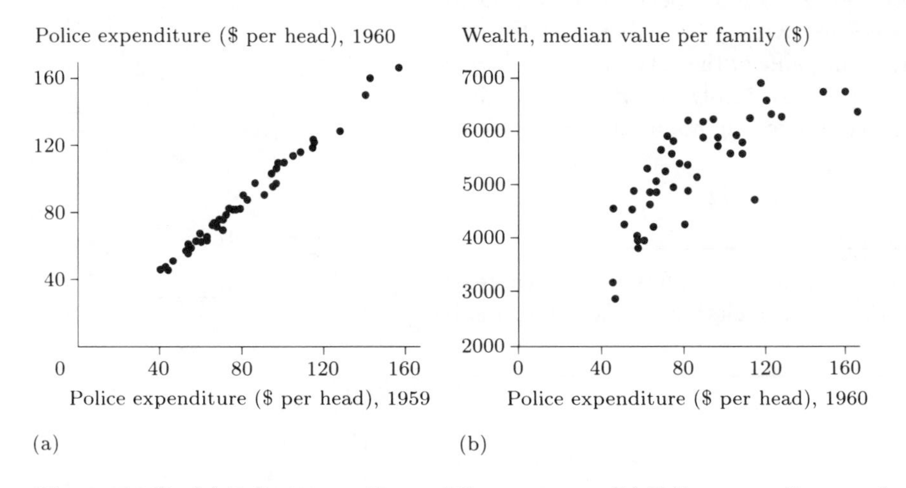

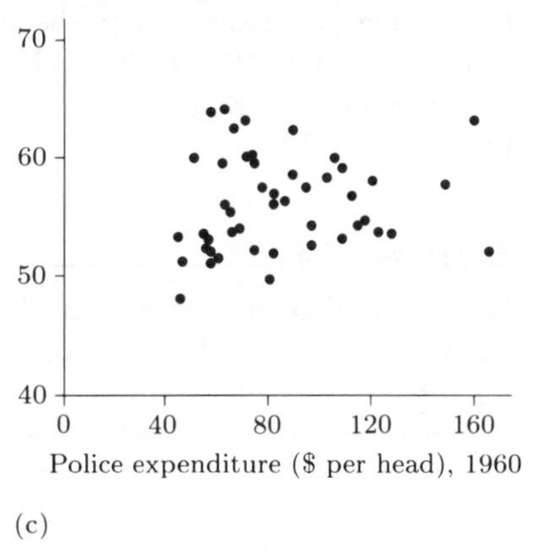

Figure 11.7 (a) Police expenditure, different years (b) Police expenditure and community wealth (c) Police expenditure and labour force participation

Vandaele, W. (1978) Participation in illegitimate activities: Erlich revisited. In Blumstein, A., Cohen, J. and Nagin, D., (eds) *Deterrence and incapacitation: estimating the effects of criminal sanctions on crime rates.* National Academy of Sciences, Washington, DC, pp. 270–335.

In each case, the two variables involved are positively related; knowing the value of one of them tells you something about the value of the other. But in Figure 11.7(a), knowing one of the police expenditure figures would tell you very accurately what the other would be. The points are not scattered very far from a straight line. By contrast, in Figure 11.7(c), knowing the police expenditure tells you very little about the labour force participation rate. The data are very scattered. Therefore, in Figure 11.7(a) we say that the two variables are strongly associated; in Figure 11.7(c) they are weakly associated; and Figure 11.7(b) comes somewhere between in terms of strength of association.

It is useful to have a summary measure of the strength of association between two random variables. Several such measures exist; one of the oldest, but still the most commonly used, is the **Pearson correlation coefficient**. As you will see later in this section, there are other correlation coefficients, but the Pearson coefficient is the most used, and it is what statisticians usually mean when they refer to a correlation coefficient without saying which one. This measure was developed by Sir Francis Galton (1822–1911), about whom

This is also known, sometimes, as the Pearson product-moment correlation coefficient. Often the word 'coefficient' is omitted, and we speak simply of the 'Pearson correlation'.

you read in *Chapter 10*, and (principally) Karl Pearson (1857–1937), who held the first post of Professor of Statistics in Britain (at University College London). Karl Pearson was the father of Egon Pearson, whom you read about in connection with hypothesis testing in *Chapter 8*.

Karl Pearson (1857–1937)

The Pearson correlation coefficient is a quantity which we shall denote by r. It takes values between -1 and $+1$. The sign of r indicates whether the relationship between the two variables involved is positive or negative. The absolute value of r, ignoring the sign, gives a measure of the strength of association between the variables. The further r is from zero, the stronger the relationship. The Pearson correlation coefficient takes the value $+1$ only if the plotted bivariate data show an exact straight-line relationship with a positive slope, and -1 if the data show an exact straight line with negative slope. Data where the two variables are unrelated have a correlation coefficient of 0. Some examples of scatter plots are given in Figure 11.8 (the data are artificial for the purposes of illustration). The three data sets depicted in Figure 11.7, for example, have the following Pearson correlation coefficients: (a) $r = 0.994$, (b) $r = 0.787$ and (c) $r = 0.121$. All are positive, reflecting the fact that in each case the two variables involved are positively related; and the stronger the relationship, the larger the value of the correlation coefficient. You should note that higher values of r do not imply anything at all about the *slope* of the straight-line fit: they say something about the *quality* of the fit.

Exercise 11.4

Based on what you have seen about values of r in different data contexts, what do you guess the value of the Pearson correlation coefficient might be for the data in Figure 11.3?

The Pearson product-moment correlation coefficient

The formula for calculating the Pearson correlation coefficient r from bivariate data $(x_1, y_1), (x_2, y_2), \ldots, (x_n, y_n)$, where the means of the x-values and the y-values are $\overline{x}$ and $\overline{y}$ and their standard deviations are s_X and s_Y, is as follows.

$$r = \frac{1}{n-1}\sum_{i=1}^{n}\left(\frac{x_i - \overline{x}}{s_X}\right)\left(\frac{y_i - \overline{y}}{s_Y}\right) \qquad (11.3)$$

$$= \frac{\frac{1}{n-1}\sum_{i=1}^{n}(x_i - \overline{x})(y_i - \overline{y})}{s_X s_Y} \qquad (11.4)$$

$$= \frac{n\sum x_i y_i - \sum x_i \sum y_i}{\sqrt{(n\sum x_i^2 - (\sum x_i)^2)(n\sum y_i^2 - (\sum y_i)^2)}} \qquad (11.5)$$

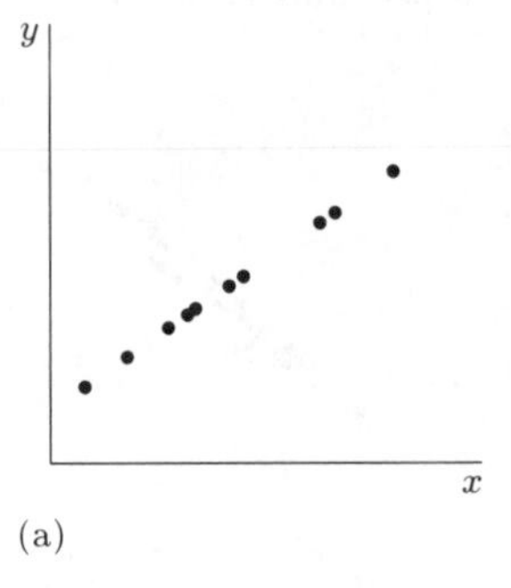

(a)

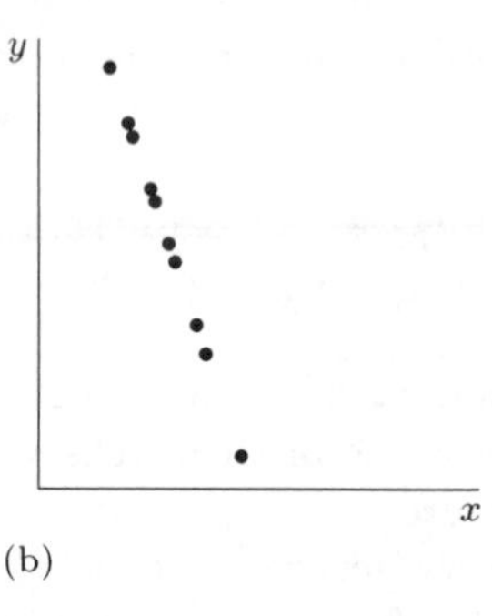

(b)

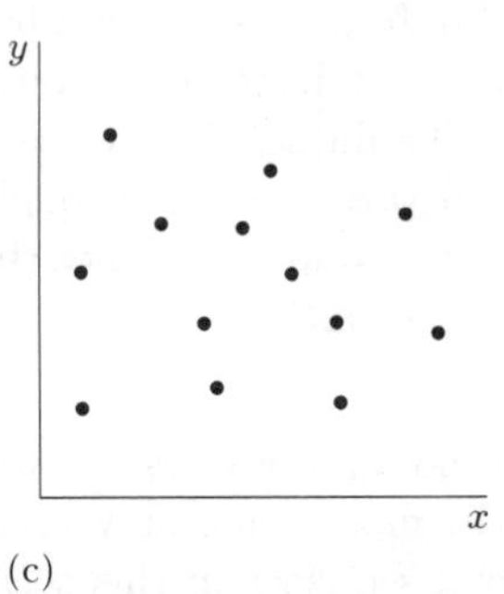

(c)

Figure 11.8 (a) $r = 1$ (b) $r = -1$ (c) $r = 0$

Of the three equivalent versions, (11.5) is the most convenient, and least prone to rounding error, if you need to calculate r using a calculator. The first formulation (11.3) gives the clearest idea of how the definition actually works. Consider the expression $(x_i - \overline{x})/s_X$. It is positive for values of x_i above their mean, and negative for values below the mean. Suppose the random variables X and Y are positively related. Then they both tend to be relatively large at the same time, and relatively small at the same time. Thus x_i and y_i are likely to be both above their mean or both below their mean for any i. If both are above their mean, then the two terms in brackets in (11.3) will be positive for that value of i, and their product will be positive, so that this data point will contribute a positive value to the sum in (11.3). If both x_i and y_i are below their means, the two terms in brackets in (11.3) will be negative, so again their product is positive, and the data point will again contribute a positive value to the sum. Since X and Y are positively related, there will be fewer data points where one of the variables is below its mean while the other is above. These points contribute a negative value to the sum in (11.3). Therefore, if X and Y are positively related, the sum in (11.3) will be positive and thus r will be positive. If X and Y are negatively related, then negative terms will dominate in the sum in (11.3) and r will turn out to be negative. The $n-1$ divisor in (11.3) is there for much the same reasons as the $n-1$ divisor in the definition of the sample variance. Expressions like $(x_i - \overline{x})/s_X$ in (11.3) include the sample standard deviation because r is intended to measure the strength of association without regard to the scales of measurement of the two variables. Thus the value of the correlation coefficient for height and weight of adults, for example, should take the same value whether weight is measured in grams or kilograms or pounds. Changing the weights from kilograms to grams will multiply both the numerator and the denominator of $(x_i - \overline{x})/s_X$ by 1000, so the overall value of this expression will not change, and hence the value of r will not change.

Formula (11.3) also shows why this is called the *product-moment* correlation coefficient. The expression in the numerator looks like that for the sample variance, one of the sample *moments*, but it involves the *product* of X and Y values.

Generally, one would use computer software to calculate correlation coefficients. However, to examine how the formulas work, let us calculate a couple of examples by hand. The easiest version of the formula for r to use for hand calculation is that given in (11.5).

Example 11.1 continued

For the data on blood pressure in Table 11.1, the necessary summary calculations are as follows (denoting systolic pressure by X and diastolic pressure by Y).

$$n = 15$$

$$\sum x_i = 210 + 169 + \cdots + 154 = 2654$$

$$\sum x_i^2 = 210^2 + 169^2 + \cdots + 154^2 = 475\,502$$

$$\sum y_i = 130 + 122 + \cdots + 100 = 1685$$

$$\sum y_i^2 = 130^2 + 122^2 + \cdots + 100^2 = 190\,817$$

$$\begin{aligned}\sum x_i y_i &= (210 \times 130) + (169 \times 122) + \cdots + (154 \times 100)\\ &= 27\,300 + 20\,618 + \cdots + 15\,400\\ &= 300\,137\end{aligned}$$

Then, using (11.5),

$$r = \frac{n\sum x_i y_i - \sum x_i \sum y_i}{\sqrt{(n\sum x_i^2 - (\sum x_i)^2)(n\sum y_i^2 - (\sum y_i)^2)}}$$

$$= \frac{15 \times 300\,137 - 2654 \times 1685}{\sqrt{(15 \times 475\,502 - 2654^2)(15 \times 190\,817 - 1685^2)}}$$

$$= \frac{30\,065}{\sqrt{88\,814 \times 23\,030}}$$

$$= 0.6648.$$

This result matches what was said in Section 11.1: these two variables are positively related. The value of the correlation coefficient is not particularly close to either 0 or 1, implying that the strength of association between these two variables is moderate. This matches the impression given by the scatter plot in Figure 11.2, where there is a moderate degree of scatter. ■

Exercise 11.5

The data in Table 11.3 were obtained in a study of a new method of measuring body composition. They give the age and body fat percentage for 14 women.

Mazess, R.B., Peppler, W.W. and Gibbons, M. (1984) Total body composition by dual-photon (^{153}Gd) absorptiometry. *American Journal of Clinical Nutrition*, **40**, 834–839.

Table 11.3 Body fat percentage and age for 14 women

Age (years)	Body fat (%)
23	27.9
39	31.4
41	25.9
49	25.2
50	31.1
53	34.7
53	42.0
54	29.1
56	32.5
57	30.3
58	33.0
58	33.8
60	41.1
61	34.5

Investigate how age and body fat percentage are related by (a) drawing a scatter plot; (b) calculating the Pearson correlation coefficient.

If you denote age by X and body fat percentage by Y your solutions will match those at the back of the book.

In Exercise 11.6 you should use your computer for the calculations.

Exercise 11.6

(a) The data in Table 11.4 are those for the heights and weights of 30 Bradford school children, illustrated in Figure 11.1.

Calculate the Pearson correlation coefficient for these data. Does the value of the coefficient match the impression of the strength of association given by the scatter plot?

Table 11.4 Heights and weights of 30 eleven-year-old schoolgirls from Heaton Middle School, Bradford

Height (cm)	Weight (kg)	Height (cm)	Weight (kg)
135	26	133	31
146	33	149	34
153	55	141	32
154	50	164	47
139	32	146	37
131	25	149	46
149	44	147	36
137	31	152	47
143	36	140	33
146	35	143	42
141	28	148	32
136	28	149	32
154	36	141	29
151	48	137	34
155	36	137	30

(b) An official investigation into differences in mortality between different occupational groups in England and Wales presented the data given in Table 11.5. They relate to male deaths in 1970–72. For each of 25 'occupational orders' (groups of occupations), the data give the 'smoking ratio', a measure of the number of cigarettes smoked on average by men in that group, and the lung cancer standardized mortality ratio (SMR), a measure of the death rate from lung cancer for men in the group. Both of these ratios are adjusted to allow for differences in the pattern of age of members of the groups, and for both, a value of 100 indicates that smoking or mortality is at the average level for England and Wales in 1970–72.

Table 11.5 Smoking ratio and SMR by occupation order

	Occupation order	Smoking ratio	Lung cancer SMR
1	Farmers, foresters, fishermen	77	84
2	Miners and quarrymen	137	116
3	Gas, coke and chemical makers	117	123
4	Glass and ceramics makers	94	128
5	Furnace, forge, foundry, rolling mill workers	116	155
6	Electrical and electronic workers	102	101
7	Engineering and allied trades not included elsewhere	111	118
8	Woodworkers	93	113
9	Leather workers	88	104
10	Textile workers	102	88
11	Clothing workers	91	104
12	Food, drink and tobacco workers	104	129
13	Paper and printing workers	107	86
14	Makers of other products	112	96
15	Construction workers	113	144
16	Painters and decorators	110	139
17	Drivers of stationary engines, cranes, etc	125	113
18	Labourers not included elsewhere	133	146
19	Transport and communications workers	115	128
20	Warehousemen, storekeepers, packers, bottlers	105	115
21	Clerical workers	87	79
22	Sales workers	91	85
23	Service, sport and recreation workers	100	120
24	Administrators and managers	76	60
25	Professional, technical workers, artists	66	51

(Extracted from the Office of Population Censuses and Surveys (1978) *Occupational mortality: the Registrar General's decennial supplement for England and Wales, 1970–72*, Series DS, No. 1, London: HMSO, p. 149.)

Draw a scatter plot of these data, calculate the Pearson correlation between the two variables, and comment on the strength of association between the two variables.

The conclusions from Exercise 11.6 can be used to illustrate a very important point about correlation. In both parts, you found a reasonably strong positive association between the variables. In part (b), this means that occupational groups where men smoke a lot also experience, on average, high mortality from lung cancer. An obvious explanation for this is that smoking causes lung cancer. But you should note that the data in Table 11.5 do not prove this causation. They merely show that high values of the two variables tend to go together, without saying anything about *why*. The data support the hypothesis that smoking causes lung cancer; but, because the analysis treated the variables on an equal footing, they support equally well the hypothesis that lung cancer causes smoking. In part (a) of Exercise 11.6, you found that tall girls tend to be relatively heavy. Again, this correlation does not establish a causal explanation, either that being tall causes girls to be heavy or that being heavy causes girls to be tall. A more reasonable explanation is that as a girl grows, this causes increases in both her weight and her height.

One should not forget the possibility of a relationship between SMR and occupation, without regard to smoking habits.

In summary, to say that there is a correlation or association between two variables X and Y is merely to say that the values of the two variables vary 'together' in some way. There can be many different explanations of why they vary together, including the following.

- Changes in X cause changes in Y.
- Changes in Y cause changes in X.
- Changes in some third variable, Z, independently cause changes in X and Y.
- The observed relationship between X and Y is just a coincidence, with no causal explanation at all.

Statisticians sometimes quote examples of pairs of variables which are correlated without there being a direct causal explanation for the relationship. For instance, there is a high positive correlation between the level of teachers' pay in the USA and the level of alcoholism; yet the alcoholism is not caused, to any great extent, by teachers who drink. In parts of Europe there is a high positive correlation between the number of nesting storks and the human birth rate; yet storks do not bring babies. Since the existence of a pattern on a scatter plot, or the value of a correlation coefficient, cannot establish which of these explanations is valid, statisticians have a slogan which you should remember.

The 'storks' data are given in Kronmal, R.A. (1993) Spurious correlation and the fallacy of the ratio standard revisited. *J. Royal Statistical Society, Series A*, **156**, 379–392.

Correlation is *not* causation.

Statisticians can be rather reticent about telling you what causation *is*!

Causation is established by other (generally non-statistical) routes; typically the aim is to carry out a study in such a way that the causal explanation in which one is interested, is the only plausible explanation for the results.

11.2.2 Care with correlation

In Section 11.1 you saw that there could be many different types of patterns in scatter plots. The Pearson correlation coefficient reduces any scatter plot to a single number. Clearly a lot of information can get lost in this process. Therefore it is hardly ever adequate simply to look at the value of the correlation coefficient in investigating the relationship between two variables. Some of the problems that arise were demonstrated very convincingly by the statistician Frank Anscombe, who invented one of the most famous sets of artificial data in statistics. It consists of four different sets of bivariate data: scatter plots of these data are given in Figure 11.9.

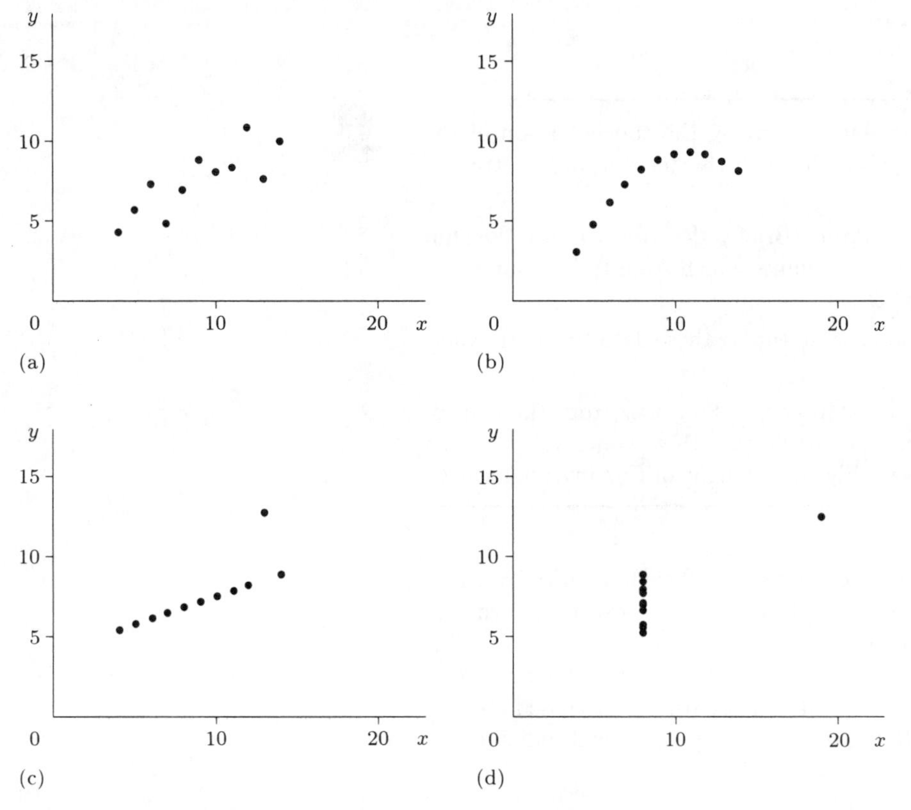

Anscombe, F.J. (1973) Graphs in statistical analysis. *American Statistician*, **27**, 17–21. The regression lines for the regression of Y on X are also the same for each of the scatter plots.

Figure 11.9 Anscombe's data sets

In each case the Pearson correlation coefficient takes the same value: $r = 0.816$. In Figure 11.9(a) this is probably a reasonable summary of the data; the two variables look reasonably strongly associated. However, the other three data sets tell a different story. In Figure 11.9(b) the variables look extremely strongly related, but it is not a straight-line relationship. For the Pearson coefficient, a 'perfect' correlation of 1 (or -1) is only found in a straight-line relationship. Therefore the Pearson correlation coefficient is an inadequate summary of this particular data set. The same is true for Figure 11.9(c), for a different reason. This time the data do lie, exactly, on a straight line—except for one outlying point. This outlier brings the correlation coefficient down; if it were omitted the Pearson correlation would be 1. In Figure 11.9(d) there is again a point that is a long way from all the others, but if this point were omitted, there would effectively be no correlation between the variables at all.

The message from Anscombe's examples is that you should always study the data before trying to interpret the value of a correlation coefficient. In fact, it is almost always best to look at a scatter plot before calculating the correlation coefficient; the scatter plot may indicate that calculating the correlation coefficient is not a sensible thing to do, or that some other analysis should be followed as well.

The Pearson correlation coefficient is most useful when the data form a more or less oval pattern on the scatter plot, as in Figure 11.9(a). It is less appropriate when the data show a curvilinear relationship. When there are points that are a long way from the rest of the data, as in Figures 11.9(c) and 11.9(d), it is often useful to calculate the correlation coefficient after omitting them from the data set.

Data that display an oval pattern of this sort are often well modelled by the bivariate normal distribution, which is described in Section 11.5.

Exercise 11.7

In *Chapter 7*, Table 7.7, you met a data set giving the concentration of the pollutant PCB in parts per million, and the shell thickness in millimetres, of 65 Anacapa pelican eggs.

(a) Produce a scatter plot of these data. Briefly describe the relationship between the variables. Are there any points which seem to be a long way from the general run of the data?

(b) Calculate the Pearson correlation coefficient for these data (using the data for all 65 eggs).

(c) Omit any data points which you think are a long way from the others. Recalculate the Pearson correlation coefficient. How would you describe the relationship between the variables in the light of this investigation?

One possibility for using the Pearson correlation coefficient with data whose scatter plot does not show a straightforward oval pattern is to transform the data in an appropriate manner in order to standardize the pattern. This method can sometimes be used to deal with data showing a curvilinear relationship (though other means of dealing with such data are discussed later in this section). It has other applications too, as Exercise 11.8 will indicate.

Exercise 11.8

In *Chapter 1*, Table 1.7, data are listed on the brain weight and body weight of 28 kinds of animal. You saw in *Chapter 1* that there were problems merely in producing a sensible plot of the data unless they were transformed first. The transformation suggested was a logarithmic transformation of both variables. Calculate the Pearson correlation coefficient for the untransformed data and for the transformed data.

The Pearson correlation for the untransformed data in Table 1.7 is so close to zero that it gives the impression that the variables are not related at all. The Pearson correlation for the transformed data (which, you might recall from *Chapter 1*, show a much more evenly-spread oval pattern on their scatter plot—see Figure 1.15) indicates that there is, in fact, a reasonably strong positive association between the variables.

The idea of transforming the data before calculating the Pearson correlation is a useful one, but it does depend on whether or not an appropriate transformation can be found. We now turn to another measure of correlation which does not depend for its usefulness on the scatter plot displaying an oval pattern.

Charles Spearman (1863–1945)

11.2.3 The Spearman rank correlation coefficient

There exist several correlation coefficients which can meaningfully be applied in a wider range of situations than can the Pearson coefficient. The first of these to be developed was published in 1904 by the British psychologist Charles Spearman (1863–1945). Spearman was one of many psychologists who, by adapting and extending existing statistical methods to make them more suitable for analysing psychological data, and by developing entirely new statistical approaches, have made important advances in statistical science. Spearman had the powerful but simple idea of replacing the original data by their ranks, and measuring the strength of association of two variables by calculating the Pearson correlation coefficient with the ranks.

Spearman, C. (1904) The proof and measurement of association between two things. *American Journal of Psychology*, **15**, 72–101.

Ranks are used in the Wilcoxon signed ranks test and the Mann–Whitney–Wilcoxon test of *Chapter 9*.

Let us see how this is done using a data set we have already met.

Example 11.4 Body fat percentage and age

In Exercise 11.5 you calculated the Pearson correlation coefficient for a data set on body fat percentage and age. The first step in calculating the Spearman correlation coefficient for these data is to find the ranks of the data on each of the variables separately. Thus, for instance, the lowest age in the data set is 23 years; this is given rank 1. The lowest body fat percentage is 25.2; this is also given rank 1. The highest age is 61 years, and since there are 14 ages in the data set, this gets rank 14; and so on. Where two data values for one of the variables are tied, they are given averaged ranks (as was done in *Chapter 9*). The resulting ranks are shown along with the original data in Table 11.6.

Table 11.6 Body fat percentage and age for 14 women, with ranks

Age (years)	Rank	Body fat (%)	Rank
23	1	27.9	3
39	2	31.4	7
41	3	25.9	2
49	4	25.2	1
50	5	31.1	6
53	$6\frac{1}{2}$	34.7	12
53	$6\frac{1}{2}$	42.0	14
54	8	29.1	4
56	9	32.5	8
57	10	30.3	5
58	$11\frac{1}{2}$	33.0	9
58	$11\frac{1}{2}$	33.8	10
60	13	41.1	13
61	14	34.5	11

The remaining calculations for the correlation use just the ranks and ignore the original data. It is fairly evident from the table, and even more obvious from a scatter plot of the ranks (see Figure 11.10) that the ranks of these two variables are positively related. That is to say, low ranks go together, and high ranks go together. This is scarcely surprising; we already saw that the two variables are positively related, which means that low values go together and high values go together.

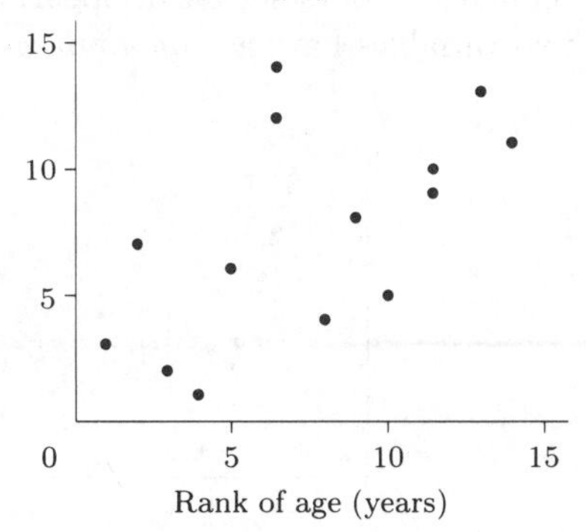

Figure 11.10 Ranked percentage body fat against ranked age

Calculating the Pearson correlation coefficient of the ranks in the usual way, we obtain the value 0.590. That is, the **Spearman rank correlation coefficient** for these data is 0.590. This value is not very different from the Pearson correlation of 0.507 for this data set, and the scatter plot of the ranks looks similar to the scatter plot of the original data. Thus the Spearman approach has not, in this case, told us anything we did not already know. But that is because these data are suitable for analysis using the Pearson correlation coefficient. The advantage of the Spearman rank correlation coefficient is that it can be used in other situations too. ■

The Spearman rank correlation coefficient is denoted r_S (S stands for Spearman). To calculate the Spearman rank correlation coefficient for a set of bivariate data, proceed as follows.

Calculation of the Spearman rank correlation coefficient

(i) Calculate the ranks for each of the variables separately (using averaged ranks for tied values);

(ii) find r_S by calculating the Pearson correlation coefficient for the ranks.

Many textbooks give a simpler formula for r_S which looks somewhat different from the Pearson formula. This formula can, however, be shown to be equivalent to the Pearson formula for data where there are no ties. If there are more than one or two ties, the simplified formula does not hold, anyway. Since you will generally be performing the calculations with your statistical software, and since all statistical packages can calculate the Pearson correlation coefficient, you are spared the details of the special Spearman formula.

Since the Spearman rank correlation coefficient is calculated using the same formula as that for the Pearson correlation coefficient, it also takes values between -1 and $+1$, with values near 0 denoting a low degree of association and values near -1 or $+1$ denoting strong association. However, the Spearman rank correlation coefficient uses a more general definition of strong association. Suppose a data set has a Spearman rank correlation coefficient of $+1$. Then the Pearson correlation coefficient for the ranks is $+1$, which means that the ranks have an exact linear (straight-line) positive relationship. The only way this can happen is when the ranks for the two variables are exactly the same: that is, the data point which has rank 1 on variable X also has rank 1 on variable Y, the data point which has rank 2 on variable X also has rank 2 on variable Y, and so on. This means that the original data points come in exactly the same order whether they are sorted in order according to the values of the variable X or the values of the variable Y. This can happen if the original variables have an exact positive linear relationship, but it can also happen if they have an exact *curvilinear* positive relationship, as shown in Figure 11.11, as long as the curve involved moves consistently upwards. Such a relationship is known as a *monotonic increasing* relationship.

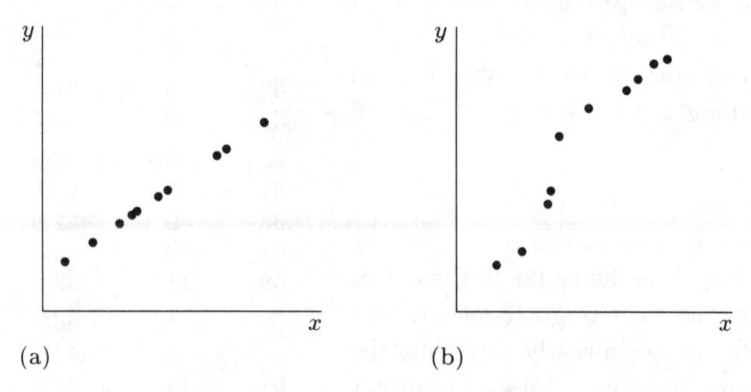

Figure 11.11 (a) $r_S = 1$, $r = 1$ (b) $r_S = 1$, $r \neq 1$

Similarly, a data set has a Spearman rank correlation coefficient of -1 if the two variables have a *monotonic decreasing* relationship (see Figure 11.12).

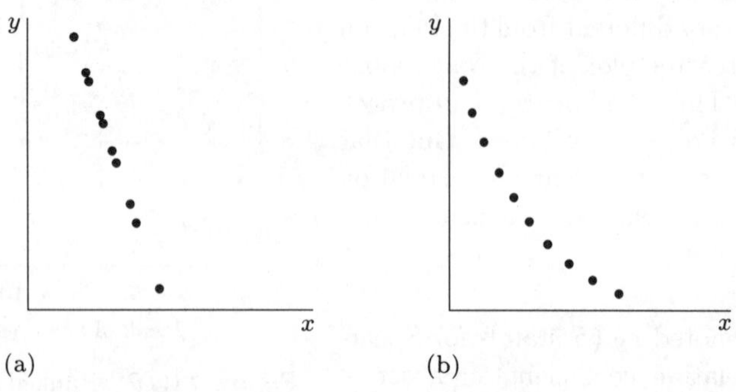

Figure 11.12 (a) $r_S = -1$, $r = -1$ (b) $r_S = -1$, $r \neq -1$

Summarizing, the Pearson correlation coefficient is a measure of strength of *linear* (that is, straight-line) association, while the Spearman rank correlation coefficient is a measure of *monotonic* (that is, always moving in a consistent direction) association.

Another advantage of the Spearman rank correlation coefficient is that it requires only the ranks of the data: in some cases, only the ranks may be available. In other cases, it might happen that the ranks are more reliable than the original data. When the data on animal body and brain weights were introduced in *Chapter 1*, various questions were raised about their accuracy. The data involve measurements for extinct animals like dinosaurs, whose body and brain weights must have been inferred from fossils, so that they are unlikely to be accurate. However, it seems more likely that the data for these animals at least come in the correct rank order.

See Table 1.7.

Exercise 11.9

(a) Calculate the value of r_S for the data on body and brain weights. What do you conclude?

(b) What is the value of the Spearman rank correlation coefficient for the log-transformed data? (You should be able to answer this without actually doing any calculations. Think about what the logarithmic transformation does to the ordering of the data.)

We have seen that the Spearman rank correlation coefficient has several advantages over the Pearson correlation coefficient. You might even be wondering why anyone bothers with the Pearson correlation coefficient. In fact, there are several reasons. First, in some circumstances, what is required is specifically a measure of the strength of linear association, and the Spearman rank correlation coefficient cannot give this. Second, there are a number of important more advanced statistical techniques which build on and use the idea of the Pearson correlation coefficient. Third, very often the aim of calculating a correlation coefficient from a sample is to make inferences about the association between the two variables concerned in the population from which the sample was drawn. We shall see some ways of doing this in Subsection 11.2.4. If, in fact, the relationship between the variables is linear, the Pearson correlation often provides more powerful inferences than does the Spearman rank correlation coefficient.

There are other measures of correlation which are beyond the scope of this course. In particular, the Kendall rank correlation coefficient measures monotonic association in rather the same way as does the Spearman rank correlation coefficient.

11.2.4 Testing correlations

So far, in this section, we have largely ignored the fact that most of our data arise as samples from some larger population. Usually the question of interest is not what the strength of association between two variables is in the sample, but what the strength of association is in the population. Just as there is a correspondence between the sample mean and the population mean, and between the sample standard deviation and the population standard deviation, so there are population analogues of both the Pearson and the Spearman correlation coefficients, and methods exist for estimating these population analogues (in fact, just as we have been doing) and for testing hypotheses about them.

The population analogue of the Pearson correlation coefficient is discussed again in Section 11.5.

In this course, we shall confine ourselves to using the sample correlation coefficients for testing the null hypothesis that, in the population, there is no association at all. That is, we shall look at methods for investigating whether or not the two variables involved are really related. Let us begin with an example.

Example 11.1 continued

The data on blood pressures can be thought of as a random sample from the population of all potential patients with moderate essential hypertension. Can we be confident that there is really a relationship between systolic and diastolic blood pressure in this population? Let us test the null hypothesis that there is no such relationship, against the two-sided alternative that there is a relationship (positive or negative).

It seems appropriate to use the value of the (sample) correlation coefficient as a test statistic. Intuitively, the further the value of r is from zero, the more evidence we have that the two variables really are related. Under the null hypothesis of no association, we would expect r to be fairly close to zero. In fact, for samples of the size we have here (sample size 15) the null distribution of R is as shown in Figure 11.13. It is symmetrical about $r = 0$. Figure 11.13 also shows the observed value, $r = 0.665$, and gives the tail area above 0.665. Since this is a two-sided test, we are interested in both tails. From Figure 11.13, it is obvious that our observed value of $r = 0.665$ is well out into the upper tail of the distribution. In fact,

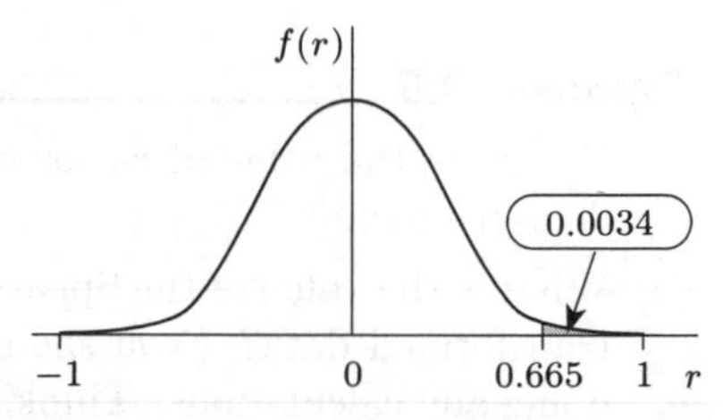

Figure 11.13 The distribution of R for samples of size 15, when there is no underlying correlation

$$SP(\text{obtained direction}) = SP(\text{opposite direction}) = 0.0034,$$

$$SP(\text{total}) = 0.0068;$$

We shall see shortly how to calculate these significance probabilities.

so there is fairly strong evidence that there really is correlation in the population. ■

This hypothesis test is based on the assumption that the data involved come from a particular distribution called the *bivariate normal distribution*, which is discussed in Section 11.5. If the null hypothesis of no association is true, this amounts to assuming that the two variables involved have independent normal distributions. More generally, if the data do not show a more or less oval pattern of points on a scatter plot, the test might possibly give misleading conclusions; but you have already seen that using the Pearson correlation coefficient can give strange answers if the scatter plot shows a strange pattern.

It is just as straightforward to test the null hypothesis of no association against a one-sided alternative, that there is a positive relationship between the two variables in the population (or that there is a negative relationship in the population). In terms of the null distribution in Figure 11.13, this amounts to finding the significance probability from one tail of the distribution rather than from both. Explore the facilities available on your computer in attempting Exercise 11.10.

Exercise 11.10

Use the data on age and body fat percentage to test the hypothesis that there is no relationship between these two variables in the population, against the one-sided alternative that the two variables have a positive relationship.

See Table 11.6.

In both *Chapter 8* and *Chapter 9*, we have had recourse to computer software when the appropriate significance tests have been too complicated or merely too time-consuming to pursue otherwise. In fact, use of the Pearson correlation coefficient to test the hypothesis of no association in a population is not at all complicated. It can be shown that when there is no association, the sampling distribution of the Pearson correlation coefficient R is most easily given in the form

$$R\sqrt{\frac{n-2}{1-R^2}} \sim t(n-2), \qquad (11.6)$$

and (11.6) suggests a convenient test statistic. For the blood pressure data ($n = 15$), we have

$$r\sqrt{\frac{n-2}{1-r^2}} = 0.664\,773\sqrt{\frac{15-2}{1-0.664\,773^2}} = 3.2085,$$

and this is at the 99.66% point of $t(13)$: hence the obtained SP of 0.0034. For the data in Exercise 11.10, your computer should have provided you with the obtained SP of 0.0323. In fact, for these data, $r = 0.506\,589$ with $n = 14$, so we should test

$$r\sqrt{\frac{n-2}{1-r^2}} = 0.506\,589\sqrt{\frac{14-2}{1-0.506\,589^2}} = 2.0354$$

against $t(12)$. This is the 96.77% quantile of $t(12)$, giving the same obtained SP of 0.0323.

Example 11.5 Testing correlations

For the data on 65 Anacapa pelican eggs discussed in Exercise 11.7, the value of the Pearson correlation coefficient r is -0.253. We can test the null hypothesis of no association between PCB concentration and shell thickness, against a two-sided alternative, by calculating the test statistic $r\sqrt{\frac{n-2}{1-r^2}}$ and comparing it against a t-distribution with $65 - 2 = 63$ degrees of freedom. The value of this quantity is

$$-0.253\sqrt{\frac{65-2}{1-(-0.253)^2}} = -2.076.$$

By looking this value up in statistical tables, or by using a computer, we find that the total SP is 0.042. There is some evidence of a relationship between the two variables, but the evidence is not strong. ■

The testing procedure for the null hypothesis of no relationship in the population, using the Spearman rank correlation coefficient as the test statistic, works in precisely the same way, except that one looks up the SP using a different computer command. For the data on body fat percentage, we saw earlier that $r_S = 0.590$ (with $n = 14$). In a test of the null hypothesis of no association against the one-sided alternative that there is a positive association, the obtained SP is 0.0144. This test provides reasonably strong evidence of a positive association between age and body fat percentage in women.

Unlike the null distribution of the Pearson correlation coefficient R given by (11.6), the null distribution of the Spearman rank correlation coefficient R_S is, in fact, extremely complicated, though it is achieved by the rather simple idea of assuming that all possible alignments of the two sets of scores are equally likely. The computations are certainly best left to a computer. For large samples, the null distribution of R_S is given by the approximation

$$R_S\sqrt{\frac{n-2}{1-R_S^2}} \approx t(n-2). \qquad (11.7)$$

Notice that the form of (11.7) is exactly that of (11.6) for the Pearson correlation coefficient.

Exercise 11.11

(a) It was stated earlier that the test for no association using the Pearson correlation coefficient was based on the assumption that the data involved followed a bivariate normal distribution; or in other words that they lay in an oval pattern on the scatter plot. When we looked at the Anacapa pelican eggs data earlier, we saw that there was some doubt about whether this was the case, and we reanalysed the data omitting one of the eggs which had a very thin shell. We found that $r = -0.157$ (with, now, $n = 64$). Using this data set with one egg omitted, test the hypothesis that there is no association between PCB concentration and thickness of shell against a two-sided alternative. What do you conclude?

(b) For the data set on body and brain weights of animals, we found that $r_S = 0.716$ (with $n = 28$). Use the large-sample approximation (even though the sample size is hardly large enough) for the distribution of R_S to test the hypothesis that there is no association between body weight and brain weight in animals, against the one-sided alternative that the two variables are positively related.

In Section 11.5 we shall return to these sorts of data, where some theoretical issues will be discussed. Meanwhile, in Sections 11.3 and 11.4, we turn to data in the form of contingency tables.

11.3 Exact tests for association in 2 x 2 contingency tables

Our attention now turns to relationships between random variables which take only a small number of discrete values—data of the sort that can be represented in *contingency tables* of the type you met in Example 11.3. In Section 11.2, we looked at measures of association for 'scatter plot' data, and then we considered methods of testing the null hypothesis of no association, using the measures of association (correlation coefficients) as test statistics. Similar concepts apply to contingency table data.

You saw in Example 11.3 that it makes sense to look for relationships or associations between the variables. There exist many different measures of association for data in contingency tables; most of these measures share some of the properties of correlation coefficients. However, in much analysis of contingency data, the emphasis is on testing for the presence of association rather than on measuring the size of the association, and the common tests of association for contingency table data do not use the common measures of association as test statistics. Therefore, in this section and the next we shall discuss only tests of association.

First, what is a contingency table? It is a table of *counts* showing the *frequencies* with which random variables take various values in a sample. The term 'contingency table' is most commonly used when the table records the values of two (or more) variables at the same time, usually (if there are two variables) with one variable corresponding to the rows of a square or rectangular table, and the other variable corresponding to the columns. The 'boxes' in the table, at the intersection of rows and columns, into which counts can fall, are usually called *cells*. In Table 11.2 the rows corresponded to the presence or absence of heart disease (2 rows) and the columns to the degree of snoring (4 columns). There were $2 \times 4 = 8$ cells in the table, which also gave information on marginal totals.

Such square or rectangular tables are often called *cross-tabulations*.

It is important to remember that contingency tables are tables of *counts*; if a table consists of percentages or proportions calculated from counts, then strictly speaking it is not a contingency table and it cannot be analysed by the methods covered here (unless it can be reverted to a table of counts).

In Example 11.3, the contingency table had two rows and four columns. The simplest two-way contingency tables have just two rows and two columns, and in this section we shall consider exact tests for these 2×2 tables. An approximate test that can be applied to larger tables as well as 2×2 tables is dealt with in Section 11.4.

In this context, an exact test is one where the SP can be calculated exactly, rather than using some approximation.

In a 2×2 contingency table, the two variables involved can each take only two values. In testing for association, as usual, the aim is to investigate whether knowing the value of one of the variables tells us anything about the value of the other. Let us see how this can be done in an example.

Example 11.6 Educational level and criminal convictions

A study was carried out on factors related to a criminal conviction after individuals had been treated for drug abuse. Sixty people who had participated in a drug rehabilitation scheme were categorized according to years of education

(15 years or less, and more than 15 years) and to whether or not they were convicted for a criminal offence after treatment. The results of the study are given in Table 11.7.

Table 11.7 Educational level and criminal convictions

Education	Convicted Yes	No	Row totals
15 years or less	16	20	36
More than 15 years	6	18	24
Column totals	22	38	60

Wilson, S. and Mandelbrote, B. (1978) Drug rehabilitation and criminality. *British Journal of Criminology*, **18**, 381–386.

It was hypothesized that those with less formal education would be more likely to face a subsequent conviction. On the face of it, the data seem to support the hypothesis. Among those less well educated, 16/36 or about 44% were convicted of a criminal offence, while among those with more education the proportion was only 6/24 or 25%. If these data represent the general state of affairs, then knowing an individual's educational history will tell you how likely it is that this person (having graduated from a drug rehabilitation programme) will face a subsequent criminal conviction. The two random variables, amount of education and conviction, are related (and not, therefore, independent).

Again, the random variables here could (and, strictly, should) be given numerical values: say, 0 and 1. But this would be to complicate matters quite unnecessarily.

However, this table includes only 60 graduates of the scheme, which is a small sample from the potential population. It might well be that this small sample does not truly represent the situation in the population. We could find out by testing the null hypothesis that there is no relationship between the two variables. An appropriate alternative hypothesis is one-sided, because the researchers' theory is that more education reduces an individual's likelihood to reoffend (or, at any rate, their likelihood to be convicted).

You do not have to learn a new test procedure to test these hypotheses. In Subsection 8.5.1 you met *Fisher's exact test*, which was used to test for differences between two binomial proportions. Here, Fisher's exact test turns out to be the appropriate test too. With the data in Table 11.7, Fisher's exact test gives an obtained SP of 0.104. There is, in fact, only very weak evidence from these data that the two variables really are related in the population. ■

It is worth exploring briefly one aspect of how this test works. When you carry out Fisher's exact test using your computer, it may well give SP values for both one-sided and two-sided tests; and you may have noticed in your work in *Chapter 8* that these values are not related in the simple kind of way that the one- and two-sided SP values are for, say, a t-test. The reason is that the distribution of the test statistic is not symmetric. In fact, the test statistic for Fisher's exact test is simply the count in one of the cells of the table, say the count in the upper left cell. The distribution of this count under the null hypothesis is found under the assumption that the marginal totals, the totals of the counts in each row and column given in the margins of the main table in Table 11.7, remain fixed. Under this assumption, considering the data in Table 11.7, the value in the top left-hand cell could be as low as 0, if none of those with more education were convicted. (It could not be larger than 22 since the total of the first column is fixed at 22 and the other count in that column could not be negative.)

Thus the null distribution of the test statistic is concentrated on the values 0, 1, 2, ..., 22, and this distribution is shown in Figure 11.14.

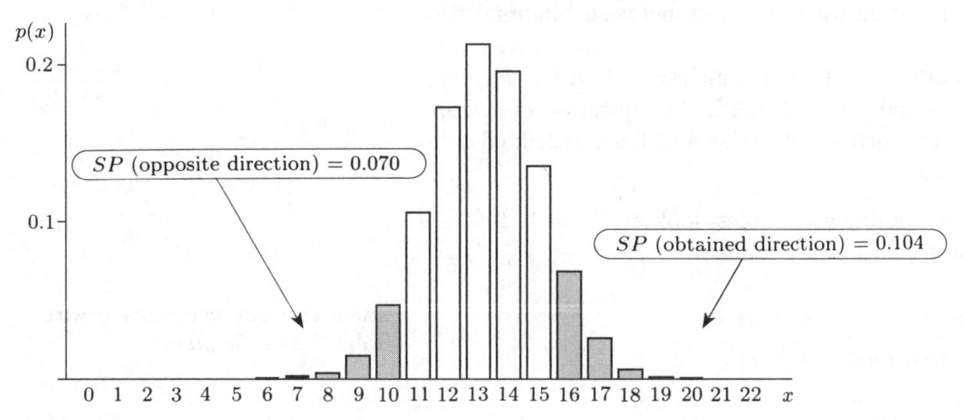

Figure 11.14

The obtained SP for our observed value in the top left-hand cell, 16, is found by adding together the probabilities for 16, 17, ..., 22: from Figure 11.14 the SP is 0.104. To calculate the total SP, we should add to 0.104 the probabilities in the left tail of the null distribution that are less than the probability for 16. Thus we must include the probabilities for 0, 1, ..., 10. Because the distribution in Figure 11.14 is not symmetric, the total SP is equal to 0.174.

In Exercise 11.12, the data are again presented as a 2×2 contingency table (with marginal totals included).

Exercise 11.12

Table 11.8 gives data from a study of 65 patients who had received sodium aurothiomalate (SA) as a treatment for rheumatoid arthritis. Of these patients, 37 had shown evidence of a toxic effect of the drug. The patients were also classified according to whether or not they had impaired sulphoxidation capacity; the researchers thought this might be linked to toxicity in some way. Use Fisher's exact test to test the null hypothesis that these two variables are not associated (against a two-sided alternative). What do you conclude? If you conclude that the variables are related, briefly describe the way in which they are related.

Ayes, R., Mitchell, S.C., Waring, R.H. *et al.* (1987) Sodium aurothiomalate toxicity and sulphoxidation capacity in rheumatoid arthritis patients. *British Journal of Rheumatology*, **26**, 197–201.

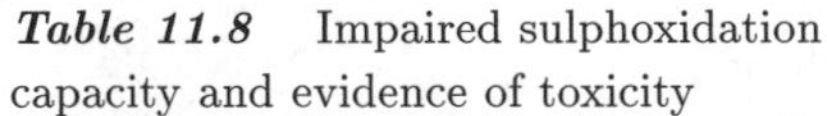

Table 11.8 Impaired sulphoxidation capacity and evidence of toxicity

Impaired sulphoxidation	Toxicity Yes	No	Row totals
Yes	30	9	39
No	7	19	26
Column totals	37	28	65

We concluded in Exercise 11.12 that patients with impaired sulphoxidation are more likely to exhibit a toxic reaction to the drug. It is important to remember that, in contingency tables just as in other kinds of data, 'correlation does not imply causation' so we cannot conclude on the basis of these data alone that the impaired sulphoxidation *causes* the toxic reaction, or vice versa.

Incidentally, it is worth examining briefly how Fisher's exact test can be used for two apparently completely different purposes: testing for association in a 2×2 contingency table and testing for differences between binomial proportions. In Exercise 8.12 you used Fisher's exact test to analyse data from an experiment where 100 male students and 105 female students asked for help; 71 of the males and 89 of the females were helped. The question of interest there was whether or not the proportions of males and females helped were the same.

To see how these data relate to contingency tables, look at Table 11.9 where they have been written out as a contingency table.

Table 11.9 can be compared with Table 2.3 of *Chapter 2*.

Table 11.9 Helping behaviour

Sex	Helped		Row totals
	Yes	No	
Male	71	29	100
Female	89	16	105
Column totals	160	45	205

This contingency table looks very like Tables 11.7 and 11.8. It does differ in one key respect, though. In Table 11.7, for instance, the researcher chose 60 people from the drug rehabilitation programme, and then classified them in two ways, by educational history and subsequent conviction record. If the researcher had chosen a different sample, the numbers in each cell of the table might have been different. In other words, the researcher did not fix either the row totals or the column totals in advance; they are all random variables that were observed during the study. Only the overall total of 60 people was fixed in advance. However, Table 11.9 is different. There, the row totals were fixed in advance, because the data come from an experiment where the researcher chose to observe exactly 100 occasions involving males and 105 involving females. The column totals are not fixed in advance, but are random variables depending on how many people happened to help overall. The data appear to have the same form in Tables 11.7 and 11.9; but the random variables involved are different. Table 11.9 does not show bivariate data because two random variables were not recorded on each occasion when help might have been offered. Two pieces of information were recorded, the sex of the person involved and whether or not help was offered; but the sex was not a random variable.

However, Fisher argued that his exact testing procedure applied in both of these different situations. You will recall that the null distribution of the test statistic was calculated on the basis that both sets of marginal totals (row and column totals) remained fixed. Without going into all the details, according to Fisher, the procedure could be applied where none of the marginal totals were fixed (testing for association, as in Table 11.7) or where one set was fixed and the other was not (testing for difference in proportions, as in Table 11.9). Most statisticians nowadays agree with Fisher about the applicability of his test.

11.4 The chi-squared test for contingency tables

Fisher's exact test requires a computer for all but the smallest data sets, and the computations involved may be too much even for some statistical software if the data set is too large. It is therefore useful to have another method of testing for association. The method you will learn about is an application of the chi-squared goodness-of-fit test that was discussed in *Chapter 9*. There, you saw that the distribution of the test statistic was approximately chi-squared, but the approximation is good provided the data set is not too small. The same applies in relation to contingency tables.

There is no simple analogue of Fisher's exact test for contingency tables that have more than two rows and more than two columns, and exact tests for such tables are not always straightforward (though they can be carried out with some statistical software). The chi-squared test, however, applies just as well to larger contingency tables as to 2×2 tables. It is worth learning about the chi-squared test for 2×2 tables, before extending the test to larger tables.

11.4.1 The chi-squared test in 2 x 2 contingency tables

Let us begin with an example that shows how a chi-squared test for association can be carried out in a 2×2 table.

Example 11.7 *Rheumatoid arthritis treatment*

Table 11.8 presented some data on reactions to a treatment for rheumatoid arthritis. How can these data be analysed using a chi-squared goodness-of-fit test? In Table 11.8 we have counts of observed frequencies in four cells (together with a number of totals). If we can construct a table of four *expected frequencies*, we can calculate the chi-squared goodness-of-fit test statistic in exactly the same way as in Section 9.2.

The null hypothesis we are testing here states that the two variables involved, toxicity and impaired sulphoxidation, are not related. The expected frequencies should be calculated on the basis that the null hypothesis is true. If it is true, then knowing whether a patient has impaired sulphoxidation will tell us nothing about the toxic effect in a patient. Altogether, according to Table 11.8, 37 patients out of 65 showed toxic effect. That is, a proportion of $37/65$ of the patients suffered toxic effects. If the null hypothesis were true, we would expect that, of the 39 patients in all who had impaired sulphoxidation, a proportion of $37/65$ of them would show toxic effect. (If the proportion differed much from $37/65$, then knowing that a person had impaired sulphoxidation would tell you something about the toxic effect in that patient; but under the null hypothesis we have assumed this does not happen.) Therefore, the expected number of patients who showed toxic effect would be $(39 \times 37)/65$ or 22.2. Out of the 26 patients who did *not* have impaired sulphoxidation the expected proportion that shows toxic effect would also be $37/65$, and the expected number of patients would be $(26 \times 37)/65$ or 14.8.

These expected frequencies are shown in Table 11.10, together with the remaining two expected frequencies. Note that each expected frequency is found by multiplying the row total for its row by the column total for its column, and then dividing by the overall total, 65.

Table 11.10 Expected values for the rheumatoid arthritis data

Impaired sulphoxidation	Toxicity Yes	No	Row totals
Yes	$(39 \times 37)/65 = 22.2$	$(39 \times 28)/65 = 16.8$	39
No	$(26 \times 37)/65 = 14.8$	$(26 \times 28)/65 = 11.2$	26
Column totals	37	28	65

The first point to note about this table is that the expected frequencies sum to the same row and column totals as the observed frequencies did. The second point is that the expected values, calculated on the basis that the two variables were independent, differ considerably from the observed counts. The differences between observed and expected values are tabulated in Table 11.11.

Table 11.11 $(O - E)$ values for the rheumatoid arthritis data

Impaired sulphoxidation	Toxicity Yes	No
Yes	$30 - 22.2 = 7.8$	$9 - 16.8 = -7.8$
No	$7 - 14.8 = -7.8$	$19 - 11.2 = 7.8$

Here, it is interesting to note that the row and column sums of the (Observed − Expected) differences are all zero (because the row and column sums were the same for the observed and expected frequencies). As a consequence, the values in Table 11.11 are all the same, apart from the signs. It is now straightforward to complete the calculation of the chi-squared test statistic. It is defined as

$$\chi^2 = \sum \frac{(O_i - E_i)^2}{E_i},$$

where the summation is over all the cells (four of them, here). Thus the value of the chi-squared test statistic is

$$\chi^2 = \frac{(7.8)^2}{22.2} + \frac{(-7.8)^2}{14.8} + \frac{(-7.8)^2}{16.8} + \frac{(7.8)^2}{11.2} = 15.905.$$

To complete the test, we need to know how many degrees of freedom are appropriate for the chi-squared distribution involved. The answer is one. This is indicated by the fact that (apart from the signs) there is only one value in the table of differences between observed and expected values. To be more formal, there are four cells, and we have estimated two parameters from the data, namely the proportion of individuals who have impaired sulphoxidation and the proportion of individuals who show toxic effect, so the number of degrees of freedom is $4 - 1 - 2 = 1$. Using a computer, the SP for an observed value of 15.905 against $\chi^2(1)$ is 0.000 067. There is extremely strong evidence that these two variables are associated, just as you found using Fisher's exact test in Exercise 11.12. ■

In general, the chi-squared goodness-of-fit test can be applied to test for association in a 2×2 contingency table as follows.

Testing for association in a 2×2 contingency table

1. Calculate a table of expected frequencies by multiplying the corresponding row and column totals together, and then dividing by the overall total.
2. Calculate the chi-squared test statistic using the formula

$$\chi^2 = \sum \frac{(O_i - E_i)^2}{E_i},$$

where the summation is over the four cells in the table.

3. Under the null hypothesis of no association, the test statistic has an approximate chi-squared distribution with 1 degree of freedom. Calculate the *SP* for the test and interpret your answer.

As with the chi-squared tests in Section 9.2, the question of the adequacy of the approximation in Step 3 arises. The same simple rule applies: the approximation is adequate if no expected frequency is less than 5, otherwise it may not be good enough.

If any expected frequency is less than 5, then the only option is an exact test, for the idea of 'pooling' cells in a 2×2 contingency table is not a meaningful one.

In practice, most statistical computer packages will carry out this test.

Note that this test, as it is presented here, is essentially two-sided; because the differences between observed and expected frequencies are squared in calculating the test statistic, their signs are ignored, and the test treats departures from the null hypothesis in either direction on the same footing. It is possible to complicate the testing procedure somewhat to give a one-sided test; we shall not do this here.

Exercise 11.13

A study was carried out in which 671 tiger beetles were classified in two ways, according to their colour pattern (bright red or not bright red) and the season in which they were found (spring or summer). The data are given in Table 11.12. Use your computer to carry out a chi-squared test for association between the two variables involved, and report your conclusions.

Sokal, R.R. and Rohlf, F.J. (1981) *Biometry*, 2nd edition, W.H. Freeman, New York, p. 745.

Table 11.12 Colour pattern and seasonal incidence of tiger beetles

Season	Colour pattern		Row totals
	Bright red	Not bright red	
Spring	302	202	504
Summer	72	95	167
Column totals	374	297	671

11.4.2 The chi-squared test in larger contingency tables

In Example 11.3 you met a contingency table which appeared (merely from a consideration of the frequencies reported) to indicate an association between snoring frequency and heart disease. How can we test whether these two variables really are associated in the population from which the sample was drawn? The examples we have analysed so far all had two rows and two columns, but the table in Example 11.3 had four columns. One answer is to use the chi-squared goodness-of-fit test again. The procedure given in Subsection 11.4.1 applies in exactly the same way to tables larger than 2×2, except that the number of degrees of freedom for the chi-squared distribution is different. For contingency tables in general, it is conventional to denote the number of rows by r, and the number of columns by c. The procedure is as follows.

Testing for association in a $r \times c$ contingency table

1 Calculate a table of expected frequencies by multiplying the corresponding row and column totals together, and then dividing by the overall total.

2 Calculate the chi-squared test statistic using the formula

$$\chi^2 = \sum \frac{(O_i - E_i)^2}{E_i},$$

where the summation is over the rc cells in the table.

3 Under the null hypothesis of no association, the test statistic has an approximate chi-squared distribution with $(r-1)(c-1)$ degrees of freedom. Calculate the SP for the test and interpret your answer.

There are $(r-1)(c-1)$ degrees of freedom because, given the marginal totals, only $(r-1)(c-1)$ cells are freely assignable.

As in previous situations, the adequacy of the approximation may not be good enough if any of the expected frequencies is less than 5.

Example 11.3 continued

We shall use the chi-squared test to test for association between snoring frequency and heart disease in the data from Table 11.2. The data (observed frequencies) were as follows.

Heart disease	Non-snorers	Occasional snorers	Snore nearly every night	Snore every night	Total
Yes	24	35	21	30	110
No	1355	603	192	224	2374
Total	1379	638	213	254	2484

The expected frequencies can be calculated as follows.

Heart disease	Non-snorers	Occasional snorers	Snore nearly every night	Snore every night	Total
Yes	$\frac{110 \times 1379}{2484}$ $= 61.067$	$\frac{110 \times 638}{2484}$ $= 28.253$	$\frac{110 \times 213}{2484}$ $= 9.432$	$\frac{110 \times 254}{2484}$ $= 11.248$	110
No	$\frac{2374 \times 1379}{2484}$ $= 1317.933$	$\frac{2374 \times 638}{2484}$ $= 609.747$	$\frac{2374 \times 213}{2484}$ $= 203.568$	$\frac{2374 \times 254}{2484}$ $= 242.752$	2374
Total	1379	638	213	254	2484

The differences $O_i - E_i$ are as follows.

Heart disease	Non-snorers	Occasional snorers	Snore nearly every night	Snore every night	Total
Yes	−37.067	6.747	11.568	18.752	0
No	37.067	−6.747	−11.568	−18.752	0
Total	0	0	0	0	0

Comparing the observed and expected frequencies, far more of the frequent snorers seem to have heart disease than would be expected under the null hypothesis of no association.

The value of the chi-squared test statistic is found from

$$\chi^2 = \sum \frac{(O_i - E_i)^2}{E_i} = \frac{(-37.067)^2}{61.067} + \cdots + \frac{(-18.752)^2}{242.752} = 72.782.$$

There are two rows and four columns in the contingency table so the number of degrees of freedom is $(2-1)(4-1) = 3$. Comparing the value of our test statistic against a chi-squared distribution with 3 degrees of freedom, the SP is very close to zero. There is very strong evidence of an association between snoring and heart disease, though because 'correlation is not causation' we cannot conclude that snoring causes heart disease, or that heart disease causes snoring. ■

The SP is about 1×10^{-15}.

Now try the following exercise.

Exercise 11.14

(a) Table 11.13 gives data on the number of failures of piston-rings in each of three legs in each of four steam-driven compressors located in the same building. The compressors have identical design and are oriented in the same way. One question of interest is whether there is an association between the leg in which a failure occurs and the compressor in which it occurs, or whether the pattern of the location of failures is the same for different compressors. Use a chi-squared test to investigate whether such an association exists, and report your conclusions.

Davies, O.L. and Goldsmith, P.L. (eds) (1972) *Statistical Methods in Research and Production*, 4th edn. Oliver and Boyd, UK, p. 324.

Table 11.13 Piston-ring failures

Compressor	Leg			Row totals
	North	Centre	South	
1	17	17	12	46
2	11	9	13	33
3	11	8	19	38
4	14	7	28	49
Column totals	53	41	72	166

(b) Some individuals are carriers of the bacterium *Streptococcus pyogenes*. To investigate whether there is a relationship between carrier status and tonsil size in schoolchildren, 1398 children were examined and classified according to their carrier status and tonsil size. The data appear in Table 11.14. Is there an association between tonsil size and carrier status? Investigate using a chi-squared test, and report your conclusions.

Krzanowski, W. (1988) *Principles of multivariate analysis.* Oxford University Press, Oxford, p. 269.

Table 11.14 Tonsil size in schoolchildren

Tonsil size	Carrier status		Row totals
	Carrier	Non-carrier	
Normal	19	497	516
Large	29	560	589
Very large	24	269	293
Column totals	72	1326	1398

The chi-squared test for contingency tables is a very useful method of analysis. Indeed, it is more widely applicable than you have yet seen. You will recall that, in 2×2 tables, Fisher's exact test could be used to compare proportions as well as to test for association. The same is true of the chi-squared test. In a 2×2 table, it can be used to compare proportions in much the same way that Fisher's test can be used. In tables with two rows and more than two columns where the column totals are fixed in advance, it can also be used to provide an overall test of the null hypothesis that several binomial proportions are all equal. In larger tables where one set of marginal totals (row or column totals) are fixed in advance, the chi-squared procedure can be used to test various different hypotheses.

However, these further uses of the chi-squared test are beyond the scope of this course. To complete our study of related variables, we now turn to a probability model for continuous bivariate data.

11.5 The bivariate normal distribution

In earlier parts of the course, when dealing with univariate data (the kind where only one random variable at a time is involved), it was often useful (and sometimes necessary) to define a probability model for the data. Often, hypothesis tests were performed or confidence intervals were calculated on the basis that the data involved came from a particular probability distribution, such as a normal distribution or a Poisson distribution. The same is true of bivariate data. It was mentioned in Section 11.2 that the test for association using the Pearson correlation coefficient as the test statistic involved an assumption that the data came from a particular probability model, the *bivariate normal distribution.* (None of the other techniques in this chapter involves the use of an explicit probability model.) The aim of this section is to show you some of the properties of the bivariate normal distribution, which is the most important probability model for continuous bivariate data, and hence to give you a flavour of what is involved in probability models for bivariate data.

The bivariate normal distribution is a model for a pair of random variables X and Y. The model defines in all respects how X and Y vary together. This involves defining the distributions of X and Y taken on their own, and you will not be surprised to learn that both these distributions are normal. But there is more to the bivariate normal distribution than that. It has to account for correlation between X and Y. This is done by defining a bivariate version of the normal probability density function.

Think of how the idea of a univariate probability density function was arrived at earlier in the course. We started with histograms, where the height of each bar in the histogram above the axis represented the frequency with which a particular range of values occurred. Many histograms had approximately the 'bell' shape characteristic of the probability density function of the normal distribution: hence its usefulness as a probability model.

To define a bivariate probability density function, we need to be able to represent how common are different pairs (x, y) of values of the two random variables involved. That is, the probability density function needs to be defined over an area of a plane, rather than simply on a line. Imagine taking a scatter plot, dividing the area up into squares, and setting up a bar on each of the squares whose height is proportional to the frequency of data in that square. The resulting object (Figure 11.15) is a bivariate version of the histogram, and in the same way that the smooth curve of the (univariate) normal probability density function looks like a smoothed and idealized histogram, so the bivariate normal probability density function is a smooth *surface* which is a smoothed and idealized version of a bivariate 'histogram' like that in Figure 11.15.

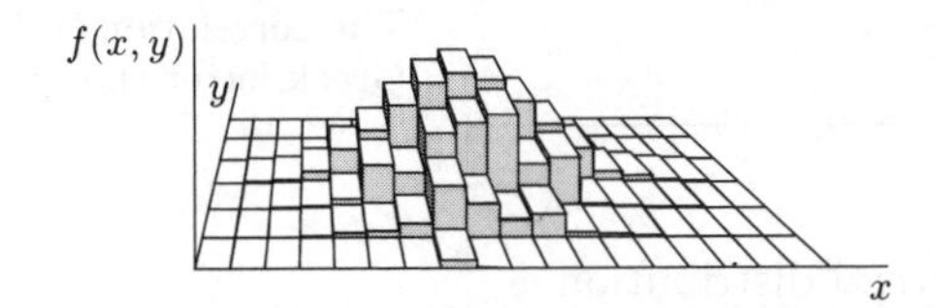

Figure 11.15 A bivariate 'histogram'

Two examples of bivariate normal probability density functions are shown in Figure 11.16. The 'contour' lines join up points of equal height, just as with contour lines on an ordinary map. In the first distribution (Figure 11.16(a)), the correlation between the two variables is zero. The 'hill' is circular in cross section. If a small sample were drawn from a population which was modelled by this distribution, a scatter plot of the sample would look like that shown in Figure 11.17; the variables are not related. In the second distribution (Figure 11.16(b)), the Pearson correlation coefficient for the two variables is high; it is 0.9. The 'hill' is a narrow ridge.

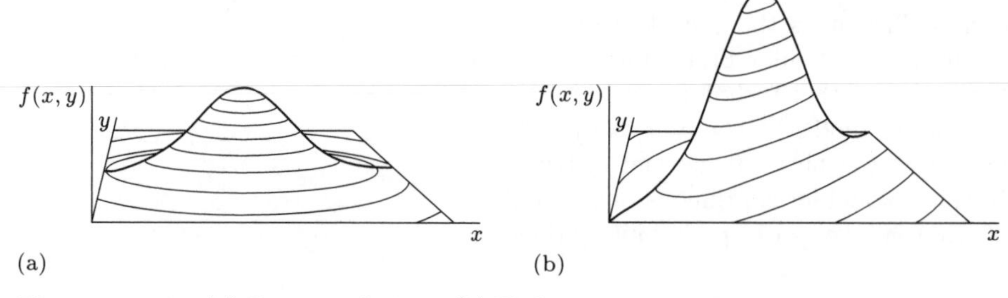

Figure 11.16 (a) Zero correlation (b) High positive correlation

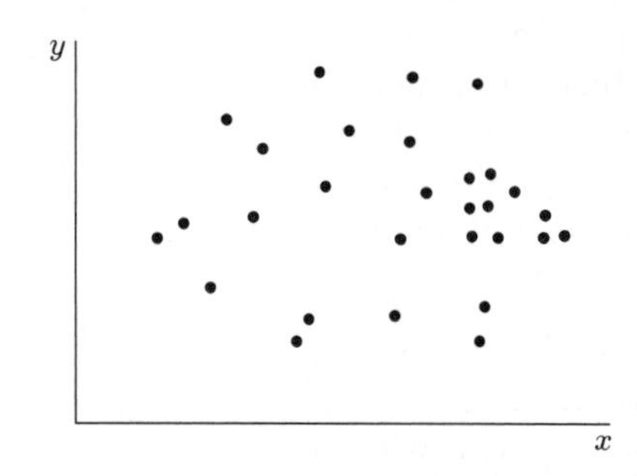

Figure 11.17 A scatter plot of observations drawn from a bivariate normal distribution with zero correlation

Exercise 11.15

Imagine that a small sample is drawn from a population following the second distribution (that with high positive correlation). Draw a rough sketch of what a scatter plot of this sample would look like.

A bivariate normal distribution for bivariate random variables X and Y is characterized by five parameters: the means of X and Y and their variances, which are denoted by μ_X, μ_Y, σ_X^2 and σ_Y^2 respectively, and the (Pearson) correlation between X and Y, which is denoted by ρ. We write

$$(X, Y) \sim N(\mu_X, \mu_Y, \sigma_X^2, \sigma_Y^2, \rho).$$

A bivariate normal distribution has several interesting properties, and these are as follows.

There is a formal definition for the correlation between random variables, involving the idea of expectation, just as there is a formal definition involving expectations for the mean and variance of a random variable. You need not worry about the details. The correlation is denoted by ρ, the Greek letter rho, pronounced 'roe'.

The bivariate normal distribution

If the random variables X and Y have a bivariate normal distribution

$$(X, Y) \sim N(\mu_X, \mu_Y, \sigma_X^2, \sigma_Y^2, \rho)$$

then

1 the distribution of X is normal with mean μ_X and variance σ_X^2, $X \sim N(\mu_X, \sigma_X^2)$, and the distribution of Y is normal with mean μ_Y and variance σ_Y^2;

2 the conditional distribution of X, given $Y = y$, is normal

$$(X|Y = y) \sim N\left(\mu_X + \rho\frac{\sigma_X}{\sigma_Y}(y - \mu_Y), \sigma_X^2(1 - \rho^2)\right)$$

(and similarly for the conditional distribution of Y, given $X = x$).

The unconditional distributions for X and Y are often referred to as the *marginal* distributions of X and Y, because they correspond to the marginal totals in a contingency table (which give total frequencies for each of the variables).

The following example gives some idea of a context where the bivariate normal model might be applied.

Example 11.8 Heights and weights of schoolgirls in Bradford

For the sample of 30 schoolchildren whose heights and weights were plotted in Figure 11.1 and listed in Table 11.4, summary sample statistics (denoting height by X and weight by Y and using the Pearson product-moment correlation coefficient) are

$$\overline{x} = 144.87,\ \overline{y} = 36.17,\ s_X = 7.61,\ s_Y = 7.70,\ r = 0.743.$$

Suppose that it is decided to model the variation in heights and weights of eleven-year-old schoolgirls in the population by a bivariate normal distribution with parameters $\mu_X = 145, \mu_Y = 36, \sigma_X^2 = 58, \sigma_Y^2 = 60, \rho = 0.75$. That is,

$$(X,Y) \sim N(145, 36, 58, 60, 0.75).$$

Then, for instance, the proportion of eleven-year-old schoolgirls in the population who are more than 140 cm tall is given by

$$P(X > 140) = 1 - \Phi\left(\frac{140 - 145}{\sqrt{58}}\right) = 0.744;$$

the proportion of schoolgirls whose weight is below 30 kg is given by

$$P(Y < 30) = \Phi\left(\frac{30 - 36}{\sqrt{60}}\right) = 0.219.$$

Considering only those schoolgirls 150 cm tall (this is taller than average: $x = 150$ while $\mu_X = 145$) the weight distribution has mean

$$\begin{aligned} E(Y|X = x = 150) &= \mu_Y + \rho\frac{\sigma_Y}{\sigma_X}(x - \mu_X) \\ &= 36 + 0.75 \times \sqrt{\frac{60}{58}} \times (150 - 145) \\ &= 36 + 3.8 \\ &= 39.8\,\text{kg}, \end{aligned}$$

which is also heavier than average.

To find, say, the proportion of eleven-year-old schoolgirls who are at the same time taller than 150 cm and heavier than 40 kg requires calculation of the probability

$$P(X > 150, Y > 40);$$

here, the only feasible approach is to use a computer programmed to return bivariate normal probabilities. The answer is 0.181. ■

As you can appreciate, to calculate all but very simple probabilities from a bivariate normal distribution requires the use of a computer, or at the very least a set of statistical tables, a calculator and plenty of time. Your statistical software may do these calculations for you.

In this section we have been able to give only a flavour of the bivariate normal distribution. There are similar analogues of the normal distribution in more than two dimensions—the *multivariate* normal distribution—but that is well beyond the scope of this course.

Summary

1. Two random variables X and Y are said to be associated (or not independent) if

$$P(X = x|Y = y) \neq P(X = x)$$

for some x and y, where the probability on the left-hand side is called a conditional probability, and is read 'the probability that $X = x$ given that $Y = y$' or 'the probability that $X = x$ conditional on $Y = y$'.

2. A measure of linear association between two variables X and Y is given by the Pearson product-moment correlation coefficient r for the sample $(x_1, y_1), (x_2, y_2), \ldots, (x_n, y_n)$, where

$$r = \frac{n\sum x_i y_i - \sum x_i \sum y_i}{\sqrt{(n\sum x_i^2 - (\sum x_i)^2)(n\sum y_i^2 - (\sum y_i)^2)}}.$$

3. A measure of monotonic association between two variables X and Y is given by the Spearman rank correlation coefficient r_S for the sample $(x_1, y_1), (x_2, y_2), \ldots, (x_n, y_n)$: here, the observations in each sample are replaced by their ranks (averaged, if necessary) and the Pearson correlation is calculated for the ranked data.

4. Correlation is not causation.

5. The null hypothesis that the underlying Pearson correlation is zero may be tested using

$$R\sqrt{\frac{n-2}{1-R^2}} \sim t(n-2).$$

6. When data are arranged as counts in a 2×2 contingency table, an exact test for no association may be provided by posing the problem as a test for equality of two proportions, and using Fisher's test.

7. When data are arranged as counts in a $r \times c$ contingency table, an approximate test for no association is provided by the test statistic

$$\chi^2 = \sum \frac{(O_i - E_i)^2}{E_i} \sim \chi^2\left((r-1)(c-1)\right),$$

where the sum is taken over all rc cells of the table, and the expected frequency E_i is given by

$$E_i = \frac{\text{row total} \times \text{column total}}{\text{overall total}}.$$

The approximation is adequate as long as every E_i is at least 5.

8. A probability model for bivariate data is given by the bivariate normal distribution

$$(X,Y) \sim N(\mu_X, \mu_Y, \sigma_X^2, \sigma_Y^2, \rho),$$

where the correlation ρ is a measure of linear association between the two variables. The marginal distributions of X and Y are given by

$$X \sim N(\mu_X, \sigma_X^2), \qquad Y \sim N(\mu_Y, \sigma_Y^2);$$

and the conditional distribution of X, given $Y = y$, is also normal:

$$(X|Y = y) \sim N\left(\mu_X + \rho\frac{\sigma_X}{\sigma_Y}(y - \mu_Y), \sigma_X^2(1 - \rho^2)\right).$$

Similarly, the conditional distribution of Y, given $X = x$, is normal.

The bivariate probability

$$P(X \leq x, Y \leq y)$$

(and variations on it) cannot be calculated as a combination of univariate normal probabilies, and requires the use of a computer.

Chapter 12
Random processes

This chapter is devoted to the idea of a random process and to models for random processes. The problems of estimating model parameters and of testing the adequacy of the fit of a hypothesized model to a set of data are addressed. As well as revisiting the familiar Bernoulli and Poisson processes, a model known as the Markov chain is introduced.

In this chapter, attention is turned to the study of *random processes*. This is a wide area of study, with many applications. We shall discuss again the familiar Bernoulli and Poisson processes, considering different situations where they provide good models for variation and other situations where they do not. Only one new model, the *Markov chain*, is introduced in any detail in this chapter (Sections 12.2 and 12.3). This is a refinement of the Bernoulli process. In Section 12.4 we revisit the Poisson process.

However, in order for you to achieve some appreciation of the wide application of the theory and practice of random processes, Section 12.1 is devoted to examples of different contexts. Here, questions are posed about random situations where some sort of random process model is an essential first step to achieving an answer. You are not expected to do any more in this section than read the examples and try to understand the important features of them. First, let us consider what is meant by the term 'random process'.

In *Chapter 2*, you rolled a die many times, recording a 1 if you obtained either a 3 or a 6 (a success) and 0 otherwise. You kept a tally of the running total of successes with each roll, and you will have ended up with a table similar to that shown in Table 2.3. Denoting by the random variable X_n the number of successes achieved in n rolls, you then calculated the proportion of successes X_n/n and plotted that proportion against n, finishing with a diagram something like that shown in Figure 2.2. For ease, that diagram is repeated here in Figure 12.1.

See Exercise 2.2.

The reason for beginning this chapter with a reminder of past activities is that it acts as a first illustration of a random process. The *sequence* of random variables

$$\{X_n; n = 1, 2, 3, \ldots\},$$

or the sequence plotted in Figure 12.1,

$$\{X_n/n; n = 1, 2, 3, \ldots\},$$

are both examples of **random processes**. You can think of a random process evolving in time: repeated observations are taken on some associated random

You now know that the distribution of the random variable X_n is binomial $B(n, \frac{1}{3})$.

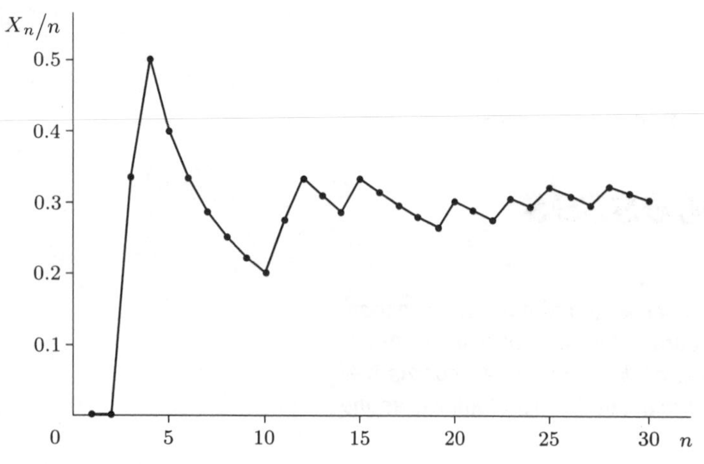

Figure 12.1 Successive calculations of X_n/n

variable, and this list of measurements constitutes what is known as a **realization** (that is, a particular case) of the random process. Notice that the order of observations is preserved and is usually important. In this example, the realization shows early fluctuations that are wide; but after a while, the process settles down (cf. Figure 2.3 which showed the first 500 observations on the random variable X_n/n). It is often the case that the long-term behaviour of a random process $\{X_n\}$ is of particular concern.

Here is another example.

Example 12.1 More traffic data

The data in Table 12.1 were collected by Professor Toby Lewis as an illustration of a random process which might possibly be modelled as a Poisson process. They show the 40 time intervals (in seconds) between 41 vehicles travelling northwards along the M1 motorway in England past a fixed point near Junction 13 in Bedfordshire on Saturday 23 March 1985. Observation commenced shortly after 10.30 pm.

Data provided by Professor T. Lewis, Centre for Statistics, University of East Anglia.

Table 12.1 Waiting times between passing traffic (seconds)

12	2	6	2	19	5	34	4
1	4	8	7	1	21	6	11
8	28	6	4	5	1	18	9
5	1	21	1	1	5	3	14
5	3	4	5	1	3	16	2

One way of viewing the data is shown in Figure 12.2. This shows the passing vehicles as 41 blobs on a time axis, assuming the first vehicle passed the observer at time 0. You have seen this type of diagram already.

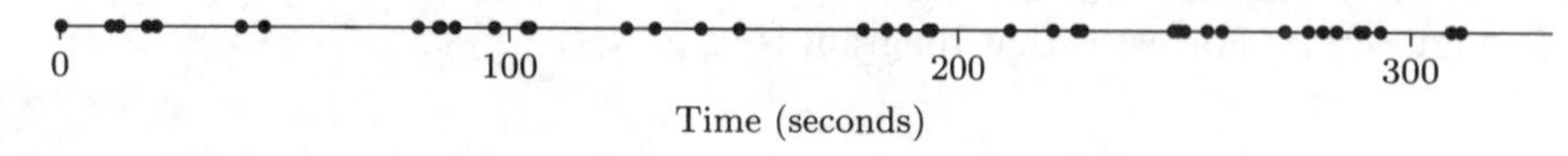

Figure 12.2 Traffic incidence, M1 motorway, late evening

A different representation is to denote by the random variable $X(t)$ the number of vehicles (after the one that initiated the sequence) to have passed the observer by time t. Thus, for instance,

$$X(10) = 0, \quad X(20) = 3, \quad X(30) = 4, \quad \ldots .$$

A graph of $X(t)$ against t is shown in Figure 12.3. It shows a *realization* of the random process

$$\{X(t); t \geq 0\}$$

and could be used in a test of the hypothesis that events (vehicles passing a fixed point) occur as a Poisson process in time.

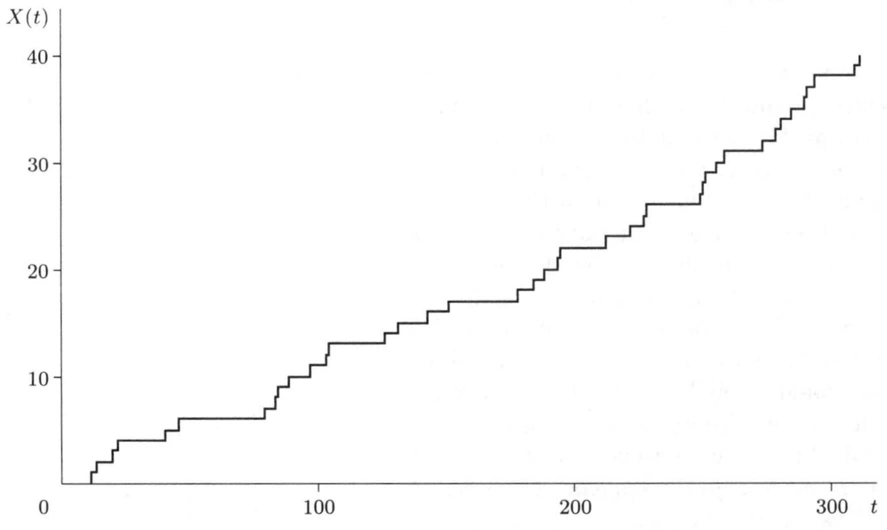

Figure 12.3 A realization of the random process $\{X(t); t \geq 0\}$

Again, the notion of ordering is important. Here, the random process is an integer-valued continuous-time process. Notice that there is no suggestion in this case that the random variable $X(t)$ settles down to some constant value. ■

In this chapter we shall consider three different models for random processes. You are already familiar with two of them: the Bernoulli and the Poisson processes. The third model is a refinement of the Bernoulli process, which is useful in some contexts when the Bernoulli process proves to be an inadequate model. We shall consider the problem of parameter estimation given data arising from realizations of random processes, and we shall see how the quality of the fit of a hypothesized model to data is tested.

These models are discussed in Sections 12.2 to 12.4, and variations of the models are briefly examined in Section 12.5. There are no exercises in Sections 12.1 or 12.5.

12.1 Examples of random processes

So far in the course we have seen examples of many kinds of data set. The first is the *random sample*. Examples of this include leaf lengths (Table 2.1), heights (Table 2.15), skull measurements (Table 3.3), waiting times (Tables 4.7 and 4.10), leech counts (Table 6.2) and wages (Table 6.9). In *Chapter 10* we looked at regression data such as height against age (Figure 10.1), boiling point of water against atmospheric pressure (Tables 10.2 and 10.4) and paper strength against hardwood content (Table 10.8). In *Chapter 11* we explored bivariate data such as systolic and diastolic blood pressure measurements (Table 11.1), snoring frequency and heart disease (Table 11.2), and height and weight (Table 11.4).

In many cases we have done more than simply scan the data: we have hypothesized models for the data and tested hypotheses about model parameters. For instance, we know how to test that a regression slope is zero; or that there is zero correlation between two responses measured on the same subject.

We have also looked at data in the context of the Poisson process. For the earthquake data (Table 4.7) we decided (Figure S9.6) that the waiting times between successive serious earthquakes world-wide could reasonably be modelled by an exponential distribution. In this case, therefore, the Poisson process should provide an adequate model for the sequence of earthquakes, at least over the years that the data collection extended. For the coal mine disaster data in Table 7.4 there was a distinct curve to the associated exponential probability plot (Figure 9.6). Thus it would not be reasonable to model the sequence of disasters as a Poisson process. (Nevertheless, the probability plot suggests some sort of systematic variation in the waiting times between disasters, and therefore the random process could surely be satisfactorily modelled without too much difficulty.) For the data on eruptions of the Old Faithful geyser (Table 3.18) we saw in Figure 3.18 that the inter-event times were not exponential: there were two pronounced modes. So the sequence of eruptions could not sensibly be modelled as a Poisson process.

Here are more examples.

Example 12.2 Times of system failures

The data in Table 12.2 give the cumulative times of failure, measured in CPU seconds and starting at time 0, of a command and control software system.

Musa, J.D., Iannino, A. and Okumoto, K. (1978) *Software reliability: measurement, prediction, application.* McGraw-Hill Book Co., New York, p. 305.

Table 12.2 Cumulative times of failure (CPU seconds)

3	33	146	227	342	351	353	444	556
571	709	759	836	860	968	1056	1726	1846
1872	1986	2311	2366	2608	2676	3098	3278	3288
4434	5034	5049	5085	5089	5089	5097	5324	5389
5565	5623	6080	6380	6477	6740	7192	7447	7644
7837	7843	7922	8738	10089	10237	10258	10491	10625
10982	11175	11411	11442	11811	12559	12559	12791	13121
13486	14708	15251	15261	15277	15806	16185	16229	16358
17168	17458	17758	18287	18568	18728	19556	20567	21012
21308	23063	24127	25910	26770	27753	28460	28493	29361
30085	32408	35338	36799	37642	37654	37915	39715	40580
42015	42045	42188	42296	42296	45406	46653	47596	48296
49171	49416	50145	52042	52489	52875	53321	53443	54433
55381	56463	56485	56560	57042	62551	62651	62661	63732
64103	64893	71043	74364	75409	76057	81542	82702	84566
88682								

Here, the random process $\{T_n; n = 1, 2, 3, \ldots\}$ gives the time of the nth failure. If the system failures occur at random, then the sequence of failures will occur as a Poisson process in time. In this case, the times between failures will constitute independent observations from an exponential distribution. An exponential probability plot of the time differences $(3, 30, 113, 81, \ldots, 4116)$ is shown in Figure 12.4.

You can see from the figure that an exponential model does not provide a good fit to the variation in the times between failures. In fact, a more

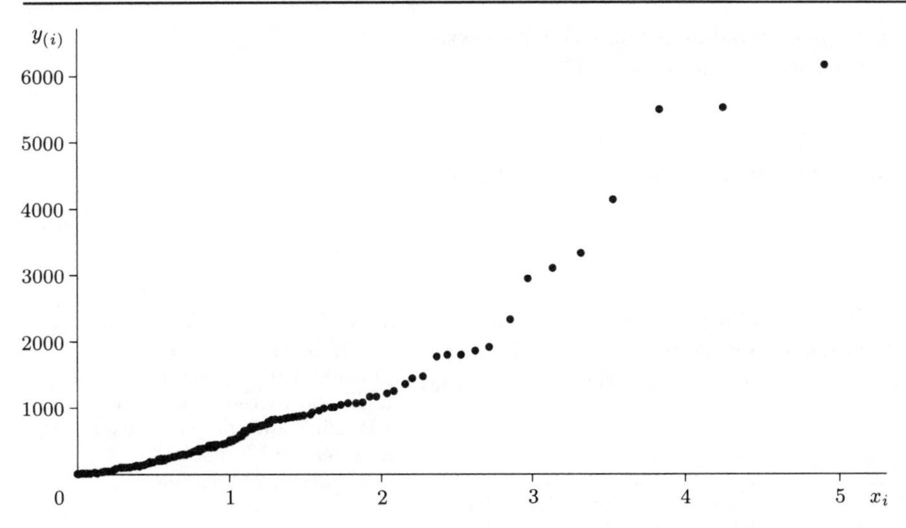

Figure 12.4 Exponential probability plot for times between failures

extended analysis of the data reveals that the failures are becoming less frequent with passing time. The system evidently becomes less prone to failure as it ages. ■

In Example 12.2 the system was monitored continuously so that the time of occurrence of each failure was recorded accurately. Sometimes it is not practicable to arrange continuous observation, and system monitoring is intermittent. An example of this is given in Example 12.3.

Example 12.3 Counts of system failures

The data in Table 12.3 relate to a DEC-20 computer which was in use at the Open University during the 1980s. They give the number of times that the computer broke down in each of 128 consecutive weeks of operation, starting in late 1983. For these purposes, a breakdown was defined to have occurred if the computer stopped running, or if it was turned off to deal with a fault. Routine maintenance was not counted as a breakdown. So Table 12.3 shows a realization of the random process $\{X_n; n = 1, 2, \ldots, 128\}$, an integer-valued random process developing in discrete time where X_n denotes the number of failures during the nth week.

The Open University (1987) M345 *Statistical Methods*, Unit 11, *Statistical Computing III*, Milton Keynes, The Open University, p. 16.

Table 12.3 Counts of failures (per week)

4	0	0	0	3	2	0	0	6	7	6	2	1	11	6	1	2	1	1
2	0	2	2	1	0	12	8	4	5	0	5	4	1	0	8	2	5	2
1	12	8	9	10	17	2	3	4	8	1	2	5	1	2	2	3	1	2
0	2	1	6	3	3	6	11	10	4	3	0	2	4	2	1	5	3	3
2	5	3	4	1	3	6	4	4	5	2	10	4	1	5	6	9	7	3
1	3	0	2	2	1	4	2	13	0	2	1	1	0	3	16	22	5	1
2	4	7	8	6	11	3	0	4	7	8	4	4	5					

Again, if breakdowns occur at random in time, then these counts may be regarded as independent observations from a Poisson distribution with constant mean. In fact, this is not borne out by the data (a chi-squared test of goodness-of-fit against the Poisson model with estimated mean 4.02 gives a SP which is essentially zero). The question therefore is: how might the incidence of breakdowns be modelled instead? (You can see occasional peaks of activity—weeks where there were many failures—occurring within the data.) ■

In the next example, this idea is explored further. The data are reported in a paper about a statistical technique called **bump-hunting**, the identification of peaks of activity.

Example 12.4 Scattering activity

A total of 25 752 scattering events called scintillations was counted from a particle-scattering experiment over 172 bins each of width 10 MeV. There are peaks of activity called *bumps* and it is of interest to identify where these bumps occur.

Good, I.J. and Gaskins, R.A. (1980) Density estimation and bump-hunting by the penalized likelihood method exemplified by scattering and meteorite data. *J. American Statistical Association*, **75**, 42–56.

Table 12.4 Scattering reactions in 172 consecutive bins

5	11	17	21	15	17	23	25	30	22	36	29	33
43	54	55	59	44	58	66	59	55	67	75	82	98
94	85	92	102	113	122	153	155	193	197	207	258	305
332	318	378	457	540	592	646	773	787	783	695	774	759
692	559	557	499	431	421	353	315	343	306	262	265	254
225	246	225	196	150	118	114	99	121	106	112	122	120
126	126	141	122	122	115	119	166	135	154	120	162	156
175	193	162	178	201	214	230	216	229	214	197	170	181
183	144	114	120	132	109	108	97	102	89	71	92	58
65	55	53	40	42	46	47	37	49	38	29	34	42
45	42	40	59	42	35	41	35	48	41	47	49	37
40	33	33	37	29	26	38	22	27	27	13	18	25
24	21	16	24	14	23	21	17	17	21	10	14	18
16	21	6										

Just glancing at the numbers, the peaks of activity do not seem consistent with independent observations from a Poisson distribution. The data in Table 12.4 become more informative if portrayed graphically, and this is done in Figure 12.5. The number of scattering reactions X_n is plotted against bin number n.

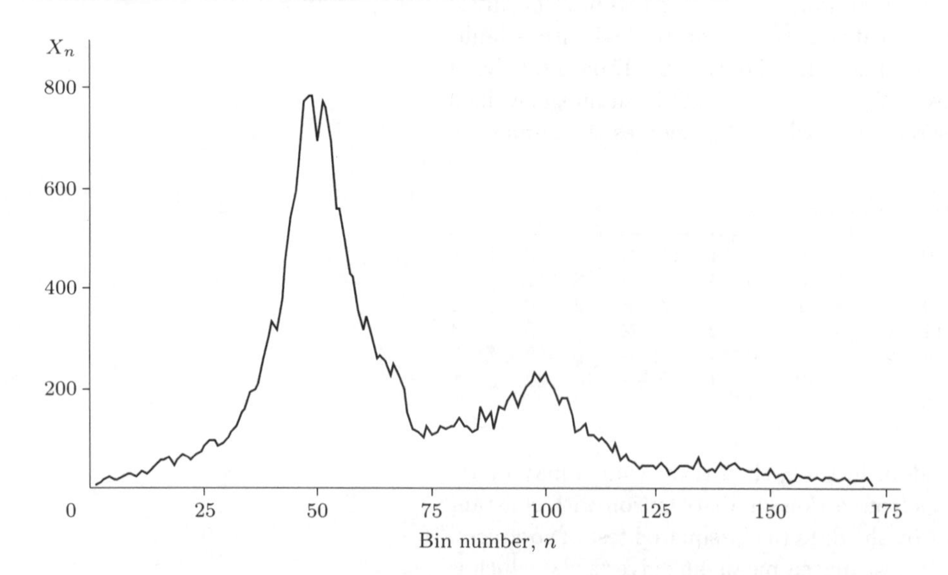

Figure 12.5 Scattering reactions in consecutive bins of constant width

The figure suggests that there are two main peaks of activity. (There are more than 30 smaller local peaks which do not appear to be significant, and are merely evidence of random variation in the data.) ■

Another example where mean incidence is not constant with passing time is provided by the road casualty data in Example 12.5. It is common to model this sort of unforeseen accident as a Poisson process in time, but, as the data show, the rate is in fact highly time-dependent.

Example 12.5 Counts of road casualties

Table 12.5 gives the number of casualties in Great Britain due to road accidents on Fridays in 1986, for each hour of the day from midnight to midnight.

Department of Transport (1987) *Road accidents in Great Britain 1986: the casualty report.* HMSO, London. Table 28.

A plot of these data (frequency of casualties against the midpoint of the associated time interval) is given in Figure 12.6.

Table 12.5 Road casualties on Fridays

Time of day	Casualties
00–01	938
01–02	621
02–03	455
03–04	207
04–05	138
05–06	215
06–07	526
07–08	1933
08–09	3377
09–10	2045
10–11	2078
11–12	2351
12–13	3015
13–14	2966
14–15	2912
15–16	4305
16–17	4923
17–18	4427
18–19	3164
19–20	2950
20–21	2601
21–22	2420
22–23	2557
23–00	4319

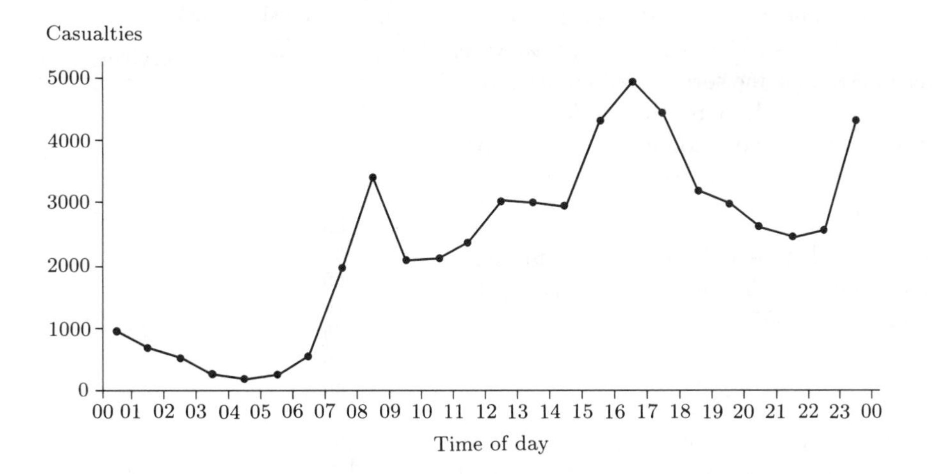

Figure 12.6 Hourly casualties on Fridays

There are two clear peaks of activity (one in the afternoon, particularly between 4 p.m. and 5 p.m., and another just before midnight). Possibly there is a peak in the early morning between 8 a.m. and 9 a.m. A formal analysis of these data would require some sort of probability model (not a Poisson process, which would provide a very bad fit) for the incidence of casualties over time. ■

There is a random process called the **simple birth process** used to model the size of growing populations. One of the assumptions of the model is that the birth rate remains constant with passing time; one of its consequences is that (apart from random variation) the size of the population increases exponentially, without limit. For most populations, therefore, this would be a useful model only in the earliest stages of population development: usually the birth rate would tail off as resources become more scarce and the population attains some kind of steady state.

One context where the simple birth process might provide a useful model is the duckweed data in *Chapter 10*, Table 10.7 and Figure 10.11. The question is: is the birth rate constant or are there signs of it tailing off with passing time? In the case of these data, we saw in Figure 10.22 that a logarithmic plot of population size against time was very suggestive of a straight line. At least over the first eight weeks of the development of the duckweed colony, there appears to be no reduction in the birth rate.

Similarly, some populations only degrade, and there are many variations on the random process known under the general term of **death process**.

Example 12.6 *Green sunfish*

A study was undertaken to investigate the resistance of green sunfish *Lepomis cyanellus* to various levels of thermal pollution. Warming of water occurs, for instance, at the seaside outflows of power stations: it is very important to discover the consequences of this for the local environment. Twenty fish were introduced into water heated at 39.7 ℃ and the numbers of survivors were recorded at pre-specified regular intervals. This followed a five-day acclimatization at 35 ℃. The data are given in Table 12.6. In fact, the experiment was repeated a number of times with water at different temperatures. The sort of question that might be asked is: how does mean survival time depend on water temperature? In order to answer this question, some sort of probability model for the declining population needs to be developed. Here, the random process $\{X_n; n = 0, 5, 10, \ldots\}$ gives the numbers of survivors counted at five-second intervals. ■

Matis, J.H. and Wehrly, T.E. (1979) Stochastic models of compartmental systems. *Biometrics*, **35**, 199–220.

Table 12.6 Survival times of green sunfish (seconds)

Time (seconds)	Survivors
0	20
⋮	⋮
45	20
50	19
55	19
60	18
65	17
70	14
75	13
80	12
85	11
90	8
95	7
100	6
105	6
110	4
115	3
120	3
125	1
130	1
135	0

Example 12.7 gives details of some very old data indeed! They come from Daniel Defoe's *Journal of the Plague Year* published in 1722, describing the outbreak of the bubonic plague in London in 1665.

Example 12.7 *The Great Plague*

Defoe used published bills of mortality as indicators of the progress of the epidemic. Up to the outbreak of the disease, the usual number of burials in London each week varied from about 240 to 300; at the end of 1664 and into 1665 it was noticed that the numbers were climbing into the 300s and during the week 17–24 January there were 474 burials. Defoe wrote: 'This last bill was really frightful, being a higher number than had been known to have been buried in one week since the preceding visitation [of the plague] of 1656.' The data in Table 12.7 give the weekly deaths from all diseases, as well as from the plague, during nine weeks from late summer into autumn 1665.

Table 12.7 Deaths in London, late 1665

Week	Of all diseases	Of the plague
Aug 8–Aug 15	5319	3880
Aug 15–Aug 22	5568	4237
Aug 22–Aug 29	7496	6102
Aug 29–Sep 5	8252	6988
Sep 5–Sep 12	7690	6544
Sep 12–Sep 19	8297	7165
Sep 19–Sep 26	6460	5533
Sep 26–Oct 3	5720	4929
Oct 3–Oct 10	5068	4327

The question arises: is there a peak of activity before which the epidemic may still be said to be taking off, and after which it may be said to be fading out?

In order for the timing of such a peak to be identified, a model for deaths due to the disease would need to be formulated. The diagram in Figure 12.7 informally identifies the peak for deaths due to the plague. However, there is a small diminution at Week 5, making conclusions not quite obvious. (For instance, it is possible that had the data been recorded midweek-to-midweek rather than weekend-to-weekend, then the numbers might tell a somewhat different story.)

The example is interesting because, at the time of writing, much current effort is devoted to Aids research, including the construction of useful models for the spread of the associated virus within and between communities. Different diseases require fundamentally different models to represent the different ways in which they may be spread. ■

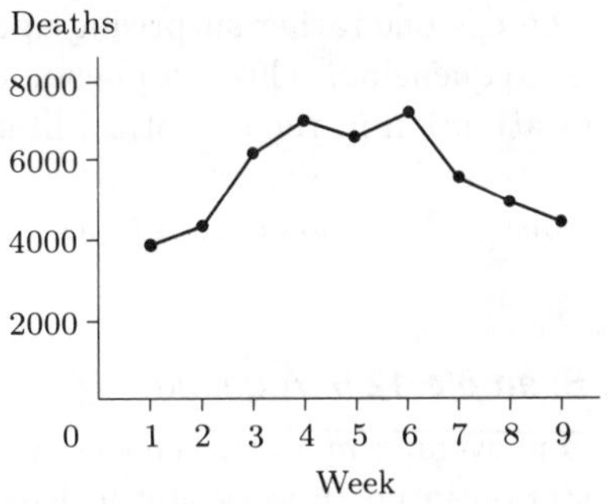

Figure 12.7 Weekly deaths due to the bubonic plague, London, 1665

For most populations, allowance has to be made in the model to reflect fluctuations in the population size due to births and deaths. Such models are called **birth-and-death processes**. Such a population is described in Example 12.8.

Example 12.8 Whooping cranes

The whooping crane (*Grus americana*) is a rare migratory bird which breeds in the north of Canada and winters in Texas. The data in Table 12.8 give annual counts of the number of whooping cranes arriving in Texas each autumn between 1938 and 1972.

Miller, R.S., Botkin, D.B. and Mendelssohn, R. (1974) The whooping crane population of North America. *Biological Conservation*, **6**, 106–111.

A plot of the fluctuating crane population $\{X(t); t = 1938, 1939, \ldots, 1972\}$ against time is given in Figure 12.8.

Table 12.8 Annual counts of the whooping crane, 1938–1972

14	22	26	15	19	21	18	17
25	31	30	34	31	25	21	24
21	28	24	26	32	33	36	38
32	33	42	44	43	48	50	56
57	56	51					

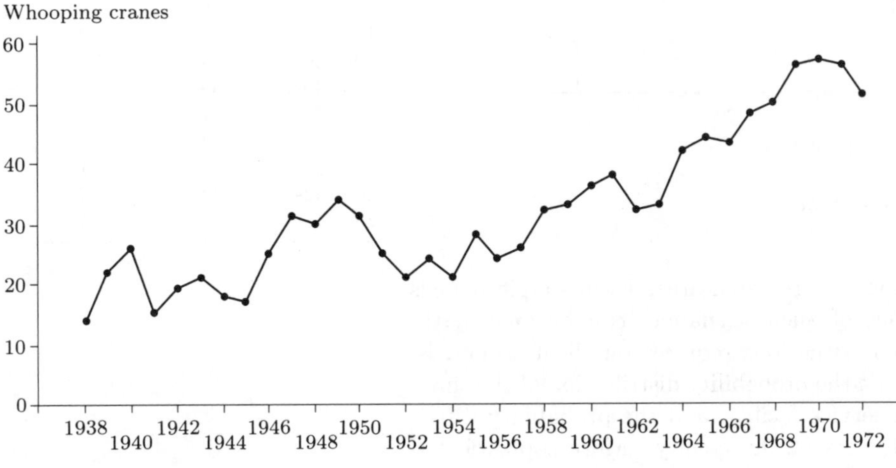

Figure 12.8 Annual counts of the whooping crane, 1938–1972

You can see from the plot that there is strong evidence of an increasing trend in the crane population. The data may be used to estimate the birth and death rates for a birth-and-death model, which may then be tested against the data. (Alternatively, perhaps, some sort of regression model might be attempted.) ■

There is one rather surprising application of the birth-and-death model: that is, to queueing. This is a phenomenon that has achieved an enormous amount of attention in the statistical literature. The approach is to think of arrivals as 'births' and departures as 'deaths' in an evolving population (which may consist of customers in a bank or at a shop, for instance).

Example 12.9 A queue

One evening in 1994, vehicles were observed arriving at and leaving a petrol station in north-west Milton Keynes, England. The station could accommodate a maximum of twelve cars, and at each service point a variety of fuels was available (leaded and unleaded petrol, and diesel fuel). When observation began there were three cars at the station. In Table 12.9 the symbol A denotes an arrival (a vehicle crossed the station entrance) and D denotes a departure (a vehicle crossed the exit line). Observation ceased after 30 minutes.

Data provided by F. Daly, The Open University.

The sequence of arrivals and departures is illustrated in Figure 12.9, where the random variable $X(t)$ (the number of vehicles in the station at time t) is plotted against passing time. The random process $\{X(t); t \geq 0\}$ is an integer-valued continuous-time random process.

Table 12.9 The number of cars at a petrol station

Time	Type	Number
0m 00s		3
2m 20s	A	4
4m 02s	D	3
4m 44s	D	2
6m 42s	D	1
7m 42s	A	2
8m 15s	D	1
10m 55s	A	2
11m 11s	A	3
14m 40s	D	2
15m 18s	A	3
16m 22s	A	4
16m 51s	A	5
19m 18s	D	4
20m 36s	D	3
20m 40s	D	2
21m 20s	A	3
22m 23s	D	2
22m 44s	D	1
24m 32s	D	0
28m 17s	A	1

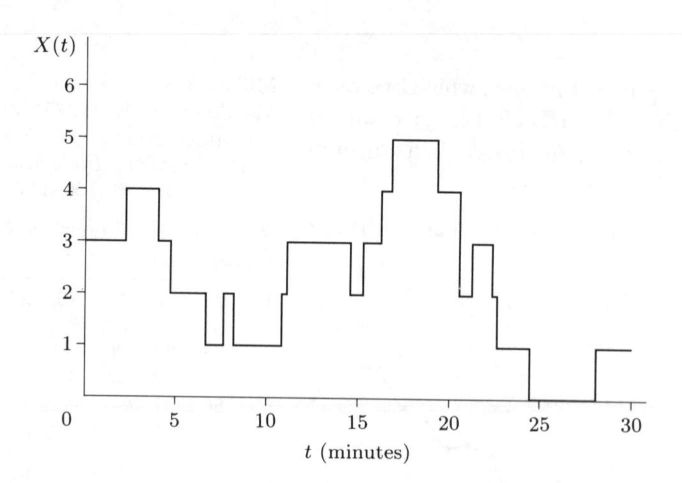

Figure 12.9 Customers at a service station

The study of queues is still an important research area. Even simple models representing only the most obvious of queue dynamics can become mathematically very awkward; and some situations require complicated models. Elements of a queueing model include the probability distribution of the times between successive arrivals at the service facility, and the probability distribution of the time spent by customers at the facility. More sophisticated models include, for instance, the possibility of 'baulking': potential customers are discouraged at the sight of a long queue and do not join it. ■

In Example 12.7 the data consisted of deaths due to the plague. It might have been interesting but not practicable to know the numbers of individuals afflicted with the disease at any one time. In modern times, this is very important—for instance, one needs to be able to forecast health demands and resources. The study of **epidemic models** is a very important part of the theory of random processes. The next example deals with a much smaller population.

Example 12.10 A smallpox epidemic

The data in Table 12.10 are from a smallpox outbreak in the community of Abakaliki in south-eastern Nigeria. There were 30 cases in a population of 120 individuals at risk. The data give the 29 inter-infection times (in days). A zero corresponds to cases appearing on the same day. Notice the long pause between the first and second reported cases.

Becker, N.G. (1983) Analysis of data from a single epidemic. *Australian Journal of Statistics*, **25**, 191–197.

Table 12.10 Waiting times between 30 cases of smallpox (days)

13	7	2	3	0	0	1	4
5	3	2	0	2	0	5	3
1	4	0	1	1	1	2	0
1	5	0	5	5			

A plot showing the progress of the disease is given in Figure 12.10.

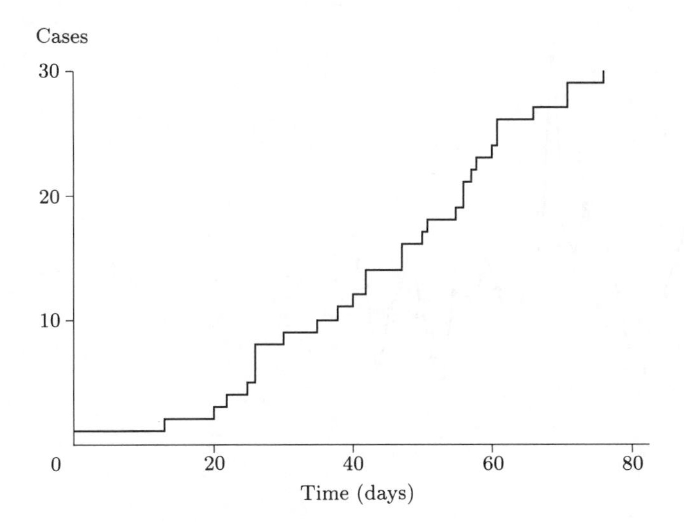

Figure 12.10 New cases of smallpox

Here, the numbers are very much smaller than they were for the plague data in Example 12.7. One feature that is common to many sorts of epidemic is that they start off slowly (there are many susceptible individuals, but rather few carriers of the disease) and end slowly (when there are many infected persons, but few survivors are available to catch the disease); the epidemic activity peaks when there are relatively large numbers of both infectives and susceptibles. If this is so, then the model must reflect this feature. Of course, different models will be required for different diseases, because different diseases often have different transmission mechanisms. ■

Here 'infectives' are those who have and who are capable of spreading the disease, and 'susceptibles' are those who have not yet contracted the disease.

Regular observations are often taken on some measure (such as unemployment, interest rates, prices) in order to identify trends or cycles. The techniques of **time series analysis** are frequently applied to such data.

Example 12.11 Sales of jeans

Data were collected on monthly sales of jeans as part of an exercise in forecasting future sales. From the data, is there evidence of trend (a steady increase or decrease in demand) or of seasonal variation (sales varying with

Conrad, S. (1989) *Assignments in Applied Statistics.* John Wiley and Sons, Chichester, p. 78.

the time of year)? If trend or seasonal variation is identified and removed, how might the remaining variation be modelled?

Table 12.11 Monthly sales of jeans in the UK, 1980–1985 (thousands)

Year	Jan	Feb	Mar	Apr	May	Jun	Jul	Aug	Sep	Oct	Nov	Dec
1980	1998	1968	1937	1827	2027	2286	2484	2266	2107	1690	1808	1927
1981	1924	1959	1889	1819	1824	1979	1919	1845	1801	1799	1952	1956
1982	1969	2044	2100	2103	2110	2375	2030	1744	1699	1591	1770	1950
1983	2149	2200	2294	2146	2241	2369	2251	2126	2000	1759	1947	2135
1984	2319	2352	2476	2296	2400	3126	2304	2190	2121	2032	2161	2289
1985	2137	2130	2154	1831	1899	2117	2266	2176	2089	1817	2162	2267

These data may be summarized in a time series diagram, as shown in Figure 12.11.

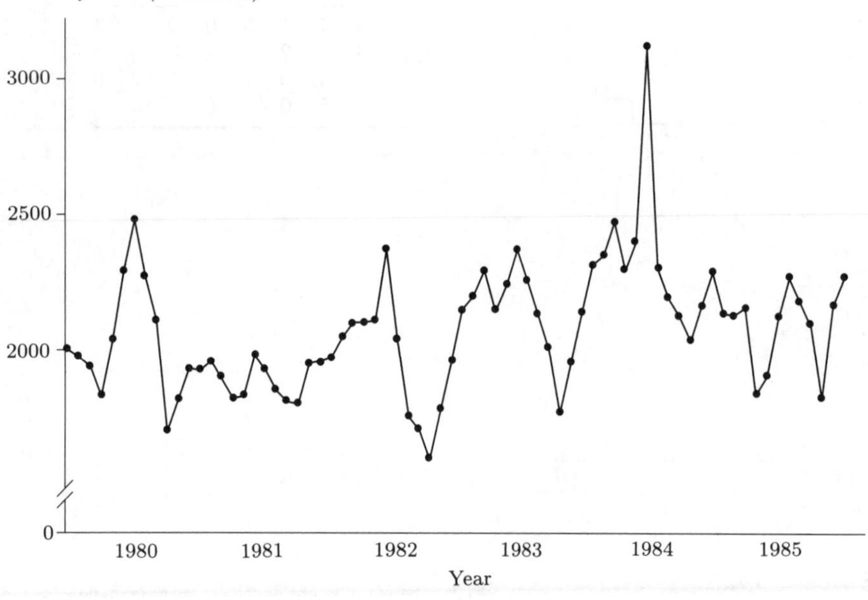

Figure 12.11 Monthly sales of jeans in the UK, 1980–1985 (thousands)

Here, a satisfactory model might be achieved simply by superimposing random normal variation $N(0, \sigma^2)$ on trend and seasonal components (if any). Then the trend component, the seasonal component and the value of the parameter σ would need to be estimated from the data. ■

Meteorology is another common context for time series data.

Example 12.12 Monthly temperatures

Average air temperatures were recorded every month for twenty years at Nottingham Castle, England. This example is similar to Example 12.11; but the time series diagram in Figure 12.12 exhibits only seasonal variation and no trend. (One might be worried if in fact there was any trend evident!)

Anderson, O.D. (1976) *Time series analysis and forecasting: the Box–Jenkins approach.* Butterworths, London and Boston, p. 166.

Table 12.12 Average monthly temperatures at Nottingham Castle, England, 1920–1939 (°F)

Year	Jan	Feb	Mar	Apr	May	Jun	Jul	Aug	Sep	Oct	Nov	Dec
1920	40.6	40.8	44.4	46.7	54.1	58.5	57.7	56.4	54.3	50.5	42.9	39.8
1921	44.2	39.8	45.1	47.0	54.1	58.7	66.3	59.9	57.0	54.2	39.7	42.8
1922	37.5	38.7	39.5	42.1	55.7	57.8	56.8	54.3	54.3	47.1	41.8	41.7
1923	41.8	40.1	42.9	45.8	49.2	52.7	64.2	59.6	54.4	49.2	36.3	37.6
1924	39.3	37.5	38.3	45.5	53.2	57.7	60.8	58.2	56.4	49.8	44.4	43.6
1925	40.0	40.5	40.8	45.1	53.8	59.4	63.5	61.0	53.0	50.0	38.1	36.3
1926	39.2	43.4	43.4	48.9	50.6	56.8	62.5	62.0	57.5	46.7	41.6	39.8
1927	39.4	38.5	45.3	47.1	51.7	55.0	60.4	60.5	54.7	50.3	42.3	35.2
1928	40.8	41.1	42.8	47.3	50.9	56.4	62.2	60.5	55.4	50.2	43.0	37.3
1929	34.8	31.3	41.0	43.9	53.1	56.9	62.5	60.3	59.8	49.2	42.9	41.9
1930	41.6	37.1	41.2	46.9	51.2	60.4	60.1	61.6	57.0	50.9	43.0	38.8
1931	37.1	38.4	38.4	46.5	53.5	58.4	60.6	58.2	53.8	46.6	45.5	40.6
1932	42.4	38.4	40.3	44.6	50.9	57.0	62.1	63.5	56.3	47.3	43.6	41.8
1933	36.2	39.3	44.5	48.7	54.2	60.8	65.5	64.9	60.1	50.2	42.1	35.8
1934	39.4	38.2	40.4	46.9	53.4	59.6	66.5	60.4	59.2	51.2	42.8	45.8
1935	40.0	42.6	43.5	47.1	50.0	60.5	64.6	64.0	56.8	48.6	44.2	36.4
1936	37.3	35.0	44.0	43.9	52.7	58.6	60.0	61.1	58.1	49.6	41.6	41.3
1937	40.8	41.0	38.4	47.4	54.1	58.6	61.4	61.8	56.3	50.9	41.4	37.1
1938	42.1	41.2	47.3	46.6	52.4	59.0	59.6	60.4	57.0	50.7	47.8	39.2
1939	39.4	40.9	42.4	47.8	52.4	58.0	60.7	61.8	58.2	46.7	46.6	37.8

The time series diagram in Figure 12.12 shows the seasonal variation quite clearly.

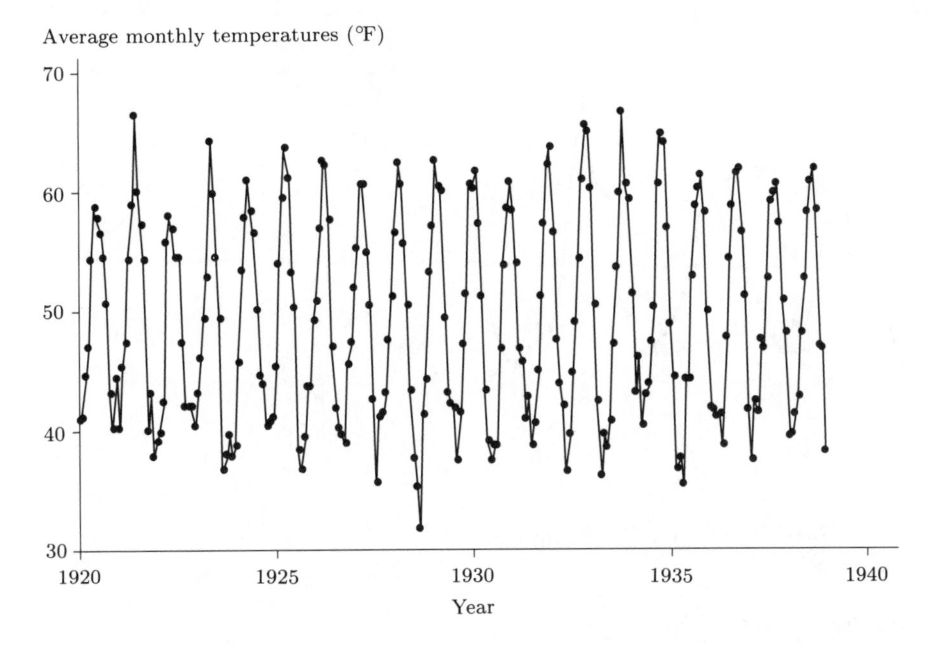

Figure 12.12 Average monthly temperatures at Nottingham Castle, England, 1920–1939 (°F)

Here, it is possible that superimposing random normal variation on seasonal components would not be enough. There is likely to be substantial correlation between measurements taken in successive months, and a good model ought to reflect this. ■

Example 12.13 is an example where one might look for trend as well as seasonal features in the data.

Example 12.13 *Death statistics, lung diseases*

The data in Table 12.13 list monthly deaths, for both sexes, from bronchitis, emphysema and asthma in the United Kingdom between 1974 and 1979. The sorts of research questions that might be posed are: what can be said about the way the numbers of deaths depend upon the time of year (seasonal variation)? What changes are evident, if any, between 1974 and 1979 (trend)?

Diggle, P.J. (1990) *Time series: a biostatistical introduction.* Oxford University Press, Oxford.

Table 12.13 Monthly deaths from lung diseases in the UK, 1974–1979

Year	Jan	Feb	Mar	Apr	May	Jun	Jul	Aug	Sep	Oct	Nov	Dec
1974	3035	2552	2704	2554	2014	1655	1721	1524	1596	2074	2199	2512
1975	2933	2889	2938	2497	1870	1726	1607	1545	1396	1787	2076	2837
1976	2787	3891	3179	2011	1636	1580	1489	1300	1356	1653	2013	2823
1977	2996	2523	2540	2520	1994	1641	1691	1479	1596	1877	2032	2484
1978	2899	2990	2890	2379	1933	1734	1617	1495	1440	1777	1970	2745
1979	2841	3535	3010	2091	1667	1589	1518	1349	1392	1619	1954	2633

The diagram in Figure 12.13 exhibits some seasonal variation but there is little evidence of substantial differences between the start and end of the series. In order to formally test hypotheses of no change, it would be necessary to formulate a mathematical model for the way in which the data are generated.

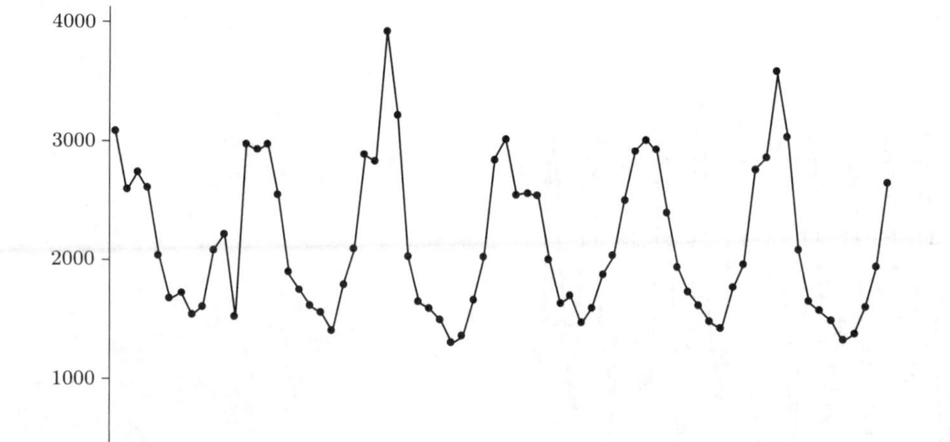

Figure 12.13 Monthly deaths from lung diseases in the UK, 1974–1979

Interestingly, most months show a decrease in the counts from 1974 to 1979. However, February shows a considerable rise. Without a model, it is not easy to determine whether or not these differences are significant. ■

Often time series exhibit cycles that are longer than a year, and for which there is not always a clear reason. A famous set of data is the so-called 'lynx data', much explored by statisticians. The data are listed in Table 12.14.

Example 12.14 The lynx data

The data in Table 12.14 were taken from the records of the Hudson's Bay Company. They give the annual number of Canadian lynx trapped in the Mackenzie River district of north-west Canada between 1821 and 1934.

Elton, C. and Nicholson, M. (1942) The ten-year cycle in numbers of the lynx in Canada. *J. Animal Ecology*, **11**, 215–244.

Table 12.14 Lynx trappings, 1821–1934

269	321	585	871	1475	2821	3928	5943	4950	2577
523	98	184	279	409	2285	2685	3409	1824	409
151	45	68	213	546	1033	2129	2536	957	361
377	225	360	731	1638	2725	2871	2119	684	299
236	245	552	1623	3311	6721	4254	687	255	473
358	784	1594	1676	2251	1426	756	299	201	229
469	736	2042	2811	4431	2511	389	73	39	49
59	188	377	1292	4031	3495	587	105	153	387
758	1307	3465	6991	6313	3794	1836	345	382	808
1388	2713	3800	3091	2985	3790	374	81	80	108
229	399	1132	2432	3574	2935	1537	529	485	662
1000	1590	2657	3396						

The counts are shown in Figure 12.14.

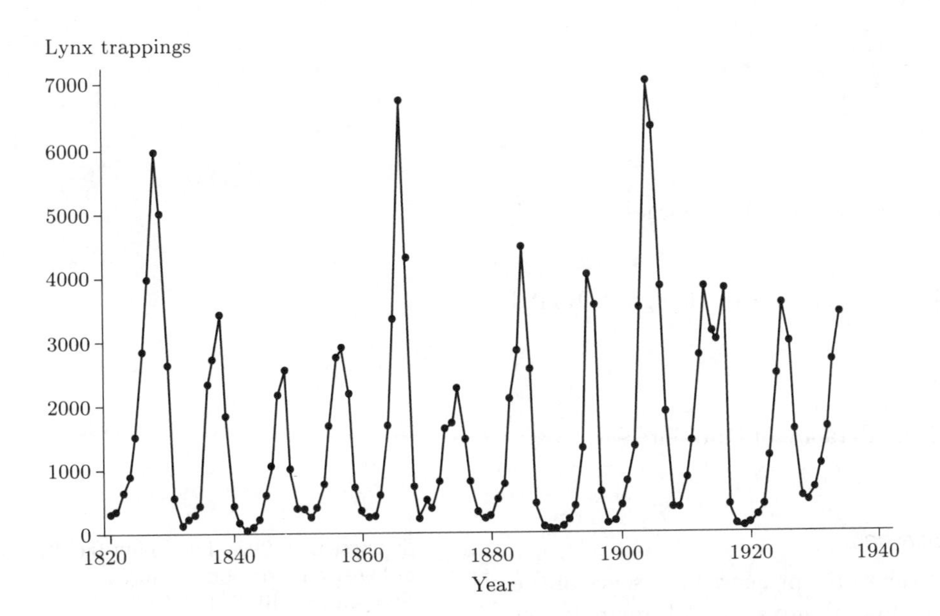

Figure 12.14 Lynx trappings, 1821–1934

There is evidence of a ten-year cycle in the population counts. Can you see any indications of a 40-year cycle? If the data set were more extended, this could be explored. ■

Recently researchers have started exploring meteorological data for evidence of long-term climatic changes. The data in Table 1.12 on annual snowfall in Buffalo, NY (USA) are an example. The 30-year period of the data in Example 12.15 is relatively brief for such purposes.

Example 12.15 March rainfall

These are data on 30 successive values of March precipitation (in inches) for Minneapolis St Paul. Again, to determine whether there are significant changes with time, one approach would be to formulate a model consistent with there being no fundamental change, and test its fit to the data.

Hinkley, D. (1977) On quick choice of power transformation. *Applied Statistics*, **26**, 67–69.

Table 12.15 March precipitation (inches)

0.77	1.74	0.81	1.20	1.95	1.20	0.47	1.43	3.37	2.20
3.00	3.09	1.51	2.10	0.52	1.62	1.31	0.32	0.59	0.81
2.81	1.87	1.18	1.35	4.75	2.48	0.96	1.89	0.90	2.05

The data are shown in Figure 12.15.

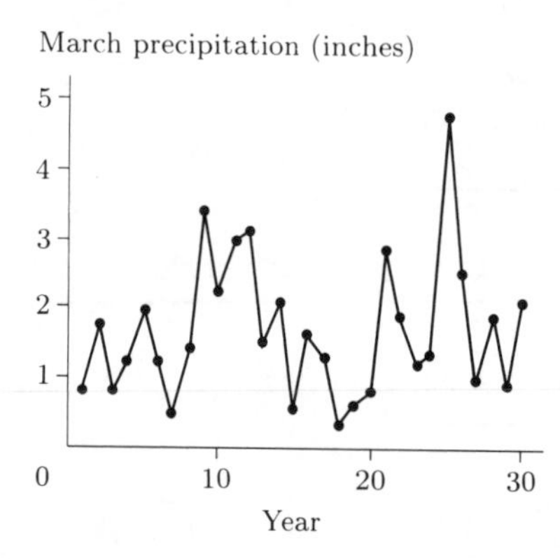

Figure 12.15 March precipitation, Minneapolis St Paul (inches)

Over this 30-year period there is little evidence of change in rainfall patterns. ■

Finally, collections of two-dimensional data are often of interest.

Example 12.16 Presence-absence data

The data illustrated in Figure 12.16 show the presence or absence of the plant *Carex arenaria* over a spatial region divided into a 24×24 square lattice. The data may be examined for evidence of clustering or competition between the plants. The plants appear to be less densely populated at the lower left of the region than elsewhere; however, for a formal test of the nature of plant coverage, a model would have to be constructed. ■

Strauss, D. (1992) The many faces of logistic regression. *American Statistician*, **46**, 321–327.

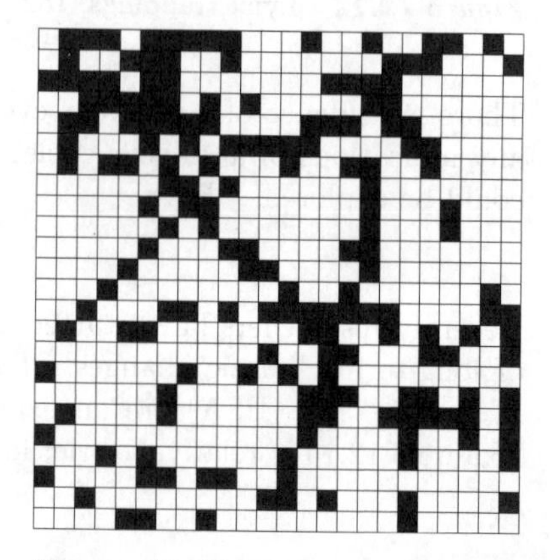

Figure 12.16

In this section a large number of different data sets have been introduced and some modelling approaches have been indicated: unfortunately, it would take another book as long as this one to describe the models and their uses in any detail! For the rest of this chapter, we shall look again at the Bernoulli and Poisson processes, and a new model, the *Markov chain*, will be introduced. The Markov chain is useful under some circumstances for modelling sequences of successes and failures, where the Bernoulli process provides an inadequate model.

12.2 The Bernoulli process as an inadequate model

The Bernoulli process was defined in *Chapter 4*. For ease, the definition is repeated here.

See page 143.

> **The Bernoulli process**
>
> A **Bernoulli process** is the name given to a *sequence of Bernoulli trials* in which
>
> (a) trials are independent;
>
> (b) the probability of success is the same from trial to trial.

Notice the notion of *ordering* here: it is crucial to the development of the ideas to follow that trials take place sequentially one after the other. Here, in Example 12.17, are two typical realizations of a Bernoulli process.

Example 12.17 Two realizations of a Bernoulli process

Each realization consists of 40 trials.

1 1 1 1 0 1 0 1 1 0 1 1 1 1 0 0 0 0 1 1 0 1 0 0 0 1 1 1 1 1 0 0 1 1 1 0 0 1 1 0

0 0 1 0 0 0 1 1 0 1 1 1 1 0 1 1 0 0 1 1 1 0 1 1 1 1 1 1 1 0 1 1 1 1 1 0 1 0 1 0

Each realization is the result of a computer simulation. In both cases the indexing parameter p for the Bernoulli process (the probability of obtaining a 1) was set equal to 0.7. ■

The Bernoulli process is an integer-valued random process evolving in discrete time. It provides a useful model for many practical contexts. However, there are several ways in which a random process, consisting of a sequence of Bernoulli trials (that is, a sequence of trials in each of which the result might be success or failure), does not constitute a Bernoulli process. For instance, the probability of success might alter from trial to trial. A situation where this is the case is described in Example 12.18.

Example 12.18 The collector's problem

Packets of breakfast cereals sometimes contain free toys, as part of an advertising promotion designed to appeal primarily to children. Often the toys form part of a family, or collection. The desire on the part of the children to complete the collection encourages purchases of the cereal.

Every time a new packet of cereal is purchased, the toy that it contains will be either new to the consumer's collection, or a duplicate. If each purchase of a new box of cereal is considered as the next in a sequence of trials, then the acquisition of a new toy may be counted as a success, while the acquisition of a duplicate is a failure. Thus each successive purchase may be regarded as a Bernoulli trial.

However, at the beginning of the promotion, the first purchase is bound to produce a new toy, for it will be the first of the collection: so, at trial number 1, the Bernoulli parameter p is equal to 1. If the number of toys in a completed collection is (say) 8, then at this stage the collector still lacks 7 toys. Assuming that no toy is any more likely to occur than any other in any given packet of cereal, then the probability of success at trial number 2 is $p = \frac{7}{8}$ (and the probability of failure is $q = \frac{1}{8}$, corresponding to the case where the second cereal packet contains the same toy as the first). The success probability $p = \frac{7}{8}$ will remain constant at all subsequent trials until a new toy is added to the collection; then there will be 6 toys missing, and the probability of success at subsequent trials changes again, from $\frac{7}{8}$ to $p = \frac{6}{8} = \frac{3}{4}$.

The assumption that no toy is any more likely to occur than any other requires that there should be no geographical bias—that the local shop does not have an unusually high incidence of toys of one type—and that no toy is deliberately produced in small numbers in order to make it a rarity.

The process continues (assuming the promotion continues long enough) until the one remaining toy of the 8 is acquired (with probability $p = \frac{1}{8}$ at each trial). Here are two realizations of the collection process. Each trial represents a purchase; a 1 indicates the acquisition of a new toy to the collection, a 0 indicates a duplicate. Necessarily, the process begins with a 1.

1 1 1 0 0 1 0 1 1 0 0 0 0 0 1 1

1 1 1 1 0 1 1 1 0 0 0 1

Actually, each of these realizations is somewhat atypical. In the first case the final toy in the collection was obtained immediately after the seventh, though in general it is likely that several duplicates would be obtained at this point. In the second case the collection was completed remarkably quickly, with seven toys of the collection acquired within the first eight purchases. In the first case it was necessary to make sixteen purchases to complete the collection; in the second case, twelve purchases were necessary. It can be shown that the expected number of purchases is 21.7. In the context of cereal purchase, this is a very large number! In practice, if the collection is sufficiently desirable, children usually increase the rate of new acquisitions by swapping their duplicates with those of others, and add to their collection in that way. ■

The derivation of this result is slightly involved.

Another example where trials remain independent but where the probability of success alters is where the value of p depends on the number of the trial.

Example 12.19 Rainy days

In *Chapter 4*, Example 4.1, where the incidence of rainy days was discussed, the remark was made that a Bernoulli process was possibly not a very good model for rain day to day, and that '... a better model would probably include some sort of mechanism that allows the parameter p to vary with the time of the year'. (That is, that the value of p, the probability of rain on any given day in the year, should vary as n, the number of the day, varies from 1 to 365.) It would not be entirely straightforward to express the dependence of p on n in mathematical terms, but the principle is clear. ■

In fact, rainfall and its incidence day to day has been studied probabilistically in some detail, as we shall see in Example 12.20.

There is one very common model for dependence from trial to trial, and that is where the probability of success at any trial depends on what occurred at the trial immediately preceding it. Here are two examples where this sort of model has proved useful in practice.

Example 12.20 The Tel Aviv rainfall study

An alternative rainfall model to that described in Example 12.19 is one where the probability of rain on any given day depends on what occurred the day before. An extensive study was carried out on the weather in Tel Aviv and the following conclusions were reached.

Gabriel, K.R. and Neumann, J. (1962) A Markov chain model for daily rainfall occurrence at Tel Aviv. *Quarterly Journal of the Royal Meteorological Society*, **88**, 90–95.

Successive days were classified as wet (outcome 1, a success) or dry (outcome 0, a failure), according to some definition. After analysing weather patterns over many days it was concluded that the probability of a day being wet, given that the previous day was wet, is 0.662; in contrast to this, the probability of a day being wet, given that the preceding day was dry, is only 0.250.

(It follows that the probability that a day is dry, given that the preceding day was dry, is $1 - 0.250 = 0.750$; and the probability that a day is dry, given the preceding day was wet, is $1 - 0.662 = 0.338$.)

We shall need a small amount of mathematical notation at this point. Denoting by the random variable X_j the weather outcome on day j, the findings of the study may be written formally as

$$P(X_{j+1} = 1|X_j = 1) = 0.662; \quad P(X_{j+1} = 1|X_j = 0) = 0.250. \qquad (12.1)$$

Remember the vertical slash notation '|' from *Chapter 11*: it is read 'given that'.

These probabilities are called **transition probabilities**. Notice that they are quite different: the weather of the day before has a substantial effect on the chances of rain on the following day. There is, or appears to be, a very strong *association* between the weather on consecutive days. For a Bernoulli process, these probabilities would be the same, equal to the parameter p—for in a Bernoulli process, success at any trial is independent of the history of the process.

The random process $\{X_j, j = 1, 2, 3, \ldots\}$ gives the sequence of wet and dry days in Tel Aviv. Any actual day-to-day data records are *realizations* of that random process. ■

Starting with a dry day on Day 1 (so $X_1 = 0$) here is a sample realization of 40 days' weather in Tel Aviv, using the probabilities given in Example 12.20. The realization was generated using a computer.

$$0000111001101011000000000111111001110000 0 \qquad (12.2)$$

You can probably appreciate that, if you did not know, it is not entirely easy to judge whether or not this sequence of trial results was generated as a Bernoulli process. In Section 12.3 we shall see some sample realizations which clearly break the assumptions of the Bernoulli process, and learn how to test the adequacy of the Bernoulli model for a given realization.

Example 12.21 provides a second context where this kind of dependence proves a more useful model than a simple Bernoulli process.

Example 12.21 The sex of children in a family

In *Chapter 3*, Example 3.13 a Bernoulli process was suggested as a useful model for the sequence of boys and girls in a family, and the value $p = 0.514$ was put forward for the probability of a boy. However, it was almost immediately remarked that '... nearly all the very considerable mass of data collected on the sex of children in families suggests that the Bernoulli model would be a very bad model to adopt ... some statistical analyses seem to show that the independence assumption of the Bernoulli model breaks down: that is, that Nature has a kind of memory, and the sex of a previous child affects to some degree the probability distribution for the sex of a subsequent child.'

One model has been developed in which transition probabilities are given for boy–boy and girl–boy sequences. The estimated probabilities (writing 1 for a boy and 0 for a girl) are (for one large data set)

$$P(X_{j+1} = 1|X_j = 1) = 0.4993; \quad P(X_{j+1} = 1|X_j = 0) = 0.5433. \qquad (12.3)$$

Edwards, A.W.F. (1960) The meaning of binomial distribution. *Nature*, **186**, 1074. For the numbers, see Daly, F. (1990) The probability of a boy: sharpening the binomial. *Mathematical Scientist*, **15**, 114–123.

These probabilities are quite close; their relative values are very interesting. A preceding girl makes a boy more likely, and vice versa. If Nature has a memory, then it can be said to operate to this effect: in the interests of achieving a balance, Nature tries to avoid long runs of either boys or girls.

In that respect, Example 12.21 is different from Example 12.20: there, wet weather leads to more wet weather, and dry weather leads to more dry weather. ■

This section ends with two brief exercises.

Exercise 12.1

For each of the following contexts, say whether you think a Bernoulli process might provide an adequate model:

(a) the order of males and females in a queue;

(b) the sequence of warm and cool days at one location;

(c) a gambler's luck (whether he wins or loses) at successive card games.

You have already seen at (12.2) a typical realization of Tel Aviv weather day to day. Exercise 12.2 involves computer simulation. You will need to understand how your computer can be made to mimic the outcome of a sequence of Bernoulli trials with altering probabilities.

Exercise 12.2

Assuming in both cases that the corresponding model is an adequate representation for the underlying probability mechanisms, use the transition probabilities given in Example 12.20 and Example 12.21 to simulate

(a) a week's weather in Tel Aviv, assuming the first day was wet;

(b) the sequence of boys and girls in a family of four children, assuming the oldest child was a girl.

Development of this model is continued in Section 12.3, where methods are explored for parameter estimation, and for testing the adequacy (or otherwise) of the fit of a Bernoulli model to data.

12.3 Markov chains

12.3.1 Definitions

The probability model for a sequence of Bernoulli trials in which the success probability at any one trial depends on the outcome of the immediately preceding trial is called a *Markov chain.*

Markov chains are named after the prominent Russian mathematician A.A. Markov (1856–1922). The term *Markov chain* actually covers rather a wide class of models, but the terminology will suffice for the process described here. In general, more than two states (0 and 1) are possible; and what has been called a Markov chain here is more precisely known as a 'two-state Markov chain'.

Markov chains

A **Markov chain** is a random process

$$\{X_n; n = 1, 2, 3, \ldots\}$$

where

(a) the random variables X_n are integer-valued (0 or 1);

(b) the transition probabilities defining the process are given by

$$P(X_{j+1} = 0|X_j = 0) = 1 - \alpha, \qquad P(X_{j+1} = 1|X_j = 0) = \alpha,$$
$$P(X_{j+1} = 0|X_j = 1) = \beta, \qquad P(X_{j+1} = 1|X_j = 1) = 1 - \beta;$$

(c) $P(X_1 = 1) = p_1$.

The transition probabilities may be conveniently written in the form of a square array as

$$M = \begin{matrix} 0 \\ 1 \end{matrix} \begin{bmatrix} 1-\alpha & \alpha \\ \beta & 1-\beta \end{bmatrix}$$

where M is called the **transition matrix**.

We need to know the value of X_1 before we can use the transition probabilities to 'drive' the rest of the process.

This particular identification of the transition probabilities may appear to be a little eccentric to you, but it turns out to be convenient.

Notice the labelling of the rows of M; it is helpful to be reminded of which probabilities correspond to which of the two outcomes, 0 and 1. A more extended notation is sometimes used: M can be written

$$M = \begin{matrix} \\ 0 \\ 1 \end{matrix} \begin{matrix} \begin{matrix} 0 & \quad 1 \end{matrix} \\ \begin{bmatrix} 1-\alpha & \alpha \\ \beta & 1-\beta \end{bmatrix} \end{matrix}.$$

with the row labels indicating the 'current' state X_j and the column labels the 'next' state X_{j+1}, respectively. In this course we shall use the less elaborate representation, that is,

$$M = \begin{matrix} 0 \\ 1 \end{matrix} \begin{bmatrix} 1-\alpha & \alpha \\ \beta & 1-\beta \end{bmatrix}.$$

Over a long period of observation, any realization of a Markov chain will exhibit a number of 0s and 1s. It is possible to show that in the long term, and with the transition probabilities given, the average proportions of 0s and 1s in a long realization will be

$$q = \frac{\beta}{\alpha + \beta}, \qquad p = \frac{\alpha}{\alpha + \beta}$$

respectively. This result is not complicated to show, but it is notationally awkward, and you need not worry about the details. A 'typical' realization of a Markov chain may therefore be achieved by taking the starting success probability for X_1, p_1, as

$$p_1 = \frac{\alpha}{\alpha + \beta}.$$

It is useful to give a name to such chains: they are called *stationary*.

A Markov chain $\{X_n, n = 1, 2, 3, \ldots\}$ with transition probability matrix

$$M = \begin{matrix} 0 \\ 1 \end{matrix} \begin{bmatrix} 1-\alpha & \alpha \\ \beta & 1-\beta \end{bmatrix}$$

and where the probability distribution of X_1 is given by

$$P(X_1 = 0) = \frac{\beta}{\alpha + \beta}, \qquad P(X_1 = 1) = \frac{\alpha}{\alpha + \beta} \tag{12.4}$$

is said to be **stationary**.

Example 12.20 continued

For the Tel Aviv rainfall study, the transition matrix M is written

$$M = \begin{matrix} 0 \\ 1 \end{matrix} \begin{bmatrix} 0.750 & 0.250 \\ 0.338 & 0.662 \end{bmatrix}$$

(where 0 indicates a dry day, 1 a wet day). In this sort of situation you might find it clearer to write

$$M = \begin{matrix} \text{Dry} \\ \text{Wet} \end{matrix} \begin{bmatrix} 0.750 & 0.250 \\ 0.338 & 0.662 \end{bmatrix}.$$

Based on these transition probabilities, the underlying proportion of wet days in Tel Aviv is given (using (12.4)) by

$$p = \frac{\alpha}{\alpha + \beta} = \frac{0.250}{0.250 + 0.338} = 0.425.$$

You could simulate a typical week's weather in Tel Aviv by taking your first day's weather from the Bernoulli distribution $X_1 \sim \text{Bernoulli}(0.425)$ and simulating the next six days' weather using M. ■

Exercise 12.3

(a) Write down the transition matrix M for sequences of girls and boys in a family using the figures given at (12.3) in Example 12.21, and estimate the underlying proportion of boys.

(b) Simulate a typical family of five children.

Finally, you should note that a Bernoulli process is a special case of a stationary Markov chain, with transition matrix

$$M = \begin{matrix} 0 \\ 1 \end{matrix} \begin{bmatrix} q & p \\ q & p \end{bmatrix}.$$

So the two rows of transition probabilities are the same.

12.3.2 Estimating transition probabilities

A question that now arises is: how do we estimate the transition matrix M from a given realization (assuming a Markov chain provides an adequate model)?

Assume that the given realization includes the outcomes of a total of $n+1$ trials; this is for ease because then there are n transitions. We need to count the frequency of each of the four possible types of transition: two types denote no change, 00 and 11, and two types denote a switch, 01 and 10.

Let us denote by n_{00} the number of 00 transitions in the sequence, by n_{01} the number of 01 transitions, by n_{10} the number of 10 transitions and by n_{11} the number of 11 transitions. (Then you will find that

$$n_{00} + n_{01} + n_{10} + n_{11} = n$$

—a useful check on your tally.) If you write these totals in matrix form as

$$N = \begin{matrix} 0 \\ 1 \end{matrix} \begin{bmatrix} n_{00} & n_{01} \\ n_{10} & n_{11} \end{bmatrix}$$

and then write in the row totals

$$N = \begin{matrix} 0 \\ 1 \end{matrix} \begin{bmatrix} n_{00} & n_{01} \\ n_{10} & n_{11} \end{bmatrix} \begin{matrix} n_{00}+n_{01} \\ n_{10}+n_{11} \end{matrix}$$

(simply to aid your arithmetic) then the estimate $\widehat{M}$ of the transition matrix M is given by

$$\widehat{M} = \begin{matrix} 0 \\ 1 \end{matrix} \begin{bmatrix} n_{00}/(n_{00}+n_{01}) & n_{01}/(n_{00}+n_{01}) \\ n_{10}/(n_{10}+n_{11}) & n_{11}/(n_{10}+n_{11}) \end{bmatrix}. \qquad (12.5)$$

It turns out that the elements of $\widehat{M}$ are the maximum likelihood estimates of the transition probabilities.

Example 12.20 continued

The original data source shows that the Tel Aviv data included altogether 2438 days (so there were 2437 transitions), with the matrix of transition frequencies given by

In fact, the data collection was not quite as simple as is suggested here, but the principles of the estimation procedure should be clear.

$$N = \begin{matrix} 0 \\ 1 \end{matrix} \begin{bmatrix} 1049 & 350 \\ 351 & 687 \end{bmatrix} \begin{matrix} 1399 \\ 1038 \end{matrix}$$

and so

$$\widehat{M} = \begin{matrix} 0 \\ 1 \end{matrix} \begin{bmatrix} 1049/1399 & 350/1399 \\ 351/1038 & 687/1038 \end{bmatrix} = \begin{matrix} 0 \\ 1 \end{matrix} \begin{bmatrix} 0.750 & 0.250 \\ 0.338 & 0.662 \end{bmatrix};$$

these estimates are the transition probabilities that have been used in all our work. ■

Example 12.22 *Counting the transitions*

For the sequence of simulated Tel Aviv weather described at (12.2) and repeated here,

0 0 0 0 1 1 1 0 0 1 1 0 1 0 1 1 0 0 0 0 0 0 0 0 1 1 1 1 1 1 0 0 1 1 1 0 0 0 0 0

the matrix of transition frequencies is given by

$$N = \begin{matrix} 0 \\ 1 \end{matrix} \begin{bmatrix} n_{00} & n_{01} \\ n_{10} & n_{11} \end{bmatrix} \begin{matrix} n_{00}+n_{01} \\ n_{10}+n_{11} \end{matrix} = \begin{matrix} 0 \\ 1 \end{matrix} \begin{bmatrix} 16 & 6 \\ 6 & 11 \end{bmatrix} \begin{matrix} 22 \\ 17 \end{matrix}.$$

(Notice that $n_{00} + n_{01} + n_{10} + n_{11} = 16 + 6 + 6 + 11 = 39$, one less than the number of trials.) The corresponding estimated transition matrix $\widehat{M}$ is given by

$$\widehat{M} = \begin{matrix} 0 \\ 1 \end{matrix} \begin{bmatrix} 16/22 & 6/22 \\ 6/17 & 11/17 \end{bmatrix} = \begin{matrix} 0 \\ 1 \end{matrix} \begin{bmatrix} 0.727 & 0.273 \\ 0.353 & 0.647 \end{bmatrix}.$$

You can compare this estimate with the matrix of transition probabilities used to drive the simulation, given by the probabilities at (12.1):

$$M = \begin{matrix} 0 \\ 1 \end{matrix} \begin{bmatrix} 0.750 & 0.250 \\ 0.338 & 0.662 \end{bmatrix}.$$

The estimate is quite good. ■

The following exercise is about estimating transition probabilities.

Exercise 12.4

For the following realizations of Markov chains, find the matrix of transition frequencies N, and hence calculate an estimate $\widehat{M}$ of the transition matrix M.

(a) 1111011100110010000001111011001010011 10

(b) 0000100110101011001000010011101000000100

(c) 1110000000000111111111111111111011110000

(d) 0000010000000000011111110000000100000011

(e) 1010010101010101010101010101001100000000100

(f) 1000101010100100001001000100010100110101

Give your estimated probabilities to three decimal places in each case.

In fact, realizations (a) and (b) in Exercise 12.4 are those of a Bernoulli process with parameter $p = 0.45$—equivalently, of a Markov chain with transition matrix

$$M = \begin{matrix} 0 \\ 1 \end{matrix} \begin{bmatrix} 0.55 & 0.45 \\ 0.55 & 0.45 \end{bmatrix}.$$

Realizations (c) and (d) were each driven by the transition matrix

$$M = \begin{matrix} 0 \\ 1 \end{matrix} \begin{bmatrix} 0.8 & 0.2 \\ 0.3 & 0.7 \end{bmatrix}.$$

Here, a 0 is very likely to be followed by another 0, and a 1 is likely to be followed by another 1. As you can see, the sample realizations exhibit long runs of 0s and 1s.

Finally, realizations (e) and (f) were generated by the transition matrix

$$M = \begin{matrix} 0 \\ 1 \end{matrix} \begin{bmatrix} 0.4 & 0.6 \\ 0.9 & 0.1 \end{bmatrix}.$$

Here, a 0 is just more likely than not to be followed by a 1, and a 1 is very likely to be followed by a 0. Both sample realizations exhibit very short runs of 0s and 1s: there is a lot of switching.

The estimates found in Exercise 12.4 show some variation, as might be expected, but in general they are reasonably close.

12.3.3 The runs test

We have identified two major types of alternative to the Bernoulli process. One corresponds to realizations exhibiting a relatively small number of long runs of 1s and 0s; and the other to realizations exhibiting a large number of short runs. A **run** is an unbroken sequence of identical outcomes. The length of a run can be just 1.

It seems intuitively reasonable that a test for the quality of the fit of a Bernoulli model to a set of trials data could therefore be developed using the observed number of runs in the data as a test statistic. An exact test based on the observed number of runs is known as the **runs test**.

Example 12.23 Counting runs

The number of runs in the realization given in Exercise 12.4(d), for instance, may be found as follows. Simply draw lines under consecutive sequences of 0s and 1s and count the total number of lines.

0 0 0 0 0 1 0 0 0 0 0 0 0 0 0 0 0 0 1 1 1 1 1 1 0 0 0 0 0 0 0 1 0 0 0 0 0 0 1 1

Figure 12.17 Counting runs

Here, there are 8 runs. Notice that in this case the matrix of transition counts is given by

$$N = \begin{matrix} 0 \\ 1 \end{matrix}\begin{bmatrix} 25 & 4 \\ 3 & 7 \end{bmatrix};$$

an alternative way to tally the number of runs is to observe that it is equal to the number of 01 switches plus the number of 10 switches plus 1: that is (denoting the number of runs by r),

$$r = n_{01} + n_{10} + 1. \qquad (12.6)$$

In this case,

$$r = 4 + 3 + 1 = 8.$$

Next, we shall need to determine whether the observed number of runs is significantly greater or less than the number that might have been expected under a Bernoulli model. ■

In Exercise 12.5 you should check that your results confirm the formula at (12.6).

Exercise 12.5

Count the number of runs in each of the realizations in Exercise 12.4(c) and (e).

The observed number of runs R in a sequence of trials may be used as the test statistic for the hypothesis that the sequence arises as a Bernoulli process. If the observed number of runs is very small or very large, then the hypothesis will be rejected. Like many of the exact tests to which you have been exposed in this course, the runs test is based on a complicated enumeration of possibilities, and only a computer renders the test feasible. You should explore the facilities available on your computer in attempting Exercise 12.6.

Exercise 12.6

For each of the six realizations given in Exercise 12.4, test whether they might reasonably be regarded as arising as a Bernoulli process.

For very long sequences of trials, the distribution of R is approximately normally distributed, with mean and variance under the null hypothesis given by

$$E(R) = \frac{2n_0n_1}{n_0+n_1} + 1, \quad V(R) = \frac{2n_0n_1(2n_0n_1 - n_0 - n_1)}{(n_0+n_1)^2(n_0+n_1-1)} \tag{12.7}$$

where n_0 is the number of 0s in the sample realization and n_1 is the number of 1s.

The Tel Aviv data provided a very long realization indeed, and this approximate test should yield useful results.

Example 12.20 continued

For the Tel Aviv data, we only have the matrix of transition counts as follows.

$$N = \begin{matrix} 0 \\ 1 \end{matrix} \begin{bmatrix} 1049 & 350 \\ 351 & 687 \end{bmatrix} \begin{matrix} 1399 \\ 1038 \end{matrix}$$

The number of runs is therefore

$$r = n_{01} + n_{10} + 1 = 350 + 351 + 1 = 702$$

(using (12.6)). We also need the overall number of 0s and 1s in the sequence of 2438 trials. In fact, it is in general not possible to deduce the values of n_0 and n_1 from the matrix N of transition frequencies, without more information. Reference to the original data gives

$$n_0 = 1400, \qquad n_1 = 1038.$$

Thus the observed value $r = 702$ of R needs to be tested against the normal distribution with mean

$$E(R) = \frac{2n_0n_1}{n_0+n_1} + 1 = \frac{2 \times 1400 \times 1038}{1400 + 1038} + 1 = 1193.12,$$

and variance

$$\begin{aligned} V(R) &= \frac{2n_0n_1\left(2n_0n_1 - n_0 - n_1\right)}{\left(n_0+n_1\right)^2\left(n_0+n_1-1\right)} \\ &= \frac{2 \times 1400 \times 1038 \times (2 \times 1400 \times 1038 - 1400 - 1038)}{2438^2 \times 2437} \\ &= 582.67, \end{aligned}$$

using (12.7). The corresponding z-score is

$$z = \frac{r - E(R)}{SD(R)} = \frac{702 - 1193.12}{\sqrt{582.67}} = -20.34,$$

to be compared against the standard normal distribution $N(0,1)$. You can see that the SP is quite negligible, and there is a strong suggestion (r much smaller than expected) that the realization exhibits long runs of dry days, and long runs of wet days, which is inconsistent with a Bernoulli model. ■

The normal approximation to the null distribution of R is useful as long as n_0 and n_1 are each about 20 or more.

Exercise 12.7

In a small study of the dynamics of aggression in young children, the following binary sequence of length 24 was observed.

1 0 1 1 1 1 0 1 1 1 0 0 0 0 1 0 0 1 1 1 0 0 0 0

Siegel, S. and Castellan, N.J. (1988) *Nonparametric Statistics for the Behavioral Sciences*, 2nd edn. McGraw-Hill Book Co., New York, p. 61.

The binary scores were derived from other aggression scores by coding 1 for high aggression, and 0 otherwise.

(a) Obtain the matrix N of transition frequencies for these data, and hence obtain an estimate $\widehat{M}$ for the transition matrix M.

(b) Find the number of runs in the data.

(c) Perform an exact test of the hypothesis that the scores could reasonably be regarded as resulting from a Bernoulli process, and state your conclusions.

(d) Notwithstanding the dearth of data, test the same hypothesis using an asymptotic test based on the normal distribution, and state your conclusions.

The runs test may not be quite appropriate here because of the precise method adopted for the coding, but this complication may be ignored.

In Section 12.4 we revisit the Poisson process as a model for random events occurring in continuous time.

12.4 The Poisson process

Earlier in the course, we met several examples of realizations of random processes that may be reasonably supposed to be well modelled by a Poisson process. These include the earthquake data of Table 4.7 (inter-event times), the nerve impulse data of Table 4.10 (inter-event times) and the coal mine disaster data of Table 7.4 (inter-event times—in fact, these do not follow an exponential distribution). The data of Table 4.4 constitute radiation counts over $7\frac{1}{2}$-second intervals. The Poisson distribution provides a good fit. (See *Chapter 9*, Example 9.4.)

In Tables 12.2 and 12.3 data are given on computer breakdowns, again exhibiting incidentally the two forms in which such data may arise: as inter-event waiting times, or as counts.

Here is another example of Poisson counts.

Example 12.24 *The Prussian horse-kick data*

This data set is one of the most well-known data collections. The data are known as Bortkewitsch's horse-kick data, after their collector. The data in Table 12.16 summarize the numbers of Prussian Militärpersonen killed by the kicks of a horse for each of 14 corps in each of 20 successive years 1875–1894. You can read down the columns to find the annual deaths in each of the 14 corps; sum across the rows for the annual death toll overall.

Ladislaus von Bortkewitsch (1868–1931) was born in St Petersburg. After studying in Russia he travelled to Germany, where later he received a Ph. D. from the University of Göttingen. The horse-kick data appear in a monograph of 1898 entitled *Das Gesetz der kleinen Zahlen*—that is, *The Law of Small Numbers*. The data are discussed in detail in Preece, D.A., Ross, G.J.S. and Kirby, S.P.J. (1988) Bortkewitsch's horse-kicks and the generalised linear model. *The Statistician*, **37**, 313–318.

Table 12.16 The Prussian horse-kick data

Year	G	1	2	3	4	5	6	7	8	9	10	11	14	15	Total
1875	0	0	0	0	0	0	0	1	1	0	0	0	1	0	3
1876	2	0	0	0	1	0	0	0	0	0	0	0	1	1	5
1877	2	0	0	0	0	0	1	1	0	0	1	0	2	0	7
1878	1	2	2	1	1	0	0	0	0	0	1	0	1	0	9
1879	0	0	0	1	1	2	2	0	1	0	0	2	1	0	10
1880	0	3	2	1	1	1	0	0	0	2	1	4	3	0	18
1881	1	0	0	2	1	0	0	1	0	1	0	0	0	0	6
1882	1	2	0	0	0	0	1	0	1	1	2	1	4	1	14
1883	0	0	1	2	0	1	2	1	0	1	0	3	0	0	11
1884	3	0	1	0	0	0	0	1	0	0	2	0	1	1	9
1885	0	0	0	0	0	0	1	0	0	2	0	1	0	1	5
1886	2	1	0	0	1	1	1	0	0	1	0	1	3	0	11
1887	1	1	2	1	0	0	3	2	1	1	0	1	2	0	15
1888	0	1	1	0	0	1	1	0	0	0	0	1	1	0	6
1889	0	0	1	1	0	1	1	0	0	1	2	2	0	2	11
1890	1	2	0	2	0	1	1	2	0	2	1	1	2	2	17
1891	0	0	0	1	1	1	0	1	1	0	3	3	1	0	12
1892	1	3	2	0	1	1	3	0	1	1	0	1	1	0	15
1893	0	1	0	0	0	1	0	2	0	0	1	3	0	0	8
1894	1	0	0	0	0	0	0	0	1	0	1	1	0	0	4
Total	16	16	12	12	8	11	17	12	7	13	15	25	24	8	196

Most often these appear as summary data. Table 12.17 shows the 196 deaths to have occurred during 280 corps-years. Here, the agreement with the Poisson distribution is moderately good.

Table 12.17 Frequency of deaths per year, 14 corps

Deaths	0	1	2	3	4	>4
Frequency	144	91	32	11	2	0

Bortkewitsch had noted that four of the corps (G, 1, 6 and 11) were atypical. If these columns are deleted, the remaining 122 deaths over 200 corps-years are summarized as shown in Table 12.18. Here, the Poisson agreement is very good indeed.

The labelling of the corps is not sequential.

Table 12.18 Frequency of deaths per year, 10 corps

Deaths	0	1	2	3	4	>4
Frequency	109	65	22	3	1	0

It is common to model the occurrence of such random haphazard accidents as deaths due to horse-kicks, as a Poisson process in time. Apart from the four corps which (for whatever reason) are atypical, this model seems to be a very useful one here. ■

12.4.1 Properties of the Poisson process

We first met the Poisson process in *Chapter 4*, Section 4.4. Events occur at random in continuous time, as portrayed in Figure 12.18.

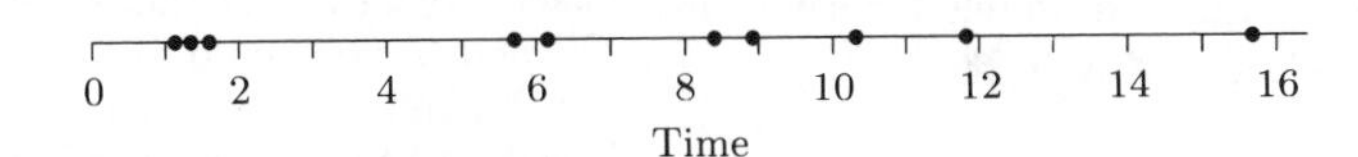

Figure 12.18 Events occurring as a Poisson process in time

This is a continuous-time process rather than a discrete-time process. Denoting by $X(t)$ the total number of events to have occurred in any time interval of duration t, then the random variable $X(t)$ has a Poisson distribution with parameter λt, where $\lambda > 0$ is the average rate of events.

By comparison, if X_n denotes the total number of successes in a sequence of n trials in a Bernoulli process, then the distribution of the random variable X_n is binomial $B(n, p)$. The distribution of the number of trials between consecutive successes in a Bernoulli process is geometric with parameter p.

The time T between events has an exponential distribution with mean $\mu = 1/\lambda$: this is written $T \sim M(\lambda)$. These characteristics may be summarized as follows.

The Poisson process

For recurrent events occurring as a Poisson process in time at average rate λ, the number of events $X(t)$ occurring in time intervals of duration t follows a Poisson distribution with mean λt: that is, $X(t) \sim \text{Poisson}(\lambda t)$. The waiting times between consecutive events are independent observations on an exponential random variable $T \sim M(\lambda)$.

Several examples of events that might reasonably be modelled as a Poisson process are listed in *Chapter 4*, Example 4.17.

12.4.2 Testing the adequacy of the fit of a Poisson process

Estimation procedures depend on the form in which the data arise. They may arise as a consequence of continuous observation, in which case they will consist of inter-event waiting times. Commonly, observation will be regular but intermittent: then the data will consist of counts of events over equal time intervals.

You are already very familiar with the estimation of parameters for exponential and Poisson models.

Example 12.25 Estimating the rate of earthquakes

For the earthquake data in Table 4.7 each time interval $t_i, i = 1, 2, \ldots, 62$ may be regarded under the Poisson process model as a data value independently chosen from the exponential distribution with parameter λ. The mean of such an exponential distribution is $1/\lambda$ (see (4.25)). So, it seems quite reasonable that a point estimate of λ should be $1/\bar{t}$, where $\bar{t} = 437$ days is the sample mean waiting time between serious earthquakes. The maximum likelihood estimate of λ (the earthquake rate) is $1/437 = 0.0023$ major earthquakes per day.

A 95% confidence interval for the mean $1/\lambda$ (again, assuming an exponential model) is given by $(346, 570)$ days. The corresponding confidence interval for the earthquake rate is $(0.0018, 0.0029)$ earthquakes per day. ■

These confidence intervals are exact. Using the methods in *Chapter 7*, you could also obtain approximate confidence intervals based on the normal distribution.

A Poisson process also forms a good probability model for certain types of traffic process. Suppose an observer is situated at a particular point at the side of a road and records the time intervals between vehicles passing that spot. A Poisson process will be a reasonable model for such data if the traffic is free-flowing. That is, each vehicle which travels along the road does so at essentially the speed its driver wishes to maintain, with no hindrance from other traffic. In saying this, many traffic situations are ruled out. For example, any road in which traffic is packed densely enough so that traffic queues form is not allowed: an adequate model for such a 'heavy-traffic' process would have to make provision for the occurrence of clusters of vehicles close together. Likewise, a main street with many intersections controlled by traffic lights also makes for traffic clumping. On the other hand, a convoy of, say, army trucks travelling long-distance may move in regular spacing, each keeping more or less the same distance from each other. Situations where free-flowing traffic can be assumed therefore reduce either to rather a quiet road where only occasional vehicles pass (singly) along, or, perhaps more usefully, to multi-lane roads such as motorways, which, when not overfull, allow vehicles to progress largely unimpeded by others.

Example 12.26 Yet more traffic data

The data in Table 12.19 are thought to correspond to free-flowing traffic. They consist of 50 vehicle inter-arrival times for vehicles passing an observation point on Burgess Road, Southampton, England, on 24 June 1981.

Data provided by Dr J.D. Griffiths of the University of Wales Institute of Science and Technology.

Table 12.19 Inter-arrival times (seconds)

13.2	1.5	1.8	7.6	8.2	10.7	9.4	8.1	1.5	2.3
3.0	1.6	4.5	7.3	9.4	13.8	30.1	1.2	0.8	6.2
7.0	7.4	6.3	1.2	4.2	4.5	2.3	1.7	2.6	3.6
7.5	10.3	3.3	28.2	23.1	1.7	2.4	3.5	16.2	1.5
12.9	5.6	4.3	13.2	3.0	5.6	4.4	14.9	1.2	4.0

The passage times are illustrated in Figure 12.19.

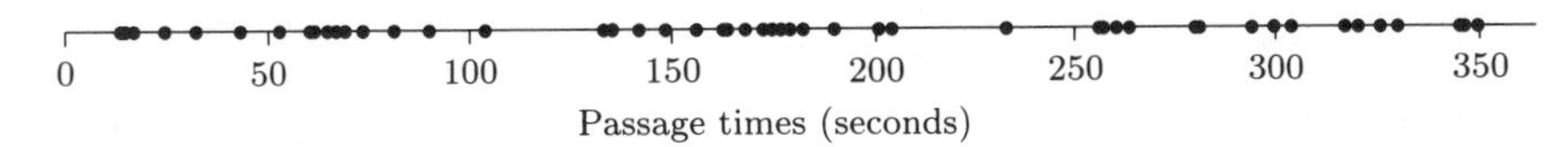

Figure 12.19 Vehicle arrival data

As we shall see, the data here can be modelled usefully by an exponential distribution. ■

Sometimes data come in the form of counts: for example, Rutherford and Geiger's radiation data in Table 4.4. If the counts arise as events in a Poisson process, these data will follow a Poisson distribution. We have seen that the Poisson model provides a good fit. On the other hand, it was remarked that for Table 12.3 in Example 12.3, the Poisson distribution did not provide a good fit to the system breakdown data summarized there.

You could confirm this lack of fit using a chi-squared test, which is an asymptotic test. In this section an exact test is described for testing the quality of the fit of a Poisson process model. The test assumes that the times of occurrence of the events are known—in other words, the test assumes continuous observation.

Now, if events occur according to a Poisson process, then over the whole interval of observation, there should be no identifiable peak of activity where there is an unusually high incidence of events. (Of course, there will be random variation exhibited in the sample realization.) Similarly, there should be no identifiable intervals over which the incidence of events is unusually sparse.

Denoting by $(0, \tau)$ the period of observation, and by

The letter τ is the lower-case Greek letter tau, pronounced 'tor'.

$$0 < t_1 < t_2 < \cdots < t_n < \tau$$

the times of occurrence of the n events observed, then it turns out that the fractions

Notice that the times $t_i, i = 1, 2, \ldots, n$, are the actual times of occurrence of events, not waiting times between events.

$$w_{(1)} = \frac{t_1}{\tau}, w_{(2)} = \frac{t_2}{\tau}, \ldots, w_{(n)} = \frac{t_n}{\tau}$$

may be regarded as a random sample from the standard uniform distribution $U(0, 1)$, if the underlying process generating the events may be modelled as a Poisson process. This is the only continuous distribution consistent with the idea of 'no preferred time of occurrence'.

A test for whether a list of times of occurrence may be regarded as arising from a Poisson process therefore reduces to testing whether a list of n numbers all between 0 and 1 may be regarded as a random sample from the standard uniform distribution $U(0, 1)$. One of the advantages of this test is that no problems of parameter estimation are raised. The test is called the **Kolmogorov test**. In general, it may be used to test data against any hypothesized continuous distribution. (For example, there is a version of the Kolmogorov test appropriate for testing normality.) We shall only use it as a test of uniformity.

The test consists of comparing the uniform cumulative distribution function

$$F(w) = w, \quad 0 \leq w \leq 1,$$

with the sample distribution function, or *empirical distribution function* obtained from the data, defined as follows.

The empirical distribution function

For any random sample $w_1, w_2, \ldots, w_n$ from a continuous distribution with unknown c.d.f. $F(w)$, an estimate $\widehat{F}(w)$ of $F(w)$ called the **empirical distribution function** may be obtained as follows.

Arrange the sample in ascending order

$$w_{(1)} \leq w_{(2)} \leq \cdots \leq w_{(n)}.$$

Then the graph of $\widehat{F}(w)$ is constructed as follows:

for all $0 < w < w_{(1)}, \widehat{F}(w) = 0$;

for all $w_{(1)} < w < w_{(2)}, \widehat{F}(w) = 1/n$;

for all $w_{(2)} < w < w_{(3)}, \widehat{F}(w) = 2/n$;

and so on; finally

for all $w_{(n-1)} < w < w_{(n)}, \widehat{F}(w) = (n-1)/n$;

for all $w_{(n)} < w < 1, \widehat{F}(w) = 1$.

At the 'jump points' $w_{(1)}, w_{(2)}, \ldots, w_{(n)}$, the flat components of the graph of $\widehat{F}(w)$ are joined by vertical components to form a 'staircase'.

The Kolmogorov test consists of superimposing on the graph of the empirical distribution function the cumulative distribution function of the hypothesized model $F(w)$ (which, in this case, is $F(w) = w$), and comparing the two.

Example 12.27 illustrates the technique. The small data set used is artificial, in the interests of simplicity.

Example 12.27 The empirical distribution function

Observation is maintained on an evolving system between times 0 and $\tau = 8$. Over that period of time, five events are observed, occurring at times 0.56, 2.40, 4.08, 4.32, and 7.60. Figure 12.20 shows the occurrence of events over time.

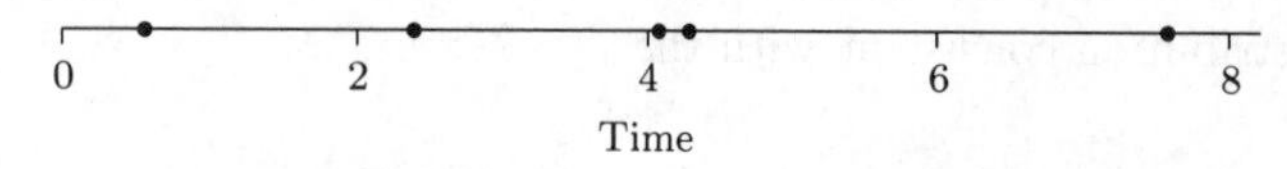

Figure 12.20 Five points, possibly occurring as a Poisson process in time

In the nature of things, the times of occurrence are already ordered. If the events occur according to a Poisson process, then the ordered fractions

$$w_{(1)} = \frac{t_1}{\tau} = \frac{0.56}{8} = 0.07,$$

$$w_{(2)} = \frac{t_2}{\tau} = \frac{2.40}{8} = 0.30,$$

$$w_{(3)} = \frac{t_3}{\tau} = \frac{4.08}{8} = 0.51,$$

$$w_{(4)} = \frac{t_4}{\tau} = \frac{4.32}{8} = 0.54,$$

$$w_{(5)} = \frac{t_5}{\tau} = \frac{7.60}{8} = 0.95,$$

may be regarded as a random sample from the standard uniform distribution $U(0,1)$. The empirical distribution function is drawn from 0 to 1 with jumps of size $1/n = 0.2$ occurring at the points 0.07, 0.30, 0.51, 0.54, 0.95. This is shown in Figure 12.21.

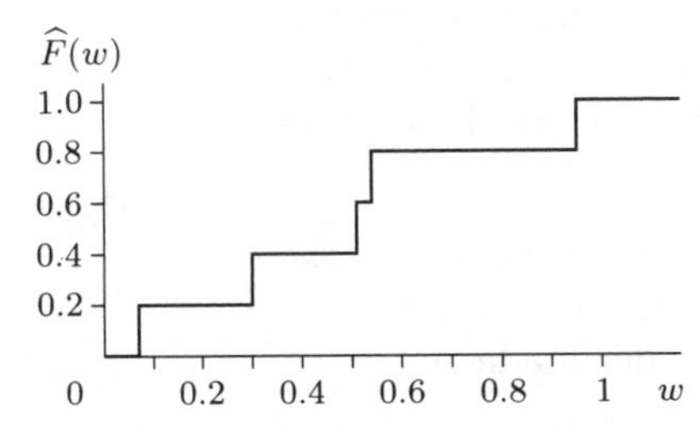

Figure 12.21 The empirical distribution function

Now the graph of the c.d.f. $F(w) = w$ of the standard uniform distribution $U(0,1)$ is superimposed on the figure. This is shown in Figure 12.22.

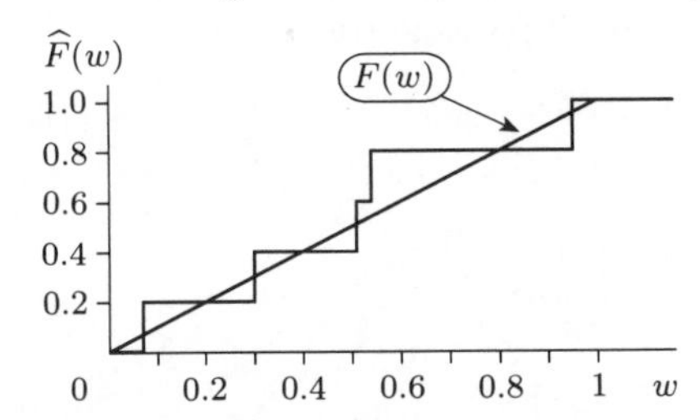

Figure 12.22 The empirical distribution function and the hypothesized distribution function

You can see in this case that there are some substantial discrepancies between the empirical distribution function $\widehat{F}(w)$ obtained for our data set and the theoretical model $F(w) = w$. Are these differences statistically significant? ■

If the hypothesis of uniformity holds, the two distribution functions will not be too different. They are equal at the point $(0,0)$ and at the point $(1,1)$. Our measure of difference between them, the Kolmogorov test statistic, is defined to be the *maximum vertical distance* between the two,

$$D = \max_{0<w<1} |\widehat{F}(w) - F(w)| = \max_{0<w<1} |\widehat{F}(w) - w|. \tag{12.8}$$

The distance D is sometimes known as the *Kolmogorov distance.*

This will always occur at one of the jump points. In Figure 12.22 the maximum difference between the two distribution functions occurs at the point $w = 0.54$, and the size of the difference is $0.8 - 0.54 = 0.26$. This is shown in Figure 12.23.

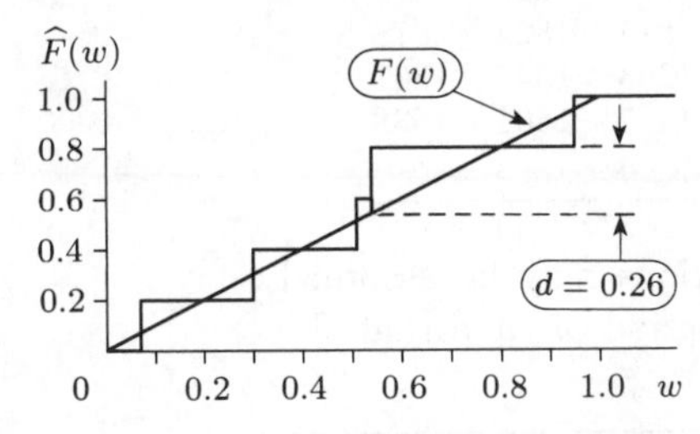

Figure 12.23 The Kolmogorov distance D: here, $d = 0.26$

So, for Example 12.27, the value of the Kolmogorov test statistic is $d = 0.26$, based on a sample of size $n = 5$. This seems quite large, but to explore whether it is significant (that is, whether the hypothesis of a Poisson process should be rejected) the null distribution of the test statistic D is required. This is extremely complicated, and it is usual at this point to employ statistical software. In fact, the associated SP here is 0.812, which is very high, reflecting the small size of the data set. There is no reason here to reject the hypothesis that the data arise from a Poisson process.

Like the χ^2 test statistic in a test of goodness-of-fit, large values of the Kolmogorov test statistic D lead to rejection of the hypothesized model; small values of D suggest that the fit is good, and that the model is adequate.

Example 12.1 continued

In this example, no value is given for τ. Observation ceased exactly with the passage of the last of the 40 cars in Table 12.1: this vehicle passed the observer after $\tau = 12 + 2 + 6 + \cdots + 16 + 2 = 312$ seconds. The first 39 passage times yield observations

$$w_{(1)} = \frac{t_1}{\tau} = \frac{12}{312} = 0.038;$$
$$w_{(2)} = \frac{t_2}{\tau} = \frac{12+2}{312} = 0.045;$$
$$w_{(3)} = \frac{t_3}{\tau} = \frac{12+2+6}{312} = 0.064;$$

and so on. The Kolmogorov test statistic takes the value $d = 0.138$ with $n = 39$. The corresponding SP is 0.412. The exponential fit seems adequate. ■

Exercise 12.8

(a) Use the Kolmogorov test against the null hypothesis of a Poisson process for the traffic data of Example 12.26 (Table 12.19). Interpret your conclusion. (Note here, again, that observation ceased only with the passage of the 50th vehicle.)

(b) Table 12.20 gives the dates (in the form of months after the start of 1851) of the major explosive volcanic eruptions in the northern hemisphere after 1851. The last recorded eruption (corresponding to $t_{56} = 1591$ in the table) occurred in July 1983. In fact, observation was continued until the start of 1985 (so $\tau = 1608$).

Solow, A.F. (1991) Exploratory analysis of occurrence of explosive volcanism. *J. American Statistical Association*, **86**, 49–54.

Table 12.20 Times of major volcanic eruptions (months)

28	38	69	226	254	265	291	318	322	367
392	394	421	451	466	587	617	622	629	675
699	721	738	745	757	796	808	814	824	878
942	1037	1129	1151	1155	1214	1231	1263	1347	1367
1377	1384	1388	1388	1410	1471	1486	1495	1501	1538
1553	1564	1565	1576	1577	1591				

Use these data to explore whether volcanic eruptions may be assumed to occur as a Poisson process (at least over the period of time that the phenomenon was researched).

12.5 Variations

This chapter concludes with some examples of chance situations where models other than those described so far would be most appropriate. The literature on the subject of model construction and testing for random processes is vast, and all we can do in an introductory text such as this is hope to obtain some notion of the wide area of application.

Example 12.28 Poisson processes with varying rate

A **time-dependent Poisson process** is a Poisson process in which the average rate of occurrence of events alters with passing time. This might be appropriate, for example, in a learning situation where the events recorded are errors—these will become less frequent as time passes. One example where such a model might be appropriate is where the times of system failures were recorded (see Table 12.2). There the remark was made that '... an exponential model does not provide a good fit to the variation in the times between failures ... the failures are becoming less frequent with passing time.'

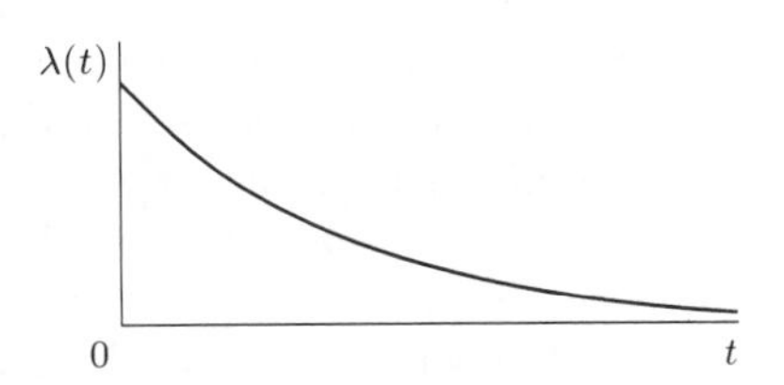

Figure 12.24 A decreasing failure rate

A sketch of a failure rate $\lambda(t)$ with approximately the right properties is shown in Figure 12.24. In this case the failure rate is always decreasing, but never attains zero. In fact, the rate may be represented by the function

$$\lambda(t) = \alpha e^{-\beta t}, \quad t \geq 0,$$

where $\alpha > 0$ and $\beta > 0$. Once a functional form for $\lambda(t)$ has been decided (linear, quadratic, trigonometric or, as in this case, exponential decay) then the parameters of the model may be estimated using maximum likelihood. If the model provides a good fit, then it can be used to forecast future failure events. For instance, the number of failures to occur during some future interval $t_1 < t < t_2$ follows a Poisson distribution whose mean is given by the integral

$$\int_{t_1}^{t_2} \lambda(t)\, dt,$$

shown in Figure 12.25. The selection of an appropriate functional form for $\lambda(t)$ is not always straightforward; but the ideas of parameter estimation and model testing are not different from those with which you are already familiar. ■

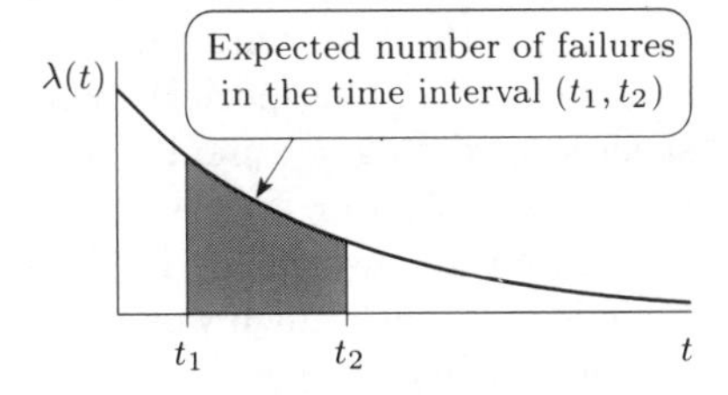

Figure 12.25 The expected number of failures in the time interval (t_1, t_2)

Example 12.29 is based on the Bernoulli process.

Example 12.29 Identifying a change in the indexing parameter

Researchers involved in a grassland experiment were interested in the presence or absence of plants in a protected habitat, year after year. For any given species, they recorded 1 if its presence was observed, 0 otherwise. This was repeated each year and so the data took the form of a sequence of 0s and 1s, possibly consistent with an underlying Bernoulli process, or a Markov chain (or neither of these). The researchers were particularly interested in one species. It certainly appeared that the plant was an intermittent visitor.

A complication was that the plant was easy to mistake: it could be missed (score 0) when it was present; and it could be 'seen' (score 1) when in fact it was not present—other plants looked very similar to it. So the data were not entirely reliable.

It became known that in reality this particular plant was present in the habitat every year until, one year, it disappeared completely. The problem was to use the sequence of recorded 0s and 1s to estimate when this had happened.

Again, this is a problem involving quite a subtle estimation procedure (not just of the change point, but also of the probabilities of error—recording a 0 when the plant was present and 1 when it was not). It is important to have in mind a useful underlying model, and to be aware of the year-to-year dependencies that need to be built in to it.

A similar but unrelated problem faced researchers studying different plants. The annual visits made by these plants to the habitat were known to be intermittent, but it was also known that after a while there was a possibility of extinction. The problem was to determine from a sequence of recent 0s what probability to attach to the possibility that extinction had taken place. Here, there were no problems of identification. Possibly the year-by-year presence or absence could be reasonably modelled as a Markov chain, for at least as long as the plant survived. ■

Example 12.30 *Competing species*

Birth-and-death models were briefly explored in Section 12.1. Biologists and ecologists often have to consider populations of two or more species living together and interacting. Sometimes these interactions are co-operative, and sometimes they are antagonistic (though not necessarily detrimental to the future existence of a species). Sometimes such communities are stable, with all species coexisting in a moderately happy association, and sometimes they are not—one species or another eventually disappears because it is not possible indefinitely to sustain life in the presence of another. The literature on the topic is vast, with much work for instance on predator-prey models and host-parasite models.

One important application is to human conflict, and there are some fascinating models to explore. ■

Example 12.31 *Epidemic models*

There are many variations on models for the spread of disease within communities. It is not just a matter of different parameters being required to reflect differences in virulence, the duration of symptoms, the length of the infectious period, and so on: different diseases are spread in entirely different ways, and the models are fundamentally different.

Populations consist of three types of individual: susceptibles (those who have not yet contracted the disease), infectives (those who have, and who are capable of spreading it) and the rest, who have recovered from the disease. Perhaps this third group are immune, or temporarily immune, to further attacks; or, as with some diseases, there is no difference between somebody who has recovered from the disease and somebody who is susceptible to it. This is, or appears to be, the case with the common cold.

Often a model developed in one context can be used in another, quite different application. Some epidemic models are appropriate to studies of the diffusion of information, where 'susceptibles' are those ignorant of certain information, 'infectives' are those who have it, and are passing it on to others, while others have known it but are no longer interested in passing it on. One interesting recent variation includes a model on drug addiction, incorporating susceptibles, addicts and dealers. ■

Billard, L. and Dayananda, P.W.A. (1988) A drug addiction model. *J. Applied Probability*, **25**, 649–662.

Summary

1. A sequence of Bernoulli trials in which trials are independent and the probability of success is the same from trial to trial is known as a Bernoulli process.

2. A Markov chain is a random process $\{X_n;\ n = 1, 2, \ldots\}$ taking values 0 or 1, in which $X_1 \sim \text{Bernoulli}(p_1)$ and thereafter transition probabilities are given by the transition matrix

$$M = \begin{matrix} 0 \\ 1 \end{matrix} \begin{bmatrix} 1-\alpha & \alpha \\ \beta & 1-\beta \end{bmatrix}.$$

3. In the long term, the proportions of 0s and 1s in a realization of a Markov chain are given by

$$q = \frac{\beta}{\alpha+\beta}, \qquad p = \frac{\alpha}{\alpha+\beta}$$

respectively.

A Markov chain in which $p_1 = \alpha/(\alpha+\beta)$ is called stationary. A Bernoulli process is a special case of a stationary Markov chain.

4. In any realization of a Markov chain the transition frequencies may be summarized through the matrix N where

$$N = \begin{matrix} 0 \\ 1 \end{matrix} \begin{bmatrix} n_{00} & n_{01} \\ n_{10} & n_{11} \end{bmatrix} \begin{matrix} n_{00}+n_{01} \\ n_{10}+n_{11} \end{matrix};$$

then the estimate $\widehat{M}$ of the transition matrix M is given by

$$\widehat{M} = \begin{matrix} 0 \\ 1 \end{matrix} \begin{bmatrix} n_{00}/(n_{00}+n_{01}) & n_{01}/(n_{00}+n_{01}) \\ n_{10}/(n_{10}+n_{11}) & n_{11}/(n_{10}+n_{11}) \end{bmatrix}.$$

5. The number of runs r in a realization of a Markov chain is given by

$$r = n_{01} + n_{10} + 1.$$

The observed value r of R may be used to test the hypothesis that a given realization is generated by a Bernoulli process. The test is exact; an approximate test is based on the asymptotic normality of R where

$$E(R) = \frac{2n_0n_1}{n_0+n_1} + 1, \qquad V(R) = \frac{2n_0n_1(2n_0n_1 - n_0 - n_1)}{(n_0+n_1)^2(n_0+n_1-1)};$$

here, n_0 is the number of 0s in the sample realization and n_1 is the number of 1s.

6. In a Poisson process the distribution of $X(t)$, the number of events in time intervals of duration t, is Poisson(λt), where λ is the average rate at which events occur. The waiting time between consecutive events follows an exponential distribution $M(\lambda)$.

7. In order to test whether events occurring at times

$$0 < t_1 < t_2 < \cdots < t_n < \tau$$

may be assumed to have been generated according to a Poisson process, the Kolmogorov test may be applied. This tests whether the observations

$$w_{(j)} = \frac{t_j}{\tau}, \quad j = 1, 2, \ldots, n$$

may be assumed to have arisen from a standard uniform distribution.

The Kolmogorov test statistic is the 'Kolmogorov distance'

$$D = \max_{0<w<1} |\widehat{F}(w) - w|.$$

The Kolmogorov test is an exact test.

Chapter 13
A Look Back

In this chapter, some of the methods and techniques studied in the course are reviewed through case studies. It is not an exhaustive review, but an attempt has been made to refer at least to the most important aspects of the 'statistical approach'. You should note particularly that no single approach is necessarily appropriate to data exploration, and that we may need to draw on more than one technique to reach a real understanding of the story behind a data set. So the studies do not constitute a chronological review of the course.

The aim of this chapter is to revise some of the more important material which has appeared in the course. However, rather than simply giving a condensed presentation of the earlier material, this revision is approached by way of a series of brief case studies. These are real questions which require the use of statistics to address them. A wide range of material is covered here, not just in the statistical techniques used but also in the nature of the examples.

In addition to providing a revision of the underlying statistical ideas, these case studies should serve both to integrate the techniques and to demonstrate that data do not always arrive in an ideal manner. A statistical analysis is more than a simple application of a single technique and sometimes there may be problems of missing or generally messy data.

Obviously there is not enough space in a single chapter to illustrate everything that has been covered in the preceding twelve, but we can hope to revisit the most important material in a way which gives a perspective on how a range of ideas and methods form a coherent structure for tackling statistical problems. As you read through the chapter, you will find that each case study is subjected to an ordered, methodical approach which you can acquire and employ for yourself.

As a general rule, the same procedures are adopted when addressing real problems.

1. Identify the question or questions that need to be answered. Generally this will be fairly clear from the start, but this is not always true. Try to pose the question in statistical terms.
2. Try to find a helpful picture of the data which, indeed, may even answer one or all of the questions, in which case you may not need to proceed any further. At all events, a picture should give some indication of an answer and should alert you to any complications which may be lying in wait.

3 Ask some fundamental questions about the data. Are the data plausibly symmetric? Are the data plausibly normal? Is there a linear relationship between any variables involved? If not, might a transformation help and, if so, which transformation is adequate for the task? Are there any possible outliers? Taking these things into account, which method should you try first?

4 Consider your likely method, or methods. Be aware of the assumptions involved and check them out. It will sometimes be necessary to try a method and then check out its validity afterwards. You may find, for instance, that you have to carry out your check by looking at residuals.

5 Ask yourself if you are satisfied that your method is justified. If not, try to modify it. For example, you might try to carry out a test based upon an assumption of normality. A preliminary check might indicate that the data are not plausibly normal because of an outlier, so you remove it and try again. Maybe the data are very skewed. Perhaps you decide that a normality assumption is not tenable, even after transformation or outlier stripping, so you elect for a non-parametric test.

6 Give a careful statement of your conclusions with a brief discussion of their relevance.

Data analysis depends, to a large extent, on a methodical approach along the lines indicated above.

Of course, there is more to statistics than merely analysing data. Data acquisition and collection are no less important and, in practice, it is vital to design a statistical experiment with the proposed techniques of analysis in mind. But this chapter is not about such considerations, important though they are. It is about analysing raw data using the methods which appear in the earlier chapters.

The course began with a discussion of the usefulness of a graphical approach to data exploration, and in the succeeding chapters you have seen literally hundreds of diagrams intended to aid insight. Often these have been used to support a more technical analysis; but it remains true that a diagram often tells you all you need to know about the structure underlying a list of numbers. And there will be some occasions where there is no obvious analytic procedure to follow, but where a diagram can at least be employed to provide a partial answer to a question, or the suggestion of an answer.

13.1 Exploiting statistical graphics

13.1.1 A question of authorship

Much statistical work has been done on the question of authorship and, indeed, some people are almost obsessive about it. Were all of the epistles written by St Paul himself? Did Shakespeare really write all of the plays attributed to him, or might Marlowe have a valid claim? The answers are often of vital importance to scholars and may well have important financial consequences. In this section, we shall look at a controversy of great concern to political historians of the USA; namely, who wrote the *Federalist* papers of 1788?

In an attempt to persuade the citizens of New York to ratify the Constitution of the United States, the *Federalist* papers were published anonymously by Alexander Hamilton, John Jay and James Madison. Seventy-seven papers appeared as letters in New York newspapers under the pseudonym 'Publius'. With eight more essays, they appeared as a book in 1788. Whilst authorship of *The Federalist* was common knowledge, no assignment of specific papers to individual authors occurred until 1807, three years after Hamilton's death. In 1818, Madison made a listing of authors, one of a variety of such lists. There is general agreement on the authorship of 70 of the papers—5 by Jay, 14 by Madison and 51 by Hamilton—but 15 remain unattributed, 12 being in dispute between Hamilton and Madison and 3 being joint works to a disputed extent.

The dispute seems to have arisen as a result of political embarrassment. Within a few years of writing, Madison and Hamilton had become bitter political enemies who sometimes contradicted their own writings, and hence neither was in any hurry to lay claim to certain tracts.

Unfortunately, the political content of the disputed papers, while it may give some indication, does not definitively resolve the problem. Both Madison and Hamilton were writing in favour of ratification, and both experienced eventual changes in political orientation. Neither does literary style help to provide an answer, since both adopted flowery, oratorical prose.

A statistical approach to this problem might well start with looking at some aspect of technical style, such as the distribution of sentence lengths. Certainly sentence length is often a characteristic of an author, but it will not distinguish reliably between authors writing in similar literary styles. Faced with this difficulty, the historian Douglas Adair opted for the frequency of occurrence of specific words. He detected a difference in choice between the words *whilst* and *while.* In the fourteen essays known to be written by Madison, *while* never occurs whereas *whilst* occurs in eight of them. So far, so good, but this is not helpful for the disputed papers since *whilst* or *while* occurs in less than half of them. In any case, how can we be sure that Madison would never use *while* here, when other of his writings have twice included the word?

But the idea of looking at keywords is a good one, provided we take common words and look at their relative frequency of use. Both authors would use most common words at the same rate, but some words may help if we can choose the right ones. Even then we need to be careful. For example, Madison uses *war* more often than Hamilton, but this could be explained by a division of labour giving him more opportunity to use it, rather than a basic predilection for the word. The answer is to restrict the analysis to non-contextual words such as *of, to, on, the, and, from,* and so on. While these are words we all use, it may turn out that rates of use differ from author to author. Table 13.1 below shows the rates for *by, from* and *to.*

You can see that low rates for *by* suggest the author Hamilton, whereas high rates for *from* indicate Madison (but there is a great deal of overlap). It would be very difficult to distinguish between authors through their use of the word *to.*

Table 13.1 Frequency distribution of rate per 1000 words in 48 Hamilton papers (H) and 50 Madison papers (M) for *by*, *from* and *to*

by			*from*			*to*		
Rate	H	M	Rate	H	M	Rate	H	M
1–2	2		1–2	3	3	23–25		3
3–4	7		3–4	15	19	26–28	2	2
5–6	12	5	5–6	21	17	29–31	2	11
7–8	18	7	7–8	9	6	32–34	4	11
9–10	4	8	9–10		1	35–37	6	7
11–12	5	16	11–12		3	38–40	10	7
13–14		6	13–14		1	41–43	8	6
15–16		5				44–46	10	1
17–18		3				47–49	3	2
						50–52	1	
						53–55	1	
						56–58	1	
Total	48	50		48	50		48	50

This is a problem in **discrimination**. Table 13.1 is based on a substantial amount of data, and could be used as the basis for probability models representing the variation in frequency of use of common words. The material whose authorship is disputed may be tested against these models.

One useful, if informal, approach is provided through the use of histograms. Figure 13.1 is a histogram showing the frequency of use of the word *to* in the twelve disputed papers.

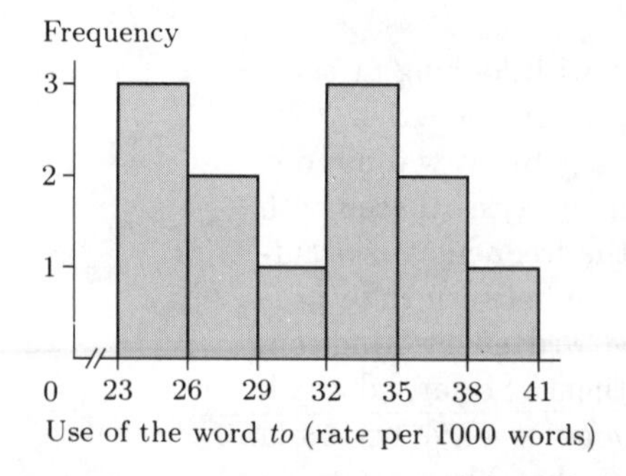

Figure 13.1 Histogram showing use of the word *to* in the disputed papers

Exercise 13.1

Use Table 13.1 to plot histograms of the frequency of the word *to* as used by Hamilton and Madison. If the disputed papers were known to be by the same author, which of the two might your histograms suggest?

As with all statistical investigations, what matters is the accumulation of evidence and the composite picture thus obtained. F. Mosteller and D.L. Wallace discuss this problem and give a wide variety of alternative analyses in their book *Applied Bayesian and Classical Inference: the case of the Federalist papers* (Springer, 1984). In one of these they chose a large number of non-contextual words and their rate of use per 1000 words is given in Table 13.2.

The words are grouped according to the degree of contextuality assessed by Mosteller and Wallace. You can see from the table that the word *upon* stands out as a very powerful discriminator. Hamilton uses *on* and *upon* equally, but Madison almost invariably uses *on*.

Table 13.2 Rate of use per 1000 words for non-contextual words

Word	Rate per 1000 words Hamilton	Madison	Word	Rate per 1000 words Hamilton	Madison
Group A			Group D		
upon	3.24	0.23	*commonly*	0.17	0.05
			consequently	0.10	0.42
Group B			*considerable(ly)*	0.37	0.17
also	0.32	0.67	*according*	0.17	0.54
an	5.95	4.58	*apt*	0.27	0.08
by	7.32	11.43			
of	64.51	57.89	Group E		
on	3.38	7.75	*direction*	0.17	0.08
there	3.20	1.33	*innovation(s)*	0.06	0.15
this	7.77	6.00	*language*	0.08	0.18
to	40.79	35.21	*vigor(ous)*	0.18	0.08
			kind	0.69	0.17
Group C			*matter(s)*	0.36	0.09
although	0.06	0.17	*particularly*	0.15	0.37
both	0.52	1.04	*probability*	0.27	0.09
enough	0.25	0.10	*work(s)*	0.13	0.27
while	0.21	0.07			
whilst	0.08	0.42			
always	0.58	0.20			
though	0.91	0.51			

In Table 13.3, the frequency distribution of use of the word *upon* shows that we can be fairly sure that Madison wrote at least eleven of the disputed papers. In fact, the one disputed paper containing *upon* is strongly classified as Madison's by other words. This analysis strongly supplements the independent opinion of historians, based on the known political beliefs of the protagonists.

Table 13.3 Frequency distribution of rate per 1000 words for *upon*

Rate per 1000 words	Hamilton	Madison	Disputed
0		41	11
0–0.4		2	
0.4–0.8		4	
0.8–1.2	2	1	1
1.2–1.6	3	2	
1.6–2.0	6		
2–3	11		
3–4	11		
4–5	10		
5–6	3		
6–7	1		
7–8	1		
Totals	48	50	12

This result is interesting in that it has been reached without the use of a formal statistical test. We have simply looked for eye-catching patterns in the data. In Table 13.3 it was not even necessary to draw histograms.

Of course, graphics can be used to good effect in regression analysis (scatter plots), hypothesis testing (comparative boxplots, for example), model fitting (histograms and probability plots) and on many other occasions. Let us move on to a review of some useful probability models.

13.2 Model fitting and testing

13.2.1 Some modelling history

Around the turn of the century, the whole notion of probability modelling was a cause of some concern to many of the more prominent statisticians. In particular, there was a lively and well-documented correspondence between W.F.R. Weldon (1860–1906), Francis Galton (1822–1911), F.Y. Edgeworth (1845–1926) and Karl Pearson (1857–1936) in which they tackled the problem of whether or not a theoretical binomial model could be said to provide an acceptable fit to observed frequencies in a dice-throwing experiment.

Weldon to Galton, 4 February 1894

There is an account of the correspondence in Pearson, E.S. (1965) Some incidents in the early history of biometry and statistics, 1890–94. *Biometrika*, **52**, 3–18.

> ... I have collected 26,306 tosses of groups of 12 dice, for use at the Royal Institution. In each group the event recorded is the number of dice with 5 or 6 points, so that the chance of success with each die is $\frac{1}{3}$ ...
>
> A certain set of 7,000 tosses, forming part of the result, was made for me by a clerk in the office of University College, whose accuracy in work of another kind I have had occasion to test by asking him to copy 24,000 numbers of 3 figures each, with excellent results.
>
> A day or so ago, Pearson wanted some records of the kind in a hurry, in order to illustrate a lecture ... I gave him the 7,000 separately from the rest, and on examination he rejects them, because he thinks the deviation from the theoretically most probable result is so great as to make the record intrinsically incredible.
>
> You will see how serious a matter this is ...
>
> Last night I saw Greenhill, whose experience in target practice at Woolwich makes him know this kind of thing statistically as well as mathematically—he is of the opinion that the record is perfectly credible, and that I have no shadow of reason to disregard it.
>
> Today I am sending it to you and Edgeworth.

In fact, there were 7006 tosses.

That is, Pearson rejected the data because they appeared to show variation not consistent with the binomial model.

Let us look at Weldon's data and carry out a preliminary analysis of the quality of fit provided by the binomial distribution $B(12, \frac{1}{3})$ using the methods you learned in *Chapter 9.* Table 13.4 gives the observed counts.

Table 13.4 Weldon's data on 7006 tosses of 12 dice

Number of 5s or 6s	Observed frequency
0	45
1	327
2	886
3	1475
4	1571
5	1404
6	787
7	367
8	112
9	29
10	2
11	1
12	0
Total	7006

The chi-squared goodness-of-fit test for comparing observed frequencies with those expected under a hypothesized model is appropriate here. For instance, the expected frequency of 0s is

$$\begin{aligned} E_0 &= 7006 \times p_0 \\ &= 7006 \times \binom{12}{0}\left(\frac{1}{3}\right)^0\left(\frac{2}{3}\right)^{12} \\ &= 7006 \times 0.007707 \\ &= 54.0 \end{aligned}$$

to one decimal place: this compares with the observed frequency of 45.

Weldon's data are summarized with an up-to-date commentary in Kemp, A.W. and Kemp, C.D. (1991) Weldon's dice data revisited. *American Statistician*, **45**, 216–222.

Exercise 13.2

Complete a table of expected frequencies assuming a binomial model $B(12, \frac{1}{3})$ and, pooling cells if necessary, calculate the value of the chi-squared test statistic

$$\chi^2 = \sum \frac{(O_i - E_i)^2}{E_i}$$

and the corresponding significance probability for the test. Assess Pearson's claim that a binomial model is inadequate in the light of your results.

Testing $\chi^2 = 15.78$ against the $\chi^2(9)$ distribution we find a SP of 0.072 and Weldon's data do not seem as surprising as Pearson thought.

Referring to the rows labelled 4 and 5 in Table 13.4, Edgeworth replied to Weldon on the subject of the comparatively large differences between observed and expected frequencies.

Edgeworth to Weldon, 7 February 1894

> The tests which I have applied to the cases with four and five dice do not yield a result which excites much suspicion. I shall be curious to know your final decision.

Edgeworth also wrote to Pearson explaining his experimental results (9 February 1894) and received an immediate reply which began as follows.

> My Dear Edgeworth,
> Probabilities are very slippery things and I may well be wrong ...

Pearson's χ^2-test was devised in 1900, some five years later. As we have seen, a χ^2-test based on 9 degrees of freedom does not indicate a discrepancy between the observed frequencies and the $B(12, \frac{1}{3})$ distribution, but the position of the two large differences next to each other should, perhaps, give cause for concern.

You may remember that Weldon's original letter mentioned 26 306 throws altogether and, perhaps, the subsample of 7006 observations is not large enough to detect a small discrepancy from a binomial model with $p = \frac{1}{3}$. Table 13.5 gives the data and an analysis for the full sample.

Notice the small discrepancy in the sum of the expected frequencies, $\sum E_i = 26\,306.02$. This sort of small rounding error is not important.

Table 13.5 Weldon's data on 26 306 tosses of 12 dice

Number of 5s or 6s	Observed frequency	Expected frequency	$O_i - E_i$	$\frac{(O_i - E_i)^2}{E_i}$
0	185	202.75	−17.75	1.55
1	1149	1216.50	−67.50	3.75
2	3265	3345.37	−80.37	1.93
3	5475	5575.61	−100.61	1.82
4	6114	6272.56	−158.56	4.01
5	5194	5018.05	175.95	6.17
6	3067	2927.20	139.80	6.68
7	1331	1254.51	76.49	4.66
8	403	392.04	10.96	0.31
9	105	87.12	17.88	3.67
10	14	13.07		
11	4	1.19	3.69	0.95
12	0	0.05		
Total	26 306	26 306.02	−0.02	35.50

Testing $\chi^2 = 35.50$ against the $\chi^2(10)$ distribution gives a SP of 0.0001. On this basis you would reject the hypothesis that the data are plausibly from a binomial distribution $B(12, \frac{1}{3})$. It seems that Pearson was right in his suspicions about these data, even if his reasons were wrong.

What is particularly noticeable is the nature of the deviations $O_i - E_i$: the signs are all negative for low counts of 5s and 6s, and positive for high counts. The dice appear to be loaded in favour of landing with 5s and 6s facing uppermost.

It seems hard to believe that the modelling assumptions which lead to a binomial model could be wrong, but you would not be entirely surprised if the assumption of the dice having equiprobable outcomes was found wanting; after all, we have all heard of loaded dice and, even though Weldon's dice are supposed to be fair, their manufacture cannot absolutely guarantee perfect symmetry of form. It is interesting to use the data to estimate the binomial parameter and use the estimate to recalculate the expected values.

The parameter turns out to be estimated by $\widehat{p} = 0.3377$. This does not look very different from $p = \frac{1}{3}$ but, using expected values calculated for $B(12, 0.3377)$, the value $\chi^2 = 8.18$ is obtained. Since we have estimated a parameter and used that estimate in calculating the expected values, we must deduct an extra degree of freedom. The SP for the test is 0.516 and we have no reason to doubt that the data are binomial. It appears that at least one of the dice was indeed not symmetric.

Try checking these calculations for yourself.

This is an important result in that it demonstrates the feasibility of detecting small effects given a large enough sample size. It is amazing that the value of χ^2 should change so much when the binomial parameter p has changed by only 0.0044.

Let us leave the concluding remarks on probability modelling to Weldon. In the following extract, he is referring to data on measurement of crabs' shells.

Weldon to Galton, 6 March 1895

> ... I am horribly afraid of pure mathematicians with no experimental training.
>
> Consider Pearson. He speaks of the curve of frontal breadths, tabulated in the report, as being a disgraceful approximation to a normal curve. I point out to him that I know of a few great causes (breakage and regeneration) which will account for these few abnormal observations ... He takes the view that the curve of frequency representing the observations must be treated as a purely geometrical figure, all the properties of which are of equal importance ...
>
> For this reason, he has fitted a 'skew' curve to my 'frontal breadths'. This skew curve fits the dozen observations at the two ends better than a normal curve, it fits the rest of the curve, including more than 90% of the observations, *worse* ...
>
> Greenhill, to whom I took my troubles, laughs at the whole thing. You know that his chief business is to teach the properties of probability surfaces to artillery officers ...
>
> The Herring, which makes a skew curve, are very heterogeneous. The mean value of the length from snout to anus, on 717 males, was widely

different from that given by 990 males—the extra 270 being obtained by opening another of the cases of herrings. I have not the figures at hand, because I sent them to Pearson, as a basis for his curve; but he says that '*the material is homogeneous, with skew variation about the mean*'. I don't believe it!

Weldon's arithmetic would seem to be adrift here.

It would have been easy for Pearson and Weldon, a mathematician and a zoologist, to drift apart, but they remained firm friends and, six years later, they were planning the first issue of *Biometrika*, now established as one of the foremost statistical journals.

Another useful discrete probability model is the Poisson distribution. This has a particular application to the occurrence of events with passing time, and whether or not they occur 'at random'.

13.2.2 The American National Hockey League

It seems likely that in many sports the number of goals scored by one team against another in each game could be modelled by a Poisson distribution. This model was suggested for goals scored in the American National Hockey League. We can begin to test this hypothesis by studying the data reproduced from Mullet's paper in Table 13.6. This shows the frequency of games in which the indicated number of home goals were scored by Boston in its matches during the 1973–1974 season.

Mullet, G.M. (1977) Simeon Poisson and the National Hockey League. *American Statistician*, **31**, 8–12.

Table 13.6 Boston's home goals in the American National Hockey League,1973–1974 season

Number of goals	Frequency
0–1	1
2	2
3	5
4	9
5	10
6	5
7	2
8	3
9	1
≥ 10	1

To test the hypothesis that these data could have arisen from a Poisson distribution we need to do two things. First, we need to fit such a distribution to the data. Second, we need to see how closely this theoretical distribution fits the data. We know that the Poisson distribution has only a single parameter, and from *Chapter 6* we know that the maximum likelihood estimate of this parameter is given by the sample mean (see Table 6.8).

However, in this case it is not possible to deduce the sample mean from Table 13.6 because of the censoring of the data. The maximum likelihood estimate $\widehat{\mu}$ of the Poisson mean μ, based on these data, requires direct calculation.

Exercise 13.3

(a) Write down the likelihood of μ for the sample, and, using your computer, show that the corresponding maximum likelihood estimate $\widehat{\mu}$ of μ is

$$\widehat{\mu} = 4.9621.$$

(b) Hence construct a table of observed and expected frequencies for the number of goals scored, and use the chi-squared test for goodness-of-fit to test the hypothesis that the Poisson distribution provides a suitable model, and interpret your findings.

Our third model for variation is the uniform distribution.

13.2.3 Deaths in the USA

Much of the time goodness-of-fit is assessed from a picture and, for continuous distributions, we have used probability plots to good effect (see *Chapter 9*). We have also used histograms to look at the general shape of the data, but we have treated them with a certain degree of caution and only used them as rough, general guides, for the sorts of reasons described in *Chapter 1*.

The data in Table 13.7 give the monthly deaths in the USA during 1966. The question of interest is whether or not the death rate is affected by the time of year. This question may be put in statistical terms by suggesting, as a null hypothesis, that deaths occur uniformly over the year and therefore testing the goodness-of-fit of a uniform distribution to these data. A preliminary look at a histogram of the data gives the impression that a uniform distribution is plausible: this is shown in Figure 13.2.

Table 13.7 Monthly deaths in USA in 1966

Month	Deaths
January	166 761
February	151 296
March	164 804
April	158 973
May	156 455
June	149 251
July	159 924
August	145 184
September	141 164
October	154 777
November	150 678
December	163 882

Figure 13.2 Histogram of deaths in the USA

However, the frequencies are of the order of 150 000 and, perhaps, one might expect the uniform fit to look almost exact with such large numbers. A chi-squared test indicates otherwise.

Table 13.8 Monthly deaths in the USA in 1966, showing expected frequencies

Month	Deaths (Observed)	Deaths (Expected)
January	166 761	158 240.1
February	151 296	142 926.5
March	164 804	158 240.1
April	158 973	153 135.5
May	156 455	158 240.1
June	149 251	153 135.5
July	159 924	158 240.1
August	145 184	158 240.1
September	141 164	153 135.5
October	154 777	158 240.1
November	150 678	153 135.1
December	163 882	158 240.1

The expected frequencies reflect the fact that there are 31 days in January, 28 in February, and so on. (1966 was not a leap year.)

The chi-squared statistic is calculated from

$$\sum_i \frac{(O_i - E_i)^2}{E_i} = 3909.8.$$

There are 11 degrees of freedom and the SP is therefore 0, which constitutes overwhelming evidence for rejection of the null hypothesis that the data are fitted by a uniform distribution or, more precisely, that deaths are distributed uniformly over the year. It appears that time of year does affect the death rate.

It is very informative in this case to consider the differences $O_i - E_i$: these are positive for winter months and negative for summer months.

It is not really surprising that the evidence for rejection is so strong. We are dealing with a very large sample indeed and intuition dictates that the larger the sample we have, the more powerful the test. Remember that power is defined as the probability of correctly rejecting the null hypothesis. Therefore, the larger the sample, the more likely we are to reject when we should reject. The histogram in Figure 13.2 has proved to be deceptive.

13.3 Comparing two populations

It is probably fair to say that one of the most important activities in statistics is to provide an answer to the question: is there a difference? In general, the question may be posed in one of two main contexts. In the first case, a sample is taken from a population and some attribute is measured. Then some treatment is administered to members of the sample, at the end of which the attribute is measured a second time. Interest centres on what difference, if any, the treatment has made to the measured attribute.

The word 'treatment' may have a very broad interpretation, depending on the context.

Possible approaches here include a t-test (which makes certain assumptions about the form of the data) or a non-parametric test such as Wilcoxon's signed rank test (which does not).

13.3.1 Pneumonia risk in smokers

As we shall see, the data in Table 13.9 are rather unusual. They are measurements of the carbon monoxide (CO) transfer factor levels in seven smokers with chickenpox who were admitted to a hospital, and which were recorded with a view to determining their risk of contracting pneumonia. The measurements were taken when the patients entered the hospital and were repeated one week later.

Table 13.9 CO transfer factor levels in smokers with chickenpox

Patient	On entry	One week later
1	40	73
2	50	52
3	56	80
4	58	85
5	60	64
6	62	63
7	66	60

Ellis, M.E., Neal, K.R. and Webb, A.K. (1987) Is smoking a risk factor for pneumonia in patients with chickenpox? *British Medical Journal*, **294**, 1002. On admission, patients were treated with intravenous acyclovir at 10 mg/kg, eight-hourly for five days. It is not recorded whether they were required to abstain from smoking.

On the face of it, this seems to be a simple data set. We are interested in the difference between the carbon monoxide transfer levels at entry and those after one week. Measurements have been repeated on the same patients, and the correct procedure is the t-test for zero difference, provided we can assume that the pairwise differences between the samples are normally distributed. We simply perform a one-sample test that the mean is zero on the differences between measurements for each individual.

Student's t-test for zero mean difference is described in *Chapter 8*.

Exercise 13.4

(a) Obtain the value of the test statistic t in a t-test for zero mean difference on the data of Table 13.9.

(b) Find the SP for your test, stating whether your test is one-sided or two-sided. Interpret your findings.

We have found little evidence to reject the null hypothesis of no difference between CO transfer factor at entry and that after one week. But look again at the data. Six out of the seven differences are positive, so the result comes as something of a surprise. With so few data points we cannot usefully draw a histogram, but we can look at boxplots.

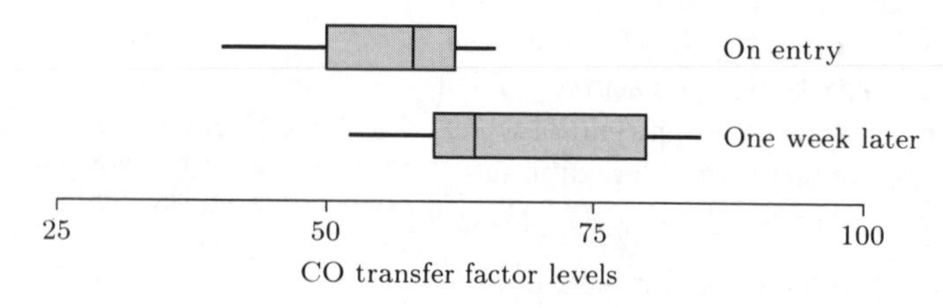

Figure 13.3 Comparative boxplots for CO transfer factors

Figure 13.3 also shows a marked difference in the boxplots, suggesting a significant difference between the CO transfer factor measurements.

In fact, the explanation of this evident contradiction is simple—we have not checked the data for normality!

Exercise 13.5

Check the data for normality by carrying out a normal probability plot for the differences in CO transfer factors. Is the assumption of normality plausible?

Clearly, the t-test was not soundly based, and its conclusions are doubtful. A test for zero difference which does not involve any distributional assumptions is the Wilcoxon signed rank test.

Exercise 13.6

Use Wilcoxon's signed rank test to investigate the hypothesis that there is no difference between CO transfer factors in smokers after a five-day acyclovir treatment.

13.3.2 Viral lesions on tobacco leaves

Bearing in mind that we should always check our test assumptions, let us consider the following problem. Two virus preparations were soaked into cheesecloth and each was rubbed onto different halves of a tobacco leaf. Numbers of local lesions appearing on each half were counted for eight leaves and these are shown in Table 13.10. Lesions appear as small, dark rings. Do the two treatments produce different effects?

Youden, W.J. and Beale, H.P. (1934) *Contributions (Boyce Thompson Institute)*, **6**, p. 437.

Table 13.10 Viral lesions on tobacco leaves

Leaf	Preparation 1	Preparation 2
1	31	18
2	20	17
3	18	14
4	17	11
5	9	10
6	8	7
7	10	5
8	7	6

A t-test for zero mean difference results in a t-value of 2.625 and, tested against $t(7)$, gives a two-sided SP of 0.034. However, we know better than to apply such a procedure blindly, so we should look first at a normal probability plot for the differences.

Exercise 13.7

(a) Check the data for normality by carrying out a normal probability plot for the differences. Is the assumption of normality plausible? If not, how do you suggest we should proceed?

(b) Proceed with a test for zero difference after implementing your suggestion.

Now this is an interesting result in that removal of the outlier, a point which, if included, is sufficiently extreme to pull the sample mean further away from zero, has reduced the SP. Intuitively this is not what we might have expected, but a moment's thought reveals why. The outlier does not simply influence the sample mean. It also has a considerable influence upon the sample standard deviation, which is used in calculating the t-statistic. In this particular case, the effect of removing the outlier is to reduce the estimated standard deviation sufficiently to increase the value of the t-statistic, and thereby reduce the SP.

In both these examples, the data were paired. In other situations, we may not wish to compare 'Before – After' differences but simply to decide whether two samples—assumed to be independent and not necessarily of the same size—may be regarded as arising from the same population.

In such a case, one might consider a two-sample t-test, which makes certain assumptions about the form of the data, or possibly a Mann–Whitney–Wilcoxon test, which is less restrictive. (And one should not forget the usefulness of a simple graphical representation such as a comparative boxplot.)

13.3.3 Survival prognosis for patients with acute myocardial infarction

Table 13.11 gives clinical data on twenty patients who were still alive a month after suffering a myocardial infarct (heart attack) and twelve who died within a month. For each patient serum urea (in grams per litre) was measured on admission to hospital and on Day 6 after admission. The questions of interest, and the reason the data were collected, were: is there, on average, a difference between the two groups on Day 0 and is there a difference on Day 6? If so, can this information be used in a predictive way, to predict the likely outcome after one month?

Albert, A. and Harris, E.K. (1987) *Multivariate Interpretation of Clinical Laboratory Data.* Marcel Dekker, New York, pp. 165–168.

Table 13.11 Serum urea (g/l) measurement on 32 patients with myocardial infarction

Group 1 Survivors		Group 2 Non-survivors	
Day 0	Day 6	Day 0	Day 6
0.55	0.58	0.39	0.63
0.31	0.54	0.34	0.46
0.58	0.72	0.54	0.74
0.48	0.52	0.38	0.60
0.32	0.38	1.04	0.68
0.32	0.46	0.45	0.94
0.42	0.66	0.25	0.79
0.40	0.42	0.34	0.56
0.44	0.80	0.26	0.52
0.25	0.61	0.35	0.69
0.30	0.36	0.22	0.90
0.47	0.58	0.42	1.94
0.40	0.62		
0.42	0.64		
0.37	0.46		
0.38	0.49		
0.20	0.32		
0.40	0.37		
0.40	0.65		
0.48	0.60		

Now let us examine the first question. One way we might seek to explore whether the groups differ on these two variables is to compare the groups' means. We learned, in *Chapter 8*, that a two sample t-test could be used for this purpose—provided certain assumptions were justified. These assumptions were that the samples were independent, the underlying populations were normal, and that the variances of the two populations were equal. We could accept some departure from the latter two assumptions, provided it was not too severe.

The first of these assumptions—the independence between the samples of surviving and non-surviving patients—seems reasonable, so we shall accept that. As to the second and third assumptions, we can obtain some insight into them from the data themselves.

We can explore informally the assumption of normality using histograms. Figure 13.4 shows histograms of the urea measurements on Day 0 and Day 6 for each group separately.

The first two of these histograms do not show any striking departure from normality—at least, bearing in mind the limitations of relatively small samples, there is no clear skewness or asymmetry.

For the non-survivors, the outlying data points in Figures 13.4(c) and 13.4(d) correspond to patient 5 (a reading of 1.04 g/l at Day 0 and 0.68 g/l at Day 6) and to patient 12 (0.42 g/l at Day 0 and 1.94 g/l at Day 6). Patient 5 was the only one whose serum urea measurement dropped from an exceptionally high reading on admission; patient 12 recorded an exceptionally high reading at Day 6.

Such isolated data points substantially different from the others are outliers, and need to be examined carefully. We need to consider whether they might have arisen due to a malfunction of the measuring instrument or perhaps a data recording or transcription error. If we do suspect such a situation then it obviously makes sense to analyse the data with these points removed. Even if we do not suspect such a situation, we can analyse the data without the offending points (as was suggested in *Chapter 9*) on the grounds that the estimates of the average for the two groups are then more robust. That is, they are less susceptible to random variation and so are more reliable. Yet other alternatives are to transform the data in some way or adopt a distribution-free procedure.

In this case let us omit the readings for the two exceptional non-surviving patients from any subsequent analysis. However, it is important to remember

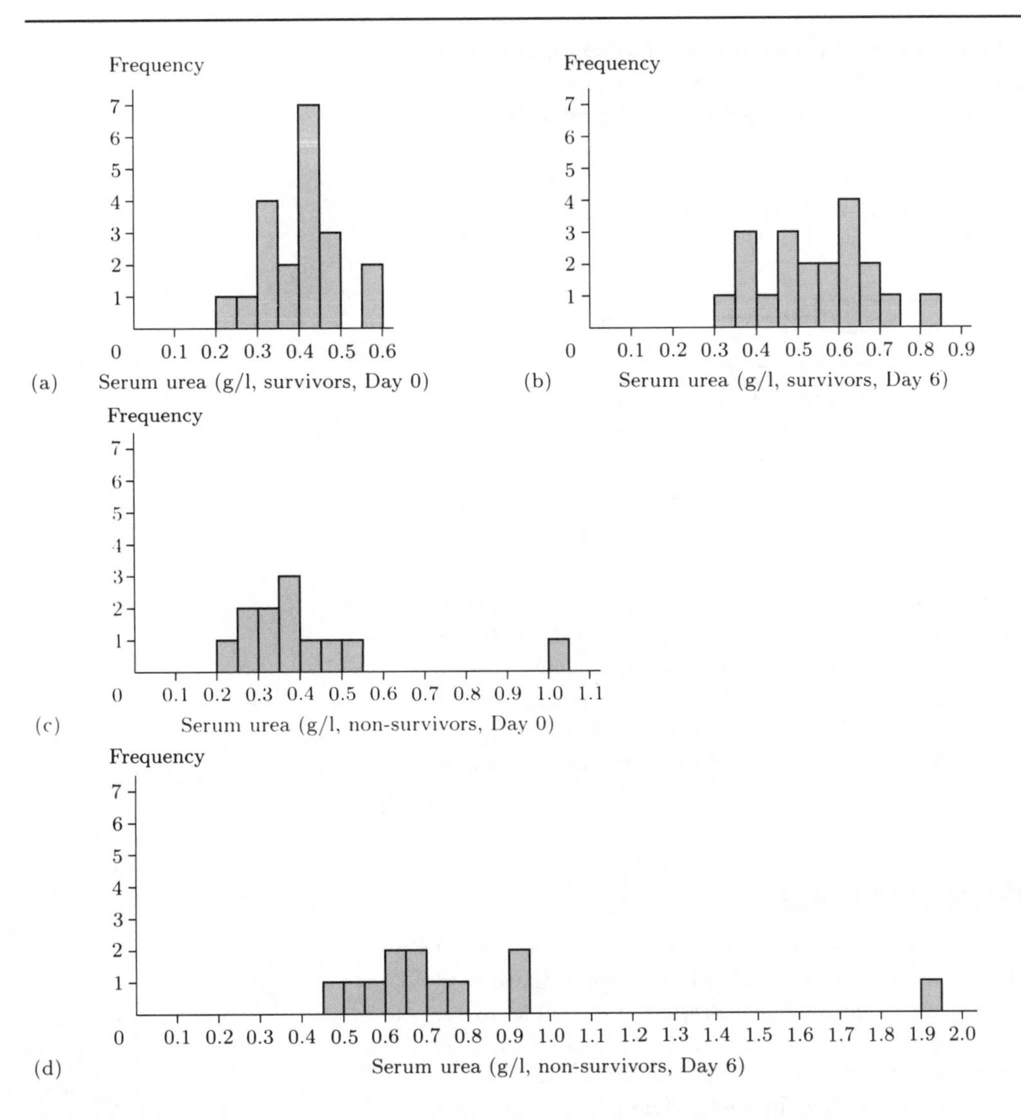

Figure 13.4 (a) Survivors at Day 0 (b) Survivors at Day 6 (c) Non-survivors at Day 0 (d) Non-survivors at Day 6

that each patient recorded exceptionally high serum urea levels at some stage following their heart attack, and that this points to a contra-indication for survival.

Exercise 13.8

(a) Use a two-sample t-test to compare the two group means at admission, omitting patients 5 and 12 from the group of non-survivors.

(b) Similarly, explore whether there is any significant difference between the underlying serum urea levels for surviving and non-surviving patients at Day 6.

So we have evidence that by the sixth day after admission there is a significant difference in serum urea levels between those patients who will survive to one month and those patients who will not.

The second question asked if it was possible to use this information to predict whether a patient was likely to survive. The non-survivors have the higher mean and, given that the distributions are roughly symmetric with equal variances (with outliers omitted), this will imply that the higher scores tend to belong to the non-survivors and the lower ones to the survivors. This is illustrated in Figure 13.5, which shows boxplots of the scores of the two groups.

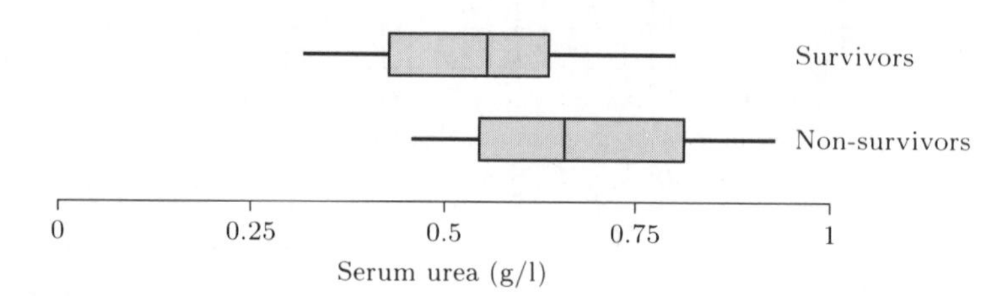

Figure 13.5 Boxplots of survivors and non-survivors on Day 6, outliers omitted

One might attempt to answer this question by choosing some threshold (say, about 0.60) and classifying a patient with serum urea concentrations below this threshold on Day 6 as 'likely to survive' and a patient with a concentration above this threshold as 'not likely to survive'. Of course, one would not get the classification correct all of the time, but it would be a start. (This idea is taken further in *Chapter 14*, where the technique of *discriminant analysis* is outlined.)

13.3.4 Diet supplements in rats

In this example straightforward comparisons of two independent groups, of the type illustrated in Subsection 13.3.3 and described in *Chapter 8*, are used to answer a more subtle question. The data are presented in Table 13.12. They show body weights in grams of rats on two diets, measured on four occasions: on starting the special diets, one week after starting, two weeks after starting and three weeks after starting the diets. There were eight rats in the first group and four rats in the second. The question the researchers were interested in was: are the rates of increase of weight the same on the two diets?

Table 13.12 Body weights (grams) of rats on two diets measured at weekly intervals

Crowder, M.J. and Hand, D.J. (1990) *Analysis of repeated measures.* Chapman and Hall, London, p. 19.

Group 1				Group 2			
Start	1 week	2 weeks	3 weeks	Start	1 week	2 weeks	3 weeks
266	265	272	278	504	507	518	525
244	238	247	245	530	543	544	559
267	264	268	269	544	553	555	548
272	274	273	275	542	550	553	569
273	276	278	280				
278	284	279	281				
271	282	281	284				
267	273	274	278				

This differs from the preceding example in a number of ways. Previously we had just two observations on each patient, and interest lay in each of these observations separately. Now we have four observations on each rat and, more importantly, we are not interested in each of them separately but in all four together: somehow we want to study the rate of growth.

Fortunately, for an individual rat we know how to do this. For an individual rat the slope of the regression line of weight on time will tell us just this—the rate of increase in weight as time changes, for that rat. We could calculate slope estimates for each rat in this way to yield, for each rat, a rate of growth. In effect we would have reduced each rat's four measurements to a single number, summarizing just that aspect of the data in which we are interested. Then we could compare the two groups using these numbers. That would certainly answer the question: do the average growth rates differ between the two groups?

Regression was introduced in *Chapter 10*.

Exercise 13.9

Use your computer to obtain slope estimates for each of the eight rats in Group 1 and for each of the four rats in Group 2. Calculate the mean and standard deviation of the growth rates in each group. Can you reach any conclusions at this stage?

Again we could consider a t-test for equal slopes in the two groups. A two-sample t-test gives a test statistic of -3.122 and a total SP of 0.011. This is highly significant. But the large difference in size between the standard deviations might make one uneasy about trusting the conclusions of this test. In addition to this, with such a small sample size we would not be able to detect departures from normality very well, so this assumption would be very much on trust.

Try checking these calculations for yourself.

Alternatively, we could adopt a distribution-free method, as outlined in *Chapter 9*.

Exercise 13.10

Perform a Mann–Whitney–Wilcoxon test on the growth rates in Table S13.5. Can you now conclude that the growth rates differ?

There is one other feature of the data that you might have noticed here: the rats in the second group were initially much heavier than those in the first group (about twice the weight). The discrepancy at the end of the experiment is similarly marked. An alternative regression model would assume that weight increase with time was multiplicative rather than additive. Straight lines fitted to the logarithms of the weights regressed against time have slopes

$$0.0158, 0.0049, 0.0037, 0.0029, 0.0083, 0.0014, 0.0137, 0.0125$$

for the eight rats in the first group, and

$$0.0144, 0.0162, 0.0026, 0.0151$$

for the four rats in the second group. The sample variances for the two groups are very similar; the total SP for a two-sided two-sample t-test of equal slope is 0.27. There is no evidence of a difference in growth rate between the two groups, using this multiplicative model.

Again, you should check these calculations if you have the time.

13.3.5 Expressed emotion

The *expressed emotion index* is a measure of the emotional climate of families with mentally ill members. Studies suggest that patients living with relatives scoring low on the expressed emotion index are less likely to relapse than those living with relatives who score high.

In a study of the relationship between expressed emotion and schizophrenia in Spain 60 patients were followed up for two years after a psychiatric evaluation. One patient dropped out of the study after twelve months, leaving only 59. Dropouts are a common problem with medical research where, as in this example, measurements take place over an extended time period. In this case we shall simply ignore the dropout—so that any inferences we make are strictly only valid for patients who do not drop out before a two-year period has elapsed. (An alternative approach, when treatments are being compared, is to regard the dropouts as treatment failures and include them as such in the analysis.)

Montero, I., Gomez-Beneyto, M., Ruiz, I., Puche, E. and Adam, A. (1992) The influence of family expressed emotion on the course of schizophrenia in a sample of Spanish patients. *British Journal of Psychiatry*, **161**, 217–222.

At the initial evaluation the families of the patients were scored on an expressed emotion scale. Table 13.13 shows the number of patients who relapsed during the two-year follow-up period for the low and high expressed emotion families.

Table 13.13 Cross-classification of high and low expressed emotion by whether the patient did or did not relapse

	Expressed emotion	
	Low	High
Relapse	17	16
No relapse	14	12

The research question of interest here is whether the proportions relapsing in the two groups are the same. Clearly they are not exactly the same. The proportion in the low expressed emotion group is $17/31 = 0.548$ while that in the high expressed emotion group is $16/28 = 0.571$. Certainly, as the theory predicted, the proportion in the low expressed emotion families is lower than that in the high expressed emotion families, but these proportions are fairly close: could the difference have arisen by chance?

You saw, in *Chapter 8*, how such a question could be answered. A suitable test is Fisher's exact test. This takes the marginals of the table as given (the row and column totals) and works out how many possible tables with these marginals would lead to a difference in proportions more extreme than that actually observed. The calculations for this test are fairly onerous, and use of a computer is really the only viable option.

Exercise 13.11

Use a computer to test the hypothesis that the proportion of relapses in a low expressed emotion group is lower than the proportion of relapses in a high expressed emotion group, using the data of Table 13.13.

In the case of the above table, the computer gives a total SP of 1 for the test of the hypothesis that there is no difference between the proportions in the populations from which the samples were taken (a two-sided test). It also gives an obtained SP of 0.534 for the (one-sided) test of the hypothesis that the low expressed emotion families have a *smaller* proportion than the high expressed emotion families (instead of merely a *different* proportion). This tells us that there is a very high probability of obtaining a table as or more extreme than the above, if the null hypothesis of identical proportions in the two populations (high and low expressed emotion) is true. In terms of the research question, we have no reason to suppose that patients belonging to families with high or low expressed emotion differ in the probability that they will relapse.

This subsection brings us rather naturally to the topic of contingency tables, a way of setting out bivariate data, and also to tests of association.

13.4 Contingency tables

Questions involving the relationships between variables are very common. They include such questions as 'Does this cause that?' 'Are these properties related?' 'Does the relationship between these change according to the level of some third variable?' and so on. The course has focused on relationships between two variables since this is the most important special case, and occurs very commonly. (*Chapter 14* briefly discusses some of the extensions which are possible when more than two variables are involved.)

As far as two variables are concerned, correlation and regression methods have been described for answering questions about the relationships between variables which are measured on a numerical scale, and contingency table and chi-squared techniques are appropriate for answering questions about the relationships between variables which have categorical scores. In this section the chi-squared test is revised.

13.4.1 Hospital treatments

The data reproduced in Table 13.14 show the results of a surgical procedure designed to improve the functioning of certain joints which have become impaired by disease. For each of five hospitals, counts are given of the numbers of patients who fall into each of three post-operative categories: no improvement, partial functional restoration and complete functional restoration. The question we would like to answer is: are the response patterns of the hospitals different? So, for example, does one hospital produce more complete functional restorations than the others?

Box, G.E.P., Hunter, W.G. and Hunter, J.S. (1978) *Statistics for Experiments*. John Wiley and Sons, New York.

Box, Hunter and Hunter take pains to point out that these categories were carefully defined in terms of measurable phenomena.

Table 13.14 Results of surgical procedure classified by hospital

	Hospital				
	A	B	C	D	E
No improvement	13	5	8	21	43
Partial functional restoration	18	10	36	56	29
Complete functional restoration	16	16	35	51	10
Total	47	31	79	128	82

Now, clearly, from looking at the table we can see that hospital D does produce more complete restorations than do the other hospitals. But this simple number does not take into account the overall number of patients seen by that hospital. Indeed, the last row of the table shows that the sample of patients seen by this hospital is greater than the sample seen in any of the other hospitals. Somehow we must make an allowance for this in comparing the patterns of results across the hospitals.

Moreover, we are just dealing with samples here. In fact, we would like our conclusions to be more general than that—we really want to make an inference to the likely outcome on future patients, not simply produce a summary of the results observed in the particular sample taken. To put it another way, one with which you should be completely familiar by now, we want to see if there are differences in the underlying population from which our sample of results was drawn. And we do this by assuming that the distributions across the hospitals are the same and exploring, under this assumption, how often one would obtain data as extreme as that observed.

One approach might be to convert the numbers to proportions. Dividing each number in a column by the column total, we can show, for each hospital, what proportion of patients had no improvement, what proportion had partial functional restoration, and so on.

The results of these calculations are shown in Table 13.15.

Table 13.15 Results of surgical procedure classified by hospital, converted into column percentages

	Hospital				
	A	B	C	D	E
No improvement	27.7	16.1	10.1	16.4	52.4
Partial functional restoration	38.3	32.3	45.6	43.8	35.4
Complete functional restoration	34.0	51.6	44.3	39.8	12.2

From this we can see that the previous exceptionally large number of patients with complete functional restoration in hospital D is, in fact, mainly a function of the large sample size assessed for this hospital: hospitals B and C both have a greater proportion of patients classed into this category in our sample.

But we are still only talking about a *sample*. Do our sample results reflect real underlying differences in proportions?

In *Chapter 11* we saw how this question could be addressed using a *chi-squared test for independence*. This tests the null hypothesis that the distributions down the columns are the same for all columns. (This is equivalent to the null hypothesis that the distributions across the rows are the same for each row, and also to the null hypothesis that the row and column classifications are independent—hence the name of the test.) That is, beginning with this null hypothesis as a basic assumption, the expected number in each cell of the table is calculated. The chi-squared test statistic is then calculated—this is a measure of the difference between the observed numbers in the cells and the expected numbers. Its size tells us how large the difference is, and its distribution under the assumption of independence can be calculated. By comparing the observed value of this test statistic with this distribution we can see how unlikely such a value would be, if the null hypothesis were true.

Exercise 13.12

Under the null hypothesis of independence, calculate expected values for each of the cells in Table 13.14. Use these to calculate a value of the chi-squared test statistic and hence obtain a significance probability for the test. Interpret your findings.

In fact, as it happens, hospital E is different from the other hospitals. It is a 'referral' hospital. Given this piece of extra information, we might ask whether the difference in outcome distributions that we have observed can, at least in part, be attributed to the difference between the referral hospital and the others—the difference between hospital E and the others. Some suggestion that this might be the case is afforded by Figure 13.6. This shows the proportions listed in Table 13.15, displayed in a more convenient way.

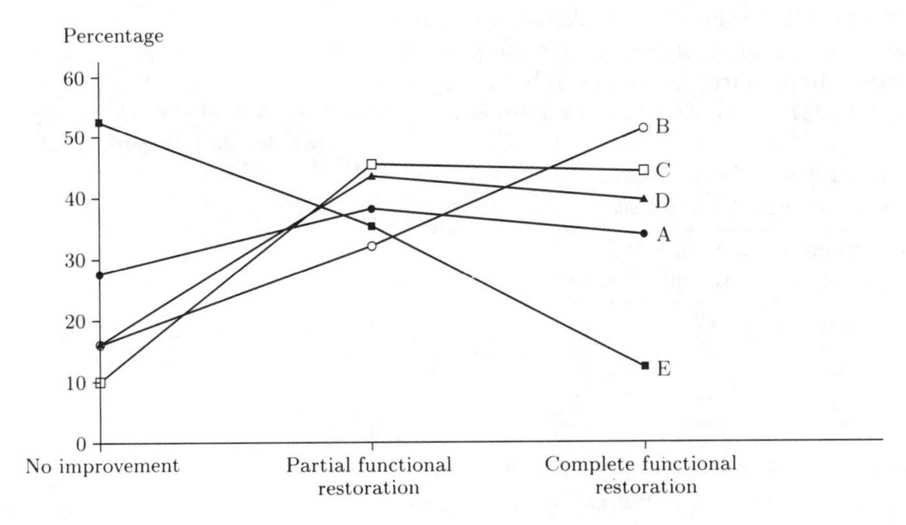

Figure 13.6 Proportions in each hospital showing each outcome

It certainly does look as if hospital E differs from the others, but again we have to ask whether the difference reflects a real underlying difference or if it could just be due to sampling variation.

The question we are now considering still involves the three outcome categories that we had before, but now it involves just two hospital categories, referral and non-referral. This means that we are interested in the condensed table shown in Table 13.16.

Table 13.16 Table showing referral and non-referral results

	Referral hospital (E)	Non-referral hospitals (A, B, C, D)
No improvement	43	47
Partial functional restoration	29	120
Complete functional restoration	10	118
Total	82	285

A chi-squared test on this table yields a test statistic value of $\chi^2 = 49.8$ and relating this to a chi-squared distribution with $(3-1)(2-1) = 2$ degrees of freedom again shows a very low significance probability $\left(SP = 1.5 \times 10^{-11}\right)$. So it does seem that hospital E differs from the others.

Box, Hunter and Hunter also conducted a test on the four non-referral hospitals, to see if any difference between them is statistically significant. It turns out not to be, with a chi-squared test statistic value of $\chi^2 = 8.3$ and, from the chi-squared distribution with 6 degrees of freedom, a SP equal to 0.22.

Thus the highly significant difference between the outcome profiles of the hospitals seems to be largely attributable to the difference between the referral and non-referral hospitals.

13.4.2 The Framingham Heart Study

The Framingham Longitudinal Study of Coronary Heart Disease was an important investigation into the factors causing heart disease. Among the variables measured were blood pressure and serum cholesterol. One question of interest is whether these are related. Reproduced below, in Table 13.17, is the distribution of these variables for 1237 subjects without heart disease.

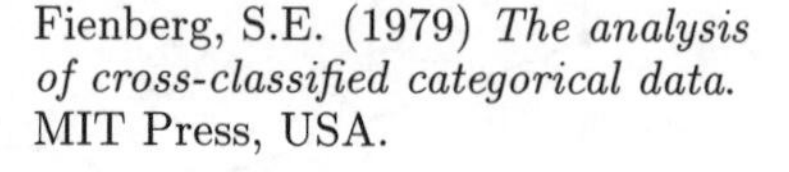
Fienberg, S.E. (1979) *The analysis of cross-classified categorical data.* MIT Press, USA.

Table 13.17 Cross-classification of systolic blood pressure by serum cholesterol for those subjects without heart disease

Serum cholesterol (mg/100cc)	Systolic blood pressure (mm/Hg) < 127	127–146	147–166	> 166
< 200	117	121	47	22
200–219	85	98	43	20
220–259	119	209	68	43
> 259	67	99	46	33

Now this example is rather different from that in Section 13.4.1. Both variables here are numerical, but they have been grouped into categories. If they had not been grouped and the original raw scores for each subject were available, then we could plot a scatter diagram and judge by eye whether the variables were related. We could also calculate a correlation coefficient. (Indeed, we could obtain a rough correlation coefficient from the data we have by assigning some numerical value to each of the levels of the two categorized variables.)

However, the correlation coefficient is only a measure of *linear* relationship between two variables. That might be useful in particular circumstances—and, in particular, it might be useful in the present case if we merely wanted to know if blood pressure and serum cholesterol increased or decreased together in the population. On the other hand, we might be interested in the more general question of whether there was any relationship between the variables, not simply a linear one. The chi-squared test addresses just that—it seeks departures from independence, without restricting them to be linear.

Also in this example, it is not so clear which set of profiles are the relevant ones. In Section 13.4.1 it was obvious that we wanted to compare column profiles. But here the real question is rather one of whether or not the variables are independent.

Exercise 13.13

Under the null hypothesis of independence calculate a value of the chi-squared test statistic for the data in Table 13.17 and hence obtain a SP for the test. Interpret your results.

13.5 Correlation

13.5.1 The dopamine hypothesis of schizophrenia

Signals in the brain are carried between cells by chemicals called neurotransmitters. These leave one cell and arrive at the receptors of another. In recent years the dopamine hypothesis of schizophrenia has postulated that in those individuals with this condition there will be an increase of a certain type of receptors called D2 dopamine receptors. This hypothesis is based on results of *post mortem* and *in vitro* studies. However, *in vivo* studies using positron emission tomography and single photon emission tomography do not report density increases of these receptors. Martinot and colleagues focused their research on the hypothesis of a link between the density of striatal D2 receptors and certain clinical dimensions of schizophrenia. To determine the former they injected a radioactive isotope into the patients and measured the ratio of striatum to cerebellum radioactivity concentration two hours after injection (this is called the S:C ratio). One measure of the latter that they used was psychomotor expressiveness which is derived as a weighted combination of the responses to a number of indicators: this is called PEF.

Martinot, J.L., Paillere-Martinot, M.L., Loc'h, C., Lecrubier, Y., Dao-Castellana, H., Aubin, F., Allilaire, J.F., Mazoyer, B., Maziere, B. and Syrota, A. (1994) Central D2 receptors and negative symptoms of schizophrenia. *British Journal of Psychiatry*, **164**, 27–34.

In fact, the PEF was the first factor which resulted when a 'principal components' analysis was performed. This is an advanced statistical technique which is outlined in *Chapter 14*.

Table 13.18 shows the results of these measurements taken on ten schizophrenic patients (seven men and three women) who formed a very homogeneous group of young, drug-free patients with a short course of illness. One reason for choosing such a homogeneous sample was that the investigators were restricted to a small sample: keeping them homogeneous eliminated superfluous variation, so making estimates more accurate. Of course, on the other hand, it makes inferences to other types of people (older patients, those taking medication, those with a long course of illness) more problematical. In an ideal world we would have a large sample randomly selected from the entire population of schizophrenic patients, but this is impracticable for several reasons, including that of cost.

Table 13.18 PEF and S:C measurements for 10 patients

PEF	S:C
48	3.32
42	3.74
44	3.70
35	3.43
36	3.65
28	4.50
30	3.85
13	4.15
22	5.06
24	4.27

The question of interest is whether there is a relationship between the two scores. A plot of the data is given in Figure 13.7. There certainly seems to

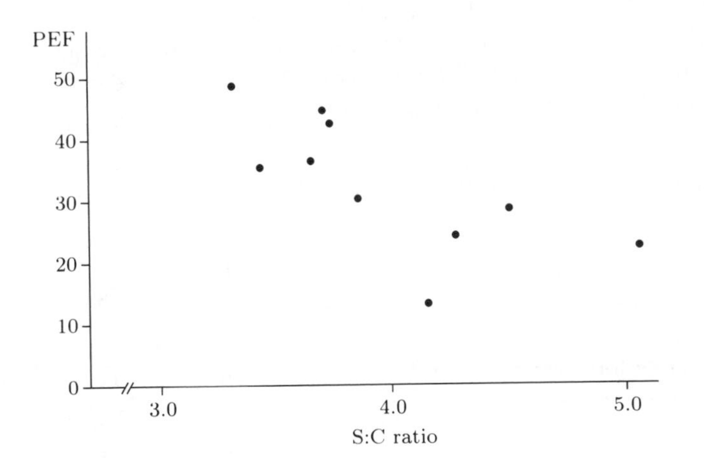

Figure 13.7 A plot of psychomotor expressiveness factor (PEF) against S:C ratio for the sample of ten schizophrenic patients

be a relationship, but how strong is it and could the apparent relationship have arisen by chance (by an accident of the way the sample of patients was drawn) from a population in which there is, in fact, no such relationship overall?

This sort of question was raised in *Chapter 11*, where the ideas of correlation were introduced. There the Pearson correlation coefficient was described as a measure of the strength of a straight-line relationship between two sets of scores. Its value for the data given in Table 13.18 is -0.72. The negative sign indicates that increasing S:C ratio is associated with decreasing PEF score (and vice versa). The absolute size (0.72) means that the relationship appears to be reasonably strong. (Whether or not a particular size of a correlation coefficient is regarded as 'large' depends very much on the field of application. In areas such as psychology and sociology, where there are typically large variances in measured values, a coefficient of 0.72 may be quite respectable. In contrast, in areas such as physics and engineering, where the variances may be small, 0.72 might be regarded as a sign of a very weak relationship. It very much depends on the question and the type of data involved.)

Now let us consider the second question. Could a correlation coefficient of -0.72 have arisen easily by chance sampling from a population in which there is no overall relationship?

Exercise 13.14

For the data in Table 13.18 with Pearson correlation -0.72, test the hypothesis that the underlying correlation is zero, and state the conclusions to be drawn from your test.

It seems unlikely that we would have obtained such a strong relationship in our sample if the population had no relationship.

13.5.2 Finger ridges of identical twins

The data in Table 13.19 are counts of the numbers of finger ridges of individuals for twelve pairs of identical twins.

Newman, H.H., Freeman, F.N. and Holzinger, K.J. (1937) *Twins.* University of Chicago Press, Chicago.

Numbers of finger ridges are evidently similar within the pairs, but vary considerably between different sets of twins. The interest lies in estimating the correlation between identical twins. A scatter plot is given in Figure 13.8. Notice, incidentally, that two of the points on the plot are superimposed at (114, 113).

The relationship looks very strong: we can proceed with calculating the sample correlation coefficient.

Table 13.19 Numbers of finger ridges of identical twins

Pair number	Twin 1	Twin 2
1	71	71
2	79	82
3	105	99
4	115	114
5	76	70
6	83	82
7	114	113
8	57	44
9	114	113
10	94	91
11	75	83
12	76	72

Exercise 13.15

Calculate the sample correlation coefficient for the data in Table 13.19. Is there high correlation between finger ridge counts for identical twins?

(Incidentally, it is not clear in Table 13.19 by what means 'Twin 1' is distinguished from 'Twin 2', but different definitions could lead to slightly different estimated correlations.)

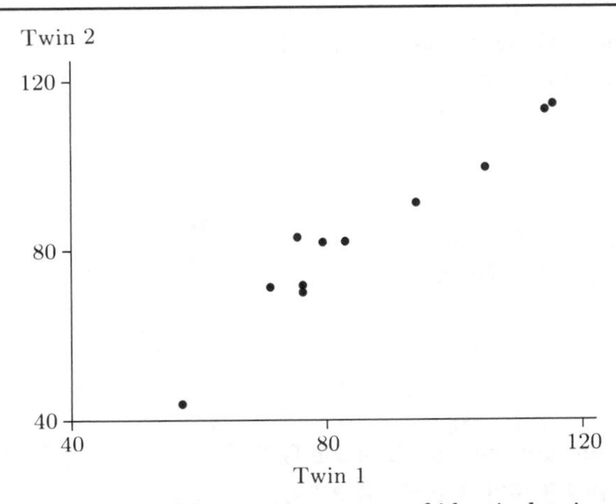

Figure 13.8 Scatter plot of finger ridge counts of identical twins

13.6 Regression analysis

Regression analysis seeks to explain, or at least to model, the way in which the distribution of a random variable Y alters with changing circumstances (such as passing time).

13.6.1 Wind speed and athletes' times

This study was designed to investigate the effect, if any, of a following wind on sprinters' running times in athletic events. It is generally accepted that a following wind aids sprinters and horizontal jumpers such as long jumpers and hurdlers: times and distances are not counted for record purposes if the following wind speed exceeds 2.0 metres per second.

Hitchcock, S. (1993) Does wind speed affect times? *Track Stats*, **31**, 19–26.

The wind speed and race time were recorded for the 21 races for the British 110 m hurdler Colin Jackson in 1990. These data are listed in Table 13.20.

A negative wind speed indicates a head wind. A scatter plot of the data is given in Figure 13.9. Here, wind speed is treated as the explanatory variable and race time as the response.

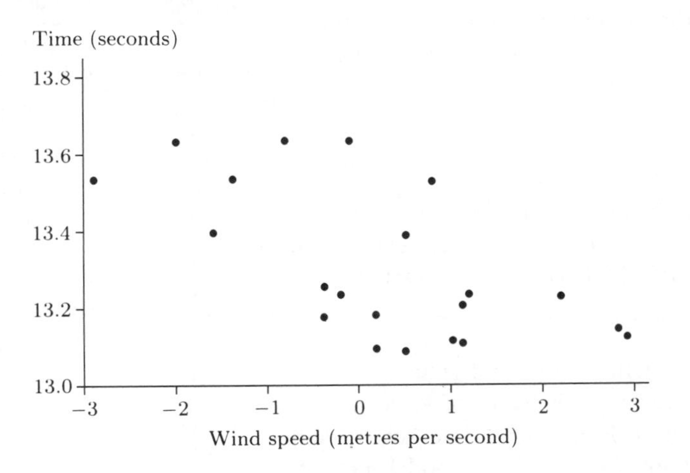

Figure 13.9 Colin Jackson's performances in 1990

Table 13.20 Wind speed and race time, 110 m hurdles

Wind speed (m/s)	Time (s)
−2.9	13.53
−2.0	13.63
−1.6	13.39
−1.4	13.53
−0.8	13.63
−0.4	13.17
−0.4	13.25
−0.2	13.23
−0.1	13.63
0.2	13.09
0.2	13.18
0.5	13.08
0.5	13.38
0.8	13.52
1.0	13.11
1.1	13.10
1.1	13.20
1.2	13.23
2.2	13.22
2.8	13.14
2.9	13.12

There is, evident in Figure 13.9, a tendency for faster times to be associated with stronger following winds, although there is considerable scatter.

Exercise 13.16

Fit a least squares straight line to the data of Figure 13.9, and comment on the usefulness of the fitted model for predicting future race times, given wind speed and direction.

The fitted model suggests that in 1990 Jackson would have run 13.32 seconds on average in windless conditions. For every metre per second following wind, he ran on average 0.085 seconds faster.

The estimated slope $\widehat{\beta} = -0.085$ seems very small in absolute terms. To test the hypothesis that wind speed has no effect on athletes' times (or, at least, on Jackson's times) one can set up a one-sided hypothesis test as follows.

$$H_0 : \beta = 0$$
$$H_1 : \beta < 0$$

Exercise 13.17

Calculate the obtained SP for the hypothesis that there is no wind speed effect, using the data provided, and interpret your findings.

13.6.2 Using X-rays to kill bacteria—Part I

The 'single hit' hypothesis of X-ray action under constant radiation fields states that bacteria have a single vital centre and that this has to be hit by an X-ray in order to kill or inactivate the bacterium. To explore this, the data in Table 13.21 were collected. These show estimates of the numbers of bacteria which survive 200 kilovolt doses of X-radiation over various periods ranging from 6 minutes to 90 minutes. If the theory is correct, then the relationship between the time of exposure, t, and the number of bacteria surviving, n_t, should have the form

$$n_t = n_0 e^{\beta t}, \qquad t \geq 0, \tag{13.1}$$

where n_0 and β are parameters describing the hypothesized relationship. The number n_0 is the number of bacteria at the start of the experiment (as can be seen by setting $t = 0$ in the model) and $\beta < 0$ is the rate of destruction of the bacteria.

Chatterjee, S. and Price, B. (1977) *Regression analysis by example*, 1st edn. John Wiley and sons, New York, p. 32.

Table 13.21 Number of surviving bacteria after exposure to X-rays for the indicated time

Time (minutes)	Estimated number surviving
6	35 500
12	21 100
18	19 700
24	16 600
30	14 200
36	10 600
42	10 400
48	6 000
54	5 600
60	3 800
66	3 600
72	3 200
78	2 100
84	1 900
90	1 500

A plot of n_t against t is shown in Figure 13.10. This has the sort of shape one would expect if the model at (13.1) were true—the value of n_t decays as t increases. And, of course, as we expect from real phenomena, the relationship is not perfect. There is some variation about the postulated model, perhaps due to measurement error or random fluctuation.

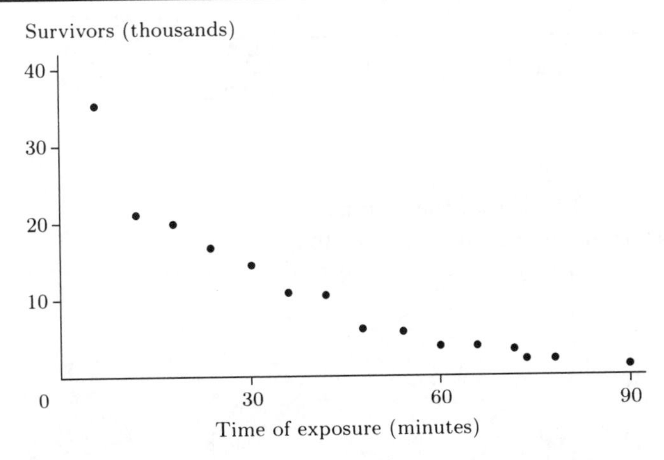

Figure 13.10 Estimated number of bacteria surviving plotted against time of exposure to X-rays

We want to go further than simply say 'it has the sort of shape we would expect'. We would like to fit our model to these data. The difficulty is that the model is not a simple linear model—it is not a straight line of the kind we have learned how to fit. We can try to transform the model so that it becomes a straight line and then fit a straight line to similarly transformed data.

Transforming the model at (13.1) by taking logarithms yields

$$\log(n_t) = \log(n_0) + \beta t \qquad t \geq 0 \tag{13.2}$$

which has exactly the linear form we want. Now we can easily fit a straight line to these data and obtain estimates of the regression slope, β, and the intercept term, $\log(n_0)$, and hence n_0.

A plot of the time of exposure against the logarithm of the number of surviving bacteria produces Figure 13.11. As can be seen, this is indeed much better modelled by a straight line—so it seems that the hypothesis, the theory that the relationship between n_t and t has the form (13.1), may well be a good model for the data.

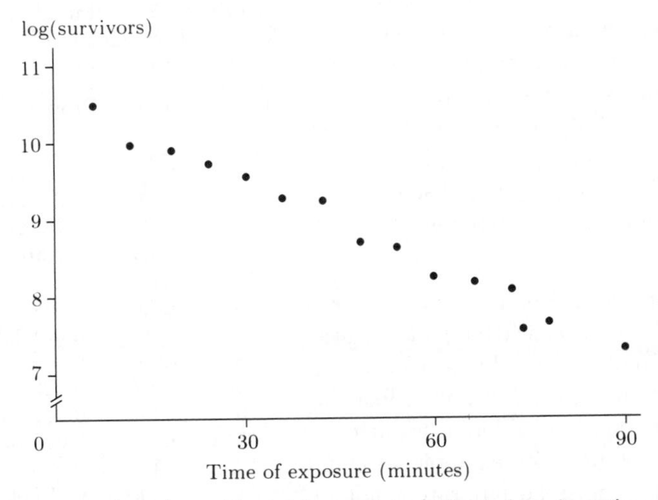

Figure 13.11 Log(number of bacteria surviving) plotted against time of exposure to X-rays

Exercise 13.18

Find the equation of the least squares regression line of $\log(n_t)$ against t, and estimate the number of bacteria at time 0.

Finally, just to check the fit of the model, Figure 13.12 shows the residuals from the fit. There seems to be no obvious pattern here—the model seems to provide a good fit.

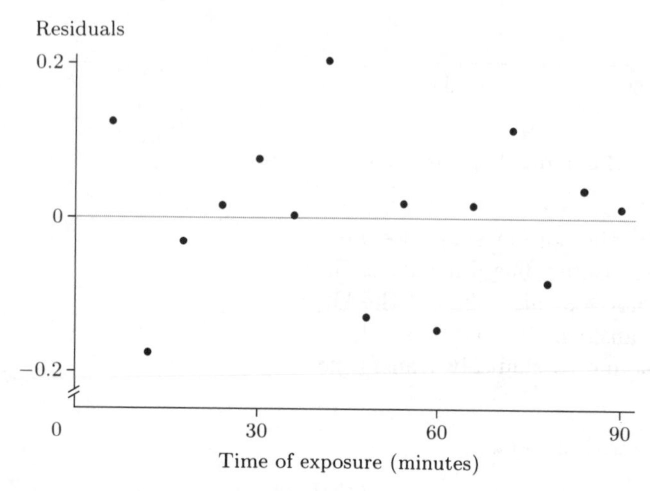

Figure 13.12 Residuals resulting from fitting a straight line to the data in Figure 13.11

13.6.3 Using X-rays to kill bacteria—Part II

A similar data set was analysed by Efron and Tibshirani (1993), and this is reproduced in Table 13.22. Here fourteen plates containing bacteria were exposed to differing doses of radiation and the proportions of bacteria which survived were measured.

Efron, B. and Tibshirani, R.J. (1993) *An introduction to the bootstrap*. Chapman and Hall, London, p. 116.

The investigator, however, was uncertain about the measurement for plate 13 —indeed, its score seems higher than one might expect by looking at neighbouring values. Following the same reasoning as in the previous study it was hypothesized that the logarithm of the number surviving might be linearly related to the dose. (In fact, Efron and Tibshirani also considered the possibility of a quadratic model—that log(proportion surviving) might decrease in a way related to the square of the dose. However, for our discussion we shall limit ourselves to a simple linear fit.) A plot of log(proportion surviving) against dose is shown in Figure 13.13.

Table 13.22 Radiation dose and proportion of bacteria surviving

Plate	Dose	Proportion surviving
1	1.175	0.440 00
2	1.175	0.550 00
3	2.350	0.160 00
4	2.350	0.130 00
5	4.700	0.040 00
6	4.700	0.019 60
7	4.700	0.061 20
8	7.050	0.005 00
9	7.050	0.003 20
10	9.400	0.001 10
11	9.400	0.000 15
12	9.400	0.000 19
13	14.100	0.007 00
14	14.100	0.000 06

It is quite apparent from Figure 13.13 that the result for plate 13 is indeed anomalous. Efron and Tibshirani comment about this scatter plot: 'Statisticians get nervous when they see one data point, especially a suspect one, dominating the answer to an important question.' Outlying points like this deserve careful study. The basic question is whether it is different for artefactual reasons, not related to the system being studied (for example, it could be different because of instrument malfunction or misread laboratory notes), or whether it represents a rare aspect of the system, so that only occasional

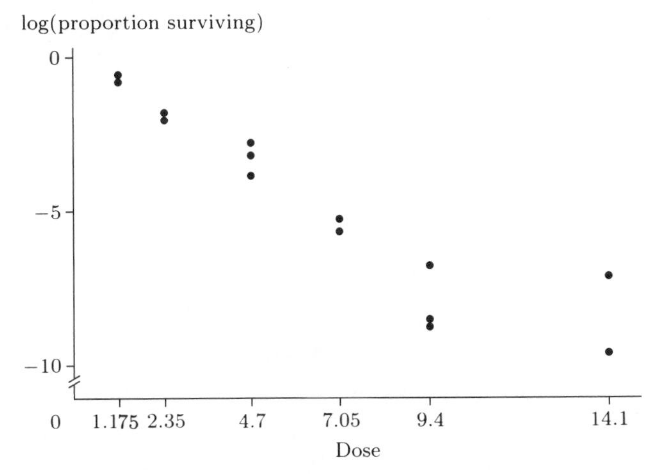

The outlying point on the right-hand side arises from plate 13, supporting the investigator's suspicions regarding the measurement for this plate.

Figure 13.13 Log(proportion surviving) plotted against dose

points would show such extreme, but perfectly valid values. It is always worth going back to the original data source to see if the question can be resolved.

In this case, given the investigators' suspicions about point 13, let us drop it from the analysis and model only the remaining data points. Doing this yields a best-fitting straight line with intercept 0.015 and slope −0.779. That is, the fitted model is

$$\log(n_d) = 0.015 - 0.779d,$$

A line constrained through the origin (corresponding to $n_0 = 1$) has slope −0.777 and equation $\log(n_d) = -0.777d$.

where d is the dose and n_d is the proportion surviving at dose d. Translating this back into the original units gives

$$n_d = 1.015e^{-0.779d}.$$

Conclusion

In the introduction, the aim of this chapter was stated as being to revise some of the material which has appeared in the course so far. Instead of giving a condensed presentation of the earlier material, we have reviewed various methods and techniques by way of case studies.

Throughout the data have been approached methodically by means of a sequence of procedures which have not always been spelled out but which have been there nevertheless. Briefly these are as follows.

1. Identify the question or questions that need to be answered, and express it in statistical terms.
2. Try to find a helpful picture for the data.
3. Ask some fundamental questions about the data and the way they seem to be distributed.
4. Be aware of the assumptions involved in any method you decide might be appropriate, and check them.
5. Ask yourself if you are satisfied that your method is justified and, if not, try to modify it.
6. Give a careful statement of your conclusions.

Chapter 14
A Look Forward

In this final chapter of the course, we look forward to some further applications of statistics, involving more advanced methods and techniques. For instance, we have already seen how to compare two independent samples, and test the hypothesis that they are drawn from populations with equal means: the technique known as analysis of variance allows us to compare more than two samples. We have seen how to represent two-variable problems through scatter diagrams, and have learned about regression analysis and correlation: multivariate problems involving several variables are less easy to represent graphically and to summarize coherently, and techniques useful here include cluster analysis, principal components analysis and multiple regression.

The aim in this course has been to provide a solid grounding in the fundamentals of statistics. There have been introduced such important notions as probability distributions as models of variation, model parameters and how to estimate them, statistical tests, correlation and regression, and many other basic statistical concepts. An attempt has been made to do this in a relaxed style, showing some of the historical context of the subject and throughout emphasizing its *applicability* by using real data which arise from real questions. Statistics is, above all else, about solving problems—it is also, however, a vast subject, and even in a book of this size we are not able to study it in depth. Moreover it is an immensely powerful subject in terms of the range and type of problems to which it can be applied. It is therefore appropriate to conclude the book with a chapter illustrating some more advanced techniques and the sorts of problems they can be used to solve.

Clearly in doing this, given the space available to us, we cannot go into too much technical detail. Therefore, what has been attempted in the sections which follow is to convey the flavour of the techniques, showing some of the questions they can be used to address, and how they can answer those questions, without labouring the theory. These descriptions are not intended to be sufficient to enable you to undertake such analyses without further instruction, but simply to convey the sorts of things that are now possible. In keeping with the overall philosophy of the course, the techniques are applied to real data sets and address real questions. And in this chapter, above all, the role of the computer is essential for the arithmetic involved. Section 14.1 looks at *analysis of variance.* This is a technique which compares several groups of scores simultaneously. At its most elementary level it can address questions such as: are there any differences between the means of the populations from which these samples are drawn? At a more sophisticated level it can tell us whether one variable influences the relationships between others.

Section 14.2 considers the class of techniques termed *cluster analysis.* Such methods are used to see if the data fall into natural groups of objects. This can be useful for scientific purposes—are there different 'kinds' of objects represented by the data? Or it can be useful for administrative purposes—do natural groups exist such that they should be treated differently?

Section 14.3 illustrates *principal components analysis.* Often objects being analysed have many measurements taken on them and it is useful to see if the natural variability between the objects can be summarized in a more efficient manner. This might be as a precursor to further analysis, or it might simply be as an end in itself, permitting, for example, a more comprehensible display of the data.

Section 14.4 gives an example of *discriminant analysis.* This is one of a class of methods for characterizing the differences between groups. There are two broad reasons for wishing to do this. One reason is simply a desire to understand the differences between groups, such as: in what ways do they differ? What are the most important differences? The other reason is to enable one to easily assign new objects to the appropriate group.

Section 14.5 demonstrates the ideas of *log-linear models.* In *Chapter 11* it was shown how two categorical variables could be examined. Log-linear models extend this to the case of more than two categorical variables, permitting different models to be fitted and explored.

Section 14.6 extends the ideas of simple regression, introduced in *Chapter 10*, to the case of *multiple regression.* This involves several explanatory variables instead of just the one introduced earlier. The extension turns out to have some interesting and unexpected properties not possessed by the single explanatory variable case.

The chapter ends with some concluding remarks about the course.

14.1 Analysis of variance

In *Chapter 8* the two-sample t-test was used to compare the means of two different groups of scores and we saw, for example, how this approach could be used to compare treatments to see which was more effective. The t-test can be generalized to permit comparisons between more than two groups. This generalization is called **analysis of variance**.

In *Chapter 8*, Table 8.10, data were given on the silver content (% Ag) of coins taken from four different mintings during the reign of King Manuel I (1143–1180). There were nine coins from the first minting, seven from the second, four from the third and seven from the fourth. The data are repeated in Table 14.1.

Table 14.1 Silver content (% Ag) of coins

1st	2nd	3rd	4th
5.9	6.9	4.9	5.3
6.8	9.0	5.5	5.6
6.4	6.6	4.6	5.5
7.0	8.1	4.5	5.1
6.6	9.3		6.2
7.7	9.2		5.8
7.2	8.6		5.8
6.9			
6.2			

What is of interest here is whether there is any significant difference in the mean silver content of the coins between different mintings. The four sample means are

$$\overline{x}_1 = 6.744, \quad \overline{x}_2 = 8.243, \quad \overline{x}_3 = 4.875, \quad \overline{x}_4 = 5.614.$$

In a hypothesis test for equality of two means, the variation in the two sample means is measured by calculating their difference $\overline{x}_1 - \overline{x}_2$. When more than two populations are involved, we cannot measure the variation in the sample mean by simply calculating a difference. Instead we take a weighted average of their squared deviations from the overall mean $\overline{x}$ of all the sample data. In this case we have $\overline{x} = 6.563$ and our measure of variation in the sample means is given by

$$\begin{aligned}&\frac{n_1(\overline{x}_1-\overline{x})^2+n_2(\overline{x}_2-\overline{x})^2+n_3(\overline{x}_3-\overline{x})^2+n_4(\overline{x}_4-\overline{x})^2}{4-1}\\&=\frac{9(6.744-6.563)^2+7(8.243-6.563)^2+4(4.875-6.563)^2+7(5.614-6.563)^2}{3}\\&=\frac{37.75}{3}\\&=12.58.\end{aligned}$$

You can see the reason for the term 'analysis of variance' in a comparison of means. The statistics involved are variances (squared differences) rather than simple differences between means. The assumptions of an analysis of variance are: first, that the samples taken from the populations under study are independent of one another; second, that the populations are normally distributed; and third, that the variances of the populations are equal. The samples are clearly independent, since the mintings are different. The sparsity of the data make any assumption of normality difficult to confirm: in fact, we saw in *Chapter 9*, Section 9.1, that normal probability plots suggest that this is a reasonable assumption, at least for the first and fourth mintings. The four sample variances are

$$s_1^2 = 0.295, \quad s_2^2 = 1.210, \quad s_3^2 = 0.203, \quad s_4^2 = 0.131,$$

and here the variation in the second coinage is substantially greater than that in any of the other three. However, the sample sizes are very small, and we have seen that when this is the case, large differences in the sample variances may not be too damaging to the conclusions of a test. An estimate of the common variance in the four populations is given by the *pooled sample variance*

$$\begin{aligned}s_p^2 &= \frac{(n_1-1)s_1^2+(n_2-1)s_2^2+(n_3-1)s_3^2+(n_4-1)s_4^2}{n_1+n_2+n_3+n_4-4}\\&=\frac{8\times 0.295+6\times 1.210+3\times 0.203+6\times 0.131}{23}\\&=\frac{11.02}{23}=0.48.\end{aligned}$$

The form of this expression for the pooled sample variance is identical to (8.7) for the two-sample case.

The test for equal population means consists of comparing the estimated variance in the means (in this case, 12.58) with the pooled sample variance (0.48). The ratio of the two estimates is 26.3, suggesting a variation in the sample means very much greater than would have been expected if the population means were equal. The distribution of this ratio under the null hypothesis of equal population means (and when the assumptions of independence, normality and equal variances are satisfied) follows a well-known form called the ***F*-distribution**. Significance probabilities are easily calculated and here, in fact, the SP of the test is just 1.3×10^{-7}. There is very considerable evidence from the analysis that the population means differ, that is, that the silver content alters significantly from coinage to coinage.

The test just performed is called a **one-way analysis of variance** because each data item was classified in one way, according to the population (minting) from which it was drawn. Here four samples were compared, but the analysis extends in an obvious way to more than four samples (and, in fact, a one-way analysis of variance on just two samples reduces exactly to the two-sample t-test with which you are familiar, because of mathematical relations existing between the t-distribution and the F-distribution).

A non-parametric version of the test, useful when the assumption of normality is very questionable, is the **Kruskal–Wallis one-way analysis of variance** in which the samples are pooled and ranked, and the ranks within each sample are then compared. This is a test for equal population medians (and again assumes equal variance within the populations).

Naturally, when there is evidence that the means of the populations are different, one would wish then to go further and identify which were 'significantly' higher than which others. To deal with this, some sort of **multiple comparisons** procedure is appropriate. (Such procedures are not always straightforward.)

The generalization of a t-test to an analysis of variance has other advantages too: in particular, we can explore the effect of *multiple factors*. For instance, in a **two-way analysis of variance** each data item is classified in two ways.

Suppose, for example, that we were interested in comparing six different prepared treatments for a particular illness. In a one-way analysis of variance we would allocate patients to six different groups, administer a different treatment to each group and explore whether there were differences in the mean response. However, suppose that, in addition to being interested in any differences between the six treatments, we were also concerned about their effects on two groups of patients: those who had the disease moderately and those who had the disease severely. We might suspect, for example, that the treatment effects differed between the two groups of patients. Perhaps it turns out that all treatments were more effective on the severely ill, or perhaps, more complicatedly, that some treatments were more effective on one patient group and some more effective on the other. We now have two *factors* which we would like to study: treatment and disease severity.

We could undertake a number of separate studies, some comparing the effects of single treatments on the two severity groups (each of these would just use a t-test since there are only two groups) and one comparing the six treatment groups. However, this would involve many tests, and would in any case not lead to a neat answer to questions about whether the pattern of responses to the six treatments differed between the two severity groups. To answer such questions we need a more global approach: we need to look at the *cross-classification* of severity by treatments. Each patient will fall into only one group according to the treatment they have received and the severity of their illness.

A cross-classification is where each data item is classified into a category according to two or more factors.

Table 14.2 shows data from an investigation into the effectiveness of different kinds of psychological treatment on the sensitivity of headache sufferers to noise. There are two groups of 22 subjects each: those suffering from a migraine headache (coded 1 in the table) and those suffering from a tension headache (coded 2).

A more detailed description is given in Hand, D.J. and Taylor, C.C. (1987) *Multivariate analysis of variance and repeated measures.* Chapman and Hall, London, p. 157.

Half of each of the two headache groups were then randomly selected to receive a treatment (coded 1 in the table). The other half (coded 2) did not receive any treatment and acted as a **control group**, so that the effectiveness of the treatment could be assessed. This means that we have a cross-classification involving two factors, headache type and treatment/control, each at two levels. There are thus four cells in the cross-classification.

Table 14.2 Relaxation training and effect of noise on headaches

Headache type	Treatment group	Score	Headache type	Treatment group	Score
1	1	5.70	2	1	2.70
1	1	5.63	2	1	4.65
1	1	4.83	2	1	5.25
1	1	3.40	2	1	8.78
1	1	15.20	2	1	3.13
1	1	1.40	2	1	3.27
1	1	4.03	2	1	7.54
1	1	6.94	2	1	5.12
1	1	0.88	2	1	2.31
1	1	2.00	2	1	1.36
1	1	1.56	2	1	1.11
1	2	2.80	2	2	2.10
1	2	2.20	2	2	1.42
1	2	1.20	2	2	4.98
1	2	1.20	2	2	3.36
1	2	0.43	2	2	2.44
1	2	1.78	2	2	3.20
1	2	11.50	2	2	1.71
1	2	0.64	2	2	1.24
1	2	0.95	2	2	1.24
1	2	0.58	2	2	2.00
1	2	0.83	2	2	4.01

Each subject then listened to a tone which gradually increased in volume. The level of volume at which the subject found the tone unpleasant was recorded, and these are the scores given in the table.

Table 14.3 shows the mean scores in each of the four groups.

Table 14.3 Mean scores for the headache treatment data

	Headache type	
	Migraine	Tension
Treatment group	4.688	4.111
Control group	2.192	2.518

Again, for legitimate application of an analysis of variance, it is necessary to make certain assumptions about the data. For example, it is assumed that the data in each group arise from a normal distribution and that the variances in the groups are equal. With only eleven cases in each group the histograms or boxplots will not give very reliable indications of shapes of distributions or sizes of standard deviations but nevertheless the boxplots in Figure 14.1 suggest that there is some skewness.

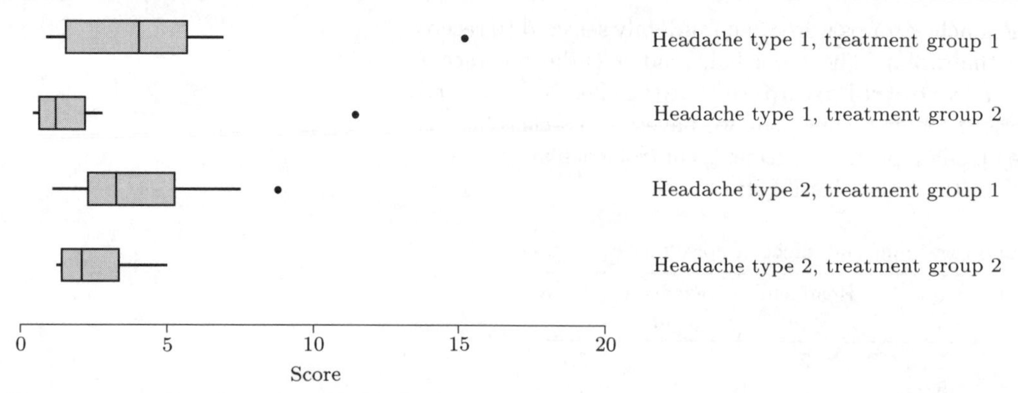

Figure 14.1 Boxplots of scores for the four groups

To reduce the skewness, logarithms of the data were taken before proceeding further. Figure 14.2 shows that skewness is improved after this and in fact a formal statistical hypothesis test showed no significant differences between the standard deviations—the differences observed could easily have arisen by chance even if the four populations in question had equal standard deviations.

Log transforms were in this case taken to base 10. The effect of the two extreme responses 15.20 and 11.50 on the skewness within groups is substantially reduced by taking logarithms. Even after this transformation, the outlier in the second treatment group remains. However, complications of a different kind ensue if the numbers of subjects in each group are not the same; and therefore this response is retained in the analysis.

The means of the transformed data are shown in Table 14.4.

Table 14.4 Mean scores for the transformed headache treatment data

	Headache type	
	Migraine	Tension
Treatment group	0.540	0.537
Control group	0.124	0.356

Obviously these means differ. The question is: do the differences reflect real differences between populations of people with the two kinds of headaches and given the two kinds of treatments? Or is it just a matter of chance that we happened to draw 44 people for whom the means differed in this way—and could they easily have been drawn from populations with identical means?

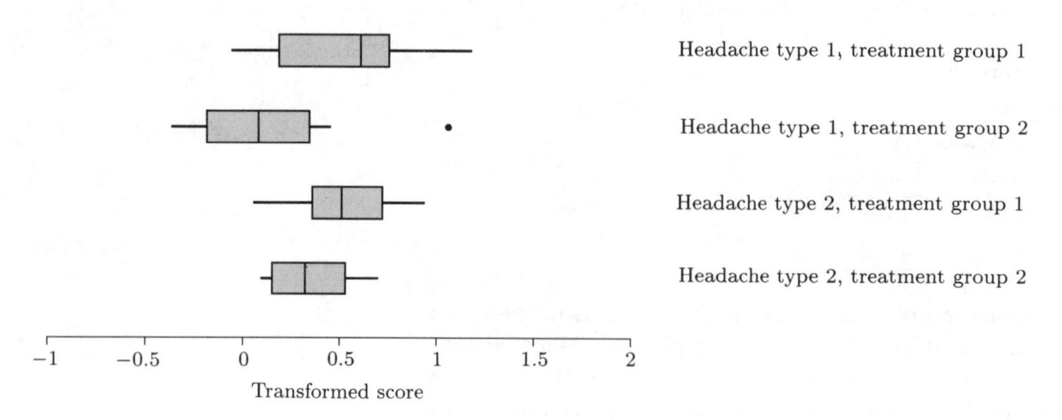

Figure 14.2 Boxplots of transformed scores for the four groups

These questions are rather general. More specifically, we want to know if there are real differences between treatments and also if there are real differences between headache types. As was noted above, we might also be interested in whether or not the pattern of responses to the treatments varies from headache type to headache type.

Let us first look at headache type. In fact, although there are only 11 subjects (patients) in each of the four groups, in each headache type there are 22 subjects. This shows an important advantage of using the cross-classification, rather than conducting separate analyses for treatment and headache type. If we performed separate analyses with 22 subjects in each group, we would need a total of $22 \times 2 = 44$ subjects for each of the two separate analyses—that is, 88 subjects in all. The cross-classification gives us 22 subjects in each group but with a total of only 44 subjects. For half the cost we have achieved the same power in our statistical analysis.

How should we proceed from here? Without going into technical details, we follow exactly the same sort of generalization as outlined for the one-way analysis of variance. That is, we calculate a measure of the variation between the means of the two headache types and compare this with the variation within groups. To do this, the 'within-group' variation is calculated as the common variance of all four of the groups in the cross-classification.

Again, the hypothesis being tested is whether the 'between-groups' variation could easily have arisen by chance from populations showing the internal variation in the data.

For the headache types the variance ratio resulting from this calculation has value 1.401. By reference to the F-distribution with appropriate degrees of freedom, the corresponding SP is found to be 0.244. This is quite a large probability, and is certainly not enough for us to reject the null hypothesis of equal mean scores with any degree of confidence. Our conclusion is thus that we have no evidence to suppose that there is a difference between headache types.

We can now do an analogous analysis for the two treatments. The results give a variance ratio of 9.49. This is substantially larger than that for headache type and, in fact, referring it to the F-distribution shows that the SP is 0.004. This is highly significant: it is very unlikely that a difference between treatment and control groups as large as this could have arisen by chance if the measurements were taken from normal populations with identical means.

Informally, our overall conclusion is thus that there is no difference in mean responses between the two headache types, but there is a difference in the mean responses to the two 'treatments' (treatment versus control). Also, looking at the table of means in Table 14.3, we see that the mean scores are large for the treatment group. This means that subjects in the treatment group could take a louder tone without finding it unpleasant: there is evidence that the treatment is effective.

A more extended analysis also showed that there is no **interaction**: that is, there is no difference between the ways the two treatments work on the two headache types.

So, analysis of variance may be regarded as a generalization of the t-test. It extends to multiple (rather than merely two) groups. It permits the effects of different factors to be disentangled. It allows one to test for interaction effects, and it provides a more powerful analysis than would be possible if the factors were analysed separately.

Remember, the tests are based on a comparison of variances—hence the name 'analysis of variance'. Note, however, that the analysis is used to test the possibility that the groups have different means.

14.2 Cluster analysis

Analysis of variance is used for answering questions about the differences between groups of subjects. The group categories are defined before the data are collected and each subject falls into just one of the categories. A common use is as a *hypothesis testing* technique, in the sense that one knows beforehand the questions or hypotheses one wants to investigate. For example, in the preceding section we wanted to know if there was a difference between the two types of headaches, if there was a difference between the two types of treatments, and also, if the response to the treatments depended on the type of headache (that is, if there was an interaction).

In contrast, **cluster analysis** is an *exploratory* technique. Cluster analysis seeks to identify natural groupings or clusterings in a collection of subjects. 'Subject' here may refer to almost anything, for instance

- people, measured on certain physiological characteristics: the aim may be to sec if there are different physical types;
- people, measured on behavioural characteristics: to see if there are characteristic patterns of behaviour;
- manufactured products, with records taken of how demand for each product varies over the course of a year: the aim could be to see if such products fall into natural groups in terms of the patterns of demand for them;
- retail outlets of a large supermarket, with interest focusing on the characteristics of the local environment: perhaps to see if the stores fall into natural groups of distinct types;
- rock samples, with measurements taken of hardness, texture, colour, frangibility, and so on: the objective may be to see if there are natural groupings;
- insects of some kind, with measurements taken of caudal width, antenna length, and so on: perhaps to find out whether in fact there is more than one species of insect represented in the sample.

The list of possibilities is endless and, while each of the above apply to very different questions, they all have a similar structure. In each of them a number of measurements is taken on a sample of subjects and interest lies in whether or not, on the basis of these measurements, the subjects fall into natural groups.

The basic approach to answering such questions is straightforward and is as follows. First, a measure of *similarity* between cases is defined using the measured variables. Then cases which are very similar are regarded as being in the same group or **cluster**, and those which are very dissimilar are regarded as being in different clusters.

This outline begs a number of questions such as: how should we measure 'similarity'? How similar is 'very' similar? These and other questions can be answered in a number of ways, but in this short outline, just two approaches will be described. Although many applications of cluster analysis involve large data sets, here an example with only 22 subjects is considered.

Table 14.5 shows data on 22 medical practices in the United Kingdom. The second column in the table (headed 'practice size') gives the number of patients registered with the practice over the six years of the study. The other columns give numbers of patients who had experienced one or more episodes of a particular psychiatric disorder. The thirteen disorders are listed at the bottom of the table.

Dunn, G. (1986) Patterns of psychiatric diagnosis in general practice: the Second National Morbidity Survey. *Psychological Medicine*, **16**, 573–581.

Table 14.5 Distribution of the thirteen most common psychiatric disorders in the UK across 22 medical practices

Practice	Practice	Disorder type												
	size	1	2	3	4	5	6	7	8	9	10	11	12	13
1	2519	221	68	83	108	70	36	58	1	17	20	7	7	6
2	1504	281	184	11	3	69	17	0	2	9	1	1	7	0
3	2161	234	206	112	55	108	63	3	28	18	2	5	3	3
4	4187	478	608	40	402	156	108	1	1	64	25	12	11	9
5	1480	76	251	173	159	23	37	4	30	14	1	9	6	3
6	2125	386	142	9	16	113	90	0	0	10	7	15	11	4
7	6514	1122	742	197	435	81	172	330	193	66	33	22	16	21
8	1820	208	398	45	14	13	35	5	1	3	4	10	12	9
9	2671	409	282	512	168	119	45	252	225	41	27	39	5	17
10	4220	314	429	105	53	65	37	20	1	29	28	23	14	11
11	2377	72	148	425	120	56	116	3	0	17	6	6	5	2
12	5009	566	305	123	174	71	99	5	16	28	15	13	9	7
13	2037	241	207	4	4	37	29	0	1	11	0	3	0	0
14	1759	390	277	56	68	160	45	17	15	26	32	23	3	2
15	1767	248	178	61	89	122	70	13	11	16	19	5	6	3
16	3443	218	185	317	198	164	75	45	89	26	21	13	22	27
17	2200	212	210	115	104	64	70	85	7	11	12	15	5	1
18	2639	280	155	286	180	53	74	36	4	27	9	16	4	12
19	1897	288	224	132	153	79	51	14	15	7	9	9	3	5
20	2278	331	251	170	439	76	84	9	24	20	8	5	5	7
21	2242	254	303	177	152	148	138	21	53	16	15	16	9	5
22	2497	330	290	186	215	119	121	21	7	16	26	13	13	9

1 = anxiety neurosis, 2 = depressive neurosis, 3 = unclassified symptoms, 4 = physical disorders of presumably psychogenic origin, 5 = insomnia, 6 = tension headache, 7 = affective psychosis, 8 = neurasthenia, 9 = enuresis, 10 = phobic neurosis, 11 = hysterical neurosis, 12 = schizophrenia, 13 = alcoholism and drug dependence

A number of questions immediately arise with these data. For example, how reliable are the data? How clear are the distinctions between the various diagnoses? To answer such questions it would be necessary to explore the way in which the data were collected, and this would be an important part of the statistical investigation: statistical techniques, no matter how sophisticated and subtle, cannot be expected to salvage a poorly designed experiment or inadequately collected data. Here, however, space limitations mean that we are forced to move on, assuming that the data are adequately reliable.

One question of interest here is whether or not the practices fall into different types. That is, whether there are categories of practice, in terms of the patterns of psychiatric disorders presented by the catchment population. This is precisely the sort of question that cluster analysis has been developed to answer.

The first thing we might do is adopt some kind of standardization to allow for the fact that the sizes of the practices range from 1480 to 6514 people. It is *pattern* of responses, not absolute numbers, that we are interested in. In Table 14.6, therefore, the numbers of patients with each disorder have been divided by the total practice size to yield proportions (thousandths). Note that these proportions do not add up to 1 since in each practice there are people who suffered from none of the disorders and also people who suffered from more than one disorder.

Table 14.6 Proportions of patients within each practice suffering from a particular disorder (thousandths)

Practice	Disorder type												
	1	2	3	4	5	6	7	8	9	10	11	12	13
1	88	27	33	43	28	14	23	0	7	8	3	3	2
2	187	122	7	2	46	11	0	1	6	1	1	5	0
3	108	95	52	25	50	29	1	13	8	1	2	1	1
4	114	145	10	96	37	26	0	0	15	6	3	3	2
5	51	170	117	107	16	25	3	20	9	1	6	4	2
6	182	67	4	8	53	42	0	0	5	3	7	5	2
7	172	114	30	67	12	26	51	30	10	5	3	2	3
8	114	219	25	8	7	19	3	1	2	2	5	7	5
9	153	106	192	63	45	17	94	84	15	10	15	2	6
10	74	102	25	13	15	9	5	0	7	7	5	3	3
11	30	62	179	50	24	49	1	0	7	3	3	2	1
12	113	61	25	35	14	20	1	3	6	3	3	2	1
13	118	102	2	2	18	14	0	0	5	0	1	0	0
14	222	157	32	39	91	26	10	9	15	18	13	2	1
15	140	101	35	50	69	40	7	6	9	11	3	3	2
16	63	54	92	58	48	22	13	26	8	6	4	6	8
17	96	95	52	47	29	32	39	3	5	5	7	2	0
18	106	59	108	68	20	28	14	2	10	3	6	2	5
19	152	118	70	81	42	27	7	8	4	5	5	2	3
20	145	110	75	193	33	37	4	11	9	4	2	2	3
21	113	135	79	68	66	62	9	24	7	7	7	4	2
22	132	116	74	86	48	48	8	3	6	10	5	5	4

Our aim now is to study the patterns of the practices to see if they fall into natural groupings of similar patterns.

First, as already noted, we have to decide what we mean by 'similar'. That is, we have to define a *distance measure* between patterns. A common distance measure is 'Euclidean distance'. This is just a generalization of the distance concept which we use in everyday life and is defined, in our situation, as follows. Suppose that we have two practices. If the first of these has proportion x_1 suffering from disorder 1, proportion x_2 from disorder 2, and so on, we can write the pattern for this case as $(x_1, x_2, x_3, \ldots, x_{13})$. Similarly, if we suppose that the second case has pattern $(y_1, y_2, y_3, \ldots, y_{13})$, then it follows that the Euclidean distance between the practices may be defined as

$$\sqrt{(x_1 - y_1)^2 + (x_2 - y_2)^2 + \cdots + (x_{13} - y_{13})^2}.$$

To take an example, consider the first two practices. Their patterns are

$$(88, 27, 33, 43, 28, 14, 23, 0, 7, 8, 3, 3, 2)$$

and

$$(187, 122, 7, 2, 46, 11, 0, 1, 6, 1, 1, 5, 0)$$

respectively. The distance between the two practices is thus

$$\begin{aligned} &\sqrt{(88-187)^2+(27-122)^2+\cdots+(2-0)^2} \\ &= \sqrt{9801+9025+\cdots+4} \\ &= 149, \end{aligned}$$

again measured in thousandths. The more alike the patterns are, then the smaller the distance between the corresponding practices will be; while the more unlike the patterns are, the larger it will be.

The geometric interpretation of this measure is that it gives the distance between a pair of points in 13-dimensional space. Some approaches to cluster analysis use the squared distance, and that is the approach adopted in what follows. This distance measure can be used to form an array showing the squared distances between each pair of medical practices, as in Table 14.7.

Table 14.7 Squared distances between each pair of medical practices ($\times 10^4$)

	1	2	3	4	5	6	7	8	9	10	11	12	13	14	15	16	17	18	19	20	21	22
1	0																					
2	221	0																				
3	71	100	0																			
4	188	151	98	0																		
5	341	454	211	171	0																	
6	145	42	93	191	524	0																
7	173	99	110	97	294	120	0															
8	398	168	186	146	252	309	212	0														
9	488	555	374	535	356	576	328	622	0													
10	73	147	39	114	232	159	160	156	528	0												
11	277	621	252	454	203	564	494	584	349	303	0											
12	27	117	35	117	297	80	106	258	488	39	318	0										
13	99	60	46	114	340	74	118	145	581	27	443	34	0									
14	395	74	197	189	474	137	138	243	485	321	732	276	223	0								
15	109	73	26	67	260	61	75	209	400	98	366	62	74	109	0							
16	64	314	69	201	179	259	222	393	310	110	110	95	185	424	127	0						
17	57	154	27	87	173	142	83	200	335	43	234	42	71	252	50	63	0					
18	80	263	75	184	174	218	157	366	255	133	118	83	181	351	113	37	58	0				
19	159	119	61	63	167	137	56	204	291	141	318	98	132	124	36	135	64	78	0			
20	355	440	307	161	221	431	218	508	449	413	472	316	435	376	235	292	259	213	129	0		
21	196	189	58	85	120	199	124	195	287	146	252	145	168	178	50	113	67	105	41	192	0	
22	152	162	59	61	137	161	83	221	301	139	263	104	149	164	36	110	54	67	10	121	21	0

From this table we can identify which are the two most similar patterns, in the squared distance sense. For instance, the most similar practices according to Table 14.7 are practices 19 and 22 (with a computed distance of 10); the next most similar are practices 21 and 22; the two least similar practices are numbers 11 and 14, with a distance of 732 between them.

One class of cluster analysis techniques identifies the two most similar patterns in this way and replaces them by their mean pattern. To do this the mean for each of the variables is calculated separately, so that a new pattern of thirteen scores results. For instance, practices 19 and 22 can be merged together, the resulting single notional practice being given a pattern

$$\begin{aligned} &\left(\tfrac{1}{2}(152+132), \tfrac{1}{2}(118+116), \ldots, \tfrac{1}{2}(3+4)\right) \\ &= \left(142, 117, \ldots, 3\tfrac{1}{2}\right). \end{aligned}$$

So, starting with 22 'clusters' each of size 1, the next stage yields 20 clusters of size 1 and one cluster of size 2.

Again the smallest distance is identified and again the closest practices are merged, to be replaced by their mean. This process continues until only a single cluster remains. By halting the process at some intermediate point a collection of clusters is produced, each containing practices that are considered 'close' to one another.

Techniques of this kind are called *hierarchical* for the reason that they hierarchically agglomerate subjects into clusters.

An alternative class of techniques is called *non-hierarchical.* Such methods begin with some initial partition of the cases into clusters and then seek to reallocate the cases to optimize some measure of cluster 'quality'. For example, a good clustering could be defined as one for which the *average* distance between practices in each cluster is small. (Obviously we need to fix the number of clusters beforehand—or else we could make this average distance zero by taking as many clusters as there are patterns!)

The ***k*-means method** of cluster analysis, uses a measure of this sort. It seeks the partition such that each practice is assigned to the cluster which has the closest mean pattern. The method works as follows.

We shall begin by supposing we want to partition our 22 practices into two clusters. That is, we want to see if the practices can be naturally partitioned into two groups of practices such that those within a group are similar and those in different groups are not similar. To initialize the process take the first two practices and regard them as cluster centres. Now sequentially assign each of the other 20 cases to that cluster which has the nearer centre. Each time an assignment is made replace the cluster centre by the mean pattern of all practices now in the cluster. This mean pattern becomes the new cluster centre and the next calculation is based on seeing which of the centres is the closer.

Once this has been done we have a partition into two clusters. However, the cluster centres will have moved around during the process because, each time a new practice was added, the cluster mean (its 'centre') was recalculated. This means that although practices were originally assigned to the cluster which had the nearer centre, they may not now be in the cluster which has the nearer centre.

To resolve this, the process is repeated, taking each practice in turn and assigning it to the cluster which has the nearer mean pattern. Again, after each such assignment, the cluster mean patterns are recalculated. (Note that this will only be necessary when a practice moves clusters. If it stays where it is—because it was already in the closer cluster—then no recalculation is necessary.)

This process is repeated, each time calculating the mean patterns and reallocating the practices, until no new reallocation of practices is required.

To follow the process by hand with thirteen variables and 22 cases would be quite unrealistic: so we use a computer. This is a good example of the impact that the computer has on statistics. Techniques like cluster analysis were impracticable before the advent of fast computing power. As a consequence the nature of statistics, the way that it is used, and the sorts of questions that can be addressed have changed dramatically.

A k-means cluster analysis on the 22 medical practices, with $k = 2$ (that is, seeking two clusters) yields

cluster 1: containing practices 5, 9, 11, 18, 16;

cluster 2: containing the remaining practices.

A k-means cluster analysis on the 22 medical practices, with $k = 3$ (that is, seeking three clusters) yields

cluster 1: containing practices 5, 9, 11, 18, 16;

cluster 2: containing practice 20 only;

cluster 3: containing the remaining practices.

Since there are three cluster centres in this analysis, and since three points define a plane, we can plot a scatter diagram of the practices in the plane defined by the cluster centres. This is shown in Figure 14.3. Cluster 2, consisting of practice 20 alone, is clearly very different from the other practices and our analysis has correctly identified this. The practices in cluster 3 form a fairly well-defined group of practices, and this is separate from the practices constituting cluster 1. This projection of the points onto the plane defined by the cluster centres shows that our analysis has been quite effective. What is also particularly clear is that it would have been virtually impossible to discern this sort of structure from Table 14.5.

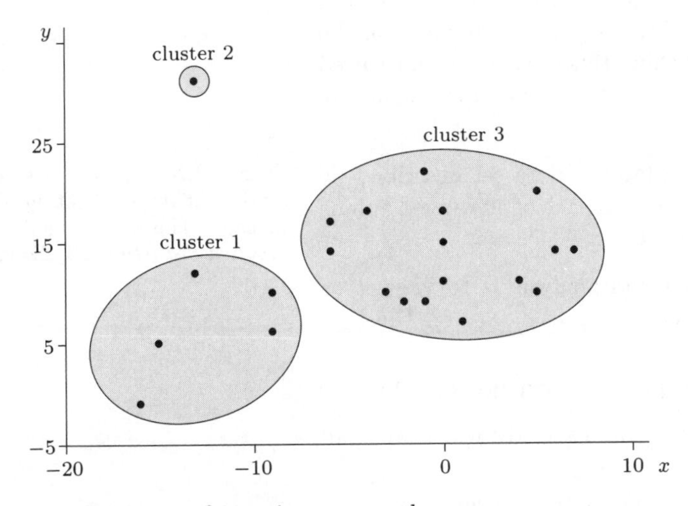

Figure 14.3 Projection of 22 points onto a plane

Even this example, though too large to be conducted by hand, is small in terms of the size of data sets on which cluster analysis is often carried out. A recent paper describes the use of cluster analysis to obtain classifications of the 108 English personal social services authorities using data from the 1979 National Housing and Dwelling Survey. Twenty variables were used in the analysis. Other quite different examples of the application of cluster analysis are its use in classifying astronomical objects.

Jolliffe, I.T., Jones, B. and Morgan, B.J.T. (1986) Comparison of cluster analyses of the English personal social services authorities. *J.Royal Statistical Society, Series A*, **149**, 253–270.

14.3 Principal components analysis

Cluster analysis is a multivariate technique—so-called because it involves multiple variables. One hopes, through using it, to be able to make some statement about natural groupings which may occur in the objects being analysed.

This section introduces another multivariate technique. Here, however, in contrast, interest is focused on the variables themselves and on the relationships between them. The technique, **principal components analysis**, can be used in a variety of ways.

For example, it can be used to see whether most of the variation between objects, even though they may be measured on many variables, can be attributed to just a few basic types of difference between them. Jolicoeur and Mosimann (1960) measured the lengths, widths, and heights of the carapaces of 24 female turtles to see whether differences could be explained simply in terms of some overall 'size' factor, or if differences in shape beyond this were also apparent. In statistical terms one will be seeking functions of the measured variables which show most variation between subjects. The aim might be simpler than in this example: one might merely wish to summarize a set of measures in the most effective way. One could, for example, take their mean, or perhaps some weighted mean. Principal components analysis can be thought of as reducing the measurements on each object to a single weighted average in such a way that these weighted averages preserve as much of the difference between the objects as possible. In effect this replaces the measured variables by a single variable, preserving as much as possible of the structure of the sample.

Jolicoeur, P. and Mosimann, J.E. (1960) Size and shape variation in the painted turtle. *Growth*, **24**, 339–354.

The data used to illustrate the technique is a famous data set described by Jeffers (1967). Nineteen measurements were taken on each of 40 winged aphids (*Alate adelges*). The data are shown in Table 14.8.

Jeffers, J.N.R. (1967) Two case studies in the application of principal component analysis. *Applied Statistics*, **16**, 225–236.

One way of thinking about principal components analysis is to view it as working in stages. Consider the measurements $x_1, x_2, \ldots, x_{19}$ on the first aphid, namely

212, 110, 75, 48, 50, 20, 20, 28, 28, 33, 3, 44, 45, 36, 70, 40, 8, 0, 30.

We can use this to form a weighted sum of this aphid's scores which we shall call y:

$$y = w_1x_1 + w_2x_2 + \cdots + w_{19}x_{19} = 212w_1 + 110w_2 + \cdots + 30w_{19},$$

with a set of weights w_1 to w_{19}. In effect this combines the aphid's scores into a single number. For any given set of weights we can do this with all of the 40 aphids, so producing 40 numbers $y_1, y_2, \ldots, y_{40}$.

Now, different sets of weights will produce different sets of 40 y-scores. These different sets will carry information about different aspects of the underlying nineteen measurements. For example, weights where $w_1 = 1$ and all the rest are 0 will simply produce a set of 40 y-scores equal to the measurements on the first variable x_1, and so will carry information about just that variable. Weights where $w_1 = w_2 = \frac{1}{2}$ and all the rest are 0 will produce a set of 40 y-scores which are the averages of the first two measurements for each insect, producing information about just these two variables.

Table 14.8 The aphid data (all lengths in hundredths of mm)

Aphid	Variable																		
	1	2	3	4	5	6	7	8	9	10	11	12	13	14	15	16	17	18	19
1	212	110	75	48	50	20	20	28	28	33	3	44	45	36	70	40	8	0	30
2	202	100	75	50	50	23	21	30	30	32	5	42	45	35	76	42	8	0	30
3	202	100	70	46	50	19	21	30	25	33	1	42	44	33	70	40	6	0	30
4	225	88	74	47	50	24	21	30	27	35	5	42	44	36	68	41	6	0	30
5	206	110	80	48	50	24	20	29	27	30	4	42	47	35	67	40	6	0	30
6	191	92	70	45	50	18	19	28	30	32	5	41	43	33	57	38	8	0	35
7	208	114	77	49	50	25	21	31	31	32	4	42	47	36	66	40	8	0	30
8	155	82	63	49	50	20	20	29	24	30	3	37	38	29	67	35	6	0	35
9	167	88	64	45	50	21	19	28	27	31	3	37	38	28	61	37	8	0	30
10	197	99	82	47	50	22	20	30	30	31	0	41	43	33	60	38	8	0	30
11	106	52	39	23	40	12	10	20	20	22	6	25	25	20	45	27	4	1	20
12	92	45	37	22	40	13	12	20	16	21	5	24	23	18	41	24	4	1	20
13	96	45	36	23	40	13	10	19	17	22	4	24	23	17	40	23	4	1	20
14	85	40	38	22	40	13	11	19	20	21	5	24	24	19	44	23	4	1	20
15	110	47	42	23	40	12	10	19	20	22	4	25	25	20	45	26	4	1	20
16	181	82	59	35	50	19	19	19	27	28	4	35	38	29	60	45	9	1	20
17	176	83	60	38	50	20	19	20	22	29	3	35	36	28	57	43	10	1	20
18	192	66	62	34	50	20	18	22	23	28	4	35	34	25	53	38	10	1	20
19	154	76	71	34	50	20	19	25	25	29	4	33	36	27	60	42	8	1	30
20	151	73	62	38	50	20	18	21	24	25	4	37	37	28	64	43	10	1	25
21	161	79	58	37	50	21	19	23	26	29	5	36	36	27	60	45	0	1	20
22	191	88	64	39	50	22	20	23	24	29	4	38	40	30	65	45	0	1	25
23	153	64	53	33	50	17	16	20	22	25	5	34	34	26	54	40	0	1	20
24	148	81	62	37	50	22	20	22	24	32	5	35	37	27	60	41	0	1	20
25	162	77	69	37	50	20	18	23	24	28	4	38	37	27	57	42	0	1	25
26	134	69	57	34	50	20	18	28	20	26	4	36	36	26	55	39	0	1	20
27	129	58	48	26	50	16	15	19	21	26	5	28	30	22	51	36	9	1	30
28	120	65	53	32	50	19	19	23	25	30	5	33	35	26	54	43	8	1	20
29	141	70	55	36	50	22	20	23	25	31	5	36	37	28	58	41	0	1	20
30	167	72	57	35	50	19	19	25	23	28	5	34	36	27	60	40	0	1	25
31	141	54	50	30	50	17	16	18	25	24	5	27	29	22	53	36	8	1	20
32	100	60	42	25	50	16	14	14	20	27	6	28	25	18	48	34	8	1	20
33	114	45	44	27	50	18	15	19	17	25	5	27	25	19	47	37	8	1	20
34	125	55	47	23	50	18	14	18	22	24	4	28	26	20	51	37	8	0	20
35	130	53	47	23	50	16	14	18	18	25	4	27	27	21	50	36	8	1	20
36	124	52	44	26	50	16	14	18	22	22	5	27	25	20	50	32	6	1	20
37	120	54	49	30	50	17	15	17	19	24	5	27	27	20	42	37	6	1	20
38	107	56	45	28	50	18	14	18	22	24	4	27	26	20	50	35	8	1	20
39	111	55	43	26	50	17	15	18	19	24	5	26	25	19	46	34	8	1	20
40	128	57	48	28	50	16	14	17	19	23	5	23	25	19	50	31	8	1	20

1 = body length, 2 = body width, 3 = forewing length, 4 = hind-wing length,
5 = number of spiracles, 6 = length antenna I, 7 = length antenna II, 8 = length antenna III,
9 = length antenna IV, 10 = length antenna V, 11 = number of antennal spines, 12 = length of tarsus,
13 = length of tibia, 14 = length of femur, 15 = rostrum length, 16 = ovipositor length,
17 = number of ovipositor spines, 18 = anal fold (present/absent), 19 = number of hind-wing hooks

To address our question—about whether the pattern of differences between the aphids can be described in simpler terms than by using all nineteen variables—we shall examine the set of weights that produce the 40 y-scores which have the most variability between them. So, for example, a set of weights for which the aphids had almost identical y-scores would tell us little about the variability between them, but a set on which they differed substantially would tell us a lot. Such a set of weights would tell us about a key aspect of the pattern of differences between the aphids.

What we are seeking is that particular set of weights which produces the maximum variability between the 40 y-scores. In order to make progress in this direction we need to clarify what we mean by 'maximum variability'. We shall use a measure that is now familiar, namely the variance of the 40 numbers. So we seek the set of weights which produces the maximum variance in the scores $y_1, y_2, \ldots, y_{40}$.

The set of weights $w_1, w_2, \ldots, w_{19}$ which achieve maximum variability in the y-scores is called the **first principal component** of the data. The corresponding y-score for each aphid is called its **score on the first principal component**.

Here, there is a small problem that you might have noticed: the variance in the y-scores can be made arbitrarily large simply by making the weights arbitrarily large. The way to overcome this is to constrain the weights in some way, and for technical reasons the constraint takes the form

$$w_1^2 + w_2^2 + \cdots + w_{19}^2 = 1.$$

It is possible to show that principal components can be calculated from the *sample correlations* calculated for each pair from the nineteen variables measured. The calculations are rather intricate, and would sensibly involve a computer.

It turns out that for the given 40 aphids, the single y-score that separates them most successfully is given by

$$\begin{aligned} y = {} & 0.253x_1 + 0.260x_2 + 0.262x_3 + 0.262x_4 + 0.159x_5 \\ & + 0.239x_6 + 0.253x_7 + 0.235x_8 + 0.240x_9 + 0.250x_{10} \\ & - 0.133x_{11} + 0.263x_{12} + 0.265x_{13} + 0.263x_{14} + 0.253x_{15} \\ & + 0.198x_{16} + 0.020x_{17} - 0.193x_{18} + 0.205x_{19}. \end{aligned}$$

You can check that $0.253^2 + 0.260^2 + \cdots + 0.205^2 = 1$.

All the x-variables (except, perhaps, x_{18}) provide some measure of the size of an aphid. This y-score may therefore be regarded as an overall measure of size for each of the 40 aphids. For instance, the size of the first aphid is given according to this definition by

$$\begin{aligned} y_1 &= 0.253 \times 212 + 0.260 \times 110 + \cdots + 0.205 \times 30 \\ &= 218.3; \end{aligned}$$

the first aphid is in this sense a little larger than the second, whose size is given by

$$\begin{aligned} y_2 &= 0.253 \times 202 + 0.260 \times 100 + \cdots + 0.205 \times 30 \\ &= 216.2. \end{aligned}$$

If all 40 y-scores are calculated, then the aphids can be ordered according to this single measurement.

This way of combining the nineteen measurements to produce a single measurement of size has other applications: for instance, a question that sometimes arises is how to combine students' scores on several different examinations into a single overall score. One way is to find the first principal component. Then the students could be ordered from 'best' to 'worst' according to their score on the first principal component, as

$$\begin{aligned}\text{Score} = 0.51 &\times \text{Mechanics}\\ + 0.37 &\times \text{Vectors}\\ + 0.35 &\times \text{Algebra}\\ + 0.45 &\times \text{Analysis}\\ + 0.53 &\times \text{Statistics}\end{aligned}$$

(say). But it is important to note that the one criterion on which the weights are calculated is that they should maximize the variation in the students' overall scores and make the differences between students as evident as possible: the weights do not necessarily reflect any perceived differences in the importance of the subjects examined. This must always be remembered when calculating principal components.

A further stage involves calculating the **second principal component**. This involves finding the weights $v_1, v_2, \ldots, v_{19}$ such that the variance of the 40 scores calculated using these weights is, in the same way as before, a maximum, subject to the constraint $v_1^2 + v_2^2 + \cdots + v_{19}^2 = 1$. However, so that the previous calculation is not merely reproduced, we also require the weights $v_1, v_2, \ldots, v_{19}$ to satisfy the additional constraint that the 40 scores along the second principal component should be *uncorrelated* with the first set of 40 scores. The exercise can be repeated to find the third and fourth, and further principal components: there are as many principal components as there are variables (in this case, 19).

It can be shown that the sum of the variances in the principal components is equal to the sum of the variances in the original variables. This means that our exercise of calculating principal components has *partitioned* the total variance in a particularly useful way. Since each principal component will have progressively less variance, the first few components show the weighted sums which are most important as they explain more of the data than the later components. In the case of the aphids, for instance, the first principal component yielding an overall measure of size for each aphid accounts for 72% of the total variation in the data. The first three components account for nearly 90% of the total variation. This means that the characteristics of each aphid can usefully be summarized in terms of just three variables rather than the original 19. Admittedly, variables such as

$$\begin{aligned}&0.253 \times \text{Body length} + 0.260 \times \text{Body width} + \cdots\\ &\quad + 0.205 \times \text{Number of hind-wing hooks}\end{aligned}$$

or (the second principal component)

$$\begin{aligned}&0.001 \times \text{Body length} - 0.055 \times \text{Body width} + \cdots\\ &\quad - 0.336 \times \text{Number of hind-wing hooks}\end{aligned}$$

do not always have a very obvious interpretation (particularly where there are minus signs) but remember, the aim of a principal components analysis is to explain as much variation as possible.

Sometimes the patterns of weights resulting from a principal components analysis turn out to have a useful interpretation. Huba et al (1981) were interested in patterns of consumption of both legal and illegal psychoactive substances. They collected data on 1634 school students in Los Angeles showing, for each of the psychoactive substances

Huba, G.J., Wingard, J.A. and Bentler, P.M. (1981) A comparison of two latent variable causal models for adolescent drug use. *Journal of Personality and Social Psychology*, **40**, 180–193.

- cigarettes
- beer
- wine
- spirits
- cocaine
- tranquillizers
- drug store medications used to get high
- heroin and other opiates
- marijuana
- hashish
- inhalents (glue, gasoline, etc.)
- hallucinogenics (LSD, mescaline, etc.)
- amphetamines/stimulants

a score of 1 (never tried), 2 (tried only once), 3 (tried a few times), 4 (tried many times), 5 (tried regularly). These scores for each of the thirteen psychoactive substances are the thirteen variables which were analysed.

The correlations between the thirteen scores, calculated from the 1634 subjects, are shown in Table 14.9.

Table 14.9 The correlations arising from an analysis of use of psychoactive substances by school students

	cigs	beer	wine	spirits	cocaine	tranq	drug	heroin	mari	hash	glue	hallu	amph
cigs	1												
beer	.447	1											
wine	.422	.619	1										
spirits	.435	.604	.583	1									
cocaine	.114	.068	.053	.115	1								
tranq	.203	.146	.139	.258	.349	1							
drug	.091	.103	.110	.122	.209	.221	1						
heroin	.082	.063	.066	.097	.321	.355	.201	1					
mari	.513	.445	.365	.482	.186	.315	.150	.154	1				
hash	.304	.318	.240	.368	.303	.377	.163	.219	.534	1			
glue	.245	.203	.183	.255	.272	.323	.310	.288	.301	.302	1		
hallu	.101	.088	.074	.139	.279	.367	.232	.320	.204	.368	.340	1	
amph	.245	.199	.184	.293	.278	.545	.232	.314	.394	.467	.392	.511	1

The correlations show us several things. First, all the correlations are positive—increased consumption of any one of the substances is associated with increased rather than decreased consumption of all of the others. However, none of the correlations is very large, so that there is not a strong dependence between any pair of substances. Also, some of the correlations are small—some pairs have only a very weak relationship.

For these data the first principal component is given by

$$\begin{aligned} &0.278 \times \text{cigarettes} + 0.286 \times \text{beer} + 0.265 \times \text{wine} + 0.318 \times \text{spirits} \\ &\quad + 0.208 \times \text{cocaine} + 0.293 \times \text{tranquillizers} + 0.176 \times \text{drug} \\ &\quad + 0.202 \times \text{heroin} + 0.339 \times \text{marijuana} + 0.329 \times \text{hashish} \\ &\quad + 0.276 \times \text{glue} + 0.248 \times \text{hallucinogenics} + 0.329 \times \text{amphetamines}. \end{aligned}$$

This component can be regarded as a measure of overall psychoactive substance usage. It says that the greatest single factor distinguishing the students (that is, maximizing the variation between them) is the overall extent to which they use the substances in question.

The second principal component is perhaps more interesting. This turns out to be

$$\begin{aligned} &0.280 \times \text{cigarettes} + 0.396 \times \text{beer} + 0.392 \times \text{wine} + 0.325 \times \text{spirits} \\ &\quad - 0.288 \times \text{cocaine} - 0.259 \times \text{tranquillizers} - 0.189 \times \text{drug} \\ &\quad - 0.315 \times \text{heroin} + 0.163 \times \text{marijuana} - 0.050 \times \text{hashish} \\ &\quad - 0.169 \times \text{glue} - 0.329 \times \text{hallucinogenics} - 0.232 \times \text{amphetamines}. \end{aligned}$$

Some of the weights are positive and others negative. In fact, if you study this second component carefully you will see that with the single exception of marijuana, all the legal drugs have positive weights and all the illegal ones have negative weights. What this means is that, having controlled for overall level of substance usage, the greatest source of difference between the students lies in their use of illegal versus legal substances. This is certainly a useful step towards explaining the differences between students' patterns of consumption in simpler terms than merely using all thirteen variables.

14.4 Discriminant analysis

Cluster analysis, described in Section 14.2, seeks to identify natural groupings within a set of objects using measurements made on those objects. In contrast, **discriminant analysis** begins with a set of objects with known group memberships and seeks to formulate some rule, in terms of measurements on the objects, so that future objects can be accurately allocated to groups. The objects may be of virtually any kind. They could be sick people, and perhaps the aim is to classify them to a disease class. Here, the rule may be based on whether or not certain symptoms are present and the values of a number of measures such as temperature, biochemical indicators, and so on. The objects could be spoken words, with the aim being automatic recognition by machine. Here, the rule will be based on the intensities of different frequencies at different times. The objects could be waveforms, such as those output by an electroencephalograph machine, where the aim is to detect anomalous shapes. Or the objects could be rocks, and the aim may be to identify the type of a new specimen.

The list is endless. Applications which have been made include disease prognosis for patients with barbiturate intoxication; diagnosis of non-toxic goitre;

diagnosis of *keratoconjunctivitis sicca*; screening for osteoporosis; prognosis in advanced breast cancer; applications in physical anthropology; machine recognition of hand-printed numerals; classification of archaeological specimens; comparing methods of preparing fish; quality rating in sheep; palaeontology.

Keratoconjunctivitis sicca is the inflammation of the conjunctiva and is associated with lacrimal deficiency.

From a discriminant analysis perspective, all such problems have a similar structure. There are two or more classes of objects. There are samples of objects from each class. For each object there is a set of measurements. The aim is to use the objects to formulate a rule such that a new object, for which the class is unknown, can be classified using only the values of the measurements on that object.

Just as there are many methods of cluster analysis, so there are many methods of discriminant analysis. The particular approach adopted in this section is *classical linear discriminant analysis*, one of the oldest methods, having been originally described by Sir Ronald Fisher in 1936. In this method one seeks a simple weighted sum of the measurements such that the groups are maximally separated on this sum. That is, we find a set of weights so that we can compute a single score for each object, much as is done in principal components analysis, but now rather than trying to maximize the variance of this score, we try to maximize the 'difference' between the *groups* of scores. More precisely, we shall choose the set of weights which maximizes the difference between the mean scores of the groups, relative to the standard deviation of scores within the groups. Once a suitable set of weights has been found, a new object can be classified simply by calculating its weighted sum and seeing which of the group mean scores it is closest to. (When we have just two groups, this reduces to defining a single threshold on the scores and classifying a new object according to whether its score is above or below the threshold.)

The method is illustrated here by applying it to data arising from a problem of identifying the population from which three historical kangaroo skull specimens arose. Measurements have been taken on the three skulls and corresponding measurements have been taken on samples from two modern kangaroo populations. The true population is known for each modern specimen. Our aim is to use the modern data to formulate a classification rule which will allow us to classify the three historical specimens.

For each of 25 male skull specimens from species *M. giganteus* (group 1) and for each of 21 male skull specimens from species *M.f. melanops* (group 2), eight measurements have been taken: basilar length, occipitonasal length, nasal length, nasal width, zygomatic width, crest width, mandible depth, and ascending ramus height. These values are shown in Table 14.10.

The first point to notice is that on each of the variables measured separately the two groups have considerable overlap. This is shown in the boxplots in Figure 14.4.

This means that none of the variables alone would not be very effective as a discriminant function: on no one of the variables could one place a threshold value such that cases with greater scores were predominantly of one class while cases with lower scores were predominantly of the other class. This, in fact, demonstrates the power of multivariate techniques such as discriminant analysis. The variables individually are not very useful at separating the classes. But what about taking them in combination? Taken together, as a multivariate set, perhaps the variables can separate the classes quite well.

Table 14.10 The kangaroo data

Group	x_1	x_2	x_3	x_4	x_5	x_6	x_7	x_8
1	1312	1445	609	241	782	153	179	591
1	1439	1503	629	222	824	141	181	643
1	1378	1464	620	233	778	144	169	610
1	1315	1367	564	207	801	116	189	594
1	1413	1500	645	247	823	120	197	654
1	1090	1195	493	189	673	188	138	476
1	1294	1421	606	226	780	149	168	578
1	1377	1504	660	240	812	128	175	628
1	1296	1439	630	215	759	151	159	578
1	1470	1563	672	231	856	103	196	683
1	1612	1699	778	263	921	86	232	772
1	1388	1500	616	220	805	107	180	652
1	1575	1655	727	271	905	82	210	712
1	1717	1821	810	284	960	104	222	731
1	1587	1711	778	279	910	81	207	692
1	1604	1770	823	272	880	57	208	713
1	1603	1703	755	268	902	81	206	754
1	1490	1599	710	278	897	115	194	688
1	1552	1540	701	238	852	82	213	722
1	1595	1709	803	255	904	83	183	701
1	1840	1907	855	308	984	84	238	795
1	1740	1817	838	281	977	121	227	770
1	1846	1893	830	288	1013	21	232	829
1	1702	1860	864	306	947	39	218	776
1	1768	1890	837	285	968	41	243	842
2	1299	1345	565	204	764	153	156	556
2	1337	1395	562	216	794	154	158	625
2	1372	1456	580	225	814	124	179	636
2	1336	1441	596	220	788	156	178	623
2	1301	1387	579	219	787	113	164	616
2	1360	1467	636	201	813	138	171	603
2	1276	1351	559	213	766	129	159	608
2	1613	1726	740	234	883	75	184	745
2	1542	1628	677	237	885	94	190	709
2	1440	1580	675	217	815	129	186	634
2	1474	1555	629	211	888	134	205	716
2	1503	1603	692	238	825	83	203	712
2	1597	1653	710	221	908	104	194	761
2	1671	1689	730	281	892	62	208	770
2	1673	1720	763	292	946	107	196	755
2	1458	1588	686	251	836	115	192	676
2	1568	1689	717	231	900	18	194	759
2	1650	1707	737	275	943	72	184	768
2	1774	1838	816	275	994	56	227	794
2	1893	1945	893	260	994	13	216	824
2	1765	1781	766	261	978	38	211	775

Andrews, D.F. and Herzberg, A.M. (1985) *Data*. Springer-Verlag, New York.

The first column gives the species group, *M. giganteus* (1) or *M.f. melanops* (2). The other eight columns are the measured variables: x_1 = basilar length, x_2 = occipitonasal length, x_3 = nasal length, x_4 = nasal width, x_5 = zygomatic width, x_6 = crest width, x_7 = mandible depth, x_8 = ascending ramus height. All measurements are in tenths of mm.

So, we begin in a way rather similar to principal components analysis. Suppose we choose a set of weights, w_i, and calculate weighted sums of the measurements for each subject. For example, for the first kangaroo skull, with measurements $(1312, 1445, \ldots, 591)$ we calculate

$$s_1 = 1312w_1 + 1445w_2 + \cdots + 591w_8.$$

Here s_1 represents the score for this skull, using these weights. Doing this for all 46 skulls gives us 46 scores, divided into two groups. Using these we can easily calculate the means of the two sets of scores and their standard deviations. (In fact, without going into details, we shall assume that the

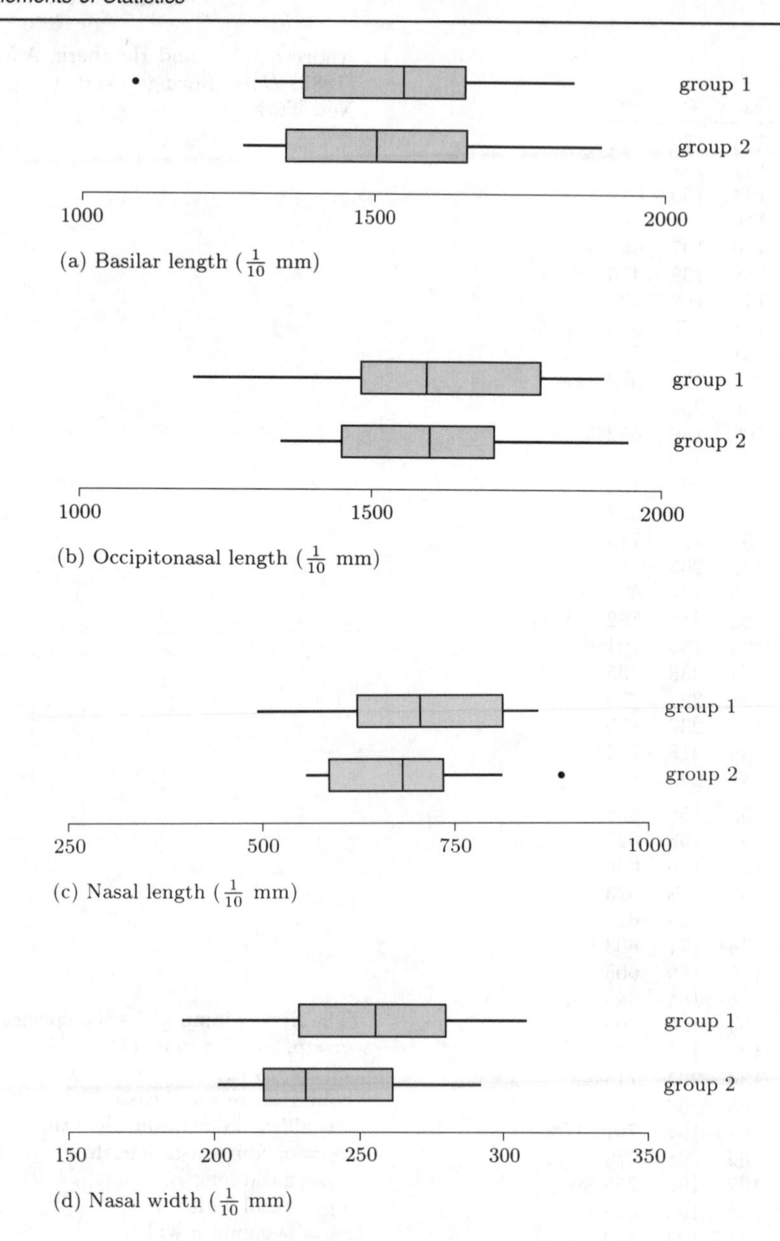

Figure 14.4 Boxplots of the eight variables, kangaroo data

standard deviations are the same in the two populations and estimate this common value—the strategy is exactly the same as that for the two sample t-test described in *Chapter 8*. Indeed, these means and the standard deviation allows us to calculate the 'distance' between the means of the two groups, taking into account their standard deviation—a straightforward t-value.)

Now, in principle we can do this for a whole range of sets of weights and choose that which leads to the greatest distance between the groups. In fact, however, with a little algebra, it is easy to find the particular set of weights which maximizes this distance without searching through such sets. In practice, of course, it is easier still, because a computer program will do the task for us.

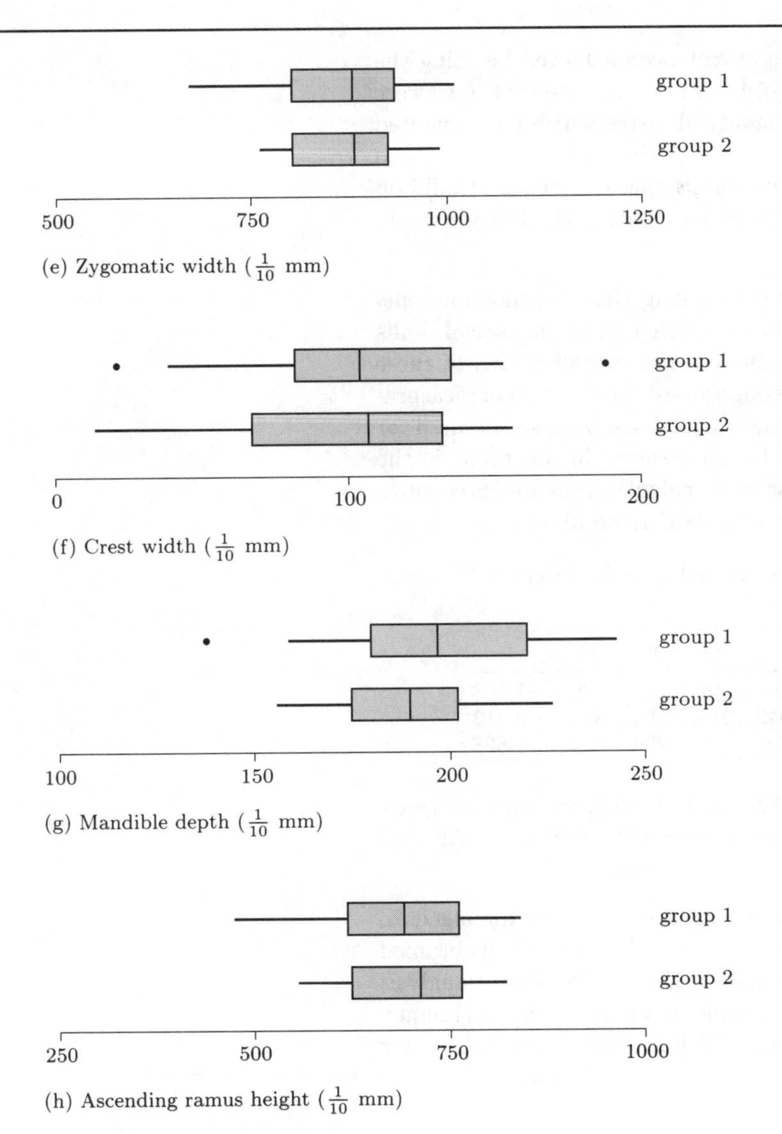

Using such a discriminant analysis program, the weighted sum of the measurements which produces a set of scores such that the means of the two groups are best separated is given by the set of weights

In this case there is no requirement for the squared weights w_i^2 to sum to one.

$$s = -0.0066x_1 - 0.0119x_2 + 0.0287x_3 + 0.0256x_4 - 0.0003x_5 - 0.0024x_6 + 0.0707x_7 - 0.0202x_8.$$

Boxplots of the scores of the cases in the two separate groups, calculated using these weights, are shown in Figure 14.5.

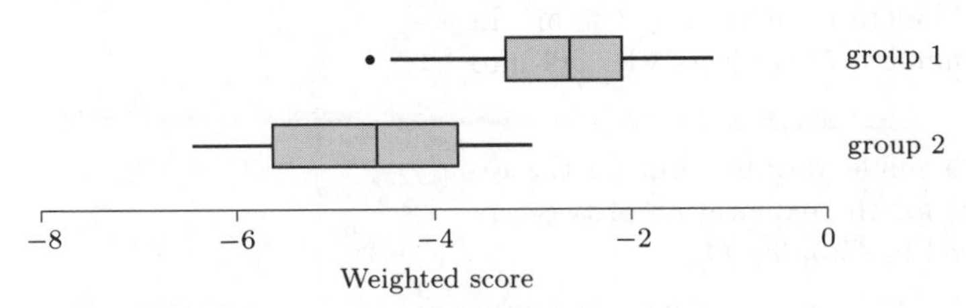

Figure 14.5 Boxplots of the data, using the weighted score s

It is clear that far superior separation has in fact been achieved by using the variables together. Moreover, from Figure 14.5 we can also see that if we were to adopt a threshold of about −3.5 and classify all cases which had weighted sums above this as belonging to the *M. giganteus* species and all with weighted sums below this as belonging to the *M.f. melanops* species, then we would do reasonably well. We would not always classify cases correctly, but we would get most of them right.

We can now classify the three historical cases using this classification rule. Unfortunately, several of the measurements are missing for the historical skulls (missing data is common with real data sets) so the overall means of these variables have been substituted for the missing values. The vectors of measurements for the three historical cases, once the means have been substituted for the missing values, are shown in Table 14.11. Also shown in this table, in the far right column, are the s-scores for these cases calculated as weighted sums of the original x-measurements, using the weights derived above.

Table 14.11 The original measured values and their final weighted sum, s

Specimen	x_1	x_2	x_3	x_4	x_5	x_6	x_7	x_8	s
A	1899	1925	905	310	1067	46	257	880	−1.568
B	1848	1609	751	287	1045	101	194	843	−6.312
C	1115	1609	699	246	756	101	204	593	1.828

Comparing the weighted sums with the threshold of −3.5, we see that specimens A and C are above the threshold and specimen B is below it. We thus classify A and C as *M. giganteus* and B as *M.f. melanops.*

In conclusion, by using all of the variables together we have managed to produce a classification rule which is superior to that which would be obtained by using the variables separately. Discriminant analysis, like cluster analysis and principal components analysis, is an example of a multivariate technique. Also like them, it is an example of a technique which relies heavily on computer power to make it practically feasible.

14.5 Log-linear modelling

In many problems data arise as counts of the frequencies with which certain categories of response occur. For example, individuals could be classified by sex (Male–Female), by response to a treatment (Healed–Healing–No change), by age (Young–Middle-aged–Old); a comparison between two stimuli may be scored on a five-point scale (Strongly prefer A–Slightly prefer A–No preference–Slightly prefer B–Strongly prefer B); the severity of a disease could be scored on a four-point scale (None–Mild–Moderate–Severe); and so on. In each of these examples the data will be the number of subjects who fall into each of the response categories.

Often interest focuses not on the scores on a single variable, but on the relationship between two variables. Techniques for the exploration of two-way tables called contingency tables were described in *Chapter 11.*

In this section, however, our main objective is to extend the problem to a third dimension, and explore the relationship between three categorical variables.

The data used to illustrate the approach are shown in Table 14.12. These arose in a study of the social origins of depression and, in particular, on the role of 'life events' in causing depression. Life events are major occurrences in one's life which, it is hypothesized, may serve to cause depression, perhaps when they occur with certain 'vulnerability factors'. Examples of such life events are bereavement, moving house, and changing jobs.

Details of the scientific background are given in Brown, G.W. and Harris, T. (1978) *Social origins of depression.* Tavistock, London.

The data set shows a three-way cross-classification of a sample of women from Camberwell in South East London. Whether or not they developed depression is cross-classified by whether or not they had experienced a life event. The table is then split according to whether or not they had an intimate supportive relationship. We shall be interested in whether or not the relationship between life events and depression is influenced by the presence of a supportive relationship.

Table 14.12 Cross-classification of depression by life event by intimacy

	Lack of intimacy Yes		Lack of intimacy No	
	Life event Yes	Life event No	Life event Yes	Life event No
Depression	24	2	9	2
No depression	52	60	79	191

We could **collapse** the three-way table and look at two-dimensional marginal cross-classifications. This is achieved simply by ignoring the third variable, so that the counts at each of its levels are added. Table 14.13 shows such a table for the cross-classification of depression by life event.

Table 14.13 Cross-classification of depression by life event

	Life event Yes	Life event No
Depression	33	4
No depression	131	251

The number 33 in the top left corner has been obtained by adding the numbers 24 and 9 in Table 14.12—collapsing across levels of 'lack of intimacy'. This new table would allow us to look at the relationship between depression and life events, simply adopting the technique of *Chapter 11.* However, this would ignore the possible effects of other factors. In particular, it would not distinguish between those women with a supportive relationship and those without. It could be the case that the relationship between depression and life events is quite different in the two categories of the third variable. Somehow we need to analyse the three-dimensional table in its entirety.

A great deal of statistics hinges around the notion of **linear models**. These are mathematical models in which the effects of different factors and the interactions between them are *additive.* For example, such a model involving two factors could be expressed as the sum of effects due to each of the factors separately. And, if it was thought that there was an interaction effect, then a model could be built which had an 'interaction term' added in.

Linear models have appealing properties—for a start, they are mathematically very tractable. Without going into details, it turns out to be possible to construct a linear model in the present context to predict the *logarithm* of the probability that an object will fall into any particular cell in the three-way table. Thus the model allows us to calculate the expected frequencies in each cell of the table under the assumption that the model is correct. Then these predicted values are compared with those observed. The difference between the fitted model and the data is given in this context by the **deviance**

$$D^2 = 2\sum O_i \log \frac{O_i}{E_i},$$

where O_i refers to the frequencies observed in the cells, E_i to the frequencies expected, and where the summation is taken over all the cells. The deviance has asymptotically a chi-squared distribution.

To test the hypothesis that the existence of a supportive relationship has no effect on the degree of association between the other variables, a model would be constructed including terms for the effects of life events, the existence of a supportive relationship, and depression; terms for the interaction of pairs of variables (the three so-called 'two-way' interaction terms); but *no* term for the effect of the existence of a supportive relationship on the degree of association between the two other variables (no three-way interaction: that is, the degree of association between depression and life events does not depend on whether there is a supportive relationship.)

For instance, existence of an interaction between life events and depression would mean that the chance of depression could depend on whether some life event had been experienced.

Log-linear models almost always require a computer to calculate the expected frequencies, since they rely on an iterative fitting process. The model fitted in this case gave a deviance of 0.4814: there are 8 cells in the table and our model involves 7 parameters (an overall mean, one for each of the three factors, and one for each of the three two-way interactions). Thus we need to compare 0.4814 with the chi-squared distribution with 1 degree of freedom. The corresponding SP is 0.488. Thus we have no reason to suppose that the three-way interaction is necessary to explain the data we have observed: we have no reason to suppose that the relationship between depression and life events is mediated by the presence of a supportive relationship.

14.6 Multiple regression analysis

Earlier in the course, in *Chapter 10*, you met simple linear regression analysis. This was a technique for constructing models to relate two variables together, with one of the variables being an 'explanatory' variable and the other a 'response' variable. Such models can be used in a number of ways. They can, for example, allow one to predict the change that one might expect in the response variable if a change is made in the explanatory variable.

In this section the simple regression model is extended to the case when there are several explanatory variables: to the case of **multiple regression**.

The data we shall use to illustrate the ideas is a subset of data which first appeared in 1932 and have since been reanalysed many times.

Woods, H., Steiner, H.H. and Starke, H.R. (1932) Effects of composition of Portland cement on heat evolved during hardening. *Industrial and Engineering Chemistry*, **24**, 1207–1212.

When cement sets, heat is generated and the aim of this study was to see how the amount of heat is influenced by the composition of the cement. Thus, the response variable was the heat evolved, measured in calories per gram of cement. The explanatory variables, summarizing the composition, are percentages of the cement by weight of three constituents: tricalcium-aluminate, tricalcium-silicate, and tetracalcium-alumino-ferrite. The data are shown in Table 14.14. Each row of this table shows the values of the explanatory variables (x_1, x_2, x_3) and the corresponding response (y) for a cement sample.

Table 14.14 Heat evolved during cement hardening (calories per gram)

Heat, y	TA, x_1	TS, x_2	TAF, x_3
78.5	7	26	6
74.3	1	29	15
104.3	11	56	8
87.6	11	31	8
95.9	7	52	6
109.2	11	55	9
102.7	3	71	17
72.5	1	31	22
93.1	2	54	18
115.9	21	47	4
83.8	1	40	23
113.3	11	66	9
109.4	10	68	8

In simple linear regression a model is constructed in which the expected value of the response variable is predicted from the sum of a constant term (which may be constrained to be zero) and some multiple of the value taken by the explanatory variable. This multiple is the slope of the regression line. It can be interpreted as showing the effect that a unit change in the value of the explanatory variable has on the response variable.

Similarly, in multiple regression, a model is constructed in which the expected value of the response variable is predicted from the sum of a constant term and multiples of the explanatory variables. Each explanatory variable will have its own multiple and these are called **regression coefficients**.

Although superficially straightforward, multiple regression in fact involves a number of subtleties. One of the chief difficulties is the question of how to interpret the regression coefficients. The regression coefficient of an explanatory variable in a multiple regression is not simply the effect that a unit change in that variable will have on the response. This would be the regression coefficient one would get from a simple linear regression involving just that one explanatory variable by itself—in other words, the slope of the regression line. Instead, the regression coefficient in a multiple regression is the effect that a unit change in this explanatory variable will produce *assuming all the other explanatory variables are held constant.* Put another way, it is the change one can uniquely attribute to the explanatory variable in question.

Notice that it will not always make sense to ask about the effect of some variable holding all others constant. In our numerical example, for instance, if we had measures of the percentages of *all* the constituents comprising the cement then the question would not be meaningful, since these would necessarily add up to 100—it would not be possible to change one without changing some others. Fortunately, such complications do not arise in the present problem and it does make sense to ask about the change in the response variable which would follow from changing each constituent without changing the others.

Our objective in the present situation is to try to build a model which allows us to relate the response variable, heat evolved, to the explanatory variables, the concentrations of the constituents. Such a model would, for example, allow us to predict the heat evolved based on the known concentrations of the constituents for a new cement sample.

The approach follows the same lines as in simple regression. There, the fitted regression line was chosen to minimize the sum of squared deviations between the predictions from the regression model and the observed values of the response variable. Exactly the same is done here, except that we have several regression coefficients to estimate.

Here, again our computer does the calculations for us. Measures of deviation other than a sum of squares could be used, but this is historically the most important and certainly by far the most widely used measure. One reason for this is its mathematical tractability. However, now that powerful computers are readily available, other measures are becoming increasingly popular. This is not the place for a comparative discussion of such measures, but it is worth noting that different measures have different properties—some, for example, are more resistant than others to outliers in the data—so that an informed choice can lead to better results.

Minimizing the sum of squared deviations

$$\sum(y_i - \widehat{y}_i)^2 = \sum(y_i - \widehat{\beta}_0 - \widehat{\beta}_1 x_1 - \widehat{\beta}_2 x_2 - \widehat{\beta}_3 x_3)^2$$

over the 13 data points gives the fitted **regression surface**

$$\widehat{y}_i = \widehat{\beta}_0 + \widehat{\beta}_1 x_1 + \widehat{\beta}_2 x_2 + \widehat{\beta}_3 x_3 = 48.19 + 1.70 x_1 + 0.66 x_2 + 0.25 x_3$$

or

$$\text{Heat evolved} = 48.19 + 1.70 \times (\%\ \text{TA}) + 0.66 \times (\%\ \text{TS}) + 0.25 \times (\%\ \text{TAF})$$

where the response is measured in calories per gram of cement.

From this it can be seen that tricalcium-aluminate is the most important of the constituents, in the sense that a 1% change in its concentration causes a 1.70 calories per gram change in the heat emitted while, for example, a 1% change in the concentration of tetracalcium-alumino-ferrite only results in a 0.25 calories per gram change in the heat emitted. Notice that we are only able to make this kind of statement about the relative importance of the explanatory variables because they are *commensurate*—they are measured in the same units (percentage points). If, in contrast, we had a regression model with one explanatory variable measured in inches and another in pounds weight, such a comparative statement would not be possible. By changing the units one could change the apparent importance of each variable.

In *Chapter 10* you learned how to test the hypothesis $H_0 : \beta = 0$ that the slope of a regression line is 0 (that is to say, that changes in the value of the explanatory variable have no effect on the response). Here, significance tests for each of the regression coefficients give rise to very small significance probabilities (you are spared the details) except in the case where the hypothesis $H_0 : \beta_3 = 0$ is tested: the SP is about 0.21. This suggests that tetracalcium-alumino-ferrite may be irrelevant to the amount of heat emitted.

Now, recall that the coefficients in a multiple regression model show the effect of each explanatory variable in a model in which the others are held constant. If a different model were to be fitted, say, one in which one of the explanatory variables was dropped, then perhaps different coefficients would be obtained for those remaining. After all, in this model with one less explanatory variable, one less variable would be held constant.

The relevance of this for our example is that the third explanatory variable, tetracalcium-alumino-ferrite, seemed not to have any effect on the response variable. Thus, it seems a little pointless including it in the model. We could fit a model involving just the other two explanatory variables.

If this is done the fitted regression surface becomes

$$y = 52.58 + 1.47x_1 + 0.66x_2,$$

or

$$\text{Heat evolved} = 52.58 + 1.47 \times (\%\ \text{TA}) + 0.66 \times (\%\ \text{TS}).$$

In this revised model all of the regression coefficients are significant; and it can be shown too that the sampling variance attaching to two of the three estimated coefficients (β_0 and β_1) has been reduced. In general, by fitting the most parsimonious model, one obtains less variable estimates of regression coefficients.

We have now achieved what we set out to do. We have constructed a model which allows us to predict the amount of heat evolved when cement sets. Moreover, as an additional bonus, we can judge the relative importance of the individual predictor variables.

In this particular example, dropping one of the explanatory variables had relatively little effect on the regression coefficients of the others. That of tricalcium-aluminate changed from 1.70 to 1.47 and that of tricalcium-silicate was unaltered. But, in fact, the changes can be dramatic—even being so much as to reverse the sign of a coefficient. The fundamental point is that in multiple regression each coefficient needs to be interpreted as the effect on the response of a unit change in its corresponding explanatory variable when the other explanatory variables which have been fitted are held constant. This will depend on what other variables have been fitted.

14.7 And beyond

The preceding six sections illustrate six statistical tools which are more advanced than those you have studied elsewhere in this course. The idea of these sections was to demonstrate the sorts of problems that could be tackled by more sophisticated methods, rather than to enable you to undertake such analyses without further instruction. But, of course, the six chosen illustrations do not go anywhere near covering the complete range of techniques available to a statistician.

However, perhaps the most important classes of topics not mentioned in the course are those of **experimental design** and **survey design**. All of the course, with the exception of just one or two passing references, has been concerned with how to analyse data once it has been collected: how to use it to address the question or questions of interest. In discussing methods of analysis, attention was drawn to any assumptions those methods may have made. For example, in discussing the two-sample t-test, it was stressed, amongst other things, that the samples had to be independently chosen from the two groups. In order for the test to be valid, the data had to be chosen such that this assumption was satisfied. But apart from such references, there has been no discussion of the best way to collect the data. The subdisciplines of statistics called experimental design and survey design deal with just such issues. They tell us the most efficient ways to collect data to answer the research questions posed, and the ways to collect data to yield the most accurate results.

A hint at the sort of ideas involved occurred in Section 14.1, when an analysis of variance arising from the cross-classification of two factors was discussed. You will recall that the data used in the example allowed us to compare the two headache types using 22 subjects in each of the two headache groups and also allowed us to compare the two treatment groups using 22 subjects in each group, but with a total of only 44 subjects, and not the 88 this would seem to imply. By controlling how the data were collected we increased the accuracy of the analysis without increasing the cost—without requiring larger samples to be collected.

In general, such tools allow great savings in resources and great increases in accuracy, and represent one of the really great contributions of statistics to scientific research.

Design represents a whole subdiscipline of statistics that we have not had space to discuss. But there are also other areas which are arguably just as important. For example, the course is based largely on current statistical practice and describes the ideas and methods underlying almost all major statistical computer packages which are currently commercially available. These, in turn, are based on a particular interpretation of probability (the *frequentist* interpretation)—a particular view of what the notion of probability means. But there are other views.

Of these the most important is undoubtedly the subjective view, leading to a school of statistics called the **Bayesian school**. In brief, this school asserts that probability has no objective meaning in the real world, but simply represents the degree of belief of the researcher that something is true. Data are collected and analysed and this leads to a modification of the researcher's beliefs in ways described by appropriate statistical techniques. One of the problems with applying such methods until recently has been the mathematical intractability of the analyses. Particular forms, often unrealistic, for the probability distributions involved had to be assumed in order that the analysis could be carried out. In recent years, however, developments have been made such that these problems no longer apply. More realistic distributional forms can be adopted and the power of the computer means that we can sidestep analytic mathematical solutions. It is certainly the case that Bayesian methods are growing in popularity and, when Bayesian software becomes readily available, they may well compete with traditional methods.

14.8 Conclusion

In the preceding sections a number of advanced statistical techniques have been presented for probing data and answering questions. Inevitably the constraints of trying to cover several such techniques in a short space mean that the impression may have been given that such techniques are isolated, bearing little relationship to one another. This is, in fact, not the case, and there are many relationships between the methods. Thus, for example, both analysis of variance and regression analysis are particular types of a more general statistical approach using a structure called a *linear model*, referred to in Section 14.5. This in turn is a particular form of a yet more abstract structure called a *generalized linear model* (of which the log-linear model is another special case). Indeed, there is an even broader generalization—to the class of *additive models*.

It is a simplifying and often misleading misconception to suppose that statistics is a collection of disconnected recipes, one to be selected and plugged in when a research question presents itself. In fact, the subject is more a language than a vocabulary, with a rich network of interconnections permitting the delicate probing of data.

The authors have been at pains to point out the central role that computers play in modern statistics. Computers mean that older statistical techniques can be performed more rapidly; they make feasible methods that had already been invented but were too time-consuming to carry out in practice; they have permitted the development of entirely new methods; and they have increased the size of problems that can be tackled, both in terms of the number of variables and the number of cases. It is in no way an exaggeration to say that the computer is the essential tool of modern statistics.

If the past three decades have seen dramatic changes, partly as a result of increasing computational power, there is no reason to suppose this will slow down or stop. Modern statistics has become a very exciting science, with an extraordinary range of tools and methods for probing and seeking structure in data, and for answering questions about the world around us. And the future promises to be even more exciting.

Solutions to the Exercises

Chapter 1

Solution 1.1

(a) Your computer may be programmed to allocate borderline cases to the next group down, or the next group up; and it may or may not manage to follow this rule consistently, depending on its handling of the numbers involved. Following a rule which says 'move borderline cases to the next group up', these are the five classifications.

(i)

1.0–1.2	1.2–1.4	1.4–1.6	1.6–1.8	1.8–2.0	2.0–2.2	2.2–2.4
6	6	4	8	4	3	4
2.4–2.6	2.6–2.8	2.8–3.0	3.0–3.2	3.2–3.4	3.4–3.6	3.6–3.8
6	3	2	2	0	1	1

(ii)

1.0–1.3	1.3–1.6	1.6–1.9	1.9–2.2	2.2–2.5
10	6	10	5	6
2.5–2.8	2.8–3.1	3.1–3.4	3.4–3.7	
7	3	1	2	

(iii)

0.8–1.1	1.1–1.4	1.4–1.7	1.7–2.0	2.0–2.3
2	10	6	10	7
2.3–2.6	2.6–2.9	2.9–3.2	3.2–3.5	3.5–3.8
6	4	3	1	1

(iv)

0.85–1.15	1.15–1.45	1.45–1.75	1.75–2.05	2.05–2.35
4	9	8	9	5
2.35–2.65	2.65–2.95	2.95–3.25	3.25–3.55	3.55–3.85
7	3	3	1	1

(v)

0.9–1.2	1.2–1.5	1.5–1.8	1.8–2.1	2.1–2.4
6	7	11	7	4
2.4–2.7	2.7–3.0	3.0–3.3	3.3–3.6	3.6–3.9
7	4	2	1	1

(b) Computer graphics: the diagrams are shown in Figures 1.9 to 1.11.

Solution 1.2

(a) Computer graphics: see Figure 1.12.

(b) Computer graphics: see Figure 1.13.

If your computer gives graphics that are text-character based (otherwise known as low-resolution graphics) then the scatter plots you obtain will not be as precise as those appearing in the text and the fitted line will not be displayed. However, the main message of the data should still be apparent.

Solution 1.3

(a) In order of decreasing brain weight to body weight ratio, the species are as follows.

Species	Body weight	Brain weight	Ratio
Rhesus Monkey	6.800	179.000	26.32
Mole	0.122	3.000	24.59
Human	62.000	1320.000	21.29
Mouse	0.023	0.400	17.39
Potar Monkey	10.000	115.000	11.50
Chimpanzee	52.160	440.000	8.44
Hamster	0.120	1.000	8.33
Cat	3.300	25.600	7.76
Rat	0.280	1.900	6.79
Mountain Beaver	1.350	8.100	6.00
Guinea Pig	1.040	5.500	5.29
Rabbit	2.500	12.100	4.84
Goat	27.660	115.000	4.16
Grey Wolf	36.330	119.500	3.29
Sheep	55.500	175.000	3.15
Donkey	187.100	419.000	2.24
Gorilla	207.000	406.000	1.96
Asian Elephant	2547.000	4603.000	1.81
Kangaroo	35.000	56.000	1.60
Jaguar	100.000	157.000	1.57
Giraffe	529.000	680.000	1.29
Horse	521.000	655.000	1.26
Pig	192.000	180.000	0.94
Cow	465.000	423.000	0.91
African Elephant	6654.000	5712.000	0.86
Triceratops	9400.000	70.000	0.007
Diplodocus	11700.000	50.000	0.004
Brachiosaurus	87000.000	154.500	0.002

(b) (i) Computer graphics: see Figure 1.14.

(ii) Computer graphics: see Figure 1.15.

Solution 1.4

There were 23 children who survived the condition. Their birth weights are 1.130, 1.410, 1.575, 1.680, 1.715, 1.720, 1.760, 1.930, 2.015, 2.040, 2.090, 2.200, 2.400, 2.550, 2.570, 2.600, 2.700, 2.830, 2.950, 3.005, 3.160, 3.400, 3.640. The median birth weight for these children is 2.200 kg (the 12th value in the sorted list).

There were 27 children who died. The sorted birth weights are 1.030, 1.050, 1.100, 1.175, 1.185, 1.225, 1.230, 1.262, 1.295, 1.300, 1.310, 1.500, 1.550, 1.600, 1.720, 1.750, 1.770, 1.820, 1.890, 1.940, 2.200, 2.270, 2.275, 2.440, 2.500, 2.560, 2.730. The middle value is the 14th (thirteen either side) so the median birth weight for these children who died is 1.600 kg.

Solution 1.5

The ordered differences are 3.8, 10.3, 11.8, 12.9, 17.5, 20.5, 20.6, 24.4, 25.3, 28.4, 30.6. The median difference is 20.5.

Solution 1.6

Once the data are entered, most computers will return the sample median at a single command. It is 79.7 inches.

Solution 1.7

(a) The mean birth weight of the 23 infants who survived SIRDS is

$$\overline{x}_S = \frac{1.130 + 1.575 + \cdots + 3.005}{23} = \frac{53.070}{23} = 2.307\,\text{kg};$$

the mean birth weight of the 27 infants who died is

$$\overline{x}_D = \frac{1.050 + 1.175 + \cdots + 2.730}{27} = \frac{45.680}{27} = 1.692\,\text{kg}.$$

The mean birth weight of the entire sample is

$$\overline{x}_T = \frac{1.130 + 1.575 + \cdots + 2.730}{50} = \frac{98.75}{50} = 1.975\,\text{kg}.$$

Notice the subscripts S, D and T used in this solution to label and distinguish the three sample means. It was not strictly necessary to do this here, since we will not be referring to these numbers again in this exercise, but it is a convenient labelling system when a statistical analysis becomes more complicated.

Solution 1.8

The mean 'After – Before' difference in Table 1.11 is

$$\overline{x} = \frac{25.3 + 20.5 + \cdots + 28.4}{11} = \frac{206.1}{11} = 18.74\,\text{pmol/l}.$$

Solution 1.9

The mean snowfall over the 63 years was 80.3 inches.

Solution 1.10

(a) The lower quartile birth weight for the 27 children who died is given by

$$q_L = x_{(\frac{1}{4}(n+1))} = x_{(7)} = 1.230\,\text{kg};$$

the upper quartile birth weight is

$$q_U = x_{(\frac{3}{4}(n+1))} = x_{(21)} = 2.200\,\text{kg}.$$

(b) For these silica data, the sample size is $n = 22$. The lower quartile is

$$q_L = x_{(\frac{1}{4}(n+1))} = x_{(\frac{23}{4})} = x_{(5\frac{3}{4})}$$

which is three-quarters of the way between $x_{(5)} = 26.39$ and $x_{(6)} = 27.08$. This is

$$q_L = 26.39 + \tfrac{3}{4}(27.08 - 26.39) = \tfrac{1}{4}(26.39) + \tfrac{3}{4}(27.08) = 26.908;$$

say, 26.9. The sample median is

$$m = x_{(\frac{1}{2}(n+1))} = x_{(\frac{23}{2})} = x_{(11\frac{1}{2})},$$

which is midway between $x_{(11)} = 28.69$ and $x_{(12)} = 29.36$. This is 29.025; say, 29.0.

The upper quartile is

$$q_U = x_{(\frac{3}{4}(n+1))} = x_{(\frac{69}{4})} = x_{(17\frac{1}{4})},$$

one-quarter of the way between $x_{(17)} = 33.28$ and $x_{(18)} = 33.40$. This is

$$q_U = 33.28 + \tfrac{1}{4}(33.40 - 33.28) = \tfrac{3}{4}(33.28) + \tfrac{1}{4}(33.40) = 33.31;$$

say, 33.3.

Solution 1.11

For the snowfall data the lower and upper quartiles are $q_L = 63.6$ inches and $q_U = 98.3$ inches respectively. The interquartile range is $q_U - q_L = 34.7$ inches.

Solution 1.12

Answering these questions might involve delving around for the instruction manual that came with your calculator! The important thing is not to use the formula—let your calculator do all the arithmetic. All you should need to do is key in the original data and then press the correct button. (There might be a choice, one of which is when the divisor in the 'standard deviation' formula is n, the other is when the divisor is $n - 1$. Remember, in this course we use the second formula.)

(a) You should have obtained $s = 8.33$, to two decimal places.

(b) The standard deviation for the silica data is $s = 4.29$.

(c) For the collapsed runners' β endorphin concentrations, $s = 98.0$.

Solution 1.13

(a) The standard deviation s is 0.66 kg.

(b) The standard deviation s is 23.7 inches.

Solution 1.14

Summary measures for this data set are

$$x_{(1)} = 23, \quad q_L = 34, \quad m = 45, \quad q_U = 62, \quad x_{(11)} = 83.$$

The sample median is $m = 45$; the sample mean is $\overline{x} = 48.4$; the sample standard deviation is 18.1. The range is $83 - 23 = 60$; the interquartile range is $62 - 34 = 28$.

Solution 1.15

The first group contains 19 completed families. Some summary statistics are

$$m = 10, \quad \overline{x} = 8.2, \quad s = 5.2, \quad \text{interquartile range} = 10.$$

For the second group of 35 completed families, summary statistics are

$$m = 4, \quad \overline{x} = 4.8, \quad s = 4.0, \quad \text{interquartile range} = 4.$$

The differences are very noticeable between the two groups. Mothers educated for the longer time period would appear to have smaller families. In each case the mean and median are of comparable size. For the smaller group, the interquartile range is much greater than the standard deviation. If the three or four very large families are removed from the second data set, the differences become even more pronounced.

Solution 1.16

(a) The five-figure summary for the silica data is given by

$$(20.77, 26.91, 29.03, 33.31, 34.82).$$

A convenient scale sufficient to cover the extent of the data is from 20 to 40. The i.q.r. is $33.31 - 26.91 = 6.40$. Then

$$q_U + \text{ i.q.r } = 33.31 + 6.40 = 39.71$$

and this exceeds the sample maximum, so the upper adjacent value is the sample maximum itself, 34.82. Also

$$q_L - \text{ i.q.r. } = 26.91 - 6.40 = 20.51.$$

This value is less than the sample minimum, so the lower adjacent value is the sample minimum itself. For these data there are no extreme values. The boxplot is shown in Figure S1.1.

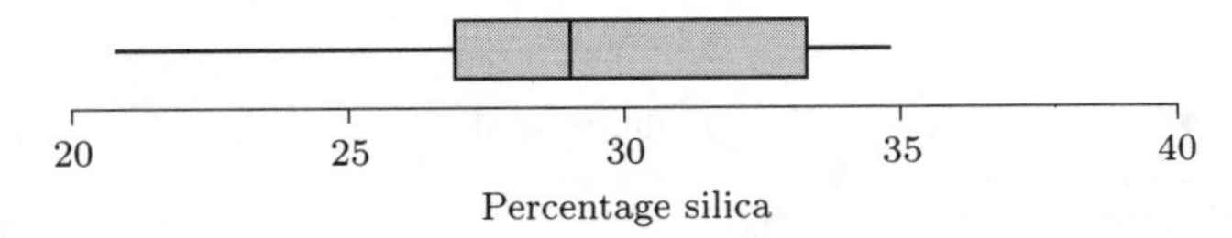

Figure S1.1

(b) For the snowfall data the lower adjacent value is 39.8; the minimum is 25.0. The upper adjacent value is equal to the maximum, 126.4. The boxplot is shown in Figure S1.2.

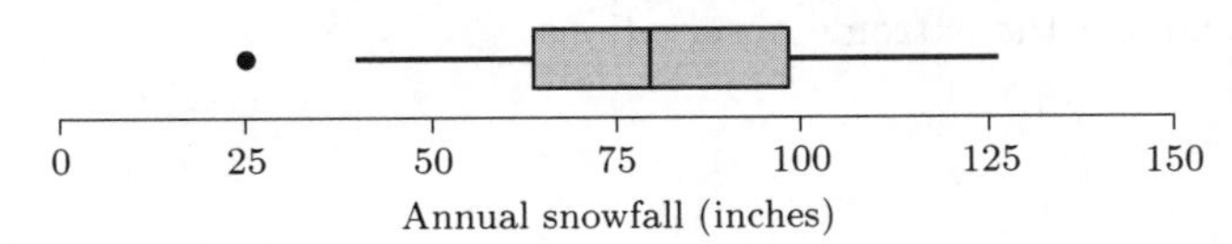

Figure S1.2

Solution 1.17

The sample skewness for the first group of mothers is -0.29.

Solution 1.18

(a) The five-figure summaries for the three groups are

normal: (14, 92, 124.5, 274.75, 655)
alloxan-diabetic: (13, 70.25,139.5, 276, 499)
insulin-treated: (18, 44, 82, 133, 465).

The normal group has one very high recording at 655; the next highest is 455, which is more consistent with the other two groups.

(b) The mean and standard deviation for each group are

normal: $\overline{x} = 186.1$, $s = 158.8$
alloxan-diabetic: $\overline{x} = 181.8$, $s = 144.8$
insulin-treated: $\overline{x} = 112.9$, $s = 105.8$.

The mean reading in the third group seems noticeably less than that for the first two groups, and has a reduced standard deviation.

(c) The sample skewness for each group is

normal:	1.47
alloxan-diabetic:	1.01
insulin-treated:	2.07.

All the samples are positively skewed: the third group has one substantial outlier at 465. Eliminating that outlier reduces the skewness to 1.02.

(d) The comparative boxplot in Figure S1.3 does not suggest any particular difference between the groups. The first two groups are substantially skewed with some evidence of extreme observations to the right; apart from three very extreme observations contributing to a high skewness, observations in the third group are more tightly clustered around the mean.

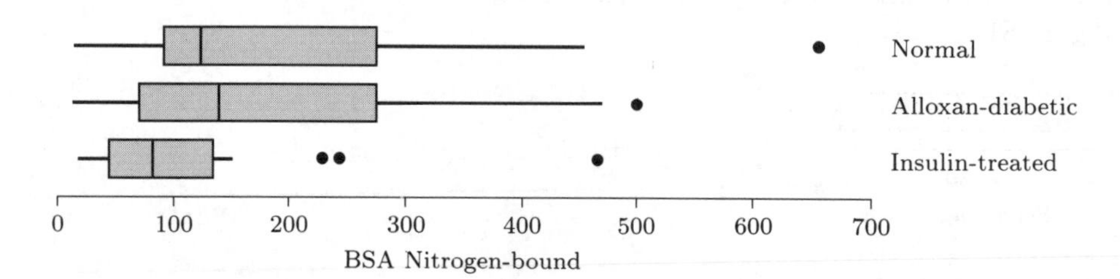

Figure S1.3

Of course, a computer makes detailed exploration of data sets relatively easy, quick and rewarding. You might find it interesting to pursue the story the data have to tell after, say, removing the extreme observations from each group.

Chapter 2

Solution 2.1

In this kind of study it is essential to state beforehand the population of interest. If this consists of rail travellers and workers then the location of the survey may be reasonable. If, on the other hand, the researcher wishes to draw some conclusions about the reading habits of the entire population of Great Britain then this sampling strategy omits, or under-represents, car users and people who never, or rarely, visit London.

A sample drawn at 9 am on a weekday will consist very largely of commuters to work, and if the researcher is interested primarily in their reading habits then the strategy will be a very useful one. On a Saturday evening there will possibly be some overrepresentation of those with the inclination, and the means, to enjoy an evening out.

Solution 2.2

This is a practical simulation. It is discussed in the text following the exercise.

Solution 2.3

A typical sequence of 40 coin tosses, and the resulting calculations and graph, follow.

Table S2.1 The results of 40 tosses of a coin

Toss number	1	2	3	4	5	6	7	8	9	10
Observed result	1	1	1	0	0	0	0	0	0	1
Total so far	1	2	3	3	3	3	3	3	3	4
Proportion (P)	1.00	1.00	1.00	0.75	0.60	0.50	0.43	0.38	0.33	0.40
Toss number	11	12	13	14	15	16	17	18	19	20
Observed result	0	1	0	1	1	0	1	0	1	1
Total so far	4	5	5	6	7	7	8	8	9	10
Proportion (P)	0.36	0.42	0.38	0.43	0.47	0.44	0.47	0.44	0.47	0.50
Toss number	21	22	23	24	25	26	27	28	29	30
Observed result	0	0	1	1	1	0	1	1	1	1
Total so far	10	10	11	12	13	13	14	15	16	17
Proportion (P)	0.48	0.45	0.48	0.50	0.52	0.50	0.52	0.54	0.55	0.57
Toss number	31	32	33	34	35	36	37	38	39	40
Observed result	1	1	1	1	0	0	1	0	0	1
Total so far	18	19	20	21	21	21	22	22	22	23
Proportion (P)	0.58	0.59	0.61	0.62	0.60	0.58	0.59	0.58	0.56	0.58

The graph of successive values of P plotted against the number of tosses is shown in Figure S2.1.

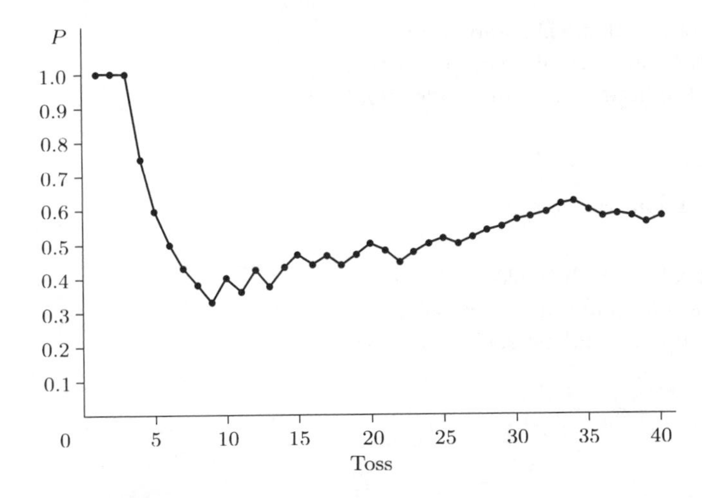

Figure S2.1 Proportion P, 40 tosses of a coin

The same phenomenon is evident here as was seen in Figures 2.2 and 2.3. In this case P seems to be tending to a value close to $\frac{1}{2}$. Did your experiment lead to similar results?

Solution 2.4

(a) The estimate of the probability that a male will be given help is

$$\frac{71}{71+29} = 0.71.$$

(b) The estimate for a female is $89/(89 + 16) = 0.85$.

(c) Since the number 0.85 is greater than the number 0.71, the experiment has provided some evidence to support the hypothesis that people are more helpful to females than to males. However, two questions arise. First, is the difference between the observed proportions sufficiently large to indicate a genuine difference in helping behaviour, or could it have arisen simply as a consequence of experimental variation when in fact there is no underlying difference in people's willingness to help others, whether male or female? Second, is the design of the experiment adequate to furnish an answer to the research question? There may have been differences (other than gender differences) between the eight students that have influenced people's responses. One matter not addressed in this exercise, but surely relevant to the investigation, is the gender of those approached.

Solution 2.5

(a) A count of yeast cells in each square is bound to result in an integer observation: you could not have 2.2 or 3.4 cells. The random variable is discrete.

(b) The data have evidently been recorded to the nearest 0.1 mm, but the actual lengths of kangaroo jawbones are not restricted in this way—within a reasonable range, any length is possible. The random variable is continuous.

(c) The lifetimes have been measured to the nearest integer and recorded as such. However, lifetime is a continuous random variable: components (in general, anyway) would not fail only 'on the hour'. A useful model would be a continuous model.

(d) Rainfall is a continuous random variable.

(e) The number of loans is an integer—the random variable measured here is discrete.

(Data might also be available on the times for which books are borrowed before they are returned. Again, this would probably be measured as integer numbers of days, even though a book could be returned at any time during a working day.)

Solution 2.6

(a) Following the same approach as that adopted in Example 2.8, we can show on a diagram the shaded region corresponding to the required proportion (or probability) $P(T > 5)$. The area of the shaded triangle is given by

$$\tfrac{1}{2} \times \text{(base)} \times \text{(height)}$$
$$= \tfrac{1}{2} \times (20 - 5) \times f(5) = \tfrac{1}{2} \times 15 \times \frac{20 - 5}{200} = 0.5625.$$

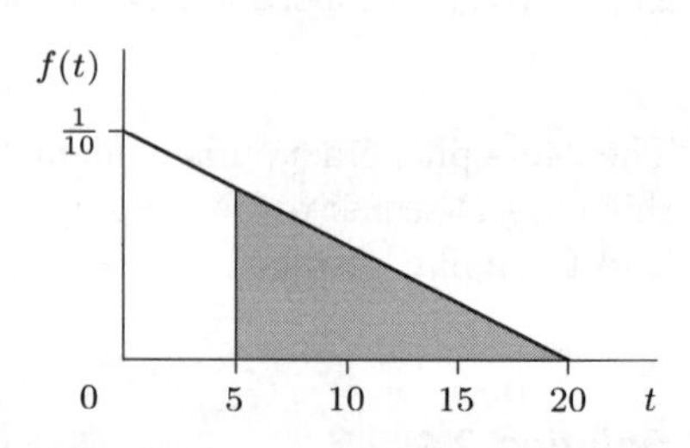

Figure S2.2 The probability $P(T > 5)$

So, according to the model, rather more than half such gaps will exceed 5 seconds. Actually the data suggest that only one-quarter might be so long: our model is showing signs that it could be improved!

(b) This part of the question asks for a general formula for the probability $P(T \leq t)$. The corresponding shaded region is shown in Figure S2.3. The area of the shaded region is given by

$$(\text{average height}) \times (\text{width})$$
$$= \tfrac{1}{2}(\text{long side} + \text{short side}) \times (\text{width}) = \tfrac{1}{2}(f(0) + f(t)) \times (t - 0)$$
$$= \tfrac{1}{2}\left(\frac{20-0}{200} + \frac{20-t}{200}\right) \times t = \frac{(40-t)t}{400} = \frac{40t - t^2}{400}.$$

This formula can now be used for all probability calculations based on this model.

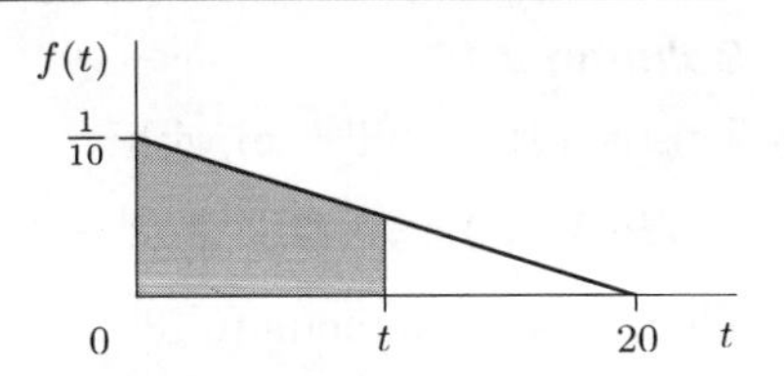

Figure S2.3 The probability $P(T \leq t)$

Solution 2.7

The probability mass function for the score on a *Double-Five* has already been established (see page 66). Summing consecutive terms gives Table S2.2.

Table S2.2 The probability distribution for a *Double-Five*

y	1	3	4	5	6
$p(y)$	$\frac{1}{6}$	$\frac{1}{6}$	$\frac{1}{6}$	$\frac{1}{3}$	$\frac{1}{6}$
$F(y)$	$\frac{1}{6}$	$\frac{1}{3}$	$\frac{1}{2}$	$\frac{5}{6}$	1

Solution 2.8

(a) Using the c.d.f.

$$P(T \leq t) = F(t) = \frac{40t - t^2}{400},$$

it follows that

$$P(T \leq 10) = F(10) = \frac{(40)(10) - 10^2}{400} = \frac{300}{400} = 0.75.$$

(b) The proportion of waiting times exceeding 5 seconds is given by $1-$ (the proportion of waiting times that are 5 seconds or less):

$$P(T > 5)$$
$$= 1 - P(T \leq 5) = 1 - F(5) = 1 - \frac{(40)(5) - 5^2}{400} = 1 - \frac{175}{400} = 0.5625$$

(see Solution 2.6(a)).

Solution 2.9

(a) The probability that a woman randomly selected from the population of 6503 has passed the menopause is

$$\frac{591}{6503} = 0.091.$$

(b) Let the random variable X take the value 1 if a woman has passed the menopause and 0 otherwise. The random variable X is Bernoulli(0.091), so

$$p(x) = (0.091)^x(0.909)^{1-x}, \quad x = 0, 1.$$

(It is very important to remember to specify the *range* of the random variable.)

Solution 2.10

This is a Bernoulli trial with

$$P(X = 0) = 0.22, \quad P(X = 1) = 0.78.$$

That is, $X \sim$ Bernoulli(0.78).

The probability mass function of X is

$$p(x) = (0.78)^x(0.22)^{1-x}, \quad x = 0, 1.$$

Solution 2.11

It is possible that the experiment will result in a sequence of 15 failures. Each of these scores 0. Then the random variable Y (the total number of successes) takes the value

$$y = 0 + 0 + \cdots + 0 = 0.$$

At the other extreme, the experiment might result in a sequence of 15 successes. Then

$$y = 1 + 1 + \cdots + 1 = 15.$$

Any sequence of failures and successes (0s and 1s) between these two extremes is possible, with y taking values $1, 2, \ldots, 14$. The range of the random variable Y is therefore

$$\{0, 1, 2, \ldots, 15\}.$$

Of course, it is unnecessary to be quite so formal. Your answer might have been a one-line statement of the range, which is all that is required.

Solution 2.12

Obviously the 100 people chosen have not been chosen independently: if one chosen person is female it very strongly influences the probability that the spouse will be male! Indeed, you can see that the distribution of the number of females is not binomial by considering the expected frequency distribution. If it was binomial there would be a non-zero probability of obtaining 0 females, 1 female and so on, up to 100 females. However, in this case you are certain to get exactly 50 females and 50 males. The probability that any other number will occur is zero.

Solution 2.13

(a) (i) The number dropping out in the placebo group is binomial $B(6, 0.14)$. The probability that all six drop out is

$$p^6 = (0.14)^6 = 7.53 \times 10^{-6}.$$

(ii) The probability that none of the six drop out is

$$(1 - p)^6 = (1 - 0.14)^6 = 0.86^6 = 0.4046.$$

(iii) The probability that exactly two drop out is

$$\binom{n}{2} p^2(1-p)^{n-2} = \binom{6}{2} (0.14)^2(0.86)^4 = 15(0.14)^2(0.86)^4 = 0.1608.$$

(b) The assumption of independence reduces, in this case, to saying that whether a patient drops out of the placebo group is unaffected by what happens to other patients in the group. Sometimes patients are unaware of others' progress in this sort of trial; but otherwise, it is at least possible that a large drop in numbers would discourage others from continuing in the study. Similarly, even in the absence of obvious beneficial effects, patients might offer mutual encouragement to persevere. In such circumstances the independence assumption breaks down.

Solution 2.14

(a) $$P(V=2) = \binom{8}{2}(0.3)^2(1-0.3)^{8-2} = \frac{8!}{2!\,6!}(0.3)^2(0.7)^6 = 28(0.3)^2(0.7)^6 = 0.2965.$$

(b) $$P(W=8) = \binom{12}{8}(0.5)^8(1-0.5)^{12-8} = \frac{12!}{8!\,4!}(0.5)^8(0.5)^4 = 495(0.5)^{12} = 0.1208.$$

(c) $$\begin{aligned} P(X>4) &= P(X=5)+P(X=6) \\ &= \binom{6}{5}(0.8)^5(0.2)^{6-5} + \binom{6}{6}(0.8)^6(0.2)^{6-6} \\ &= 6(0.8)^5(0.2)+(0.8)^6 = 0.6554. \end{aligned}$$

(d) $$\begin{aligned} P(Y\le 2) &= P(Y=0)+P(Y=1)+P(Y=2) \\ &= \binom{6}{0}\left(\frac{1}{3}\right)^0\left(\frac{2}{3}\right)^{6-0} + \binom{6}{1}\left(\frac{1}{3}\right)^1\left(\frac{2}{3}\right)^{6-1} + \binom{6}{2}\left(\frac{1}{3}\right)^2\left(\frac{2}{3}\right)^{6-2} \\ &= \left(\frac{2}{3}\right)^6 + 6\left(\frac{1}{3}\right)\left(\frac{2}{3}\right)^5 + 15\left(\frac{1}{3}\right)^2\left(\frac{2}{3}\right)^4 = 0.6804. \end{aligned}$$

(e) Writing

$$P(Z\le 7) = P(Z=0)+P(Z=1)+\cdots+P(Z=7)$$

involves calculating eight probabilities and adding them together. It is easier to say

$$\begin{aligned} P(Z\le 7) &= 1-P(Z\ge 8) \\ &= 1-[P(Z=8)+P(Z=9)+P(Z=10)] \\ &= 1-\left[\binom{10}{8}\left(\frac{1}{4}\right)^8\left(\frac{3}{4}\right)^2 + \binom{10}{9}\left(\frac{1}{4}\right)^9\left(\frac{3}{4}\right) + \binom{10}{10}\left(\frac{1}{4}\right)^{10}\left(\frac{3}{4}\right)^0\right] \\ &= 1-\left[45\left(\tfrac{1}{4}\right)^8\left(\tfrac{3}{4}\right)^2 + 10\left(\tfrac{1}{4}\right)^9\left(\tfrac{3}{4}\right) + \left(\tfrac{1}{4}\right)^{10}\right] \\ &= 1-0.000\,416 = 0.999\,584. \end{aligned}$$

(Actually, it is even easier to use your computer for binomial probability calculations.)

Solution 2.15

(a) The distribution of wrinkled yellow peas amongst a 'family' of eight is $B(8, \frac{3}{16})$.

(b) The probability that all eight are wrinkled and yellow is

$$\left(\tfrac{3}{16}\right)^8 = 1.53 \times 10^{-6}.$$

(c) The distribution of wrinkled green peas amongst eight offspring is binomial $B\left(8, \frac{1}{16}\right)$. The probability that there are no wrinkled green peas is

$$\binom{8}{0}\left(\tfrac{1}{16}\right)^0\left(1-\tfrac{1}{16}\right)^{8-0} = \left(\tfrac{15}{16}\right)^8 = 0.597.$$

Solution 2.16

You should find that your computer gives you the following answers. (These answers are accurate to six decimal places.)

(a) 0.200 121 (b) 0.068 892 (c) 0.998 736 (d) 0.338 529

(e) If four dice are rolled simultaneously, then the number of 6s to appear is a binomial random variable $M \sim B\left(4, \frac{1}{6}\right)$. The probability of getting at least one 6 is

$$P(M \geq 1) = 1 - P(M = 0) = 1 - \left(\tfrac{5}{6}\right)^4 = 0.5177.$$

If two dice are rolled, the probability of getting a double-6 is $\frac{1}{6} \times \frac{1}{6} = \frac{1}{36}$. The number of double-6s in twenty-four such rolls is a binomial random variable $N \sim B\left(24, \frac{1}{36}\right)$. The probability of getting at least one double-6 is

$$P(N \geq 1) = 1 - P(N = 0) = 1 - \left(\tfrac{35}{36}\right)^{24} = 0.4914.$$

So it is the first event of the two that is the more probable.

(f) If X is $B(365, 0.3)$ then

$$P(X \geq 100) = 0.8738.$$

(This would be very time-consuming to calculate other than with a computer.) In answering this question the assumption has been made that rain occurs independently from day to day; this is a rather questionable assumption.

Solution 2.17

(a) A histogram of the data looks like the following. The sample mean and standard deviation are:

$$\overline{x} = 18.11\,\text{mm}, \quad s = 8.602\,\text{mm}.$$

The average book width appears to be about 18.11 mm, so for 5152 books the required shelving would be 5152 × 18.11 mm = 93.3 m.

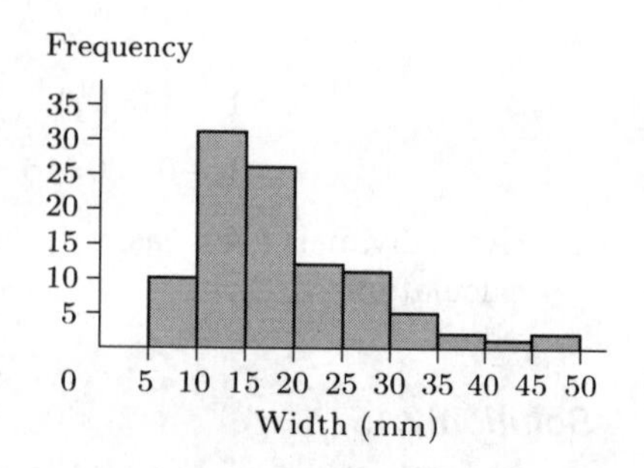

Figure S2.4 Widths of 100 books

(b) This is a somewhat subjective judgement, since no formal tests have been developed for a 'bell-shaped' appearance, or lack of it. The histogram suggests the data are rather skewed. It is worth observing that the width of the widest book in the sample is about 3.5 standard deviations above the mean; the narrowest book measures only 1.5 standard deviations below the mean.

Solution 2.18

(a) You might have obtained a sequence of 0s and 1s as follows.

0 0 1 1 0 0 0 0 0 0

The number of 1s in the ten trials is 2. A single observation from $B(10, 0.2)$ was then obtained: it was 3. The sum of ten independent Bernoulli random variables Bernoulli(0.2) is binomial $B(10, 0.2)$. The two observations, 2 and 3, are independent observations, each from $B(10, 0.2)$.

(b) Figure S2.5 shows three bar charts similar to those you might have obtained.

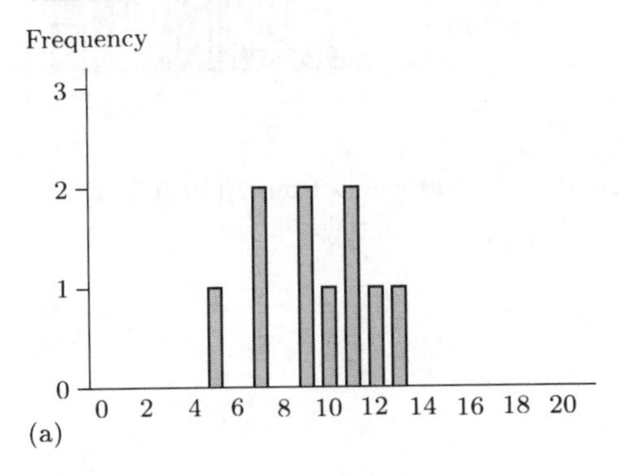

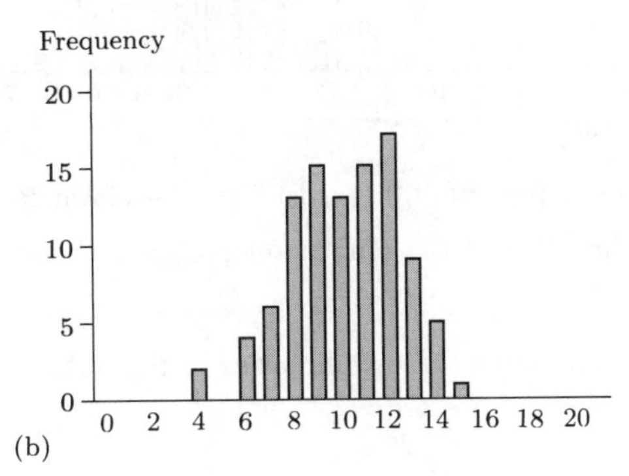

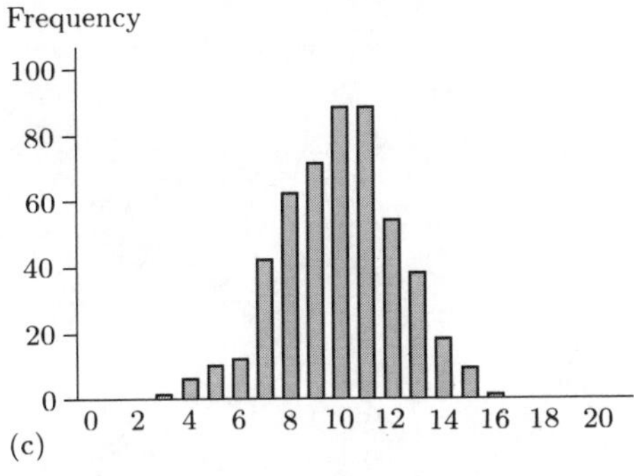

Figure S2.5 (a) 10 values from $B(20, 0.5)$ (b) 100 values from $B(20, 0.5)$ (c) 500 values from $B(20, 0.5)$

Notice that, as the sample size increases, the bar charts for the observed frequencies become less jagged. Even in the case of a sample of size 100, however, the bar chart can be very irregular: this is bimodal. When the sample is of size 500, the observed frequencies are very suggestive of the underlying probability distribution, whose probability mass function is shown in Figure S2.6.

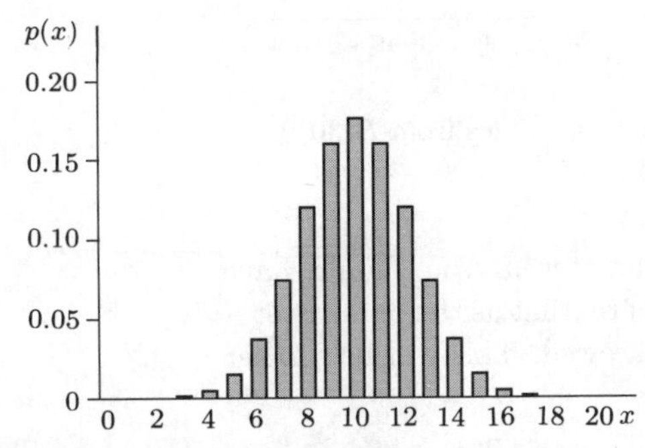

Figure S2.6 The binomial probability distribution $B(20, 0.5)$

(c) Here are three typical bar charts.

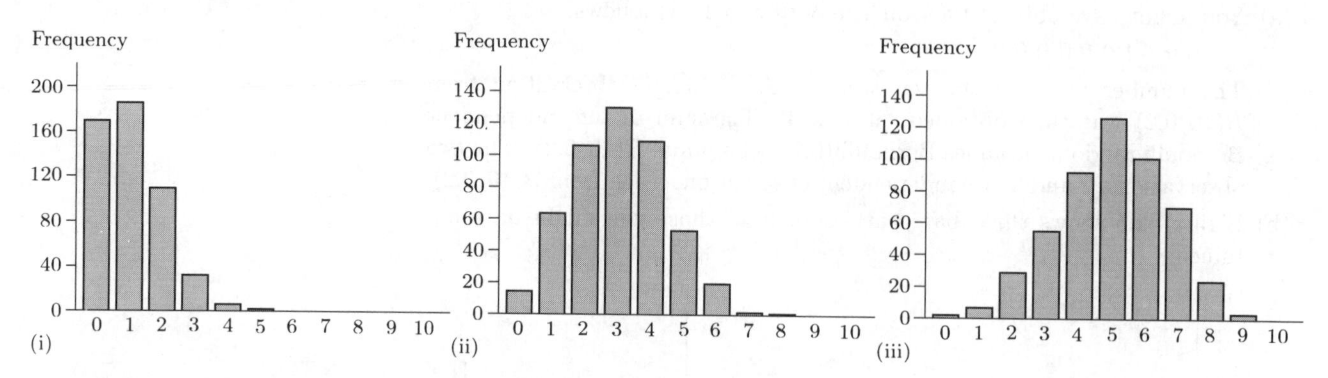

Figure S2.7 (i) 500 values from $B(10, 0.1)$ (ii) 500 values from $B(10, 0.3)$ (iii) 500 values from $B(10, 0.5)$

You can see that the value of the parameter p affects the skewed nature of the sample data.

(d) The following diagrams show three summaries of the data.

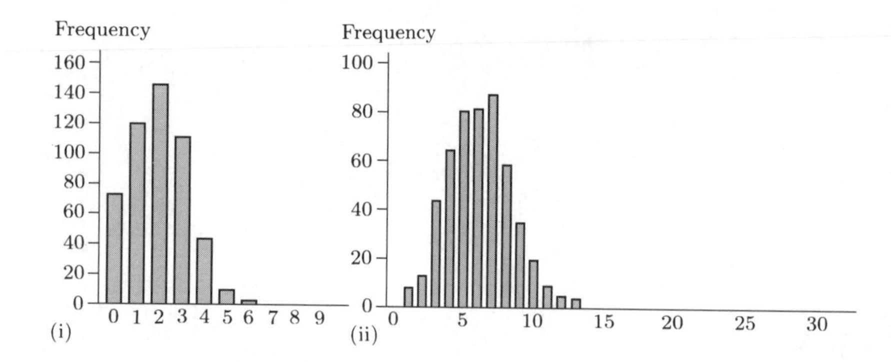

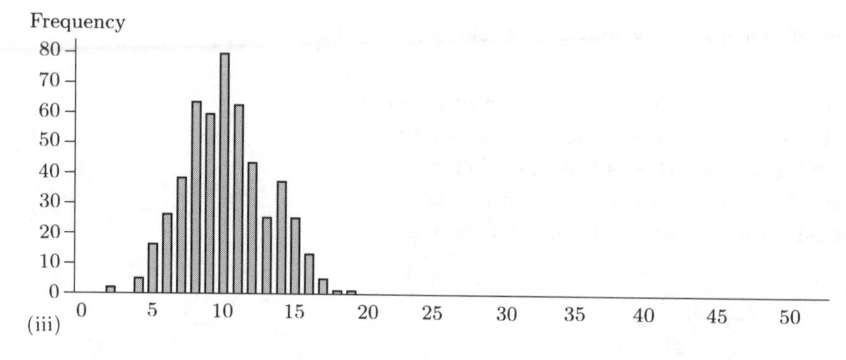

Figure S2.8 (i) 500 values from $B(10, 0.2)$ (ii) 500 values from $B(30, 0.2)$ (iii) 500 values from $B(50, 0.2)$

Even for a value as low as 0.2 for the parameter p, you should have observed from your data, rather as is evident here, that as the parameter n increases the sample histograms become less skewed. This will be further discussed in *Chapter 5*.

Solution 2.19

Out of interest, this experiment was repeated three times, thus obtaining the frequencies in Table S2.3.

Table S2.3 'Opening the bag' three times

Number of defective fuses	Frequency		
0	95	94	93
1	5	4	6
2	0	2	1
3	0	0	0
4	0	0	0
5	0	0	0
6	0	0	0
7	0	0	0
8	0	0	0

You can see that there is some variation in the results here.

Solution 2.20

(a) In 6 rolls of the die, the following results were obtained.

Table S2.4 Rolling a die 6 times

Roll number	1	2	3	4	5	6
Frequency	0	0	1	1	1	3
Relative frequency	0	0	$\frac{1}{6}$	$\frac{1}{6}$	$\frac{1}{6}$	$\frac{1}{2}$

You can see that the sample relative frequencies are widely disparate and do not always constitute very good estimates of the theoretical probabilities: in all cases, these are $\frac{1}{6} = 0.1667$.

(b) In 600 rolls, the following frequencies were obtained.

Table S2.5 Rolling a die 600 times

Roll number	1	2	3	4	5	6
Frequency	80	95	101	111	97	116
Relative frequency	0.1333	0.1583	0.1683	0.1850	0.1617	0.1933

Even in a sample as large as 600, probability estimates can be quite wrong in the second decimal place! But these are generally more consistent and closer to the theoretical values than is the case with the sample of just 6 rolls.

Solution 2.21

One hundred observations on the binomial distribution $B(33, 0.1)$ were generated. Three observations were 8 or more, giving an estimated probability of 0.03 that a sample as extreme as that reported could occur. For interest, the number of left-handed people in each of a 100 groups of 33 individuals was counted. The frequencies were as listed in Table S2.6.

Table S2.6 Left-handedness in 100 groups of 33 individuals

Number of left-handed people	Frequency
0	4
1	9
2	20
3	26
4	23
5	12
6	3
7	0
8	2
9	1
⋮	⋮

Actually, if X is binomial $B(33, 0.1)$, then

$$P(X \geq 8) = 0.014.$$

This makes it seem very unlikely that the circumstance observed could have arisen by mere chance.

Solution 2.22

(a) If the random variable V follows a triangular distribution with parameter 60, then the c.d.f. of V is given by

$$F(v) = P(V \le v) = 1 - \left(1 - \frac{v}{60}\right)^2, \quad 0 \le v \le 60.$$

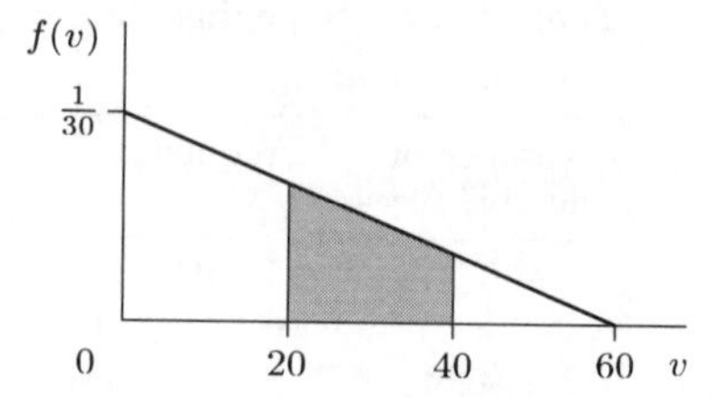

Figure S2.9 The probability $P(20 \le V \le 40)$

Then (either directly from your computer, or by using this formula together with your calculator) the following values will be obtained.

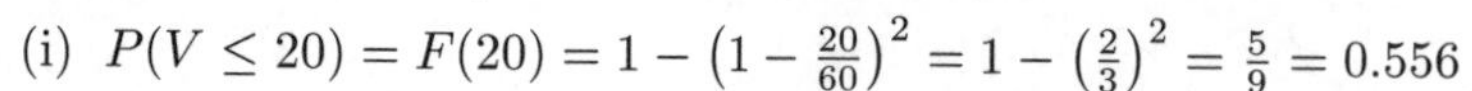

(i) $P(V \le 20) = F(20) = 1 - \left(1 - \frac{20}{60}\right)^2 = 1 - \left(\frac{2}{3}\right)^2 = \frac{5}{9} = 0.556$

(ii) $P(V > 40) = 1 - F(40) = \left(1 - \frac{40}{60}\right)^2 = \left(\frac{1}{3}\right)^2 = \frac{1}{9} = 0.111$

(iii) The probability $P(20 \le V \le 40)$ is equal to the area of the shaded region in Figure S2.9. It is given by

$$P(V \le 40) - P(V \le 20) = F(40) - F(20) = \frac{8}{9} - \frac{5}{9} = \frac{1}{3} = 0.333$$

(using your answers to parts (i) and (ii)).

(b) (i) A histogram of these data is shown in Figure S2.10.

(ii) The data are skewed, with long waiting times apparently less likely than shorter waiting times. The sample is very small, but in the absence of more elaborate models to consider, the triangular model is a reasonable first attempt. The longest waiting time observed in the sample is 171 hours. Any number higher than this would be a reasonable guess at the model parameter—say, 172 or 180 or even 200, without going too high (300, perhaps). Try 180.

(iii) With θ set equal to 180, and denoting by W the waiting time (in hours), then

$$P(W > 100) = 1 - F(100) = \left(1 - \frac{100}{180}\right)^2 = 0.198.$$

In the sample of 40 there are 5 waiting times longer than 100 hours (102, 116.5, 122, 144, 171), so the sample-based estimate for the proportion of waiting times exceeding 100 hours is $\frac{5}{40} = 0.125$.

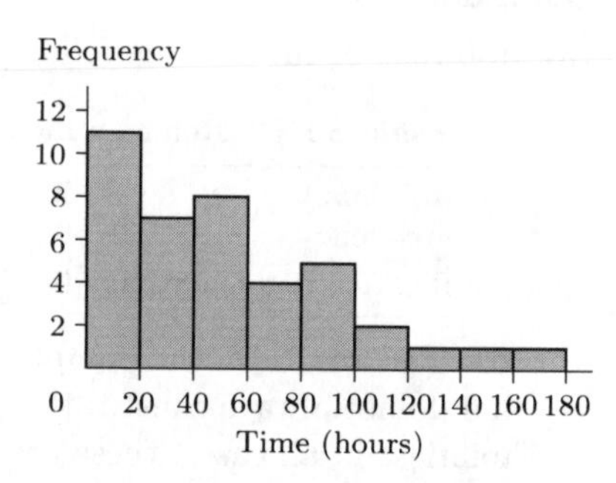

Figure S2.10 Waiting times between admissions

Chapter 3

Solution 3.1

The mean score on a *Double-Five* is given by

$$\mu = 1 \times \tfrac{1}{6} + 3 \times \tfrac{1}{6} + 4 \times \tfrac{1}{6} + 5 \times \tfrac{1}{3} + 6 \times \tfrac{1}{6} = 4.$$

Hence an effect of replacing the 2-face of a fair die by a second 5 is to increase the mean from its value of 3.5 (see Example 3.3) to 4.

Solution 3.2

From the given probability distribution of X, the mean number of members of the family to catch the disease is

$$\begin{aligned}\mu &= 1 \times \tfrac{3}{90} + 2 \times \tfrac{8}{90} + 3 \times \tfrac{15}{90} + 4 \times \tfrac{20}{90} + 5 \times \tfrac{24}{90} + 6 \times \tfrac{20}{90} \\ &= \tfrac{1}{90}(3 + 16 + 45 + 80 + 120 + 120) = \tfrac{384}{90} = 4.3.\end{aligned}$$

Solution 3.3

(a) For a fair coin, $P(\text{Heads}) = p(1) = \frac{1}{2}$. So $p = \frac{1}{2}$ and the mean of the Bernoulli random variable is p, i.e. $\frac{1}{2}$.

(b) As in *Chapter 2*, Exercise 2.2, $p(1) = P(3 \text{ or } 6) = \frac{1}{3}$. Thus $\mu = p = \frac{1}{3}$.

Solution 3.4

The expected value of Y is given by

$$E(Y) = \sum_{y=0}^{2} y p(y)$$
$$= 0p(0) + 1p(1) + 2p(2) = p^3(1-p)^2\left(2p^2 + 2p + 5\right) + 2p(1-p)^6.$$

(a) When $p = 0.1$,

$$E(Y) = 0.1^3 \times 0.9^2 \times 5.22 + 2 \times 0.1 \times 0.9^6 = 0.1105.$$

(b) When $p = 0.4$,

$$E(Y) = 0.4^3 \times 0.6^2 \times 6.12 + 2 \times 0.4 \times 0.6^6 = 0.1783.$$

(c) When $p = 0.6$,

$$E(Y) = 0.6^3 \times 0.4^2 \times 6.92 + 2 \times 0.6 \times 0.4^6 = 0.2441.$$

(d) When $p = 0.8$,

$$E(Y) = 0.8^3 \times 0.2^2 \times 7.88 + 2 \times 0.8 \times 0.2^6 = 0.1615.$$

You can see that when the chain is very fragile or very robust, the expected number of quads is low; only for intermediate p is the expected number of quads more than about 0.2.

Solution 3.5

(a) In one experiment the results in Table S3.1 were obtained. The sample mean is 5.63.

(b) The mean of the first sample drawn in an experiment was 6.861. Together with nine other samples, the complete list of sample means is shown in Table S3.2.

(c) In one experiment the following results were obtained: (i) 9.974; (ii) 97.26; (iii) 198.5.

(d) These findings suggest that the mean of the Triangular(θ) distribution is $\frac{1}{3}\theta$.

Table S3.1

9.72	3.37	12.99	6.92	1.35
2.38	2.08	8.75	7.79	0.95

Table S3.2

6.861	6.468	6.532	6.713	6.667
6.628	6.744	6.586	6.808	6.671

Solution 3.6

Using the information given, the probability required is

$$P(T > \mu) = P\left(T > \tfrac{1}{3}\theta\right) = 1 - P\left(T \le \tfrac{1}{3}\theta\right) = 1 - F\left(\tfrac{1}{3}\theta\right)$$
$$= 1 - \left[1 - \left(1 - \frac{\theta/3}{\theta}\right)^2\right] = \left(1 - \tfrac{1}{3}\right)^2 = \tfrac{4}{9}.$$

So in any collection of traffic waiting times (assuming the triangular model to be an adequate representation of the variation in waiting times) we might expect just under half the waiting times to be longer than average. Notice that this result holds irrespective of the actual value of the parameter θ.

Solution 3.7

The probability distribution for the *Double-Five* outcome is shown in Table 3.4.

The population mean is 4 (see solution to Exercise 3.1).

The calculation of the variance is as follows.

j	1	3	4	5	6
$j-\mu$	-3	-1	0	1	2
$(j-\mu)^2$	9	1	0	1	4

$$\begin{aligned}\sigma^2 &= \sum_j (j-\mu)^2 p(j) \\ &= 9 \times \tfrac{1}{6} + 1 \times \tfrac{1}{6} + 0 \times \tfrac{1}{6} + 1 \times \tfrac{1}{3} + 4 \times \tfrac{1}{6} = 2.67\end{aligned}$$

The variance of the score on a fair die is 2.92. So, while the mean score on a *Double-Five* is greater than that on a fair die, the variance of the *Double-Five* outcome is smaller. This is not unreasonable since, by replacing the 2 by another 5, one can intuitively expect a little more 'consistency', that is, less variability, in the outcomes.

Solution 3.8

To check for independence, we shall work out $p_{X,Y}(x, y)$ assuming independence, and compare the outcome with Table 3.8. For instance, $p_{X,Y}(0, -1)$ would be the product $p_X(0)p_Y(-1) = 0.4 \times 0.3 = 0.12$, $p_{X,Y}(2, -1)$ would be the product $p_X(2)p_Y(-1) = 0.2 \times 0.3 = 0.06$, and so on. In this way, we produce Table S3.3 of the joint p.m.f. of X and Y *under independence*.

Table S3.3 The joint p.m.f. of X and Y under independence

x	0	1	2
$y = -1$	$0.4 \times 0.3 = 0.12$	$0.4 \times 0.3 = 0.12$	$0.2 \times 0.3 = 0.06$
$y = 1$	$0.4 \times 0.7 = 0.28$	$0.4 \times 0.7 = 0.28$	$0.2 \times 0.7 = 0.14$

These values are shown more clearly in Table S3.4.

Table S3.4

		x		
		0	1	2
y	-1	0.12	0.12	0.06
	1	0.28	0.28	0.14

These values are not the same as those in Table 3.8. For instance, under independence we would require $p_{X,Y}(1, 1)$ to equal 0.28, whereas $p_{X,Y}(1, 1)$ is 0.30. Hence X and Y are *not* independent.

Solution 3.9

(a) The random variable N takes the value 1 if the first trial results in a 'success': $P(N = 1) = p$.

(b) Success occurs for the first time only at the second trial if initially there is a failure, followed immediately by a success: $P(N = 2) = qp$.

(c) Here, there are two failures followed by a success: $P(N = 3) = q^2p$.

(d) A clear pattern is emerging. The random variable N takes the value n only if $(n - 1)$ failures are followed at the nth trial by a success:

$$P(N = n) = q^{n-1}p.$$

(e) The range of possible values N can take is $1, 2, 3, \ldots$, the set of positive integers (which you might also know as the set of natural numbers).

Solution 3.10

(a) The proportion of families comprising at least 4 children is found from $P(N \geq 4) = 1 - P(N \leq 3)$.

$$\begin{aligned} 1 - P(N \leq 3) &= 1 - (p(1) + p(2) + p(3)) = 1 - (p + qp + q^2p) \\ &= 1 - (0.514)(1 + 0.486 + 0.486^2) = 1 - (0.514)(1.722) \\ &= 1 - 0.885 = 0.115. \end{aligned}$$

(b) Denoting by 'success' the identification of a defective chip, $p = 0.012$. The size of the inspector's sample of chips is a random variable N where $N \sim G(0.012)$. Then

$$\begin{aligned} P(N < 6) = P(N \leq 5) &= p + qp + q^2p + q^3p + q^4p \\ &= (0.012)(1 + 0.988 + 0.988^2 + 0.988^3 + 0.988^4) \\ &= (0.012)(4.8814) = 0.0586, \end{aligned}$$

so about 6% of daily visits involve a halt in production.

Solution 3.11

In this case, the random variable N follows a geometric distribution with parameter $p = 0.02$. So

$$P(N > 20) = q^{20} = (0.98)^{20} = 0.668.$$

The probability that the inspector will have to examine at least 50 chips is

$$P(N \geq 50) = P(N > 49) = q^{49} = (0.98)^{49} = 0.372.$$

Notice that it is much easier to use the formula $P(N > n) = q^n$ to calculate tail probabilities for the geometric distribution than to add successive terms of the probability function as in Solution 3.10.

Solution 3.12

(a) 2 seems intuitively correct.

(b) If the probability of throwing a 5 is $\frac{1}{3}$, this suggests that the average number of throws necessary to achieve a 5 will be 3.

(c) 6.

(d) By the same argument, guess $\mu = 1/p$.

Solution 3.13

The number N of rolls necessary to start playing is a geometric random variable with parameter $p = 1/6$.

(a) $P(N = 1) = p = 1/6 = 0.167$.

(b) $P(N = 2) = qp = 5/36 = 0.139$; $P(N = 3) = q^2p = 25/216 = 0.116$.

(c) The probability that at least six rolls will be necessary to get started is given by $P(N \geq 6) = P(N > 5) = q^5 = 3125/7776 = 0.402$.

(d) The expected number of rolls for a geometric random variable is $1/p$; which is 6 in this case. The standard deviation is $\sqrt{q}/p = 6\sqrt{5/6} = 5.48$.

Solution 3.14

Your results should not be too different from the following, which were obtained on a computer.

(a) A frequency table for the 1200 rolls summarizes the data as follows.

Outcome	1	2	3	4	5	6
Frequency	195	202	227	208	181	187

(b) The corresponding bar chart is shown in Figure S3.1. The bar chart shows some departures from the theoretical expected frequencies (200 in each of the six cases): these departures may be ascribed to random variation.

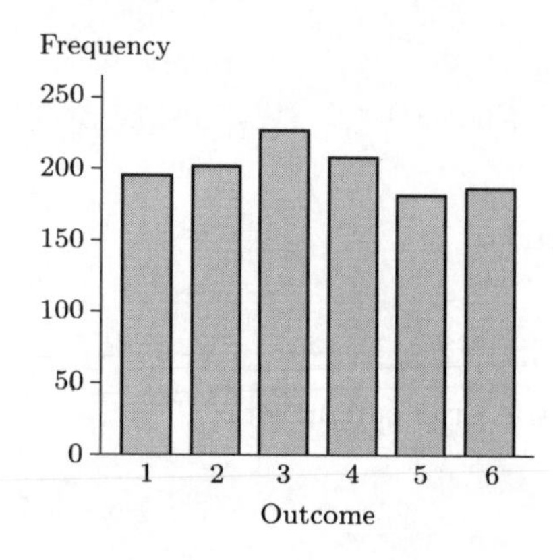

Figure S3.1 Bar chart for 1200 rolls of the die

(c) The computer gave

$$\overline{x} = 3.45, \quad s^2 = 2.798\,08,$$

so $s = 1.67$.

This may be compared with the theoretical sample moments for a discrete uniform distribution:

$$\mu = \tfrac{1}{2}(n+1) = \tfrac{1}{2}(6+1) = 3.5, \quad \sigma^2 = \tfrac{1}{12}\left(n^2-1\right) = \tfrac{1}{12}(36-1) = 2.917,$$

so $\sigma = 1.71$.

The sample gave results that were on average slightly lower than the theoretical scores, and that are slightly less dispersed. These differences are scarcely perceptible and can be ascribed to random variation.

Solution 3.15

A sketch of the p.d.f. of X when $X \sim U(a, b)$ is shown in Figure S3.2.

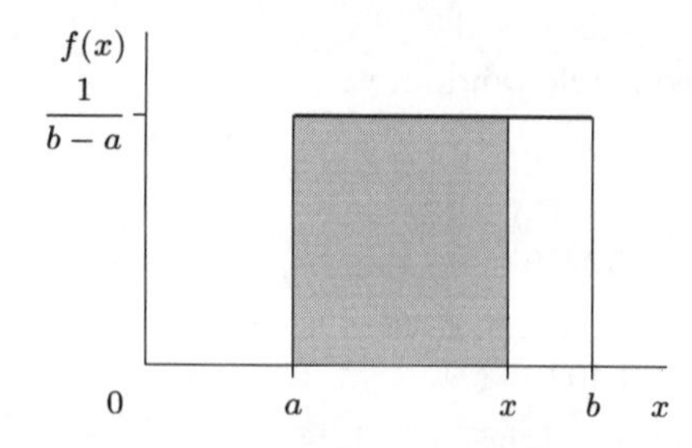

Figure S3.2 The p.d.f. of X, $X \sim U(a, b)$

(a) By symmetry, the mean of X is $\mu = \frac{1}{2}(a + b)$.

(b) The probability $P(X \leq x)$ is equal to the area of the shaded rectangle in the diagram. This is $(x-a) \times \frac{1}{b-a}$. So the c.d.f. of X is given by

$$F(x) = \frac{x-a}{b-a}, \quad a \leq x \leq b.$$

Solution 3.16

The formula for the variance of a continuous uniform random variable $U(a,b)$ is

$$\sigma^2 = \frac{(b-a)^2}{12}.$$

For the standard continuous uniform distribution $U(0,1)$, $a=0$ and $b=1$, so the variance is

$$\sigma^2 = \tfrac{1}{12} = 0.083,$$

and the standard deviation is

$$\sigma = \sqrt{\tfrac{1}{12}} = 0.289.$$

Solution 3.17

(a) From Solution 3.16, the c.d.f. of the $U(a,b)$ distribution is

$$F(x) = \frac{x-a}{b-a}, \quad a \leq x \leq b.$$

To solve $F(m) = \frac{1}{2}$, we need to solve the equation

$$\frac{m-a}{b-a} = \frac{1}{2}$$

or

$$m - a = \frac{b-a}{2}.$$

This gives

$$m = \frac{a+b}{2},$$

and is the median of the $U(a,b)$ distribution. You might recall that this is also the value of the mean of the $U(a,b)$ distribution, and follows immediately from a symmetry argument.

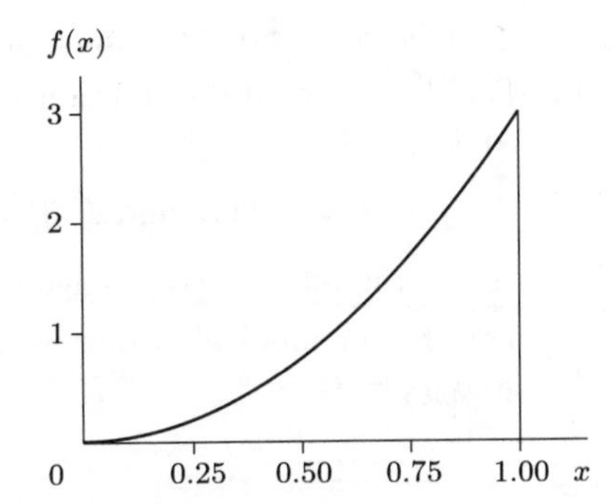

Figure S3.3 $f(x) = 3x^2$, $0 \leq x \leq 1$

(b) (i) The density function $f(x) = 3x^2$, $0 \leq x \leq 1$, is shown in Figure S3.3.

(ii) The mean and median are shown in Figure S3.4.

(iii) From $F(x) = x^3$, it follows that the median is the solution of the equation

$$x^3 = \tfrac{1}{2}.$$

This is

$$m = \left(\tfrac{1}{2}\right)^{1/3} = 0.794.$$

The mean $\mu = 0.75$ and the median $m = 0.794$ are shown in Figure S3.4.

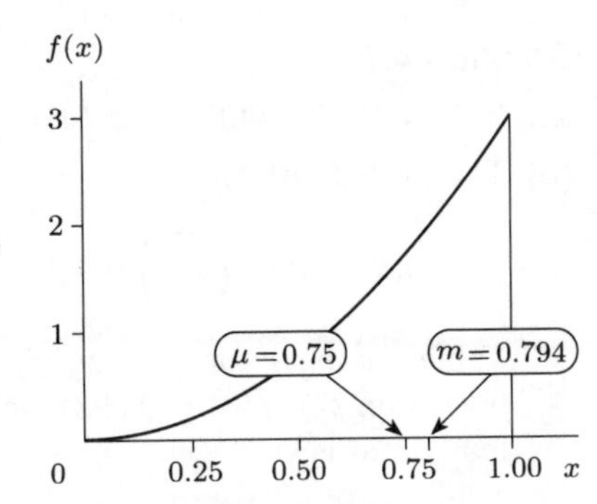

Figure S3.4 The mean and median of X

Solution 3.18

(a) The c.d.f. of this distribution is

$$F(x) = x^3, \quad 0 \le x \le 1.$$

To obtain the interquartile range, we need both q_U and q_L. To obtain q_L, we solve

$$F(q_L) = q_L^3 = \tfrac{1}{4},$$

and obtain

$$q_L = \left(\tfrac{1}{4}\right)^{1/3} = 0.630.$$

Likewise,

$$F(q_U) = q_U^3 = \tfrac{3}{4},$$

hence

$$q_U = \left(\tfrac{3}{4}\right)^{1/3} = 0.909.$$

So, the interquartile range is

$$q_U - q_L = 0.909 - 0.630 = 0.279.$$

Solution 3.19

(a) For the binomial distribution $B(10, 0.5)$, $F(4) = 0.3770$, $F(5) = 0.6230$, so the median is 5.

(b) For the binomial distribution $B(17, 0.7)$, $F(11) = 0.4032$, $F(12) = 0.6113$, so the median is 12.

(c) For the binomial distribution $B(2, 0.5)$, $F(1) = 0.75$, therefore the upper quartile is 1. (So is the median!)

(d) For the binomial distribution $B(19, 0.25)$, $F(5) = 0.6678$, $F(6) = 0.8251$, so $q_{0.75} = 6$.

Since $F(2) = 0.1113$ and $F(3) = 0.2631$, $q_{0.25} = 3$.

Hence the interquartile range is $q_{0.75} - q_{0.25} = 6 - 3 = 3$.

(e) For the binomial distribution $B(15, 0.4)$, $F(7) = 0.7869$, $F(8) = 0.9050$, so $q_{0.85} = 8$.

Chapter 4

Solution 4.1

(a) If $X \sim B(50, 1/40)$, then $P(X = 0) = (39/40)^{50} = (0.975)^{50} = 0.2820$.

(b) The probability that the cyclist gets wet twice is

$$P(X = 2) = \binom{50}{2}(0.025)^2(0.975)^{48} = 0.2271.$$

(c) Values of $p(x)$ for $x = 0, 1, 2, 3$, are $p(0) = 0.2820$, $p(1) = 0.3615$, $p(2) = 0.2271$, $p(3) = 0.0932$; so the probability that she gets wet at least four times is

$$\begin{aligned} 1 - (p(0) + p(1) + p(2) + p(3)) &= 1 - (0.2820 + 0.3615 + 0.2271 + 0.0932) \\ &= 1 - 0.9638 = 0.0362. \end{aligned}$$

Solution 4.2

When $X \sim B(60, 1/48)$, $P(X = 0) = (47/48)^{60} = 0.2827$. Rounding to, say, $(0.979)^{60} = 0.28$ would induce rather serious rounding errors. Continuing in this way, obtain the table of probabilities as follows.

	$P(X=0)$	$P(X=1)$	$P(X=2)$	$P(X=3)$	$P(X \geq 4)$
$B(50, 1/40)$	0.2820	0.3615	0.2271	0.0932	0.0362
$B(60, 1/48)$	0.2827	0.3610	0.2266	0.0932	0.0365

The last value was obtained by subtraction. In fact, if you use a computer you would find that the probability $P(X \geq 4)$ when X is $B(60, 1/48)$ is 0.0366, to 4 decimal places.

Solution 4.3

(a) In this case $X \sim B(360, 0.01)$.

(b) Including also the probabilities calculated in the text for $B(320, 0.011\,25)$, the results are as listed in the table below.

	$P(X=0)$	$P(X=1)$	$P(X=2)$	$P(X=3)$	$P(X \geq 4)$
$B(320, 0.01125)$	0.0268	0.0975	0.1769	0.2134	0.4854
$B(360, 0.01)$	0.0268	0.0976	0.1769	0.2133	0.4854

In this case the results are close, identical to three decimal places. (Again, the last column was found by subtraction. To 4 decimal places, when $X \sim B(320, 0.011\,25)$, the probability $P(X \geq 4)$ is 0.4855.)

Solution 4.4

Using the given recursion,

(a) $p_X(1) = \frac{\mu}{1} p_X(0) = \frac{\mu}{1} e^{-\mu} = \mu e^{-\mu}$,

(b) $p_X(2) = \frac{\mu}{2} p_X(1) = \frac{\mu}{2} \mu e^{-\mu} = \frac{\mu^2}{2!} e^{-\mu}$,

(c) $p_X(3) = \frac{\mu}{3} p_X(2) = \frac{\mu}{3} \frac{\mu^2}{2!} e^{-\mu} = \frac{\mu^3}{3!} e^{-\mu}$,

where the notation $k!$ means the number $1 \times 2 \times \ldots \times k$.

(d) There is an evident pattern developing here: a general formula for the probability $p_X(x)$ is

$$p_X(x) = \frac{\mu^x}{x!} e^{-\mu}.$$

Solution 4.5

The completed table is as follows.

	$P(X=0)$	$P(X=1)$	$P(X=2)$	$P(X=3)$	$P(X \geq 4)$
$B(320, 0.01125)$	0.0268	0.0975	0.1769	0.2134	0.4854
$B(360, 0.01)$	0.0268	0.0976	0.1769	0.2133	0.4854
Poisson(3.6)	0.0273	0.0984	0.1771	0.2125	0.4847

(Probabilities in the last column are found by subtraction: to 4 decimal places, the probability $P(X \geq 4)$ when X is Poisson(3.6) is 0.4848.)

Solution 4.6

(a) The exact probability distribution for the number of defectives in a box is $B(50, 0.05)$ which (unless you have access to very extensive tables!) will need calculation on a machine as follows (i.e. recursively, retaining displayed values on the machine):

$$p_X(0) = (0.95)^{50} = 0.0769$$

$$p_X(1) = 50 \times \frac{0.05}{0.95} \times p_X(0) = 0.2025$$

$$p_X(2) = \frac{49}{2} \times \frac{0.05}{0.95} \times p_X(1) = 0.2611$$

$$p_X(3) = \frac{48}{3} \times \frac{0.05}{0.95} \times p_X(2) = 0.2199$$

$$p_X(4) = \frac{47}{4} \times \frac{0.05}{0.95} \times p_X(3) = 0.1360$$

and, by subtraction,

$$P(X > 4) = 1 - (0.0769 + 0.2025 + \cdots + 0.1360) = 0.1036.$$

(b) The approximating probability distribution is Poisson(2.5). The probabilities are shown for comparison in the following table.

	$P(X = 0)$	$P(X = 1)$	$P(X = 2)$	$P(X = 3)$	$P(X = 4)$	$P(X \geq 5)$
$B(50, 0.05)$	0.0769	0.2025	0.2611	0.2199	0.1360	0.1036
Poisson(2.5)	0.0821	0.2052	0.2565	0.2138	0.1336	0.1088

(c) The probabilities are 'similar', but are not really very close—certainly, not as close as in some previous exercises and examples. The parameter $p = 0.05$ is at the limit of our 'rule' for when the approximation will be useful (and, in some previous examples, n has been counted in hundreds, not in tens).

Solution 4.7

(a) You should have observed something like the following. The computer gave the random sample

9 7 6 8 6.

The sample mean is

$$\frac{9 + 7 + 6 + 8 + 6}{5} = \frac{36}{5} = 7.2,$$

resulting in an estimate of 7.2 for the population mean μ (usually unknown, but in this case known to be equal to 8).

(b) From 100 repetitions of this experiment, the observed sample means ranged from as low as 4.9 to as high as 11.6, with frequencies as follows.

[4, 5)	1
[5, 6)	12
[6, 7)	14
[7, 8)	25
[8, 9)	25
[9, 10)	19
[10, 11)	3
[11, 12)	1

(c) A histogram of the distribution of sample means is shown in Figure S4.2. The data vector had mean 7.96 and variance 1.9.

(d) Repeating the experiment for samples of size 50 gave the following results. Observed sample means ranged from 6.90 to 9.46, with frequencies

$[6.5, 7)$	2
$[7, 7.5)$	10
$[7.5, 8)$	41
$[8, 8.5)$	39
$[8.5, 9)$	7
$[9, 9.5)$	1

and corresponding histogram as shown in Figure S4.1. The data vector had mean 7.9824 and variance 0.2. What has happened is that the sample means based on samples of size 50 (rather than 5) are much more contracted about the value $\mu = 8$. A single experiment based on a sample of size 50 is likely to give an estimate of μ that is closer to 8 than it would have been in the case of an experiment based on a sample of size 5.

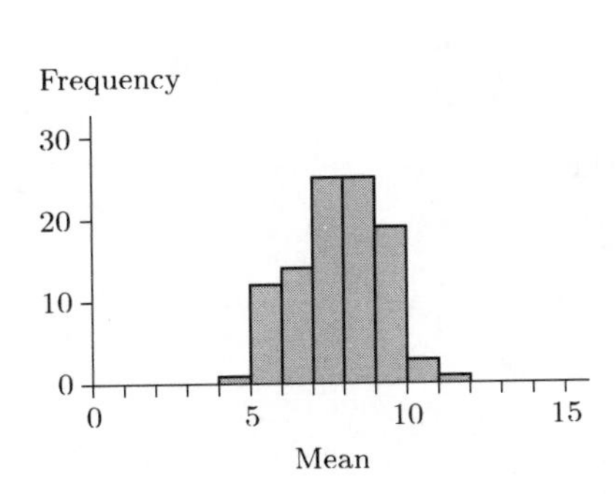

Figure S4.1

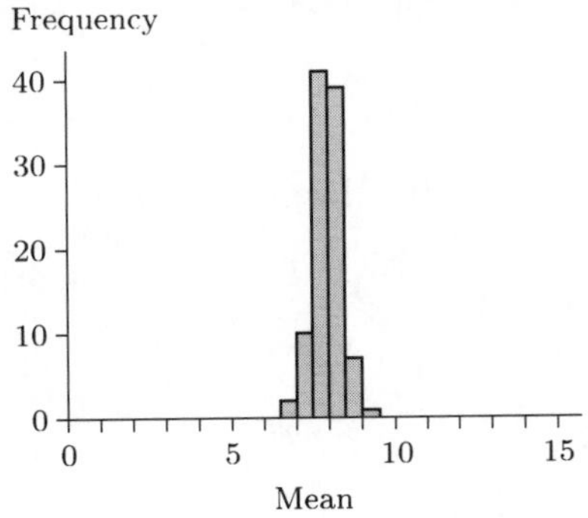

Figure S4.2

Solution 4.8

(a) If X (chest circumference measured in inches) has mean 40, then the random variable $Y = 2.54X$ (chest circumference measured in cm) has mean

$$E(Y) = E(2.54X) = 2.54E(X) = 2.54 \times 40 = 101.6.$$

(Here, the formula $E(aX + b) = aE(X) + b$ is used, with $a = 2.54$ and $b = 0$.)

(b) If X (water temperature measured in degrees Celsius) has mean 26, then the random variable $Y = 1.8X + 32$ (water temperature measured in °F) has mean

$$E(Y) = E(1.8X + 32) = 1.8E(X) + 32 = 1.8 \times 26 + 32 = 78.8.$$

Solution 4.9

If X (finger length in cm) has mean 11.55 and standard deviation 0.55, and if the random variable finger length (measured in inches) is denoted by Y, then $Y = X/2.54$, hence

$$\mu_Y = \frac{\mu_X}{2.54} = \frac{11.55}{2.54} = 4.55, \qquad \sigma_Y = \frac{\sigma_X}{2.54} = \frac{0.55}{2.54} = 0.22.$$

Solution 4.10

The probability distribution for the outcome of throws of a *Double-Five* is as follows.

x	1	3	4	5	6
$p(x)$	$\frac{1}{6}$	$\frac{1}{6}$	$\frac{1}{6}$	$\frac{1}{3}$	$\frac{1}{6}$

The expected value of X^2 is given by

$$\begin{aligned}E(X^2) &= 1^2 \times \tfrac{1}{6} + 3^2 \times \tfrac{1}{6} + 4^2 \times \tfrac{1}{6} + 5^2 \times \tfrac{1}{3} + 6^2 \times \tfrac{1}{6}\\ &= 1 \times \tfrac{1}{6} + 9 \times \tfrac{1}{6} + 16 \times \tfrac{1}{6} + 25 \times \tfrac{1}{3} + 36 \times \tfrac{1}{6}\\ &= \tfrac{1}{6} + \tfrac{9}{6} + \tfrac{16}{6} + \tfrac{25}{3} + \tfrac{36}{6}\\ &= \tfrac{112}{6},\end{aligned}$$

and so from the formula (4.16) the variance of X is given by

$$V(X) = E(X^2) - (E(X))^2 = \frac{112}{6} - 4^2 = \frac{16}{6} = 2.67$$

as before.

Solution 4.11

If X is binomial with parameters $n = 4$, $p = 0.4$ then according to (4.17) the mean of X is

$$E(X) = np = 4 \times 0.4 = 1.6$$

and the variance of X is

$$V(X) = npq = 4 \times 0.4 \times 0.6 = 0.96.$$

From the individual probabilities for X, it follows that

$$\begin{aligned}E(X^2) &= 0^2 \times 0.1296 + 1^2 \times 0.3456 + \cdots + 4^2 \times 0.0256\\ &= 0 + 0.3456 + 1.3824 + 1.3824 + 0.4096 = 3.52,\end{aligned}$$

and so

$$V(X) = E(X^2) - (E(X))^2 = 3.52 - 1.6^2 = 3.52 - 2.56 = 0.96,$$

confirming the result obtained previously.

Solution 4.12

A time interval of four years includes one leap year—1461 days altogether. The probability of a lull exceeding 1461 days is

$$\left(1 - \frac{62}{27\,107}\right)^{1461} \simeq 0.0352;$$

so, in a list of 62 waiting times one might expect about two of them to exceed 1461 days. In this case there were exactly two such lulls, one of which lasted 1617 days, and the other, already identified, was of 1901 days' duration.

Solution 4.13

Set the parameter λ equal to $1/437$.

(a) A time interval of three years including one leap year will last 1096 days altogether. The probability that no earthquake occurs during this interval is

$$P(T > t) = e^{-\lambda t} = e^{-1096/437} = e^{-2.508} = 0.0814.$$

(b) The equation $F(x) = \frac{1}{2}$ may be written

$$1 - e^{-x/437} = \tfrac{1}{2},$$

or

$$e^{-x/437} = \tfrac{1}{2},$$

or

$$x = -437 \log \tfrac{1}{2} = 437 \log 2 = 303 \text{ days}.$$

(c) The proportion of waiting times lasting longer than expected is

$$P(T > 437) = e^{-437/437} = e^{-1} = 0.368;$$

thus just over one-third of waiting times are longer than average!

Solution 4.14

If $X \sim \text{Poisson}(8.35)$ then $p(0) = 0.0002$, $p(1) = 0.0020$, $p(2) = 0.0082$ and $p(3) = 0.0229$. So,

(a) the probability of exactly two earthquakes is 0.0082;

(b) the probability that there will be at least four earthquakes is

$$1 - (0.0002 + 0.0020 + 0.0082 + 0.0229) = 1 - 0.0333 = 0.9667.$$

Solution 4.15

The general median waiting time is the solution of the equation

$$F(x) = 1 - e^{-\lambda x} = \tfrac{1}{2},$$

or

$$x = \frac{-\log \frac{1}{2}}{\lambda} = \frac{\log 2}{\lambda} = \mu_T \times \log 2 = 0.6931\mu_T,$$

where μ_T is the mean waiting time. So for an exponential random variable the median is approximately 70% of the mean.

Solution 4.16

(a) You will probably have got something not too different to this. The simulation can be shown on a table as follows. There are 7300 days in twenty years, so the simulation has to be extended up to or beyond 7300 days. When we start, we do not know how many random numbers that will take, so we just have to keep going. Waiting times are drawn from the exponential distribution with mean 437.

The 16th earthquake happened just after the twenty years time limit. A diagram of the incidence of earthquakes with passing time is shown in Figure S4.3.

Earthquake	Waiting time	Cumulative time
1	695.4	695.4
2	279.2	974.6
3	685.2	1659.8
4	420.8	2080.6
5	758.3	2838.9
6	385.3	3224.2
7	156.0	3380.2
8	681.7	4061.9
9	334.5	4396.4
10	370.1	4766.5
11	557.7	5324.2
12	381.2	5705.4
13	901.8	6607.2
14	179.0	6786.2
15	92.9	6879.1
16	851.6	7730.7
17		
18		

0 2000 4000 6000 8000

Sixteen earthquakes, time in days

Figure S4.3 Incidence of earthquakes (simulated)

(b) There are 15 earthquakes counted in the simulation. The expected number was

$$\lambda t = \left(\frac{1}{437 \text{ days}}\right) \times (7300 \text{ days}) = 16.7.$$

The number of earthquakes is an observation on a Poisson random variable with mean 16.7. The median of the Poisson(16.7) distribution is 17. (For, if $X \sim \text{Poisson}(16.7)$, then $F(16) = P(X \leq 16) = 0.4969$, while $F(17) = 0.5929$.)

Solution 4.17

(a) A histogram of the data is given in Figure S4.4. The data are very skewed and suggest that an exponential model might be plausible.

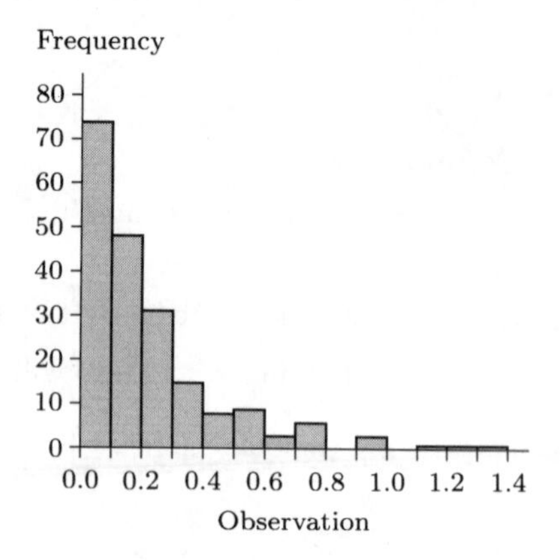

Figure S4.4

(b) (i) The sample mean is 0.224 and the sample median is 0.15. So the sample median is about 67% of the sample mean, mimicking corresponding properties of the exponential distribution (69%).

(ii) The sample standard deviation is 0.235 which is close to the sample mean. (For the exponential distribution, the mean and standard deviation are equal.)

(c) The c.d.f. of the exponential distribution is given by

$$F(x) = 1 - e^{-\lambda t} = 1 - e^{-t/\mu}, \quad t \geq 0$$

and so the lower quartile is the solution of the equation

$$F(x) = 1 - e^{-x/\mu} = 0.25.$$

That is,

$$e^{-x/\mu} = 0.75,$$

so

$$x/\mu = -\log 0.75;$$

so

$$q_L = -\mu \log 0.75 = 0.29\mu.$$

Similarly,

$$q_U = 1.39\mu.$$

For these data, the sample lower quartile is 0.06, which is 0.27 times the sample mean, and the sample upper quartile is 0.29, which is 1.29 times the sample mean. The similarity to corresponding properties of exponential distribution is fairly marked.

(d) During the first quarter-minute there were 81 pulses; during the second there were 53.

(e) It looks as though it might be reasonable to model the incidence of nerve pulses as a Poisson process, with mean waiting time between pulses estimated by the sample mean

$$\bar{t} = 0.2244 \text{ seconds.}$$

Then the pulse rate λ may be estimated by

$$\frac{1}{\bar{t}} = 4.456 \text{ per second.}$$

Over quarter-minute (15-second) intervals the expected number of pulses is

$$(4.456 \text{ per second}) \times (15 \text{ seconds}) = 66.8,$$

and so our two observations 81 and 53 are observations on the Poisson distribution with mean 66.8.

Solution 4.18

(a) Your simulation may have given something like the following. Twenty observations from the Poisson distribution Poisson(3.2) were

4	6	2	3	3	2	8	1	3	4
0	4	4	4	6	3	4	3	7	2

with frequencies as follows.

Count	Frequency
0	1
1	1
2	3
3	5
4	6
5	0
6	2
7	1
8	1

For a random sample of size 50 a typical frequency table is given by

Count	Frequency
0	3
1	8
2	11
3	13
4	6
5	4
6	4
7	1

and for a random sample of size 100 the frequencies are as follows.

Count	Frequency
0	6
1	13
2	26
3	16
4	19
5	13
6	4
7	1
8	2

(b) For a sample of size 1000 the sample relative frequencies were as shown below. These may be compared with the probability mass function

$$p(n) = \frac{e^{-3.2}3.2^n}{n!}, \quad n = 0, 1, 2, 3, \ldots .$$

Count	Frequency	Relative frequency	Probability
0	31	0.031	0.0408
1	146	0.146	0.1304
2	194	0.194	0.2087
3	236	0.236	0.2226
4	168	0.168	0.1781
5	124	0.124	0.1140
6	61	0.061	0.0608
7	25	0.025	0.0278
8	11	0.011	0.0111
9	2	0.002	0.0040
10	2	0.002	0.0013
11	0	0.000	0.0004
12	0	0.000	0.0001
13	0	0.000	0.0000

(Notice the small rounding error in the assessment of the probabilities in the fourth column. They add to 1.0001.)

Solution 4.19

(a) One simulation gave $x = 59$ for the number of males, and therefore (by subtraction) 41 females.

(b) The number of colour-deficient males present is therefore a random observation from $B(59, 0.06)$: this simulation gave $y_1 = 3$. The number of colour-deficient females is a random observation from $B(41, 0.004)$. This simulation gave $y_2 = 0$.

(c) The resulting observation on the random variable W is

$$w = y_1 + y_2 = 3 + 0 = 3.$$

(d) The expected number of males is 50, equal to the expected number of females. Intuitively, the expected number of colour-deficient males is $50 \times 0.06 = 3$; the expected number of colour-deficient females is $50 \times 0.004 = 0.2$. The expected number of colour-deficient people is $3 + 0.2 = 3.2$. This result is, as it happens, correct, though quite difficult to confirm formally: no attempt will be made to do so here.

(e) Repeating the exercise gave a data vector of 1000 observations on W with the following frequencies.

Count	Frequency
0	30
1	137
2	186
3	243
4	182
5	115
6	63
7	28
8	12
9	3
10	0
11	0
12	1

This data set has mean $\overline{w} = 3.25$ and standard deviation $s = 1.758$.

Solution 4.20

(a) If the mean arrival rate is $\lambda = 12$ claims per week this is equivalent to

$$\frac{12}{7 \times 24} = \frac{1}{14}$$

claims per hour. So the mean waiting time between claim arrivals is 14 hours. By adding together 20 successive observations from the exponential distribution with mean 14, the twenty arrival times may be simulated. You might have got something like the following.

Claim number	Waiting time	Arrival time	Approximation
1	4.0	4.0	4 am, Mon
2	13.2	17.2	5 pm, Mon
3	3.3	20.5	9 pm, Mon
4	44.3	64.8	5 pm, Wed
5	17.3	82.1	10 am, Thu
6	6.0	88.1	4 pm, Thu
7	4.7	92.8	9 pm, Thu
8	4.0	96.8	1 am, Fri
9	3.2	100.0	4 am, Fri
10	11.7	111.7	4 pm, Fri
11	25.5	137.2	5 pm, Sat
12	33.3	170.5	3 am, Mon
13	1.3	171.8	4 am, Mon
14	0.5	172.3	4 am, Mon
15	4.9	177.2	9 am, Mon
16	2.7	179.9	12 noon, Mon
17	5.5	185.4	5 pm, Mon
18	3.7	189.1	9 pm, Mon
19	30.7	219.8	4 am, Wed
20	3.6	223.4	7 am, Wed

(b) Ten weeks of simulated claims gave 8 claims in the first week, 18 in the second and 14 in the third. You should have observed a continuing sequence with similar numbers. These are all observations on a Poisson random variable with mean 14.

Chapter 5

Solution 5.1

In Figure 5.4(a), $\mu = 100$; it looks as though $\mu + 3\sigma$ is about 150; so σ is about 17. In Figure 5.4(b), $\mu = 100$ and $\mu + 3\sigma$ is about 115: therefore, the standard deviation σ looks to be about 5. In Figure 5.4(c), $\mu = 72$ and σ is a little more than 1; and in Figure 5.4(d), $\mu = 1.00$ and σ is about 0.05.

Solution 5.2

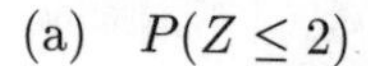

(a) $P(Z \leq 2)$

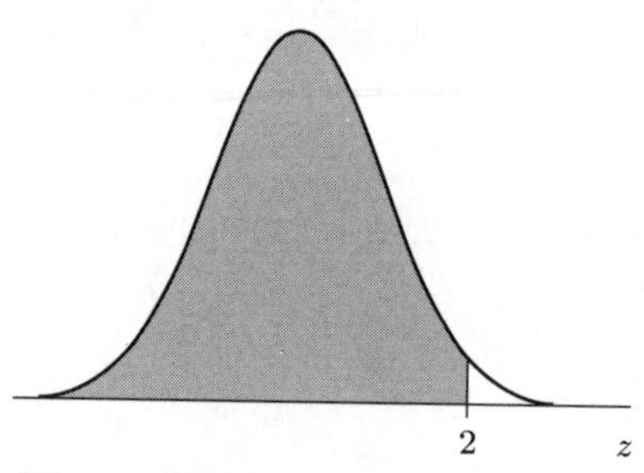

Figure S5.1

(b) $P(Z > 1)$

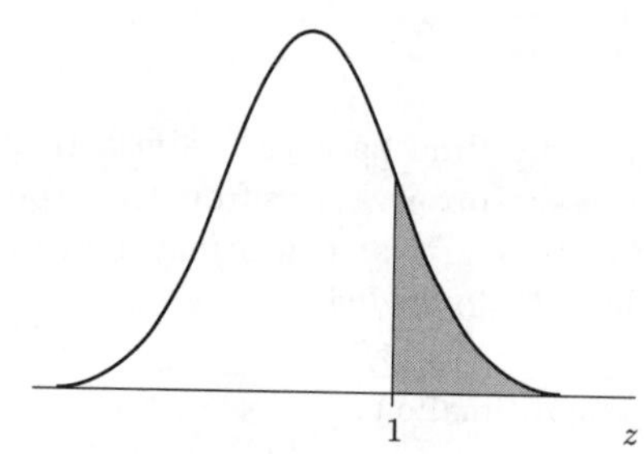

Figure S5.2

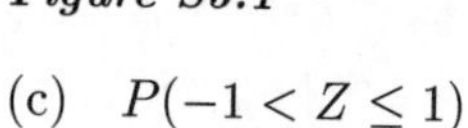

(c) $P(-1 < Z \leq 1)$

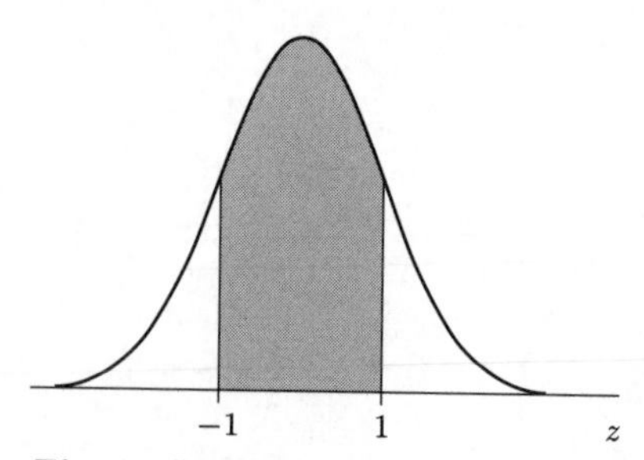

Figure S5.3

(d) $P(Z \leq -2)$

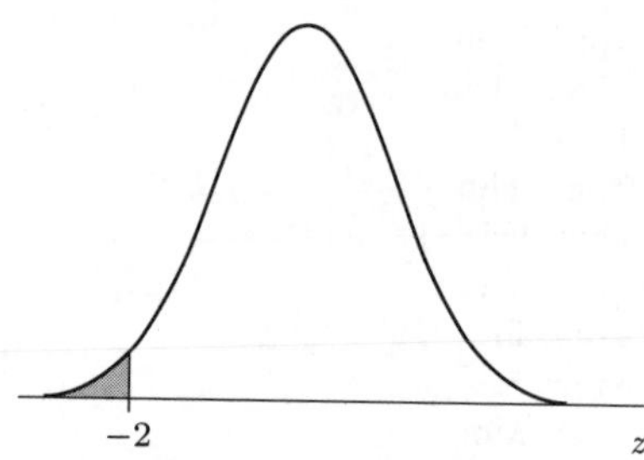

Figure S5.4

Solution 5.3

(a) Writing $X \sim N(2.60, 0.33^2)$, where X is the enzyme level present in individuals suffering from acute viral hepatitis, the proportion of sufferers whose measured enzyme level exceeds 3.00 is given by

$$P(X > 3.00) = P\left(Z > \frac{3.00 - 2.60}{0.33}\right)$$

(writing $z = (x - \mu)/\sigma$). This probability reduces to

$$P\left(Z > \frac{0.40}{0.33}\right) = P(Z > 1.21)$$

and is represented by the shaded area in Figure S5.5.

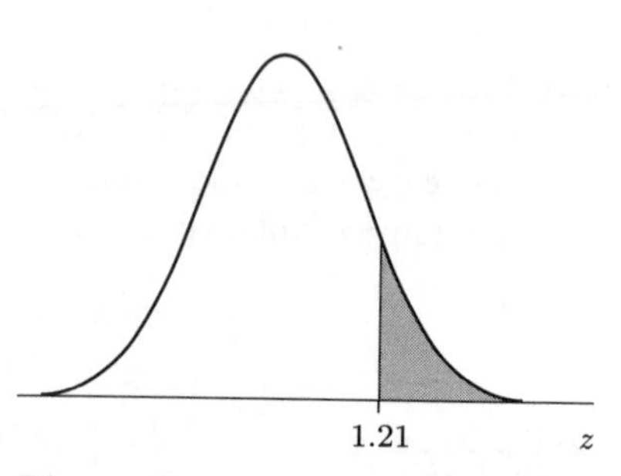

Figure S5.5

(b) Writing $Y \sim N(2.65, 0.44^2)$, where Y is the enzyme level in individuals suffering from aggressive chronic hepatitis, the proportion required is given by the probability

$$\begin{aligned} P(Y < 1.50) &= P\left(Z < \frac{1.50 - 2.65}{0.44}\right) \\ &= P\left(Z < -\frac{1.15}{0.44}\right) \\ &= P(Z < -2.61); \end{aligned}$$

this (quite small) proportion is given by the shaded area in Figure S5.6.

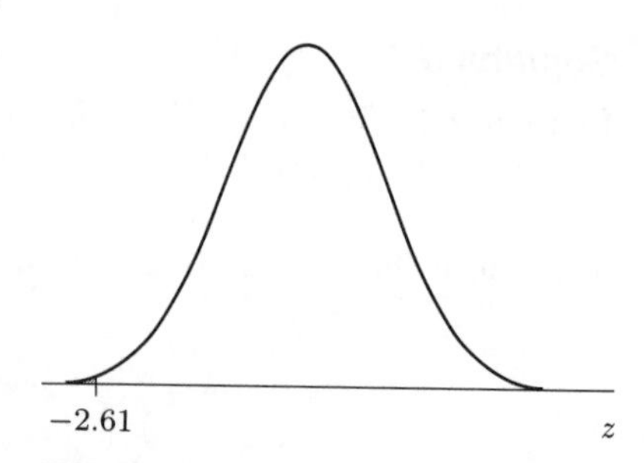

Figure S5.6

(c) The sample mean and sample standard deviation are

$$\overline{x} = 1.194 \quad \text{and} \quad s = 0.290.$$

The lower extreme (0.8 mm) may be written in standardized form as

$$z = \frac{x - \mu}{\sigma} = \frac{0.8 - 1.194}{0.290} = -1.36;$$

and the upper extreme (1.2 mm) as

$$z = \frac{x - \mu}{\sigma} = \frac{1.2 - 1.194}{0.290} = 0.02.$$

The proportion of ball-bearings whose diameter is between 0.8 mm and 1.2 mm can be shown on a sketch of the standard normal density as in Figure S5.7.

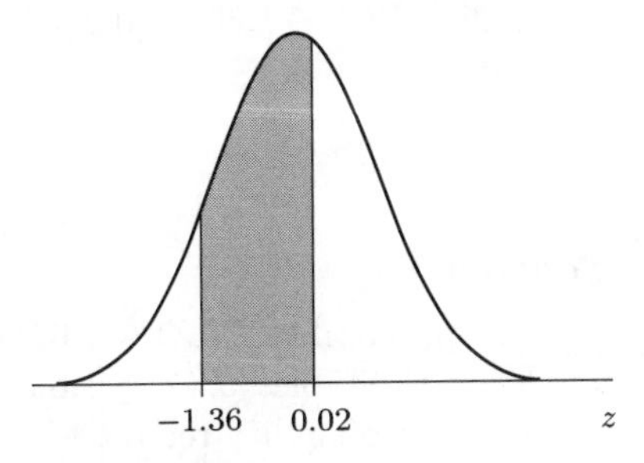

Figure S5.7

Solution 5.4

(a) The probability $P(Z \leq 1.00)$ is to be found in the row for $z = 1.0$ and in the column headed 0: this gives $P(Z \leq 1.00) = 0.8413$. This is shown in Figure S5.8.

(b) The probability $P(Z \leq 1.96)$ is given in the row for $z = 1.9$ and in the column headed 6: $P(Z \leq 1.96) = 0.9750$. This is illustrated in Figure S5.9.

(c) The probability $P(Z \leq 2.25)$ is to be found in the row for $z = 2.2$ and in the column headed 5: that is, $P(Z \leq 2.25) = 0.9878$. This probability is given by the shaded area in Figure S5.10.

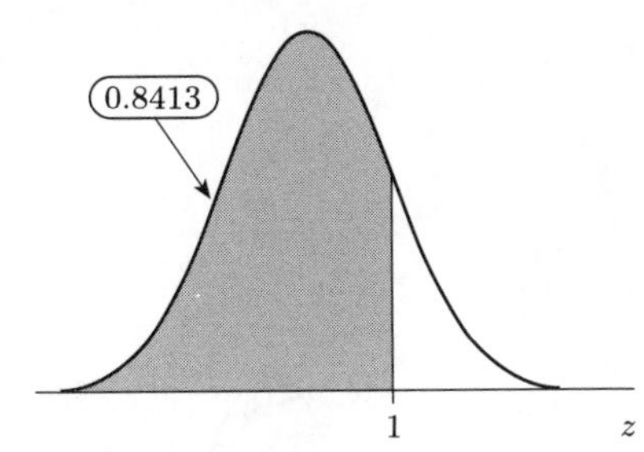

Figure S5.8

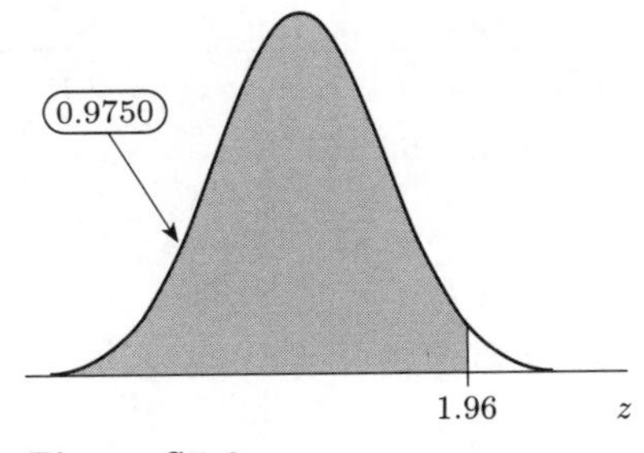

Figure S5.9

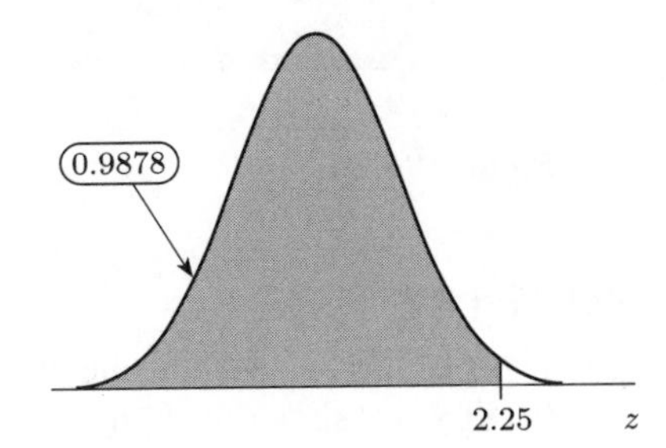

Figure S5.10

Solution 5.5

(a) First, sketch the standard normal density, showing the critical points $z = -1.33$ and $z = 2.50$. From the tables, $P(Z \leq 2.50) = 0.9938$ and so $P(Z > 2.50) = 0.0062$; by symmetry, $P(Z \leq -1.33) = P(Z \geq 1.33) = 1 - 0.9082 = 0.0918$. By subtraction, the probability required is

$$1 - 0.0062 - 0.0918 = 0.9020.$$

(b) From the tables, $P(Z \geq 3.00) = 1 - 0.9987 = 0.0013$. By symmetry,

$$P(-3.00 \leq Z \leq 3.00) = 1 - 0.0013 - 0.0013 = 0.9974.$$

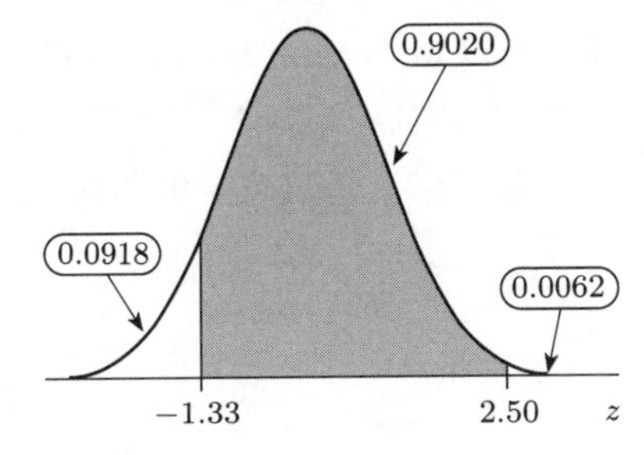

Figure S5.11

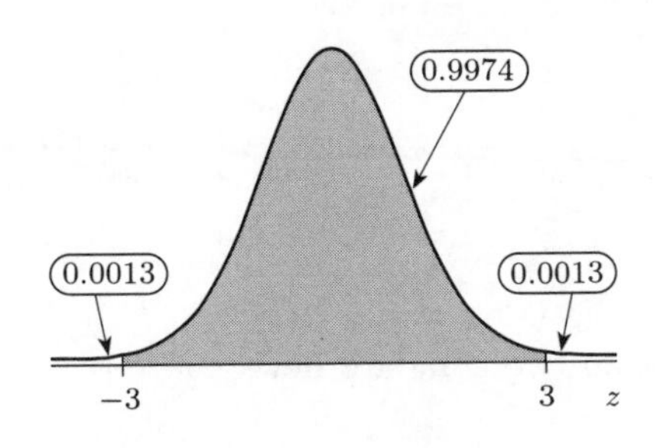

Figure S5.12

(c) First, sketch the standard normal density, showing the critical points $z = 0.50$ and $z = 1.50$. The probability $P(Z \leq 0.50)$ is 0.6915; the probability $P(Z \leq 1.50)$ is 0.9332. By subtraction, therefore, the probability required is

$$P(0.50 \leq Z \leq 1.50) = 0.9332 - 0.6915 = 0.2417.$$

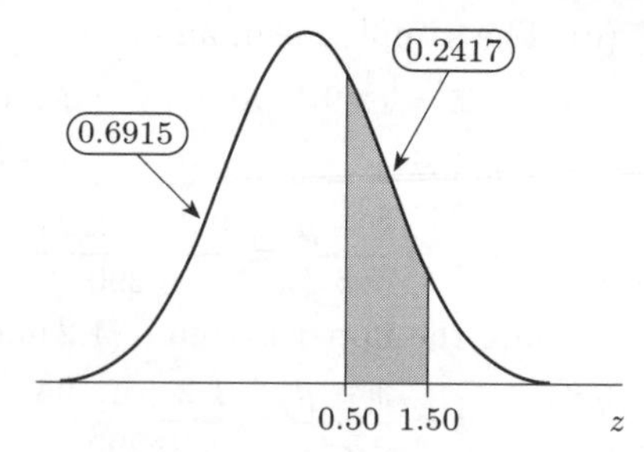

Figure S5.13

Solution 5.6

(a) The probability $P(|Z| \leq 1.62)$ is given by the shaded area in Figure S5.14. From the tables, $P(Z \geq 1.62) = 1 - P(Z \leq 1.62) = 1 - 0.9474 = 0.0526$, so the probability required is

$$P(|Z| \leq 1.62) = 1 - 0.0526 - 0.0526 = 0.8948.$$

(b) The probability $P(|Z| \geq 2.45)$ is given by the sum of the two shaded areas in Figure S5.15. From the tables, $P(Z \geq 2.45) = 1 - P(Z \leq 2.45) = 1 - 0.9929 = 0.0071$, so the total probability is $2 \times 0.0071 = 0.0142$.

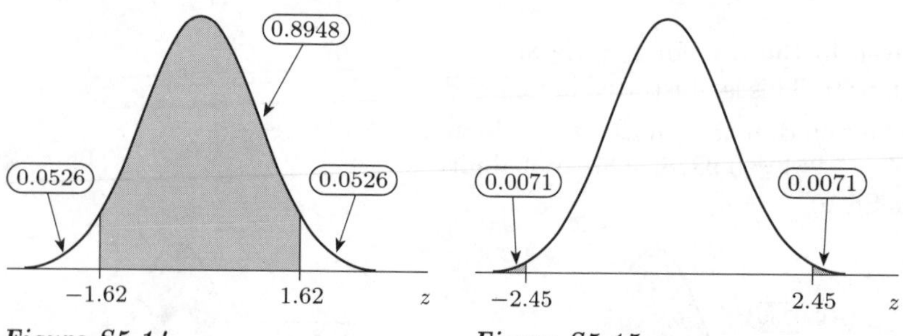

Figure S5.14

Figure S5.15

Solution 5.7

(a) The proportion within one standard deviation of the mean is given by $P(|Z| \leq 1)$, shown in Figure S5.16. Since

$$P(Z > 1) = 1 - P(Z \leq 1) = 1 - 0.8413 = 0.1587,$$

the answer required is $1 - 0.1587 - 0.1587 = 0.6826$: that is, nearly 70% of a normal population are within one standard deviation of the mean.

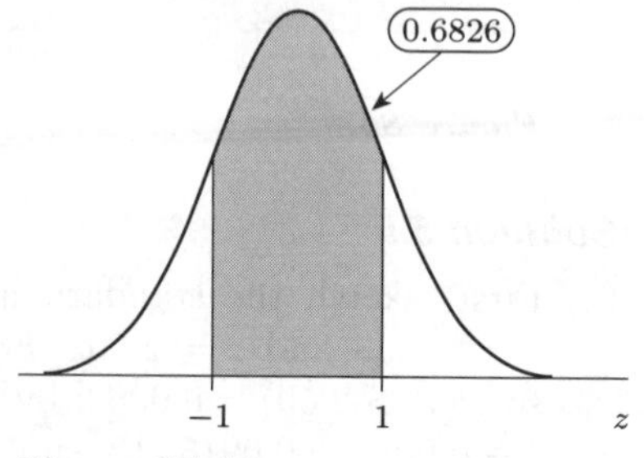

Figure S5.16

(b) Here we require the probability $P(|Z| > 2)$. Since

$$P(Z > 2) = 1 - P(Z \leq 2) = 1 - 0.9772 = 0.0228,$$

this is $2 \times 0.0228 = 0.0456$.

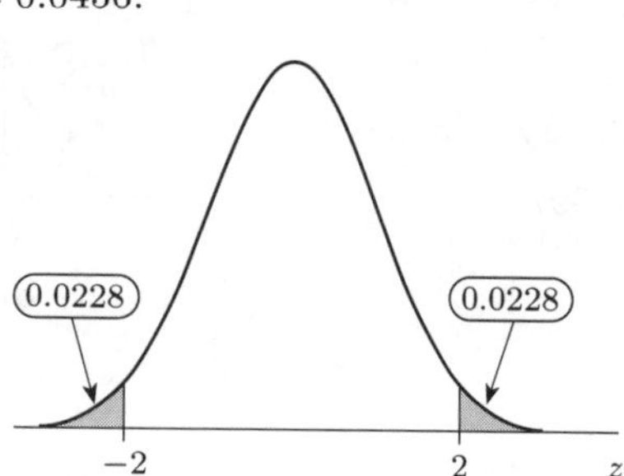

Figure S5.17

Less than 5% of a normal population are more than two standard deviations away from the mean.

Solution 5.8

We are told that $X \sim N(40, 4)$ ($\mu = 40, \sigma^2 = 4$; so $\sigma = 2$). The probability required is

$$P(37 \leq X \leq 42) = P\left(\frac{37-40}{2} \leq Z \leq \frac{42-40}{2}\right) = P(-1.50 \leq Z \leq 1.00)$$

shown in Figure S5.18. The probability required is

$$P(-1.50 \leq Z \leq 1.00) = 0.7745.$$

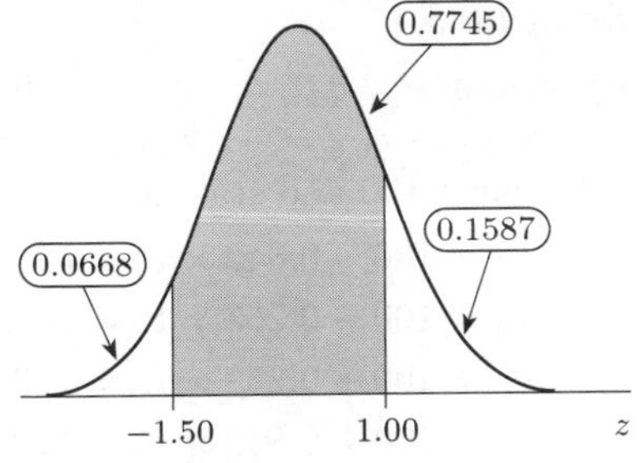

Figure S5.18

Solution 5.9

Writing $A \sim N(0, 2.75)$, the probability required is $P(0 < A < 2)$.

Subtracting the mean μ and dividing by the standard deviation σ, using $\mu = 0$ and $\sigma = \sqrt{2.75}$, this may be rewritten in terms of Z as

$$P\left(\frac{0-0}{\sqrt{2.75}} < Z < \frac{2-0}{\sqrt{2.75}}\right) = P(0 < Z < 1.21).$$

From the tables, $\Phi(1.21) = 0.8869$, so the probability required is

$$0.8869 - 0.5 = 0.3869.$$

Solution 5.10

(a) If T is $N(315, 17\,161)$ then a sketch of the distribution of T is given in Figure S5.19.

(b) Standardizing gives

$$P(T < 300) = P\left(Z < \frac{300-315}{131}\right) = P(Z < -0.11) = 0.4562.$$

This is shown in Figure S5.20.

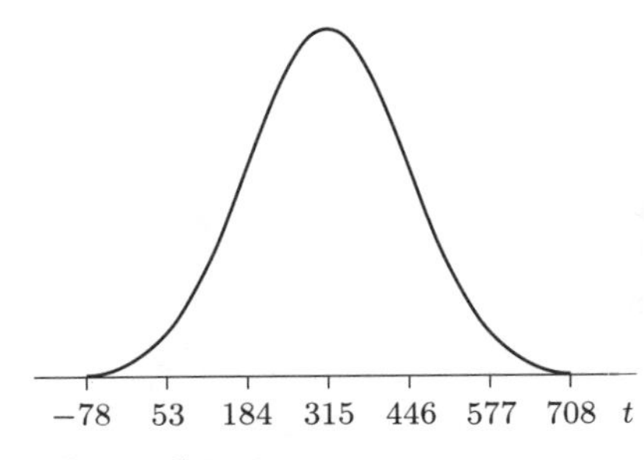

Figure S5.19

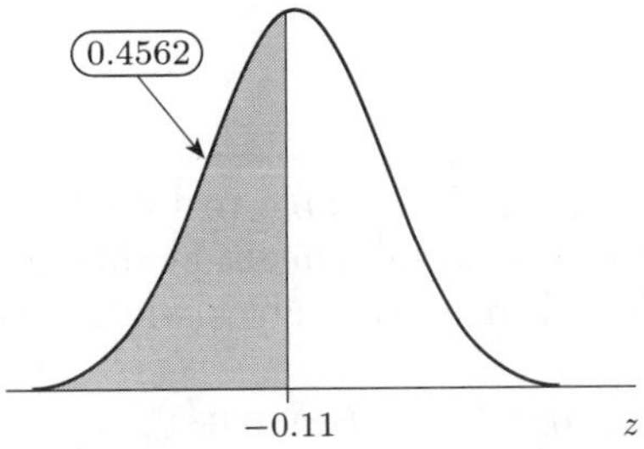

Figure S5.20

(c) $$P(300 \leq T \leq 500) = P\left(\frac{300-315}{131} \leq Z \leq \frac{500-315}{131}\right) = P(-0.11 \leq Z \leq 1.41).$$

This is shown in Figure S5.21.The area of the shaded region is 0.4645.

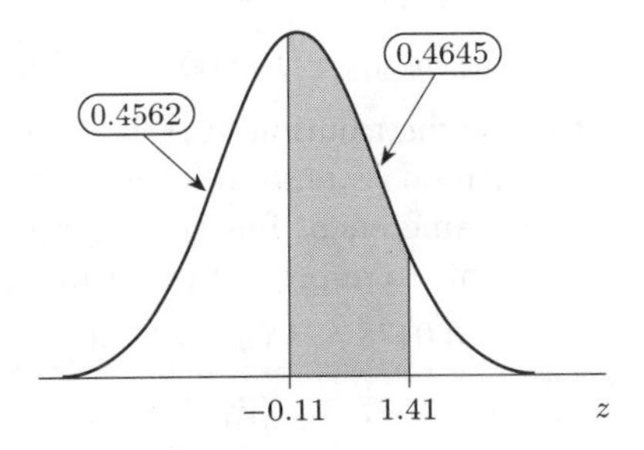

Figure S5.21

(d) First, we need

$$P(T > 500) = P(Z > 1.41) = 1 - 0.9207 = 0.0793.$$

The number of smokers with a nicotine level higher than 500 in a sample of 20 smokers has a binomial distribution $B(20, 0.0793)$. The probability that at most one has a nicotine level higher than 500 is

$$p_0 + p_1 = (0.9207)^{20} + 20(0.9207)^{19}(0.0793) = 0.1916 + 0.3300 = 0.52.$$

Solution 5.11

By symmetry, $q_{0.2} = -q_{0.8} = -0.842$ for the standard normal distribution, and $q_{0.4} = -q_{0.6} = -0.253$. Assuming IQ scores to be normally distributed with mean 100 and standard deviation 15, then

$$q_{0.2} = 100 - 0.842 \times 15 = 87.4$$
$$q_{0.4} = 100 - 0.253 \times 15 = 96.2$$
$$q_{0.6} = 100 + 0.253 \times 15 = 103.8$$
$$q_{0.8} = 100 + 0.842 \times 15 = 112.6$$

and these quantiles are illustrated in Figure S5.22.

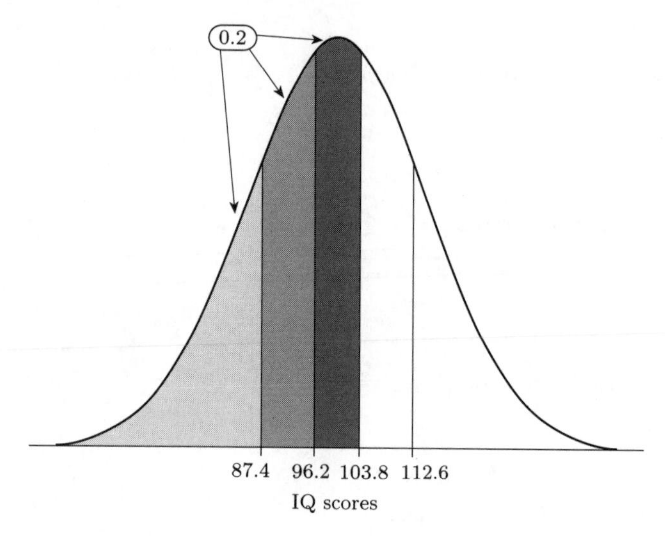

Figure S5.22

Solution 5.12

(a) Most statistical computer programs should be able to furnish standard normal probabilities and quantiles. The answers might be different in the fourth decimal place to those furnished by the tables when other than simple calculations are made.

(i) 0.0446 (ii) 0.9641 (iii) 0.9579 (iv) 0.0643

(v) $q_{0.10} = -q_{0.90} = -1.2816$ (vi) $q_{0.95} = 1.6449$

(vii) $q_{0.975} = 1.9600$ (viii) $q_{0.99} = 2.3263$

(b) The distribution of X is normal $N(100, 225)$. Most computers should return non-standard normal probabilities routinely, taking the distribution parameters as function arguments, and insulating users from the requirements to re-present problems in terms of the standard normal distribution.

(i) 0.0478 (ii) 0.1613 (iii) 100 (iv) 119.2 (v) 80.8

(c) (i) 0.1587 (ii) 166.22 cm

(iii) The first quartile is $q_L = 155.95$; the third quartile is $q_U = 164.05$; the interquartile range is given by the difference $q_U - q_L = 8.1$ cm.

(iv) 0.3023

(d) (i) 0.1514 (ii) 530.48 (iii) 0.6379

(iv) This question asks 'What proportion of smokers have nicotine levels within 100 units of the mean of 315?'. Formally,

$$P(|T - 315| \leq 100) = P(-100 \leq T - 315 \leq 100) = P(215 \leq T \leq 415)$$

which is 0.5548.

(v) $q_{0.80} = 425.25$

(vi) The range of levels is that covered by the interval $(q_{0.04}, q_{0.96})$ allowing 4% either side. This is $(85.7, 544.3)$.

(vii)

$$P(215 < T < 300) + P(350 < T < 400)$$
$$= 0.231\,794 + 0.136\,451 = 0.3682.$$

Solution 5.13

Your solution might have gone something like the following.

(a) The first sample of size 5 from Poisson(8) consisted of the list 6, 7, 3, 8, 4. This data set has mean $\overline{x}_5 = \frac{1}{5}(6 + 7 + 3 + 8 + 4) = 5.6$. When a vector of 100 observations on $\overline{X}_5$ was obtained, the observed frequencies of different observations were as follows.

[5, 6)	4
[6, 7)	25
[7, 8)	27
[8, 9)	28
[9, 10)	10
[10, 11)	6

So there were 90 observed in [6, 10).

(b) The 100 observations on $\overline{X}_{20}$ were distributed as follows. (Your results will be somewhat different.)

[6, 7)	8
[7, 8)	42
[8, 9)	43
[9, 10)	6
[10, 11)	1

So all the observations but one were in [6, 10), and 85 of the 100 were in [7, 9).

(c) All the 100 observations were in [7, 9).

(d) The larger the sample size, the less widely scattered around the population mean $\mu = 8$ the observed sample means were. In non-technical language, 'larger samples resulted in sample means that were more precise estimates of the population mean'.

Solution 5.14

The exponential distribution is very skewed, and you might have expected more scatter in the observations. This was apparent in the distributions of the sample means. For samples of size 5 the following observations were obtained on $\overline{X}_5$. (Remember, $\overline{X}_5$ estimates the population mean, $\mu = 8$.)

$[0, 5)$	18
$[5, 10)$	56
$[10, 15)$	22
$[15, 20)$	3
$[20, 25)$	1

The largest observation was $\overline{x}_5 = 21.42$. Nevertheless, it is interesting to observe that the distribution of observations on $\overline{X}_5$ peaks not at the origin but somewhere between 5 and 10.

For samples of size 20 the following distribution of observations on $\overline{X}_{20}$ was obtained.

$[4, 6)$	9
$[6, 8)$	37
$[8, 10)$	39
$[10, 12)$	12
$[12, 14)$	3

These observations are peaked around the point 8.

Finally, for samples of size 80 the following observations on $\overline{X}_{80}$ were obtained.

$[6, 7)$	11
$[7, 8)$	42
$[8, 9)$	33
$[9, 10)$	12
$[10, 11)$	2

Solution 5.15

(a) (i) The following typical 100 observations on $\overline{X}_2$ resulted in a histogram almost as skewed as the distribution of X itself.

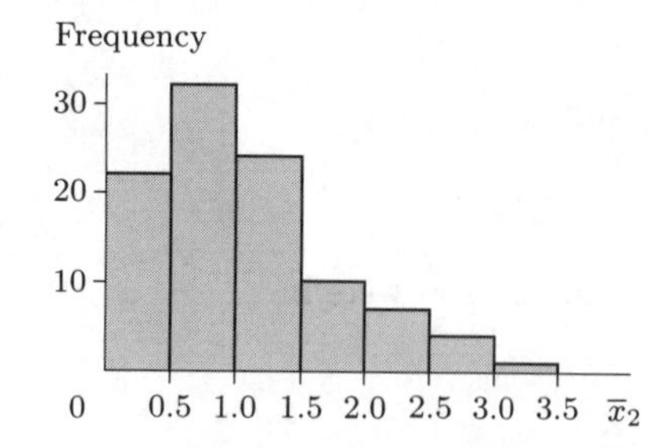

Figure S5.23

(ii) The histogram for 100 observations on $\overline{X}_{30}$ is given in Figure S5.24.

(iii) The histogram of part (ii) is evidently more symmetric than that of part (i), and it appears that a normal density might provide a usable approximation to the distribution of $\overline{X}_{30}$.

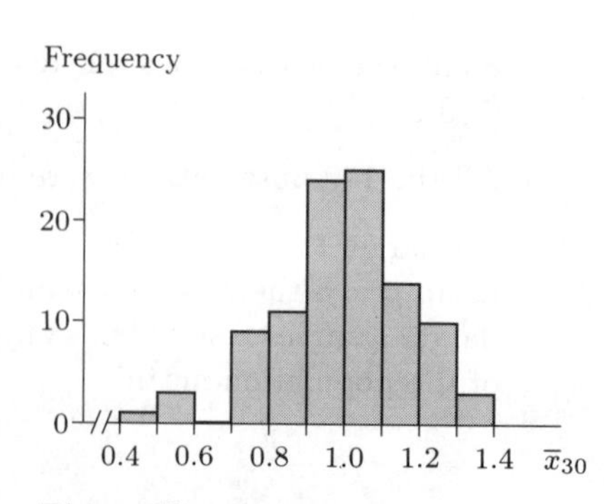

Figure S5.24

(b) (i) The simulation of 100 observations on $\overline{X}_2$ resulted in a histogram that was roughly triangular over $[0, 2]$, and very different to that obtained at part (a)(i).

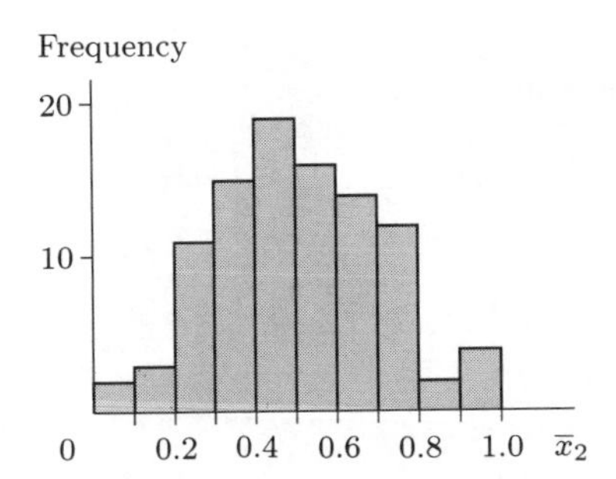

Figure S5.25

(ii) For means of samples of size 30, the following histogram was obtained.

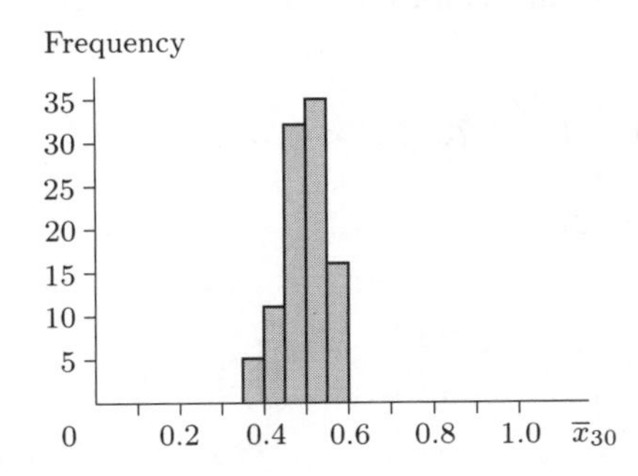

Figure S5.26

(iii) The differences between the histograms of parts (b)(i) and (b)(ii) are not so apparent. The distribution in both cases peaks at 1 and is roughly symmetric. However, notice that for samples of size 30 there is much less variation in the distribution of the sample mean.

Solution 5.16

It does not matter that there is no proposed model for the duration of individual patients' visits: the numbers $\mu = 20$ minutes and $\sigma = 15$ minutes provide sufficient information. By the central limit theorem, the total time T required of the dentist by the 12 patients is approximately normally distributed with mean

$$12 \times 20 = 240$$

and variance

$$12 \times 15^2 = 2700.$$

The probability that T will be less than 3 hours (that is, 180 minutes) is

$$P(T < 180) = P\left(Z < \frac{180 - 240}{\sqrt{2700}}\right) = P(Z < -1.15)$$

or $1 - 0.8749 = 0.1251$ from the tables. So she will only be free at 12 with probability about $\frac{1}{8}$.

(If you used a computer for the normal probability without the intermediate calculation $Z = -1.15$, you would have obtained the solution 0.1241.)

Solution 5.17

If the error in any individual transaction is written W, then $W \sim U(-\frac{1}{2}, \frac{1}{2})$. The mean of W is $E(W) = 0$, by symmetry, and the variance of W is $V(W) = \frac{1}{12}\left(\frac{1}{2} - (-\frac{1}{2})\right)^2 = \frac{1}{12}$.

The accumulated error in 400 transactions is given by the sum

$$S = W_1 + W_2 + \cdots + W_{400}.$$

By the central limit theorem, S has mean

$$\mu = 400 \times 0 = 0,$$

and variance

$$\sigma^2 = \frac{400}{12} = 33.333,$$

and is approximately normally distributed.

The probability that after 400 transactions her estimate of her bank balance is less than ten pounds in error is

$$P(-10 < S < 10) = P\left(\frac{-10-0}{\sqrt{33.333}} < Z < \frac{10-0}{\sqrt{33.333}}\right) = P(-1.73 < Z < 1.73).$$

This probability is given by the shaded area in Figure S5.27.

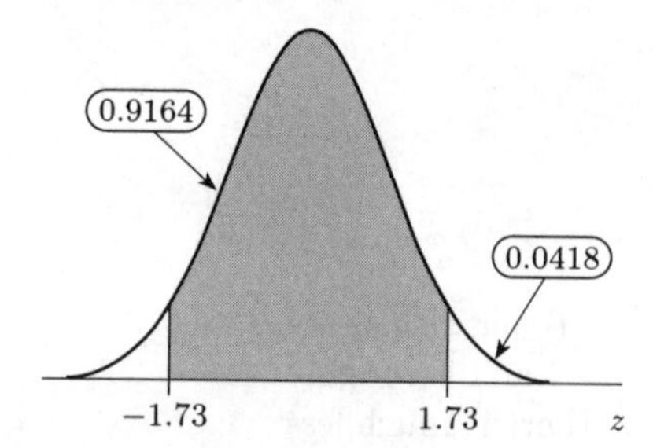

Figure S5.27

Since $P(Z > 1.73) = 1 - 0.9582 = 0.0418$, the probability required is

$$1 - 2 \times 0.0418 = 0.9164$$

(or rather more than 90%).

Solution 5.18

The probability

$$P(12 \leq X \leq 15)$$

can be approximated by the probability

$$P\left(11\tfrac{1}{2} \leq Y \leq 15\tfrac{1}{2}\right),$$

where $Y \sim N(8, 4)$; this is the same as

$$P\left(\frac{11\frac{1}{2} - 8}{2} \leq Z \leq \frac{15\frac{1}{2} - 8}{2}\right) = P(1.75 \leq Z \leq 3.75)$$

and (from the tables) this is $0.9999 - 0.9599 = 0.0400$. The actual binomial probability is 0.0384.

Solution 5.19

(a) (i) 0.164 538 (ii) 0.182 820 (iii) 0.165 408 (iv) 0.124 056

(b) 0.472 284

(c) Since $np = \frac{25}{4} = 6.25$, $npq = \frac{75}{16} = 4.6875$, then $\mu = 6.25$, $\sigma^2 = 4.6875$.

(d) (i) $P(X = 6) \simeq P\left(5\frac{1}{2} \leq Y \leq 6\frac{1}{2}\right) = 0.181\,446$. This may be compared with the binomial probability 0.182 820.

(ii) $P(X = 7) \simeq P\left(6\frac{1}{2} \leq Y \leq 7\frac{1}{2}\right) = 0.172\,185$. This may be compared with the exact binomial probability, which is 0.165 408.

(iii) $P(X = 8) \simeq P\left(7\frac{1}{2} \leq Y \leq 8\frac{1}{2}\right) = 0.132\,503$. The exact binomial probability is 0.124 056.

(iv) $P(6 \leq X \leq 8) \simeq P\left(5\frac{1}{2} \leq Y \leq 8\frac{1}{2}\right) = 0.486\,134$. The corresponding binomial probability is 0.472 284.

Solution 5.20

A computer gives the probability $P(30 \leq X \leq 45)$ when $X \sim \text{Poisson}(40)$ as 0.766 421. The central limit theorem permits the approximation

$$P(30 \leq X \leq 45) = P\left(29\tfrac{1}{2} \leq Y \leq 45\tfrac{1}{2}\right)$$

where Y is normal with mean $\mu = 40$ and variance $\sigma^2 = 40$. The right-hand side is given by

$$P\left(\frac{29\frac{1}{2} - 40}{\sqrt{40}} \leq Z \leq \frac{45\frac{1}{2} - 40}{\sqrt{40}}\right)$$
$$= P(-1.66 \leq Z \leq 0.87) = 0.8078 - (1 - 0.9515) = 0.7593$$

from the tables. Directly from a computer, we would obtain 0.759 310. This approximation is reasonable.

Chapter 6

Solution 6.1

For any random sample taken from a population with mean μ and variance σ^2, the sample mean $\overline{X}$ has mean and variance

$$E(\overline{X}) = \mu, \qquad V(\overline{X}) = \frac{\sigma^2}{n},$$

where n is the sample size. When the population is Poisson, the variance σ^2 is equal to the mean μ, so

$$E(\overline{X}) = \mu, \qquad V(\overline{X}) = \frac{\mu}{n}.$$

The random variable $\overline{X}_{(1)}$ was based on samples of size 103, the random variable $\overline{X}_{(2)}$ on samples of size 48. So

$$E(\overline{X}_{(1)}) = E(\overline{X}_{(2)}) = \mu;$$

but

$$V(\overline{X}_{(1)}) = \frac{\mu}{103}, \qquad V(\overline{X}_{(2)}) = \frac{\mu}{48}.$$

The larger the sample taken, the smaller the variance in the sample mean.

Solution 6.2

The mean of the random variable Y_i is given by

$$E(Y_i) = E(\alpha + \beta x_i + W_i),$$

which is of the form $E(\text{constant} + W_i)$, since α, β and x_i are all constants. This is therefore

$$E(Y_i) = \alpha + \beta x_i + E(W_i),$$

and from our assumption that W_i has mean 0 it follows that

$$E(Y_i) = \alpha + \beta x_i,$$

for all $i = 1, 2, \ldots, 6$.

Similarly, the variance of Y_i is given by

$$V(Y_i) = V(\text{constant} + W_i) = V(W_i) = \sigma^2,$$

for all $i = 1, 2, \ldots, 6$.

Solution 6.3

(a) The midpoint of the coordinates (x_1, Y_1) and (x_2, Y_2) is the point $\left(\frac{1}{2}(x_1 + x_2), \frac{1}{2}(Y_1 + Y_2)\right)$. The midpoint of the coordinates (x_5, Y_5) and (x_6, Y_6) is the point $\left(\frac{1}{2}(x_5 + x_6), \frac{1}{2}(Y_5 + Y_6)\right)$. The slope of the line joining the two midpoints is

$$\widehat{\beta}_2 = \frac{\frac{1}{2}(Y_5 + Y_6) - \frac{1}{2}(Y_1 + Y_2)}{\frac{1}{2}(x_5 + x_6) - \frac{1}{2}(x_1 + x_2)} = \frac{Y_5 + Y_6 - Y_1 - Y_2}{x_5 + x_6 - x_1 - x_2}.$$

The centre of gravity of the points (x_1, Y_1), (x_2, Y_2) and (x_3, Y_3) is $\left(\frac{1}{3}(x_1 + x_2 + x_3), \frac{1}{3}(Y_1 + Y_2 + Y_3)\right)$. The centre of gravity of the points (x_4, Y_4), (x_5, Y_5) and (x_6, Y_6) is $\left(\frac{1}{3}(x_4 + x_5 + x_6), \frac{1}{3}(Y_4 + Y_5 + Y_6)\right)$. The slope of the line joining the two centres of gravity is

$$\widehat{\beta}_3 = \frac{\frac{1}{3}(Y_4 + Y_5 + Y_6) - \frac{1}{3}(Y_1 + Y_2 + Y_3)}{\frac{1}{3}(x_4 + x_5 + x_6) - \frac{1}{3}(x_1 + x_2 + x_3)} = \frac{Y_4 + Y_5 + Y_6 - Y_1 - Y_2 - Y_3}{x_4 + x_5 + x_6 - x_1 - x_2 - x_3}.$$

(b) Consequently

$$\begin{aligned} E(\widehat{\beta}_2) = E\left(\frac{Y_5 + Y_6 - Y_1 - Y_2}{x_5 + x_6 - x_1 - x_2}\right) &= \frac{1}{x_5 + x_6 - x_1 - x_2} E(Y_5 + Y_6 - Y_1 - Y_2) \\ &= \frac{1}{x_5 + x_6 - x_1 - x_2}((\alpha + \beta x_5) + (\alpha + \beta x_6) - (\alpha + \beta x_1) - (\alpha + \beta x_2)) \\ &= \frac{1}{x_5 + x_6 - x_1 - x_2}(\beta x_5 + \beta x_6 - \beta x_1 - \beta x_2) = \beta; \end{aligned}$$

and $E(\widehat{\beta}_3)$ reduces to β in a similar way.

Solution 6.4

The variance of the second estimator $\widehat{\beta}_2$ is given by

$$\begin{aligned}
V(\widehat{\beta}_2) &= V\left(\frac{Y_5+Y_6-Y_1-Y_2}{x_5+x_6-x_1-x_2}\right) \\
&= \frac{1}{(x_5+x_6-x_1-x_2)^2} V(Y_5+Y_6-Y_1-Y_2) \\
&= \frac{1}{(x_5+x_6-x_1-x_2)^2} (V(Y_5)+V(Y_6)+V(Y_1)+V(Y_2)) \\
&= \frac{4\sigma^2}{8^2} = 0.0625\sigma^2.
\end{aligned}$$

The variance of $\widehat{\beta}_3$ is

$$\begin{aligned}
V(\widehat{\beta}_3) &= V\left(\frac{Y_4+Y_5+Y_6-Y_1-Y_2-Y_3}{x_4+x_5+x_6-x_1-x_2-x_3}\right) \\
&= \frac{1}{(x_4+x_5+x_6-x_1-x_2-x_3)^2} V(Y_4+Y_5+Y_6-Y_1-Y_2-Y_3) \\
&= \frac{1}{(x_4+x_5+x_6-x_1-x_2-x_3)^2} (V(Y_4)+V(Y_5)+V(Y_6)+V(Y_1)+V(Y_2)+V(Y_3)) \\
&= \frac{6\sigma^2}{9^2} = 0.0741\sigma^2.
\end{aligned}$$

Solution 6.5

(a) There is one unknown parameter here, μ, which is also the mean of the Poisson distribution. Matching sample and population moments gives the estimate $\widehat{\mu} = \overline{x}$; the corresponding estimator for μ is $\widehat{\mu} = \overline{X}$.

(b) In this case the population mean is $1/p$. Matching moments gives $\overline{X} = 1/\widehat{p}$; so $\widehat{p} = 1/\overline{X}$.

(c) Here, there are two unknown parameters, so we shall need to use two sample moments. These are $\overline{X}$, the sample mean, and S^2, the sample variance. Notice the use of the upper-case letter S, implying that like the sample mean, the sample variance is a random variable. Matching moments gives $\widehat{\mu} = \overline{X}$, $\widehat{\sigma}^2 = S^2$.

(d) The mean of the exponential distribution is $1/\lambda$: matching moments gives $\overline{X} = 1/\widehat{\lambda}$; so $\widehat{\lambda} = 1/\overline{X}$.

(e) There is one unknown parameter here. Matching the sample mean to the binomial mean gives $\overline{X} = m\widehat{p}$, so

$$\widehat{p} = \frac{\overline{X}}{m} = \frac{X_1+X_2+\cdots+X_n}{mn}.$$

(This was the 'intuitive' estimate $\widehat{p}$ of p that was used in Example 6.1.)

Solution 6.6

(a) You might have obtained something like this: the 1000 samples of size 2 may be represented as a rectangular array

$$\begin{bmatrix} 0.156 & 0.093 \\ 0.183 & 0.203 \\ 0.066 & 0.168 \\ & \vdots & \\ 0.679 & 0.218 \end{bmatrix}$$

with sample means

$$\begin{bmatrix} 0.124 \\ 0.193 \\ 0.117 \\ \vdots \\ 0.449 \end{bmatrix}.$$

Taking reciprocals gives

$$\begin{bmatrix} 8.05 \\ 5.19 \\ 8.56 \\ \vdots \\ 2.23 \end{bmatrix}$$

which is a data vector of 1000 different independent estimates of λ.

(b) The mean of this data vector is 9.20, close to twice the true value of λ.

(For interest, the experiment was repeated four more times, resulting in four further estimates 11.4, 11.1, 11.8 and 9.6.)

Solution 6.7

(a) The method of moments says, simply, that $\widehat{\mu} = \overline{X}$.

(b) For any random sample, the sample mean $\overline{X}$ has expectation μ, the population mean, so in this case

$$E(\widehat{\mu}) = E(\overline{X}) = \mu;$$

it follows that $\widehat{\mu}$ is unbiased for μ.

(c) Using the same set of 1000 random samples of size 2 as was used in Solution 6.6, our data vector of 1000 different independent estimates of μ is

$$\begin{bmatrix} 0.124 \\ 0.193 \\ 0.117 \\ \vdots \\ 0.449 \end{bmatrix}.$$

(d) This data vector has mean 0.202, close to the true value $\mu = 0.2$.

(For interest, the experiment was repeated four more times, resulting in estimates 0.202 (again), 0.196, 0.195 and 0.201.)

Solution 6.8

The average run length is the sample mean

$$\overline{x} = \frac{1 \times 71 + 2 \times 28 + 3 \times 5 + 4 \times 2 + 5 \times 2 + 6 \times 1}{71 + 28 + 5 + 2 + 2 + 1} = \frac{166}{109} = 1.523;$$

assuming a geometric model with mean $1/p$, the moment estimate of p is

$$\widehat{p} = \frac{1}{\overline{x}} = \frac{109}{166} = 0.657.$$

Solution 6.9

(a) The mean of these data is $\overline{t} = 437.21$, and so the moment estimator for the exponential parameter λ is

$$\widehat{\lambda} = 1/\overline{t} = 0.0023.$$

The units of $\widehat{\lambda}$ are 'earthquakes per day'. We know that the moment estimator is biased:

$$E(\widehat{\lambda}) = \lambda\left(1 + \frac{1}{n-1}\right).$$

However, in this case $n = 62$. The moment estimator may be expected to overestimate the true value of λ by a factor $1 + 1/61 = 1.016$, which is very small.

(b) The moment estimate of μ is $\widehat{\mu} = \overline{t} = 437.21$; the estimator $\widehat{\mu}$ is unbiased. The units of $\widehat{\mu}$ are 'days between earthquakes'.

Solution 6.10

In this case $n = 3$ and $x_{\max} = 13.1$, so

$$\widehat{\theta} = \left(1 + \tfrac{1}{n}\right) x_{\max} = \tfrac{4}{3}(13.1) = 17.5.$$

Solution 6.11

(a) The mean of the Pareto$(100, \theta)$ probability distribution is

$$\mu = \frac{100\theta}{\theta - 1}.$$

(b) The method of moments says that the moment estimator of θ for a sample from the Pareto distribution where $K = 100$ may be found using

$$\overline{X} = \frac{100\widehat{\theta}}{\widehat{\theta} - 1};$$

so

$$\widehat{\theta}\overline{X} - \overline{X} = 100\widehat{\theta};$$

thus, finally,

$$\widehat{\theta} = \frac{\overline{X}}{\overline{X} - 100}.$$

In this case the sample total is $\sum x_i = 3624$ and so the sample mean is $\overline{x} = 3624/30 = 120.8$. The moment estimate of θ is

$$\widehat{\theta} = \frac{\overline{x}}{\overline{x} - 100} = \frac{120.8}{20.8} = 5.81.$$

Solution 6.12

The maximum likelihood estimate of θ is (after a little calculation) $\widehat{\theta}_{ML} = 5.59$; this is to be compared with the moment estimate $\widehat{\theta}_{MM} = 5.81$ obtained in the previous exercise. Numerically there is little to choose between either estimate. We know that maximum likelihood estimators possess good properties. (However, $\widehat{\theta}_{MM}$ was in this case a great deal easier to obtain, and to calculate.)

Solution 6.13

(a) The total number of mice tested is $12 + 20 + \cdots + 20 = 505$; the total number afflicted is $0 + 0 + \cdots + 4 = 43$. The maximum likelihood estimate of p is $\widehat{p} = 43/505 = 0.085$.

(b) The total number of normal *Drosophila* is 419; the total number of vestigial *Drosophila* is 68. The maximum likelihood estimate of the proportion normal is given by

$$\widehat{p} = \frac{419}{419 + 68} = \frac{419}{487} = 0.86.$$

Solution 6.14

The likelihood for the sample observed is

$$\left(\tfrac{1}{2}(1-r)\right)^{147} \left(\tfrac{1}{2}r\right)^{65} \left(\tfrac{1}{2}r\right)^{58} \left(\tfrac{1}{2}(1-r)\right)^{133}$$

$$= \left(\tfrac{1}{2}(1-r)\right)^{147+133} \left(\tfrac{1}{2}r\right)^{65+58} = \left(\tfrac{1}{2}(1-r)\right)^{280} \left(\tfrac{1}{2}r\right)^{123} = \frac{(1-r)^{280}r^{123}}{2^{403}}.$$

This expression is maximized where $(1-r)^{280}r^{123}$ is maximized. This occurs at $\widehat{r} = 0.3052$.

You might have found this using numerical procedures or—perhaps not quite so precisely—by scanning a graph of the function $(1-r)^{280}r^{123}$. Differentiation gives the exact fractional answer, $\widehat{r} = 123/403$.

Solution 6.15

The likelihood of p for the sample of 1469 cars is given by

$$p_1^{902} p_2^{403} p_3^{106} p_4^{38} p_5^{16} P(X \geq 6)^4,$$

where $p_j = P(X = j) = (1-p)^{j-1}p, \quad j = 1, 2, \ldots$. This is

$$p^{902}((1-p)p)^{403} \left((1-p)^2p\right)^{106} \left((1-p)^3p\right)^{38} \left((1-p)^4p\right)^{16} \left((1-p)^5\right)^4$$

$$= p^{1465}(1-p)^{813}.$$

This is maximized at $\widehat{p} = 1465/2278 = 0.6431$. (The exact fraction 1465/2278 was found using differentiation—numerical and graphical techniques should provide an answer close to 0.6431.)

Notice that for these data the sample mean is at least $2282/1469 = 1.553$ (that is, the sample mean if all four of the fullest cars only contained six passengers); so the maximum likelihood estimate for p is going to be just under $1469/2282 = 0.6437$, as indeed it is. This small calculation is a useful check on your answer. Notice that the censoring has not in fact influenced the calculation unduly.

Solution 6.16

For these data (and assuming a Poisson model) the likelihood is given by

$$p_0^{11} p_1^{37} p_2^{64} p_3^{55} p_4^{37} p_5^{24} P_6^{12},$$

where p_j is the probability $P(X = j)$ when X is Poisson(μ), and where P_j is the probability $P(X \geq j)$. The likelihood is maximized at $\widehat{\mu} = 2.819$. (Again, the sample mean assuming at most 6 colonies per quadrat would be $670/240 = 2.792$. The estimate is not very different for the censored data.)

Solution 6.17

The average time between pulses is given by the sample mean $\bar{t} = 0.2244$. The units are hundredths of a second. Consequently, the maximum likelihood estimate of the pulse rate (per second) is $100/0.2244 = 446$.

Solution 6.18

All that is required here is the sample mean $\widehat{\mu} = \overline{x} = 159.8$ (measurements in cm).

Solution 6.19

(a) In one simulation the observations

$$x_1 = 96.59, \qquad x_2 = 99.87, \qquad x_3 = 107.15$$

were obtained, having sample variance $s^2 = 29.2$, which is fairly close to 25. However, this sequence was immediately followed by the sample

$$x_1 = 100.82, \qquad x_2 = 99.30, \qquad x_3 = 100.91,$$

having sample variance $s^2 = 0.82$, which is very far from 25!

(b) Your collected samples may have looked something like

$$\begin{bmatrix} 108.08 & 96.18 & 89.85 \\ 102.19 & 97.58 & 106.97 \\ 98.52 & 99.23 & 96.88 \\ & \vdots & \\ 106.01 & 95.45 & 96.58 \end{bmatrix}$$

with sample variances

$$\begin{bmatrix} 85.68 \\ 22.04 \\ 1.44 \\ \vdots \\ 33.60 \end{bmatrix}.$$

(i) The mean of this vector of sample variances is 24.95 which is very close to 25; but you can see from the four elements listed that the variation is very considerable. The highest sample variance recorded in the 100 samples was 133.05; the lowest was 0.397.

(ii) The variation in the sample variances is evident from a frequency table, and from the histogram shown in Figure S6.1.

Observation	Frequency
0–20	48
20–40	32
40–60	15
60–80	1
80–100	2
100–120	1
120–140	1

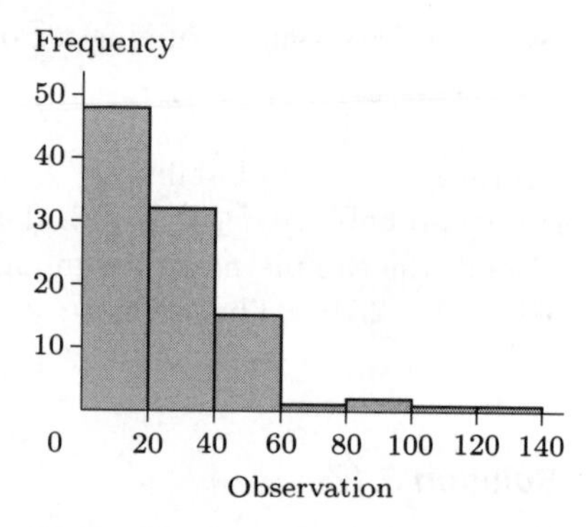

Figure S6.1

(iii) The variance in the recorded sample variances is 524.3!

Your results will probably have shown similarly gross variation in the observed sample variances.

Solution 6.20

(a) The mean of the 100 observed sample variances, based on samples of size 10 was, in one particular experiment, 24.22.

(b) The results are summarized below. A histogram is given in Figure S6.2.

Observation	Frequency
0–10	10
10–20	27
20–30	36
30–40	15
40–50	10
50–60	2

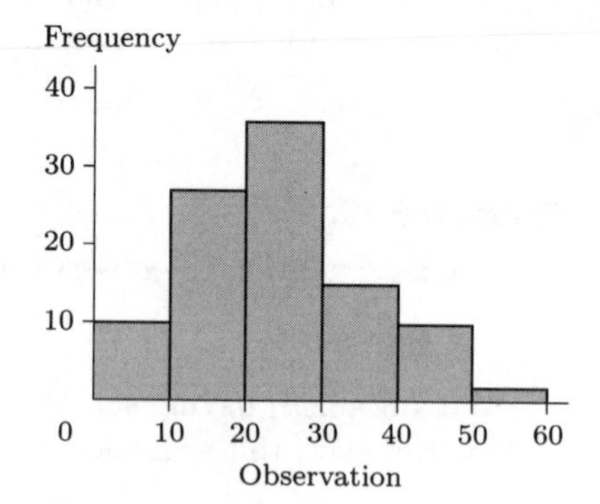

Figure S6.2

The distribution is very much less dispersed.

(c) The variance in the sample variances is now only 128. (The maximum observation was 55.6, the minimum was 3.49.)

Solution 6.21

(a) (i) $P(-1 \leq Z \leq 1) = 0.6827.$ (ii) $P(-\sqrt{2} \leq Z \leq \sqrt{2}) = 0.8427.$

(iii) $P(-\sqrt{3} \leq Z \leq \sqrt{3}) = 0.9167.$ (iv) $P(-2 < Z \leq 2) = 0.9545.$

(b) (i)
$$P(W \leq 1) = P(Z^2 \leq 1) \quad \text{(by definition)}$$
$$= P(-1 \leq Z \leq 1) = 0.6827.$$

(ii) $P(W \leq 2) = P(-\sqrt{2} \leq Z \leq \sqrt{2}) = 0.8427.$

(iii) $P(W \leq 3) = 0.9167.$

(iv) $P(W \leq 4) = 0.9545.$

(c) In one particular run, the following results were obtained:

(i) proportion less than 1 = 677/1000 = 0.677;

(ii) proportion less than 2 = 831/1000 = 0.831;

(iii) proportion less than 3 = 908/1000 = 0.908;

(iv) proportion less than 4 = 955/1000 = 0.955.

(d) The corresponding histogram is shown in Figure S6.3. Notice that this histogram is exceedingly skewed.

(e) From this simulation, an estimate of μ_W is

$$\overline{w} = 1.032,$$

and an estimate for σ_W^2 is

$$s_W^2 = 2.203.$$

(For interest, the experiment was repeated a further four times. Estimates of μ_W and σ_W^2 were

$$1.009, 2.287; \quad 0.975, 2.033; \quad 1.001, 1.782; \quad 1.044, 1.982.)$$

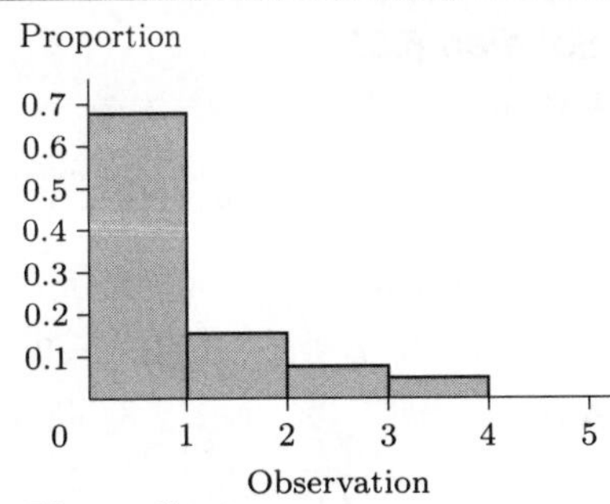

Figure S6.3

Solution 6.22

If Z^2 has mean 1 and variance 2, then the sum of r independent observations on Z^2,

$$W = Z_1^2 + Z_2^2 + \cdots + Z_r^2,$$

will have mean and variance

$$E(W) = r, \qquad V(W) = 2r.$$

Solution 6.23

(a) 0.5697 (b) 0.1303 (c) 0.0471 (d) 0.0518

Solution 6.24

(a) 4.594 (b) 15.507 (c) 11.651 (d) 18.338 (e) 36.191

The last three results are summarized in Figure S6.4.

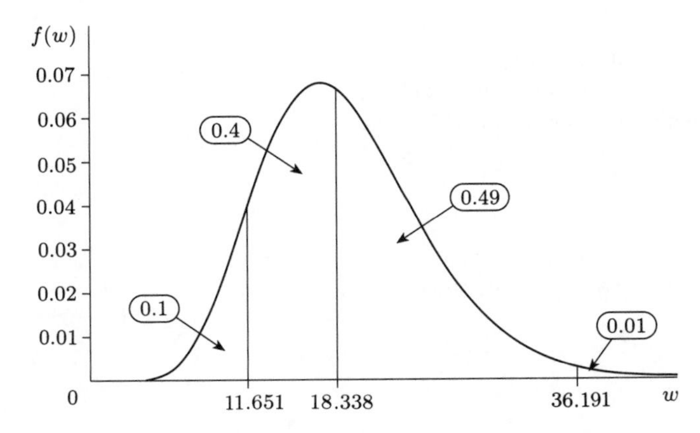

Figure S6.4

Solution 6.25

You should have obtained the following quantiles from the table.

(a) 13.091 (b) 4.168 (c) 11.340 (d) 24.769 (e) 3.841

Solution 6.26

Writing

$$W = \frac{(n-1)S^2}{\sigma^2} \sim \chi^2(n-1),$$

it follows that

$$E(W) = n-1, \quad V(W) = 2(n-1).$$

Also,

$$S^2 = \frac{\sigma^2}{n-1}W$$

so

$$E(S^2) = E\left(\frac{\sigma^2}{n-1}W\right) = \frac{\sigma^2}{n-1}E(W) = \frac{\sigma^2}{n-1}(n-1) = \sigma^2;$$

and

$$V(S^2) = V\left(\frac{\sigma^2}{n-1}W\right) = \frac{\sigma^4}{(n-1)^2}V(W) = \frac{\sigma^4}{(n-1)^2}2(n-1) = \frac{2\sigma^4}{n-1}.$$

Chapter 7

Solution 7.1

The lower confidence limit for μ is found by solving the equation

$$P(T \geq t) = e^{-t/\mu} = 0.05$$

or

$$\frac{t}{\mu} = -\log 0.05$$

or

$$\mu = \frac{t}{-\log 0.05} = \frac{157}{2.996} = 52.4 \text{ days}$$

(to 3 significant figures). The confidence statement may be expressed thus: 'A 90% confidence interval for the mean number of days between disasters, assuming an exponential model and based on the single observation 157 days, is from $\mu_- = 52.4$ days to $\mu_+ = 3060$ days.'

Solution 7.2

To obtain the upper confidence limit p_+ for p in this case, it is necessary to solve the equation

$$P(N \geq 13) = 0.025$$

or

$$(1-p)^{12} = 0.025.$$

This has solution

$$p_+ = 1 - 0.025^{1/12} = 1 - 0.735 = 0.265.$$

Consequently, a 95% confidence interval for p, assuming a geometric model and based on the single observation 13, is given by

$$(p_-, p_+) = (0.002, 0.265).$$

Notice that the confidence interval contains the maximum likelihood estimate $\widehat{p} = 0.077$. The approach has usefully provided a range of plausible values for p.

Solution 7.3

To obtain a 90% confidence interval for p, it is necessary to solve the two equations

$$P(N \leq 13) = 0.05 \quad \text{and} \quad P(N \geq 13) = 0.05.$$

The first may be written

$$1 - (1-p)^{13} = 0.05$$

and has solution

$$p_- = 1 - (1 - 0.05)^{1/13} = 1 - 0.996 = 0.004.$$

The second may be written

$$(1-p)^{12} = 0.05$$

and has solution

$$p_+ = 1 - 0.05^{1/12} = 1 - 0.779 = 0.221.$$

Thus a 90% confidence interval for p is given by

$$(p_-, p_+) = (0.004, 0.221).$$

This interval is narrower than the 95% confidence interval (0.002, 0.265) and in this sense is more useful; but less confidence may be attached to it. The only way to reduce the width of a confidence interval while maintaining a high confidence level is to increase the sample size.

Solution 7.4

The confidence level required is 99%, so $\frac{1}{2}\alpha = 0.005$. Writing

$$P(N \leq n) = 1 - \left(1 - \frac{1}{\mu}\right)^n = \tfrac{1}{2}\alpha,$$

we need to solve, in this case, the equation

$$1 - \left(1 - \frac{1}{\mu}\right)^4 = 0.005$$

for μ. From

$$\left(1 - \frac{1}{\mu}\right)^4 = 1 - 0.005 = 0.995$$

it follows that

$$\mu = \frac{1}{1 - 0.995^{1/4}} = 798.5,$$

and this is the upper confidence limit μ_+. Similarly, writing

$$P(N \geq n) = \left(1 - \frac{1}{\mu}\right)^{n-1} = \tfrac{1}{2}\alpha,$$

we need to solve, in this case, the equation

$$\left(1 - \frac{1}{\mu}\right)^3 = 0.005.$$

This has solution

$$\mu_- = \frac{1}{1 - 0.005^{1/3}} = 1.206.$$

So a 99% confidence interval for the average length of runs of diseased trees, based on the observation 4 and assuming a geometric model, is given by

$$(\mu_-, \mu_+) = (1.206, 798.5).$$

Notice the width of this confidence interval, and particularly the extent of the upper confidence limit! This is due to the inherent skewed nature of the geometric distribution, but also to the dearth of data. The only way to reduce the width of the confidence interval is to collect more data.

Solution 7.5

If T has a triangular density with parameter θ ($T \sim$ Triangular(θ)) then the c.d.f. of T is given by

$$F(t) = P(T \leq t) = 1 - \left(1 - \frac{t}{\theta}\right)^2, \quad 0 \leq t \leq \theta.$$

(a) Writing

$$P(T \leq t) = 1 - \left(1 - \frac{t}{\theta}\right)^2 = \tfrac{1}{2}\alpha,$$

it follows that

$$\left(1 - \frac{t}{\theta}\right)^2 = 1 - \tfrac{1}{2}\alpha$$

or

$$\frac{t}{\theta} = 1 - \sqrt{1 - \tfrac{1}{2}\alpha};$$

so, finally,

$$\theta = \frac{t}{1 - \sqrt{1 - \frac{1}{2}\alpha}}.$$

This is the upper confidence limit θ_+: it is high values of θ that render low values of t unlikely.

The lower confidence limit θ_- is found by solving the equation

$$P(T \geq t) = \left(1 - \frac{t}{\theta}\right)^2 = \tfrac{1}{2}\alpha$$

for θ; writing

$$\frac{t}{\theta} = 1 - \sqrt{\tfrac{1}{2}\alpha}$$

it follows that

$$\theta_- = \frac{t}{1 - \sqrt{\frac{1}{2}\alpha}}.$$

So a $100(1-\alpha)\%$ confidence interval for the triangular parameter θ, based on a single observation t, is given by

$$(\theta_-, \theta_+) = \left(\frac{t}{1-\sqrt{\frac{1}{2}\alpha}}, \frac{t}{1-\sqrt{1-\frac{1}{2}\alpha}} \right).$$

(b) For instance, if $t = 5$, then a 95% confidence interval ($\frac{1}{2}\alpha = 0.025$) is given by

$$(\theta_-, \theta_+) = \left(\frac{5}{1-\sqrt{0.025}}, \frac{5}{1-\sqrt{0.975}} \right) = (5.94, 397).$$

Again, the confidence interval is extremely wide. But it makes sense: the parameter θ specifies the right-hand edge of the range of T. If the value $t = 5$ has been observed, the value of θ must be at least 5.

Solution 7.6

First, a model is required. Assuming the unknown number of *Firefly* dinghies manufactured to date to be equal to θ then, in the absence of any information to the contrary, we could assume that any one of the dinghies is as likely to have been observed as any other. That is, denoting by X the sail number observed, the random variable X has a discrete uniform distribution

$$P(X = x) = \frac{1}{\theta}, \quad x = 1, 2, 3, \ldots, \theta.$$

Then X has c.d.f.

$$P(X \leq x) = \frac{x}{\theta}, \quad x = 1, 2, 3, \ldots, \theta.$$

The confidence level required is 90%: so $\frac{1}{2}\alpha = 0.05$. Writing

$$P(X \leq 3433) = \frac{3433}{\theta} = 0.05,$$

we obtain the upper confidence limit $\theta_+ = 3433/0.05 = 68\,660$.

Now, the probability $P(X \geq 3433)$ is given by

$$P(X \geq 3433) = 1 - P(X \leq 3432) = 1 - \frac{3432}{\theta},$$

and so the lower confidence limit θ_- for θ is given by the solution of the equation

$$P(X \geq 3433) = 1 - \frac{3432}{\theta} = 0.05;$$

this solution is

$$\theta_- = \frac{3432}{0.95} = 3612.6.$$

The unknown number θ is indubitably an integer. Erring a little on the safe side, we can conclude from the one sighting made (3433) that a 90% confidence interval for the number θ of *Firefly* dinghies manufactured to date is given by

$$(\theta_-, \theta_+) = (3612, 68\,660).$$

Again, the interval is so wide as to be of questionable use. We shall see in Subsection 7.2.5 the very useful consequences of taking a larger sample.

Solution 7.7

(a) (i) A 90% confidence interval for p, based on observing 4 successes in 11 trials, is given by

$$(p_-, p_+) = (0.1351, 0.6502).$$

(ii) The corresponding 95% confidence interval is

$$(p_-, p_+) = (0.1093, 0.6921).$$

(b) Confidence intervals based on observing 8 successes in 22 trials are

$$90\%: \quad (p_-, p_+) = (0.1956, 0.5609);$$
$$95\%: \quad (p_-, p_+) = (0.1720, 0.5934).$$

In both cases, the larger sample size has led to narrower confidence intervals. The reason is the increase in information. Just as larger samples lead to reduced variation in parameter estimates, so they permit narrower (more precise) confidence intervals.

(c) A 99% confidence interval for p based on observing 4 successes in 5 trials is given by

$$(p_-, p_+) = (0.1851, 0.9990).$$

(d) In one experiment (your results might have been similar) the sequence of 10 observations on $B(20, 0.3)$ is

1 9 8 7 3 9 6 4 8 5.

The corresponding confidence limits for p are

$$1: \quad (p_-, p_+) = (0.0026, 0.2161);$$
$$9: \quad (p_-, p_+) = (0.2587, 0.6531);$$
$$8: \quad (p_-, p_+) = (0.2171, 0.6064);$$
$$7: \quad (p_-, p_+) = (0.1773, 0.5580);$$
$$3: \quad (p_-, p_+) = (0.0422, 0.3437);$$
$$9: \quad (p_-, p_+) = (0.2587, 0.6531);$$
$$6: \quad (p_-, p_+) = (0.1396, 0.5078);$$
$$4: \quad (p_-, p_+) = (0.0714, 0.4010);$$
$$8: \quad (p_-, p_+) = (0.2171, 0.6064);$$
$$5: \quad (p_-, p_+) = (0.1041, 0.4556).$$

Of these ten intervals, only the first one does not contain the known value of p, 0.3. Remember the interpretation of a confidence interval—in *repeated experiments*, a proportion $100(1 - \alpha)\%$ of confidence intervals obtained may be expected to contain the (usually) unknown value of the parameter.

Here, the confidence level set was 90%; and, as it happened, exactly nine out of the ten calculated intervals contained the known parameter value $p = 0.3$—just as expected. What happened in your experiment?

The observation '1 success in 20 trials' is so low, that it reduces our confidence that the underlying success probability is, or could be, as high as 0.3.

(Incidentally, an observed success count as high as 10 in 20 would have resulted in the confidence interval

$$(p_-, p_+) = (0.3020, 0.6980),$$

which does not contain the value $p = 0.3$, either.)

For discrete random variables, the calculations underlying the construction of confidence intervals conceal an interesting feature, exemplified here. Only the confidence intervals for $x = 3, 4, \ldots, 9$ contain the value 0.3: other values of x are 'too low' or 'too high'. If X is binomial $B(20, 0.3)$, then $P(3 \leq X \leq 9) = 0.9166 > 0.90$. That is, an average of about 92% (more than 90%) of confidence intervals generated in this way will contain the (usually unknown) parameter value. The procedure is 'conservative'.

Solution 7.8

(a) (i) A 90% confidence interval for a Poisson mean μ, based on the single observation 3, is given by

$$(\mu_-, \mu_+) = (0.8177, 7.754).$$

(ii) The corresponding 95% confidence interval is

$$(\mu_-, \mu_+) = (0.6187, 8.767).$$

(b) (i) An estimate of the mean underlying accident rate μ is given by the sample mean

$$\widehat{\mu} = \overline{x} = \frac{4+4+3+0+5+3+2}{7} = \frac{21}{7} = 3.0.$$

So the estimate of μ is the same as it was in part (a), but this time it is based on seven observations rather than on one.

(ii) Confidence intervals for μ based on these data are given by

$$90\%: \quad (\mu_-, \mu_+) = (2.010, 4.320);$$
$$95\%: \quad (\mu_-, \mu_+) = (1.857, 4.586).$$

Notice that the increased information has resulted in narrower confidence intervals.

(c) (i) The estimated mean annual accident rate for girls of this age is

$$\widehat{\mu} = \frac{20}{6} = 3.333.$$

(ii) In this case we are only told the sample total $t = 20$; but we know that the random variable T, on which t is a single observation, is Poisson(6μ). All that is required is that we obtain confidence limits for the mean of T, and then divide these limits by 6. This approach gives

$$90\%: \quad (\mu_-, \mu_+) = (2.209, 4.844);$$
$$95\%: \quad (\mu_-, \mu_+) = (2.036, 5.148).$$

Solution 7.9

(a) The mean of the 62 time intervals is 437.21 days, about 14 months. Assuming that times between earthquakes are exponentially distributed, a 90% confidence interval for the mean time interval between serious earthquakes world-wide is given by

$$(\mu_-, \mu_+) = (359.06, 546.06),$$

or from about twelve months to eighteen months.

(b) (i) The mean waiting time between vehicles is

$$\widehat{\mu} = \frac{212}{50} = 4.24 \text{ seconds},$$

so the estimated traffic rate is

$$\widehat{\lambda} = \frac{1}{\mu} = 0.2358 \text{ vehicles per second} = 14.15 \text{ vehicles per minute}.$$

(ii) A 90% confidence interval for the mean traffic rate is given by

$$(\lambda_-, \lambda_+) = (11.03, 17.60).$$

(c) When this experiment was tried the ten resulting confidence intervals were as follows. (The sample mean is also shown in brackets.)

0.7821 (1.090) 1.645
0.9361 (1.305) 1.969
0.7864 (1.096) 1.654
0.7149 (0.997) 1.504
0.6848 (0.955) 1.440
0.8603 (1.199) 1.810
0.7142 (0.996) 1.502
0.8163 (1.138) 1.717
0.7145 (0.996) 1.503
0.6423 (0.895) 1.351

Interestingly, an eleventh experiment gave

0.4114 (0.5735) 0.8654

which does not contain the number 1.0, but the sample had extraordinarily low numbers in it. They were

0.11 0.75 0.80 0.07 0.09 1.54 0.54 0.34 0.15 1.67
0.01 0.60 0.36 0.66 0.72 0.44 0.57 0.06 1.86 0.13.

Note that only two of these numbers exceed the mean, 1.

A twelfth experiment gave

1.152 (1.606) 2.423

which also does not contain the number 1; in this case the numbers sampled from $M(1)$ were unusually high.

0.91 0.13 3.71 1.23 0.56 2.45 0.03 2.51 0.42 2.09
0.56 1.09 1.05 3.13 4.29 1.96 2.10 1.59 0.35 1.95

Solution 7.10

(a) One simulation gave the following ten observations on N.

20 6 3 7 4 2 8 10 3 15

The corresponding 90% confidence interval for p based on these data is

$$(p_-, p_+) = (0.0712, 0.1951).$$

The width of the confidence interval is $0.1951 - 0.0712 = 0.124$. It contains both the values $p = 1/10$ and $p = 1/6$.

(b) For interest, this part of the exercise was performed three times. The corresponding confidence intervals were

$$(0.0890, 0.1217), \quad \text{width} = 0.033;$$
$$(0.0825, 0.1130), \quad \text{width} = 0.030;$$
$$(0.0940, 0.1284), \quad \text{width} = 0.034.$$

In all three cases the confidence interval contained the value $p = 1/10$ and *not* the value $p = 1/6$, providing quite strong evidence that the die is loaded (as we know; but you can imagine that there are circumstances where one might not know this, but merely suspect it).

(The experiment was repeated on the computer a further 997 times, making a total of 1000 experiments altogether. The number 1/10 was included in the resulting confidence intervals 892 times, roughly as expected—90% of 1000 is 900. The number 1/6 was not included in any of the intervals.)

Solution 7.11

Either from tables or your computer, you should find (a) $t = 1.699$; (b) $t = 1.697$; (c) $t = -3.365$; (d) $t = 2.262$.

Solution 7.12

For the five data points the sample mean and sample standard deviation are $\overline{x} = 3.118$ and $s = 0.155$. For a confidence level of 95%, the corresponding critical t-value is obtained from the t-distribution $t(4)$: it is the 97.5% point of $t(4)$, $q_{0.975} = 2.776$. This is shown in Figure S7.1. A 95% confidence interval for μ, the mean coal consumption in pounds per draw-bar horse-power hour, is given by

$$\begin{aligned}(\mu_-, \mu_+) &= \left(\overline{x} - \frac{ts}{\sqrt{n}}, \overline{x} + \frac{ts}{\sqrt{n}}\right)\\ &= \left(3.118 - \frac{2.776 \times 0.155}{\sqrt{5}}, 3.118 + \frac{2.776 \times 0.155}{\sqrt{5}}\right)\\ &= (3.118 - 0.192, 3.118 + 0.192) = (2.93, 3.31).\end{aligned}$$

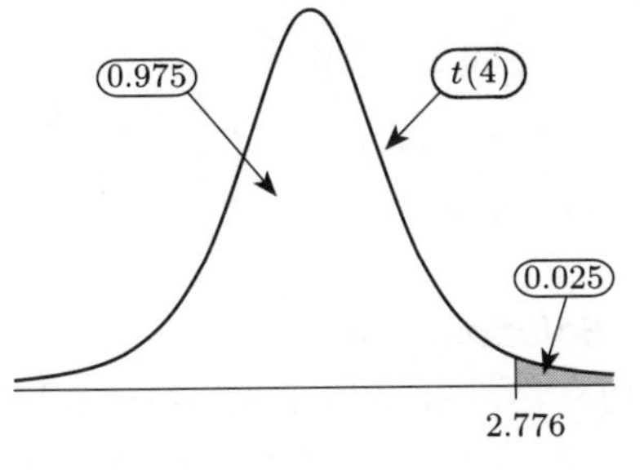

Figure S7.1 Critical values from $t(4)$

Solution 7.13

(a) For the data

1.2 2.4 1.3 1.3 0 1.0 1.8 0.8 4.6 1.4

the sample mean is $\overline{x} = 1.58$ and the sample standard deviation is $s = 1.23$. A 95% confidence interval is required for the mean difference μ between the two treatments. Reference to tables of $t(9)$ gives the critical value $t = 2.262$. Thus the corresponding confidence interval is

$$\begin{aligned}(\mu_-, \mu_+) &= \left(\overline{x} - \frac{ts}{\sqrt{n}}, \overline{x} + \frac{ts}{\sqrt{n}}\right)\\ &= \left(1.58 - \frac{2.262 \times 1.23}{\sqrt{10}}, 1.58 + \frac{2.262 \times 1.23}{\sqrt{10}}\right)\\ &= (1.58 - 0.88, 1.58 + 0.88) = (0.70, 2.46).\end{aligned}$$

(b) Since the number 0 (indicating no difference between the two treatments) is not included in this confidence interval, there is considerable evidence that the treatment L-hyoscyamine hydrobromide is a more effective hypnotic than the alternative.

Solution 7.14

This is a 'self-help' exercise included to encourage you to use your computer.

Solution 7.15

In this case the sample size was $n = 8$; the sample variance was $s^2 = 37.75$.

(a) For a 90% confidence interval we will require the 5% and the 95% points of $\chi^2(7)$. These are $c_L = 2.167$ and $c_U = 14.067$ respectively. Consequently, a 90% confidence interval for the population variance σ^2, based on the data provided, is

$$(\sigma_-^2, \sigma_+^2) = \left(\frac{7 \times 37.75}{14.067}, \frac{7 \times 37.75}{2.167}\right) = (18.79, 121.9).$$

(b) For a 95% confidence interval we need

$$c_L = q_{0.025} = 1.690, \quad c_U = q_{0.975} = 16.013.$$

The interval is

$$(\sigma_-^2, \sigma_+^2) = \left(\frac{7 \times 37.75}{16.013}, \frac{7 \times 37.75}{1.69}\right) = (16.50, 156.4).$$

Solution 7.16

(a) An estimate for σ is given by the sample standard deviation

$$s = \sqrt{37.75} = 6.1.$$

(b) We found in Solution 7.15(a) that a 90% confidence interval for the population variance σ^2 is given by (18.79, 121.9). Taking square roots, the corresponding 90% confidence interval for the population standard deviation σ is given by

$$(\sigma_-, \sigma_+) = (4.3, 11.0).$$

Solution 7.17

For the earthquake data, the mean time between quakes is given by $\widehat{\mu} = \overline{x} = 437.21$ days. There were 62 waiting times recorded; to establish a 95% confidence interval for the mean time between serious earthquakes world-wide, it is necessary to use the fact that the 97.5% point of the standard normal distribution is $z = 1.96$. The confidence interval required is

$$\begin{aligned}(\mu_-, \mu_+) &= \left(\frac{\widehat{\mu}}{1 + z/\sqrt{n}}, \frac{\widehat{\mu}}{1 - z/\sqrt{n}}\right) \\ &= \left(\frac{437.21}{1 + 1.96/\sqrt{62}}, \frac{437.21}{1 - 1.96/\sqrt{62}}\right) \\ &= (350.1, 582.1).\end{aligned}$$

This is an interval of between about twelve and nineteen months. Here, it is assumed that the sum of 62 exponential waiting times is approximately normally distributed.

Solution 7.18

(a) For the Kwinana traffic data the mean waiting time between vehicles was $\widehat{\mu} = 4.24$ seconds (corresponding to a traffic rate of $\widehat{\lambda} = 14.15$ vehicles per minute); the sample size was $n = 50$. Assuming the sample total to be approximately normally distributed, a 90% confidence interval for the mean waiting time (taking $z = 1.645$) is given by

$$\begin{aligned}(\mu_-, \mu_+) &= \left(\frac{\widehat{\mu}}{1+z/\sqrt{n}}, \frac{\widehat{\mu}}{1-z/\sqrt{n}}\right)\\ &= \left(\frac{4.24}{1+1.645/\sqrt{50}}, \frac{4.24}{1-1.645/\sqrt{50}}\right)\\ &= (3.44, 5.53).\end{aligned}$$

By taking reciprocals, confidence limits for the mean traffic rate are $\lambda_- = 1/\mu_+ = 0.18$ vehicles per second, or 10.9 vehicles per minute; and $\lambda_+ = 1/\mu_- = 0.29$ vehicles per second, or 17.4 vehicles per minute.

(b) The confidence interval in the solution to Exercise 7.9 was

$$(\lambda_-, \lambda_+) = (11.03, 17.60).$$

The differences, which are very slight, are due to the approximation induced by assuming a normal distribution for the sample total.

Solution 7.19

The stated constraints imply that the following inequalities must hold:

$$\frac{\widehat{\mu}}{1+z/\sqrt{n}} \geq 0.97\widehat{\mu} \quad \text{and} \quad \frac{\widehat{\mu}}{1-z/\sqrt{n}} \leq 1.03\widehat{\mu}$$

where $z = 1.96$. The first inequality gives

$$\frac{1}{1+z/\sqrt{n}} \geq 0.97,$$

so

$$1 + \frac{z}{\sqrt{n}} \leq \frac{1}{0.97}.$$

This gives

$$\frac{z}{\sqrt{n}} \leq \frac{0.03}{0.97},$$

so

$$\sqrt{n} \geq \frac{97z}{3} = \frac{97 \times 1.96}{3};$$

that is, $n \geq 4016.2$. The second inequality gives

$$\frac{1}{1-z/\sqrt{n}} \leq 1.03,$$

so

$$\frac{z}{\sqrt{n}} \leq \frac{0.03}{1.03}.$$

This gives

$$\sqrt{n} \geq \frac{103z}{3} = \frac{103 \times 1.96}{3};$$

that is, $n \geq 4528.4$. For both inequalities to hold, the sample size must be 4529 or more. This is an extremely large sample!

Solution 7.20

The sample mean is

$$\frac{0 \times 101 + 1 \times 143 + 2 \times 120 + \cdots + 10 \times 2}{101 + 143 + 120 + \cdots + 2} = \frac{1522}{621} = 2.451.$$

An approximate 95% confidence interval for the underlying mean accident rate over the whole of the eight-year period (assuming a Poisson model) is

$$\begin{aligned}(\mu_-, \mu_+) &= \left(\widehat{\mu} - z\sqrt{\frac{\widehat{\mu}}{n}}, \widehat{\mu} + z\sqrt{\frac{\widehat{\mu}}{n}}\right) \\ &= \left(2.451 - 1.96\sqrt{\frac{2.451}{621}}, 2.451 + 1.96\sqrt{\frac{2.451}{621}}\right) \\ &= (2.451 - 0.123, 2.451 + 0.123) \\ &= (2.33, 2.57).\end{aligned}$$

Solution 7.21

There were 109 runs observed. The mean length of a run was

$$\widehat{\mu} = 166/109 = 1.523.$$

For a 99% confidence interval the 99.5% point of Z is required. This is $z = 2.576$. The resulting approximate 99% confidence interval for the mean length of runs of diseased trees is

$$\begin{aligned}(\mu_-, \mu_+) &= \left(\widehat{\mu} - z\sqrt{\frac{\widehat{\mu}(\widehat{\mu}-1)}{n}}, \widehat{\mu} + z\sqrt{\frac{\widehat{\mu}(\widehat{\mu}-1)}{n}}\right) \\ &= \left(1.523 - 2.58\sqrt{\frac{1.523 \times 0.523}{109}}, 1.523 + 2.58\sqrt{\frac{1.523 \times 0.523}{109}}\right) \\ &= (1.523 - 0.221, 1.523 + 0.221) \\ &= (1.30, 1.74).\end{aligned}$$

Here, it is assumed that the sum of 109 geometric run lengths is approximately normally distributed.

Solution 7.22

(a) In this case, the estimate of p is $\widehat{p} = 286/5387 = 0.0531$. An approximate 90% confidence interval for the underlying proportion of red-haired children in Scotland is

$$\begin{aligned}(p_-, p_+) &= \left(\widehat{p} - z\sqrt{\frac{\widehat{p}(1-\widehat{p})}{n}}, \widehat{p} + z\sqrt{\frac{\widehat{p}(1-\widehat{p})}{n}}\right) \\ &= \left(0.0531 - 1.645\sqrt{\frac{0.0531 \times 0.9469}{5387}}, 0.0531 + 1.645\sqrt{\frac{0.0531 \times 0.9469}{5387}}\right) \\ &= (0.0531 - 0.0050, 0.0531 + 0.0050) \\ &= (0.048, 0.058).\end{aligned}$$

Here (as well as the usual assumption of normality) it has been assumed that the school children of Caithness are typical of all those in Scotland, which may not necessarily be the case.

(b) Here, the estimated proportion of fair-haired children who are blue-eyed is $\widehat{p} = 1368/5789 = 0.236\,31$. An approximate 95% confidence interval for this proportion is

$$
\begin{aligned}
(p_-, p_+) &= \left(\widehat{p} - z\sqrt{\frac{\widehat{p}(1-\widehat{p})}{n}}, \widehat{p} + z\sqrt{\frac{\widehat{p}(1-\widehat{p})}{n}}\right) \\
&= \left(0.236\,31 - 1.96\sqrt{\frac{0.236\,31 \times 0.763\,69}{5789}}, 0.236\,31 - 1.96\sqrt{\frac{0.236\,31 \times 0.763\,69}{5789}}\right) \\
&= (0.236\,31 - 0.010\,94, 0.236\,31 + 0.010\,94) \\
&= (0.225, 0.247).
\end{aligned}
$$

Again, it has been assumed here that Aberdeen school children are representative of all school children in Scotland.

Solution 7.23

For these data the sample mean is $\overline{x} = 1.992$ and the sample standard deviation is $s = 1.394$. A 90% confidence interval for the average number of books borrowed in a year is given by

$$
\begin{aligned}
(\mu_-, \mu_+) &= \left(\overline{x} - z\frac{s}{\sqrt{n}}, \overline{x} + z\frac{s}{\sqrt{n}}\right) \\
&= \left(1.992 - 1.645 \times \frac{1.394}{\sqrt{122}}, 1.992 + 1.645 \times \frac{1.394}{\sqrt{122}}\right) \\
&= (1.992 - 0.208, 1.992 + 0.208) \\
&= (1.78, 2.20).
\end{aligned}
$$

Solution 7.24

For these data the sample mean is $\overline{x} = 0.3163$ and the sample standard deviation is $s = 0.0805$. Making no assumptions at all about the distribution of eggshell thicknesses for Anacapa pelicans, a 95% confidence interval for the mean thickness is given by

$$
\begin{aligned}
(\mu_-, \mu_+) &= \left(\overline{x} - z\frac{s}{\sqrt{n}}, \overline{x} + z\frac{s}{\sqrt{n}}\right) \\
&= \left(0.3163 - 1.96 \times \frac{0.0805}{\sqrt{65}}, 0.3163 + 1.96 \times \frac{0.0805}{\sqrt{65}}\right) \\
&= (0.3163 - 0.020, 0.3163 + 0.020) \\
&= (0.30, 0.34).
\end{aligned}
$$

Chapter 8

Solution 8.1

For this test the null hypothesis is

$$H_0 : p = \tfrac{2}{3};$$

the alternative hypothesis is

$$H_1 : p \neq \tfrac{2}{3}.$$

An appropriate model for the number of 1s in a sequence of 25 trials is binomial $B(25, p)$. In this experiment the number of 1s observed is 10.

(a) A 95% confidence interval for p, based on observing 10 successes in 25 trials is given by

$$(p_-, p_+) = (0.2113, 0.6133).$$

(b) The hypothesized value $p_0 = \frac{2}{3} = 0.6667$ is not contained in this confidence interval. The conclusions of the test may be stated as follows.

On the basis of these data, there is evidence at the significance level 0.05 to reject the null hypothesis $p = \frac{2}{3}$ in favour of the alternative hypothesis that p differs from $\frac{2}{3}$. (In fact, there is evidence from the sample that $p < \frac{2}{3}$.)

(c) A 99% confidence interval for p is given by

$$(p_-, p_+) = (0.1679, 0.6702).$$

The interval contains the hypothesized value $p_0 = \frac{2}{3}$: at the 1% level of significance there is no evidence, from these data, to reject the hypothesis.

Solution 8.2

The Kwinana Freeway data consist of 50 observations on times assumed to be exponentially distributed. The mean waiting time μ is unknown. The hypothesis that the mean traffic flow rate is 10 vehicles per minute is equivalent to hypothesizing a mean waiting time of $\frac{1}{10}$ minute, or 6 seconds:

$$H_0 : \mu = 6.$$

An appropriate alternative hypothesis is furnished by

$$H_1 : \mu \neq 6.$$

Confidence intervals at levels 90%, 95% and 99% for μ, based on the data, are

$$90\% : \quad (3.41, 5.44);$$
$$95\% : \quad (3.27, 5.71);$$
$$99\% : \quad (3.02, 6.30).$$

Only the last of these contains the hypothesized value $\mu_0 = 6$. One may conclude the test as follows. Based on these data, the hypothesis that the mean traffic flow rate is 10 vehicles per minute is rejected at the 5% level of significance; at the 1% level the evidence is insufficient to reject the hypothesis.

There are two additional points to notice here. The first is that no mention is made of the conclusion of the test at the 10% significance level: this is because rejection at 5% implies rejection at 10%. Second, at some significance level between 1% and 5% it is clear that the hypothesized value μ_0 will itself be at the very boundary of the decision rule. This idea will be explored in Section 8.3.

Solution 8.3

Large-sample confidence intervals for the Bernoulli parameter p, based on approximate normal distribution theory, are of the form

$$(p_-, p_+) = \left(\widehat{p} - z\sqrt{\frac{\widehat{p}(1-\widehat{p})}{n}}, \widehat{p} + z\sqrt{\frac{\widehat{p}(1-\widehat{p})}{n}}\right)$$

(a) In this case $n = 1064$ and $\widehat{p} = 787/1064 = 0.740$. A specified 10% significance level for the test implies a 90% confidence level; this in turn implies $z = q_{0.95} = 1.645$, the 95% quantile of the standard normal distribution. The confidence interval required is given by

$$(p_-, p_+) = \left(0.740 - 1.645\sqrt{\frac{0.740 \times 0.260}{1064}}, 0.740 + 1.645\sqrt{\frac{0.740 \times 0.260}{1064}}\right)$$
$$= (0.740 - 0.022, 0.740 + 0.022) = (0.718, 0.762).$$

The confidence interval contains the hypothesized value $p_0 = 0.75$. Thus, at the 10% level of significance, these data offer no evidence to reject the hypothesis that the proportion of yellow peas is equal to $\frac{3}{4}$.

(b) Here, $n = 100$ and $\widehat{p} = \frac{60}{100} = 0.6$. A 95% confidence interval for p is given by

$$(p_-, p_+) = \left(0.6 - 1.96\sqrt{\frac{0.6 \times 0.4}{100}}, 0.6 + 1.96\sqrt{\frac{0.6 \times 0.4}{100}}\right)$$
$$= (0.6 - 0.096, 0.6 + 0.096) = (0.504, 0.696).$$

The hypothesized value $p_0 = \frac{2}{3} = 0.667$ is contained in the interval: on the basis of these data, there is no evidence at this level to reject H_0.

Solution 8.4

The observed mean ratio is

$$\overline{r} = \frac{0.693 + 0.662 + \cdots + 0.933}{20} = 0.6605.$$

The sample standard deviation is $s = 0.0925$. Consequently, the observed value t of the test statistic T under the null hypothesis is

$$t = \frac{\overline{r} - \mu}{s/\sqrt{n}} = \frac{0.6605 - 0.618}{0.0925/\sqrt{20}} = 2.055.$$

To work out the rejection region for the test, we need the 2.5% and 97.5% quantiles for $t(19)$. These are

$$q_{0.025} = -2.093, \qquad q_{0.975} = 2.093.$$

The rejection region is shown in Figure S8.1, together with the observed value t of T.

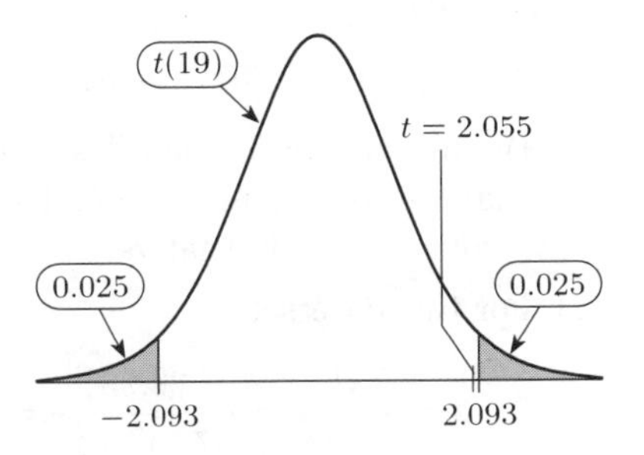

Figure S8.1

As you can see, the observed value $t = 2.055$ is very close to the boundary of the rejection region (suggesting Shoshoni rectangles are somewhat 'square'); but strictly according to the predetermined significance level, there is insufficient evidence, on the basis of these data, to reject the null hypothesis that $\mu = 0.618$.

Solution 8.5

The null hypothesis here is

$$H_0 : \mu = 0$$

and, as for L-hyoscyamine hydrobromide, the alternative hypothesis is

$$H_1 : \mu > 0.$$

Assuming a normal model for the variation in sleep gain after a dose of D-hyoscyamine hydrobromide, then under the null hypothesis

$$T = \frac{\overline{D}}{S/\sqrt{n}} \sim t(n-1).$$

To determine the rejection region at level 0.05, the critical quantile is $q_{0.95} = 1.833$. The observed value t of the test statistic T is

$$t = \frac{\overline{d}}{s/\sqrt{n}} = \frac{0.75}{1.79/\sqrt{10}} = 1.33$$

which is less than $q_{0.95} = 1.833$, and therefore falls outside the rejection region. On the basis of these data, and at this level, there is no reason to suspect that the hypnotic D-hyoscyamine hydrobromide has any measurable effect in prolonging sleep.

Solution 8.6

(a) The null and alternative hypotheses are

$$H_0 : \mu = 0, \qquad H_1 : \mu \neq 0,$$

where μ is the mean difference between the heights of a pair of cross-fertilized and self-fertilized plants whose parents were grown from the same seed. An appropriate test statistic is

$$T = \frac{\overline{D}}{S/\sqrt{n}}$$

with null distribution $t(n-1)$.

You will notice here the very precise statement of the meaning of the parameter μ. Sometimes it is important to be pedantic in this way.

(b) The test is two-sided at 10%: the rejection region is defined by the boundary points

$$q_{0.05} = -1.761; \qquad q_{0.95} = 1.761,$$

the 5% and 95% quantiles of $t(14)$. If the observed value t of T is less than -1.761 or more than 1.761, the null hypothesis will be rejected in favour of the alternative.

(c) For this data set

$$t = \frac{\overline{d}}{s/\sqrt{n}} = \frac{20.93}{37.74/\sqrt{15}} = 2.15.$$

This exceeds $q_{0.95} = 1.761$, so the hypothesis of zero difference is rejected in favour of the alternative hypothesis that there is a difference in the mean height of cross-fertilized and self-fertilized plants. (In fact, on the basis of the data, it appears that cross-fertilized plants are taller than self-fertilized plants.)

Solution 8.7

(a) Assume in this case that the test statistic is $X \sim \text{Poisson}(\mu)$. Under the null hypothesis, the distribution of X is Poisson(3.0). We require to find values x_1 and x_2 such that, as closely as can be attained,

$$P(X \leq x_1) \simeq 0.05, \quad P(X \geq x_2) \simeq 0.05.$$

This is shown in Figure S8.2.

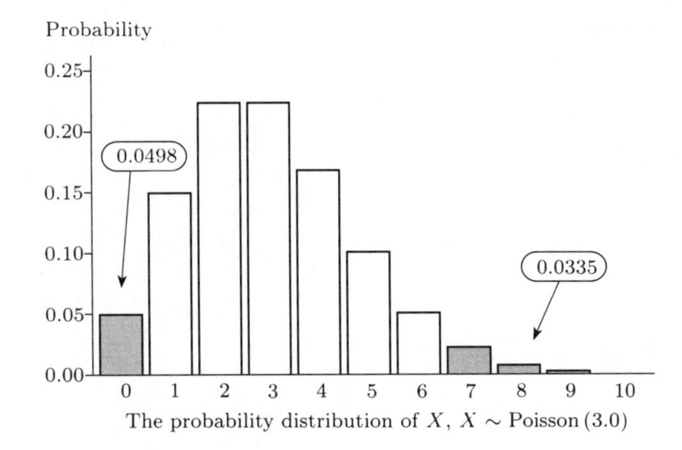

Figure S8.2 Identifying the rejection region

In this case

$$P(X \leq 0) = 0.0498 \text{ and } P(X \geq 7) = 0.0335$$

so the rejection region for the test, based on the single observation x, is

$$x = 0 \text{ or } x \geq 7.$$

Otherwise the hypothesis $H_0 : \mu = 3$ is not rejected. The actual level of this test is $0.0498 + 0.0335 = 0.0833$.

(b) A sensible test statistic is the sample total $V = X_1 + X_2 + \cdots + X_5$, which under the null hypothesis has a Poisson distribution with mean 15. In this case

$$P(V \leq 8) = 0.0374 \text{ and } P(V \geq 22) = 0.0531$$

and the null hypothesis will not be rejected if the sample total is between 9 and 21 inclusive. The level of the test is $0.0374 + 0.0531 = 0.0905$.

(c) For a sample of size 10 the sample total $W = X_1 + X_2 + \cdots + X_{10}$ has the null distribution Poisson(30). Useful quantiles are given by

$$P(W \leq 21) = 0.0544 \text{ and } P(W \geq 40) = 0.0463,$$

and so the rejection region for the test based on the sample total W is

$$w \leq 21 \text{ or } w \geq 40.$$

The level of the test is $0.0544 + 0.0463 = 0.1007$.

Solution 8.8

(a) The forecast relative frequency of light-greys is $\frac{1}{8}$: if this is true, the distribution of the number of light-greys in samples of size 18 is binomial $N \sim B(18, \frac{1}{8})$.

(b) The probability of observing four light-greys is

$$p_N(4) = 0.1152.$$

The diagram shows all the values in the range of the null distribution $B(18, \frac{1}{8})$ that are as extreme as that observed.

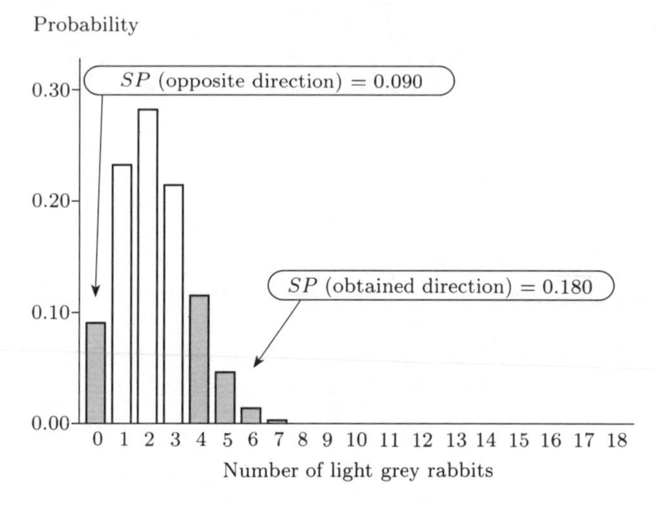

Figure S8.3 Counts as extreme as 4 from $B(18, \frac{1}{8})$

The SP may be calculated as follows.

$$SP(\text{obtained direction}) = P(N \geq 4) = 0.180$$
$$SP(\text{opposite direction}) = P(N = 0) = 0.090$$
$$SP(\text{total}) = 0.270$$

This completes the assessment of the null hypothesis in the light of the sample. There is no substantial evidence that the null hypothesis is flawed: were it true, more than a quarter of future samples would, in fact, offer less support for it than did the sample collected.

Solution 8.9

If the number of insects caught in a trap is denoted by $N \sim \text{Poisson}(\mu)$, then the total number caught in 33 traps is

$$T = N_1 + N_2 + \cdots + N_{33} \sim \text{Poisson}(33\mu).$$

Under the null hypothesis $H_0 : \mu = 1$, the distribution of T is Poisson(33).

In fact, the total number of insects counted was

$$\begin{aligned} t &= 0 \times 10 + 1 \times 9 + 2 \times 5 + 3 \times 5 + 4 \times 1 + 5 \times 2 + 6 \times 1 \\ &= 0 + 9 + 10 + 15 + 4 + 10 + 6 = 54. \end{aligned}$$

Counts more extreme than $t = 54$ are shown in Figure S8.4.

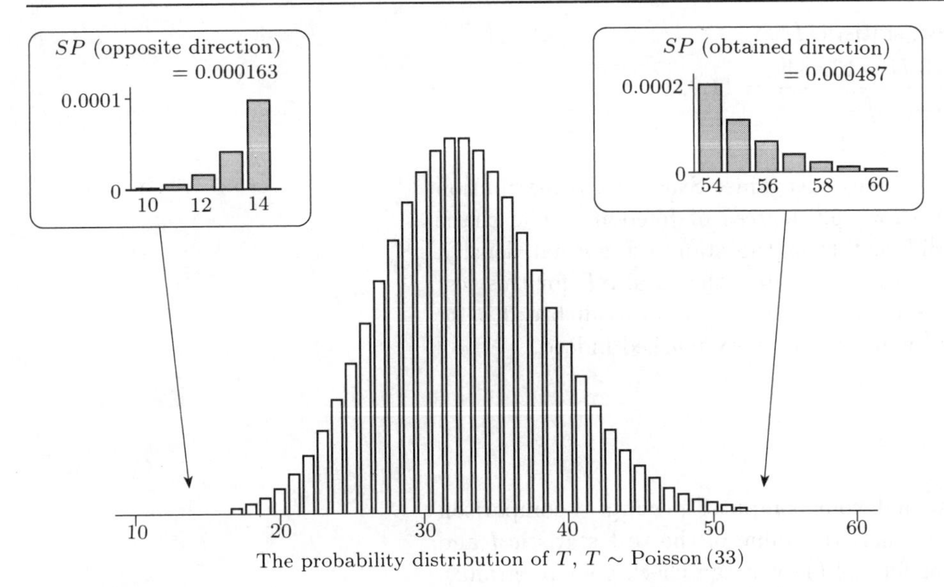

Figure S8.4 Counts more extreme than $t = 54$ when $T \sim \text{Poisson}(33)$

The significance probabilities for the test are given by

$$SP(\text{obtained direction}) = P(T \geq 54) = 0.000\,487$$
$$SP(\text{opposite direction}) = P(T \leq 14) = 0.000\,163$$
$$SP(\text{total}) = 0.000\,487 + 0.000\,163 = 0.000\,650.$$

These significance probabilities are extremely small, offering strong evidence that the hypothesis $\mu = 1$ is false. The value obtained suggests that in fact μ is rather greater than 1.

The observed sample mean catch is $54/33 = 1.64$, which does not at first glance appear to be so very different from the hypothesized value. Evidently the difference is very considerable.

Solution 8.10

(a) Neither the boxplots nor the histograms suggest that either sample is skewed, or that the variances are substantially different. The histograms suggest that a normal model would be an extremely useful representation of the variability in the measurements.

(b) For the Etruscan skulls,

$$s_1^2 = 35.65;$$

for the modern Italian skulls,

$$s_2^2 = 33.06.$$

These are remarkably close: there is no question of having to forgo the test on the grounds that the test assumptions are not satisfied.

(c) Further summary statistics are

$$n_1 = 84, \qquad \overline{x}_1 = 143.77; \qquad n_2 = 70, \qquad \overline{x}_2 = 132.44.$$

The pooled estimate of the variance σ^2 is given by

$$s_p^2 = \frac{(n_1 - 1)s_1^2 + (n_2 - 1)s_2^2}{n_1 + n_2 - 2} = \frac{83 \times 35.65 + 69 \times 33.06}{84 + 70 - 2} = 34.47.$$

The observed value of the test statistic T is

$$t = \frac{\overline{x}_1 - \overline{x}_2}{s_p\sqrt{\dfrac{1}{n_1} + \dfrac{1}{n_2}}} = \frac{143.77 - 132.44}{\sqrt{34.47}\sqrt{\dfrac{1}{84} + \dfrac{1}{70}}} = 11.92.$$

The test statistic needs to be assessed against Student's t-distribution with $n_1 + n_2 - 2 = 84 + 70 - 2 = 152$ degrees of freedom. The shape of $t(152)$ is not markedly different from the standard normal density. Without recourse to tables, or to a computer, the total SP for this test is very close to 0: differences in the mean maximum skull breadth for Etruscans and the modern Italian male are very marked indeed.

Solution 8.11

It may be that at a single command your computer permits a comparison of the two diets returning, for instance, the value of the test statistic t and associated SP. The details are as follows (so you can check your program!). For those rats given the restricted diet,

$$n_1 = 106, \quad \overline{x}_1 = 968.745, \quad s_1^2 = 80\,985.7;$$

and for those rats given the *ad libitum* diet,

$$n_2 = 89, \quad \overline{x}_2 = 684.011, \quad s_2^2 = 17\,978.6.$$

Then the pooled estimate of σ^2 is given by

$$s_p^2 = \frac{(n_1 - 1)s_1^2 + (n_2 - 1)s_2^2}{n_1 + n_2 - 2} = \frac{105 \times 80\,985.7 + 88 \times 17\,978.6}{106 + 89 - 2} = 52\,257.1.$$

This estimate of σ^2 is rather different to either of s_1^2 or s_2^2; and, in fact, the ratio of the sample variances is given by

$$s_1^2/s_2^2 = 4.5$$

which exceeds 3, and so suggests that the assumption of equal variances underlying the two-sample t-test is untenable. This suggestion may be confirmed formally using an appropriate test: in fact, there turns out to be considerable evidence that the variances are different.

Both samples are also considerably skewed. The software on your computer may blindly perform a two-sample t-test if you ask it to do so, with or without a warning message that the assumptions of the test may be seriously adrift. If so, then the resulting value of the test statistic T is

$$t = \frac{\overline{x}_1 - \overline{x}_2}{s_p\sqrt{\frac{1}{n_1} + \frac{1}{n_2}}} = \frac{968.745 - 684.011}{\sqrt{52\,257.1}\sqrt{\frac{1}{106} + \frac{1}{89}}} = 8.66.$$

Against the null distribution $t(193)$ (or, essentially, against the normal distribution $N(0,1)$) the SP is negligible. Assuming the t-test to be viable, there is very considerable evidence that the mean lifelengths under the different diet regimes are different (and that, in fact, a restricted diet leads to increased longevity).

The question of whether the t-test is a valid procedure in this case is a real one, and also a worrying one: however, lacking any other procedure for comparing two means, the t-test has provided some 'feel' for the extent to which the data indicate a difference between the two diet regimes. Other tests for a comparison of two means, with less restrictive assumptions, are available in the statistical literature.

Solution 8.12

(a) The data $\widehat{p}_1 = r_1/n_1 = 71/100 = 0.71$ and $\widehat{p}_2 = r_2/n_2 = 89/105 = 0.85$ suggest that females are more likely to be successful in a request for help. Fisher's exact test applied to the data $r_1 = 71, n_1 = 100, r_2 = 89$, $n_2 = 105$, yields the results

$$SP(\text{obtained direction}) = 0.013$$
$$SP(\text{opposite direction}) = 0.006$$
$$SP(\text{total}) = 0.019.$$

The total SP is less than 2%; in a directed test of the hypothesis $H_0 : p_1 = p_2$ against the alternative $H_1 : p_1 < p_2$ the SP is 1.3%. There is strong evidence of a difference in proportions, and that females are more likely to be given help when it is requested.

(b) In this case the proportions to be compared are $\widehat{p}_1 = r_1/n_1 = 8/20 = 0.40$ and $\widehat{p}_2 = r_2/n_2 = 11/20 = 0.55$. A reasonable null hypothesis might be

$$H_0 : p_1 = p_2.$$

No particular suggestion has been offered that certain types of brain damage might reduce a person's facility in handling syllogisms: we might write

$$H_1 : p_1 \neq p_2.$$

However, the data suggest the possibility that p_1 is less than p_2. In fact, Fisher's test gives

$$SP(\text{obtained direction}) = 0.264$$
$$SP(\text{opposite direction}) = 0.264$$
$$SP(\text{total}) = 0.527.$$

There is no serious evidence (whatever alternative hypothesis one might care to suggest) on the basis of these data that the individuals tested showed a significantly different capability.

Solution 8.13

(a) In this case the data available to test the null hypothesis $H_0 : \mu_1 = \mu_2$ (assuming a Poisson model) are

$$n_1 = n_2 = 1, \quad t = 3 + 6 = 9$$

and consequently the aim is to test the observed value $t_1^* = 3$ against the binomial distribution $B(t, n_1/(n_1 + n_2))$ or $B(9, \frac{1}{2})$. By the symmetry of the binomial distribution

$$SP(\text{obtained direction}) = SP(\text{opposite direction})$$
$$= p(3) + p(2) + p(1) + p(0) = 0.254$$

and so

$$SP(\text{total}) = 2 \times 0.254 = 0.508.$$

The SPs in either direction are not small: there is no strong evidence to reject the hypothesis that the two underlying mean densities are equal.

(b) In this case

$$n_1 = 4, \quad n_2 = 8;$$
$$t_1 = 77 + 61 + 157 + 52 = 347;$$
$$t_2 = 17 + 31 + 87 + 16 + 18 + 26 + 77 + 20 = 292;$$
$$t = t_1 + t_2 = 639.$$

In fact $\widehat{\mu}_1$ (the estimate of the mean plant density under Treatment 1) is 347/4=86.75; and $\widehat{\mu}_2$ is 292/8=36.50. There seems to be a considerable difference here. Formally, we need to test the observed value $t_1^* = 347$ against the binomial distribution $B(639, \frac{1}{3})$. The null distribution peaks at the mean $(\frac{1}{3} \times 639 = 213)$ and so fairly evidently

$$SP(\text{obtained direction}) = P(T_1^* \geq 347) \simeq 0;$$

this was calculated using a computer.

Working out the SP in the opposite direction strictly involves a scan of the binomial distribution $B(639, \frac{1}{3})$ to ascertain those values in the range occurring with probability less than $p(347)$.

A normal approximation has $T_1^* \simeq N(213, 142)$ (for the binomial mean is $np = 639 \times \frac{1}{3} = 213$; the variance is $npq = 639 \times \frac{1}{3} \times \frac{2}{3} = 142$). The SP in the obtained direction is again

$$P(T_1^* \geq 347);$$

using a continuity correction this is approximately

$$1 - \Phi\left(\frac{346\frac{1}{2} - 213}{\sqrt{142}}\right) = 1 - \Phi(11.2) \simeq 0;$$

by the symmetry of the normal distribution the SP in the opposite direction is the same as the SP in the obtained direction. There is very considerable evidence that the mean plant densities under the two treatments differ and, in fact, that Treatment 1 leads to a higher density than Treatment 2.

Chapter 9

Solution 9.1

Since $n = 7$, we determine the points for which $\Phi(x_i) = i/8$.

i	$y_{(i)}$	$i/8$	x_i
1	5.1	0.125	−1.150
2	5.3	0.250	−0.674
3	5.5	0.375	−0.319
4	5.6	0.500	0.000
5	5.8	0.625	0.319
6	5.8	0.750	0.674
7	6.2	0.875	1.150

A probability plot for the points $y_{(i)}$ against x_i is given in Figure S9.1.

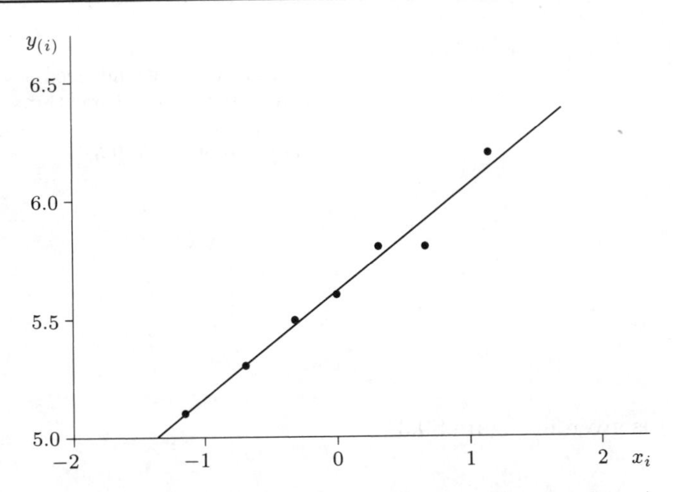

Figure S9.1 Silver content against normal scores (fourth coinage)

A straight line fits the points quite closely and we can conclude that the normal distribution provides an adequate model for the variation in the data.

Solution 9.2

The differences (in microns) are given in Table S9.1.

Table S9.1 Corneal thickness in patients with glaucoma (microns)

Patient	1	2	3	4	5	6	7	8
Glaucomatous eye	488	478	480	426	440	410	458	460
Normal eye	484	478	492	444	436	398	464	476
Difference	4	0	−12	−18	4	12	−6	−16

Since $n = 8$, we determine the points for which $\Phi(x_i) = i/9$. The points $y_{(i)}$ and x_i are shown in the table in the margin. The points $(x_i, y_{(i)})$ are shown plotted in Figure S9.2.

i	$y_{(i)}$	$i/9$	x_i
1	−18	1/9	−1.221
2	−16	2/9	−0.765
3	−12	3/9	−0.431
4	−6	4/9	−0.140
5	0	5/9	0.140
6	4	6/9	0.431
7	4	7/9	0.765
8	12	8/9	1.221

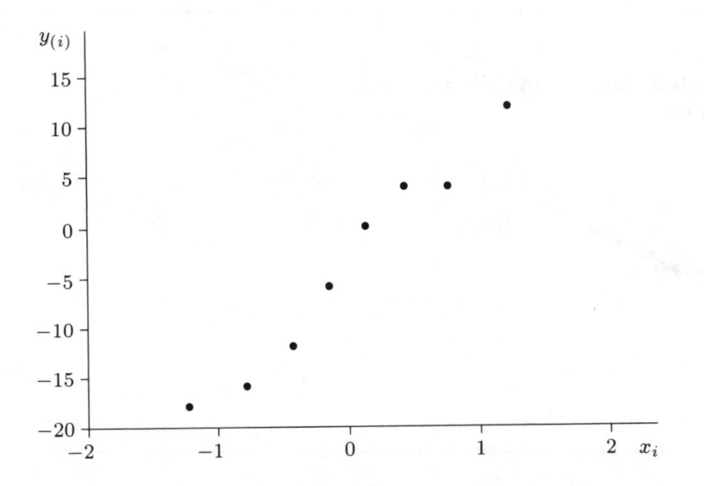

Figure S9.2 Corneal thickness differences against normal scores

The points do not appear to lie on a straight line and the evidence in favour of a normal modelling distribution for the differences in corneal thickness is not strong. However, there is no systematic pattern to the points.

Solution 9.3

Since $n = 20$, we plot $y_{(i)}$ against $x_i = -\log((21 - i)/21)$.

Normally you would need a calculator for the logarithms, but in this case the x_is are the same as they are in Table 9.5.

i	$y_{(i)}$	x_i	i	$y_{(i)}$	x_i
1	1.45	0.049	11	3.87	0.742
2	1.67	0.100	12	4.33	0.847
3	1.90	0.154	13	5.35	0.965
4	2.02	0.211	14	5.72	1.099
5	2.32	0.272	15	6.48	1.253
6	2.35	0.336	16	6.90	1.435
7	2.43	0.405	17	8.68	1.658
8	2.47	0.480	18	9.47	1.946
9	2.57	0.560	19	10.00	2.351
10	3.33	0.647	20	10.93	3.045

The exponential probability plot ($y_{(i)}$ against x_i) is given in Figure S9.3.

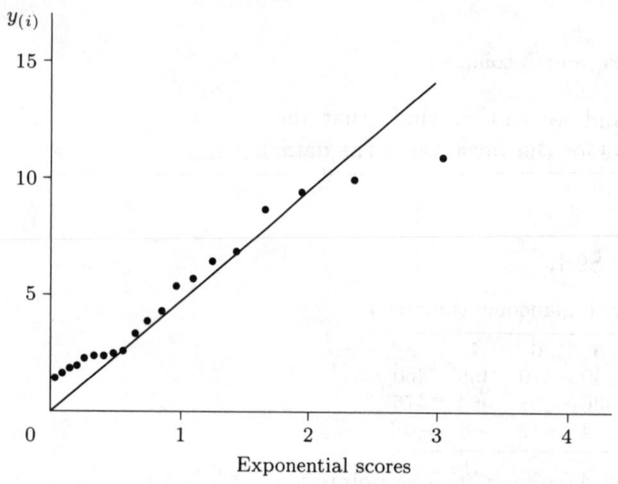

Figure S9.3 Unpleasant memory recall times against exponential scores

Remember that for an exponential probability plot, the fitted straight line must pass through the origin.

The points do not lie on a straight line and the evidence does not support an exponential modelling distribution for the variation in recall times.

Solution 9.4

The normal probability plot, together with a fitted straight line, for the data on the 84 Etruscan skulls is shown in Figure S9.4.

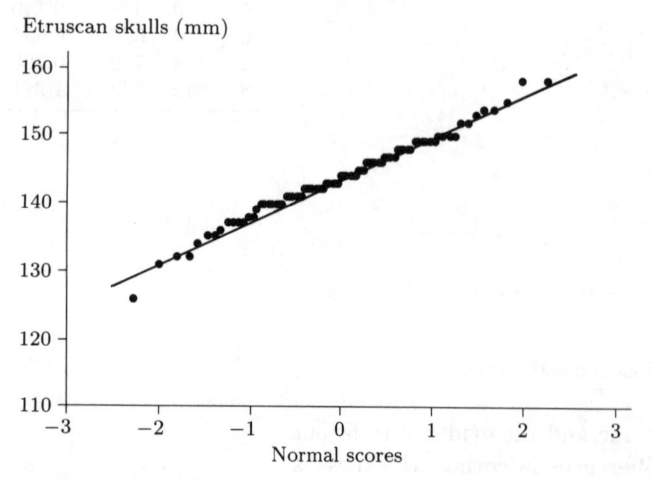

Figure S9.4 Normal probability plot, Etruscan skulls

The fit appears to be very good, supporting an earlier assertion that the variation in the data may plausibly be modelled by a normal distribution. Notice also that the fitted line has intercept at about 144 ($\overline{x}_E = 143.8$) and slope about 6 ($s_E = 5.97$).

The corresponding plot for the 70 modern Italian skulls is given in Figure S9.5. Again, a straight line appears to fit the points very well (though there is a departure from the fitted line at both extremes—small skulls are surprisingly small and large skulls are surprisingly large). The fitted line in the diagram has intercept at about 132 ($\overline{x}_I = 132.4$) and slope again about 6 ($s_I = 5.75$).

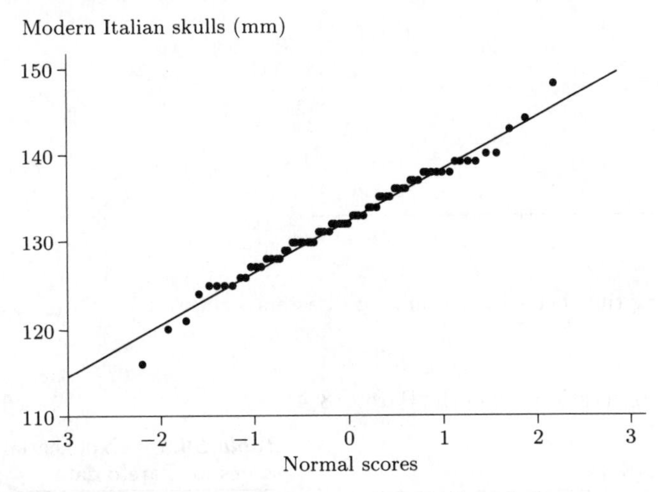

Figure S9.5 Normal probability plot, modern Italian skulls

Solution 9.5

(a) The exponential probability plot for the 62 waiting times between earthquakes, together with a fitted straight line through the origin, is given in Figure S9.6. The fit looks very good. (The slope of the line appears to be about 450: the sample mean for the data set is 437.)

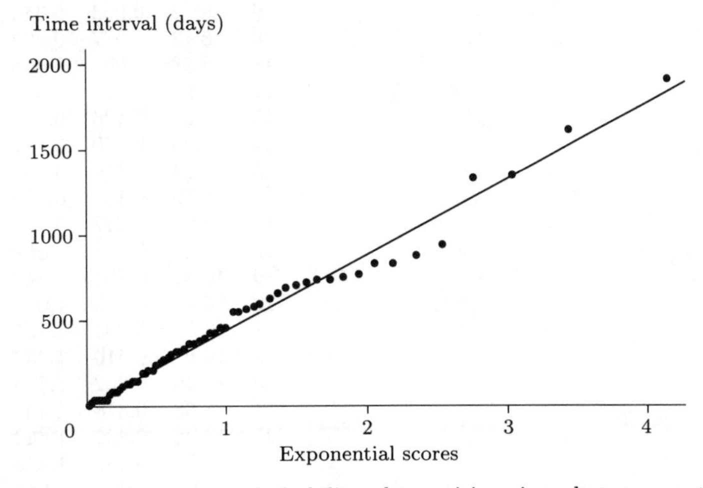

Figure S9.6 Exponential probability plot, waiting times between earthquakes

(b) In this case (see Figure S9.7) there is a clear departure from linearity, suggesting that the variation in the waiting times between successive coal-mining disasters is other than exponential. Nevertheless, the plot does suggest some sort of systematic variation in these data.

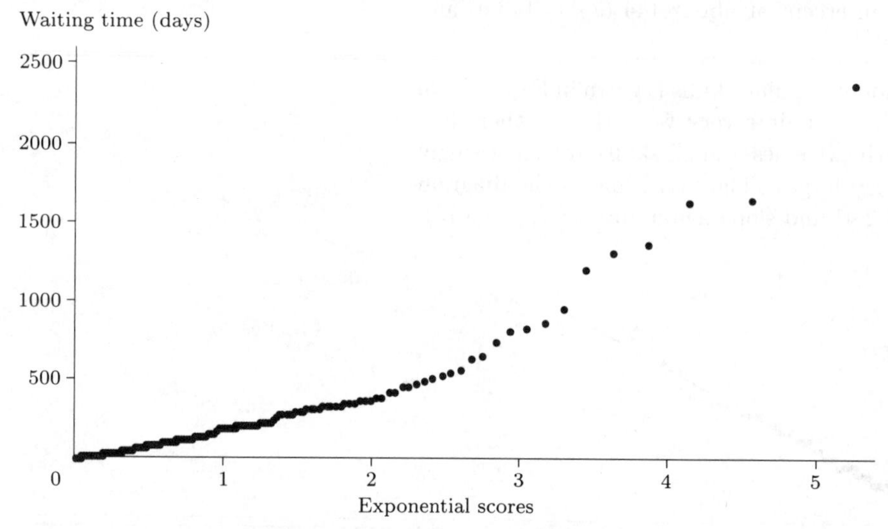

Figure S9.7 Exponential probability plot, waiting time between coal-mining disasters

Solution 9.6

Your computer may have provided you with 30 observations from the Pareto(8, 4) distribution similar to those shown in Table S9.2.

Table S9.2 Thirty observations from Pareto(8, 4)

8.18	10.91	8.73	8.17	12.90	8.88	10.31	8.19	9.59	13.73
11.68	8.52	13.14	10.10	8.44	10.72	8.18	8.12	8.33	9.20
8.78	10.41	11.49	9.54	12.55	12.28	17.26	9.07	8.05	9.99

Writing $W \sim \text{Pareto}(8, \theta)$, then the random variable $Y = \log(W/8)$ has an exponential distribution. The ordered sample $y_{(1)}, y_{(2)}, \ldots, y_{(30)}$ and the associated exponential scores $x_1, x_2, \ldots, x_{30}$ are shown in Table S9.3.

The plot of the points y_i against x_i, together with a fitted straight line through the origin, is given in Figure S9.8.

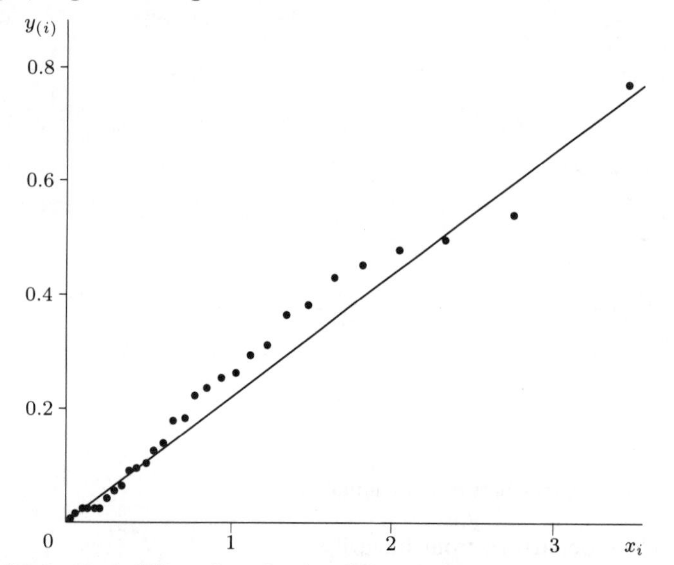

Figure S9.8 Probability plot, simulated Pareto data

Table S9.3 Exponential scores for Pareto data

i	$w_{(i)}$	$y_{(i)}$	x_i
1	8.05	0.006	0.033
2	8.12	0.015	0.067
3	8.17	0.021	0.102
4	8.18	0.022	0.138
5	8.18	0.022	0.176
6	8.19	0.023	0.215
7	8.33	0.040	0.256
8	8.44	0.054	0.298
9	8.52	0.063	0.343
10	8.73	0.087	0.389
11	8.78	0.093	0.438
12	8.88	0.104	0.490
13	9.07	0.126	0.544
14	9.20	0.140	0.601
15	9.54	0.176	0.661
16	9.59	0.181	0.726
17	9.99	0.222	0.795
18	10.10	0.233	0.869
19	10.31	0.254	0.949
20	10.41	0.263	1.036
21	10.72	0.293	1.131
22	10.91	0.310	1.237
23	11.49	0.362	1.355
24	11.68	0.378	1.488
25	12.28	0.429	1.642
26	12.55	0.450	1.825
27	12.90	0.478	2.048
28	13.14	0.496	2.335
29	13.73	0.540	2.741
30	17.26	0.769	3.434

If the maximum point of the 30 is ignored, there is a pronounced curve to the probability plot, suggesting in this case some systematic error with the Pareto generator.

Did your plot provide a more convincing straight line?

Solution 9.7

(a) There are four categories with expected frequencies

$$E_i = n\theta_i = 290\theta_i, \quad i = 1, 2, 3, 4,$$

where

$$\theta_1 = \tfrac{9}{16},\ \theta_2 = \tfrac{3}{16},\ \theta_3 = \tfrac{3}{16},\ \theta_4 = \tfrac{1}{16}.$$

A table for calculating the chi-squared test statistic is given in Table S9.4.

Table S9.4 *Pharbitis nil*, simple theory

i	O_i	E_i	$(O_i - E_i)$	$(O_i - E_i)^2/E_i$
1	187	163.125	23.875	3.49
2	35	54.375	−19.375	6.90
3	37	54.375	−17.375	5.55
4	31	18.125	12.875	9.15

The chi-squared value for the test is

$$\chi^2 = \sum_{i=1}^{4} \frac{(O_i - E_i)^2}{E_i} = 3.49 + 6.90 + 5.55 + 9.15 = 25.09.$$

Measured against $\chi^2(3)$ ($k = 4$; no model parameters were estimated from the data), this gives a SP of about 0.000 015. This is exceptionally small: there is very considerable evidence that the simple theory is flawed.

(b) Allowing for genetic linkage, then the expected frequencies are considerably changed, as shown in Table S9.5.

For instance,
$$E_1 = n\theta_1 = 290 \times 0.6209 = 180.061.$$

Table S9.5 *Pharbitis nil*, genetic linkage

i	O_i	E_i	$(O_i - E_i)$	$(O_i - E_i)^2/E_i$
1	187	180.061	6.939	0.27
2	35	37.439	−2.439	0.16
3	37	37.439	−0.439	0.01
4	31	35.061	−4.061	0.47

The chi-squared value for this test is

$$\chi^2 = \sum_{i=1}^{4} \frac{(O_i - E_i)^2}{E_i} = 0.27 + 0.16 + 0.01 + 0.47 = 0.91.$$

Measured against $\chi^2(2)$ ($k = 4$; one model parameter was estimated from the data so $p = 1$), this gives a SP of 0.63. There is no evidence to reject the genetic linkage.

Solution 9.8

Pielou assumed a geometric modelling distribution with parameter $p = 0.657$ for the following data; expected frequencies obtained by multiplying the hypothesized probability mass function by 109 are included in Table S9.6. For instance, $E_4 = 109\theta_4 = 109(0.343)^3(0.657) = 2.89$.

Run lengths of diseased trees in an infected plantation

Run length	1	2	3	4	5	6	> 6
Observed number of runs	71	28	5	2	2	1	0
Estimated number of runs	71.61	24.56	8.43	2.89	0.99	0.34	0.18

Pooling runs of length 3 or more and performing the chi-squared test calculation gives Table S9.6.

In fact, pooling runs of length 4 or more gives an expected frequency of $2.89 + 0.99 + 0.34 + 0.18 = 4.40$ which is less than 5; but the conclusions of the chi-squared test would not be seriously adrift.

Table S9.6 Diseased trees: testing a geometric fit

Run length	O_i	E_i	$O_i - E_i$	$(O_i - E_i)^2/E_i$
1	71	71.61	-0.61	0.005
2	28	24.56	3.44	0.482
≥ 3	10	12.83	-2.83	0.624

The test statistic is

$$\chi^2 = \sum_{i=1}^{3} \frac{(O_i - E_i)^2}{E_i} = 0.005 + 0.482 + 0.624 = 1.1.$$

One parameter ($p = 0.657$) was estimated from these data, so the chi-squared null distribution has $(3 - 1 - 1) = 1$ degree of freedom for a SP of 0.29. The geometric distribution is not rejected as a model and this confirms that Pielou's assumptions were not unreasonable.

Solution 9.9

The suggested grouping gives the following observed frequencies.

Level	< 150	$150 - 250$	$250 - 350$	$350 - 450$	≥ 450
Observed frequency	4	14	19	7	11

Working to full computer accuracy, the expected frequencies and chi-squared test calculations are as shown in Table S9.7, using $\overline{x} = 314.91$ and $s = 131.16$.

Table S9.7

O_i	E_i	$(O_i - E_i)$	$(O_i - E_i)^2/E_i$
4	5.74	-1.74	0.53
14	11.33	2.67	0.63
19	16.23	2.77	0.47
7	13.37	-6.37	3.03
11	8.33	2.67	0.86

The value of the test statistic is

$$\chi^2 = \sum_{i=1}^{5} \frac{(O_i - E_i)^2}{E_i} = 0.53 + 0.63 + \cdots + 0.86 = 5.52.$$

Two parameters were estimated from these data, so the chi-squared distribution has $(5 - 2 - 1) = 2$ degrees of freedom for a SP of 0.061. There is some evidence for rejection of the null hypothesis that the data are fitted by a normal distribution.

Solution 9.10

The details of observed and expected frequencies, and of the chi-squared test calculations are given in Table S9.8, using $\overline{x} = 1369.1$ and $s = 693.7$.

Table S9.8 Chi-squared calculations, rainfall data

Rainfall	O_i	E_i	$(O_i - E_i)$	$(O_i - E_i)^2/E_i$
< 600	5	6.29	−1.29	0.26
600–1000	14	7.69	6.31	5.18
1000–1400	10	10.36	−0.36	0.01
1400–1800	8	10.10	−2.10	0.44
1800–2200	5	7.13	−2.13	0.64
≥ 2200	5	5.43	−0.43	0.03

The value of the chi-squared test statistic is

$$\chi^2 = \sum_{i=1}^{6} \frac{(O_i - E_i)^2}{E_i} = 0.26 + 5.18 + \cdots + 0.03 = 6.56.$$

Two parameters were estimated from these data, so the chi-squared distribution has $(6 - 2 - 1) = 3$ degrees of freedom for a SP of 0.09. There is insufficient evidence for rejection of the null hypothesis that the data are fitted by a normal distribution, although the fit is not good and one should have reservations about it.

Solution 9.11

The sample skewness of birth weights for the group of children who survived is 0.229; and that for the group of children who died is 0.491.

The skewness of the group of children who died is rather higher than one would like and it is worth trying to transform the data. One possible transformation is to take logarithms: this reduces the sample skewnesses to −0.291 and 0.177 respectively.

It is worth noticing that this transformation not only gives skewnesses for the two groups that are similar in size and opposite in sign but also gives approximately equal variances to the two groups.

Solution 9.12

A two-sample t-test for equal means may now be carried out. This gives a t-statistic of −3.67 on 48 degrees of freedom, with a total SP equal to 0.0006. It may be concluded that there is a significant difference, provided the residuals are plausibly normal. Subtract 0.482 from the transformed data of the first group (that is, those who died) and subtract 0.795 from the transformed data of the second group (survivors); pool the data and construct a normal probability plot. This is shown in Figure S9.9.

It is possible to fit an acceptable straight line and there is strong evidence for rejection of the null hypothesis that there is no difference between the groups.

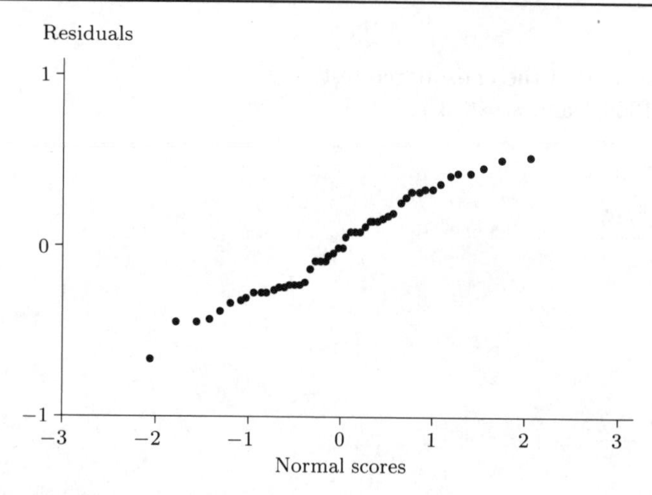

Figure S9.9 Residuals against normal scores

Solution 9.13

The analysis is very similar to that in Example 9.8. It turns out that the same transformation works well.

As before, we can carry out a two-sample t-test for equal means. This gives a t-statistic of 2.46 on 52 degrees of freedom, which has a SP equal to 0.017. It may be concluded that there is evidence of a significant difference, provided the residuals are plausibly normal. Subtract 4.565 from the transformed data of the group with less formal education and subtract 3.029 from the transformed data of the group with more formal education, pool the data and construct a normal probability plot. This is shown in Figure S9.10.

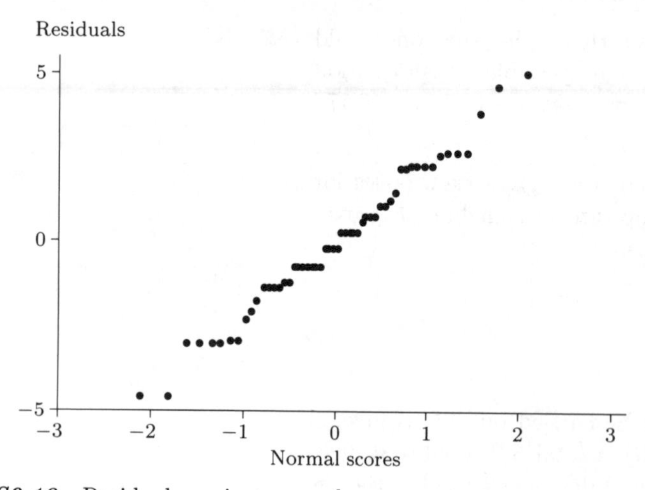

Figure S9.10 Residuals against normal scores

This does not suggest a good straight line, although it may be just about acceptable. This casts doubt on our assumption of normality.

Solution 9.14

(a) Subtracting 0.618 from each entry in Table 8.4 and allocating signed ranks produces the following table.

Difference	0.075	0.044	0.072	−0.012	−0.048	0.131	0.054	0.010	−0.009	0.226
Sign	+	+	+	−	−	+	+	+	−	+
Rank	17	10	16	$5\frac{1}{2}$	11	18	14	4	3	19
Difference	0.036	−0.003	0.050	−0.017	−0.042	0.052	−0.012	−0.007	−0.065	0.315
Sign	+	−	+	−	−	+	−	−	−	+
Rank	8	1	12	7	9	13	$5\frac{1}{2}$	2	15	20

There are no 0s and there are only two tied differences.

(b) The sums of the ranks are 151 for the positive differences and 59 for the negative differences, thus the Wilcoxon signed rank statistic is 151. This gives

$$SP(\text{obtained direction}) = SP(\text{opposite direction}) = 0.044;$$

$$SP(\text{total}) = 0.088.$$

There is some evidence for rejection of the null hypothesis of zero difference; in other words, there is some evidence that the rectangles do not conform to the Greek standard.

(c) A t-test for zero difference gives a total SP of 0.054, which may be interpreted as giving some evidence, although not strong evidence, for rejection of the hypothesis. There must be doubt about such a result because of the lack of normality of the data.

Solution 9.15

The sample size is 20, so that

$$E(S) = \frac{n(n+1)}{4} = \frac{20 \times 21}{4} = 105,$$

$$V(S) = \frac{n(n+1)(2n+1)}{24} = \frac{20 \times 21 \times 41}{24} = 717.5.$$

Therefore, we have

$$z = \frac{151 - 105}{\sqrt{717.5}} = 1.717$$

which is the 0.957 quantile of the standard normal distribution. The SP is therefore 0.086, which is very close to that given by the exact test in Exercise 9.14.

Solution 9.16

The ranks are as follows.

Pleasant memory	Rank	Unpleasant memory	Rank
1.07	1	1.45	5
1.17	2	1.67	7
1.22	3	1.90	8
1.42	4	2.02	10
1.63	6	2.32	$12\frac{1}{2}$
1.98	9	2.35	14
2.12	11	2.43	15
2.32	$12\frac{1}{2}$	2.47	16
2.56	17	2.57	18
2.70	19	3.33	25
2.93	20	3.87	27
2.97	21	4.33	28
3.03	22	5.35	31
3.15	23	5.72	33
3.22	24	6.48	35
3.42	26	6.90	36
4.63	29	8.68	37
4.70	30	9.47	38
5.55	32	10.00	39
6.17	34	10.93	40
	$345\frac{1}{2}$		$474\frac{1}{2}$

Labelling the pleasant memory recall times as group A, $u_A = 345\frac{1}{2}$. An exact test gives

$$SP(\text{total}) = 0.082.$$

Alternatively, using a normal approximation,

$$E(U_A) = \frac{n_A(n_A + n_B + 1)}{2} = \frac{20 \times 41}{2} = 410,$$

$$V(U_A) = \frac{n_A n_B(n_A + n_B + 1)}{12} = \frac{20 \times 20 \times 41}{12} = 1366.667$$

giving

$$z = \frac{345.5 - 410}{\sqrt{1366.667}} = -1.745.$$

Normal tables give a total SP of 0.081 for the two-sided test. Therefore, the conclusion is that the evidence for rejection of the null hypothesis, namely that memory recall times are different for pleasant and unpleasant memories, is not very strong.

Solution 9.17

The data are rather interesting. A naive estimate of the age of the site is the sample mean, 2622 years. However, Figure S9.11 shows (a) a boxplot for these data; (b) a normal probability plot; and (c) a normal probability plot with the single high outlier (3433 years) removed.

A straight line fits the seven points in Figure S9.11(c) very well. A good estimate of the age of the site would seem to be provided by the mean of the trimmed sample, 2506 years, nearly 1000 years less than the untrimmed estimate.

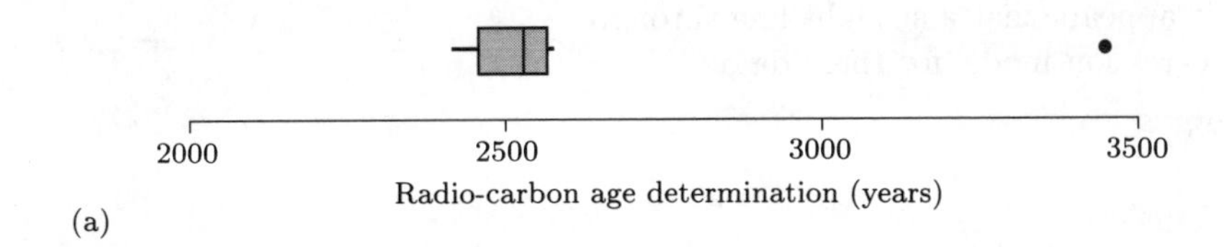

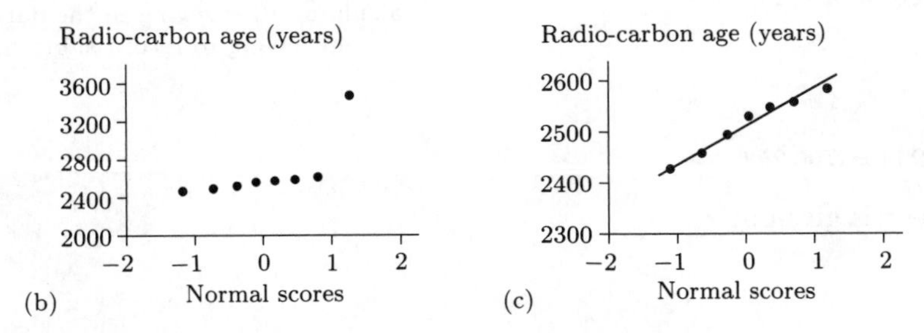

Figure S9.11 (a) Boxplot (b) Normal probability plot (c) Trimmed normal probability plot

Chapter 10

Solution 10.1

(a) The problem has been set up as a prediction problem, with bracket weight the explanatory variable (x) and beetle count the response variable (y). A scatter plot of beetle count against bracket weight is given in Figure S10.1.

Figure S10.1 also shows the fitted straight line found in part (b).

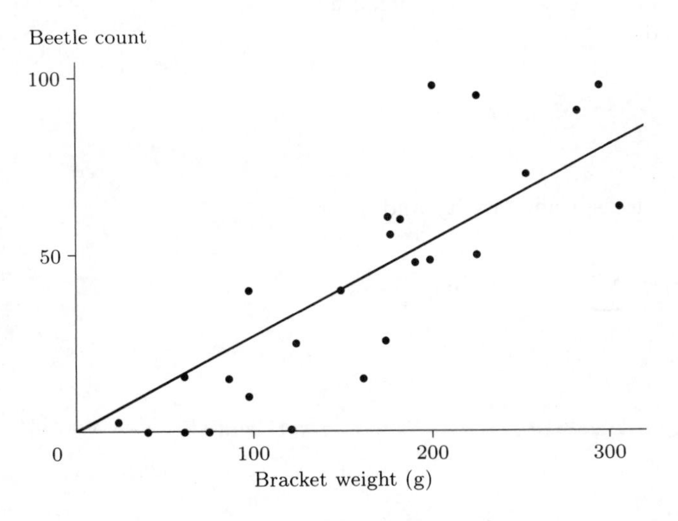

Figure S10.1 Beetle count against bracket weight

(b) From the scatter plot in part (a), it appears that a straight line through the origin would provide a useful regression model for these data:

$$Y_i = \gamma x_i + W_i, \quad i = 1, 2, \ldots, 25.$$

Using

$$\sum x_i y_i = 62 \times 16 + 226 \times 50 + \cdots + 162 \times 15$$
$$= 992 + 11\,300 + \cdots + 2430 = 219\,817$$

You will probably be able to obtain these sums directly from your calculator after keying in the data, without having to record subtotals.

and

$$\sum x_i^2 = 62^2 + 226^2 + \cdots + 162^2$$
$$= 3844 + 51\,076 + \cdots + 26\,244 = 796\,253,$$

the least squares estimate for the slope γ is given by

$$\widehat{\gamma} = \frac{\sum x_i y_i}{\sum x_i^2} = \frac{219\,817}{796\,253} = 0.276.$$

(c) For a fungus weighing 240 g, the predicted beetle count is

$$240\widehat{\gamma} = 240 \times 0.276 = 66.3 \text{ beetles;}$$

say, 66 beetles.

(d) There are two useful representations for the residual sum of squares (see (10.4)). In either case we need the result $\sum y_i^2 = 68\,918$. Either say

$$\sum (y_i - \widehat{y}_i)^2 = \sum y_i^2 - \frac{\sum (x_i y_i)^2}{\sum x_i^2} = 68\,918 - \frac{219\,817^2}{796\,253} = 8234.4$$

or say

$$\sum (y_i - \widehat{y}_i)^2 = \sum y_i^2 - \widehat{\gamma}^2 \sum x_i^2 = 68\,918 - (0.276)^2(796\,253) = 8262.6.$$

Notice that the rounding error induced by using $\widehat{\gamma}$ at the calculator keypad (0.276 instead of 0.276 064 266) has been considerable.

A properly programmed computer will give you all these answers, including the scatter plot, from a few keyboard commands.

Solution 10.2

The summary statistics in this case (writing x for specific gravity and y for strength) are

$$n = 10, \quad \sum x_i = 4.951, \quad \sum y_i = 118.77, \quad \sum x_i^2 = 2.488\,995,$$
$$\sum x_i y_i = 59.211\,61,$$

so the slope estimate is

$$\widehat{\beta} = \frac{10 \times 59.211\,61 - 4.951 \times 118.77}{10 \times 2.488\,995 - 4.951^2}$$
$$= 10.8220,$$

and the estimate of the constant term is

$$\widehat{\alpha} = \overline{y} - \widehat{\beta}\overline{x} = \tfrac{1}{10}(118.77 - 10.8220 \times 4.951) = 6.5190.$$

So the fitted model is

$$y = 6.52 + 10.82x$$

or

$$\text{Strength} = 6.52 + 10.82 \times \text{Specific gravity}.$$

Solution 10.3

For the finger-tapping data (tapping frequency y against caffeine dose x), the summary statistics are

$$n = 30, \quad \sum x_i = 3000, \quad \sum y_i = 7395, \quad \sum x_i^2 = 500\,000,$$

$$\sum x_i y_i = 743\,000,$$

so the slope estimate is

$$\widehat{\beta} = \frac{30 \times 743\,000 - 3\,000 \times 7395}{30 \times 500\,000 - 3000^2}$$

$$= \frac{105\,000}{6\,000\,000} = 0.0175,$$

and the constant term is estimated by

$$\widehat{\alpha} = \overline{y} - \widehat{\beta}\overline{x}$$

$$= \tfrac{1}{30}(7395 - 0.0175 \times 3000)$$

$$= 244.75.$$

So the fitted model is

$$y = 244.75 + 0.0175x$$

or

$$\text{Tapping frequency} = 244.75 + 0.0175 \times \text{Caffeine dose}.$$

Solution 10.4

(a) With the relevant data keyed in (or the appropriate data file accessed), most statistical software would provide the equation of the fitted straight line at a single command. For Forbes' data, the estimators are $\widehat{\alpha} = 155.30$ and $\widehat{\beta} = 1.90$, so the equation of the fitted line is given by

$$\text{Boiling point} = 155.30 + 1.90 \times \text{Atmospheric pressure},$$

where temperature is measured in °F and atmospheric pressure in inches Hg.

(b) For Hooker's data, $\widehat{\alpha} = 146.67$, $\widehat{\beta} = 2.25$; the fitted line has equation

$$\text{Boiling point} = 146.67 + 2.25 \times \text{Atmospheric pressure}.$$

Solution 10.5

(a) The problem here was to predict morbidity rates from mortality rates, so in the following scatter diagram (see Figure S10.2) morbidity rates (y) are plotted against mortality rates (x).

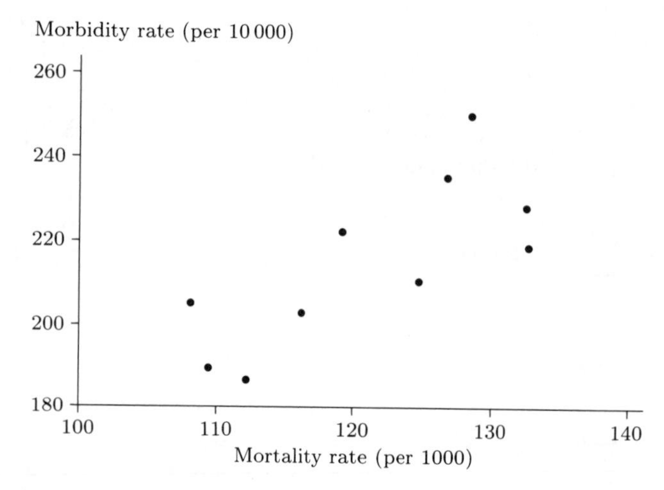

Figure S10.2 Morbidity rates against mortality rates

(b) The points are suggestive of a useful straight line fit. From $\widehat{\alpha} = 16.5478$ and $\widehat{\beta} = 1.6371$ it follows that the least squares regression line equation is given by

$$\text{Morbidity rate (per 1000)} = 16.55 + 1.64 \times \text{Mortality rate (per 10 000)}.$$

Solution 10.6

(a) A heavy car uses a lot of fuel because it is heavy: it is possible that under unusual circumstances one might wish to predict kerb weight from fuel consumption figures, but in general the problem would be to estimate fuel consumption, given the size of the car. Therefore plot consumption (miles per gallon, y) against kerb weight (kilograms, x). The appropriate scatter diagram is shown in Figure S10.3.

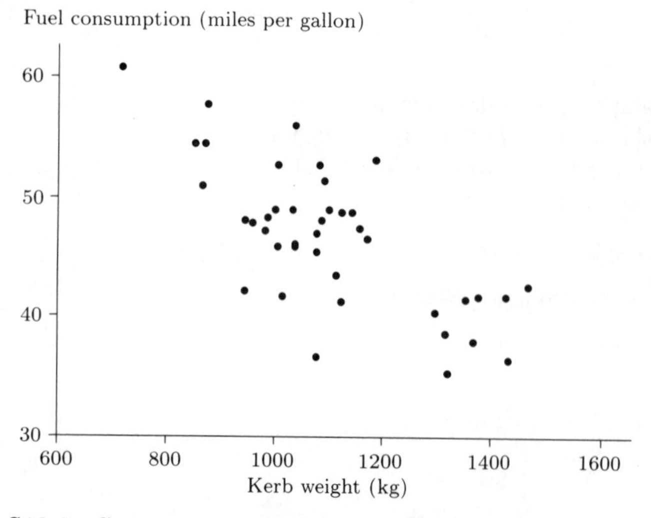

Figure S10.3 Consumption against kerb weight

(b) There is a lot of scatter: there is less evidence of an underlying 'formula' relating fuel consumption to kerb weight than there is in some other contexts. However, there is a pronounced downward trend. The least squares regression line has equation

$$\text{Fuel consumption} = 73.48 - 0.024 \times \text{Kerb weight},$$

where fuel consumption is measured in miles per gallon, and kerb weight in kilograms.

Solution 10.7

A plot of temperature against chirping frequency for these data is given in Figure S10.4.

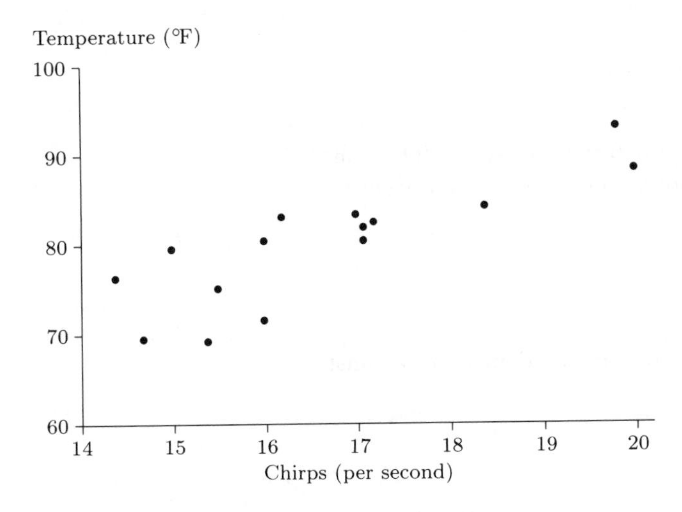

Figure S10.4 Temperature against chirping frequency

The least squares fitted line through the scattered points has equation

$$\text{Temperature} = 25.23 + 3.29 \times \text{Chirping frequency},$$

where temperature is measured in °F and chirps are counted every second. If $x_0 = 18$ chirps per second, the corresponding estimate of temperature, y_0, is given by

$$y_0 = 25.23 + 3.29x_0 = 25.23 + 3.29 \times 18 = 84.5\,^\circ\text{F}.$$

Solution 10.8

Taking examination score as the response variable (Y) and time taken as the explanatory variable (x), then we fit the model

$$Y_i = \alpha + \beta x_i + W_i, \quad i = 1, 2, \ldots, 134,$$

where the random terms W_i are independent and normally distributed random variables with mean 0 and variance σ^2. The proposition that x is valueless as a predictor for Y is covered by the hypothesis

$$H_0 : \beta = 0.$$

For these data,

$$\widehat{\alpha} = 56.7333, \qquad \widehat{\beta} = -0.0012,$$

and the residual sum of squares is

$$\sum(y_i - \widehat{y}_i)^2 = \sum(y_i - \overline{y})^2 - \widehat{\beta}^2 \sum(x_i - \overline{x})^2 = 12\,809.38 - 41.82 = 12\,767.56.$$

So our estimator of variance is given by

$$s^2 = \frac{\sum(y_i - \widehat{y})^2}{n-2} = \frac{12\,767.56}{132} = 96.724.$$

The observed value of the test statistic

$$\frac{\widehat{\beta} - 0}{s/\sqrt{\sum(x_i - \overline{x})^2}} = \frac{-0.0012 - 0}{\sqrt{96.724}/\sqrt{28\,651\,067.56}} = -0.658$$

is at the 26% quantile of $t(132)$. So we have

$$SP(\text{obtained direction}) = SP(\text{opposite direction}) = 0.26;$$
$$SP(\text{total}) = 0.52.$$

The total SP is not small: the hypothesis that $\beta = 0$ is not rejected. This confirms our suspicion that examination time is not a good predictor for eventual score.

Solution 10.9

(a) A scatter diagram for percentage of non-contaminated peanuts against aflatoxin level is given in Figure S10.5.

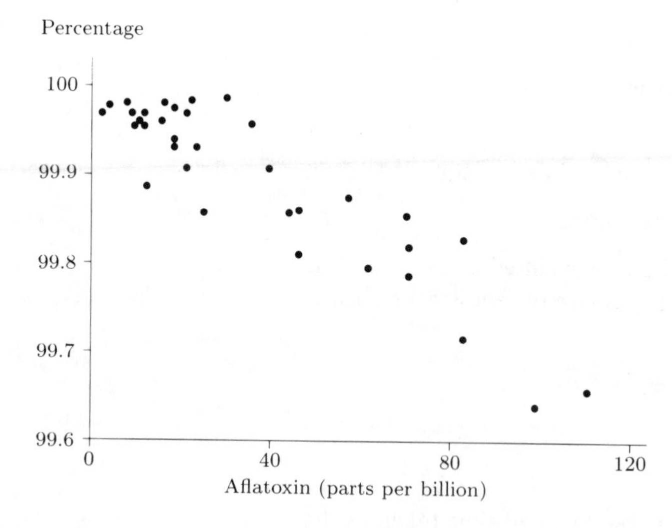

Figure S10.5 Percentage of non-contaminated against aflatoxin level

(b) The least squares regression line has equation

$$\text{Percentage non-contaminated} = 100.002 - 0.003 \times \text{Aflatoxin level}$$

and the scatter plot suggests that a straight line would be a good fit to the data.

However, on the grounds of these results, and on the basis of what we know of the problem—that zero aflatoxin level would indicate 100% non-contamination—a better model would be to fit the constrained line

$$\text{Percentage non-contaminated} = 100 - \gamma \times \text{Aflatoxin level},$$

or, equivalently

$$y = 100 - \gamma x.$$

This model has one parameter: the equivalent hypothesis under test is $H_0 : \gamma = 0$.

(c) Proceeding, however, with the two-parameter model, the residual sum of squares is

$$\sum(y_i - \widehat{y}_i)^2 = \sum(y_i - \overline{y})^2 - \widehat{\beta}^2 \sum(x_i - \overline{x})^2 = 0.288\,645 - 0.239\,150$$
$$= 0.049\,495.$$

Our estimate of variance is

$$s^2 = \frac{0.049\,495}{32} = 0.001\,547,$$

and our test statistic is

$$\frac{\widehat{\beta} - 0}{s/\sqrt{\sum(x_i - \overline{x})^2}} = \frac{-0.003 - 0}{\sqrt{0.001\,547}/\sqrt{28\,367.7}} = -12.434.$$

The SP is negligible. Despite the slope estimate being a small number in absolute terms, it represents a significant downward trend (as indicated by the scatter plot in part (a)).

Solution 10.10

In this context, breathing resistance is the response variable (y) and height is the explanatory variable (x). A scatter plot of the data (see Figure S10.6) was not asked for in this question, but it is always a useful first step. (Some would say it is an essential first step of a regression analysis.)

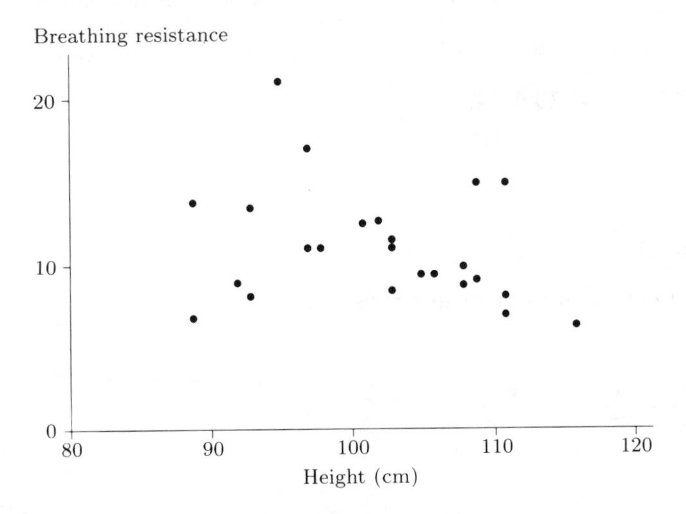

Figure S10.6 Breathing resistance against height

The fitted line through the scattered points has equation

$$y = 23.807 - 0.125x$$

and so the predicted breathing resistance for a child $x_0 = 100$ cm tall is 11.346. From the data given in Table 10.15, the sample size is $n = 24$, the sample mean is $\overline{x} = 102.042$, $\sum(x_i - \overline{x})^2 = 1352.96$ and

$$\begin{aligned}\sum(y_i - \widehat{y}_i)^2 &= \sum(y_i - \overline{y})^2 - \widehat{\beta}^2 \sum(x_i - \overline{x})^2 \\ &= 284.64 - (0.125)^2(1352.96) = 263.6.\end{aligned}$$

Using a computer package, intermediate results are not really necessary: the residual sum of squares is a standard summary statistic in the context of regression.

So our estimate of σ^2 is

$$s^2 = \frac{263.6}{22} = 11.98.$$

The 97.5% quantile of $t(22)$ is 2.074, and consequently the 95% confidence interval for the mean breathing resistance for children 100 cm tall, based on these data, is

$$\begin{aligned}&\left(\widehat{\alpha} + \widehat{\beta}x_0 \pm q_{0.975}s\sqrt{\frac{(x_0 - \overline{x})^2}{\sum(x_i - \overline{x})^2} + \frac{1}{n}}\right) \\ &= \left(11.307 \pm 2.074\sqrt{11.98}\sqrt{\frac{(-2.042)^2}{1352.96} + \frac{1}{24}}\right) \\ &= (11.307 \pm 1.519) = (9.8, 12.8).\end{aligned}$$

The units of measurement are those for breathing resistance, not stated in the original table.

Solution 10.11

The important measures obtained from the data in Table 10.16 are

If all your calculations are done on a computer, these intermediate results are unnecessary.

$$n = 42, \quad \overline{x} = 124.31, \quad \sum(x_i - \overline{x})^2 = 7058.98, \quad \widehat{\alpha} = 27.516,$$

$$\widehat{\beta} = -0.136\,06, \quad \sum(y_i - \widehat{y}_i)^2 = 796.055, \quad s^2 = 19.9014.$$

Also, the 97.5% quantile of $t(40)$ is 2.021.

The predicted mean when x_0 is 100 is

$$\widehat{\alpha} + \widehat{\beta}x_0 = 27.516 - 0.136\,06 \times 100 = 13.91,$$

and the 95% prediction interval for this particular child's breathing resistance is given by

$$\begin{aligned}&\left(13.91 \pm 2.021\sqrt{19.9014}\sqrt{\frac{(-24.31)^2}{7058.98} + \frac{1}{42} + 1}\right) \\ &= (13.91 \pm 9.49) = (4.4, 23.4).\end{aligned}$$

The prediction interval is extremely wide: the reason for this is that there is a great deal of scatter in the data.

Solution 10.12

From the data given in Table 10.13

$$n = 15, \quad \overline{x} = 16.653, \quad \sum(x_i - \overline{x})^2 = 40.557, \quad \widehat{\alpha} = 25.2323,$$

$$\widehat{\beta} = 3.2911, \quad \sum(y_i - \widehat{y}_i)^2 = 190.547, \quad s^2 = 14.6575.$$

Some of these were calculated in Exercise 10.7, though not to as many places of decimals.

Also, the 99.5% quantile of $t(13)$ is 3.012. The predicted mean is

$$\widehat{\alpha} + \widehat{\beta}x_0 = 84.472.$$

The 99% prediction interval is

$$\left(84.472 \pm 3.012\sqrt{14.6575}\sqrt{\frac{(18 - 16.653)^2}{40.557} + \frac{1}{15} + 1}\right)$$

$$= (84.472 \pm 12.157) = (72.3, 96.6).$$

Solution 10.13

From the data given in Table 10.12,

$$n = 42, \quad \overline{x} = 1104.69, \quad \sum(x_i - \overline{x})^2 = 1\,229\,650.98, \quad \widehat{\alpha} = 73.480,$$

$$\widehat{\beta} = -0.0242, \quad \sum(y_i - \widehat{y}_i)^2 = 689.33, \quad s^2 = 17.2333.$$

Some of these were calculated in Exercise 10.6, though less precisely.

The 97.5% quantile of $t(40)$ is 2.021. The predicted mean consumption is

$$\widehat{\alpha} + \widehat{\beta}x_0 = 47.344.$$

The 95% prediction interval is

$$\left(47.354 \pm 2.021\sqrt{17.2333}\sqrt{\frac{(1080 - 1104.69)^2}{1\,229\,650.98} + \frac{1}{42} + 1}\right)$$

$$= (47.354 \pm 8.491) = (38.9, 55.8).$$

Chapter 11

Solution 11.1

In scatter plot (a) the variables are negatively related.

In scatter plot (b) the variables are positively related.

In scatter plot (c) the variables do not appear to be related at all. Knowing the value of one of them tells you nothing about the value of the other.

Solution 11.2

(a) The random variables X and Y are related, because the conditional probability that $Y = 10$ given $X = 4$ is not the same as the unconditional probability that $Y = 10$.

(b) In this case it is not possible to say whether W and Z are related. The question does not give enough information. Knowing that $W = 4$ tells us nothing new about the probability that Z took the value 5. However, we do not know what would happen if we knew, say, that W took the value 6. Would that change the probability distribution of Z? If knowing the values of W does not change the probability distribution of Z, then W and Z are not related, but otherwise they are related.

Solution 11.3

(a) Altogether, out of 2484 people, 110 provided a value of Yes for the random variable X. Thus an estimate for the probability $P(X = \text{Yes})$ is 110/2484 or 0.044.

(b) There are 254 people in Table 11.2 for whom the value of the random variable Y is Snore every night. Of these, 30 provided a value of Yes for X (heart disease). Thus an estimate for the conditional probability $P(X = \text{Yes}|Y = \text{Snore every night})$ is 30/254 or 0.118. This is getting on for three times the unconditional probability that $X = \text{Yes}$. That is, knowing that someone snores every night tells you something about how likely it is that they have heart disease: snorers are more likely to have it. (Note that this does not tell you that snoring causes heart disease, or for that matter that heart disease causes snoring. More on this point is discussed in Section 11.3.)

This could be put more formally, if you prefer. Unless for some reason these estimates are very inaccurate, it appears that

$$P(X = \text{Yes}|Y = \text{Snore every night}) \neq P(X = \text{Yes});$$

therefore, using (11.1), X and Y are related.

Solution 11.4

Your answer may be as follows. The relationship is negative, therefore the Pearson correlation r will be negative, and the amount of scatter in the data in Figure 11.3 is not too different from that in Figure 11.7(b). The value of r for Figure 11.7(b) is 0.787, so the value of r for Figure 11.3 might be around -0.7 or -0.8. Perhaps a guess of anything between -0.6 and -0.9 would be reasonable. In fact, the value of r for Figure 11.3 is -0.767.

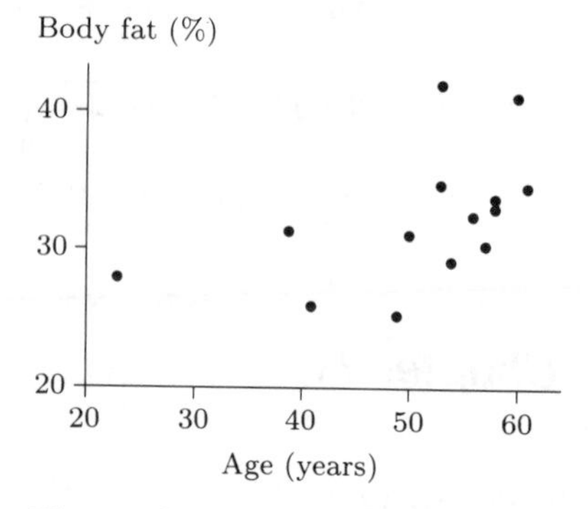

Figure S11.1 Body fat percentage against age

Solution 11.5

(a) The scatter plot is as shown in Figure S11.1. This clearly shows that the two variables are positively related, though the association between them is not particularly strong.

(b) For these data,

$$n = 14, \quad \sum x_i = 712, \quad \sum y_i = 452.5, \quad \sum x_i^2 = 37\,600,$$
$$\sum y_i^2 = 14\,937.57, \quad \sum x_i y_i = 23\,346.5.$$

So using (11.5),

$$r = \frac{14 \times 23\,346.5 - 712 \times 452.5}{\sqrt{(14 \times 37\,600 - 712^2)(14 \times 14\,937.57 - 452.5^2)}}$$
$$= \frac{4671}{\sqrt{19\,456 \times 4369.73}} = 0.507.$$

This value indicates a moderate positive association between the two variables.

Solution 11.6

(a) The correlation is 0.743. This matches the impression given by the scatter plot of a reasonably strong positive relationship between the variables.

If you want to check your computer output, the exact value of r using (11.5) is

$$r = \frac{37\,880}{\sqrt{2\,600\,135\,980}}.$$

(b) The scatter plot for these data is as shown in Figure S11.2. This shows a reasonably strong positive association between the variables. The Pearson correlation coefficient is 0.716, which gives the same impression of the strength of the relationship. The exact value of r is

$$r = \frac{193\,000}{\sqrt{72\,610\,213\,900}}.$$

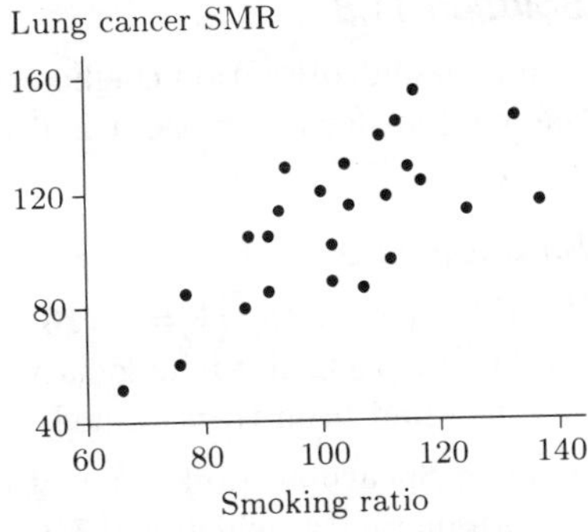

Figure S11.2 Lung cancer SMR against smoking ratio

Solution 11.7

(a) The scatter plot is as shown in Figure S11.3.

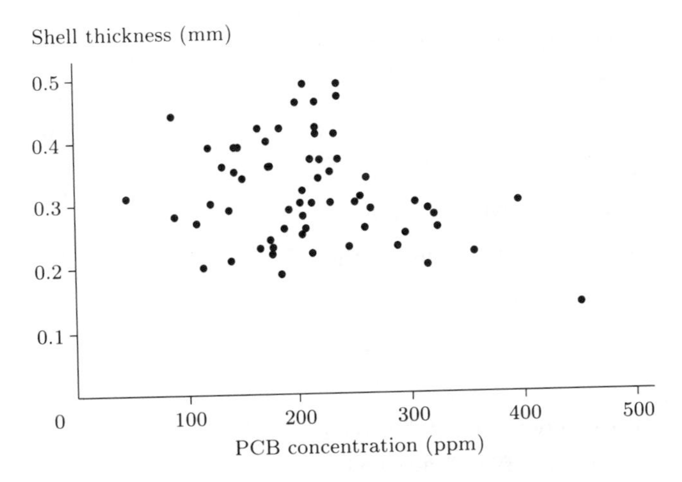

Figure S11.3 Shell thickness against PCB concentration

There seems to be a rather weak negative relationship between the two variables; the more the PCB, the thinner the shell. But remember that we cannot conclude from these data that the PCB causes the shells to become thin. (However, there is evidence from other sources that this causal explanation is true.)

Several points appear to be some distance from the main pattern; but the most obvious is the point at the bottom right. It is the thinnest shell with the highest concentration of PCB.

(b) The Pearson correlation coefficient for the full data set is -0.253. This is in line with the interpretation (a weak negative relationship) in part (a).

(c) Omitting the most extreme point at the bottom right from the calculation, the Pearson correlation becomes -0.157. This value is considerably nearer 0 than the correlation coefficient for the full data set. This indicates that much of the impression that these two variables are negatively correlated stems from this single egg, which seems to be rather atypical.

Solution 11.8

The Pearson correlation coefficient for the untransformed data is -0.005. For the log transformed data, it is 0.779.

Solution 11.9

(a) For these data, $r_S = 0.716$. This fairly large value corresponds to the fairly strong linear association seen in the scatter plot of the data after they had been transformed.

(b) The Spearman rank correlation coefficient would be the same as for the original data, which is 0.716. This is because the logarithmic transformation does not change the order in which the data come; consequently, the ranks of the log-transformed data are the same as the ranks of the original data. (If you do not believe this, check it!)

Solution 11.10

The computed SP is

$$SP(\text{obtained direction}) = 0.0323;$$

so there is moderate evidence of a positive relationship between these two variables in the population.

Solution 11.11

(a) In this case you should calculate $r\sqrt{\frac{n-2}{1-r^2}}$ and compare it against a t-distribution with $64 - 2 = 62$ degrees of freedom. The value of this quantity is

$$-0.157\sqrt{\frac{64-2}{1-(-0.157)^2}} = -1.256.$$

Hence the total SP is 0.214. There is no evidence of a relationship between the variables. Our previous impression, that the apparent relationship between the variables depends on the single extreme egg, is confirmed.

(b) The calculations are much the same as for part (a). You need to calculate the test statistic and compare it against a t-distribution on $28 - 2 = 26$ degrees of freedom. The value of this quantity is

$$r_S\sqrt{\frac{n-2}{1-r_S^2}} = 0.716\sqrt{\frac{28-2}{1-0.716^2}} = 5.232.$$

It is not really necessary to turn on your computer or open your statistical tables to see that the obtained SP is approximately zero. There is very strong evidence that the two variables are related.

Solution 11.12

Using Fisher's exact test, $SP = 0.0001$. There is very strong evidence that the two variables are related. It is clear from the original table that the relationship works in such a way that patients with impaired sulphoxidation are more likely to exhibit a toxic reaction to the drug.

Solution 11.13

The expected frequencies are as follows.

Season	Colour pattern		Row totals
	Bright red	Not bright red	
Spring	280.918	223.082	504
Summer	93.082	73.918	167
Column totals	374	297	671

In all four cells, the absolute difference between observed and expected frequencies is 21.082. This gives

$$\chi^2 = \frac{21.082^2}{280.918} + \frac{21.082^2}{223.082} + \frac{21.082^2}{93.082} + \frac{21.082^2}{73.918} = 14.36$$

and the SP is 0.000 15. There is strong evidence of association between season and colour pattern. Comparing the expected and observed frequencies, we see that there are more bright red beetles in spring and fewer in summer than is expected under the null hypothesis of no association.

Solution 11.14

(a) Looking at the observed and expected frequencies, Compressor 1 seems to fail relatively frequently in the centre leg and relatively rarely in the south leg, while Compressor 4 fails relatively rarely in the centre leg and relatively frequently in the south leg. However, the value of the chi-squared test statistic is 11.72, and since there are 4 rows and 3 columns, the number of degrees of freedom is $(4-1)(3-1) = 6$. The significance probability is $SP = 0.068$. There is only very weak evidence of association between the two variables: the data are consistent with the null hypothesis that the pattern of location of failures is the same in all four compressors.

(b) The value of the chi-squared test statistic is 7.885. The number of degrees of freedom is $(3-1)(2-1) = 2$, and the significance probability is $SP = 0.019$. There is fairly strong evidence of a relationship between tonsil size and carrier status. Comparing the observed and expected frequencies, it appears that carriers are less likely than non-carriers to have normal tonsil size, and carriers are more likely than non-carriers to have very large tonsils. In short, on average, carriers have larger tonsils than do non-carriers.

Solution 11.15

Your scatter plot should look like the scatter plot in Figure S11.4.

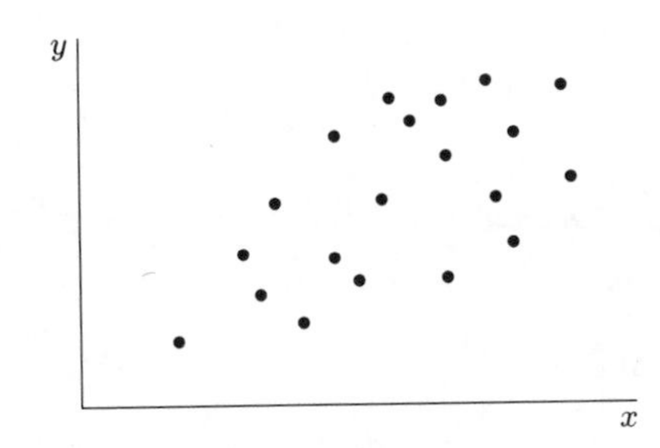

Figure S11.4

Chapter 12

Solution 12.1

(a) The adequacy of the Bernoulli model would rather depend on what service was being provided. For instance, queues at a bank or a post office might consist largely of individuals who have arrived at the service point independently of one another, and alone. On the other hand, queues at a cinema box office will often include male–female pairs, implying a strong dependence between consecutive individuals.

(b) Rather as for sequences of wet and dry days, it is likely that there will be some noticeable association between the characteristics of consecutive days, and a Bernoulli model would not be appropriate.

(c) Some card-players strenuously maintain a belief in runs of luck (good and bad) and if the phenomenon exists, then the Bernoulli process will not be an appropriate model here. For games involving some skill, it is probable that players perform better on some occasions than on others, over relatively long periods. For games involving chance alone there may be some association (the order of cards dealt) but it would be difficult to demonstrate and very hard to quantify.

Solution 12.2

(a) Using the probabilities at (12.1), three different weekly weather sequences (starting with a wet day) were simulated. They were as follows.

1 1 0 0 0 0 0

1 0 0 1 1 0 0

1 0 0 0 0 0 0

You should have generated a sequence similar to these.

(b) Four different families of size 4 were generated (starting with a girl). They were as follows.

0 1 0 1

0 1 1 1

0 1 0 1

0 1 1 1

Again, you should have generated a similar sequence.

Solution 12.3

(a) Using (12.3), the transition matrix M is given by

$$M = \begin{matrix} 0 \\ 1 \end{matrix} \begin{bmatrix} 0.4567 & 0.5433 \\ 0.5007 & 0.4993 \end{bmatrix},$$

and the overall probability of a boy is given by

$$p = \frac{\alpha}{\alpha + \beta} = \frac{0.5433}{0.5433 + 0.5007} = 0.5204.$$

(b) Three typical families of 5 children were generated. They were

00110
01011
11011.

Solution 12.4

(a) The matrix of transition frequencies is

$$N = \begin{matrix} 0 \\ 1 \end{matrix} \begin{bmatrix} 9 & 8 \\ 9 & 13 \end{bmatrix} \begin{matrix} 17 \\ 22 \end{matrix};$$

the corresponding matrix of estimated transition probabilities is

$$\widehat{M} = \begin{matrix} 0 \\ 1 \end{matrix} \begin{bmatrix} 9/17 & 8/17 \\ 9/22 & 13/22 \end{bmatrix} = \begin{matrix} 0 \\ 1 \end{matrix} \begin{bmatrix} 0.529 & 0.471 \\ 0.409 & 0.591 \end{bmatrix}.$$

(b) $$N = \begin{matrix} 0 \\ 1 \end{matrix} \begin{bmatrix} 15 & 10 \\ 10 & 4 \end{bmatrix} \begin{matrix} 25 \\ 14 \end{matrix}; \quad \widehat{M} = \begin{matrix} 0 \\ 1 \end{matrix} \begin{bmatrix} 15/25 & 10/25 \\ 10/14 & 4/14 \end{bmatrix} = \begin{matrix} 0 \\ 1 \end{matrix} \begin{bmatrix} 0.600 & 0.400 \\ 0.714 & 0.286 \end{bmatrix}.$$

(c) $$N = \begin{matrix} 0 \\ 1 \end{matrix} \begin{bmatrix} 12 & 2 \\ 3 & 22 \end{bmatrix} \begin{matrix} 14 \\ 25 \end{matrix}; \quad \widehat{M} = \begin{matrix} 0 \\ 1 \end{matrix} \begin{bmatrix} 12/14 & 2/14 \\ 3/25 & 22/25 \end{bmatrix} = \begin{matrix} 0 \\ 1 \end{matrix} \begin{bmatrix} 0.857 & 0.143 \\ 0.120 & 0.880 \end{bmatrix}.$$

(d) $$N = \begin{matrix} 0 \\ 1 \end{matrix} \begin{bmatrix} 25 & 4 \\ 3 & 7 \end{bmatrix} \begin{matrix} 29 \\ 10 \end{matrix}; \quad \widehat{M} = \begin{matrix} 0 \\ 1 \end{matrix} \begin{bmatrix} 25/29 & 4/29 \\ 3/10 & 7/10 \end{bmatrix} = \begin{matrix} 0 \\ 1 \end{matrix} \begin{bmatrix} 0.862 & 0.138 \\ 0.300 & 0.700 \end{bmatrix}.$$

(e) $$N = \begin{matrix} 0 \\ 1 \end{matrix} \begin{bmatrix} 9 & 14 \\ 15 & 1 \end{bmatrix} \begin{matrix} 23 \\ 16 \end{matrix}; \quad \widehat{M} = \begin{matrix} 0 \\ 1 \end{matrix} \begin{bmatrix} 9/23 & 14/23 \\ 15/16 & 1/16 \end{bmatrix} = \begin{matrix} 0 \\ 1 \end{matrix} \begin{bmatrix} 0.391 & 0.609 \\ 0.938 & 0.063 \end{bmatrix}.$$

(f) $$N = \begin{matrix} 0 \\ 1 \end{matrix} \begin{bmatrix} 12 & 13 \\ 13 & 1 \end{bmatrix} \begin{matrix} 25 \\ 14 \end{matrix}; \quad \widehat{M} = \begin{matrix} 0 \\ 1 \end{matrix} \begin{bmatrix} 12/25 & 13/25 \\ 13/14 & 1/14 \end{bmatrix} = \begin{matrix} 0 \\ 1 \end{matrix} \begin{bmatrix} 0.480 & 0.520 \\ 0.929 & 0.071 \end{bmatrix}.$$

Solution 12.5

For Exercise 12.4(c) the matrix of transition frequencies is

$$N = \begin{matrix} 0 \\ 1 \end{matrix} \begin{bmatrix} 12 & 2 \\ 3 & 22 \end{bmatrix} \begin{matrix} 14 \\ 25 \end{matrix};$$

the number of runs is

$$r = 2 + 3 + 1 = 6.$$

For Exercise 12.4(e) the matrix of transition frequencies is

$$N = \begin{matrix} 0 \\ 1 \end{matrix} \begin{bmatrix} 9 & 14 \\ 15 & 1 \end{bmatrix} \begin{matrix} 23 \\ 16 \end{matrix};$$

the number of runs is

$$r = 14 + 15 + 1 = 30.$$

Solution 12.6

(a) The total SP is 0.421; there is no evidence to reject a Bernoulli model here.

(b) The total SP is 0.853; there is no evidence to reject a Bernoulli model here.

(c) The total SP is 1.4×10^{-6}, which is very small indeed. The Bernoulli model is firmly rejected. (In fact, the realization shows a very small number of long runs.)

(d) The total SP is 0.004; the realization shows a small number of long runs inconsistent with a Bernoulli model.

(e) The SP is very small (0.001); but here there are many short runs inconsistent with a Bernoulli model.

(f) The total SP is 0.016; there is evidence to reject the Bernoulli model in the light of many short runs.

Solution 12.7

(a) The matrix of transition frequencies is

$$N = \begin{matrix} 0 \\ 1 \end{matrix} \begin{bmatrix} 7 & 4 \\ 5 & 7 \end{bmatrix} \begin{matrix} 11 \\ 12 \end{matrix};$$

the corresponding matrix of estimated transition probabilities is

$$\widehat{M} = \begin{matrix} 0 \\ 1 \end{matrix} \begin{bmatrix} 7/11 & 4/11 \\ 5/12 & 7/12 \end{bmatrix} = \begin{matrix} 0 \\ 1 \end{matrix} \begin{bmatrix} 0.636 & 0.364 \\ 0.417 & 0.583 \end{bmatrix}$$

(b) The number of runs in the data is

$$r = n_{01} + n_{10} + 1 = 4 + 5 + 1 = 10.$$

(c) The total SP against a hypothesized Bernoulli model is $SP = 0.301$. There is no evidence to reject a Bernoulli model for the sequence of 0s and 1s here.

(d) In this case, $n_0 = n_1 = 12$ and so from (12.7)

$$E(R) = \frac{2 \times 12 \times 12}{12 + 12} + 1 = 13, \qquad V(R) = \frac{288 \times 264}{24^2 \times 23} = 5.739.$$

The corresponding z-score is

$$z = \frac{r - E(R)}{SD(R)} = \frac{10 - 13}{\sqrt{5.739}} = -1.252;$$

the corresponding SP is $2 \times \Phi(-1.252) = 0.211$. The number of runs is not noticeably extreme (high or low). Again, there is no evidence to reject the Bernoulli model here.

Solution 12.8

(a) The accumulated times are given in Table S12.1.

Table S12.1

13.2	14.7	16.5	24.1	32.3
43.0	52.4	60.5	62.0	64.3
67.3	68.9	73.4	80.7	90.1
103.9	134.0	135.2	136.0	142.2
149.2	156.6	162.9	164.1	168.3
172.8	175.1	176.8	179.4	183.0
190.5	200.8	204.1	232.3	255.4
257.1	259.5	263.0	279.2	280.7
293.6	299.2	303.5	316.7	319.7
325.3	329.7	344.6	345.8	349.8

Observation ceased with the passing of the 50th vehicle at time $\tau = 349.8$. This leaves 49 vehicle observations:

$$w_1 = \frac{t_1}{\tau} = \frac{13.2}{349.8} = 0.038, \quad w_2 = \frac{t_2}{\tau} = \frac{14.7}{349.8} = 0.042, \quad \ldots,$$

$$w_{49} = \frac{t_{49}}{\tau} = \frac{345.8}{349.8} = 0.989;$$

the observed value of the Kolmogorov test statistic D is $d = 0.090$ with $n = 49$. The corresponding SP exceeds 0.2. We can conclude that the Poisson process is a reasonable model for the traffic flow along Burgess Road during the observation period.

(b) Here τ is known to be 1608; we therefore have 56 observations

$$w_1 = \frac{t_1}{\tau} = \frac{28}{1608} = 0.017, \quad w_2 = \frac{t_2}{\tau} = \frac{38}{1608} = 0.024, \quad \ldots,$$

$$w_{56} = \frac{t_{56}}{\tau} = \frac{1591}{1608} = 0.989;$$

the observed value of D is $d = 0.159$ with $n = 56$. The corresponding SP is between 0.1 and 0.2: there is some small evidence that volcano eruptions are not well fitted by a Poisson process.

Chapter 13

Solution 13.1

Use of the word *to* provides possibly the most difficult discriminant of the three: *by* would have been easier! But we are obliged to use the data available, not the data we wish had been available. Histograms showing Hamilton's and Madison's use of the word *to* are given in Figure S13.1.

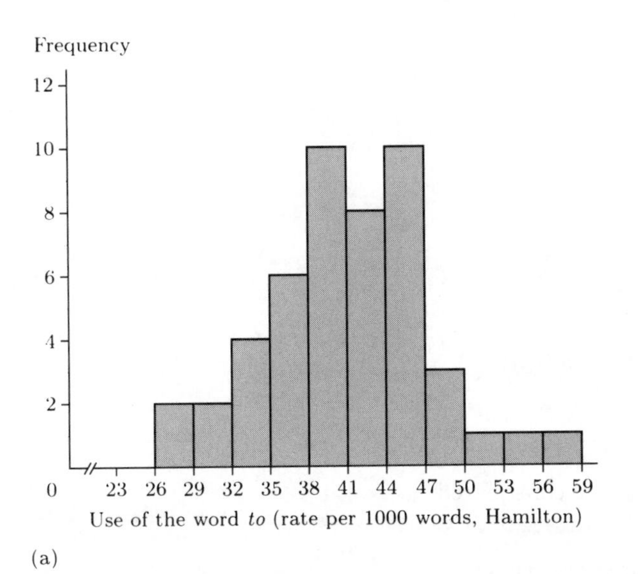

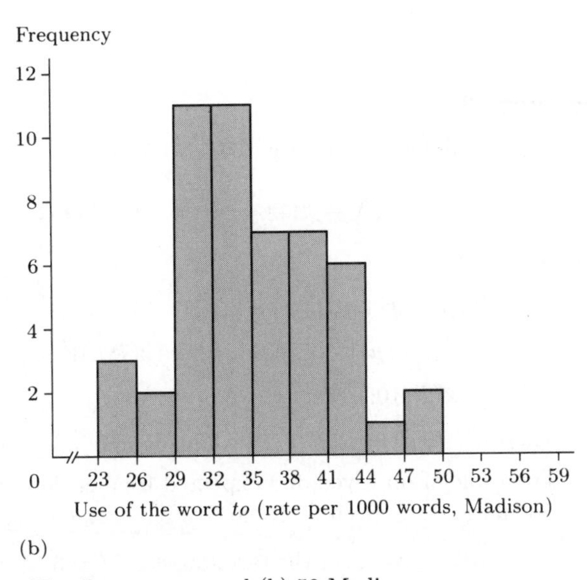

Figure S13.1 Distribution of rates of occurrence of *to* in (a) 48 Hamilton papers and (b) 50 Madison papers

If we were sure that all twelve disputed papers were by the same author, then by comparison with Figure 13.1, the histograms might possibly suggest Madison as the author.

Solution 13.2

The completed table of observed and expected frequencies, showing the chi-squared calculations, is given in Table S13.1. The last four cells were pooled to ensure that all expected frequencies were at least 5.

Table S13.1 Weldon's data: chi-squared calculations

Number of 5s or 6s	Observed frequency	Expected frequency	$O_i - E_i$	$\frac{(O_i - E_i)^2}{E_i}$
0	45	54.0	−9.0	1.50
1	327	324.0	3.0	0.03
2	886	891.0	−5.0	0.03
3	1475	1484.9	−9.9	0.07
4	1571	1670.6	−99.6	5.94
5	1404	1336.4	67.6	3.42
6	787	779.6	7.4	0.07
7	367	334.1	32.9	3.24
8	112	104.4	7.6	0.55
9	29	23.2		
10	2	3.5	5.0	0.93
11	1	0.3		
12	0	0.0		
Total	7006	7006	0	15.78

The observed chi-squared value

$$\chi^2 = \sum \frac{(O_i - E_i)^2}{E_i} = 1.50 + 0.03 + \cdots + 0.93 = 15.78$$

is compared against the χ^2 distribution with $10 - 1 = 9$ degrees of freedom. (There are ten cells; the binomial model $B(12, \frac{1}{3})$ was specified completely, with no parameters requiring estimation.) The corresponding SP is 0.072. There is some evidence of a poor binomial fit, but the evidence is not overwhelming—Pearson's objections seem somewhat exaggerated.

Solution 13.3

(a) The likelihood of μ for the data, writing

$$p(x;\mu) = P(X = x) = \frac{e^{-\mu}\mu^x}{x!}, \quad x = 0, 1, 2, \ldots$$

is given by

$$(p(0;\mu) + p(1;\mu))^1 \times (p(2;\mu))^2 \times (p(3;\mu))^5 \times (p(4;\mu))^9$$
$$\times (p(5;\mu))^{10} \times (p(6;\mu))^5 \times (p(7;\mu))^2 \times (p(8;\mu))^3 \times (p(9;\mu))^1$$
$$\times (P(10;\mu))^1$$

where $P(10;\mu) = P(X \geq 10)$. This is maximized at $\mu = \widehat{\mu} = 4.9621$.

(b) The table of observed and expected frequencies, showing the chi-squared calculations, is given in Table S13.2.

The observed value of the test statistic $\chi^2 = 3.51$ compared against $\chi^2(4)$ (six cells, one parameter estimated from the data) gives a SP of 0.476. There is no evidence to reject the hypothesis of a Poisson model for the goal frequency. It is reasonable to suppose that the incidence of goals occurs at random during the course of play.

Table S13.2

Number of goals	Observed frequency	Expected frequency	$O_i - E_i$	$\frac{(O_i - E_i)^2}{E_i}$
0 – 1	1	1.63	−1.99	0.79
2	2	3.36		
3	5	5.56	−0.56	0.06
4	9	6.89	2.11	0.65
5	10	6.84	3.16	1.46
6	5	5.66	−0.66	0.08
7	2	4.01	−2.06	0.47
8	3	2.49		
9	1	1.37		
≥ 10	1	1.19		
Total	39	39	0	3.51

Solution 13.4

(a) The seven differences are

$$33, 2, 24, 27, 4, 1, -6$$

with mean $\overline{d} = 12.143$ and standard deviation $s = 15.378$. The corresponding value of test statistic t in a test of zero mean difference is

$$t = \frac{\overline{d}}{s/\sqrt{n}} = 2.089.$$

(b) The t-distribution against which the test statistic is compared has 6 degrees of freedom, and $P(T_6 > 2.089) = 0.041$.

No indication has been given whether there should be an expected increase or decrease in the CO transfer factors. For a two-sided test, the corresponding SP is 0.082. This provides little evidence that there is any significant difference between CO transfer factors in smokers at entry and one week later.

Solution 13.5

The normal probability plot is shown in Figure S13.2.

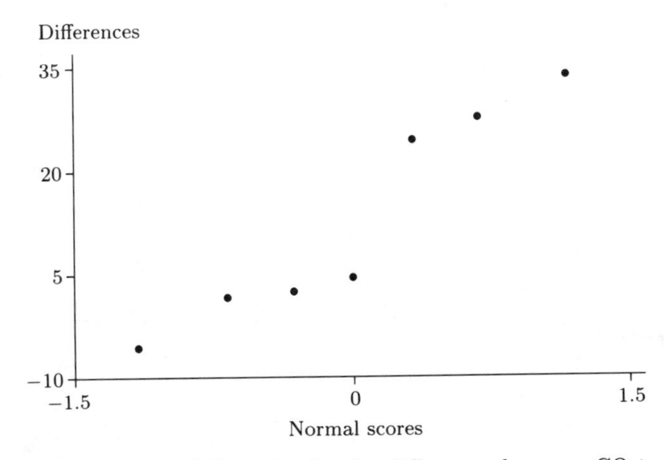

Figure S13.2 Normal probability plot for the differences between CO transfer factors

The plot shows that the data are split into two smaller groups—a group of four where there is little change in the transfer factor and a group of three where the change is comparatively large. It is clear that the normality assumption is not tenable.

Solution 13.6

Wilcoxon's signed rank test involves ranking the absolute differences, as shown below.

Table S13.3 Ranked differences

Patient	Entry	One week	Difference	Rank
1	40	73	33	7
2	50	52	2	2
3	56	80	24	5
4	58	85	27	6
5	60	64	4	3
6	62	63	1	1
7	66	60	−6	4

The sum over the positive ranks is 24 and the sum over the negative ranks is 4. The observed value of the test statistic is $w_+ = 24$. Using a computer, the total SP is 0.109, and there is no reason to reject the null hypothesis of no difference between the CO transfer factors.

Using a normal approximation, the corresponding z-score is

$$z = \frac{w_+ - E(W_+)}{SD(W_+)} = \frac{24 - 14}{\sqrt{35}} = 1.690,$$

and since $P(Z > z) = 0.0455$, the total SP is 0.091.

Solution 13.7

(a) The normal probability plot for the differences is shown in Figure S13.3.

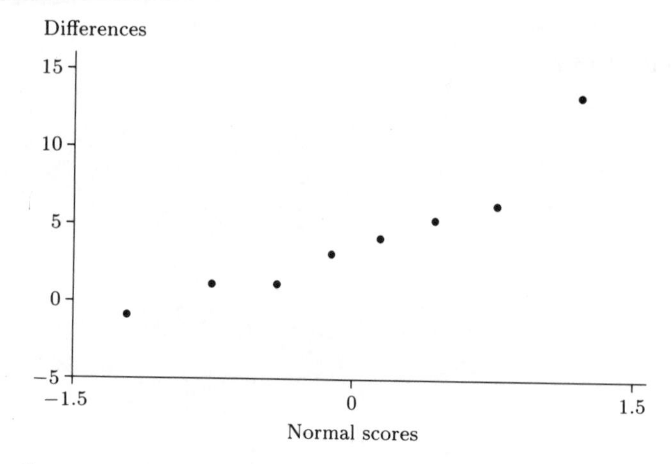

Figure S13.3 Normal probability plot

Clearly there is an outlier and this is confirmed in Table S13.4, which shows the differences. Leaf 1 is atypical.

Table S13.4 Viral lesions on tobacco leaves

Leaf	Preparation 1	Preparation 2	Difference
1	31	18	13
2	20	17	3
3	18	14	4
4	17	11	6
5	9	10	−1
6	8	7	1
7	10	5	5
8	7	6	1

However, the other seven points appear to lie on a straight line and removal of the outlier should leave data which are plausibly normal.

(b) A t-test on the seven points gives a t-value of 2.875 which, tested against $t(6)$, results in a total SP of 0.028. There is evidence for rejecting the null hypothesis, and for concluding that the two preparations have a significantly different effect. (The experimental results suggest that the first preparation leads to a substantially higher average number of lesions.)

Solution 13.8

(a) First, let us consider the two groups at Day 0. Sample statistics are

$$n_1 = 20, \quad \overline{x}_1 = 0.3945, \quad s_1^2 = 0.0089,$$
$$n_2 = 10, \quad \overline{x}_2 = 0.3520, \quad s_2^2 = 0.0093.$$

Notice that the sample variances are very close, well within a factor of 3 of one another. There is no evidence that the assumption of equal variances for the two populations is broken. The pooled sample variance is given by

$$s_p^2 = \frac{(n_1 - 1)s_1^2 + (n_2 - 1)s_2^2}{n_1 + n_2 - 2} = 0.009\,031,$$

and the value of the test statistic t for a test of equal population means is

$$t = \frac{\overline{x}_1 - \overline{x}_2}{s_p\sqrt{\dfrac{1}{n_1} + \dfrac{1}{n_2}}} = 1.155.$$

Compared against Student's t-distribution with $n_1 + n_2 - 2 = 28$ degrees of freedom, this gives a total SP of $2 \times 0.129 = 0.258$.

There is no evidence that there is any difference between the mean urea levels at admission for the two groups of patients.

(b) For the two groups at Day 6, the analysis proceeds as follows. Sample statistics are

$$n_1 = 20, \quad \overline{x}_1 = 0.5390, \quad s_1^2 = 0.0168,$$
$$n_2 = 10, \quad \overline{x}_2 = 0.6830, \quad s_2^2 = 0.0254.$$

Again, the sample variances do not suggest any significant difference between the two population variances.

The pooled sample variance is

$$s_p^2 = 0.0196,$$

and the value of the test statistic t is

$$t = \frac{\overline{x}_1 - \overline{x}_2}{s_p\sqrt{\dfrac{1}{n_1} + \dfrac{1}{n_2}}} = -2.655.$$

Compared against $t(28)$, this gives a total SP of $2 \times 0.0065 = 0.013$.

In this case, there is substantial evidence that by Day 6 after admission there is a significant difference between mean serum urea levels for the surviving and non-surviving patients.

Solution 13.9

The regression coefficients, computed separately for each rat, appear in Table S13.5. In fact, the difference between the groups is striking—the second group have the larger growth rates. On the other hand, there are only four values in the second group, and one of these is similar in size to the values from the first group. So, with such small sample sizes, could these results have arisen easily by chance?

Table S13.5 Regression slopes for each rat separately

Group 1:	4.3	1.2	1.0	0.8	2.3	0.4	3.8	3.4
Group 2:	7.4	8.8	1.4	8.4				

The means for the first and second groups are respectively 2.15 and 6.50 and their respective standard deviations are 1.514 and 3.451. At first glance there seems to be a difference, but we should not jump to conclusions at this stage.

Solution 13.10

Group A:	4.3	1.2	1.0	0.8	2.3	0.4	3.8	3.4
Rank	9	4	3	2	6	1	8	7
Group B:	7.4	8.8	1.4	8.4				
Rank	10	12	5	11				

In this solution the groups have been relabelled A and B to achieve a convenient notation, consistent with previous work.

The value of the Mann–Whitney–Wilcoxon test statistic is $u_A = 40$ with a computed total SP of 0.048. There is some evidence that the rates of growth in the two groups differ.

(The normal approximation gives

$$z = \frac{u_A - E(U_A)}{SD(U_A)} = \frac{40 - 52}{\sqrt{34.667}} = -2.038$$

with a total SP of 0.042.)

Solution 13.11

The result of performing Fisher's exact test on the two sample proportions 17/31 and 16/28 is

$$SP(\text{obtained direction}) = 0.534$$

using a one-tailed test. For a two-sided test exploring merely whether there is a significant difference, the total SP is 1. There is no evidence for a difference in either direction.

Solution 13.12

The expected values for each cell are shown in brackets. For instance, the expected value in the top left-hand cell is found from

$$\frac{90 \times 47}{367} = 11.5.$$

	Hospital					Total
	A	B	C	D	E	
No improvement	13 (11.5)	5 (7.6)	8 (19.4)	21 (31.4)	43 (20.1)	90
Partial	18 (19.1)	10 (12.6)	36 (32.1)	56 (52.0)	29 (33.3)	149
Complete	16 (16.4)	16 (10.8)	35 (27.5)	51 (44.6)	10 (28.6)	128
Total	47	31	79	128	82	367

The chi-squared test statistic is calculated from

$$\chi^2 = \sum \frac{(O_i - E_i)^2}{E_i} = 56.7,$$

summed over all 15 cells. For the chi-squared test of independence, we need the number of degrees of freedom for the null distribution. This parameter is calculated as $(r-1)(c-1)$, where r is the number of rows in the table and c the number of columns. In our case these are 3 and 5 respectively, so that the distribution we need is the chi-squared distribution with $(3-1)(5-1) = 8$ degrees of freedom.

The probability of obtaining a value as high as 56.7 from a chi-squared distribution with 8 degrees of freedom is very low indeed, about 2×10^{-9}. So we conclude from this test that there are real differences between the distributions of outcomes across the hospitals.

Solution 13.13

The chi-squared test statistic is 20.85, and relating this to a chi-squared distribution with $(4-1)(4-1) = 9$ degrees of freedom gives a SP equal to 0.013. There is thus a low probability that such a distribution would be obtained by chance if the two variables really were independent. We conclude that serum cholesterol level and systolic blood pressure are associated.

Solution 13.14

Depending on your software, you might be able to calculate directly that for a bivariate sample of size 10 with an underlying correlation of zero, the probability of obtaining a sample correlation of $r = -0.72$ or less is

$$P(R \leq -0.72) = P(R \geq 0.72) = 0.0098$$

and so the total SP for the test is $2 \times 0.0098 = 0.02$.

This offers considerable evidence that there is an underlying association between PEF and the S:C ratio; in fact, there is evidence that there is a negative association between the two measures.

(Alternatively, use the fact that

$$R\sqrt{\frac{n-2}{1-R^2}} \sim t_{(n-2)}$$

and compare

$$t = r\sqrt{\frac{n-2}{1-r^2}} = -0.72\sqrt{\frac{10-2}{1-(-0.72)^2}} = -2.910$$

against $t(8)$. Again, the SP for the test is given by $2 \times 0.0098 = 0.02$.)

Solution 13.15

The correlation coefficient is given by $r = 0.971$. This is a high value and we see that the number of finger ridges in identical twins are highly correlated. The SP for a test that the underlying correlation is zero is given by 1.5×10^{-7}: this is very low!

Solution 13.16

Writing x for the explanatory variable (wind speed) and y for the response (race time), summary statistics are

$$n = 21, \quad \sum x_i = 4.7, \quad \sum y_i = 279.36, \quad \sum x_i^2 = 45.11,$$
$$\sum x_i y_i = 58.796,$$

and the estimated slope is

$$\widehat{\beta} = \frac{n\sum x_i y_i - \sum x_i \sum y_i}{n\sum x_i^2 - (\sum x_i)^2} = \frac{21 \times 58.796 - 4.7 \times 279.36}{21 \times 45.11 - (4.7)^2}$$
$$= \frac{-78.276}{925.22} = -0.084\,602\,6.$$

Also,

$$\widehat{\alpha} = \overline{y} - \widehat{\beta}\overline{x} = 13.321\,792.$$

Consequently, the fitted regression model is

$$\text{Race time} = 13.32 - 0.085 \times \text{Wind speed},$$

where race time is measured in seconds and wind speed in metres per second. This reflects the fact, suggested in the scatter plot of the data, that stronger following winds tend to lead to reduced race times. However, the model should not be extrapolated too far. For instance, the current world record for 110 m Hurdles (men) is 12.91 s (held by Jackson). The model suggests that with wind speeds much above 4.9 m/s, he would routinely race inside world record times! In fact, some hurdlers are hampered by severe following winds: it gets them too close to the next hurdle to jump, and therefore destroys their rhythm.

Solution 13.17

The sampling distribution of the estimator, assuming the scatter to be normally distributed about the model $y = \alpha + \beta x$, is given by

$$\frac{\widehat{\beta} - \beta}{S/\sqrt{\sum(x_i - \overline{x})^2}} \sim t_{(n-2)}.$$

Under the null hypothesis $H_0 : \beta = 0$, the value of the test statistic is

$$\frac{\widehat{\beta}}{s/\sqrt{\sum(x_i - \overline{x})^2}}$$

where $\widehat{\beta} = -0.085$, $\quad s^2 = \dfrac{\sum(y_i - \widehat{y})^2}{n-2} = \dfrac{0.4665}{19} = 0.024\,55,$

and $\sum(x_i - \overline{x})^2 = 44.058\,095$. So

$$t = -\frac{0.085}{\sqrt{0.024\,55}\,/\sqrt{44.058\,095}} = -3.584.$$

The obtained SP for the test is $P(T_{19} \leq -3.584) = 0.001$. There is very considerable evidence from these data to reject the hypothesis that, in fact, $\beta = 0$, and so wind speed effect is significant.

Solution 13.18

Writing $y = \log(n_t)$, $x = t$, then the regression line of y on x has slope

$$\widehat{\beta} = \frac{n\sum xy - \sum x \sum y}{n\sum x^2 - (\sum x)^2} = -0.0364$$

and intercept

$$\widehat{\alpha} = 10.578.$$

The fitted model is therefore

$$y = \widehat{\alpha} + \widehat{\beta}x = 10.578 - 0.0364x$$

or

$$\log(n_t) = 10.578 - 0.0364t$$

or, taking exponentials,

$$n_t = e^{10.578 - 0.0364t} = 39\,270e^{-0.0364t}.$$

The estimated value of n_0 is 39 270 (or, say, 40 000).

Appendix

Table A1 Random digits

00	51996	67729	79247	41569	33165	35451	23991	58948	35972	79025
01	42589	94219	48403	95170	51525	26521	19879	88310	06468	69670
02	47158	09333	20386	98024	56877	74668	69637	20418	43374	13817
03	24299	45224	77139	55789	13580	88149	39939	48812	81717	59789
04	89780	97312	96536	41006	38359	10485	47297	14435	13116	30340
05	34818	50722	68004	73739	22459	86729	30045	81460	48440	05831
06	58691	21970	21706	66206	61987	04895	17174	19145	34750	27306
07	05854	36875	73598	74348	92678	59310	86612	55945	66336	79130
08	88701	39905	74970	28201	58425	64949	59834	10334	23308	88985
09	32190	87180	00048	35539	88002	79496	46392	78480	34534	11088
10	76323	55586	81840	53045	87354	88903	54436	23499	90535	80748
11	78081	60126	89511	74863	37984	94185	16085	52489	79069	46885
12	90370	02201	46991	30198	06715	30131	26103	21861	47840	52797
13	29793	92737	53256	40922	70214	57666	36685	54891	78546	79872
14	86488	65850	83262	02569	40204	92446	70258	05162	08948	04650
15	54259	39910	33839	78858	36960	88771	51258	11678	44321	63761
16	90881	60632	21639	09218	99920	56005	12957	12618	43513	77511
17	83146	91326	87784	04430	77922	96620	46118	94261	94202	43525
18	65591	21105	61573	40528	21033	67926	79719	51285	57556	51047
19	46315	01019	53430	52773	78977	36208	78401	21227	39427	47685
20	51874	60786	76136	01981	79965	65102	53480	24424	25663	21500
21	76325	00394	91369	48736	88814	80774	86803	15586	76164	26888
22	76888	57236	05793	12186	79565	71263	12415	88798	72284	20594
23	83892	49337	09661	90125	11556	60473	20335	71856	24650	47294
24	57027	65058	03251	93347	12224	95951	75882	12911	08275	11412
25	26491	47289	13263	46917	19336	02069	26016	45865	19210	35623
26	48630	06084	93016	23591	77042	22990	51693	90767	20940	56019
27	85059	68964	70918	53925	16336	41504	04625	41846	99061	20570
28	05898	04239	52072	05896	35551	25514	07630	85771	47374	07537
29	31975	43807	85551	23751	76974	92537	49677	44965	37367	85647
30	19073	25910	18699	86575	99921	53081	88838	54490	38442	94343
31	04661	54191	69904	93133	15357	28526	59079	82950	76625	61952
32	10024	48420	51286	70040	23930	46131	10686	05596	68401	01470
33	56539	26854	82381	19159	82920	18466	77083	26012	84265	30123
34	15701	04626	74015	71536	27491	65959	79442	18296	85352	08189
35	52557	40974	54757	54926	68073	98814	55199	49034	10210	73231
36	12194	76804	51429	90336	32874	49375	05322	20554	97481	02246
37	02503	48204	05664	58186	27985	58688	28998	30263	22634	22274
38	63964	53059	81216	87634	07009	46984	80855	30533	19662	07691
39	10501	74442	63688	88339	77187	59025	28555	19827	03346	33567

This table was calculated using SC — Statistical Calculator v.321.10, 1994

Table A2 Probabilities for the standard normal distribution $\Phi(z) = P(Z \leq z)$

z	0	1	2	3	4	5	6	7	8	9
0.0	0.5000	0.5040	0.5080	0.5120	0.5160	0.5199	0.5239	0.5279	0.5319	0.5359
0.1	0.5398	0.5438	0.5478	0.5517	0.5557	0.5596	0.5636	0.5675	0.5714	0.5753
0.2	0.5793	0.5832	0.5871	0.5910	0.5948	0.5987	0.6026	0.6064	0.6103	0.6141
0.3	0.6179	0.6217	0.6255	0.6293	0.6331	0.6368	0.6406	0.6443	0.6480	0.6517
0.4	0.6554	0.6591	0.6628	0.6664	0.6700	0.6736	0.6772	0.6808	0.6844	0.6879
0.5	0.6915	0.6950	0.6985	0.7019	0.7054	0.7088	0.7123	0.7157	0.7190	0.7224
0.6	0.7257	0.7291	0.7324	0.7357	0.7389	0.7422	0.7454	0.7486	0.7517	0.7549
0.7	0.7580	0.7611	0.7642	0.7673	0.7704	0.7734	0.7764	0.7794	0.7823	0.7852
0.8	0.7881	0.7910	0.7939	0.7967	0.7995	0.8023	0.8051	0.8078	0.8106	0.8133
0.9	0.8159	0.8186	0.8212	0.8238	0.8264	0.8289	0.8315	0.8340	0.8365	0.8389
1.0	0.8413	0.8438	0.8461	0.8485	0.8508	0.8531	0.8554	0.8577	0.8599	0.8621
1.1	0.8643	0.8665	0.8686	0.8708	0.8729	0.8749	0.8770	0.8790	0.8810	0.8830
1.2	0.8849	0.8869	0.8888	0.8907	0.8925	0.8944	0.8962	0.8980	0.8997	0.9015
1.3	0.9032	0.9049	0.9066	0.9082	0.9099	0.9115	0.9131	0.9147	0.9162	0.9177
1.4	0.9192	0.9207	0.9222	0.9236	0.9251	0.9265	0.9279	0.9292	0.9306	0.9319
1.5	0.9332	0.9345	0.9357	0.9370	0.9382	0.9394	0.9406	0.9418	0.9429	0.9441
1.6	0.9452	0.9463	0.9474	0.9484	0.9495	0.9505	0.9515	0.9525	0.9535	0.9545
1.7	0.9554	0.9564	0.9573	0.9582	0.9591	0.9599	0.9608	0.9616	0.9625	0.9633
1.8	0.9641	0.9649	0.9656	0.9664	0.9671	0.9678	0.9686	0.9693	0.9699	0.9706
1.9	0.9713	0.9719	0.9726	0.9732	0.9738	0.9744	0.9750	0.9756	0.9761	0.9767
2.0	0.9772	0.9778	0.9783	0.9788	0.9793	0.9798	0.9803	0.9808	0.9812	0.9817
2.1	0.9821	0.9826	0.9830	0.9834	0.9838	0.9842	0.9846	0.9850	0.9854	0.9857
2.2	0.9861	0.9864	0.9868	0.9871	0.9875	0.9878	0.9881	0.9884	0.9887	0.9890
2.3	0.9893	0.9896	0.9898	0.9901	0.9904	0.9906	0.9909	0.9911	0.9913	0.9916
2.4	0.9918	0.9920	0.9922	0.9925	0.9927	0.9929	0.9931	0.9932	0.9934	0.9936
2.5	0.9938	0.9940	0.9941	0.9943	0.9945	0.9946	0.9948	0.9949	0.9951	0.9952
2.6	0.9953	0.9955	0.9956	0.9957	0.9959	0.9960	0.9961	0.9962	0.9963	0.9964
2.7	0.9965	0.9966	0.9967	0.9968	0.9969	0.9970	0.9971	0.9972	0.9973	0.9974
2.8	0.9974	0.9975	0.9976	0.9977	0.9977	0.9978	0.9979	0.9979	0.9980	0.9981
2.9	0.9981	0.9982	0.9982	0.9983	0.9984	0.9984	0.9985	0.9985	0.9986	0.9986
3.0	0.9987	0.9987	0.9987	0.9988	0.9988	0.9989	0.9989	0.9989	0.9990	0.9990
3.1	0.9990	0.9991	0.9991	0.9991	0.9992	0.9992	0.9992	0.9992	0.9993	0.9993
3.2	0.9993	0.9993	0.9994	0.9994	0.9994	0.9994	0.9994	0.9995	0.9995	0.9995
3.3	0.9995	0.9995	0.9995	0.9996	0.9996	0.9996	0.9996	0.9996	0.9996	0.9997
3.4	0.9997	0.9997	0.9997	0.9997	0.9997	0.9997	0.9997	0.9997	0.9997	0.9998
3.5	0.9998	0.9998	0.9998	0.9998	0.9998	0.9998	0.9998	0.9998	0.9998	0.9998
3.6	0.9998	0.9998	0.9999	0.9999	0.9999	0.9999	0.9999	0.9999	0.9999	0.9999
3.7	0.9999	0.9999	0.9999	0.9999	0.9999	0.9999	0.9999	0.9999	0.9999	0.9999
3.8	0.9999	0.9999	0.9999	0.9999	0.9999	0.9999	0.9999	0.9999	0.9999	0.9999
3.9	1.0000	1.0000	1.0000	1.0000	1.0000	1.0000	1.0000	1.0000	1.0000	1.0000
4.0	1.0000	1.0000	1.0000	1.0000	1.0000	1.0000	1.0000	1.0000	1.0000	1.0000

This table was calculated using SC — Statistical Calculator v.321.10, 1994

Table A3 Standard normal quantiles

α	q_α	α	q_α	α	q_α	α	q_α
0.50	0.0	0.67	0.4399	0.84	0.9945	0.955	1.695
0.51	0.02507	0.68	0.4677	0.85	1.036	0.960	1.751
0.52	0.05015	0.69	0.4959	0.86	1.080	0.965	1.812
0.53	0.07527	0.70	0.5244	0.87	1.126	0.970	1.881
0.54	0.1004	0.71	0.5534	0.88	1.175	0.975	1.960
0.55	0.1257	0.72	0.5828	0.89	1.227	0.980	2.054
0.56	0.1510	0.73	0.6128	0.90	1.282	0.985	2.170
0.57	0.1764	0.74	0.6433	0.905	1.311	0.990	2.326
0.58	0.2019	0.75	0.6745	0.910	1.341	0.991	2.366
0.59	0.2275	0.76	0.7063	0.915	1.372	0.992	2.409
0.60	0.2533	0.77	0.7388	0.920	1.405	0.993	2.457
0.61	0.2793	0.78	0.7722	0.925	1.440	0.994	2.512
0.62	0.3055	0.79	0.8064	0.930	1.476	0.995	2.576
0.63	0.3319	0.80	0.8416	0.935	1.514	0.996	2.652
0.64	0.3585	0.81	0.8779	0.940	1.555	0.997	2.748
0.65	0.3853	0.82	0.9154	0.945	1.598	0.998	2.878
0.66	0.4125	0.83	0.9542	0.950	1.645	0.999	3.090

This table was calculated using SC — Statistical Calculator v.321.10, 1994

Table A4 Standard normal random numbers — $Z \sim N(0, 1)$

00	0.46	−0.64	−1.05	−1.66	−0.55	−0.18	0.06	0.46	−1.94	−0.16
01	0.41	1.08	−0.22	−0.40	−0.60	−0.45	0.46	1.51	0.71	1.25
02	1.24	0.44	−1.71	0.98	−1.17	−0.33	1.41	−0.57	0.53	0.57
03	−1.09	0.93	1.08	−1.61	−0.60	−0.02	−1.20	0.14	0.05	−0.46
04	0.34	0.50	−0.43	−0.61	−0.22	−1.67	−1.39	−0.35	0.19	2.24
05	0.27	−1.11	−0.21	−0.15	1.72	0.78	−0.48	0.48	−1.02	−0.56
06	−0.19	−0.35	0.50	−1.78	0.47	0.59	1.22	0.76	0.59	−0.60
07	−2.27	0.86	−0.89	0.26	−0.66	0.71	0.31	−0.53	−0.72	0.36
08	−0.20	−0.10	−0.80	−1.26	−1.08	0.60	−1.13	0.39	0.63	−1.32
09	0.30	0.05	0.67	−0.95	0.41	0.09	0.06	2.34	0.31	−0.56
10	−0.97	−0.08	−0.11	0.17	0.01	1.20	−1.75	−0.69	0.36	1.17
11	−0.10	−0.66	−0.80	−1.59	0.08	0.95	0.31	2.23	−2.35	−0.40
12	−1.38	−0.05	−0.97	−1.26	0.36	0.04	−0.06	0.70	0.65	1.71
13	−3.81	−0.26	−0.68	−0.49	0.12	−0.08	1.17	0.91	−0.64	−1.66
14	0.66	−2.38	−2.41	2.11	0.48	0.41	−1.90	0.18	−0.81	2.83
15	−1.58	0.89	1.66	−1.34	1.11	−1.50	−0.66	−1.52	−2.10	−0.90
16	0.01	−1.04	0.30	−1.17	−0.62	−0.63	1.16	−0.28	−1.41	−0.07
17	0.63	−0.30	−0.52	0.45	0.10	−0.86	−0.33	0.97	−0.37	0.35
18	−0.05	0.88	0.71	−2.44	0.38	1.49	−0.59	0.65	0.56	−0.76
19	0.01	−0.19	0.91	0.25	0.59	−0.77	1.37	−0.66	0.96	−0.87
20	0.12	−1.69	−1.97	0.44	0.17	0.18	−2.13	−1.59	0.53	−1.16
21	0.24	−0.52	1.66	−0.95	−2.34	−1.18	1.58	−1.14	−1.53	1.09
22	0.34	0.92	0.73	−0.30	−0.77	−1.53	1.32	0.68	0.74	0.46
23	1.12	1.12	0.87	0.36	−2.55	1.55	−0.52	−1.21	1.13	0.71
24	0.88	0.57	1.02	−2.40	−0.49	0.55	0.55	−0.09	1.70	−0.41
25	−0.74	0.87	0.69	1.35	1.54	0.75	0.87	−0.96	−1.81	0.79
26	−1.32	−0.28	−0.05	−0.72	−0.27	1.55	1.86	0.77	−0.02	0.03
27	−1.65	−0.55	0.55	−1.28	0.38	0.17	−0.02	−0.20	−0.97	0.63
28	−0.11	0.93	0.66	−0.06	−0.43	−0.82	−2.03	2.48	−1.20	1.80
29	0.93	0.01	1.54	2.18	1.24	2.98	−2.02	0.54	0.65	−0.75
30	−0.19	0.08	0.07	0.73	−0.39	−0.52	−0.62	−0.19	0.60	1.41
31	0.20	−0.26	0.95	0.15	0.51	0.98	−0.08	1.51	0.24	−0.86
32	1.51	−0.41	1.15	−0.26	−0.37	−1.02	0.13	0.78	0.76	0.02
33	−1.11	−1.10	0.04	−0.14	−2.72	0.64	1.69	−1.55	−0.97	0.18
34	0.10	0.74	−0.25	1.04	0.20	−0.34	−0.04	−1.46	0.63	0.29
35	−1.16	−1.37	0.10	−0.25	−0.23	0.65	−0.11	0.07	−0.06	1.3
36	0.01	−0.57	−0.08	−0.91	0.45	0.26	−0.54	−0.93	0.90	−0.18
37	−0.93	−0.12	−1.28	0.23	−0.42	0.09	−1.02	−0.52	−0.95	−0.78
38	0.20	0.02	−0.81	−1.38	0.14	−1.84	−0.67	−1.10	−0.35	−1.93
39	2.11	−1.15	1.22	0.68	−0.93	−0.26	−1.31	−0.70	−0.74	−0.60
40	−1.22	1.46	0.34	−0.42	−1.84	−0.82	0.37	−0.71	−1.50	0.12
41	−0.29	−1.58	−1.36	1.02	−0.24	−0.58	−0.85	1.19	0.16	1.37
42	0.19	−0.37	1.61	0.62	0.53	−1.11	0.45	−0.04	0.19	0.14
43	1.36	1.33	1.04	−0.95	1.00	0.18	−0.16	0.67	0.30	0.06
44	0.13	−0.86	1.08	−0.96	0.41	0.70	−0.24	0.22	−0.61	−0.22
45	−0.16	−0.42	1.75	0.73	1.28	−0.14	1.25	−0.28	0.92	0.94
46	0.81	0.66	0.14	−0.91	0.47	1.94	1.78	2.24	0.22	−0.59
47	0.94	−1.96	−1.15	0.17	0.25	−0.91	0.64	0.19	−1.35	0.87
48	0.18	0.96	−1.06	0.91	−0.14	−0.90	0.59	0.24	1.69	−0.43
49	−1.09	−0.22	−0.06	0.39	−0.84	−0.69	0.59	−0.90	0.25	0.31

This table was calculated using SC — Statistical Calculator v.321.10, 1994

Table A5 t-quantiles

df	0.90	0.95	0.975	0.99	0.995	0.999
1	3.078	6.314	12.71	31.82	63.66	318.3
2	1.886	2.920	4.303	6.965	9.925	22.33
3	1.638	2.353	3.182	4.541	5.841	10.21
4	1.533	2.132	2.776	3.747	4.604	7.173
5	1.476	2.015	2.571	3.365	4.032	5.893
6	1.440	1.943	2.447	3.143	3.707	5.208
7	1.415	1.895	2.365	2.998	3.499	4.785
8	1.397	1.860	2.306	2.896	3.355	4.501
9	1.383	1.833	2.262	2.821	3.250	4.297
10	1.372	1.812	2.228	2.764	3.169	4.144
11	1.363	1.796	2.201	2.718	3.106	4.025
12	1.356	1.782	2.179	2.681	3.055	3.930
13	1.350	1.771	2.160	2.650	3.012	3.852
14	1.345	1.761	2.145	2.624	2.977	3.787
15	1.341	1.753	2.131	2.602	2.947	3.733
16	1.337	1.746	2.120	2.583	2.921	3.686
17	1.333	1.740	2.110	2.567	2.898	3.646
18	1.330	1.734	2.101	2.552	2.878	3.610
19	1.328	1.729	2.093	2.539	2.861	3.579
20	1.325	1.725	2.086	2.528	2.845	3.552
21	1.323	1.721	2.080	2.518	2.831	3.527
22	1.321	1.717	2.074	2.508	2.819	3.505
23	1.319	1.714	2.069	2.500	2.807	3.485
24	1.318	1.711	2.064	2.492	2.797	3.467
25	1.316	1.708	2.060	2.485	2.787	3.450
26	1.315	1.706	2.056	2.479	2.779	3.435
27	1.314	1.703	2.052	2.473	2.771	3.421
28	1.313	1.701	2.048	2.467	2.763	3.408
29	1.311	1.699	2.045	2.462	2.756	3.396
30	1.310	1.697	2.042	2.457	2.750	3.385
31	1.309	1.696	2.040	2.453	2.744	3.375
32	1.309	1.694	2.037	2.449	2.738	3.365
33	1.308	1.692	2.035	2.445	2.733	3.356
34	1.307	1.691	2.032	2.441	2.728	3.348
35	1.306	1.690	2.030	2.438	2.724	3.340
36	1.306	1.688	2.028	2.434	2.719	3.333
37	1.305	1.687	2.026	2.431	2.715	3.326
38	1.304	1.686	2.024	2.429	2.712	3.319
39	1.304	1.685	2.023	2.426	2.708	3.313
40	1.303	1.684	2.021	2.423	2.704	3.307
45	1.301	1.679	2.014	2.412	2.690	3.281
50	1.299	1.676	2.009	2.403	2.678	3.261
55	1.297	1.673	2.004	2.396	2.668	3.245
60	1.296	1.671	2.000	2.390	2.660	3.232
65	1.295	1.669	1.997	2.385	2.654	3.220
70	1.294	1.667	1.994	2.381	2.648	3.211
75	1.293	1.665	1.992	2.377	2.643	3.202
80	1.292	1.664	1.990	2.374	2.639	3.195
85	1.292	1.663	1.988	2.371	2.635	3.189
90	1.291	1.662	1.987	2.368	2.632	3.183
100	1.290	1.660	1.984	2.364	2.626	3.174

This table was calculated using SC — Statistical Calculator v.321.10, 1994

Table A6 χ^2 quantiles

df	0.005	0.01	0.025	0.05	0.1	0.5	0.9	0.95	0.975	0.99	0.995
1	.00004	0.0001	0.0009	.0039	0.016	0.455	2.71	3.84	5.02	6.63	7.88
2	0.010	0.020	0.051	0.103	0.211	1.39	4.61	5.99	7.38	9.21	10.60
3	0.072	0.115	0.216	0.352	0.584	2.37	6.25	7.81	9.35	11.34	12.84
4	0.207	0.297	0.484	0.711	1.06	3.36	7.78	9.49	11.14	13.28	14.86
5	0.412	0.554	0.831	1.14	1.61	4.35	9.24	11.07	12.83	15.09	16.75
6	0.676	0.872	1.24	1.64	2.20	5.35	10.64	12.59	14.45	16.81	18.55
7	0.989	1.24	1.69	2.17	2.83	6.35	12.02	14.07	16.01	18.48	20.28
8	1.34	1.65	2.18	2.73	3.49	7.34	13.36	15.51	17.53	20.09	21.95
9	1.73	2.09	2.70	3.33	4.17	8.34	14.68	16.92	19.02	21.67	23.59
10	2.16	2.56	3.25	3.94	4.87	9.34	15.99	18.31	20.48	23.21	25.19
11	2.60	3.05	3.82	4.57	5.58	10.34	17.28	19.68	21.92	24.72	26.76
12	3.07	3.57	4.40	5.23	6.30	11.34	18.55	21.03	23.34	26.22	28.30
13	3.57	4.11	5.01	5.89	7.04	12.34	19.81	22.36	24.74	27.69	29.82
14	4.07	4.66	5.63	6.57	7.79	13.34	21.06	23.68	26.12	29.14	31.32
15	4.60	5.23	6.26	7.26	8.55	14.34	22.31	25.00	27.49	30.58	32.80
16	5.14	5.81	6.91	7.96	9.31	15.34	23.54	26.30	28.85	32.00	34.27
17	5.70	6.41	7.56	8.67	10.09	16.34	24.77	27.59	30.19	33.41	35.72
18	6.26	7.01	8.23	9.39	10.86	17.34	25.99	28.87	31.53	34.81	37.16
19	6.84	7.63	8.91	10.12	11.65	18.34	27.20	30.14	32.85	36.19	38.58
20	7.43	8.26	9.59	10.85	12.44	19.34	28.41	31.41	34.17	37.57	40.00
21	8.03	8.90	10.28	11.59	13.24	20.34	29.62	32.67	35.48	38.93	41.40
22	8.64	9.54	10.98	12.34	14.04	21.34	30.81	33.92	36.78	40.29	42.80
23	9.26	10.20	11.69	13.09	14.85	22.34	32.01	35.17	38.08	41.64	44.18
24	9.89	10.86	12.40	13.85	15.66	23.34	33.20	36.42	39.36	42.98	45.56
25	10.52	11.52	13.12	14.61	16.47	24.34	34.38	37.65	40.65	44.31	46.93
26	11.16	12.20	13.84	15.38	17.29	25.34	35.56	38.89	41.92	45.64	48.29
27	11.81	12.88	14.57	16.15	18.11	26.34	36.74	40.11	43.19	46.96	49.64
28	12.46	13.56	15.31	16.93	18.94	27.34	37.92	41.34	44.46	48.28	50.99
29	13.12	14.26	16.05	17.71	19.77	28.34	39.09	42.56	45.72	49.59	52.34
30	13.79	14.95	16.79	18.49	20.60	29.34	40.26	43.77	46.98	50.89	53.67
31	14.46	15.66	17.54	19.28	21.43	30.34	41.42	44.99	48.23	52.19	55.00
32	15.13	16.36	18.29	20.07	22.27	31.34	42.58	46.19	49.48	53.49	56.33
33	15.82	17.07	19.05	20.87	23.11	32.34	43.75	47.40	50.73	54.78	57.65
34	16.50	17.70	19.81	21.66	23.95	33.34	44.90	48.60	51.97	56.06	58.96
35	17.19	18.51	20.57	22.47	24.80	34.34	46.06	49.80	53.20	57.34	60.27
36	17.89	19.23	21.34	23.27	25.64	35.34	47.21	51.00	54.44	58.62	61.58
37	18.59	19.96	22.11	24.07	26.49	36.34	48.36	52.19	55.67	59.89	62.88
38	19.29	20.69	22.88	24.88	27.34	37.34	49.51	53.38	56.90	61.16	64.18
39	20.00	21.43	23.65	25.70	28.20	38.34	50.66	54.57	58.12	62.43	65.48
40	20.71	22.16	24.43	26.51	29.05	39.34	51.81	55.76	59.34	63.69	66.77
45	24.31	25.90	28.37	30.61	33.35	44.34	57.51	61.66	65.41	69.96	73.17
50	27.99	29.71	32.36	34.76	37.69	49.33	63.17	67.50	71.42	76.15	79.49
55	31.73	33.57	36.40	38.96	42.06	54.33	68.80	73.31	77.38	82.29	85.75
60	35.53	37.48	40.48	43.19	46.46	59.33	74.40	79.08	83.30	88.38	91.95
65	39.38	41.44	44.60	47.45	50.88	64.33	79.97	84.82	89.18	94.42	98.11
70	43.28	45.44	48.76	51.74	55.33	69.33	85.53	90.53	95.02	100.4	104.2
75	47.21	49.48	52.94	56.05	59.79	74.33	91.06	96.22	100.8	106.4	110.3
80	51.17	53.54	57.15	60.39	64.28	79.33	96.58	101.9	106.6	112.3	116.3
85	55.17	57.63	61.39	64.75	68.78	84.33	102.1	107.5	112.4	118.2	122.3
90	59.20	61.75	65.65	69.13	73.29	89.33	107.6	113.1	118.1	124.1	128.3
100	67.33	70.06	74.22	77.93	82.36	99.33	118.5	124.3	129.6	135.8	140.2

This table was calculated using SC — Statistical Calculator v.321.10, 1994

Index of data sets

Index of examples

General index